Cancer Care in the United Arab Emirates

Humaid O. Al-Shamsi
Editor

Cancer Care in the United Arab Emirates

Editor
Humaid O. Al-Shamsi
Burjeel Cancer Institute
Burjeel Medical City, Burjeel Holdings
Abu Dhabi, United Arab Emirates

Ras Al Khaimah Medical and Health Sciences University
Ras Al Khaimah, United Arab Emirates

Gulf Medical University
Ajman, United Arab Emirates

Emirates Oncology Society
Emirates Medical Association
Dubai, United Arab Emirates

College of Medicine
University of Sharjah
Sharjah, United Arab Emirates

Gulf Cancer Society
Alsafa, Kuwait

This book is an open access publication.

ISBN 978-981-99-6793-3   ISBN 978-981-99-6794-0   (eBook)
https://doi.org/10.1007/978-981-99-6794-0

Emirates Oncology Society

This Springer imprint is published by the registered company Springer Nature Singapore Pte Ltd.
The registered company address is: 152 Beach Road, #21-01/04 Gateway East, Singapore 189721, Singapore

Paper in this product is recyclable.

Foreword

The United Arab Emirates (UAE), established in 1971, has undergone an unparalleled transformation over the past 50 years. The country has experienced rapid growth in its native and expatriate populations, with an enormous and steady influx of tourists. Likewise, the UAE's healthcare delivery system has undergone incredible growth and modernization, becoming a regional beacon for contemporary, patient-centric healthcare.

A key milestone in the UAE's healthcare evolution is the establishment of infrastructure supporting world-class medical facilities and care providers. From state-of-the-art hospitals to specialized clinics, these facilities are equipped with cutting-edge medical technology and are increasingly staffed by highly skilled professionals from across the globe. Among the significant advancements over the past five decades in the UAE has been the evolution of cancer care. However, until now, there has been no comprehensive documentation encapsulating this development.

Cancer Care in the United Arab Emirates fills this gap. This unprecedented and inclusive reference seeks to fully characterize the landscape of cancer care in the UAE and is poised to become an invaluable resource for those seeking to understand the current state of cancer care across the seven emirates.

The book embarks on a comprehensive analysis of cancer and its management through the UAE's relatively short history. Included are summaries of key regulatory policies regarding medical education and training, cancer control, and care delivery. It also describes the impressive expansion of innovative diagnostic and treatment modalities, with a growing emphasis on precision medicine. In doing so, the book's authors can celebrate a number of notable achievements while acknowledging persistent challenges and unmet needs. Authored by an esteemed group of experts from various arenas, it brings together perspectives from oncology specialists, researchers, healthcare policymakers, patient advocates, and other key stakeholders involved in the cancer care delivery system.

In addition to addressing general aspects of cancer care, this pioneering book sheds light on unique factors and challenges specific to the UAE. It explores cultural and societal impacts, healthcare infrastructure, regulatory frameworks, and the incorporation of technological advancements in cancer care. The critical role of patient support organizations, patient empowerment, and community engagement in reducing the current burden of cancer is also examined in detail.

Each chapter of *Cancer Care in the United Arab Emirates* provides valuable insights into the challenges faced by different disciplines involved in cancer care and offers a roadmap for transformative change. By presenting an overarching vision and strategic outlook for the future, it catalyzes positive change in cancer prevention, control, and care across the UAE.

I wholeheartedly commend Professor Humaid O. Al-Shamsi for his remarkable vision and effort in compiling this book over the past few years, making it broadly accessible and ensuring widespread benefit.

Endorsed by prestigious organizations such as the Emirates Oncology Society, Emirates Medical Association, and the Gulf Cancer Society, *Cancer Care in the United Arab Emirates* stands as a testament to its credibility and relevance. It is a tour de force and will undoubtedly play a pivotal role in reducing the burden of cancer with its emphasis on prevention, screening, and enhanced cancer care. Furthermore, with its call for collaboration, it is certain to facilitate better outcomes for the country and the wider region.

In summary, Cancer Care in the United Arab Emirates heralds a new dawn in our collective understanding of cancer care in the UAE. As we navigate the complex systems of care required to promote outstanding cancer care, this book offers guidance, inspiration, and the knowledge required to make meaningful strides. It is an invitation to all stakeholders to come together, learn, and contribute to the noble cause of defeating cancer. Here's to a future where we rise to the challenge, foster collaborations, and continue our journey towards a healthier and cancer-free UAE.

July 15, 2024

Robert A. Wolff
Clinical and Educational Affairs,
Division of Cancer Medicine,
Department of Gastrointestinal Medical Oncology
The University of Texas, MD Anderson Cancer Center
Houston, TX, USA

Acknowledgment

Dedicated to my Creator, who may accept this work for His honorable sake.

To my father and idol, the late Sheikh Zayed bin Sultan Al Nahyan, the founder and first president of the United Arab Emirates, may Allah have mercy on him, for his love for this land and his belief in his people (us).

To the late Sheikh Hamdan bin Rashid Al Maktoum, the father of the Emirates Medical Association, for his support of sciences and research in the UAE and the region.

To my country and my leaders, whom I do not have enough words to thank.

To the UAE armed forces that have taken care of and supported me during my 20 years of medical education and training, believing, and trusting in me.

To my late father, may Allah Almighty have mercy on him who always blamed me for my long travel and absence away from him.

To my dear mother, who has planted the love of science and medicine in me since childhood. I hope I am today the man you dreamed I would be.

To my dear wife and best friend, Khadija, the love of my life, who has been there for me at all times.

To my kids, Sarah, Abdullah, Muhammad, Noor, Aseel, and Sama, who are my source of strength and inspiration. Forgive me if I'm away from you due to my work and my national mission to fight cancer.

To my brother, Hassan, and my sisters, Fatima, Eng. Nabila, Salma, and Dr. Nada, for being away from them for many years and for their support and prayers for me always.

To Prof. Parveen Wasi, my internal medicine program director during my residency at McMaster University in Canada, for her support and encouragement to pursue clinical research.

To Prof. Robert Wolff, Prof. Bindi Dhesy, and Prof. Peter Ellis for their support and mentorship during my oncology fellowship and early career in oncology.

To Dr. Mouza AlSharhan, the Past President of the Emirates Medical Association, for her support for the Emirates Oncology Society over the years.

To Dr. Shamsheer Vayalil for his continuous support to transform cancer care in the UAE and the region through the Burjeel Holdings oncology program.

To my dear friends: Dr. Sherif Almarzooqi, Dr. Mitref Altunaiji, Dr. Abdullah Alqimzi, Dr. Khaled Alfarsi, Dr. Uthman Alao, Dr. Ibrahim Abu Ghida, and Tariq Elfikki, for being there for me in my ups and downs.

To my friend and role model in science, medicine, and clinical research, Prof. Waleed Al-Hazzani, it is hard to follow your steps. You set the bar very high in clinical research for all of us.

To my past, current, and future teachers, and to everyone who supported me throughout my life journey.

To all my past and future patients: I promise you that I will take care of you as I take care of my own family.

For humanity, believing in the words of Allah Almighty: "We have sent you nothing but a mercy for the world."

To cancer, which taught me a lot and which we will eliminate, Allah willing.

Sharjah, UAE   Humaid O. Al-Shamsi
July 15, 2024

Contents

About the Editor

Prof. Humaid Obaid Al-Shamsi is the Chief Executive Officer of Burjeel Cancer Institute in Abu Dhabi, UAE, President of the Emirates Oncology Society, Lead of the Gulf Cancer Society, Full Professor of Oncology at the Ras Al Khaimah Medical and Health Sciences University, Ras Al Khaimah, UAE, and an Adjunct Professor of Oncology at the College of Medicine, University of Sharjah. He is the first Emirati to be promoted as a professor in oncology in the UAE. He is also the Chairman for Colorectal Cancer in the MENA region, appointed by the prestigious National Comprehensive Cancer Network®. He is also the only member of Lung Cancer Policy Network in the MENA region that aims to advance lung cancer research and screening globally. He is the Chairman of the Oncology and Hematology Fellowship Training Program for the National Institute for Health Specialties in the United Arab Emirates. He is the only member in GCC in the WIN Consortium which is comprised of organizations representing all stakeholders in personalized cancer medicine globally.

He is board-certified in both internal medicine and oncology from the UK, USA (ABIM), and Canada (FRCPC). He has also been awarded the FRCP (London) in 2023 and FRCP (Glasgow) in 2024. He is the only physician in the UAE with a subspecialty fellowship certification and training in gastrointestinal oncology and the first Emirati to train and complete a clinical post-doctoral fellowship in palliative care. He was an assistant professor at the University of Texas MD Anderson Cancer Center between 2014 and 2017. He has published more than 140 peer-reviewed articles in JAMA Oncology, Lancet Oncology, The Oncologist, BMC Cancer, and many others. His area of expertise includes precision oncology and cancer care in the UAE. In 2016, he published with his group from MD Anderson the JCO paper describing a new distinct subgroup of CRC, NON V600 BRAF-mutated CRC. In 2022, he published the first book about cancer research in the UAE and also the first book about cancer in the Arab world, both of which were launched at Dubai Expo 2020. *Cancer in the Arab World* has been downloaded more than 450,000 times in its first 18 months of publication and is the ultimate source of cancer data in the Arab region. He also published the first comprehensive book about cancer care in the UAE which is the first book in UAE history to document the cancer care in the UAE with many topics addressed for the first time, e.g., neuroendocrine tumors in the UAE. He is passionate about advancing cancer care in the

UAE and the GCC and has made significant contributions to cancer awareness and early detection for the public using social media platforms. He is considered as the most followed oncologist in the world with over 300,000 subscribers across his social media platforms (Instagram, Twitter, LinkedIn, and TikTok). In 2022, he was awarded the prestigious Feigenbaum Leadership Excellence Award from Sheikh Hamdan Smart University for his exceptional leadership and research and the Sharjah Award for Volunteering. He was also named the Researcher of the Year in the UAE in 2020 and 2021 by the Emirates Oncology Society.

In May 2024, HH Sheikh Mansour bin Zayed Al Nahyan, Vice President of the United Arab Emirates, awarded him the first place in the Emirati Talent Competitiveness Council (NAFIS) program for outstanding leadership in the private sector across all business and medical disciplines. Beside his clinical and administrative duties, he is engaged in education and various levels of research training for medical trainees to enhance their clinical and research skills. His mission is to advance cancer care in the UAE and the MENA region and make cancer care accessible to everyone in need around the globe.

Introduction

Humaid O. Al-Shamsi

1.1 Introduction

During my personal experience with cancer, I consistently sought a reliable source of current information on cancer care in the United Arab Emirates (UAE). Regrettably, I encountered difficulties in finding the specific information I sought. Recognizing the pressing need for a comprehensive reference on cancer care in the UAE, I have resolved to transform this aspiration into a tangible resource that will endure for years to come, serving the welfare of my country and fellow citizens.

To begin my endeavor, I embarked on a comprehensive examination of existing literature pertaining to cancer care in the UAE. I diligently explored various sources, including official health websites within the country as well as reputable platforms like PubMed and Google Scholar. Additionally, I reached out to prominent healthcare organizations such as the UAE Ministry of Health and Prevention (MOHAP), the Dubai Health Authority (DHA), and the Abu Dhabi Department of Health (DOH), seeking relevant data and accessible resources.

The second part was to get the experts in each topic and oncology subspecialty to contribute to this project, and I have received an overwhelmingly positive response to participate and contribute from all the cancer institutions in the UAE. I have also provided the opportunity for all oncologists and hematologists to suggest

H. O. Al-Shamsi (✉)
Burjeel Cancer Institute, Burjeel Medical City, Burjeel Holdings, Abu Dhabi, United Arab Emirates

Ras Al Khaimah Medical and Health Sciences University, Ras Al Khaimah, United Arab Emirates

Gulf Medical University, Ajman, United Arab Emirates

Emirates Oncology Society, Emirates Medical Association, Dubai, United Arab Emirates

College of Medicine, University of Sharjah, Sharjah, United Arab Emirates

Gulf Cancer Society, Alsafa, Kuwait
e-mail: alshamsi@burjeel.com; humaid.al-shamsi@medportal.ca

© The Author(s) 2024
H. O. Al-Shamsi (ed.), *Cancer Care in the United Arab Emirates*,
https://doi.org/10.1007/978-981-99-6794-0_1

topics and chapters. This made me confident that we are covering as many oncology topics as possible in the UAE without missing any major topics or issues.

The objective of this book is to offer a comprehensive and profound examination of the demographic, economic, and prevailing trends in oncology practice that have an impact on cancer care in the UAE. Its purpose is to chart a path for healthcare providers, regulators, decision-makers, patients, and all other relevant parties to propel the progress of cancer care in the UAE. By providing valuable insights, this book aims to guide and inform stakeholders in their efforts to enhance cancer care within the country.

Within the pages of this book, you will discover a thorough exploration of the evolution of cancer care in the UAE, along with a detailed analysis of the national cancer incidence. Dedicated chapters are devoted to addressing the most prevalent types of cancer. Furthermore, the book elucidates the current challenges faced in the UAE, some of which are unique to the region, while also highlighting opportunities for improvement. The author presents recommendations and proposes a comprehensive cancer control plan intended for all stakeholders, policymakers, healthcare providers in the oncology field, and the UAE community as a whole. The ultimate goal is to deliver cutting-edge, high-quality cancer care to our population and effectively manage the anticipated rise in the cancer burden in the coming years and decades.

We have also addressed many topics that have not been addressed before, like non-governmental organizations' (NGO) role in cancer care in the UAE, clinical cancer research in the UAE, basic cancer research in the UAE, oncology and hematology fellowship training in the UAE, oncology nursing in the UAE, genomic medicine in cancer care in the UAE, genetic testing for cancer in the UAE, oncofertility in the UAE, psycho-oncology in the UAE, traditional, complementary, and integrative medicine and cancer care in the UAE, artificial intelligence (AI) in oncology in the UAE, geriatric oncology in the UAE, cancer survivorship programs in the UAE, suggested quality control measures for cancer care in the UAE and many others.

1.2   The UAE

The UAE was established on December 2, 1971, and is located in Asia in the southeast of the Arabian Peninsula on the Arabian Gulf and Gulf of Oman. It is a federation of seven states (Emirates) composed of Abu Dhabi, the largest state in size (which serves as the capital); Dubai, which is considered the trade hub; Sharjah, the cultural hub; Ajman; Ras Al Khaimah; Fujairah; and Umm Al-Quwain (the last five Emirates/States are known collectively as the Northern Emirates, which we will refer to throughout this book) (Fig. 1.1). The UAE's population statistics from 2010 to 2020 are given in Appendix 1 (Source: Federal Competitiveness and Statistics Centre).

In order to grasp the UAE's healthcare system, including cancer care, it is crucial to have an understanding of the demographic makeup of the UAE population. As of 2013, the estimated population of the UAE stood at 8.6 million. It is worth

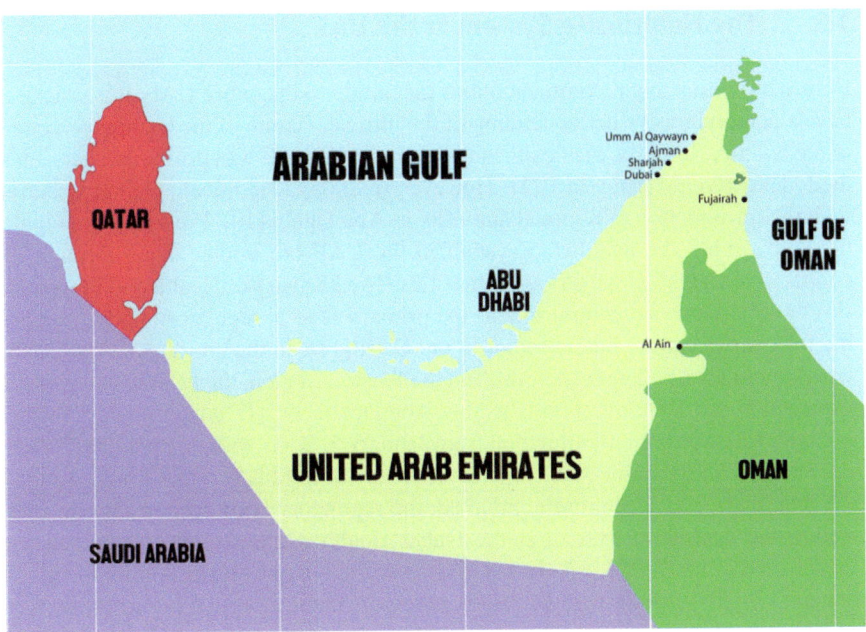

Fig. 1.1 The UAE map

noting that the most recent official census was conducted in 2005 [1, 2]. Although the UAE is relatively small in terms of land area, it possesses the world's third-largest reserves of conventional oil and the fifth-largest reserves of natural gas [3]. The UAE's 2015 gross domestic product (GDP) per capita ranked in the 95th percentile globally [4, 5]. In 2000, the World Health Organization ranked the UAE's healthcare system as the 27th best globally [6]. Healthcare services in the UAE are financed through government-funded health insurance plans available to all UAE citizens, private insurance options, or self-payment for legal residents [7]. Government-funded health coverage is provided to all UAE citizens, and the extent of coverage varies depending on the Emirate of residence. Additionally, according to UAE immigration regulations, all expatriates are required to have at least basic health insurance.

Cancer continues to be a significant public health concern in the UAE, resulting in considerable loss of life and health complications that incur substantial expenses for the UAE healthcare system. It ranks as the third-highest cause of death in the country, following cardiovascular diseases and injuries, and contributed to approximately 10% of all deaths in the UAE in 2010 [8]. The UAE government is committed to decreasing cancer-related fatalities. Decreasing the mortality rate caused by cancer is a significant benchmark within the UAE's national agenda's "Pillar of World-Class Healthcare" [9].

1.3   The Healthcare System in the UAE

Before providing any recommendations that apply to the entire UAE, it is crucial to have a comprehensive understanding of the intricate nature of the healthcare system in the country. Initially, the healthcare system in the UAE was primarily overseen at the federal level, starting in 1971. However, in 2005, a significant change occurred with the introduction of regional authority in Abu Dhabi [10]. Presently, the regulations governing the healthcare system in the UAE are implemented through the cooperative endeavors of various authorities. The Ministry of Health and Prevention (MOHAP) assumes a central role as the primary federal regulatory body responsible for overseeing the health sector in the UAE [11]. In the emirate of Abu Dhabi, which includes Abu Dhabi, Al-Ain, and the Western region, the healthcare system is overseen by the Department of Health—Abu Dhabi (DOH), which was established in 2007. It is worth mentioning that from January 1, 2005, to January 1, 2007, DOH operated under the name "the General Authority of Health Services for the Emirate of Abu Dhabi" [12]. Similarly, in Dubai, the regulatory body responsible for overseeing the healthcare system is the Dubai Health Authority (DHA), which was established in June 2007 [13]. In a similar vein, the Sharjah Health Authority (SHA), founded in 2010, along with MOHAP, is responsible for regulating health services in Sharjah. However, the remaining Northern Emirates, namely Ajman, Fujairah, Ras Al Khaimah, and Umm Al-Quwain, have their health services solely regulated by MOHAP.

1.4   History of Cancer Care in the UAE

From the 1940s to the 1970s, there was a lack of comprehensive records regarding cancer care in the UAE. The initial scientific documentation and publication on cancer care in the country appeared in the medical literature from Al-Qassimi Hospital, located in Sharjah. This report, published in 1981, consisted of a case series encompassing five instances of hepatocellular carcinoma (liver cancer) [14]. The inaugural cancer care center introduced in the UAE emerged as Tawam Hospital, situated in Al-Ain City within the Emirate of Abu Dhabi, commencing operations in September 1979. Tawam Hospital proudly held the distinction of being the nation's pioneer facility to provide both chemotherapy and radiation therapy services [14]. In February 1983, Tawam Hospital was designated as the official tertiary referral hospital for cancer cases in the United Arab Emirates [14]. In the past, individuals from all over the UAE would journey to Tawam Hospital to receive specialized cancer treatment.

In August 1983, Al Mafraq Hospital, situated 35 km away from Abu Dhabi, commenced operations and offered specialized medical care encompassing various fields, including oncology services [14]. By September 1983, Al Mafraq Hospital had established both medical oncology and radiation services. Nevertheless, the radiation service was discontinued in 2007, while the medical oncology service persisted until November 2019. At that time, it was relocated to the recently inaugurated Sheikh Shakhbout Medical City (SSMC) [15]. Sheikh Khalifa Medical City (SKMC) in Abu Dhabi, formed in 2005 through the consolidation of various publicly operated healthcare organizations on Abu Dhabi island, has been delivering specialized oncology and hematology services for both adults and children since its inception. SKMC had one of the most extensive leukemia programs in the UAE until it transferred this service to Sheikh Shakhbout Medical City in 2020. In 2020, SKMC ceased providing adult oncology services, but it has continued to offer pediatric oncology services up to the present time (February 2023).

Established in 1983, Dubai Hospital stands as the third hospital funded by the government. It offers a comprehensive array of medical and surgical oncology services for adults, as well as pediatric oncology and hematology services. However, radiation therapy services are presently not offered at Dubai Hospital. The hospital is equipped with a renowned nuclear medicine department that delivers a variety of diagnostic and therapeutic services. Additionally, in 2018, the hospital introduced PET/CT as part of its imaging capabilities.

Cleveland Clinic Abu Dhabi (CCAD), managed by Mubadala, is a leading multispecialty hospital with 364 beds. Affiliated with the Cleveland Clinic Foundation in the USA, CCAD has been serving the public in Abu Dhabi since May 2015. The hospital offers a wide range of surgical and medical oncology services. Additionally, in December 2022, CCAD expanded its offerings to include radiation oncology services.

The private sector in the UAE has displayed reluctance to offer oncology care services due to several factors. These include a shortage of specialized personnel and equipment, as well as the potential high costs associated with capital investment. Another contributing factor was the provision of free cancer treatment to all UAE residents, irrespective of citizenship or residency status, until 2007. Visitors to the country who had been diagnosed with cancer were also eligible for complimentary cancer treatment, leading to a significant influx of patients from Asian countries seeking free cancer care in the UAE until 2007. In 2005, the first privately owned outpatient facility providing chemotherapy was established at the American Hospital in Dubai. Subsequently, in 2011, radiation therapy services were added. The Gulf International Cancer Center (GICC) in Abu Dhabi commenced its operations in 2007, offering both chemotherapy and radiation therapy. Notably, the center introduced the first PET/CT scanner in the UAE in 2009 [16]. The American Hospital in Dubai introduced the second PET/CT scanner in the UAE. In 2013, the Tawam Molecular Imaging Centre, which is privately owned and unrelated to the previously mentioned Tawam Hospital, became the third facility in the UAE to offer PET scans [17]. The center referred to is presently recognized as the Cleveland Clinic Abu Dhabi—Al Ain. Following the closure of the radiation service at Al Mafraq, and in addition to Tawam Hospital and GICC, the American Hospital in Dubai emerged as the third facility to offer radiation treatment in the UAE. Subsequently, the government-funded Sheikh Khalifa Specialty Hospital in Ras Al Khaimah, which opened in February 2015, became the fourth provider of radiation therapy in the country. Notably, it was the first and only radiation facility available in the

Northern Emirates. In 2016, Mediclinic City Hospital became the fifth provider of radiation therapy in the UAE, introducing the country's first stereotactic body radiotherapy (SBRT). Advanced Oncology Cancer Center, a specialized private cancer center in Dubai, commenced offering radiation therapy services in 2019. Other notable private oncology hospitals in the UAE include Al Zahra Hospital and Mediclinic City Hospital in Dubai, as well as Zulekha Hospitals in Dubai and Sharjah [18]. Several private hospitals have established radiation facilities in recent years. These include Burjeel Medical City in Abu Dhabi, which opened its facility in February 2021; Mediclinic Abu Dhabi in 2022; Saudi-German Hospital Dubai in 2021; and Neurospinal Hospital Dubai in the summer of 2020.

Burjeel Medical City, which opened in October 2020, became one of the main cancer centers in the UAE with unique services like the Brainlab© radiation facility and adult and pediatric bone marrow transplantation [BMT]. Burjeel Medical City is the only center in the UAE that provides adult and pediatric allogenic transplantation. The center has completed more than 55 BMTs for both adults and pediatrics in its first 15 months of operation. Burjeel Medical City is the most publishing clinical cancer research center in the UAE.

The primary cancer center catering to the Northern Emirates region is the Sheikh Khalifa Specialty Hospital in Ras Al Khaimah, offering a semi-comprehensive range of cancer services. Sharjah University Hospital, on the other hand, provides cancer screening and treatment, including surgery and chemotherapy, but lacks a radiation treatment facility.

The Al Jalila Foundation, an esteemed member of the Sheikh Mohammed Bin Rashid Al Maktoum Global Initiatives, is in the process of establishing a groundbreaking cancer charity hospital in Dubai called the "Hamdan Bin Rashid Cancer Hospital." This institution aims to unite top-notch experts to oversee the prevention, diagnosis, and treatment of cancer within a single facility. Named after the late Sheikh Hamdan Bin Rashid Al Maktoum, the hospital will provide a comprehensive range of services, including outpatient, ambulatory, and diagnostic care, as well as inpatient and surgical treatments. Patient well-being will be prioritized in an environment known for its personalized and compassionate approach. The hospital will extend its services to patients from all across the UAE, offering medical treatments that are either free or highly subsidized to alleviate the financial burden faced by individuals unable to afford quality healthcare. The Al Jalila Foundation has committed an investment of AED 1.2 billion towards this innovative project, which will be the region's first fully modular-built hospital. Anticipated to open in 2026, this all-in-one cancer care facility will encompass prevention, diagnosis, and treatment services, with the capacity to serve up to 30,000 patients annually [19].

1.5 Oncology Manpower in the UAE

In 2018, the UAE had a total of 66 registered medical and radiation oncologists. The Department of Health (DOH) had the highest number of oncologists, with 34 medical and radiation oncologists. The Dubai Health Authority (DHA) had 26 oncologists, while the Ministry of Health and Prevention (MOHAP) had 6 oncologists [20]. In the UAE, the ratio of oncologists to the population is 0.6 per 100,000 people. This figure is relatively low when compared to developed nations such as Canada, where there were 1.6 oncologists per 100,000 individuals in 2016, and the United States, which had 4 oncologists per 100,000 population in the same year [21]. In 2010, Switzerland had a rate of 3.3 oncologists per 100,000 population, while the United Kingdom had a rate of 3.6 oncologists per 100,000 population [22]. The rate of oncologists per 100,000 in the UAE is higher than in other developing countries; for example, Turkey has 0.4 [22] and India has 0.0001 [23], most likely due to the large populations of these countries.

The UAE represents a diverse and multicultural society, with over 190 nationalities residing within its borders. Physicians seeking licensure in the UAE undergo a meticulous evaluation process before obtaining their license. Medical oncologists, radiation oncologists, and surgical oncologists originate from various backgrounds, including Europe, North America, South Asia (Pakistan and India), South America, Australia, Africa, and more. This diversity presents a challenge in delivering standardized, high-quality cancer care.

1.6 Cancer Registry in the UAE

In the UAE, there are an estimated 4500 new cases of cancer reported each year. However, accurately describing and reporting cancer incidence in the country requires addressing the challenges and limitations posed by the current fragmented and multiple tumor registries. The first tumor registry in the UAE was established in 1983 as a hospital-based registry at Tawam Hospital, which serves as the official cancer tertiary referral hospital in the country. Mr. Antony D. R. Beal, a radiation physicist, spearheaded the establishment of this registry. It contained valuable information regarding cancer occurrences in the UAE, and the data were presented at the inaugural UAE Cancer Congress in 1985 [14]. The tumor registry at Tawam Hospital did not include data on cases treated at other healthcare facilities across the country, excluding patients who had received cancer treatment at hospitals like Al Mafraq Hospital. In these hospitals, cancer data was being reported to the Ministry of Health and Prevention (MOHAP) [14]. Under the supervision of the Ministry of Health and Prevention (MOHAP), the first official report on cancer incidence was published in 2002 by Dr. Falah Al-Khatib and colleagues at Tawam Hospital [24]. The collected data from the UAE was included in the regional Gulf Countries Cancer Registry, which was centered in Riyadh at King Faisal Hospital. This registry encompassed all the countries in the Gulf Cooperation Council (GCC). The findings from this collective effort were published in a paper titled "Cancer incidence for common cancers in Gulf Cooperation Council countries during 1998–2001" [25].

In 2011, the Department of Health (DOH) released its inaugural cancer statistics specifically for the Emirate of Abu Dhabi, which included the data from Tawam Hospital [26]. The available data at that time only presented information on cancer mortality, lacking any data on cancer incidence. According to the report, cancer was responsible for 14% of all deaths in the Emirate in 2011, with lung cancer accounting for 14.8% of male mortality and breast cancer contributing to 27% of female mortality. In 2012, the Department of Health (DOH) initiated the Abu Dhabi Central Cancer Registry (ADCCR) with the objective of monitoring cancer incidence rates, assessing the effectiveness of cancer screening programs, identifying risk factors for cancer, and planning interventions to control and prevent cancer. The first comprehensive data on cancer incidence in the Emirate of Abu Dhabi was published in 2012. The report indicated a total of 1729 cancer cases that year, with 28% of cases involving UAE citizens and 72% involving expatriates. Among males, hematological malignancies were the most common cancer (29%), while breast cancer was the most prevalent among females (26%). The cancer mortality rate was recorded at 12.9% [27].

Dubai Hospital has been collecting tumor registry data since 2001; however, this data has not been published separately. Instead, the data is reported annually to the Ministry of Health and Prevention (MOHAP). In 2014, the MOHAP National Cancer Registry was established as the official entity responsible for collecting cancer data from hospitals across the United Arab Emirates. It became mandatory for both private and government hospitals to report all malignant cases to the registry.

The Ministry of Health and Prevention (MOHAP) endeavored to establish comprehensive and accurate national disease registries that would provide access to medical information across the entire country while ensuring the confidentiality of data. As part of this initiative, the United Arab Emirates National Cancer Registry was established under the authority of MOHAP, following the directive of the UAE Cabinet and the Minister of Health and Prevention. The National Cancer Registry operates as a population-based cancer registry for the United Arab Emirates. Its primary purpose is to systematically gather, store, summarize, analyze, and disseminate information concerning individuals diagnosed and/or treated for cancer within the UAE. The registry plays a crucial role in monitoring cancer incidence, identifying patterns and trends over time, and offering insights into the prevalence of cancer in the country. This data is invaluable for planning cancer services, implementing cancer screening programs, and conducting cancer research initiatives. The Cancer Registry operates as part of the National Disease Registries and operates under the jurisdiction of the Statistics and Research Center. Annually, the UAE National Cancer Registry produces a report on cancer incidence and mortality, which is available both online and in print, starting in 2014. The registry's ultimate goal is to uncover disease trends and facilitate the study of disease distribution in different regions of the country [28].

The registry follows the guidelines recommended by the International Agency for Research on Cancer (IARC) to collect data on malignant neoplasms. Data is sourced from various channels, including:

(a) The central cancer registry at the Department of Health (DOH): This registry, known for its high level of expertise, serves as a centralized repository for cancer data in Abu Dhabi.
(b) The central cancer registry at the Dubai Health Authority (DHA): Similar to the DOH registry, this centrally located registry in the DHA is also highly qualified.
(c) Hospital admissions and medical records departments: Data is obtained from these departments in public, private, and university hospitals throughout the UAE. The data is classified using the International Classification of Diseases and the International Classification of Diseases for Oncology.
(d) Notifications by medical professionals: Information is received through notifications made by healthcare professionals.
(e) Reports from pathology laboratories: Data is collected from reports generated by pathology laboratories.
(f) Mandatory reporting: Starting in 2013, various sources, such as mortality data, medical treatment abroad, and other notifications, were made compulsory for reporting.

By leveraging these multiple sources, the registry ensures a comprehensive collection of data on cancer cases in the UAE, following the established guidelines set forth by the IARC.

According to GLOBOCAN, the projected growth of all malignancies in the UAE is expected to increase from 4.81k in 2020 to 15.9K by the year 2040 [29]. The substantial projected increase mentioned might be an overestimation and could exceed the actual rise in cancer incidence by 2040 (Fig. 1.2).

Within the UAE, there are numerous prospects for enhancing cancer care. In this publication, we have identified some of these opportunities, addressed challenges, and presented practical solutions. Our recommendations align with the UAE government's vision of reducing cancer mortality rates and providing optimal healthcare for the population. Throughout the book, we extensively discuss these recommendations, covering areas such as enhancing data collection and cancer reporting practices throughout the UAE, as well as promoting and facilitating cancer research specific to the UAE.

We propose a unified, multidisciplinary approach to clinical decision-making, based on evidence-based medicine. Additionally, we advocate for improving existing cancer screening programs to instill greater trust in the general public and increase screening uptake. The UAE Cancer Care Task Force remains committed to closely collaborating with health authorities and policymakers to advance cancer care across the UAE.

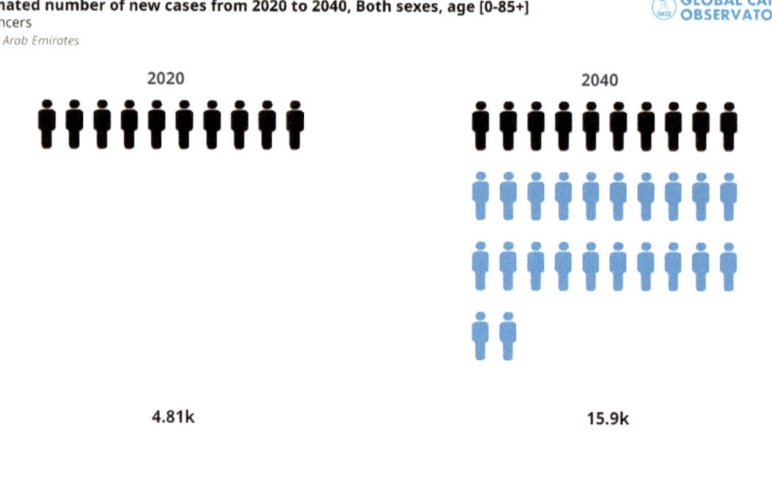

Fig. 1.2 Estimated number of new cases from 2020 to 2040, both sexes and age [0–85+] (Source: Used with permission from International Agency of Research on Cancer (IARC)/World Health Organization (WHO))

Table 1.1 Number of population in the UAE by gender, 2010–2020

المجموع	إناث	ذكور	السنة
Total	Female	Male	Year
82,64,070	21,02,250	61,61,820	2010
83,94,019	21,49,356	62,44,663	2011
85,26,425	21,97,598	63,28,827	2012
86,61,345	22,47,008	64,14,337	2013
87,98,841	22,97,617	65,01,224	2014
89,38,974	23,49,456	65,89,518	2015
91,21,167	28,22,873	62,98,294	2016
93,04,277	28,88,335	64,15,942	2017
93,66,828	30,69,166	62,97,662	2018
95,03,738	32,01,014	63,02,724	2019
92,82,410	28,13,950	64,68,460	2020
			المصدر: المركز الاتحادي للتنافسية والإحصاء

Source: Federal Competitiveness and Statistics Centre

Conflicts of Interest The author has no conflict of interest to declare.

Appendix

See Tables 1.1 and 1.2.

Table 1.2 Population and demographic indicators, 2020

Item	المعدل Rate	البيان
Sex ratio	229.9	نسبة النوع
Life expectancy at birth (in years)	79.7	لعمر المتوقع عند الميلاد (بالسنوات)
Life expectancy at birth for males (in years)	78.0	العمر المتوقع عند الميلاد للذكور (بالسنوات)
Life expectancy at birth for females (in years)	81.4	(العمر المتوقع عند الميلاد للإناث (بالسنوات)
Median age (in years)	32.8	العمر الوسيط (بالسنوات)
Median age for males (in years)	33.5	العمر الوسيط للذكور (بالسنوات)
Median age for females (in years)	31.2	العمر الوسيط للإناث (بالسنوات)
Age dependency ratio	20.0	معدل الاعالة العمرية
Children dependency ratio	17.9	معدل إعالة الصغار
Elderly dependency ratio	2.1	معدل إعالة كبار السن
Population density (per km^2)	130.7	(الكثافة السكانية (لكل كم^2)
Average annual population growth rate, 2010–2020	1.2%	متوسط معدل النمو السكاني السنوي 2010–2020
Average annual population growth rate, 2015–2020	0.8%	متوسط معدل النمو السكاني السنوي 2015–2020
Average annual population growth rate, 2019–2020	−2.3%	متوسط معدل النمو السكاني السنوي 2019–2020
		المصدر: المركز الاتحادي للتنافسية والإحصاء

Source: Federal Competitiveness and Statistics Centre

References

1. Blair I, Sharif AA. Population structure and the burden of disease in the United Arab Emirates. J Epidemiol Glob Health. 2012;2:61–71.
2. United Arab Emirates National Bureau of Statistics 2013.
3. Magazine Tf. The rising Northern Emirates, special report. 2016. pp. 31–32.
4. John I. UAE's per capita GDP stays high. Khaleej Times. 2016.
5. Country classification by the Department of Economic and Social Affairs of the United Nations Secretariat. http://www.un.org/en/development/desa/policy/wesp/wesp_current/2014wesp_country_classification.pdf. Accessed 7 Feb 2023.
6. https://www.who.int/news/item/07-02-2000-world-health-organization-assesses-the-world's-health-systems.
7. Hamidi S, Shaban S, Mahate AA, et al. Health insurance reform and the development of health insurance plans: the case of the Emirates of Abu Dhabi, UAE. J Health Care Finance. 2014;40:47–66.
8. Health Statistics. Health Authority Abu Dhabi. 2011. https://www.haad.ae/HAAD/LinkClick. aspx?fileticket=c-lGoRRszqc=. Accessed 26 Feb 2023.
9. The official portal of the UAE government. Cancer. https://government.ae/en/information-and-services/health-and-fitness/special-health-issues/cancer-. Accessed 26 Mar 2023.
10. Babar ZUD. Pharmaceutical policy in countries with developing healthcare systems. In: Babar ZUD, editor. Pharmaceutical policy in countries with developing healthcare systems. Springer. 2017. pp. 369–370.
11. Health authorities. The offical portal of UAE governemnt. https://government.ae/en/information-and-services/health-and-fitness/health-authorities. Accessed 7 Jan 2023.
12. General background of HAAD. By Health Authority—Abu Dhabi (HAAD). https://www.haad. ae/HAAD/LinkClick.aspx?fileticket=PjQdcYm0zT8%3D&tabid=790. Accessed 7 Jan 2023.

13. About us. By The Dubai Health Authority (DHA). https://www.dha.gov.ae/en/Pages/AboutUs. aspx. Accessed 7 Jan 2023.
14. Tadmouri GO, Al-Sharhan M. Cancers in the United Arab Emirates, genetic disorders in the Arab World. pp. 59–61.
15. https://gulfnews.com/uae/health/741-bed-hospital-to-open-in-abu-dhabi-1.66698698. Accessed 18 Feb 2023.
16. Gulf International Cancer Center. http://www.gulficc.com/about.html. Accessed 26 Sep 2023.
17. http://www.mubadala.com/en/news/mubadala-healthcare-officially-opens-tawam-molecular-imaging-centre. Accessed 26 Feb 2023.
18. https://azhd.ae/services-2/cancer-and-oncology-center-in-dubai/. Accessed 15 Feb 2023.
19. https://www.aljalilafoundation.ae/what-we-do/treatment/hamdan-bin-rashid-cancer-hospital/ (Accessed on 08 Apr 2024).
20. Ministry of Health UAE wide physician census 2016. 2016.
21. American Society of Clinical Oncology. The state of cancer care in America, 2017: a report by the American Society of Clinical Oncology. J Oncol Pract. 2017;13:e353–94.
22. Physicians by medical speciality. https://knoema.com/pnilhuf/physicians-by-medical-speciality?tsId=1002380. Accessed 7 Jan 2023.
23. India has 1.8 mn cancer patients but only one oncologist to treat every 2,000. http://www.business-standard.com/article/current-affairs/india-has-1-8-mn-cancer-patients-but-only-one-oncologist-to-treat-every-2-000-114052401140_1.html. Accessed 7 Jan 2023.
24. UAE Ministry of Health. Cancer incidence report: UAE (1998–2002). Abu Dhabi: UAE Ministry of Health. 2002.
25. Al-Hamdan N, Ravichandran K, Al-Sayyad J, et al. Incidence of cancer in gulf cooperation council countries, 1998-2001. East Mediterr Health J. 2009;15:600–11.
26. Health Statistics. Department of Health 2011. Abu Dhabi: Health Authority. 2011.
27. Health Statistics. Department of Health Abu Dhabi. 2012.
28. MCCR-01-344 Cancer Notification Policy—United Arab Emirates Ministry of Health Central Cancer Registry (MCCR). 2014.
29. GLOBOCAN database (March 2023). https://gco.iarc.fr/today/home. GLOBOCAN database (September 2018). https://gco.iarc.fr/today/home. Accessed 22 Mar 2023.

Prof. Humaid Obaid Al-Shamsi is the Chief Executive Officer of Burjeel Cancer Institute in Abu Dhabi, UAE, President of the Emirates Oncology Society, Lead of the Gulf Cancer Society, Full Professor of Oncology at the Ras Al Khaimah Medical and Health Sciences University, Ras Al Khaimah, UAE, and an Adjunct Professor of Oncology at the College of Medicine, University of Sharjah. He is the first Emirati to be promoted as a professor in oncology in the UAE. He is also the Chairman for Colorectal Cancer in the MENA region, appointed by the prestigious National Comprehensive Cancer Network®. He is also the only member of Lung Cancer Policy Network in the MENA region that aims to advance lung cancer research and screening globally. He is the Chairman of the Oncology and Hematology Fellowship Training Program for the National Institute for Health Specialties in the United Arab Emirates. He is the only member in GCC in the WIN Consortium which is comprised of organizations representing all stakeholders in personalized cancer medicine globally.

He is board-certified in both internal medicine and oncology from the UK, USA (ABIM), and Canada (FRCPC). He has also been awarded the FRCP (London) in 2023 and FRCP (Glasgow) in 2024. He is the only physician in the UAE with a subspecialty fellowship certification and training in gastrointestinal oncology and the first Emirati to train and complete a clinical post-doctoral fellowship in palliative care. He was an assistant professor at the University of Texas MD Anderson Cancer Center between 2014 and 2017. He has published more than 140 peer-reviewed articles in JAMA Oncology, Lancet Oncology, The Oncologist, BMC Cancer, and many others. His area of expertise includes precision oncology and cancer care in the UAE. In 2016, he published with his group from MD Anderson the JCO paper describing a new distinct subgroup of CRC, NON V600 BRAF-mutated CRC. In 2022, he published the first book about cancer research in the UAE and also the first book about cancer in the Arab world, both of which were launched at Dubai Expo 2020. *Cancer in the Arab World* has been downloaded more than 450,000 times in its first 18 months of publication and is the ultimate source of cancer data in the Arab region. He also published the first comprehensive book about cancer care in the UAE which is the first book in UAE history to document the cancer care in the UAE with many topics addressed for the first time, e.g., neuroendocrine tumors in the UAE. He is passionate about advancing cancer care in the UAE and the GCC and has made significant contributions to cancer awareness and early detection for the public using social media platforms. He is considered as the most followed oncologist in the world with over 300,000 subscribers across his social media platforms (Instagram, Twitter, LinkedIn, and TikTok). In 2022, he was awarded the prestigious Feigenbaum Leadership Excellence Award from Sheikh Hamdan Smart University for his exceptional leadership and research and the Sharjah Award for Volunteering. He was also named the Researcher of the Year in the UAE in 2020 and 2021 by the Emirates Oncology Society.

In May 2024, HH Sheikh Mansour bin Zayed Al Nahyan, Vice President of the United Arab Emirates, awarded him the first place in UAE Nafis program for outstanding leadership in private sector across all business and medical disciplines. Beside his clinical and administrative duties, he is engaged in education and various levels of research training for medical trainees to enhance their clinical and research skills. His mission is to advance cancer care in the UAE and the MENA region and make cancer care accessible to everyone in need around the globe.

Cancer Care in the UAE

2

Humaid O. Al-Shamsi ⓘ and Amin M. Abyad

2.1 Introduction

Cancer remains a significant health concern in the United Arab Emirates (UAE), resulting in substantial illness and loss of life. It represents approximately 8.2% of all fatalities in the UAE, positioning it as the fifth leading cause of death in 2021 [1]. The UAE has made a dedicated commitment to decrease cancer-related deaths by approximately 30% before the year 2030. The reduction of cancer mortality is a significant benchmark within the UAE national plan, aligning with the objective of achieving a "Pillar of World-Class Healthcare".

There is a lack of comprehensive data regarding the state of cancer care in the UAE. This data is crucial for recognizing deficiencies and enhancing the provision of cancer care in the country [2–4]. This review offers the most extensive and up-to-date information on cancer care, addressing various previously unexplored subjects such as psycho-oncology, onco-fertility, oncology medical tourism, cancer education and training, precision oncology cancer

H. O. Al-Shamsi (✉)
Burjeel Cancer Institute, Burjeel Medical City, Burjeel Holdings, Abu Dhabi, United Arab Emirates

Ras Al Khaimah Medical and Health Sciences University, Ras Al Khaimah, United Arab Emirates

Gulf Medical University, Ajman, United Arab Emirates

Emirates Oncology Society, Emirates Medical Association, Dubai, United Arab Emirates

College of Medicine, University of Sharjah, Sharjah, United Arab Emirates

Gulf Cancer Society, Alsafa, Kuwait
e-mail: alshamsi@burjeel.com; humaid.al-shamsi@medportal.ca

A. M. Abyad
Emirates Oncology Society, Emirates Medical Association, Dubai, United Arab Emirates

Burjeel Medical City, Abu Dhabi, United Arab Emirates

© The Author(s) 2024
H. O. Al-Shamsi (ed.), *Cancer Care in the United Arab Emirates*,
https://doi.org/10.1007/978-981-99-6794-0_2

research in the UAE, cancer survivorship, oncology nursing, cancer support programs, and more. In conclusion, the review will provide recommendations to different stakeholders, including regulators and policymakers, payers, and the UAE oncology community, regarding the provision and future planning of top-notch cancer care.

2.2 History of Cancer Care in the UAE

We have presented a concise summary of the historical development of cancer care in the UAE, which can be traced back to 1981, when the inaugural specialized cancer center, Tawam Hospital in Al Ain, was established [2]. Over the past decade, there has been a consistent and notable advancement in cancer care in the UAE. Key achievements during this period include the integration of cutting-edge technologies like cyberknife radiation and the initiation of hematopoietic stem cell transplantation (HSCT) programs [5, 6], along with a significant rise in the number of cancer centers. Figure 2.1 summarizes the UAE oncology landscape outline.

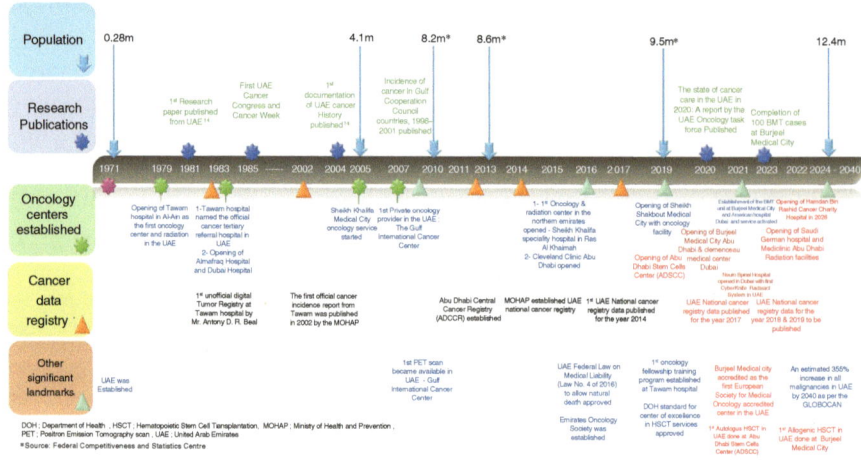

Fig. 2.1 UAE oncology landscape timeline as of July 2024

2.3   Cancer Incidence in the UAE in 2021

According to the UAE National Cancer Registry (UAE-NCR), a total of 5830 new cases of cancer, including both malignant and in-situ cases, were reported during the period from January 1st to December 31st 2021 [1]. Out of the total reported cases, 5612 (96%) were invasive cancers, while 218 (4%) were classified as in situ cases (as shown in Appendix 1). Women were more affected by cancer, accounting for 3210 (55.1%) cases, compared to 2620 (44.9%) cases in males. These statistics include both UAE citizens and expatriates. Among UAE citizens, there were 1493 new cases reported, with 1431 (95.8%) categorized as malignant and 62 (4.2%) as in situ cases. Similarly, among non-UAE citizens, there were 4337 new cancer cases, with 4181 (96.4%) classified as malignant and 156 (3.6%) as in situ cases. The crude incidence rate for both genders combined was 60.5 per 100,000. Notably, a higher incidence of cancer was observed in females compared to males. The crude incidence rate for females was significantly higher at 108.7 per 100,000, whereas for males, it was 39.5 per 100,000. The overall age-standardized incidence rate (ASR) for both genders was recorded as 107.8 per 100,000. Breast, thyroid, colorectal, leukemia, and skin were identified as the most prevalent cancers among both males and females, as indicated in Table 2.1. Among males, the top-ranked cancers were colorectal, prostate, leukemia, thyroid, and skin, as shown in Table 2.2. On the other hand, among females, the leading cancers were breast, thyroid, colorectal, uterus, and cervix uteri,

Table 2.1 Top ten most common malignant primary sites among UAE population in 2021

Primary site ICD-10	Number of malignant cancer cases 2021	%
Breast	1139	20.3
Thyroid	595	10.6
Colorectal	532	9.5
Leukemia	304	5.4
Skin (Carcinoma)	273	4.9
Prostate	251	4.5
Bronchus and Lung	231	4.1
Non-Hodgkin lymphoma	228	4.1
Uterus	173	3.1
Lip, Oral cavity and pharynx	154	2.7

Source: Ministry of Health and Prevention, Statistics and Research Center, National Disease Registry—UAE National Cancer Registry Report, 2021

Table 2.2 Top ten most common malignant primary sites among females and males in 2021

Primary site	%	Primary site	%
Breast	36.9	Colorectal	12.5
Thyroid	13.8	Prostate	9.8
Colorectal	7.0	Leukemia	8.2
Uterus	5.7	Thyroid	6.8
Cervix uteri	4.6	Skin (Carcinoma)	6.4
Skin (Carcinoma)	3.6	Bronchus and Lung	6.3
Ovary	3.5	Non-Hodgkin lymphoma	5.6
Leukemia	3.1	Lip, Oral cavity and pharynx	4.6
Non-Hodgkin lymphoma	2.8	Kidney & Renal pelvis	4.2
Bronchus and Lung	2.3	Urinary bladder	4.0

Source: Ministry of Health and Prevention, Statistics and Research Center, National Disease Registry—UAE National Cancer Registry Report, 2021

respectively, as expected, as displayed in Table 2.2. In 2021, a total of 154 children aged 0–14 years were diagnosed with new invasive cases of cancer across the UAE, with 45.4% being female and 54.5% being male. These cases accounted for approximately 2.7% of all registered malignant cases. The most frequently encountered cancers in boys and girls were leukemia, brain & CNS, non-Hodgkin's lymphoma, Kidney and Renal pelvis, and bone and articular cartilage. Cancer ranked as the fifth most common cause of death among the UAE population. In 2021, there were 975 deaths from cancer (506 males and 469 females), representing 8.2% of all deaths, irrespective of gender, nationality, or cancer type, as illustrated in Table 2.3. This corresponds to an estimated age-standardized cancer mortality rate of 29.6 deaths per 100,000 for both genders combined. Malignant neoplasm of colon emerged as the leading cause of cancer-related deaths, accounting for an average of 11.9% of annual cancer mortality. Malignant neoplasm of trachea, bronchus & lung ranked as the second most common cause of cancer-related deaths in both genders, while malignant neoplasm of breast stood as the third most common cause of cancer mortality for both males and females, as depicted in Table 2.3 [1]. In our previous report, the young population in the UAE indicated a higher incidence rate of breast and colorectal cancer [98, 99].

Table 2.3 Distribution of malignant cancer deaths by type of cancer in the UAE in 2021

Underlying cause of death	%
Malignant neoplasm of colon	11.49
Malignant neoplasm of trachea, bronchus and lung	9.85
Malignant neoplasm of breast	9.64
Leukemia	4.92
Malignant neoplasm of stomach	4.31
Malignant neoplasm of cervix uteri	1.33
Malignant neoplasm of rectum	1.33

Source: Ministry of Health and Prevention, Statistics and Research Center, National Disease Registry—UAE National Cancer Registry Report, 2021

2.4   Cancer Screening Programs in the UAE

The United Arab Emirates (UAE) has achieved notable advancements in cancer screening programs. In 2014, the Ministry of Health and Prevention (MOHAP), in collaboration with various governmental and private healthcare sectors and international experts, introduced national cancer screening guidelines for breast, cervical, and colorectal cancers. These guidelines were subsequently updated in 2018 and 2023. However, there is a lack of officially published data on the participation rate in cancer screening. Unofficial reports suggest that only around 25% of the eligible population undergoes breast and colorectal cancer screenings, indicating low utilization rates [7–12]. There is a significant requirement for the implementation of a comprehensive nationwide screening program in the UAE, which should encompass breast, colorectal, cervical, lung, and prostate cancers. This program should adopt a unified approach and utilize a call and recall system, replacing the existing screening methods [10, 13, 14]. It is important to investigate the disparity in cancer screening rates between citizens and non-citizens, as the data show that 73% of all cancer cases occur in non-citizens, whereas only 27% of cases are reported in UAE citizens [15]. Consequently, the majority of cancer care resources are currently allocated to non-citizens. It is crucial to address the social obstacles that hinder cancer screening in an appropriate manner [12, 16–18]. The adoption of universally approved blood-based cancer screening tools, such as Epi proColon, should be taken into consideration as a less invasive alternative that has the potential to be more socially acceptable for screening purposes [19]. We should contemplate active involvement in clinical trials for blood-based cancer screening methods within our population [20–22].

2.5   Modifiable Cancer Risk Factors in the UAE Population

2.5.1   Obesity/Fast Food/Smoking/HPV/HBV

The relatively elevated occurrence of obesity among the young population in the UAE aligns with other neighboring Arab nations in the Middle East and is strongly associated with the excessive consumption of unhealthy diets that are high in calories and low in nutritional value [23]. As mentioned earlier, the UAE has been making endeavors to manage modifiable risk factors for cancer, including smoking and various forms of obesity. These factors have been prioritized on the active agenda of the UAE government, with initiatives in place to regulate and control them [2].

Tobacco control stands as a significant focal point for health authorities in the UAE. Following the UAE's endorsement of the WHO Framework Convention on Tobacco Control in November 2005, the Ministry of Health devised an encompassing strategy to combat this widespread issue through the National Tobacco Control Program. Notably, the UAE Ministry of Industry and Advanced Technology (MOIAT) became the first among GCC countries to enforce a compulsory standard, known as UAE.S 5030:2018, governing electronic nicotine products to align them with traditional tobacco products. Additionally, the UAE government implemented a 100% taxation policy on tobacco products.

Due to the prevalence of obesity as a public health issue, the Ministry of Health and Prevention (MOHAP) initiated a national strategy to address childhood obesity. In 2017, an excise tax was implemented throughout the UAE, which was further escalated in 2019. This tax structure included a 50% levy on carbonated and sweetened beverages, a 100% levy on energy drinks, as well as a 100% levy on tobacco products and associated items.

Additionally, we suggest the implementation of calorie labeling for all fast food items and the restriction of advertising for fast food restaurants, as the growing prevalence of such advertising is increasingly noticeable both online and in public spaces [24].

Cervix uteri ranks as the fifth most prevalent cancer among females in the UAE [1]. Due to this fact, the UAE took the lead in the Middle East and North Africa (MENA) region by initiating an extensive human papillomavirus (HPV) vaccination campaign. In 2008, the Abu Dhabi Department of Health (DOH) introduced HPV vaccination for all eligible schoolgirls attending public and private schools. Subsequently, in 2013, the vaccination program was expanded to include all females between the ages of 15 and 26, encompassing both UAE nationals and non-nationals. The Dubai Health Authority (DHA) also advises vaccination for all eligible girls aged 11–12 years [2]. In 2018, the inclusion of HPV vaccination in the national vaccination program was implemented to encompass the entire UAE population. The vaccination campaign aimed to target schoolgirls between the ages of 13 and 14, administering 2 doses, while girls aged 15 and above up to 26 years received 3 doses. As of 2022, the coverage rate for the 2-dose vaccination was reported at 82%

[26]. Our suggestion is to improve public, parental, and adolescent understanding and awareness regarding HPV infection and vaccination. This focus aims to minimize misconceptions and the associated stigma related to HPV vaccination [2, 27].

Since 1991, the Hepatitis B vaccine has been implemented and required for all eligible female citizens in the UAE [28]. Since 2006, it has been mandatory for all expatriates arriving in the UAE for employment to undergo testing for hepatitis B and C infections. If an individual is found to be non-immune to Hepatitis B (HBV), they are required to receive the HBV vaccination [2].

2.6   Established Comprehensive Cancer Center

There are over 30 oncology centers and clinics scattered throughout the UAE. In previous discussions, we have provided a detailed overview of the significant achievements and developments in establishing various cancer centers in the country [2]. A formal definition for a "comprehensive cancer center (CCC)" is currently not available in the UAE. However, the Department of Health (DOH) has recently issued general standards for "centers of excellence" in Abu Dhabi, which encompass various healthcare facilities and are not exclusively applicable to oncology centers [29].

The comprehensive cancer center (CCC) should serve as a convenient and comprehensive destination for fulfilling all cancer care requirements [30]. According to our perspective, for a facility to be recognized as a comprehensive cancer center (CCC), it should offer a range of services, including medical oncology for both adult and pediatric patients, hematology, surgical oncology, radiation oncology, nuclear medicine, and palliative care [30]. Presently, there are four centers that fulfill the criteria to be classified as comprehensive cancer centers (CCCs). However, there are notable cancer centers, such as Dubai Hospital (DH), Saudi German Hospital Dubai, Sheikh Shakhbout Medical City (SSMC), Cleveland Clinic Abu Dhabi (CCAD), and Mediclinic Hospital Abu Dhabi, that do not meet the CCC criteria. A list of these centers and hospitals providing oncology services across the UAE can be found in Appendix 2, arranged alphabetically.

The Al Jalila Foundation, an organization affiliated with Sheikh Mohammed Bin Rashid Al Maktoum Global Initiatives, is currently in the process of establishing the first comprehensive cancer hospital in Dubai, named the "Hamdan Bin Rashid Cancer Hospital." This charitable hospital aims to bring together leading experts and pioneers in the field of cancer care, with a focus on the comprehensive diagnosis and management of various cancer cases. The hospital will offer a range of services, including outpatient ambulatory care, inpatient care, and surgical services. It aims to provide advanced technology and diagnostic services within a compassionate environment that prioritizes personalized patient care. The hospital is named in honor of the late Sheikh Hamdan Bin Rashid Al Maktoum, in recognition of his significant contributions to improving healthcare in the UAE. As a charitable institution, the hospital will accept patients from across the UAE, regardless of their place of residence. The medical services provided will be predominantly free or heavily subsidized, aiming to alleviate the financial burden for patients who are unable to afford high-quality healthcare. The Al Jalila Foundation has invested AED 1.2 billion in constructing the region's first fully modular-built hospital. The planned opening is set for 2026. The hospital aims to provide a comprehensive and integrated approach to cancer care, encompassing prevention, diagnosis, and evidence-based management, with a capacity to treat approximately 30,000 patients annually [31].

2.7   Oncology Manpower in the UAE

According to the author's knowledge, as of August 2022, there were approximately 100 specialized physicians practicing in the field of oncology in the UAE. This includes medical oncologists, malignant hematology physicians, and radiation oncologists. These oncologists are predominantly located in the emirates of Abu Dhabi and Dubai. They come from diverse backgrounds, with a significant number hailing from Arab countries such as Lebanon, Syria, and Jordan, while others originate from the UK, USA, and various other western countries. Many oncologists are choosing to relocate to the UAE due to its reputation for better remuneration and the high quality of life it offers. These professionals have received training from various sources, with a considerable portion holding certifications from Canada and other western countries, including board certifications from the American and UK medical boards. There are also oncologists who have received their training in their home countries, including Syria, Jordan, and Lebanon. The local UAE oncologists represent only a small percentage, approximately 10–12, of the total oncologists practicing in the UAE. Most local oncologists from the UAE have pursued their training in Canada and the United States, with many having completed advanced fellowship programs at renowned institutions like MD Anderson Cancer Centers. There is a need to foster the interest of UAE medical graduates in pursuing careers in oncology by raising awareness about the significance of this specialty among medical students and early-career physicians.

2.8   Advanced Surgical Oncology and Robotic Surgeries

In the UAE, surgical oncology has been a part of medical practice for many years. However, it is noteworthy that a significant number of general surgeons performing oncology procedures have not undergone formal training in surgical oncology. Over the past decade, there has been a gradual increase in the presence of formally trained surgical oncologists in the UAE. Extensive research indicates that a surgeon's training significantly impacts the outcomes of cancer patients. In a comprehensive analysis of 27 studies focusing on surgeon training, specialization, and associated outcomes, it was consistently observed that well-trained and specialized surgical oncologists achieved better surgical outcomes for cancer patients compared to non-specialized general surgeons [32]. In recent times, the UAE has experienced an expansion in the availability of robotic surgery technology, although only a limited number of cancer centers currently provide this advanced surgical modality.

One of the persistent challenges in the field of surgical oncology in the UAE relates to the types of surgeries performed for cancer patients. A conflict of interest and competition, often driven by financial considerations due to involvement in private practices, has led to a situation where surgeons without a background in surgical oncology may hesitate to refer patients to specialized oncology surgeons for surgical procedures. Moreover, surgeries performed by non-specialized surgeons are typically not discussed during multidisciplinary team meetings (MDT) or tumor boards. Consequently, this can result in inappropriate surgical management, particularly in complex cases such as colorectal and breast cancers, where patients may benefit from neoadjuvant therapy to enhance surgical and long-term outcomes.

The regulatory bodies should take steps to tackle this concern by enforcing the requirement for multidisciplinary team (MDT) evaluation and approval of all cancer cases that are planned for surgical interventions. It is crucial that even insurance approval for cancer surgeries is contingent upon MDT discussions, with the exception of emergency surgeries, which should undergo auditing for surgeons who frequently perform cancer emergency procedures. Additionally, the utilization of robotic surgery should be limited to surgeons who have received adequate and appropriate surgical training [33]. This precaution is necessary as some surgeons may have undergone only brief and simplified courses, falsely presenting themselves as proficient robotic surgeons, thereby impacting the appropriate selection of surgery for cancer patients.

2.9   Radiation Therapy in the UAE

Presently, radiation therapy services are available at approximately ten centers located across the UAE. Among these centers, four are situated in Abu Dhabi and Al Ain, five are in Dubai, and one is in Ras Al-Khaimah, serving as the sole radiation therapy center in the Northern Emirates. These facilities are equipped with a total of seven linear accelerators (LINACs), one tomotherapy unit, and two brachytherapy units in the Abu Dhabi/Al Ain region. In Ras Al-Khaimah, there are currently two linear accelerators and one ViewRay MR Linac, while Dubai houses four linear accelerators, one tomotherapy machine, one cyberknife machine, and two brachytherapy units. The LINACs used in the UAE are mostly of recent generations, with Elekta Versa HD™ and Varian—TrueBeam® being the most commonly utilized models. Advanced radiation therapy techniques, including complex intensity modulated radiation therapy (IMRT), volumetric modulated arc therapy (VMAT), and RAPID-ARC assisted by advanced image guidance radiation (IGRT), are employed across all these sites. Cyberknife radiosurgery was introduced by the Neurospinal Hospital in Dubai in 2022 [5]. This year (2022), Burjeel Medical City in the UAE has introduced Novalis' comprehensive system, which includes Brain Lab and Elements software. This advanced tool enables the delivery of highly precise radiotherapy and stereotactic radiosurgery (SRS) for the treatment of various conditions [34]. In the Middle East, Sheikh Khalifa Hospital in Ras Al-Khaimah stands as the sole provider of ViewRay—MRIdian, which is the first Food and Drug Administration (FDA)-cleared MRI-Guided Radiation Therapy system in the region. This innovative system combines the guidance of magnetic resonance imaging with adaptive radiation therapy, allowing for enhanced precision and accuracy in radiation treatments [35].

The field of radiation therapy in the UAE is experiencing rapid growth and achieving notable advancements in terms of technology and treatment outcomes. Additionally, there are plans for at least four more centers to introduce radiation oncology services in the coming months, incorporating advanced technologies like the Gamma knife and artificial adaptive planning LINACS. Furthermore, there are intentions to establish a proton therapy service at one of the centers; however, previous endeavors to introduce proton therapy to the UAE faced obstacles due to high costs and a limited number of patients who require this particular form of treatment [36]. This demonstrates the positive trend of growth and expansion in the number of radiotherapy providers. However, we strongly advocate for implementing stricter regulations and limitations on the establishment of additional radiation centers. This approach is crucial to uphold quality control over existing practices and to preserve the higher quality typically associated with centers that handle a higher volume of cases.

To enhance radiation therapy in the UAE, it is crucial to establish an independent and impartial regulatory body responsible for implementing a comprehensive quality control program across all radiation facilities in the country. Furthermore, a shift in the payment model is necessary, moving away from the "per-fraction" approach and transitioning toward a "per-site" model. This change aims to prevent unnecessary clinical and financial burdens on patients and healthcare systems while ensuring the delivery of high-quality and effective therapy. Adopting an episode-based payment approach, which may involve fewer treatment sessions, is expected to reduce travel time, minimize treatment side effects, decrease the duration of hospital stays, and provide patients with more free time to engage in social activities, ultimately enhancing their overall quality of life [37].

2.10 Hematopoietic Stem Cell Transplantation

Hematopoietic stem cell transplantation (HSCT) is widely recognized as a standard and highly successful treatment approach, offering the potential to cure and even save lives for various benign and malignant hematologic diseases, immune disorders, and solid tumors. According to reports, between 2016 and 2018, a total of 164 pediatric and 161 adult patients who were UAE citizens underwent HSCT outside of their country [6]. Every year, around 200 patients, both citizens and non-citizens, in the UAE are in need of hematopoietic stem cell transplantation (HSCT).

In 2019, the Abu Dhabi Stem Cell Centre established the first hematopoietic stem cell transplantation (HSCT) service in the UAE. From that time until August 2022, they successfully conducted 11 autologous non-cryopreservation HSCTs for low-risk cases. In October 2021, Burjeel Medical City introduced the largest and most comprehensive HSCT unit in the UAE, catering to both adults and pediatrics. This marked the introduction of cryopreserved HSCT in the UAE, with 14 autologous cases utilizing cryopreservation reported. Additionally, they achieved a significant milestone by completing the first pediatric allogeneic HSCT, encompassing five cases. The American Hospital Dubai (AHD) is the third healthcare provider in the UAE offering HSCT services. They initiated their program in December 2021 and have successfully performed nine autologous HSCT cases. It is important to note that all the mentioned providers are private healthcare institutions. Sheikh Shakhbout Medical City (SSMC), a public hospital, has announced plans to commence HSCT services, with an expected launch of their program in late 2022 or early 2023.

Considering the relatively small number of potential HSCT candidates in the UAE each year, it is of utmost importance to establish a national and trustworthy center of excellence for HSCT. This approach would involve consolidating the number of providers and concentrating expertise in a single reference center that serves the entire UAE. In addition, we strongly recommend the establishment of a National Marrow Donor Program in the UAE. This program would facilitate lifesaving transplants for patients who lack a matched related donor within their family, ensuring access to suitable donors and improving the chances of successful transplant outcomes.

At present, CAR-T cell therapy is not available in the UAE. However, Burjeel Medical City has outlined its intentions to become the pioneering center in the country to offer this treatment modality. In the coming 24 months, they have plans to introduce CAR-T cell therapy as part of their expanding gene and cellular therapy programs.

2.11   Gynecology Oncology

In 2021, women's cancers ranked one of the most frequently observed cancers in the UAE. Specifically, there were 173 reported cases of uterus cancer, 141 cases of cervix uteri, and 108 cases of ovarian cancer, totaling 422 cases of women's cancers. This accounts for approximately 7.5% of all malignancies reported in the UAE during that year [1]. It is worth noting that many cases of women's cancers are managed by general gynecologists and general surgeons, as dedicated gynecology oncology units are currently unavailable in the UAE. Furthermore, the UAE has a scarcity of trained gynecology oncology surgeons, with fewer than ten available. There are currently no laws or regulations limiting the role of general surgeons or general gynecologists in the management of the more complicated gynecology oncology cases. As mentioned earlier, the involvement of general surgeons in managing these cases has been demonstrated to result in less than ideal clinical and oncological outcomes [38]. Therefore, we advocate for the establishment of dedicated gynecology oncology units throughout the UAE, with mandatory referrals to these units, and for general surgeons and/or gynecologists to refrain from performing complex gynecology oncology surgeries.

Efforts must be made to effectively address the low participation in cervical cancer screening and promptly identify and tackle the obstacles preventing its uptake. The inclusion of the HPV vaccine in the UAE's national vaccination program for girls is a positive step, and its successful implementation is crucial. Additionally, it is important to note that gynecology oncology fellowship programs are currently lacking in the UAE. Therefore, we strongly recommend the establishment of these programs to provide clinicians with comprehensive and advanced training in this specialized field.

2.12 Pediatric Oncology

Among children aged 5–14 years old, cancer is the second most common cause of mortality, following accidents [39]. The occurrence of pediatric cancers varies significantly across various countries globally, accounting for approximately 0.5–4.6% of all cancer cases. The overall incidence rates worldwide typically range between 50 and 200 cases per one million children [40]. Although childhood cancer has a relatively small number of newly diagnosed cases and cancer-related deaths, it imposes a significant burden of disease.

Based on data from the UAE National Cancer Registry (NCR) [1] in 2021, there were 154 newly diagnosed cases of cancer among children aged 0–14 years in the UAE. This represents approximately 2.74% of all reported malignant cases, with a slight majority of 54.5% being male and 45.4% being females. The data reveal that the highest number of childhood cancer cases occurred in the age group of 0–4 years, accounting for 72 cases or 46.8% of the total. The age group of 10–14 years followed with 43 cases, making up 27.9% of the cases. The age group of 5–9 years had the fewest reported cancer cases among the pediatric population.

The data represent that the most commonly reported cancer was leukemia (42.9%), followed by brain and CNS tumors (14.9%), non-Hodgkin lymphoma (8.4%), Kidney & Renal pelvis (6.5%), and bone and articular cartilage (4.5%) [1] (Table 2.4).

In the UAE, there are seven hospitals offering pediatric hematology-oncology services, with an approximate count of 28–30 specialized physicians in this field.

Table 2.4 Distribution of top five pediatric cancer cases by primary sites in the UAE, 2021

Primary sites ICD-10	Number of cancer cases	%
Leukemia	66	42.9
Brain and CNS	23	14.9
Non-Hodgkin lymphoma	13	8.4
Kidney and Renal pelvis	10	6.5
Bone and articular cartilage	7	4.5

Source: Ministry of Health and Prevention, Statistics and Research Center, National Disease Registry—UAE National Cancer Registry Report, 2021

Table 2.5 List of hospitals providing pediatric oncology in the UAE

Hospital	City	
Burjeel Medical City	Abu Dhabi	Private hospital
Tawam Hospital	Alain	Public hospital
Dubai Hospital	Dubai	Public hospital
American Hospital, Dubai	Dubai	Private hospital
NMC Hospital, Abu Dhabi	Abu Dhabi	Private hospital
Sheikh Khalifa Medical City	Abu Dhabi	Public hospital
Mediclinic City Hospital	Dubai	Private hospital

Public hospitals such as Tawam Hospital, Sheikh Khalifa Medical City (SKMC), and Dubai Hospital are among the facilities providing these services and experiencing the highest influx of patients. It is important to mention that currently, there are no pediatric hematology-oncology services available in the Northern Emirates region (Table 2.5).

Burjeel Medical City, located in Abu Dhabi, is the sole hospital in the UAE offering a pediatric bone marrow transplant (BMT) program. This program has successfully conducted five allogeneic transplants, marking a significant milestone as the first instance of such procedures in the UAE. The first transplant took place in April 2022 [41].

Pediatric hematology and oncology services in the UAE are currently spread across multiple providers, serving a comparatively smaller patient population. To enhance patient outcomes, it is recommended to consolidate and centralize pediatric hematology-oncology services in the UAE [42].

2.13 Palliative and Supportive Care

A report titled "Palliative Care in the United Arab Emirates, a Desperate Need" was published in 2018 [43]. Palliative and supportive care have developed markedly in the UAE over the last few years, starting with only two centers providing palliative care and now including four centers with palliative care services across the country. The palliative care program at Tawam Hospital was established in 2007 with the aim of providing support to oncology patients receiving treatment at the hospital. To this day, it remains the sole palliative care program funded by the government in the UAE. Additionally, the American Hospital in Dubai initiated its own palliative care program in the latter part of 2014 [43]. Mediclinic Hospital, Dubai, started its service in 2019. Recently, Burjeel Medical City (BMC) in Abu Dhabi was the latest hospital to join, and it became the first palliative care program in Abu Dhabi City since March 2020. In May 2022, Dr. Neil A. Nijhawan, the founding director of the palliative care program at Burjeel Medical City in Abu Dhabi, became the first representative of the UAE in the World Health Organization's (WHO) palliative care network and was appointed by the WHO as an expert member of the Eastern Mediterranean Regional Office's (EMRO) Expert Network on Palliative Care.

The Emirates Medical Association (EMA) has recently granted approval for the establishment of the palliative and supportive care working group, which will operate under the Emirates Oncology Society (EOS). The launch of this group is scheduled for late 2022. This significant development aims to increase public awareness of palliative care and foster its implementation throughout the UAE. The working group will actively engage with stakeholders and regulatory bodies to advocate for the specialty. Education and training initiatives will also be central to the group's objectives.

Enhancing the provision of palliative care in the UAE demands a comprehensive and well-structured approach that goes beyond simply adopting the Western model of supportive and palliative care. Our previous publication highlighted recommendations for improving palliative care, and we have further refined and categorized these recommendations as follows [44]:

- Implementation of a nationwide strategy for palliative and supportive care as an integral component of the UAE's cancer control strategy.
- Ensuring the availability of necessary pain and palliative care medications at all healthcare levels, including injectable opioids, morphine pumps, and devices for patients receiving end-of-life care at home.
- Providing essential palliative care training to healthcare professionals who are not specialized in the field, emphasizing pain management and the effective treatment of distressing symptoms. Additionally, incorporating palliative and supportive care into the undergraduate curricula of medical and nursing students.
- Strengthening the capabilities and support for various essential members of the multidisciplinary team involved in palliative care, such as clinical nurse specialists, imams, and chaplains, through training and ongoing development (palliative care clinical governance).
- Regularly reviewing and updating the Allow Natural Death (AND) policy from 2016 to incorporate proactive advance care planning and treatment de-escalation plans [44].

2.14   Cancer Survivorship Program

Cancer survivorship represents a crucial aspect of cancer care that is currently in its early stages of development [45]. It encompasses both short-term and long-term aspects of care, which involve monitoring and anticipating treatment complications, assessing the risk of cancer recurrence, addressing potential increased risks of secondary malignancies, ensuring adherence to recommended adjuvant therapies, and promoting lifestyle modifications such as weight management, increased physical activity, and exercise [45, 46]. Cancer survivorship programs in the country are still in the early stages of development, with only two centers, Tawam Hospital and Burjeel Medical City in Abu Dhabi and Al Ain, offering comprehensive programs for cancer survivors. However, there is a clear need for such programs in other emirates as well. Existing programs should also facilitate knowledge sharing to improve practices and address challenges. It is crucial to raise awareness among stakeholders about the significance of these programs in ensuring seamless care for cancer patients throughout their survivorship journey [47].

2.15 Onco-fertility

With the increasing number of younger adults with cancer, which is considered a global phenomenon [48], as outlined earlier, the UAE is witnessing a rise in the number of individuals diagnosed with cancer during their reproductive years, emphasizing the growing importance of fertility preservation for these young patients [49].

A notable disparity has been identified between international practice guidelines and the current implementation of fertility preservation in Arab countries, including the UAE. Several obstacles hinder the optimal delivery of these services, such as the absence of certain advanced techniques, insufficient physician training or awareness, and the absence of dedicated fertility teams or clinics within cancer centers [50].

The Fakih IVF Fertility Center and Al Ain Fertility Center are the leading institutions in onco-fertility in the UAE. They specialize in cryopreservation techniques, such as freezing ova or embryos, as well as ovarian tissue freezing for female patients. For young male patients with cancer, the available options include sperm freezing and/or testicular tissue freezing. Additionally, various procedures like ovarian transposition, fertility-sparing surgery, and hormonal ovarian suppression are readily accessible in cancer centers throughout the UAE.

Physician awareness and attitudes towards fertility preservation options for young adults with cancer, as well as limited insurance coverage, present significant challenges. Many insurance policies do not cover fertility preservation procedures, which is particularly problematic for expat patients. On the other hand, UAE citizens have insurance coverage for fertility preservation services.

The availability of pre-implantation genetic diagnosis, which involves testing embryos or oocytes for genetic defects prior to implantation, such as BRCA testing for carriers, is currently limited in the country. Patients in need of this procedure are often referred to highly specialized fertility centers abroad.

To enhance access to fertility preservation for cancer patients in the UAE, comprehensive workshops targeting healthcare providers in oncology and involving various stakeholders, such as providers, regulators, and patient advocacy groups, are essential. These workshops should aim to address the needs of all indicated cancer patients, irrespective of their insurance coverage or nationality.

2.16 Psycho-oncology

Patients who have been recently diagnosed with cancer or have experienced disease recurrence face the possibility of developing various emotional challenges, including anxiety, depression, adjustment disorder, and a decrease in self-confidence. Studies suggest that around 50% of cancer patients encounter emotional difficulties to some extent. When these difficulties become severe, they can impede the patient's capacity to effectively manage the impact of cancer, its associated symptoms, and treatment-related complications. Individuals with an increased likelihood of experiencing psychiatric illness and depression are often those diagnosed with advanced cancer, dealing with poorly managed symptoms (particularly pain), having a history of previous mental health issues, and facing additional life stressors simultaneously [51]. Psycho-oncology plays a significant role in addressing the social, behavioral, and psychological well-being of cancer patients. This specialized field focuses on two key psychological aspects of cancer. Firstly, it addresses the psychological responses of patients, their families, and caregivers at every stage of the disease. Secondly, it explores the factors that can impact the disease process, encompassing psychological, behavioral, and social factors [52].

Psycho-oncology is a developing field in the UAE, and currently, there is a lack of specifically trained and certified specialists in this area. However, many psychiatrists and psychologists within cancer centers provide support to patients. To enhance support for cancer patients, there is a growing initiative to establish dedicated psycho-oncology clinics within UAE cancer centers. Important actions to improve psycho-oncology in the UAE include attracting specialized physicians, offering scholarships for psychiatry trainees to pursue psycho-oncology fellowships and advanced training, and collaborating closely with regulators and healthcare providers. These steps are crucial in addressing the existing gap and advancing the field of psycho-oncology throughout the UAE.

2.17 Cancer Support Programs and Support Groups

Cancer support programs aim to provide financial assistance to cancer patients, while support groups offer psychological support and guidance. However, a common challenge with financial support programs is their limited budgetary allocation for each patient, often falling short of covering the full cost of treatment. Exceptions exist for complex cancer cases, with higher coverage options available. One notable example is the weekly TV show "Alam W Amal" on Sharjah TV, which provides coverage of up to 180,000 dirhams per case. Additionally, many pharmaceutical companies offer patient support programs specifically for the cancer drugs they manufacture, which can be beneficial for patients requiring expensive medications. Numerous support programs are available to address the emotional and physical well-being of cancer patients. Details of the current cancer support programs and support groups in the UAE can be found in Table 2.6.

Table 2.6 Cancer support programs and cancer support groups in the UAE

Cancer support programs	Funded by	Location	Service provided
Friends of Cancer Patients (FoCP)	Charity	Sharjah	Financial support
Cancer Patient Support Program (BASMAH)-ISAHD	Dubai government	Dubai	Oncology medication support
Cancer Patient Care Society RAHMA	Charity	Abu Dhabi	Financial support
Emirates Cancer Society (previously known as Moazzara)	Charity	Al Ain	Financial support
UAE access programs – Axios International	Medications support	Dubai	Oncology medication support
Various Charity Organizations in the UAE	Treatment cost coverage (cost allowed per case varies)	Across the UAE	Financial support
Brest Friends	N/A	Dubai	Psychological support
Majlis Al Amal-Al Jalila Foundation	N/A	Dubai	Psychological support For female cancer patients
The Cancer Majlis	N/A	Dubai	Psychological support
Bosom Buddies	N/A	Abu Dhabi	Psychological support

2.18 Genetic Testing and Counseling

Access to genetic counseling services in the UAE is extremely limited, with a lack of dedicated genetic counselors in most centers. As a result, counseling and testing are typically conducted by treating oncologists, leading to significant variation in knowledge, attitude, and skills in this specialized area. Moreover, many health insurance policies exclude genetic testing from their coverage, making it challenging for most patients in the UAE to access these services. The few centers that do provide genetic counseling and testing experience long appointment waiting times. Additionally, many tests are not covered by insurance, and when they are, samples are sent overseas for testing, resulting in lengthy turnaround times of 4–6 weeks. To overcome these barriers, several pharmaceutical companies offer free genetic testing for specific indications, such as BRCA testing for breast cancer patients. Our recommendation is to expand genetic counseling services in comprehensive cancer centers throughout the UAE, facilitated by experienced genetic counselors. Regulators and stakeholders should collaborate to ensure the availability and accessibility of genetic testing for all cancer patients, regardless of their insurance coverage, aligning with international guidelines and best practices in cancer care [53, 54].

2.19   Precision Oncology in the UAE

Precision oncology, which involves the use of advanced molecular profiling of tumors to identify specific genetic changes that can be targeted, is a rapidly evolving discipline that has become an integral part of clinical practice [55]. Targetable biomarkers found in non-small cell lung cancer serve as a compelling illustration of the significance of precision oncology at various stages, including initial diagnosis, treatment, and disease progression. Treating patients with lung cancer without incorporating precision oncology and assessing mutations is now deemed suboptimal [56].

Many oncologists in the UAE make use of widely accessible Next Generation Sequencing (NGS) testing and other precision medicine tools. Various commercial testing programs, including Illumina Inc. foundation medicine and The Guardant360®, are available, but all testing is conducted abroad as no testing is currently performed within the UAE. These tests are generally not covered by insurance providers in the UAE, except for a small percentage of premium private insurance plans, which only benefit a limited number of cancer patients requiring NGS testing. The cost of commercially available NGS testing panels remains a significant barrier for many patients seeking more comprehensive tumor profiling. The UAE would greatly benefit from the availability of a more affordable and reliable testing platform. While some small multinational companies offer more cost-effective NGS testing options in the UAE, further data is necessary to validate their reliability and consider them as viable alternatives.

The Oncotype DX Breast Recurrence Score® Test is a commercially available test in the UAE that examines the activity of 21 genes to predict tumor behavior and chemotherapy response. This test assists clinicians in estimating the risk of breast cancer recurrence in certain early breast cancer cases. Despite being recommended by international guidelines, most insurance companies do not provide coverage for this test. It is often excluded from health insurance policies due to being categorized as a genetic test, even though it is not strictly a genetic test. Similarly, MammaPrint® is a diagnostic test that analyzes 70 key genes associated with breast cancer recurrence. While some insurance companies have recently started covering this test, it is important for insurance coverage to align with international guidelines and recommendations to ensure these tests are covered for all appropriate cases of breast cancer [57].

The lack of insurance coverage poses a significant challenge in the field of precision oncology, particularly for non-FDA-approved drugs targeting specific genetic mutations. These drugs are generally not covered by insurance companies, making it difficult to obtain approval for such cases, although government-funded insurance may provide some level of coverage. In light of this, the molecular tumor board plays a crucial role in providing individualized recommendations for each patient. However, currently, there is no molecular tumor board established in the UAE where complex cases can be discussed and experiences can be shared. Most oncologists rely on their own expertise and institutional tumor boards to make clinical decisions, while some seek guidance from international institutions with specialized molecular tumor boards abroad. The establishment of a UAE-wide molecular tumor board is necessary to enhance precision oncology practice and ultimately improve outcomes for cancer patients.

In May 2021, Burjeel Medical City introduced the first specialized clinic for precision oncology. This clinic focuses on utilizing the latest tools in precision medicine to provide personalized treatment recommendations for patients.

Launched in July 2021, the Emirati Genome Program is a government initiative aimed at sequencing the genetic information of the UAE population. Its objective is to provide personalized and preventive healthcare by analyzing genetic data. The response from the general Emirati population has been predominantly positive, expressing optimism about the potential benefits of this program and the associated biobank for biomedical research [58].

The primary objective of the program is to equip clinicians, healthcare professionals, and decision-makers with accurate population-based data. This information will enable them to offer precise diagnoses, treatment options, and personalized preventive programs that align with the specific genetic characteristics of the population. By analyzing individuals' unique genetic profiles, the program aims to anticipate and prevent genetic diseases more effectively. It also seeks to facilitate the implementation of innovative therapies for rare and chronic conditions, including cancer [59].

Advancing precision oncology in the UAE is a complex endeavor that necessitates collaboration among various stakeholders and the development of national-level policies. This collaborative effort involves healthcare providers, payers, policymakers, and patient advocacy groups. In order to enhance the affordability and effectiveness of precision oncology, it is crucial to establish standardized and interoperable protocols for NGS testing and indications. Adequate infrastructure, funding, data management, and research initiatives must also be in place to support these efforts and improve outcomes for cancer patients in the UAE [60].

2.20 Pathology, Molecular, and Cytogenetics Testing

Prior to the initiation of cancer treatment, it is necessary to examine a diagnostic tumor tissue sample in a pathology laboratory. This evaluation is conducted by a pathologist to confirm the type of malignancy and provide essential prognostic factors that guide the subsequent treatment approach [61]. The quality of pathology reports in the UAE can vary significantly based on the pathologist's background and experience. Many laboratories in the country are accredited by the College of American Pathologists (CAP), which mandates the use of tumor checklists for reporting cancer histopathology. These checklists ensure that a minimum set of essential pathology data is included in a synoptic report. Accredited laboratories are also expected to have measures in place to ensure physician competency and to benchmark performance in report categories such as cytology, cervical smears, and thyroid FNAs. Comprehensive Cancer Centers (CCCs) tend to prioritize standardized and more reliable pathology reporting compared to smaller clinics and peripheral healthcare facilities. The presence of multidisciplinary tumor (MDT) boards, which provide the pathologist with relevant clinical information, can improve performance. International collaborations have published audited data on pathology discrepancies and revisions of primary diagnoses.

In the UAE, there are experts specializing in various subspecialties of histopathology, often grouped within one or two larger organizations. However, the provision of organ-specific expert practices is currently decentralized, with multiple hospitals and laboratories offering these services. The number of organ-specific specialists holding board certification in their subspecialties is limited, primarily to American Boards (such as hematopathology, cytopathology, dermatopathology, and neuropathology), as well as the Royal College UK. One significant unmet need is the availability and accessibility of molecular and cytogenetic testing in the UAE. While there are some exceptions, many advanced tests are typically sent outside of the country to international centers in the USA and Europe. This leads to delays in diagnosis and treatment initiation, sometimes ranging from 2 to 4 weeks in certain cases. To address this, there is a need to establish centralized pathology reporting for complex suspected cancer cases. The use of digital pathology can facilitate second-look reviews from specialized cancer centers abroad. A second pathology reading is often required to confirm malignancies and is expected in accredited laboratories. Incorporating second pathology reviews into cancer treatment programs contributes to higher quality. While dramatic changes in diagnoses are rare (usually around 3%), even subsidiary changes can have clinical significance [62, 63]. Extensive testing, including molecular data, is becoming increasingly necessary for the diagnosis and classification of soft tissue and brain tumors. Integrated reports that provide clinically significant information and tumor classification are essential in these cases. We believe that the advancements in human epidermal growth factor receptor 2 (HER2) testing and its impact on therapy require a systematic and obligatory second review. Inconsistencies in HER2 testing can result in significant errors in clinical management, emphasizing the importance of a thorough and reliable evaluation process [64, 65].

An examination of malpractice data in the United States often highlights breast, thyroid, and melanoma fine needle aspiration (FNA) as the pathology reports that carry the highest risk of legal action. Analyzing national data can help identify areas with high risk and prioritize them for improvement and mitigation through second opinions or expert reviews [64–66].

The following are the recommendations for improving pathology, molecular, and cytogenetic testing:

1. As digital pathology becomes more prevalent, pathologists in the UAE now have access to a broader network of experts, leading to faster report generation. It is recommended to have a second pathologist, preferably with organ-specific expertise or specialization, review the initial diagnosis of malignancy.
2. All high-risk and organ-specific pathologies should undergo review by specialized experts who have received appropriate training. It is essential to establish or identify centers of excellence for specific types of malignancies to ensure that

a sufficient number of cases are handled by specialists in those particular organs. For instance, hematopathologists should handle lymphomas and leukemias, cytopathologists should focus on thyroid FNAs, dermatopathologists should specialize in melanomas, and neuropathologists should specialize in brain tumors [57, 59, 60]. Similarly, a restricted number of central and reference laboratories should provide treatment-specific tests like HER2 FISH and NGS.

3. The Department of Health Abu Dhabi and/or Dubai Health Authority (DHA) should oversee and assess standard quality measures and the competence of pathologists through monitoring and audits.
4. As we enhance our research capabilities, it is important to consider implementing a centralized pathology review process.
5. The UAE should promote the establishment of biobanking facilities to facilitate the collection of data and support future research endeavors.

2.21 Cancer Drug Availability and Cost

Cancer medications, including the latest approved drugs, are widely accessible in the UAE. The approval process in the UAE has been remarkably swift, often granting approval shortly after FDA approval. As an example, the drug "Sotorasib" for lung cancer was swiftly approved in the UAE, making it the second country to do so after the USA [67]. Additional challenges arise when the parent company fails to register certain drugs, leading to a prolonged and costly procurement process lasting 4–6 weeks.

Clinical evidence and real-world data have demonstrated the effectiveness and safety of oncology biosimilar drugs [68]. In response to the rising cost of cancer drugs globally, the utilization of biosimilars has become prevalent to mitigate the escalating expenses. Several biosimilars have been approved and are accessible in the UAE, although their usage varies among cancer centers and physicians. We strongly advocate for the incorporation of approved biosimilar cancer drugs, when suitable, into the cancer center's formulary.

2.22 Oncology Nursing

The scarcity of oncology nurses in the UAE is an ongoing issue, and the COVID-19 pandemic has further intensified this shortage [69]. Similar to other countries, the UAE also faces significant shortages of well-trained and experienced oncology nursing staff, as seen in neighboring countries. Oncology nurses in the UAE primarily come from Jordan, the Philippines, India, and Lebanon. These nurses often rotate between different cancer centers within the UAE due to better salary prospects. In an effort to address this shortage, the UAE has recently eliminated the requirement of 2 years of experience for nursing licensure, aiming to attract more nurses to join the workforce in the country.

The Emirates Oncology Nursing Society (EOHNS) serves as the official organization representing oncology nurses in the UAE. One of its primary focuses is encouraging participation in continuing medical education (CME) activities. The EOHNS aims to foster a sense of community among cancer nurses in the UAE, working together to enhance nursing care for cancer patients and their families. This is achieved through the development of nursing leaders and the promotion of nurses' roles in cancer care, ultimately shaping the future of oncology nursing. Currently, there is a lack of structured training programs for oncology nursing in the UAE. It is recommended to establish a dedicated training program for oncology nursing to address the existing shortage of nurses in this field in the UAE.

The role of a nurse practitioner is not widely recognized or established in the UAE. Implementing this role may present challenges due to current medical practice laws and cultural attitudes among patients and their families. In the UAE, there is a prevailing expectation for direct care from physicians rather than nurses. Additionally, there is an abundance of oncologists in the UAE, which may impact the demand for nurse practitioners compared to countries like the US, where there is a shortage of oncologists and a greater emphasis on the role of nurse practitioners.

The EOHNS has taken the lead in promoting nursing research and evidence-based practice in oncology nursing, aligning with its mission to ensure excellence in cancer care and research for cancer patients in the UAE. The society has actively advocated for oncology nursing research and has dedicated specific tracks in its annual conference for nursing research and evidence-based practice. One of the key focuses is the advancement of the advanced practice role in oncology nursing within the UAE. Furthermore, efforts are being made to enhance oncology nursing research and evidence-based practice through initiatives led by SEHA. SEHA plays a crucial role in defining and addressing research-related issues, attracting skilled research nurses, supporting research activities, and raising the profile of nursing research in Abu Dhabi. The nursing research committee within SEHA has implemented a comprehensive program to facilitate education on evidence-based practice and research processes for registered nurses, with a particular emphasis on oncology nursing research studies.

We suggest promoting the active involvement of oncology nurses in cancer research through the provision of research training programs and incentives for career advancement throughout the UAE.

2.23   Artificial Intelligence and Cancer Care in the UAE

The UAE has embraced the use of artificial intelligence (AI) in cancer care ahead of other countries in the region. In 2016, the IBM Watson oncology program was implemented as a pilot project at Tawam Hospital with the aim of assisting clinicians in their daily management of cancer cases. However, the project was later discontinued after IBM halted the Watson program due to the AI technology not meeting the anticipated outcomes.

Table 2.7 The current AI uses in clinical practice in the UAE

AI technology	Facility	Year	Format	Status
IBM™ Watson Oncology—Pilot	SEHA—Tawam Hospital	2016	Clinical decision in oncology	Suspended
AI-enabled digital mammography system, Lunit INSIGHT MMG lung cancer screening-Coreline-Medical AI solutions	International Radiology Centre—Sharjah Commercial	2021	AI-enabled independent reader for breast cancer screening and lung cancer screening	Active
Prognica Labs	Dubai Commercial	2021	Prognica Labs uses artificial intelligence to detect masses in mammography screenings	Active
Mammography Intelligent Assessment (Mia)™	UAE Commercial	2021	First and only AI-enabled independent reader for breast cancer screening to be commercially available in the UAE	Active
The GI Genius™ intelligent endoscopy module	Sheikh Shakhbout Medical City Abu Dhabi	2021	Is the first-to-market, computer-aided polyp detection system powered by AI	Active
Khalifa University researchers	Research Abu Dhabi	2021	To identify cancer in tissue samples, which could speed up diagnosis and improve outcomes in patients with colorectal cancer	Active
DoH—Abu Dhabi	Research Abu Dhabi	2022	First personalized precision medicine for oncology in collaboration with Mubadala Health, Cleveland Clinic Abu Dhabi, NYU Abu Dhabi, Mohamed bin Zayed University of Artificial Intelligence and G42 Healthcare	Active
Mohamed bin Zayed University of Artificial Intelligence team	Research Abu Dhabi	2022	AI tool for better diagnosis and treatment of pancreatic cancer	Active

At present, several AI platforms are being employed in the UAE, primarily to assist in the diagnosis of cancer through imaging techniques such as breast and lung cancer screening. The current utilization of AI in clinical practice in the UAE is outlined in Table 2.7.

The UAE has made significant contributions to AI and cancer research. A collaborative team of researchers from New York University (NYU) in the United States and NYU Abu Dhabi has successfully developed and presented a novel AI system that aids in the identification of breast cancer using ultrasound images. This system demonstrates accuracy comparable to that of a radiologist and has been created as a decision-support tool for healthcare professionals. The research, published in the journal *Nature*, reveals that the system is designed to assist radiologists in reducing false-positive results and unnecessary biopsy rates, all while maintaining a high level of sensitivity [70, 71]. Results from a team of researchers at Mohamed bin Zayed University for AI have been published, showcasing their progress in the development of machine learning algorithms capable of predicting cancer type classification using multi-omics data [72].

2.24 COVID-19 and Cancer Care in the UAE

The UAE demonstrated exceptional management of the COVID-19 pandemic, achieving low mortality rates and conducting the world's first phase 3 clinical trial for a COVID-19 vaccine. They achieved a remarkable vaccination rate of 100% among eligible subjects and ensured the continued management of acute and chronic medical conditions throughout the pandemic. Furthermore, the UAE played a vital role in providing crucial medical aid, ventilators, testing kits, personal protective equipment (PPE), and supplies to 135 countries worldwide. As a result, the UAE has been recognized as one of the top countries to live in during the pandemic [73].

During the initial phase of the pandemic, a significant number of cancer patients returned to the UAE from other countries, and appropriate arrangements were made to ensure the continuation of their treatment. In contrast to many other countries where cancer services were disrupted, the UAE successfully maintained uninterrupted cancer care. Vital cancer treatment components such as chemotherapy infusion centers, radiation therapy facilities, oncological surgeries, and outpatient oncology clinics continued to operate seamlessly across the entire UAE [74]. The successful continuity of cancer care during the pandemic can be attributed to the effective communication efforts led by the Emirates Oncology Society. Through educational webinars and awareness campaigns, the Society encouraged oncologists and healthcare providers to persist in delivering cancer care services despite the challenges posed by the pandemic [75].

The UAE has played a pivotal role in advancing both cancer care and COVID-19 research and publication. Early on in the pandemic, the country published the first international recommendations for cancer care during this challenging period. These recommendations have gained significant recognition, with over 650 citations to date. In fact, they were honored as the Publication of the Year (2020) by the renowned Oncologist journal in the United States [76, 77]. It ranked among the highly cited articles from the Arab World during the pandemic [78]. Another significant publication was the pioneering study on pre-chemotherapy COVID-19 screening, which gained considerable attention and was featured as a trending article upon its publication in JAMA Oncology [79, 80]. Additional research studies from the UAE were also disseminated in both regional and global scientific journals [81–86].

2.25 Research

We conducted a comprehensive and systematic literature search to locate publications related to breast cancer originating from the UAE. On August 8, 2021, we utilized PubMed and employed the search terms "breast" AND "Cancer* OR Oncol* OR malignant* OR tumor OR tumor" AND "emirates OR UAE." This search yielded a total of 203 journal publications authored by individuals from the UAE, with the earliest publication dating back to 2001. The majority of the publications consisted of basic science/translational studies (45.8%) or observational studies (26.1%), while 40 publications (19.1%) were non-data-driven, such as reviews, consensus statements, and editorials. Furthermore, we identified six clinical trials within this dataset (Fig. 2.2). Notably, among the 163 data-driven publications, only 62 (38%) were conducted within the UAE.

Conversely, the remaining studies were carried out in foreign locations, with a significant portion of the authors holding affiliations both within and outside the UAE. This indicates a substantial level of international collaboration among researchers [87].

In general, the UAE has been steadily progressing in cancer research, although there are still many gaps in evidence. The country's academic institutions and research programs dedicated to cellular and molecular studies have been growing, contributing to the advancement of both basic and translational research in the field of cancer. Initially, observational studies have focused on fundamental

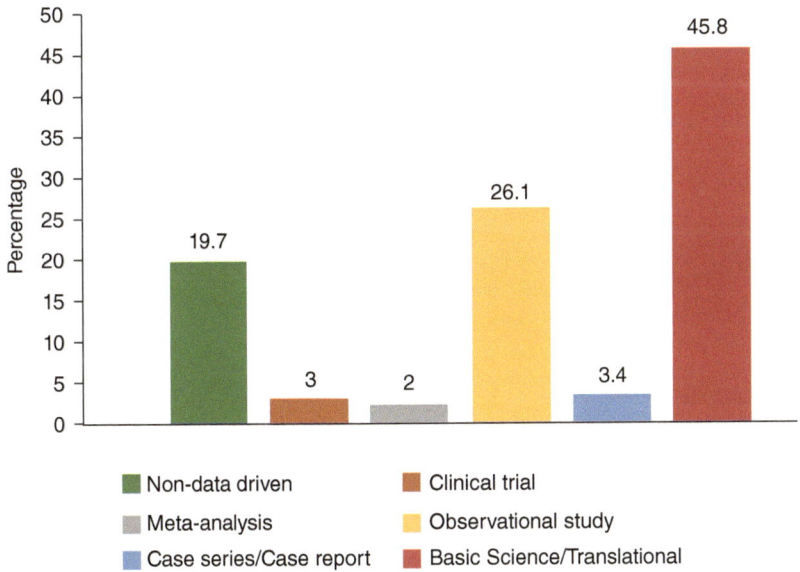

Fig. 2.2 Cancer research output in the UAE by the type of publication over a 20-year period from 2001 to 2021

epidemiological aspects and screening parameters. Moving forward, it is important to prioritize the expansion of national cancer registries and the collection of longitudinal data on clinically relevant variables. This will help improve the molecular and clinical understanding of different types of cancer in the UAE and provide valuable information on survival rates, which can inform therapeutic strategies. However, there is a noticeable lack of therapeutic clinical trials, which is a shared challenge faced by neighboring countries.

Previous efforts to launch large, randomized trials have been limited and faced obstacles, particularly with participant enrollment. This highlights the importance of increasing public and healthcare provider awareness regarding the significance of clinical trials in providing access to cutting-edge cancer treatments.

Given the progressive and adaptable resource and regulatory infrastructures in place, ongoing efforts are being made to enhance collaborations with clinical trial sponsors, with the aim of facilitating interventional studies within the country that align with international requirements and standards [88].

The chart presented in Fig. 2.3 provides the ranking of cancer centers according to their engagement in cancer research activities during the year 2021 [89].

Fig. 2.3 The ranking of cancer centers based on their cancer research output in 2021

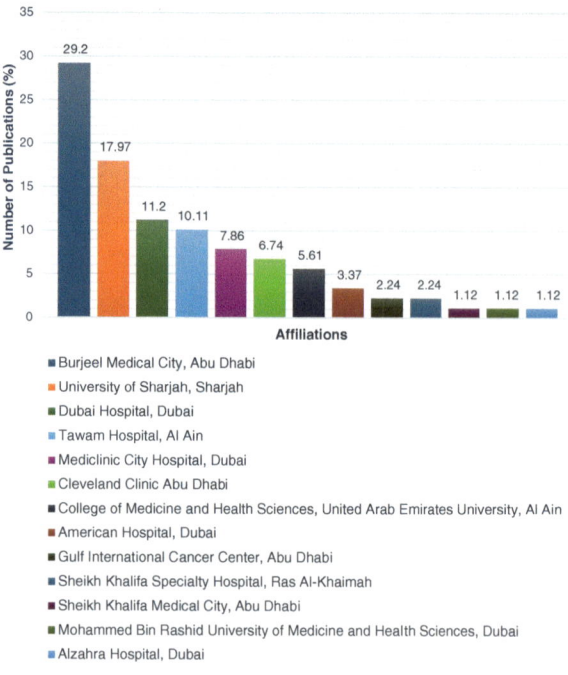

- Burjeel Medical City, Abu Dhabi
- University of Sharjah, Sharjah
- Dubai Hospital, Dubai
- Tawam Hospital, Al Ain
- Mediclinic City Hospital, Dubai
- Cleveland Clinic Abu Dhabi
- College of Medicine and Health Sciences, United Arab Emirates University, Al Ain
- American Hospital, Dubai
- Gulf International Cancer Center, Abu Dhabi
- Sheikh Khalifa Specialty Hospital, Ras Al-Khaimah
- Sheikh Khalifa Medical City, Abu Dhabi
- Mohammed Bin Rashid University of Medicine and Health Sciences, Dubai
- Alzahra Hospital, Dubai

2.26   Education and Training

At present, there is a lack of well-structured and comprehensive fellowship training programs in oncology and hematology in the UAE. The only fellowship program that has received accreditation from the Accreditation Council for Graduate Medical Education-International (ACGME-I) in the UAE is the medical oncology fellowship training program, which commenced in August 2019. Since its inception, three fellows have joined the program, but there have been no graduates announced thus far. The program has a duration of 3 years and focuses solely on medical oncology, excluding hematology. As for hematology, the only fellowship training program in this field was initiated in Dubai in 2020 and currently stands as the sole hematology fellowship training program available in the UAE.

The National Institute for Health Specialties (NIHS) was established as a national institution with the objective of organizing, regulating, and leading the professional development of the healthcare workforce, particularly in advanced specialty training. It operates under the umbrella of the United Arab Emirates University, the country's primary national higher education institute, and is governed by a Board of Directors chaired by the Ministry of Education in the UAE. The NIHS has formed an oncology committee comprising 12 members, currently chaired by Professor Humaid Al-Shamsi. This committee, established in May 2022, is tasked with developing the NIHS Program Requirements for Specialty Education in Hematology and Medical Oncology, which will lead to the Emirati Board in Hematology and Medical Oncology. Graduates of these programs will hold Tier I qualifications and serve as consultants in the UAE.

At present, there is a lack of oncology residency training programs specifically designed for nurses in the UAE. There is a pressing and unfulfilled requirement to establish comprehensive and organized oncology training programs that cater to the needs of both physicians and nurses, given the global scarcity of adequately trained oncology nursing professionals [69].

2.27   Emirates Oncology Society

The Emirates Oncology Society (EOS) serves as the official governing entity that represents oncology healthcare providers in the United Arab Emirates (UAE) [90]. The Emirates Oncology Society (EOS) operates as part of the Emirates Medical Association (EMA), which was founded in 1981. The EMA is a non-profit organization that unites healthcare professionals from different specialties who fulfill the membership requirements outlined in its regulations. Under the purview of the Ministry of Social Affairs, the EMA is headquartered in Dubai and takes on the responsibility of organizing and conducting scientific training programs, conferences, and events. It also collaborates with various healthcare organizations to promote collaboration and advancement in the field of medicine.

The Emirates Oncology Society (EOS) was initially founded in 2016, but it became fully operational in 2020. The society comprises over 80 physicians specializing in various areas of oncology, including medical, surgical, and radiation oncology. Membership in EOS requires physicians to hold an active UAE license in the field of oncology.

In the past 3 years, the Emirates Oncology Society (EOS) has published over 50 scholarly articles, establishing itself as the most prolific scientific society in the country. The society has also organized a wide range of scientific activities, delivering over 200 continuing medical education (CME) hours to healthcare professionals and the general public. One of the notable events organized by EOS is the annual conference held in September. Additionally, EOS has introduced the groundbreaking EOS Annual Award, which recognizes individuals and organizations making significant positive contributions at the national, regional, and global levels. This includes prestigious accolades such as the EOS Lifetime Achievement Awards.

The Emirates Oncology Society (EOS) plays a crucial role as a trusted advisor to healthcare regulatory bodies. Expert members of the society have been providing valuable guidance and counsel to the Ministry of Health and Prevention, UAE, and the Dubai Health Authorities. Their expertise and insights contribute to the development of best practices aimed at improving the quality of cancer care. Furthermore, selected members of EOS actively participate as members of the national committee for cancer control, further enhancing their influence in shaping cancer control strategies at the national level.

2.28 Oncology Continuing Medical Education (CME) in the UAE

Numerous organizations, including the pharmaceutical industry, host a wide range of oncology educational events throughout the UAE year-round. These educational activities aim to enhance knowledge and expertise in the field of oncology. Examples of such events include the International Oncology Conference initiated by VPS Healthcare and the International Oncology Conference at Tawam Hospital, both established in 2011. In Dubai, the Excellence in Oncology Care event was launched in 2010, while the American Hospital in Dubai organized the International Oncology Summit, which began in 2019. As mentioned earlier, the Emirates Oncology Society annual conference was introduced in 2020, further contributing to the educational landscape. Additionally, the pharmaceutical industry plays an active role in delivering continuing medical education (CME) programs, featuring renowned oncologists from leading clinical and research institutions worldwide.

2.29 Cancer Medical Tourism in the UAE

The UAE is a sought-after healthcare destination, particularly for patients from Africa and Iraq. According to the World Health Organization (WHO), the UAE's healthcare system is ranked 27th globally [81, 91]. The UAE government has introduced medical tourism portals that enable international patients to access a comprehensive range of tourism services and schedule medical procedures. These portals provide various services, including visa issuance, appointment booking, accommodation, transportation, and social and recreational activities [82].

The UAE's prime geographical location is one of the key factors that contribute to its appeal as a prominent destination for medical tourism. Additionally, the presence of internationally recognized centers and hospitals, healthcare professionals with extensive training and experience in the US and Europe, and the availability of cutting-edge cancer drugs and treatment approaches not widely accessible in many countries make the UAE an attractive choice for numerous cancer patients. However, the high cost of medical services in the UAE poses a significant challenge, as it tends to be more expensive compared to other countries renowned for cancer medical tourism, such as India. Further endeavors are necessary to enhance the affordability and allure of cancer medical tourism, particularly for specialized treatments like bone marrow transplantations.

2.30   Government-Funded Cancer Care Medical Tourism Outside the UAE

In 2013, the UAE government allocated a budget of more than $163 million US dollars for various purposes, including government-funded cancer care, medical tourism conducted outside the UAE, and medical treatment provided abroad [92]. There is no publicly available official data regarding the specific stages or types of cancer cases treated abroad. The top five destinations for cancer care medical tourism from the UAE are the United States of America, Germany, Singapore, South Korea, and Thailand [93, 94]. A study analyzed administrative data from the Dubai Health Authority (DHA) for UAE nationals who sought medical treatment abroad between 2009 and 2016. The dataset included information from 6557 UAE nationals. The primary medical travel destinations were Germany (46%), the UK (19%), and Thailand (14%). The most common intended medical specialties were orthopedic surgery (13%), oncology (13%), and neurosurgery (10%). After adjusting for other factors, oncology had the highest expected number of patient trips, with an incidence rate ratio (IRR) of 1.34 (95% CI: 1.24–1.44) [94].

Multiple entities, including health authorities such as the Department of Health AD, Dubai Health Authority, and the Ministry of Health and Prevention, as well as Presidential affairs offices, armed forces, police, and charitable organizations, provide sponsorship for cancer care abroad. These entities may offer financial support or assistance for individuals seeking treatment outside of the UAE [2]. The specific requirements and procedures for sponsorship vary among different agencies that provide sponsorship for cancer treatment. One key requirement is that the individual seeking sponsorship must be a UAE citizen. However, there may be exceptions in certain cases for non-UAE citizens when the required treatment is not available in the UAE. Despite the availability of cancer treatment options within the UAE, many patients still choose to seek treatment abroad. Currently, there are no established criteria or guidelines for these entities/agencies to determine the selection of patients for treatment abroad [2].

During an internal evaluation conducted at an oncology center outside the UAE, a review was conducted on 273 patients who sought to travel abroad for treatment between January and September 2017. The findings revealed that 86% of these referrals were deemed unnecessary from a clinical perspective, as comparable oncology services were already accessible within the UAE [2]. This evaluation took place prior to the establishment of multiple bone marrow transplantation programs, which was a significant factor for these cases since many of them required such treatment [6]. Based on our expertise, we believe that over 95% of cancer cases can be appropriately managed and treated within the UAE.

As mentioned in the previous section discussing cancer care medical tourism in the UAE, despite the presence of well-established cancer care centers, there are still factors that drive individuals to seek treatment abroad [95]:

- Patients may reject the initial diagnosis and seek a second opinion to validate the diagnosis.
- Patients and their families may believe that more advanced treatment options for their cancer are available abroad, including new drugs, technologies, and expertise that are not accessible locally.
- Patients and their families may receive generous government funding while receiving treatment abroad, including full paid leave during sickness. In contrast, patients receiving treatment within the UAE do not receive similar flexible sick leave benefits. The same applies to family members, who may be granted companion leave while accompanying the patient abroad, which is not the case if the patient is treated within the UAE.

Seeking cancer care abroad is not a viable and sustainable solution in the long run. It is crucial to conduct targeted research to understand the reasons behind the preference for overseas treatment and identify the barriers that prevent patients from receiving local treatment. By gaining this insight, we can provide recommendations to reduce the unnecessary demand for cancer care abroad. As previously discussed in our earlier report, it is recommended to limit overseas treatment to complex cancer cases that require advanced treatment options not available in the UAE. Building public confidence in local cancer care is of utmost importance, and it requires dedicated efforts such as national outreach programs and active involvement from regulatory bodies and sponsoring agencies involved in facilitating medical travel abroad.

2.31 National Cancer Control Plan

The establishment of the National Cancer Control Committee took place in 2017, accompanied by the launch of a National Cancer Control Plan that aligns with the UAE National Agenda for 2021 and the WHO Cancer Control Plan. This comprehensive plan is built upon key pillars including leadership and governance, prevention and awareness, early detection, capacity building, treatment, palliative care, research, and surveillance. In 2022, the committee and plan were updated based on reliable data obtained from the National Cancer Registry. The vision of the national plan is to reduce cancer mortality and morbidity, while improving the survival rate among the population of the UAE. The mission is focused on saving lives and alleviating suffering by prioritizing cancer prevention, early detection, and providing the best possible curative and palliative care. Currently, efforts are underway to develop a national surveillance monitoring framework in collaboration with the World Health Organization (WHO).

2.32 Integrative Oncology

Integrative oncology refers to a patient-centered approach in cancer care that is supported by evidence and incorporates various mind-body practices, natural products, and lifestyle modifications from diverse traditions alongside conventional cancer treatments [96]. Integrative oncology is an emerging field in the UAE, and currently, no cancer centers in the country offer this service as part of their practice. However, despite the ongoing debates surrounding this approach, some cancer patients from the UAE choose to seek integrative cancer treatment abroad. In the UAE, several centers utilize complementary medicine, specifically hemopathy, in the treatment of cancer patients. To ensure the appropriate use of integrative oncology and prevent its misuse in cancer care, proper control and monitoring measures are necessary. The addition of the integrative oncology specialty to the existing cancer care landscape in the UAE would be a significant development.

2.33 Major Recommendations for the Advancement of Cancer Care in the UAE

In Fig. 2.4, we have highlighted the key recommendations to advance cancer care in the UAE to a higher level.

These initiatives will encompass the following:

- Implementation of a National Cancer Control Plan (CCP): A well-executed and comprehensive cancer control plan necessitates precise data collection and analysis, a dependable cancer registry, and regular monitoring and evaluation.
- Establishment of a National Cancer Care Agency: This agency will oversee all aspects of cancer care, from prevention and screening to accurate diagnosis and the establishment of clear management guidelines. Its aim is to ensure that cancer patients throughout the country receive appropriate, safe, evidence-based, and effective care.
- Designation of central comprehensive tertiary oncology referral facilities as Centers of Excellence
- Establishing Centers of Excellence for cancer care that meet rigorous criteria to enhance the outcomes of cancer patients in the UAE.
- Enhancing and facilitating the availability of a comprehensive national cancer registry and improving reporting mechanisms.

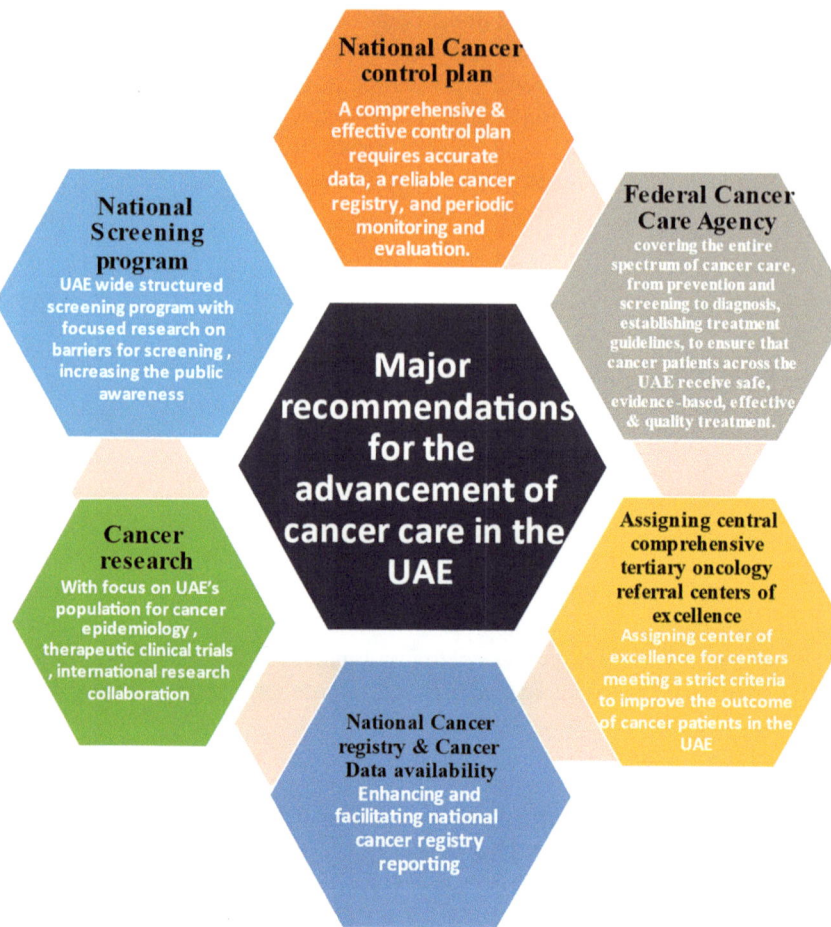

Fig. 2.4 Major recommendations for the advancement of cancer care in the UAE

- Promoting cancer research focused on the UAE's population, including studies on cancer epidemiology, therapeutic clinical trials, and international research collaborations.
- Implementing a well-structured national screening program across the UAE, which includes targeted research to identify barriers to screening and increase public awareness.

2.34   Conclusion

In this review, our objective was to provide an overview of the current state of cancer care in the UAE. We highlighted the rapid advancements in cancer care within the country, featuring cutting-edge cancer centers. However, to further improve cancer care, it is crucial to implement quality control measures under the guidance of regulatory bodies throughout the UAE. The establishment of a federal cancer agency is recommended to ensure specialized oversight and governance of cancer care. It is imperative to enhance the national UAE cancer control program, focusing on refining early detection, cancer screening, and facilitating timely referrals to cancer networks. Education and training programs are necessary to address the global shortage of oncology healthcare providers, particularly oncology nurses. Additionally, the activation of survivorship programs that address onco-fertility, psycho-oncology, and precision oncology and foster UAE-specific cancer research through regional and international collaborations is essential for the next phase of cancer care in the UAE.

Conflicts of Interest   The authors have no conflict of interest to declare.

Appendix

Appendix 1 Number of Cancer Cases Among the UAE Population According to Primary Site, Gender, and Nationality in 2021

Primary site ICD-10	Non-Emirati			Emirati			Grand Total
	Female	Male	Total	Female	Male	Total	Total
(C00-C96) All invasive cancers (malignant cases)	**2237**	**1944**	**4181**	**822**	**609**	**1431**	**5612**
C00-C14 lip, oral cavity and pharynx	27	97	124	10	20	30	154
C15 esophagus	3	15	18	5	4	9	27
C16 stomach	25	79	104	14	16	30	134
C17 small intestine	6	14	20	2	4	6	26
C18–C21 colorectal	132	240	372	81	79	160	532
C22 liver and intrahepatic bile ducts	22	62	84	11	19	30	114
C23, C24 gallbladder, other and unspecified part of biliary tract	15	19	34	8	4	12	46
C25 pancreas	29	50	79	12	19	31	110
C26 other and ill-defined digestive organs	1	5	6	1	1	2	8
C30, C31 nasal cavity, middle ear, accessory sinuses	2	9	11		1	1	12
C32 larynx		17	17		12	12	29
C34 bronchus and lung	53	118	171	17	43	60	231
C37 thymus	3	4	7		3	3	10
C38 heart, mediastinum, and pleura		6	6				6
C40–C41 bone and articular cartilage	2	20	22	5	7	12	34
C43 skin melanoma	18	30	48	2	1	3	51
C44 skin (carcinoma)	94	155	249	15	9	24	273
C45 mesothelioma	4	1	5	1		1	6
C46 Kaposi sarcoma		2	2		1	1	3
C48 retroperitoneum and peritoneum	5	7	12	5	1	6	18

Primary site ICD-10	Non-Emirati			Emirati			Grand Total
	Female	Male	Total	Female	Male	Total	
C49 connective and soft tissue	7	23	30	9	8	17	47
C50 breast	915	6	921	213	5	218	1139
C51 vulva	2		2	1		1	3
C52 vagina	2		2	1		1	3
C53 cervix uteri	118		118	23		23	141
C54–C55 uterus	113		113	60		60	173
C56 ovary	85		85	23		23	108
C57 other and unspecified female genital organs	5		5	1		1	6
C58 placenta	4		4				4
C61 prostate		180	180		71	71	251
C62 testis		45	45		15	15	60
C64–C65 kidney and renal pelvis	29	80	109	14	28	42	151
C66, C68 ureter and other urinary organs	1	2	3		2	2	5
C67 urinary bladder	12	65	77	11	38	49	126
C69 eye	3	1	4		1	1	5
C70–C72 brain and CNS	28	77	105	22	22	44	149
C73 thyroid	266	136	402	155	38	193	595
C74–C75 other endocrine glands	4	5	9	1	1	2	11
C76–C80 unknown or unspecified sites	17	21	38	11	12	23	61
C81 Hodgkin's lymphoma	26	33	59	13	19	32	91
C82–C85, C96 Non-Hodgkin lymphoma	53	98	151	32	45	77	228
C88, C90 multiple myeloma	22	45	67	13	9	22	89
C91–C95 leukemia	69	163	232	26	46	72	304
Other hematopoietic malignancies	15	13	28	3	5	8	36
Other malignancy		1	1	1		1	2
(D00-D09) Non-invasive cancers (in-situ cases)	**107**	**49**	**156**	**44**	**18**	**62**	**218**
D01 carcinoma in situ of other and unspecified digestive organs	2	3	5	3	1	4	9
D02 carcinoma in situ of middle are and respiratory system		1	1				1
D03 melanoma in situ	3	4	7				7
D04 carcinoma in situ of skin		1	1	3	1	4	5
D05 carcinoma in situ of breast	45	3	48	20		20	68
D06 carcinoma in situ of cervix uteri	51		51	17		17	68
D07 carcinoma in situ of other and unspecified genital organs	2	9	11	1	2	3	14
D09 carcinoma in situ of other and unspecified sites	4	28	32		14	14	46
Grand total—invasive and non-invasive	**2344**	**1993**	**4337**	**866**	**627**	**1493**	**5830**

Source: Ministry of Health and Prevention, Statistics and Research Center, National Disease Registry—UAE National Cancer Registry Report, 2021

Appendix 2 Cancer Centers and Hospitals in the UAE in Alphabetical Order

Cancer center	Location	Facility available								
		Medical oncology and infusion unit	Acute hematology service	Bone marrow transplantation unit	Pediatric oncology	Radiation	Surgical oncology	Nuclear medicine / PET imaging	Palliative care unit	Research unit
Abu Dhabi Stem Cells Center	Abu Dhabi	No	No	Yes	No	No	No	No	No	No
Advanced Care Oncology Center	Dubai	Yes	No	No	No	No	No	No	No	No
American Hospital Dubai	Dubai	Yes	Yes	Yes	Yes	Yes	Yes	Yes	Yes	No
Al Zahra Hospital Dubai	Dubai	Yes	No	No	No	No	Yes	No	No	No
Aster Hospital	Dubai	Yes	No	No	No	No	Yes	No	No	No
Burjeel Hospital	Abu Dhabi	Yes	No	No	No	No	Yes	No	No	No
Burjeel Hospital for Advanced Surgery Dubai	Dubai	Yes	No	No	No	No	Yes	No	No	No
Burjeel Medical City	Abu Dhabi	Yes	Yes	Yes	Yes	Yes	Yes	Yes	Yes	Yes
Burjeel Day Surgery Center, Al Reem Island	Abu Dhabi	Yes	No	No	No	No	No	No	No	No
Burjeel Royal Hospital	Alain	Yes	No	No	No	No	Yes	No	No	No
Burjeel Specialty Hospital	Sharjah	Yes	No	No	No	No	Yes	No	No	No
Canadian Hospital	Dubai	Yes	No	No	No	No	Yes	No	No	No
Cleveland Clinic Abu Dhabi	Abu Dhabi	Yes	No	No	No	No	Yes	Yesa (Off-site)	No	No
Clemenceau Medical Center	Dubai	Yes	No	No	Yes	No	Yes	Yes	No	Yes
Dubai Hospital	Dubai	Yes	Yes	No	Yes	No	Yes	Yes	No	Yes
Gulf International Cancer Center	Abu Dhabi	Yes	No	No	No	Yes	No	Yes	No	No
Lifeline Hospital VPS	Abu Dhabi	Yes	No	No	No	No	No	No	No	No

Cancer center	Location	Facility available								
		Medical oncology and infusion unit	Acute hematology service	Bone marrow transplantation unit	Pediatric oncology	Radiation oncology	Surgical oncology	Nuclear medicine / PET imaging	Palliative care unit	Research unit
King's College Hospital London, Dubai	Dubai	Yes	Yes	No	No	No	Yes	No	No	No
Medcare hospital	Sharjah	Yes	No	No	No	No	Yes	No	No	No
Mediclinic Airport Road	Abu Dhabi	Yes	Yes	No	No	Yes	Yes	No	No	No
Mediclinic City Hospital	Dubai	Yes	Yes	No	Yes	Yes	Yes	Yes	Yes	Yes
Neuro Spinal Hospital	Dubai	Yes	No	No	No	Yes	Yes	No	No	No
NMC Royal Hospital Sharjah	Sharjah	Yes	No	No	No	No	Yes	No	No	No
NMC Specialty Hospital	Abu Dhabi	Yes	No	No	Yes	No	Yes	No	No	No
Saudi German Hospital Dubai	Dubai	Yes	No	No	No	Yes	Yes	No	No	No
Saudi German Hospital Ajman	Ajman	Yes	No	No	No	No	Yes	No	No	No
Sharjah University Hospital	Sharjah	Yes	No	No	No	No	Yes	No	No	Yes
Sheikh Khalifa Specialty Hospital	Rasal-Khaimah	Yes	No	No	No	Yes	Yes	Yes	No	No
Sheikh Shakhbout Medical City	Abu Dhabi	Yes	Yes	No	No	No	Yes	No	No	No
Tawam Hospital	Alain	Yes	Yes	No	Yes	Yes	Yes	No[a]	Yes	Yes
Yas Clinic	Abu Dhabi	Yes	No	No	No	No	No	No	No	No
Zulekha Hospital Sharjah	Sharjah	Yes	No	No	No	No	Yes	No	No	No

[a] Cleveland Clinic Abu Dhabi's (CCAD) Positron emission tomography (PET) scanner is located in Al Ain city (150 km away from CCAD main campus) near the main building of Tawam Hospital

References

1. Cancer incidence in United Arab Emirates, Annual Report of the UAE - National Cancer Registry, 2021. Statistics and Research Center, Ministry of Health and Prevention.
2. Al-Shamsi H, et al. The state of cancer care in the United Arab Emirates in 2020: challenges and recommendations, a report by the United Arab Emirates Oncology Task Force. Gulf J Oncolog. 2020;1(32):71–87.
3. Abu-Gheida IH, et al. General oncology care in the UAE. In: Al-Shamsi HO, et al., editors. Cancer in the Arab world. Singapore: Springer Singapore; 2022. p. 301–19.
4. Abu-Gheida I, Nijwahan N, Al-Shamsi HO. Oncology care in the UAE. In: Laher I, editor. Handbook of healthcare in the Arab world. Cham: Springer International Publishing; 2020. p. 1–18.
5. Shanbhag NM, et al. Meningioma treated with hypofractionated stereotactic radiotherapy using CyberKnife®: first in the United Arab Emirates. Cureus. 2022;14(2):e21821.
6. Al-Shamsi HO, et al. Establishment of the first comprehensive adult and pediatric hematopoietic stem cell transplant unit in the United Arab Emirates: rising to the challenge. Clin Pract. 2022;12(1):84–90.
7. https://www.doh.gov.ae/en/news/Early-Cancer-Diagnosis-Saves-Lives. Accessed 9 Aug 2022.
8. https://mohap.gov.ae/en/about-us/projects-and-initiatives/itmenan. Accessed 9 Aug 2022.
9. https://www.khaleejtimes.com/health/uae-new-screening-facility-launched-for-early-detection-of-diseases. Accessed 9 Aug 2022.
10. https://www.bayut.com/mybayut/department-of-health-abu-dhabi/. Accessed 9 Aug 2022.
11. Cancer Patient Support Program (BASMAH). https://www.isahd.ae/content/docs/PD%20 PSP%20Program_Basmah%20Guidelines_24-11-2020.pdf. Accessed 6 Aug 2022.
12. Almansoori A, Alzaabi M, Alketbi L. Colorectal cancer screening in ambulatory healthcare service clinics in Abu Dhabi, United Arab Emirates in 2015–2016. BMC Cancer. 2021;21(1):897.
13. Buehler SK, Parsons WL. Effectiveness of a call/recall system in improving compliance with cervical cancer screening: a randomized controlled trial. CMAJ. 1997;157(5):521–6.
14. Abdul Rashid RM, et al. Is the phone call the most effective method for recall in cervical cancer screening?—results from a randomised control trial. Asian Pac J Cancer Prev. 2013;14(10):5901–4.
15. United Arab Emirates Ministry of Health and Prevention. Cancer incidence in United Arab Emirates. Annual report of the UAE National Cancer Registry 2017. https://www.mohap. gov.ae/Files/MOH_OpenData/1585/CANCER%20INCIDENCE%20IN%20UNITED%20 ARAB%20EMIRATES%20ANNUAL%20REPORT%20OF%20THE%20UAE%20-%20 NATIONAL%20CANCER%20REGISTRY%20-%202017.pdf. Accessed 20 May 2022.
16. Sabih WK, et al. Barriers to breast cancer screening and treatment among women in Emirate of Abu Dhabi. Ethn Dis. 2012;22(2):148–54.
17. Khan S, Woolhead G. Perspectives on cervical cancer screening among educated Muslim women in Dubai (the UAE): a qualitative study. BMC Womens Health. 2015;15:90.
18. Al Abdouli L, et al. Colorectal cancer risk awareness and screening uptake among adults in the United Arab Emirates. Asian Pac J Cancer Prev. 2018;19(8):2343–9.
19. Parikh RB, Prasad V. Blood-based screening for colon cancer: a disruptive innovation or simply a disruption? JAMA. 2016;315(23):2519–20.
20. https://www.aacr.org/blog/2016/04/26/fda-approval-epi-pro-colon-colorectal-cancer/. Accessed 9 Aug 2022.
21. https://www.nhs-galleri.org/. Accessed 9 Aug 2022.
22. https://clinicaltrials.gov/ct2/show/NCT04213326. Accessed 9 Aug 2022.
23. Radwan H, Ballout RA. The epidemiology and economic burden of obesity and related cardio-metabolic disorders in the United Arab Emirates: a systematic review and qualitative synthesis. J Obes. 2018;2018:2185942.

24. Alkharfy KM. Food advertisements: to ban or not to ban? Ann Saudi Med. 2011;31(6):567–8.
25. Abdalla M, et al. Sleep-disordered breathing in children and adolescents seeking paediatric dental care in Dubai, UAE. Eur Arch Paediatr Dent. 2022;23:485.
26. Al-Shamsi HO. The state of cancer care in the United Arab Emirates in 2022. Clin Pract. 2022;12(6):955–85. https://doi.org/10.3390/clinpract12060101. PMID: 36547109; PMCID: PMC9777273.
27. Elbarazi I, et al. A content analysis of Arabic and English newspapers before, during, and after the human papillomavirus vaccination campaign in the United Arab Emirates. Front Public Health. 2016;4:176.
28. Al Awaidy, S.T.; Ezzikouri, S. Moving towards hepatitis B elimination in Gulf Health Council states: From commitment to action. J. Infect. Public Health 2020, 13, 221–227
29. https://www.doh.gov.ae/-/media/F27DED3FBBA340A58EA4FD891EE39215.ashx. Accessed 9 Aug 2022.
30. Grosso D, Aljurf M, Gergis U. Building a Comprehensive Cancer Center: overall structure. In: Aljurf M, et al., editors. The Comprehensive Cancer Center: development, integration, and implementation. Cham: Springer International Publishing; 2022. p. 3–13.
31. https://www.aljalilafoundation.ae/what-we-do/treatment/hamdan-bin-rashid-cancer-hospital/. (Accessed on 07 June 2024)
32. Bilimoria KY, et al. Effect of surgeon training, specialization, and experience on outcomes for cancer surgery: a systematic review of the literature. Ann Surg Oncol. 2009;16(7):1799–808.
33. Chen IHA, et al. Evolving robotic surgery training and improving patient safety, with the integration of novel technologies. World J Urol. 2021;39(8):2883–93.
34. https://gulfnews.com/uae/health/burjeel-medical-city-introduces-4d-radiation-radiosurgery-to-treat-cancer-1.88281497. Accessed 8 Aug 2022.
35. https://viewray.com/find-mridian-mri-guided-radiation-therapy/. Accessed 8 Aug 2022.
36. https://www.laingbuissonnews.com/imtj/news-imtj/first-abu-dhabi-proton-centre/. Accessed 9 Aug 2022.
37. Kehoe G, et al. Innovation in payment for radiotherapy: the radiation oncology model. JCO Oncol Pract. 2021;17(12):e786–92.
38. Minig L, Padilla-Iserte P, Zorrero C. The relevance of gynecologic oncologists to provide high-quality of care to women with gynecological cancer. Front Oncol. 2015;5:308.
39. Ward E, et al. Childhood and adolescent cancer statistics, 2014. CA Cancer J Clin. 2014;64(2):83–103.
40. Sminkey L. Cancer: international childhood cancer 2017. [Serial on the internet]. http://www.who.int/cancer/en/.
41. https://www.wam.ae/en/details/1395303041032. Accessed 9 Aug 2022.
42. https://www.cancer.org/cancer/cancer-in-children/how-are-childhood-cancers-treated.html. Accessed 9 Aug 2022.
43. Al-Shamsi HO, Tareen M. Palliative care in the United Arab Emirates, a desperate need. Palliat Med Care. 2018;5(3):1–4. https://doi.org/10.15226/2374-8362/5/3/00161.
44. Nijhawan NA. An update on the state of palliative care development in the United Arab Emirates. Palliat Med Hosp Care Open J. 2022;8(2):27–9. https://doi.org/10.17140/PMHCOJ-8-149.
45. Bodai BI, Tuso P. Breast cancer survivorship: a comprehensive review of long-term medical issues and lifestyle recommendations. Perm J. 2015;19(2):48–79.
46. Nardin S, et al. Breast cancer survivorship, quality of life, and late toxicities. Front Oncol. 2020;10:864.
47. Jefford M, et al. Improved models of care for cancer survivors. Lancet. 2022;399(10334):1551–60.
48. di Martino E, et al. Incidence trends for twelve cancers in younger adults—a rapid review. Br J Cancer. 2022;126(10):1374–86.
49. Oktay K, Harvey BE, Loren AW. Fertility preservation in patients with cancer: ASCO clinical practice guideline update summary. J Oncol Pract. 2018;14(6):381–5.
50. Kassem L, et al. Awareness and practices of Arab oncologists towards oncofertility in young women with cancer. Ecancermedicalscience. 2022;16:1388.

51. Management of psychiatric disorders in patients with cancer. https://www.uptodate.com/contents/management-of-psychiatric-disorders-in-patients-with-cancer?search=Psycho%20Oncology&source=search_result&selectedTitle=1~5&usage_type=default&display_rank=1#H1. Accessed 10 Aug 2022.
52. Psycho-Oncology Journal. https://onlinelibrary.wiley.com/journal/10991611. Accessed 10 Aug 2022.
53. Darabi S, Braxton DR, Homer J, et al. Precision medicine, genetics and genomics in a community cancer clinic. J Clin Oncol. 2020;38(15 suppl):abstr e13511. https://doi.org/10.1200/JCO.2020.38.15_suppl.e13511.
54. Robson ME, et al. American Society of Clinical Oncology policy statement update: genetic and genomic testing for cancer susceptibility. J Clin Oncol. 2015;33(31):3660–7.
55. Schwartzberg L, et al. Precision oncology: who, how, what, when, and when not? Am Soc Clin Oncol Educ Book. 2017;37:160–9.
56. Tafe LJ. Non-small cell lung cancer as a precision oncology paradigm: emerging targets and tumor mutational burden (TMB). Adv Anat Pathol. 2020;27(1):3–10.
57. Andre F, et al. Biomarkers for adjuvant endocrine and chemotherapy in early-stage breast cancer: ASCO guideline update. J Clin Oncol. 2022;40(16):1816–37.
58. https://jscholarship.library.jhu.edu/handle/1774.2/66617. Accessed 9 Aug 2022.
59. https://u.ae/en/information-and-services/health-and-fitness/research-in-the-field-of-health. Accessed 9 Aug 2022.
60. Horgan D, et al. An index of barriers for the implementation of personalised medicine and pharmacogenomics in Europe. Public Health Genomics. 2014;17(5–6):287–98.
61. Anglade F, Milner DA Jr, Brock JE. Can pathology diagnostic services for cancer be stratified and serve global health? Cancer. 2020;126(S10):2431–8.
62. Khan FA, Maqbool S, Elkhoury M, Habermann TM, Dufan T, Siddiqi AH, Han A, Sayed K, Ahmed Y, Kantawala K, Hamza M, Sonal A, Ahmed S, Yang C, Abdelrahman D, AHD-MDT Members, Mayo eTumor Board Members. Impact of international collaboration utilizing E-consult and E-tumor board in the multidisciplinary management of cancer patients: a study from Mayo Clinic Care Network (MCCN) members. J Clin Oncol. 2019;37(15_suppl):e18261.
63. Habermann TM, Khurana A. Analysis and impact of a multidisciplinary lymphoma virtual tumor board. Leuk Lymphoma. 2020;61(14):3351–9.
64. Peck M, et al. Review of diagnostic error in anatomical pathology and the role and value of second opinions in error prevention. J Clin Pathol. 2018;71(11):995–1000.
65. Bailey GE, et al. The value of second-opinion consultation in nongynecologic cytopathology. Am J Clin Pathol. 2022;157(5):724–30.
66. Farooq A, et al. Assessing the value of second opinion pathology review. Int J Qual Health Care. 2021;33(1):mzab032.
67. http://wam.ae/en/details/1395302945069. Accessed 9 Aug 2022.
68. Peeters M, et al. Biosimilars in an era of rising oncology treatment options. Future Oncol. 2021;17(29):3881–92.
69. Young A, Samadi M. The global power of oncology nurses in low- and middle-income countries. Asia Pac J Oncol Nurs. 2022;9(3):131–2.
70. Makino T, et al. Differences between human and machine perception in medical diagnosis. Sci Rep. 2022;12(1):6877.
71. https://www.nyu.edu/about/news-publications/news/2022/april/radiologists%2D%2Dai-systems-show-differences-in-breast-cancer-scree.html. Accessed 24 Oct 2022.
72. https://mbzuai.ac.ae/news/dedicated-to-ai-cancer-solutions/. Accessed 24 Oct 2022.
73. https://www.uae-embassy.org/sites/default/files/inline-files/UAE%20COVID-19%20Global%20Response-3August2021.pdf. Accessed 10 Aug 2022.
74. Richards M, et al. The impact of the COVID-19 pandemic on cancer care. Nat Cancer. 2020;1(6):565–7.
75. https://www.fiercepharma.com/marketing/astrazeneca-debuts-new-normal-same-cancer-awareness-campaign. Accessed 24 Oct 2022.

76. Al-Shamsi HO, et al. A practical approach to the management of cancer patients during the novel coronavirus disease 2019 (COVID-19) pandemic: an international collaborative group. Oncologist. 2020;25(6):e936–45.
77. https://theoncologist.onlinelibrary.wiley.com/journal/1549490x/homepage/anniversaryretrospective. Accessed 9 Aug 2022.
78. Zyoud SEH. The Arab region's contribution to global COVID-19 research: bibliometric and visualization analysis. Glob Health. 2021;17(1):31.
79. Al-Shamsi HO, Coomes EA, Alrawi S. Screening for COVID-19 in asymptomatic patients with cancer in a hospital in the United Arab Emirates. JAMA Oncol. 2020;6(10):1627–8.
80. Al-Shamsi HO, et al. Serial screening for COVID-19 in asymptomatic patients receiving anti-cancer therapy in the United Arab Emirates. JAMA Oncol. 2021;7(1):129–31.
81. Blair I, Sharif A. Health and health systems performance in the United Arab Emirates. World Hosp Health Serv. 2013;49(4):12–7. PMID: 24683809.
82. Abou Ghayda R, et al. The global case fatality rate of coronavirus disease 2019 by continents and national income: A meta-analysis. J Med Virol. 2022;94(6):2402–13.
83. Benbrahim Z, et al. National approaches to managing cancer care: responses of countries in the MENA region to the COVID-19 pandemic. Ecancermedicalscience. 2021;15:1189.
84. Coomes EA, Al-Shamsi HO. Evolution of cancer care in response to the COVID-19 pandemic. Oncologist. 2020;25(9):e1426–7.
85. Tashkandi E, Zeeneldin A. Virtual management of patients with cancer during the COVID-19 pandemic: web-based questionnaire study. J Med Internet Res. 2020;22(6):e19691.
86. Iskanderian RR, et al. Outcomes and impact of a universal COVID-19 screening protocol for asymptomatic oncology patients. Gulf J Oncolog. 2020;1(34):7–12.
87. Lewison G, et al. Cancer research in the 57 Organisation of Islamic Cooperation (OIC) countries, 2008-2017. Ecancermedicalscience. 2020;14:1094.
88. Al-Shamsi HO, Abyad AM, Rafii S. A proposal for a National Cancer Control Plan for the UAE: 2022-2026. Clin Pract. 2022;12(1):118–32.
89. https://eos-uae.com/official-ranking-for-cancer-research-by-institution-in-uae-in-2021-by-eos/. Accessed 10 Aug 2022.
90. https://eos-uae.com/. Accessed 9 Aug 2022.
91. https://www.who.int/news/item/07-02-2000-world-health-organization-assesses-the-world's-health-systems.
92. Al-Marzooq F, et al. Supragingival microbiome alternations as a consequence of smoking different tobacco types and its relation to dental caries. Sci Rep. 2022;12(1):2861.
93. Sahoo S. Rising numbers in outbound medical tourism. http://www.thenational.ae/business/industry-insights/tourism/rising-numbers-in-outbound-medical-tourism. Accessed 26 June 2017.
94. Alnakhi WK, et al. Treatment destinations and visit frequencies for patients seeking medical treatment overseas from the United Arab Emirates: results from Dubai Health Authority reporting during 2009-2016. Trop Dis Travel Med Vaccines. 2019;5:10.
95. Al-Shamsi HO, Al-Hajeili M, Alrawi S. Chasing the cure around the globe: medical tourism for cancer care from developing countries. J Glob Oncol. 2018;4:1–3.
96. Witt CM, et al. A comprehensive definition for integrative oncology. J Natl Cancer Inst Monogr. 2017;2017(52)
97. Al-Shamsi HO, et al. Early onset colorectal cancer in the United Arab Emirates, where do we stand? Acta Sci Med Sci. 2020;4(11):24–7.
98. Al-Shamsi HO, Alrawi S. Breast cancer screening in the United Arab Emirates: is it time to call for a screening at an earlier age? J Cancer Prev Curr Res. 2018;9(3):123–6. https://doi.org/10.15406/jcpcr.2018.09.00334.

Prof. Humaid Obaid Al-Shamsi is the Chief Executive Officer of Burjeel Cancer Institute in Abu Dhabi, UAE, President of the Emirates Oncology Society, Lead of the Gulf Cancer Society, Full Professor of Oncology at the Ras Al Khaimah Medical and Health Sciences University, Ras Al Khaimah, UAE, and an Adjunct Professor of Oncology at the College of Medicine, University of Sharjah. He is the first Emirati to be promoted as a professor in oncology in the UAE. He is also the Chairman for Colorectal Cancer in the MENA region, appointed by the prestigious National Comprehensive Cancer Network®. He is also the only member of Lung Cancer Policy Network in the MENA region that aims to advance lung cancer research and screening globally. He is the Chairman of the Oncology and Hematology Fellowship Training Program for the National Institute for Health Specialties in the United Arab Emirates. He is the only member in GCC in the WIN Consortium which is comprised of organizations representing all stakeholders in personalized cancer medicine globally.

He is board-certified in both internal medicine and oncology from the UK, USA (ABIM), and Canada (FRCPC). He has also been awarded the FRCP (London) in 2023 and FRCP (Glasgow) in 2024. He is the only physician in the UAE with a subspecialty fellowship certification and training in gastrointestinal oncology and the first Emirati to train and complete a clinical post-doctoral fellowship in palliative care. He was an assistant professor at the University of Texas MD Anderson Cancer Center between 2014 and 2017. He has published more than 140 peer-reviewed articles in JAMA Oncology, Lancet Oncology, The Oncologist, BMC Cancer, and many others. His area of expertise includes precision oncology and cancer care in the UAE. In 2016, he published with his group from MD Anderson the JCO paper describing a new distinct subgroup of CRC, NON V600 BRAF-mutated CRC. In 2022, he published the first book about cancer research in the UAE and also the first book about cancer in the Arab world, both of which were launched at Dubai Expo 2020. *Cancer in the Arab World* has been downloaded more than 450,000 times in its first 18 months of publication and is the ultimate source of cancer data in the Arab region. He also published the first comprehensive book about cancer care in the UAE which is the first book in UAE history to document the cancer care in the UAE with many topics addressed for the first time, e.g., neuroendocrine tumors in the UAE. He is passionate about advancing cancer care in the UAE and the GCC and has made significant contributions to cancer awareness and early detection for the public using social media platforms. He is considered as the most followed oncologist in the world with over 300,000 subscribers across his social media platforms (Instagram, Twitter, LinkedIn, and TikTok). In 2022, he was awarded the prestigious Feigenbaum Leadership Excellence Award from Sheikh Hamdan Smart University for his exceptional leadership and research and the Sharjah Award for Volunteering. He was also named the Researcher of the Year in the UAE in 2020 and 2021 by the Emirates Oncology Society.

In May 2024, HH Sheikh Mansour bin Zayed Al Nahyan, Vice President of the United Arab Emirates, awarded him the first place in UAE Nafis program for outstanding leadership in private sector across all business and medical disciplines. Beside his clinical and administrative duties, he is engaged in education and various levels of research training for medical trainees to enhance their clinical and research skills. His mission is to advance cancer care in the UAE and the MENA region and make cancer care accessible to everyone in need around the globe.

Dr. Amin M. Abyad earned his medical degree, Bachelor in Medicine and Surgery (MBChB), from Beirut Arab University. After completing his internship, he joined the Internal Medicine Residency at Makassed Hospital in Beirut, Lebanon, which is affiliated with the American University of Beirut Medical Center (AUBMC). Dr. Amin was appointed as Chief Resident of Internal Medicine (2017–2018). Then Dr. Amin started his fellowship in hematology and medical oncology at Makassed Hospital, where he received intensive training in hematology and medical oncology. Dr. Amin joined Burjeel Medical City in July 2021. Dr. Amin is highly interested in malignant hematology and solid malignancies. He has been highly involved in clinical research, being involved in multiple research projects and publishing in multiple peer-reviewed journals. Dr. Abyad believes in patient-centered care, trying to enhance patient outcomes through the application of the latest evidence-based practice and personalized medicine.

Alya Zaid Harbi, Buthaina Abdulla Bin Belaila,
Wael Shelpai, and Hira Abdul Razzak

3.1 Background

The United Arab Emirates (UAE), a member of the Gulf Cooperation Council (GCC) and the Arab world, boasts an exceptional healthcare system, particularly renowned for its oncology care. This includes unique integration and alliances with the public and private healthcare sectors within the country.

Established with the objective of offering guidance and oversight in the field of healthcare, the Ministry of Health and Prevention (MOHAP) serves as a governmental entity. Our main focus is on fostering the overall health and prosperity of individuals, aiming to facilitate a state of well-being for all. The Statistics and Research Center houses the UAE National Cancer Registry (UAE-NCR), which offers valuable tools and information for cancer surveillance systems. This crucial initiative is dedicated to safeguarding our society from the detrimental impact of cancer by monitoring and addressing its health risks.

Consequently, cancer registration and surveillance form the fundamental pillars of the UAE National Cancer Registry's mission. In this chapter, we provide a concise overview of the UAE-NCR's methodology for cancer registration and surveillance, aimed at monitoring and analyzing cancer-related outcomes. The UAE National Cancer Registry meticulously collects and ensures the quality of cancer data in the United Arab Emirates. This endeavor is regarded as a valuable opportunity to gain insights into the prevailing landscape and is anticipated to drive future advancements and transformations.

A. Z. Harbi · H. A. Razzak
Statistic and Research Center, Dubai, United Arab Emirates

Ministry of Health and Prevention, Dubai, United Arab Emirates
e-mail: alya.harbi@mohap.gov.ae; Hira.AbdulRazzak@mohap.gov.ae

B. A. B. Belaila · W. Shelpai (✉)
Ministry of Health and Prevention, Dubai, United Arab Emirates
e-mail: Buthaina.Abdulla@mohap.gov.ae; Wael.shelpai@mohap.gov.ae

© The Author(s) 2024
H. O. Al-Shamsi (ed.), *Cancer Care in the United Arab Emirates*,
https://doi.org/10.1007/978-981-99-6794-0_3

The present chapter has been collaboratively developed by the UAE National Cancer Registry Section, a division of the Statistics and Research Center under the Ministry of Health and Prevention (MOHAP). The NCD and Mental Health Sections have also provided valuable support throughout the process. The team has actively engaged in discussions and continuous improvement efforts concerning cancer registration activities undertaken by the Statistics and Research Center—UAE National Cancer Registry Section. This collective endeavor aims to prioritize the resolution of critical issues and ensure comprehensive coverage of the pertinent aspects.

3.2   Introduction

Noncommunicable diseases (NCDs) are the primary cause of mortality worldwide, resulting in 41 million fatalities annually, which corresponds to 74% of all global deaths. Each year, approximately 17 million individuals die from NCDs prior to reaching the age of 70, with 85% of these premature deaths taking place in low- and middle-income nations. Among NCD-related deaths, cardiovascular diseases claim the most lives, accounting for 17.9 million fatalities on an annual basis, followed by cancer (9.3 million), chronic respiratory diseases (4.1 million), and diabetes (2.0 million, such as kidney disease fatalities associated with diabetes). Collectively, these four categories contribute to 80% of all premature deaths attributed to NCDs.

The risk of NCD-related mortality is elevated by factors such as physical inactivity, tobacco use, excessive alcohol consumption, and unhealthy eating habits. Essential elements in addressing NCDs include early detection, screening, treatment, and the provision of palliative care [1].

In the fight against the burden imposed by NCDs, nations have pledged their commitment to attain global NCD targets, including Sustainable Development Goal 3.4. This target aims to reduce premature mortality caused by noncommunicable diseases by one-third by the year 2030 [2].

Accurate and up-to-date cancer incidence and mortality data play a critical role in the development and evaluation of cancer control programs. To meet this need, a population-based cancer registry systematically gathers comprehensive cancer information. The collection of high-quality cancer data over time serves as a catalyst for transforming healthcare services, leading to enhanced patient outcomes.

An organization dedicated to storing, collecting, analyzing, and interpreting data on individuals affected by cancer is known as a cancer registry. A population-based cancer registry acquires data from various healthcare providers within a specific geographic area and can be utilized to illustrate trends in cancer occurrence at a variety of locations over time or among different population groups.

Cancer registry data holds the potential to provide valuable information for evaluating the long-term impact of various treatment methods and the effectiveness of early detection initiatives like colorectal screening or mammography. Furthermore, these data can be utilized in epidemiological research aimed at identifying the underlying factors responsible for the development of cancer. Information on cancer incidence and mortality, as well as changing trends, is a significant component in the monitoring and planning of programs for early detection, prevention of cancer, and treatment.

The UAE National Disease Registry (UAE-NDR) is central to public health and healthcare in the UAE. The UAE-NDR at the Ministry of Health and Prevention includes data on all diseases that have a high impact on society, as the UAE-National Cancer Registry (UAE-NCR) is one of the largest, most advanced, and most complex cancer data curation services.

The UAE National Cancer Registry (UAE-NCR) carries out a systematic process of gathering, storing, summarizing, analyzing, and disseminating data concerning individuals diagnosed with cancer and receiving treatment within the UAE. The main objectives of the UAE-NCR are to ascertain cancer statistics on a national level in the UAE, furnish decision-makers and researchers with trustworthy data, monitor the effectiveness of cancer screening and early detection initiatives, and plan for cancer services and cancer control.

The UAE National Cancer Registry (UAE-NCR) is an integral component of the UAE National Disease Registries and operates under the supervision of the Statistics and Research Center at the Ministry of Health and Prevention (MOHAP). UAE-NCR offers comprehensive and reliable data services that encompass the entire cancer journey for all individuals diagnosed with cancer in the UAE. These services ensure the provision of high-quality, carefully validated information.

MOHAP has a qualified cadre that has acquired the international accreditation to register and manage the cancer data according to international standards, like Certified Tumor Registrars (CTRs), clinical coders, biostatisticians, and epidemiologists, as well as the registration data is utilized to generate various outputs, such as official statistics and data tools.

3.3   UAE National Cancer Control Program

The UAE National Cancer Registry annual report relies on the incidence data and cancer mortality statistical data provided by the UAE-NCR as the foundation for all the cited cancer statistics. The crucial role of these surveillance activities, combined with the data gathered through the screening program, the national health survey, and other initiatives, cannot be overstated in the context of cancer control programs.

The UAE National Cancer Control Committee comprises members from key stakeholders involved in cancer control within the UAE. These members are individuals officially appointed through the minister's decision. The committee meets regularly with different stakeholders.

According to the World Health Organization (WHO) guidelines, the UAE National Cancer Control Program supports a national agenda to lessen cancer mortality and incidence, increase the survival and quality of life of cancer patients, and develop strategic methods to minimize or prevent the cancer impact in the UAE.

Utilize data to determine and track the extent of the cancer burden in the UAE, prioritize effective strategies for cancer control, formulate cancer plans, and implement them. Additionally, use this data, in conjunction with screening and other cancer-related measures, to assess and monitor cancer programs. For instance, information extracted from the UAE National Cancer Registry reveals that breast, colorectal, and thyroid cancer are prevalent types of cancer in the UAE, with cancer ranking as the third leading cause of mortality. The UAE National Cancer Control Program has used this information to develop a plan, in collaboration with local health authorities, to educate and empower UAE citizens in the prevention and control of breast and colorectal cancer. As well as developing a UAE lung and prostate cancer screening plan, UAE National Cancer Registry data have been utilized to demonstrate that breast, colorectal, thyroid, prostate, and lung cancer are the most common cancers across the UAE. This information has been used to develop a prostate and lung cancer screening plan in association with local health authorities.

The UAE National Cancer Control Committee is currently reviewing and updating the National Strategy for Cancer Control and its Action Plan, which have existed since 2017.

3.4   UAE National Cancer Registry (UAE-NCR)

Cancer poses a significant health challenge in both developed and developing nations, with the UAE experiencing it as the third most common cause of death.

Cancer registration involves an ongoing, organized procedure that entails collecting, storing, analyzing, interpreting, and reporting data related to the occurrence and attributes of cancer.

Population-based cancer registries serve as a distinctive and invaluable information resource for cancer control programs. These data aid in the provision of financial and human resources for efficient healthcare planning, in addition to the development of prevention and early detection initiatives. Furthermore, cancer registries play a vital role in providing critical data to determine the occurrence and frequency of cancers in a specific population.

The establishment of the Gulf Centre for Cancer Registration (GCCR) aimed to establish a comprehensive database on cancer incidence for the GCC countries, namely the United Arab Emirates (UAE), Bahrain, Saudi Arabia, Oman, Qatar, and Kuwait. The GCCR commenced the collection and accumulation of data for this purpose on January 1, 1998, where the Gulf Center for Cancer Registration is provided with the required data periodically. The main goal of the GCCR is to gather and categorize data on all instances of cancer, enabling the production of statistical information regarding cancer occurrence within a specific population. This initiative serves to offer technical assistance for early detection and screening initiatives as well as facilitate epidemiological research on cancer. Ultimately, the aim is to establish a framework for evaluating and managing the impact of cancer on the communities residing in the GCC states.

In 2012, the Department of Health (DOH) in the Emirate of Abu Dhabi initiated the establishment of the Abu Dhabi Central Cancer Registry, a proficient and centralized cancer registry within the DOH. This registry serves as a comprehensive database encompassing all cancer-related information in Abu Dhabi. Similarly, the Emirate of Dubai has a Dubai Health Authority (DHA) cancer registry and various hospital-based cancer registries.

In 2014, a comprehensive UAE National Cancer Registry (UAE-NCR) was established by the Statistics and Research Center and works under the jurisdiction of the Ministry of Health and Prevention (MOHAP). Its primary responsibilities include assessing the population-based incidence of cancer in the UAE, gauging the magnitude of the cancer burden, conducting epidemiological studies, promoting early detection methods, and implementing cancer screening initiatives.

The UAE-NCR is responsible for collecting, processing, and analyzing complex data on patients who are diagnosed with or treated for the condition of cancer in the UAE. The data are systematically gathered from various sources and entities throughout the UAE, including the Abu Dhabi Central Cancer Registry at the Department of Health Registry, the cancer registry of the Dubai Health Authority, cancer registries based on private hospitals in Dubai, medical facilities in the Northern Emirates, certified records of malignancies from public and private hospitals, pathology laboratory reports, and mortality data. This comprehensive approach ensures that individuals diagnosed with and/or receiving cancer treatment within the UAE are included in the data collection process. Consequently, the data compiled by the UAE-NCR encompasses all individuals in the country who have undergone a cancer diagnosis or treatment. The collected data encompasses various aspects, including demographic information, diagnostic details, cancer type, staging, direct treatment methods, and follow-up data. The UAE-NCR obtains this information through either the medical record concept or electronic reporting at the level of the medical facility. All incident cases from central registries, public and private hospitals, primary healthcare centers (PHC), clinics, pathology laboratories, treatment facilities, and other medical establishments are mandated to report to the UAE-NCR. The central registry plays a vital role in consolidating the data from the cancer registry, creating a unified record for each individual cancer case, and providing a subset of this data to the UAE-NCR on an annual basis. Collaborating with local health authorities, the UAE-NCR works toward ensuring the coherence and comparability of cancer incidence and mortality data.

The data obtained from the UAE-NCR plays a crucial role in shaping cancer care service plans and oncology research programs, as well as the development of future initiatives, including advancements in screening programs. Each year, a report is published based on the data collected by the registry, focusing on in situ and invasive neoplasms, and adhering to international standards. In 2013 and 2020, the Prime Minister's Office issued a circular to all health authorities and medical facilities to report all cancer cases according to an updated form to be submitted to the

UAE-NCR in MOHAP. The UAE-NCR is an associate member of the International Association of Cancer Registries (IACR).

3.5   Coverage of the Cancer Registry

Data for the cancer registry is gathered from both public and private healthcare facilities throughout the UAE, enabling the UAE-NCR to compile individual records that encompass different levels of geographical coverage. This coverage spans from Abu Dhabi to Dubai and extends to the Northern Emirates of the UAE.

3.6   Legal Framework for Cancer Registration and Surveillance

The cancer is notifiable by law, and the notification of cancer cases is done by all healthcare providers—public and private—and other related entities.

In 2013 and 2020, the UAE Prime Minister's Office approved a cancer registration policy to obligate notification of cancer cases by all public and private facilities across the UAE.

3.7   UAE-NCR Staff (Personnel)

At the level of technical expertise, the expertise and technical competence of the registry staff significantly impact the quality of the cancer registry data. MOHAP hired qualified cadres that had acquired international accreditation to register and manage the cancer registry data according to international standards, like CTRs, clinical coders, biostatisticians, data scientists, and epidemiologists.

3.8   Registry Data Linkages

The UAE-NCR also links data with national databases for the purpose of supplementing and improving the quality of the data, and it regularly links cancer data with UAE death databases and other data sources, such as treatment abroad and health insurance claim files.

3.9   Methodology and Standard Operating Procedures

The UAE National Cancer Registry (UAE-NCR) under MOHAP acquires comprehensive demographic, diagnostic, cancer, staging, treatment, and follow-up data for all cancer cases diagnosed and/or treated within the UAE. The collection of this information aligns with internationally recognized registration protocols, utilizing

ICD coding and TNM staging standards. Annually, all invasive and in situ cases diagnosed in public or private medical facilities are notified and registered with the UAE-NCR.

Two data collection methods are available:

3.9.1   Active Method

Representatives from public and private hospitals, as well as NGOs, serve as key contacts responsible for gathering cancer data from patient medical records, the Health Information Management System, and pathology reports. After completing a standardized form, the collected information is submitted to the UAE-NCR.

Meanwhile, the UAE-NCR staff conducts regular visits to the medical treatment abroad department at MOHAP to collect and abstract registry data.

3.9.2   Passive Method

Designated representatives from various stakeholders and medical facilities throughout the UAE gather cancer data from patient files, the Health Information Management System, and pathology reports. They then proceed to fill out a standardized form and submit the collected information to the UAE-NCR.

The Department of Health in Abu Dhabi supplies mortality data specific to Abu Dhabi, while MOHAP provides mortality data for the other Emirates.

In cases where TNM information is available in the patient's medical file, information on the extent of the disease is collected as part of the TNM staging system. The SEER summary stage is abstracted.

The abstraction form (data collection form) is distributed to the main public hospitals that have oncological services.

The unique identification (Emirates ID) number is available and used for the deduplication procedures as well as by comparing the demographic information, such as patient name and date of birth, through cancer registry software.

3.10   Reporting Sources and Data Processing

The UAE National Cancer Registry (UAE-NCR) includes the following reporting sources: the DOH central cancer registry, the DHA cancer registry, public and private hospitals, private physician clinics, public and private laboratories, death notifications, and medical treatment abroad.

Ensuring the participation of all public and private hospitals, including both inpatient and outpatient clinics, within the reporting region responsible for diagnosing and/or treating cancer is crucial for maintaining comprehensive reporting. All data provided for this report underwent coding according to ICD-10-CM and

ICD-O-3, which was subsequently converted to ICD-10-CM for analysis and report generation.

To maintain data comparability, all cases reported to the UAE-NCR must adhere to the rules and recommendations set forth by the International Agency for Research on Cancer (IARC).

A thorough examination of all pertinent details pertaining to new cases is conducted to identify potential duplications using a master index. Subsequently, the clinical data is verified by the Cancer Tumor Registrar (CTR) and a team of proficient registry staff.

The collected data serve various purposes, including monitoring incidence trends, facilitating research endeavors, supporting planning initiatives, and evaluating the quality of cancer care facilities.

3.11 Reportable List and International Standards

The registry encompasses all neoplasms classified in the International Classification of Diseases for Oncology, Third Edition (ICD-O-3), specifically those with a behavior code of 2 or 3, as well as malignant and in situ cases in the ICD-10 CM. It is mandatory to report all diagnosed reportable tumors in individuals residing in the UAE by the reference date specified by the registry.

If a term in ICD-O-3 with a behavior code of "0" or "1" is confirmed as in situ ("2") or malignant ("3") by a pathologist, the case is considered reportable. The multiple primary rules are employed to determine the number of primary tumors a patient has, providing guidance and standardization in this process.

Additionally, the histology rules, which contain specific instructions for coding histology, are applied to ensure consistent reporting among all participants.

3.12 Data Items, Data Management, and Quality Control Procedures

Cancer has been designated as a mandatory notifiable disease through a ministerial decree, guaranteeing the opportunity for inclusive data gathering. The UAE-NCR endeavors to achieve unrestricted access to cancer data from all medical establishments, both public and private, across the UAE.

A cancer registry or cancer surveillance system can gather numerous data items, and the UAE-NCR determines the necessary elements for data collection. Each data item on the sample data collection form, as displayed in Table 3.1, is accompanied by a comprehensive description, specific codes, and definitions.

Every data item relating to the patient is collected and updated. The registry registers all new cases of cancer diagnosed in the UAE. Multiple data sources had aided in optimizing collection completeness; however, this could lead to the problem of multiple patient notifications. This problem was addressed by cross-checking the Emirates ID number, names, age, gender, date of birth, and address, which is a good

Table 3.1 UAE National
Cancer Registry Data Items

UAE-NCR data item
Facility name
Facility license number
Facility referred from
Facility referred to
Date case reported
Case reported by
First name
Middle name
Last name
Date of birth
Gender
Nationality
Marital status
Occupation
Emirates ID
Medical record number
Address emirate—current
Address city—current
Smoking status
Alcohol status
Family history of cancer
Basis of diagnosis
Date of first contact
Date of initial diagnosis
Age at diagnosis
Comorbidities code
Comorbidities description
Primary site ICD10 code
Primary site ICD10 description
Primary site ICDO 3
Primary site ICDO 3 description
Histology ICDO-3 codes
Histology description
Behavior code
Behavior description
Sequence number
Grade
Grade description
Laterality code
Laterality description
Date of surgical diagnostic procedure
SEER summary stage
Clinical T
Clinical N
Clinical M
Clinical stage group

Table 3.1 (continued)

AJCC edition
Pathological T
Pathological N
Pathological M
Pathological group
Other staging system—name
Other staging system—value
Date of start of treatment
Surgery (Y/N)
Rx date—surgery
Rx FIN—surgery
Radiotherapy (Y/N)
Rx start date—radiotherapy
Chemotherapy (Y/N)
Rx start date—chemotherapy
Rx protocol—chemotherapy
Hormonal therapy (Y/N)
Rx start date—hormonal therapy
Rx FIN—hormone therapy
Immunotherapy (Y/N)
Rx start date—immunotherapy
Rx FIN—immunotherapy
Hematologic, transplant, and endocrine procedure
Hematologic, transplant, and endocrine procedure date
Rx FIN—hematologic, transplant, and endocrine procedure
Hematologic, transplant, and endocrine procedure type
Systematic/surgery sequence
Other treatment
Other treatment—specify
Other treatment date
Palliative care
Palliative care date
Rx FIN—palliative care
Rx code—palliative
Date of last contact
Date of death
Vital status
Underlying cause of death
Cause of death

Source: Ministry of Health and Preventions (MOHAP), Statistics and Research Center—UAE National Cancer Registry

quality indicator as well as demonstrating good coverage along with the extensiveness of cancer cases in the UAE. The Emirates identification card number serves as a distinct identifier assigned to both UAE citizens and non-UAE citizens individually.

Following a thorough review and filtering of the received cancer data, updates were made to ensure the exclusion of any duplicate or previously registered cases. Every endeavor is made to ensure the completeness of all variables. In cases where information is found to be incomplete, notification forms with insufficient details are returned to the respective data providers for further clarification. Upon completion, the forms are returned to the registry.

All the revised information obtained in electronic format, whether through passive or active means, was entered into both the enterprise data warehouse (EDW) and the disease registry databases. The electronic data preserved in the cancer registry databases is imperiled by ongoing quality control. Quality control procedures and activities involving data checks for accuracy, completeness, consistency, validity, uniqueness, and timeliness are implemented.

Internal consistency and edit checks are performed during data entry into the EDW and disease registry system.

Collaboration between the UAE-NCR and local health authorities is established to guarantee the coherence and consistency of cancer incidence and mortality data. As part of this effort, the annual cancer registry data is published on the MOHAP's website as open data [3]. The website serves as a platform for accessing the official UAE cancer statistics as well as the annual reports of the UAE cancer registry. Moreover, the UAE cancer registry data is designed to be comparable with that of other GCC countries and their international counterparts.

3.13   Improving the Quality of Cancer Incidence and Mortality Data

Several workshops and trainings have been done in collaboration with the WHO Regional Office for the Eastern Mediterranean (WHO-EMRO), the International Agency for Research on Cancer (IARC), and the Union for International Cancer Control (UICC) to raise the quality of cancer incidence and mortality data collected at the hospital level. For example:

The Statistics and Research Center—UAE National Cancer Registry, in the UAE, conducted several workshops and trainings for cancer registrars and countries about cancer registration, cancer data quality, and cancer staging. In addition to that, the department conducted several trainings for physicians about the mechanism of writing the causes of death according to international standards and provided instructions on how to complete and file death certificates.

3.14   UAE National Cancer Registry Software

To alleviate the challenges posed by a multifaceted system for data collection, the UAE-NCR has implemented multiple surveillance and informatics initiatives. These initiatives aim to automate developments and facilitate the electronic exchange of data for cancer and other disease reporting. For example, the development of an enterprise data warehouse (EDW), which is a web-based repository to automate data collection and facilitate data cleaning, analytic data extraction, and writing reports, will improve cancer registration and its quality on a short-term basis.

3.15   Dissemination and Use of Data and Reports from the UAE National Cancer Registry

Ensuring the confidentiality of data is a key priority for the UAE-NCR, as it plays a central role in releasing data for clinical purposes, research, and healthcare planning. The registry has established robust procedures for data release that guarantee the preservation of confidentiality.

Unless mandated by law or with the explicit written consent of the healthcare provider or facility, no identifying information regarding an individual healthcare provider or facility will be disclosed. Furthermore, specific patient information will not be furnished to individuals (patients) unless otherwise stipulated by law.

All requests for data should be directed to the Statistics and Research Director, the UAE-NCR manager, or another designated member of the registry staff who is authorized to respond. The Statistics and Research Center has a data request form available for researchers, registry staff, and other individuals to utilize.

This form serves as internal documentation for data requests, ensuring the documentation of all information inquiries, aiding in staff effort monitoring, and facilitating the preparation of periodic summary reports on data requests.

To request the release of statistical cancer registry data, individuals can submit formal requests via email to SARC.Request@mohap.gov.ae. The data will be prepared for the statistical staff's review. All correspondence and the cancer registry data are carefully documented and filed for reference, facilitating the generation of summary tabulations and routine reports.

The release of UAE-NCR data is contingent upon its utilization solely for medical purposes. These permissible medical purposes include surveillance, clinical audit, cancer service evaluation, and ethically approved research.

It is important to note that any information that can potentially identify an individual will not be disclosed. Comprehensive information regarding all the data released by the registry is made available on MOHAP's website through open data initiatives. This website serves as a platform for hosting the official UAE cancer statistics and annual reports from the UAE cancer registry.

3.16   Confidentiality Procedures and Data Security

Maintaining confidentiality is an obligation upheld by the UAE National Cancer Registry, ensuring that any information that could potentially identify a patient, healthcare professional, or medical facility remains confidential. Stringent confidentiality procedures are implemented and upheld throughout all stages of registry operations, aiming to:

- Safeguard the privacy of each individual patient.
- Ensure the confidentiality of the facilities that report the cases, ensuring their privacy is protected.
- Offer public reassurance that the data will be handled responsibly and will not be subject to misuse.

Access to identifiable data is strictly limited to a selected number of UAE-NCR staff members, and such access is only granted when necessary to uphold data quality or investigate specific incidents. These registry specialists undergo certification, possess extensive training, and work in secure environments to ensure the utmost confidentiality and security of the data.

Whenever feasible, the tasks are conducted using de-identified data, wherein any direct identifiers of individuals are removed. In cases where an analyst necessitates access to patient-identifiable information, such as for record accuracy verification, they must provide a justifiable reason and obtain special permission.

All data collected by UAE-NCR is stored on secure servers within the MOHAP network, and MOHAP is legally obligated to ensure proper storage and usage of information in accordance with the Data Protection Act. A daily backup of the UAE-NCR database is required to be performed at the end of each day.

3.17   Cancer Statistics and Annual Reports in the UAE

The UAE-NCR serves as the sole means to monitor the incidence of cancer and its various types. It provides valuable insights into the annual number of cancer diagnoses, the prevalence of cancer within the population, and survival rates for different types of cancer. By consistently tracking these statistics over time, UAE-NCR helps determine whether the incidence of cancer is on the rise or decline, as well as the overall progress in extending the lifespan of individuals affected by cancer.

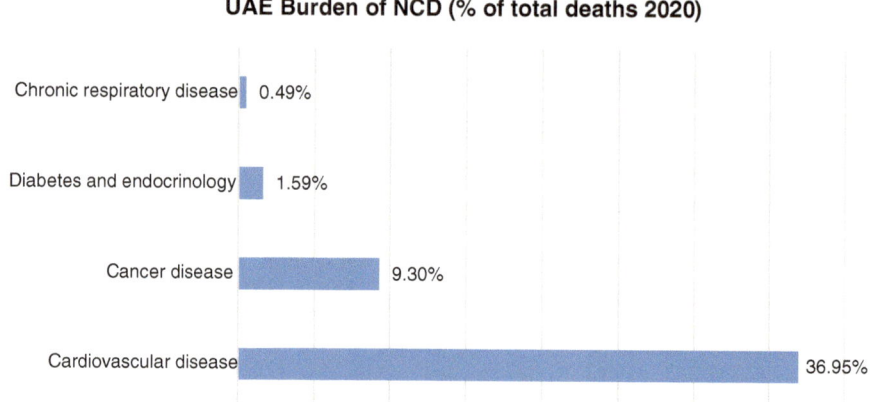

UAE Burden of NCD (% of total deaths 2020)

Chronic respiratory disease — 0.49%

Diabetes and endocrinology — 1.59%

Cancer disease — 9.30%

Cardiovascular disease — 36.95%

Fig. 3.1 UAE's burden of NCDs by MOHAP in 2020 (mortality rate) (Source: Ministry of Health and Prevention (MOHAP), Statistics and Research Center)

All indicators related to cancer are calculated annually, such as age-standardized mortality rates, crude mortality rates, crude and age-standardized incidence rates, and other indicators.

MOHAP shows the UAE burden of NCDs (mortality ratio) in 2020, which shows the percentage of total deaths in 2020 due to cancer was 9.3% (Fig. 3.1).

According to the recent UAE cancer registry annual report 2021, there were a total of 5830 newly diagnosed cancer cases (malignant and in-situ) in the UAE (total population 2020 = 9,282,410). Among these cases, 5612 (96%) were malignant. There were 2553 (45%) males and 3059 (55%) females among them. In males, the most common cancers were colorectal (12.5%), followed by prostate (9.8%), leukemia (8.2%), thyroid (6.8%), and skin (carcinoma) (6.4%). In females, the most common cancers were breast (36.9%), thyroid (13.8%), colorectal (7.0%), uterus (5.7%), and cervix uteri (4.6%) (Table 3.2) [4].

Based on the latest data available from the UAE-NCR, covering the timeframe of January 1 to December 31, 2021, a total of 5830 newly diagnosed cancer cases (malignant and in-situ) were recorded in the UAE. Among these cases, 5612 (96%) were classified as malignant, indicating active cancer, while 218 (4%) were categorized as in situ cases, representing early-stage cancer. The data reveal that cancer was more prevalent among females, with 3210 cases (55.1%) reported, compared to 2620 cases (44.9%) among males. For more detailed information, please refer to Table 3.3 [4].

In the UAE citizens, there were 1493 newly diagnosed cases of cancer (malignant and in-situ), with 1431 (95.8%) classified as malignant and 62 (4.2%) categorized as in situ cases. Among non-UAE citizens, a total of 4337 cancer cases were newly diagnosed, with 4181 (96.4%) being malignant and 156 (3.6%)

Table 3.2 The top ten most prevalent malignant primary sites among both females and males were determined as the highest occurrences in 2021 [4]

Primary site ICD-10 Female	%	Primary site ICD-10 Male	%
C50 Breast	36.9	C18–C21 Colorectal	12.5
C73 Thyroid	13.8	C61 Prostate	9.8
C18–C21 Colorectal	7	C91-C95 Leukemia	8.2
C54–C55 Uterus	5.7	C73 Thyroid	6.8
C53 Cervix uteri	4.6	C44 Skin (Carcinoma)	6.4
C44 Skin (Carcinoma)	3.6	C34 Bronchus and Lung	6.3
C56 Ovary	3.5	C82-C85, C96 Non-Hodgkin lymphoma	5.6
C91-C95 Leukemia	3.1	C00-C14 Lip, Oral cavity & pharynx	4.6
C82-C85, C96 Non-Hodgkin lymphoma	2.8	C64-C65 Kidney & Renal pelvis	4.2
C34 Bronchus and lung	2.3	C67 Urinary bladder	4

Source: Ministry of Health and Preventions (MOHAP), Statistics and Research Center, UAE National Disease Registries—UAE National Cancer Registry Report 2021 [4]

categorized as in situ cases. For more detailed information, please refer to Table 3.3 [4].

On the whole, the crude incidence rate of cancer for both genders in the UAE was 60.5 per 100,000 individuals. The data clearly showed a higher incidence of cancer among females, with a crude incidence rate of 108.7 per 100,000, compared to males, with a rate of 39.5 per 100,000.

Furthermore, the overall age-standardized incidence rate (ASR) for both genders was calculated to be 107.8 per 100,000 individuals, for females 149.4 per 100,000 females, and for males 96.6 per 100,000 males.

Breast, thyroid, colorectal, leukemia, and skin (carcinoma) were identified as the most common cancers among both genders in the UAE. Among males, the top-ranked cancers were colorectal, prostate, leukemia, thyroid, and skin (carcinoma). The leading cancers among females were breast, thyroid, colorectal, uterus, and cervix uteri.

According to IARC's GLOBOCAN [5], in 2022, there were an estimated 5 526 new cancer cases (2 607 in males and 2 919 in females) and 2 283 cancer deaths (1 283 in males and 1 000 in females) over a total population of approximately ten million inhabitants. Females are more affected because of the high burden of breast cancer (39.1%), which ranks first, followed by the thyroid (11.5%) and the colorectum (7.0%). The three most frequent cancer sites for males are the colorectum (13.3%), prostate (11.9%), and bladder (7.3%).

Table 3.3 Number of cancer cases among the UAE population based on primary site, nationality, and gender in 2021

Primary site ICD-10	Non-Emirati			Emirati			Grand total
	Female	Male	Total	Female	Male	Total	
(C00–C96) all invasive cancers (malignant cases)	**2237**	**1944**	**4181**	**822**	**609**	**1431**	**5612**
C00–C14 lip, oral cavity & pharynx	27	97	124	10	20	30	154
C15 esophagus	3	15	18	5	4	9	27
C16 stomach	25	79	104	14	16	30	134
C17 small intestine	6	14	20	2	4	6	26
C18–C21 colorectal	132	240	372	81	79	160	532
C22 liver and intrahepatic bile ducts	22	62	84	11	19	30	114
C23, C24 gallbladder, other and unspecified part of biliary tract	15	19	34	8	4	12	46
C25 pancreas	29	50	79	12	19	31	110
C26 other and ill-defined digestive organs	1	5	6	1	1	2	8
C30, C31 nasal cavity, middle ear, accessory sinuses	2	9	11		1	1	12
C32 larynx		17	17		12	12	29
C34 bronchus and Lung	53	118	171	17	43	60	231
C37 thymus	3	4	7		3	3	10
C38 heart, mediastinum, and pleura		6	6				6
C40–C41 bone and articular cartilage	2	20	22	5	7	12	34
C43 skin melanoma	18	30	48	2	1	3	51
C44 skin (Carcinoma)	94	155	249	15	9	24	273
C45 mesothelioma	4	1	5	1		1	6
C46 Kaposi sarcoma		2	2		1	1	3
C48 retroperitoneum and peritoneum	5	7	12	5	1	6	18
C49 connective and soft tissue	7	23	30	9	8	17	47
C50 breast	915	6	921	213	5	218	1139
C51 vulva	2		2	1		1	3
C52 vagina	2		2	1		1	3

Primary site ICD-10	Non-Emirati			Emirati			Grand total
	Female	Male	Total	Female	Male	Total	
C53 cervix uteri	118		118	23		23	141
C54–C55 uterus	113		113	60		60	173
C56 ovary	85		85	23		23	108
C57 other and unspecified female genital organs	5		5	1		1	6
C58 placenta	4		4				4
C61 prostate		180	180		71	71	251
C62 testis		45	45		15	15	60
C64–C65 kidney & renal pelvis	29	80	109	14	28	42	151
C66, C68 ureter and other urinary organs	1	2	3		2	2	5
C67 urinary bladder	12	65	77	11	38	49	126
C69 eye	3	1	4		1	1	5
C70–C72 brain & CNS	28	77	105	22	22	44	149
C73 thyroid	266	136	402	155	38	193	595
C74–C75 other endocrine glands	4	5	9	1	1	2	11
C76–C80 unknown or unspecified sites	17	21	38	11	12	23	61
C81 Hodgkin's lymphoma	26	33	59	13	19	32	91
C82–C85, C96 Non-Hodgkin lymphoma	53	98	151	32	45	77	228
C88, C90 Multiple myeloma	22	45	67	13	9	22	89
C91–C95 Leukemia	69	163	232	26	46	72	304
Other hematopoietic malignancies	15	13	28	3	5	8	36
Other Malignancy		1	1	1		1	2
(D00–D09) non-invasive cancers (in-situ cases)	**107**	**49**	**156**	**44**	**18**	**62**	**218**
D01 Carcinoma in situ of other and unspecified digestive organs	2	3	5	3	1	4	9
D02 Carcinoma in situ of middle are and respiratory system		1	1				1
D03 Melanoma in situ	3	4	7				7
D04 Carcinoma in situ of skin		1	1	3	1	4	5
D05 Carcinoma in situ of breast	45	3	48	20		20	68

Table 3.3 (continued)

Primary site ICD-10	Non-Emirati			Emirati			Grand total
	Female	Male	Total	Female	Male	Total	
D06 Carcinoma in situ of cervix uteri	51		**51**	17		**17**	**68**
D07 Carcinoma in situ of other and unspecified genital organs	2	9	**11**	1	2	**3**	**14**
D09 Carcinoma in situ of other and unspecified sites	4	28	**32**		14	**14**	**46**
Grand total—invasive and non-invasive	**2344**	**1993**	**4337**	**866**	**627**	**1493**	**5830**

Source: Ministry of Health and Preventions (MOHAP), Statistics and Research Center, UAE National Disease Registries—UAE National Cancer Registry Report 2021

3.18   Pediatric Cancer Cases

In 2021, there were 154 newly diagnosed cases of cancer in children aged 0–14 years in the UAE, with 45% being females and 55% being males. These cases accounted for roughly 2.7% of all malignant cases registered.

The most common cancers among pediatric patients were leukemia, brain and CNS, non-Hodgkin lymphoma, kidney & renal pelvis, and bone and articular cartilage.

3.19   Cancer Mortality

In 2021, cancer was identified as the fifth leading cause of death in the UAE.

The total number of cancer-related deaths was 975, with 506 in males and 469 in females. These deaths represented 8.2% of all deaths, irrespective of nationality, gender, or type of cancer.

The estimated age-standardized rate of mortality for both genders in 2021 was 29.6 deaths per 100,000 population. Among all cancer-related deaths, colon cancer accounted for the highest percentage, with an average of 11.49% per year. Trachea, bronchus & lung ranked as the second most common cause of cancer death in both males and females. Breast cancer was the third-most common cause of cancer death.

3.20   Conclusion

In this chapter, we aim to give a full picture of cancer registration and surveillance in the UAE, describe the UAE-NCR's approach to cancer registration and surveillance to monitor cancer-related outcomes, and give some examples of how the UAE National Cancer Registry can improve cancer care in the UAE.

The UAE National Cancer Registry, as a population-based cancer registry, plays an important role in the planning, operation, and evaluation of the UAE national cancer prevention and control program, not only to articulate the disease burden, trends, and geographical comparisons but also to evaluate the quality of cancer care.

Conflict of Interest   The authors have no conflicts of interest to disclose.

References

1. World Health Organization. Noncommunicable diseases. [Online]. https://www.who.int/news-room/fact-sheets/detail/noncommunicable-diseases.
2. World Health Organization. Countries to act on noncommunicable diseases but need to speed up efforts to meet global commitments. [Online]. https://www.who.int/news/item/18-07-2016-countries-start-to-act-on-noncommunicable-diseases-but-need-to-speed-up-efforts-to-meet-global-commitments. Accessed 7 July 2022.
3. MOHAP's website, open data. https://mohap.gov.ae/en/open-data/mohap-open-data.

4. UAE-NCR annual report 2021. https://mohap.gov.ae/assets/download/ab960117/CANCER%20INCIDENCE%20IN%20UNITED%20ARAB%20EMIRATES%20ANNUAL%20REPORT%20OF%20THE%20UAE%20-%202021.pdf.aspx
5. IARC's GLOBOCAN, in 2022. https://gco.iarc.who.int/media/globocan/factsheets/populations/784-united-arab-emirates-fact-sheet.pdf

Dr. Alya Zaid Harbi is a director of statistics and research within the Ministry of Health and Prevention-UAE and a PhD holder in the field of computer science, specialized in data science. She plays a critical role in illustrating the health of the UAE population through the management of health data and information across the country. She launched and completed one of the biggest health studies in the UAE that is related to the National Health Survey, where she worked in collaboration with different organizations and stakeholders to make it successful. Moreover, she is leading the first National Cancer Registry project in the UAE and is responsible for making the UAE a leader in that field. Dr. Alya is currently establishing the enterprise data warehouse and nationwide disease registries for all UAE health data, which is an integrated information technology solution that will help MOHAP collect, aggregate, analyze, and represent data to help in decision-making and refining policies.

Dr. Buthaina Abdulla Bin Belaila is a consultant family physician and Head of Non-Communicable Disease and Mental Health at MOHAP. A leader in the field of health care systems in the UAE with extensive experience in non-communicable diseases. She is a graduate of the UAE Faculty of Medicine and the UAE Women Leadership Program. She played a critical role in implementing the WHO ISH/CVD/risk assessment tool and integrating NCD services in all PHCs in 2015. She has participated in multiple international publications on NCD. Dr. Buthaina led the development of the master national plan for NCDs in 2017 and the national action plans for diabetes and obesity, cardiovascular disease, and cancer control for 2023–2026, which have made a significant impact on the health and well-being of the people of the UAE.

Wael Shelpai is a Certified Tumor Registrar (CTR®) and (ODS) from NCRA, USA. He is the GICR—IARC/WHO Regional Expert and Trainer in Cancer Registration and Cancer Registry Data Quality for the MENA Region.

He is an expert on national disease registries, Ministry of Health and Prevention, UAE. He has received training in disease-cancer registration, epidemiology, and survival analysis from several international organizations, like IARC/WHO—France, WIA, India, etc. He is affiliated with several professional committees and associations, and he has delivered international cancer registration and cancer registry data quality trainings and workshops. He is the author of several publications and has presented abstracts at national and international conferences. Since 2006, he has been an active member of the NCRA (National Cancer Registry Association) in the USA.

Hira Abdul Razzak is an accomplished Health Research Specialist at the Ministry of Health and Prevention in the UAE. With an MSc in Genetics and a gold medal from the University of Karachi, she has over 12 years of experience conducting scientific research.

Previously serving as a Research Manager, she has engaged in collaborative endeavors with renowned research networks and faculty members possessing diverse backgrounds at local, national, and international levels. Through these collaborations, she has showcased her proficiency in conducting systematic reviews and national surveys, authoring books, extracting and monitoring data, designing studies, analyzing and disseminating research findings, and maintaining stringent quality control measures. Moreover, she has played a crucial role in spearheading the digital transformation of research services, recognizing the potential of technology to revolutionize the field. She believes in utilizing the collective power of her profession, passion, and knowledge to create meaningful narratives and scholarly articles to benefit the people she serves. She has embarked on a journey to engineer positive change in the research culture as well as in the realms of human health, behavior, and well-being.

Her research interests primarily revolve around non-communicable and communicable diseases, mental disorders, and genetic diseases, reflecting her dedication to advancing health care in the country. Colleagues readily acknowledge her as an innovative specialist, renowned for her unwavering capability to consistently generate new and groundbreaking ideas that can be relied upon. Overall, her academic achievements, extensive experience, and contributions in the field position her as a respected figure in the field of medical research.

Cancer Prevention, Screening, and Early Detection in the UAE

4

Saeed Rafii and Humaid O. Al-Shamsi

4.1 Cancer Incidence in the UAE

The United Arab Emirates National Cancer Registry (UAE-NCR) 2021 has reported that between January 1 and December 31, 2021, 5830 patients were diagnosed with either malignant or in situ cancer, of which 5612 (96%) were malignant and 218 (4%) were in situ cases [1].

GLOBOCAN, the International Agency for Research on Cancer, predicts 4807 newly diagnosed cancers in the UAE by 2020 [2]. Cancer incidence in the UAE has risen in the past decade. According to the World Health Organization (WHO), the number of breast cancer cases has almost doubled between 2012 (568 cases) and 2018 (1054 cases). It is estimated that the number of breast cancer cases will almost triple and increase to 2993 by 2040. Additionally, WHO has forecasted a steep and alarming increase in lung cancer, from 190 new cases in 2018 to 1020 new cases by 2040 (Fig. 4.1). Furthermore, the probability of premature death from cancer per year in the UAE has increased from 3.29% in 2000 to 3.93% in

S. Rafii
Department of Oncology, Mediclinic City Hospital, Dubai, United Arab Emirates

Emirates Oncology Society, Emirates Medical Association, Dubai, United Arab Emirates
e-mail: saeed.rafii@mediclinic.ae

H. O. Al-Shamsi (✉)
Burjeel Cancer Institute, Burjeel Medical City, Burjeel Holdings, Abu Dhabi, United Arab Emirates

Ras Al Khaimah Medical and Health Sciences University, Ras Al Khaimah, United Arab Emirates

Gulf Medical University, Ajman, United Arab Emirates

Emirates Oncology Society, Emirates Medical Association, Dubai, United Arab Emirates

College of Medicine, University of Sharjah, Sharjah, United Arab Emirates

Gulf Cancer Society, Alsafa, Kuwait
e-mail: alshamsi@burjeel.com; humaid.al-shamsi@medportal.ca

© The Author(s) 2024
H. O. Al-Shamsi (ed.), *Cancer Care in the United Arab Emirates*,
https://doi.org/10.1007/978-981-99-6794-0_4

79

Estimated past and future trends in total cases per year (breast and lung)

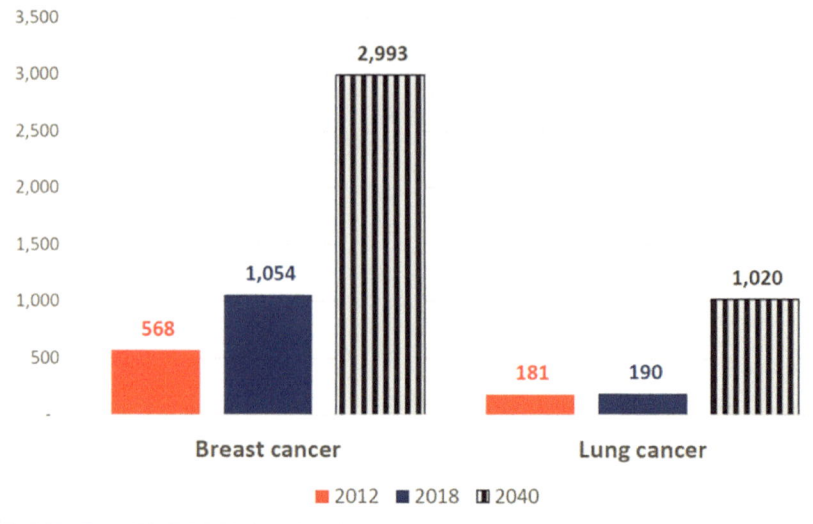

Fig. 4.1 An estimated past and future trend in total breast and lung cancer cases per year [3]. (Used with permission from the World Health Organization)

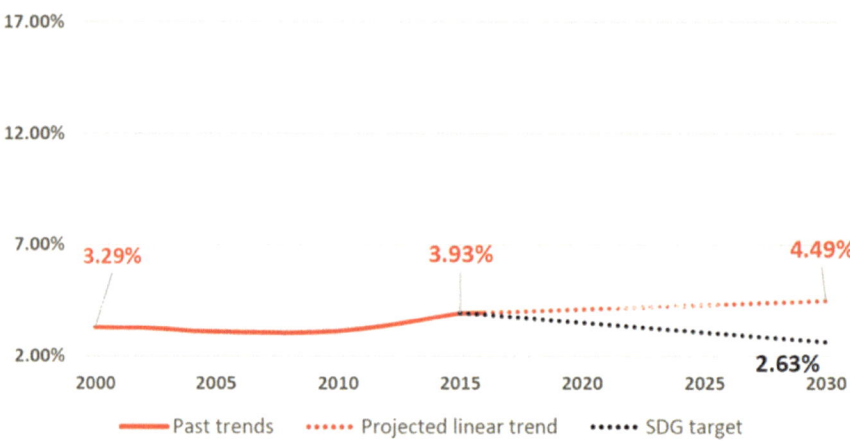

Fig. 4.2 Probability of premature death from cancer per pear between 2000 and 2030 [3]. (Used with permission from the World Health Organization)

2015 [3]. If the current trend continues, it is projected that by 2030 the probability of premature death from cancer will have increased to 4.49% (Fig. 4.2). These statistics highlight the importance of public awareness and a national cancer prevention strategy.

4.2   Cancer Risk Factors in the UAE

Identifiable and modifiable risk factors for developing cancer include obesity, smoking, industrialism, sedentary life, and high-risk viral infections such as the human papillomavirus (HPV). HPV and hepatitis are among the most common factors in the UAE, similar to other parts of the world. However, some of the risk factors, such as the increased risk of obesity, smoking, and sedentary life, are increasingly alarming. The most common population-attributable fractions are thought to be tobacco use (11.6%), infections (11.9%), obesity (4.6%), UV-light (3.1%), alcohol (2.6%), and occupational risks (1.6%) [3].

4.2.1   Obesity and Physical Exercises

The prevalence of obesity in the UAE is rising due to increased consumption of fast food and reduced physical activity. The UAE climate, particularly in late spring and summer, may limit outdoor activities, forcing many people to stay indoors or use large shopping malls, which in turn encourages the consumption of fast food in large food courts. In the past few years, initiatives such as 30 minutes of activities for 30 days, led by His Highness Sheikh Hamdan, Crown Prince of Dubai, the Dubai Marathon etc., have tried to encourage the public to have more outdoor activities and exercises. The high level of safety and security in the UAE encourages women to take part in physical activities, yet female obesity remains a concerning issue. The UAE government has initiated multiple programs for controlling cancer-risk factors, including obesity. The National Nutrition Strategy 2022–2030 has recommended the implementation of the specifications related to trans fatty acids and the implementation of a reformulation program to reduce salt in processed foods. In 2017, an excise tax of 50% was applied to soft drinks and 100% to energy drinks [4].

In 2019, the excise tax was applied to 50% for sweetened beverages [5].

4.2.2   Smoking

Many forms of active smoking, including habitual or occasional smoking, are increasingly observed in modern Arab culture and in the UAE. In particular, smoking shisha with flavored tobacco has become popular among youth [6]. The UAE Federal Law No. 15 of 2009 forbids the sale of tobacco products to those under 18, smoking in private cars when a child under the age of 12 is present, smoking in houses of worship, educational institutions (such as universities and schools), health and sports facilities, the sale of sweets that resemble tobacco products, automatic vending equipment and devices for tobacco distribution inside the country, tobacco advertisement, and smoking in closed public spaces [7]. It is encouraging that many restaurants and businesses have adopted the policy of allocating designated smoking areas. Nonetheless, both active and passive smoking are of concern. The UAE National Tobacco Control Committee intends to outlaw smoking in public places entirely [8]. Excise tax was introduced across the UAE in 2017; in 2019, the tax for tobacco products and electronic smoking devices increased from 50 to 100% [5].

4.2.3   HPV Infection and Cervical Cancer in the UAE

According to the UAE-NCR, cervix uteri is ranked the fifth-most common cancer among females in the UAE in 2021, with 141(4.6%) malignant cases. This is also the sixth most common cause of cancer death, with an estimated average of 13 (1.33%) [1]. Although there is a paucity of data regarding the rate of HPV infection in the UAE, it is estimated that between 2.3 and 2.5% of the general female population carry HPV 16 and 18 variants, which are the most common causes of invasive cervical cancer [9]. Women aged 25–65 who are residing in the UAE are eligible for cervical cancer screening that includes, according to the National Cancer Screening Guidelines, a PAP smear and an HPV test as co-testing, and it is recommended that the screening test be repeated every 3 years for women aged 25–29 years, every 5 years for women 30–65 years, and annually for women who are immune-compromised due to disease or medication [10].

In 2008, Abu Dhabi's Health Authority (HAAD) launched an HPV vaccination program for girls entering grades 11 and 12. Abu Dhabi is the first state within the Middle East or Arab nations to introduce HPV vaccination in the public sector [11].

In 2018, the Department of Health (DOH) announced that the early detection of cervical cancer campaign had resulted in a significant decrease in the late-stage diagnosis of cervical cancer rate from 30.3% in 2012 to 14.8% in 2015. We have previously recommended improving public, parental, and adolescent education about HPV vaccination, specifically to address misconceptions and fears surrounding HPV vaccination [12].

4.2.4   Hepatitis Infection and Hepatocellular Cancer Risk in the UAE

The UAE mandated hepatitis B vaccination in 1991 and introduced compulsory hepatitis B and C testing for all new immigrants since 2006, followed by a mandatory HBV vaccination in cases of non-immunity [13, 14]. It is estimated that the national prevalence is between 1.0 and 1.5% in the Emirati general population, translating to 12,000 and 18,000 cases [13]. A study looking at the burden of virus-associated liver cancer in the Arab world from 1990 to 2010 reported in 2015 that the age-standardized death rate for HBV-associated HCC in the UAE increased by approximately 10% between 1990 and 2010 to 3.2 per 100,000 males and 1.2 per 100,000 females in 2010 [15].

Given that it takes decades to observe the effect of HBV vaccination on the reduction of HCC, there is currently no data available in the UAE to compare the prevalence of HCC before and after the HBV vaccination mandate.

4.3   Cancer Screening Programs in the UAE

The national cancer screening program is outlined in Table 4.1 [16].

Table 4.1 Recommendations for regular cancer screening by the national cancer screening program

Cancer type	Primary population group	Screening test and frequency
Colon and rectum cancer	Men and women Age: 40–75 years	Colonoscopy, every 10 years Or Stool test, every 2 years
Breast cancer	Women Age: 40 years and above	Mammogram Every 2 year
Cervical cancer	Women Age: 25–65 years	PAP smear test Every 3–5 years

Source: The UAE government portal, updated 24 October 2018

4.4 Current Status of Cancer Screening in the UAE

Despite the existence of a cancer screening program and the technology needed to deliver cancer screening, the rate and coverage of national cancer screening are currently suboptimal. No official published data on cancer screening uptake is available, although it is estimated that the uptake rate is generally low. One study reported that of the 45,147 UAE nationals in the Emirates of Abu Dhabi eligible for screening in 2015–2016, only 23.5% were screened [17]. When applied to the general eligible population, regardless of nationality, we estimate that cancer screening uptake will be lower than this. Multiple factors are responsible for the current low coverage and uptake of cancer screening in the UAE, which have been summarized in Table 4.2 [12].

We have already proposed a UAE-wide national screening program that includes breast, colorectal, cervical, lung, and prostate cancer based on a call-and-recall system, and reformatory actions needed to improve nationwide cancer screening [12, 18].

4.4.1 Available Screening Techniques in the UAE

There is a wide range of radiological and non-radiological screening tests available both in the public and private sectors in the UAE.

According to a WHO report, in 2020, the following radiology equipment was available in the UAE per 100,000 population: mammography (129.6), CT scanners (422.8), MRI (373.9), and PET CT scanners (12.7) [3]. It is likely that this statistic has improved as many more medical facilities have been established in the past few years.

In addition to radiological screening, there are a few non-invasive and blood-based early cancer tests, such as those being introduced in the UAE commercially, e.g., hPG80® and Epiprocolon®.

Table 4.2 Causes for low uptake of cancer screening and reformatory actions needed to improve a nationwide cancer screening program in the UAE [12]

Cause/factor	Root causes	Reformatory actions
Access to the service	• Difficulty in accessing services • Cost • The location of examination centers • Health insurance does not include screening • Population growth and change in the population pyramid • Recommendations and guidelines do not include the younger population	• Mass education/campaign on screening and early detection • Address the target groups at educational institutes, media, etc. • Launch mobile screening services, such as mammograms, cervical examinations, stool examinations (FIT test), and medical advice • Covering the cost through health insurance or delivering it for free • Develop discounted rates for screening packages • Creating a unified cost and quality standard for cancer screening • Establish a specialized center for cancer detection
Mechanisms and quality control of cancer screening services	• The need for a unified national program for cancer screening • Poor compliance in the application of the national screening guidelines • The absence of a dedicated team to monitor the quality of services • Lack of human resources for auditing, and lack of audits to assess the quality of services	• Activating a national program for early detection of cancer that includes a central call system and text messaging to call the target high-risk groups • Establishing a mechanism and unified targets to measure the coverage rate • Linking the Emirates ID to the cancer screening record • Commit all service providers to achieve the target percentage • Activating the national registry for screening and early detection • Integrating primary care and screening • Building capacity and logistical resources • Assigning coordinators for quality assurance • Establishing screening services • Establishing a quality assurance and monitoring department for screening services • Integrating screening and early detection into insurance coverage • Monitoring the impact of screening on outcome mortality

Table 4.2 (continued)

Cause/factor	Root causes	Reformatory actions
Community awareness of the importance of early examination and detection of cancer	• Lack of awareness campaigns • Lack of education and understanding of cancer • Lack of integration of information about cancer into the educational curriculum • Lack of availability of smart awareness applications • Lack of awareness about early detection programs • Lack of awareness of global awareness days for different types of cancer	• Conducting a cancer awareness measurement assessment by questionnaire in the community • Intensifying awareness campaigns on the importance of early detection • Information about the availability and locations of services • Involving prominent and famous figures to disseminate awareness • Emphasize the importance of early detection of cancer • Develop a public information website • Incorporating cancer information into educational programs and campaigns in universities and schools • Include representatives from the relevant authorities • Conducting free campaigns to detect cancer during ambulatory care • Smart awareness campaigns and applications to help community members make the right decisions and get early detection tests

End-user factors influencing the screening and detection service	1. Emotional factors
	  • Fear
	  • Shyness
	  • Anxiety about the outcome
	  • Hospital admission
	2. Practical factors
	  • Absence from work
	  • Transportation
	3. Understanding and education
	  • Lack of knowledge
	  • Misconceptions and stigma
	  • Confidentiality of the information
	  • Awareness of the importance of examination
	4. Cultural differences in society affect concepts of prevention and early detection	

Capacity building in human resources and logistic support	• Lack of competencies • Inadequate distribution of human resources • Lack of logistical resources and modern equipment	• Increasing the budget allocation for early detection programs • Increasing the logistical and human resources to increase the coverage rate through training and manpower • Linking the annual performance of general practitioners in primary healthcare centers • Hosting independent international experts regularly to evaluate the program and staff • Raising the efficiency level of employees through training courses

4.5   Specific Cancer Screening Initiatives

4.5.1   Breast Cancer Screening

Breast cancer awareness and screening are more developed in the UAE compared with other cancers. Most major hospitals offer mammograms and breast ultrasound scans. Breast cancer screening is also accepted by many major insurance companies. In 2011, an initiative called "Pink Caravan" was established by Sheikha Jawaher Bint Mohammed Al Qasimi to raise awareness about breast cancer screening. Since its foundation and until 2020, the caravan has travelled 1953 kilometers and provided 20,794 free mammograms and 3584 US scans to women across the UAE, leading to the diagnosis of 80 breast cancer cases [19].

4.5.2   Colon Cancer Screening

According to the national guidelines, all members of the public over the age of 40 are advised to have an annual stool fecal immunochemical test as well as a colonoscopy every 10 years. At-risk populations, such as those with hereditary susceptibility genes, family history, and patients with inflammatory bowel disease, need to undergo colonoscopies once a year or every 5 years, at the discretion of their physicians [20, 21].

Data from the DOH indicates that between 2012 and 2019, 42% of the estimated 845 patients who were diagnosed with colorectal cancer in Abu Dhabi were detected as a result of routine screening [20].

The UAE Ministry of Health and Prevention (MOHAP) has also issued a national guideline for colorectal cancer screening and diagnosis [21]. However, there is no official data on the coverage and uptake of colorectal cancer screening.

4.5.3   Prostate Cancer Screening

There is no official nationwide prostate cancer screening program in the UAE. Most major hospitals offer prostate cancer screening to men as part of a health package with reduced costs. In their most expensive insurance plans, some insurance companies provide free annual prostate cancer screenings to all men over the age of 45.

4.5.4   Lung Cancer Screening

There is no national lung cancer screening program in the UAE at the moment. Despite the availability of low-dose CT scans, general awareness and lung cancer screening are low throughout the UAE [22]. To the best of our knowledge, lung cancer screening is not used routinely throughout the UAE, although the DOH in Abu Dhabi has a published guideline for lung cancer screening [23].

4.6   Challenges in Cancer Screening in the UAE

The UAE has experienced a rapid expansion in its economy and population in the past 20 years. Its unique population is comprised of a majority of immigrants from across the globe with diverse backgrounds. Such a rapid expansion in the population has resulted in an increase in non-communicable diseases, including cancer. Due to the historical nature of short-term placement in the country, the UAE healthcare system has not fully integrated non-UAE residents into cancer screening programs. Additionally, the cost of cancer screening has not been fully covered by some insurance companies, and access to cancer screening is not equally available to all at-risk groups. Even when structured cancer screening is available (see Sects. 4.3 and 4.4), the uptake and compliance with such programs may be unsatisfactorily low. Many factors, such as cultural barriers, education and awareness, access to healthcare, and the unequal distribution of specialized healthcare facilities, contribute to the lack of effective cancer screening in the country. This may impact the late-stage diagnosis of cancer and, subsequently, the higher risk of cancer-related death, despite the advanced healthcare system.

4.7   Conclusion and Recommendations

Much has been done in the past few years by the UAE government to reduce the burden of cancer in the country, including raising public awareness, encouraging physical activities, passing laws in order to reduce obesity, and reducing smoking. However, more needs to be done in order to have effective cancer screening programs that are accessible and effective, result in early-stage cancer diagnosis, and subsequently reduce cancer-related mortalities in the country.

We recommend focusing on educating the public and enhancing awareness about cancer and its risk factors. It is important to address cultural misconceptions and beliefs about cancer. In parallel, governmental legislation is imperative to target known causes of cancer.

It is important to recognize that reducing cancer-related mortality is not possible without an effective early cancer detection program. Therefore, we encourage legislators and health authorities to plan and implement a nationwide call/recall cancer screening program for the most common cancers, such as breast, colorectal, lung, and prostate. Guidelines must take into account the young population of the UAE and tailor screening recommendations that reflect the age of diagnosis of the most common cancers in the country. Such a program should also address the referral pathway by which patients are appropriately referred to highly specialized centers.

We have previously proposed an all-inclusive cancer control program to formulate strategy, implement prevention, and enforce comprehensive cancer management through a collaborative process between government organizations, the community, and non-government organizations [12]. Given the current increasing trend in cancer incidence and cancer mortality rate, it is urgent to formulate an action plan with an emphasis on preventing cancers or detecting cases early at a curable stage in order to control the rising incidence rate at its current level or, ideally, to reduce it. It takes a concerted effort from all stakeholder groups in order to set a series of targets that result in significant risk reduction by increasing early detection, improving treatment, and enhancing survivorship [12].

Conflict of Interest   The authors have no conflict of interest to declare.

References

1. Cancer incidence in United Arab Emirates, Annual Report of the UAE - National Cancer Registry - 2021. Statistics and Research Center, Ministry of Health and Prevention. Accessed on 25Mar2024.
2. https://gco.iarc.fr/today/data/factsheets/populations/784-united-arab-emirates-fact-sheets.pdf. Accessed 19 Mar 2023.
3. https://cdn.who.int/media/docs/default-source/country-profiles/cancer/are-2020.pdf?sfvrsn =c932b567_2&download=true.
4. https://u.ae/en/about-the-uae/strategies-initiatives-and-awards/strategies-plans-and-visions/ health/national-nutrition-strategy-2030#:~:text=The%20National%20Nutrition%20 Strategy%202030,nutrition%20to%20all%20age%20groups.
5. https://tax.gov.ae/DataFolder/Files/Pdf/Taxable%20Person%20Guide%20for%20Excise%20 Tax%20-%20Sept%202019%20Latest%20ver.pdf.
6. Razzak HA, Harbi A, Ahli S. Tobacco smoking prevalence, health risk, and cessation in the UAE. Oman Med J. 2020;35(4):e165. https://doi.org/10.5001/omj.2020.107. PMID: 32904941; PMCID: PMC7462068.
7. https://mohap.gov.ae/app_content/legislations/php-law-en-32/mobile/index.html.
8. https://u.ae/en/information-and-services/health-and-fitness/tobacco-provisions,   https://hpv-centre.net/statistics/reports/ARE_FS.pdf.
9. Ortashi O, Abdalla D. Colposcopic and histological outcome of atypical squamous cells of undetermined significance and atypical squamous cell of undetermined significance cannot exclude high-grade in women screened for cervical cancer. Asian Pac J Cancer Prev. 2019;20(9):2579–82.   https://doi.org/10.31557/APJCP.2019.20.9.2579.   PMID: 31554348; PMCID: PMC6976842.
10. https://www.doh.gov.ae/-/media/51BDF280150B4AD481064B8E945BDB1D.ashx.
11. Al Shdefat S, Al Awar S, Osman N, Khair H, Sallam G, Elbiss H. Health Care System View of Human Papilloma Virus (HPV) Vaccine Acceptability by Emirati Men. Comput Math Methods Med. 2022 Jan 28;2022:8294058. https://doi.org/10.1155/2022/8294058. PMID: 35126638; PMCID: PMC8816567.
12. Al-Shamsi HO, Abyad AM, Rafii S. A proposal for a national cancer control plan for the UAE: 2022-2026. Clin Pract. 2022;12(1):118–32. https://doi.org/10.3390/clinpract12010016. PMID: 35200267; PMCID: PMC8870909.
13. https://doi.org/10.33590/emjhepatol/19-00061.
14. Abu-Gheida IH, Nijhawan N, Al-Awadhi A, Al-Shamsi HO. General oncology care in the UAE. In: Al-Shamsi HO, Abu-Gheida IH, Iqbal F, Al-Awadhi A, editors. Cancer in the Arab world. Singapore: Springer; 2022. https://doi.org/10.1007/978-981-16-7945-2_19.
15. Khan G, Hashim MJ. Burden of virus-associated liver cancer in the Arab world, 1990-2010. Asian Pac J Cancer Prev. 2015;16(1):265–70. https://doi.org/10.7314/apjcp.2015.16.1.265. PMID: 25640363.
16. https://u.ae/en/information-and-services/health-and-fitness/chronic-diseases-and-natural-disorders/cancer. Accessed 10 Dec 2022.
17. Almansoori A, Alzaabi M, Alketbi L. Colorectal cancer screening in ambulatory healthcare service clinics in Abu Dhabi, United Arab Emirates in 2015–2016. BMC Cancer. 2021;21:897. https://doi.org/10.1186/s12885-021-08623-9.
18. Al-Shamsi HO. The state of cancer care in the United Arab Emirates in 2022. Clin Pract. 2022;12:955–85. https://doi.org/10.3390/clinpract12060101.
19. https://www.focp.ae/our-programs/womens-health/.

20. https://www.doh.gov.ae/en/news/ADPHC-revises-recommendations-for-early-colorectal-cancer-screening#:~:text=According%20to%20the%20new%20recommendations,a%20colonoscopy%20every%20ten%20years.
21. https://www.isahd.ae/content/docs/MOHAP%20Guidelines%20For%20Colorectal%20Cancer%20Screening.pdf.
22. Al-Shamsi H, Darr H, Abu-Gheida I, Ansari J, McManus MC, Jaafar H, Tirmazy SH, Elkhoury M, Azribi F, Jelovac D, et al. The state of cancer care in the United Arab Emirates in 2020: challenges and recommendations, a report by the United Arab Emirates Oncology Task Force. Gulf J Oncolog. 2020;1:71–87.
23. DOH lung cancer screening specifications. Department of Health. 2018. www.doh.gov.ae.

Dr. Saeed Rafii is a board-certified consultant medical oncologist. After completion of his primary medical degree, he was trained in internal medicine, followed by subspecialty training in medical oncology in two of the most prestigious cancer hospitals in the UK, Queen Elizabeth Hospital Birmingham and the Royal Marsden Hospital, London. He then completed a clinical fellowship in early-phase clinical trials at the Royal Marsden Hospital, London, and received his CCT (certificate of completion of training) from the UK General Medical Council. Dr. Rafii was subsequently appointed as an associate professor and consultant in medical oncology at the University of Manchester and the Christie Hospital, where he helped to establish and expand the Experimental Cancer Medicine Centre. He then moved to University College London Hospital and the Oxford Cancer Network as a consultant medical oncologist.

He has extensive expertise in clinical trials and has been chief, principal, or co-investigator on over 100 early and late-phase oncology clinical trials. Dr. Rafii also holds a PhD and a postdoctoral fellowship in molecular cancer genetics. He is a member of the Royal College of Physicians of the UK, the European Society of Medical Oncology (ESMO), and the American Association for Clinical Oncology (ASCO). In 2018, he was elected as Fellow of the Royal College of Physicians of UK (FRCP) for his outstanding medical and research activities.

Prof. Humaid Obaid Al-Shamsi is the Chief Executive Officer of Burjeel Cancer Institute in Abu Dhabi, UAE, President of the Emirates Oncology Society, Lead of the Gulf Cancer Society, Full Professor of Oncology at the Ras Al Khaimah Medical and Health Sciences University, Ras Al Khaimah, UAE, and an Adjunct Professor of Oncology at the College of Medicine, University of Sharjah. He is the first Emirati to be promoted as a professor in oncology in the UAE. He is also the Chairman for Colorectal Cancer in the MENA region, appointed by the prestigious National Comprehensive Cancer Network®. He is also the only member of Lung Cancer Policy Network in the MENA region that aims to advance lung cancer research and screening globally. He is the Chairman of the Oncology and Hematology Fellowship Training Program for the National Institute for Health Specialties in the United Arab Emirates. He is the only member in GCC in the WIN Consortium which is comprised of organizations representing all stakeholders in personalized cancer medicine globally.

He is board-certified in both internal medicine and oncology from the UK, USA (ABIM), and Canada (FRCPC). He has also been awarded the FRCP (London) in 2023 and FRCP (Glasgow) in 2024. He is the only physician in the UAE with a subspecialty fellowship certification and training in gastrointestinal oncology and the first Emirati to train and complete a clinical post-doctoral fellowship in palliative care. He was an assistant professor at the University of Texas MD Anderson Cancer Center between 2014 and 2017. He has published more than 140 peer-reviewed articles in JAMA Oncology, Lancet Oncology, The Oncologist, BMC Cancer, and many others. His area of expertise includes precision oncology and cancer care in the UAE. In 2016, he published with his group from MD Anderson the JCO paper describing a new distinct subgroup of CRC, NON V600 BRAF-mutated CRC. In 2022, he published the first book about cancer research in the UAE and also the first book about cancer in the Arab world, both of which were launched at Dubai Expo 2020. *Cancer in the Arab World* has been downloaded more than 450,000 times in its first 18 months of publication and is the ultimate source of cancer data in the Arab region. He also published the first comprehensive book about cancer care in the UAE which is the first book in UAE history to document the cancer care in the UAE with many topics addressed for the first time, e.g., neuroendocrine tumors in the UAE. He is passionate about advancing cancer care in the UAE and the GCC and has made significant contributions to cancer awareness and early detection for the public using social media platforms. He is considered as the most followed oncologist in the world with over 300,000 subscribers across his social media platforms (Instagram, Twitter, LinkedIn, and TikTok). In 2022, he was awarded the prestigious Feigenbaum Leadership Excellence Award from Sheikh Hamdan Smart University for his exceptional leadership and research and the Sharjah Award for Volunteering. He was also named the Researcher of the Year in the UAE in 2020 and 2021 by the Emirates Oncology Society.

In May 2024, HH Sheikh Mansour bin Zayed Al Nahyan, Vice President of the United Arab Emirates, awarded him the first place in UAE Nafis program for outstanding leadership in private sector across all business and medical disciplines. Beside his clinical and administrative duties, he is engaged in education and various levels of research training for medical trainees to enhance their clinical and research skills. His mission is to advance cancer care in the UAE and the MENA region and make cancer care accessible to everyone in need around the globe.

A Proposal for Cancer Control Plan in the UAE

Humaid O. Al-Shamsi ⓘ and Amin M. Abyad

Abbreviations

EMRO	Eastern Mediterranean Region Office
ICER	Institute for Clinical and Economic Review
KPI	Key performance indicators
MDT	Multidisciplinary team
MENA	Middle East and North Africa
NCD	Noncommunicable diseases
NICE	National Institute for Health and Care Excellence
UAE	United Arab Emirates
WHO	World Health Organization

H. O. Al-Shamsi (✉)
Burjeel Cancer Institute, Burjeel Medical City, Burjeel Holdings, Abu Dhabi,
United Arab Emirates

Ras Al Khaimah Medical and Health Sciences University, Ras Al Khaimah, United Arab Emirates

Gulf Medical University, Ajman, United Arab Emirates

Emirates Oncology Society, Emirates Medical Association, Dubai, United Arab Emirates

College of Medicine, University of Sharjah, Sharjah, United Arab Emirates

Gulf Cancer Society, Alsafa, Kuwait
e-mail: alshamsi@burjeel.com; humaid.al-shamsi@medportal.ca

A. M. Abyad
Emirates Oncology Society, Emirates Medical Association, Dubai, United Arab Emirates

Burjeel Medical City, Abu Dhabi, United Arab Emirates

© The Author(s) 2024
H. O. Al-Shamsi (ed.), *Cancer Care in the United Arab Emirates*,
https://doi.org/10.1007/978-981-99-6794-0_5

5.1 Introduction

Cancer poses a significant worldwide health challenge, with nearly nine million lives lost to malignant diseases across the globe in 2015. It stands as the second most prevalent cause of death internationally, responsible for approximately one-sixth of all fatalities in present times [1]. The Eastern Mediterranean Region is renowned for its notable occurrence of cardiovascular ailments [2].

Within the swiftly expanding population of the Middle East, despite significant advancements in life expectancy, the occurrence of cancer is projected to double over the next two decades, aligning with the global trend of rising cancer rates [3]. The transition toward a more westernized lifestyle plays a significant role in driving this transformation. The adoption of westernized lifestyle choices is a major factor contributing to this shift.

Based on the aforementioned information, the incidence of cancer in our region is projected to experience a twofold increase within the next two decades, making it the region with the highest expected rise among all World Health Organization (WHO) regions. By the year 2030, it is expected that the number of cancer-related deaths will reach 652,097, which is a significant rise from the 367,441 reported in 2012 [4]. These projections stem from the combined influence of population growth and aging, the cumulative impact of heightened exposure to cancer risk factors such as smoking, dietary shifts, lifestyle changes, and the exacerbation of environmental pollution. These factors contribute to a substantial increase in the burden of cancer, placing a significant strain on the healthcare system.

In the UAE population, cancer mortality has been the fifth leading cause of death in 2021 [5].

In a significant number of cases, cancer can be prevented, making it a highly preventable disease. Timely diagnosis plays a crucial role in successful treatment outcomes. Even when diagnosed at later stages, it is possible to manage cancer-related symptoms, slow down disease progression, improve the quality of life, and provide support to patients and their caregivers throughout their journey. It is essential to develop and implement effective cancer control plans to address these needs. Many countries have ongoing cancer control programs at different levels. However, in several other countries, either no effective cancer control plan exists, the existing plan is outdated, or it is not effectively implemented. Therefore, there is a need for a comprehensive plan that is efficiently executed, regularly monitored, and subject to necessary modifications and updates based on local needs and requirements [1, 2, 4].

Cancer management now requires a comprehensive global approach that ensures equal and uniform access to care. The UAE National Cancer Strategy is designed to prioritize various initiatives aimed at reducing the impact of cancer in the UAE and facilitating the provision of optimal cancer care for patients and their caregivers. This strategy aims to enhance cancer care efforts in the UAE, aligning them with international guidelines and the most effective evidence-based practices while adapting them to suit the country's specific needs [3, 5, 7].

5.2   The Cancer Burden

During the previous century, the economy of the UAE experienced rapid growth, positioning it as one of the world's swiftest-growing economies [8, 9]. The remarkable economic growth witnessed in the UAE resulted in various transformations in the population's economy, sociodemographics, and way of life. Concurrently, there has been an epidemiological rise in the prevalence of noncommunicable diseases (NCDs), with cancer being particularly noteworthy. Epidemiological studies conducted in the UAE have linked this increase in cancer rates to various risk factors, including reduced physical activity and sedentary lifestyles [10–12], the consumption of high-calorie and low-nutrient meals [13], the obesity pandemic [12, 13], a rise in the number of individuals who smoke [14], and higher levels of air pollution [15]. Cancer diagnostic tools and procedures have witnessed advancements, leading to improved detection capabilities. In 2016, noncommunicable diseases (NCDs) accounted for 77.3% of all reported deaths. The top two main causes of death included cardiovascular diseases and malignancy. Approximately 6% of all deaths are attributed to infectious, maternal, perinatal, and nutritional conditions, while chronic respiratory diseases and diabetes account for approximately 5% each [9, 16].

5.2.1   Cancer Incidence Rate

In 2021, a total of 5830 cases were diagnosed in the UAE, with 218 cases (4%) classified as in situ cancer and 5612 cases (96%) classified as malignant. Among these cases, there were 2620 (44.9%) male patients and 3210 (55.1%) female patients. Due to the UAE's diverse population, a significant proportion of the reported cancer cases were individuals who had migrated to the country. The distribution of cancer cases by nationality is not evenly distributed, with 866 female cases and 627 male cases reported among UAE citizens, while non-UAE citizens accounted for 2344 female cases and 1993 male cases [5] (Table 5.1).

5.2.2   The Most Common Cancer Types

From January 1 to December 31, 2021, the UAE National Cancer Registry (UAE-NCR) reported a total of 5830 newly diagnosed cancer cases, comprising both malignant and in situ cases. Among these, 5612 cases (96%) were classified as malignant, while 218 cases (4%) were in situ. Overall, cancer affected more

Table 5.1 The most common primary malignant tumors in the UAE for both genders in 2021

Primary site	%	Primary site	%
Breast	36.9	Colorectal	12.5
Thyroid	13.8	Prostate	9.8
Colorectal	7.0	Leukemia	8.2
Uterus	5.7	Thyroid	6.8
Cervix uteri	4.6	Skin (Carcinoma)	6.4
Skin (Carcinoma)	3.6	Bronchus and Lung	6.3
Ovary	3.5	Non-Hodgkin lymphoma	5.6
Leukemia	3.1	Lip, Oral cavity & pharynx	4.6
Non-Hodgkin lymphoma	2.8	Kidney & Renal pelvis	4.2
Bronchus and Lung	2.3	Urinary bladder	4.0

Source: Ministry of Health and Prevention, Statistics and Research Center, National Disease Registry—UAE National Cancer Registry Report, 2021

women than men, with 3210 (55.1%) females and 2620 (44.9%) males being diagnosed. Among UAE citizens, there were 1493 newly diagnosed cancer cases, with 1431 cases (95.8%) being malignant and 62 cases (4.2%) being in situ. Similarly, among non-UAE citizens, there were 4337 newly diagnosed cancer cases, with 4181 cases (96.4%) being malignant and 156 cases (3.6%) being in situ. The overall crude incidence rate for both genders was 60.5 per 100,000 population. The incidence rate was higher for females, with a rate of 108.7 per 100,000, compared to males at 39.5 per 100,000. The overall age-standardized incidence rate (ASR) for both genders was 107.8 per 100,000. The most common cancers among both genders were breast, thyroid, colorectal, leukemia, and skin. Among males, colorectal, prostate, leukemia, thyroid, and skin were the top-ranked cancers, while among females, breast, thyroid, colorectal, uterus, and cervix uteri were the most prevalent. In 2021, there were 154 children between the ages of 0–14 diagnosed with new cancer in the UAE, with 45% being females and 55% being males. These cases constituted approximately 2.7% of all registered malignant cases. The most common cancers in boys and girls were leukemia, brain and CNS, non-Hodgkin lymphoma, kidney & renal pelvis, and bone and articular cancers [5].

5.2.3   Cancer Mortality

In 2021, cancer was the cause of 975 deaths, accounting for 8.2% of all reported fatalities. This corresponds to an estimated age-standardized mortality rate of 29.6 deaths per 100,000 population for both genders. Among the leading causes of cancer-related deaths, colon cancer ranked first, responsible for 11.49% of the deaths, followed by trachea, bronchus, and lung cancer (9.85% deaths), and breast (9.64% deaths) [5].

5.3   Cancer Prevention, Screening, and Early Detection

Around one-third of cancers can be prevented through lifestyle modifications, genetic testing for high-risk individuals, and vaccination. Despite the relatively low reported incidence of cancer in the UAE, it is crucial to prioritize enhancing outcomes for diagnosed and treated cancer patients. Cancer screening and early detection have the most significant influence on patient outcomes. Early detection plays a vital role in increasing the likelihood of successful treatment and positive results. However, the 2017 data indicate that the global population coverage rate for early cancer screening among the targeted age groups is lower (less than 70%) than the targets set by the World Health Organization [1, 2, 4, 5, 17].

The factors contributing to the low rates of cancer screening in the UAE population, as well as the suggested actions to enhance these rates, are outlined in Table 5.2 [18].

Table 5.2 Root causes for low cancer screening and proposed actions [18]

Cause/factor	Root causes	Reformatory actions
Access to the service	• Nonaccessibility to service • Financial inability • Examination center location • Screening is not covered by health insurance • Demographic change and population growth/change in the population pyramid • Younger populations are not addressed in recommendations or local guidelines	• Public education and media campaigns on early detection and cancer screening • Utilizing media and other tools to address specific target groups • Launch and use mobile screening apps for services such as mammography, pap smear, and stool occult blood (FIT test) • Free screening campaigns or insurance coverage for screening tools • Creation of screening packages with a discounted price • Unifying the cost and quality of screening tools • Establishment of a specialized cancer screening and detection center
Quality control mechanisms for cancer screening services	• Poor compliance with the adopted national screening guidelines and standards • The absence of a team dedicated to assessing the quality of services • Deficiency in human resources for monitoring and a lack of auditing to assess the quality of services	• Using a national program for early detection of cancer that utilizes smart technology, such as an automated system for text messaging, to notify the target high-risk population • Establishing a clear mechanism and identifying targets to monitor patients' coverage rates • Cancer screening records are being linked to Emirates ID • Committing all service providers to the achievement of the target numbers • Activation of the national cancer registry for screening and early case detection • More integration of screening into primary care • Expanding logistical resources and capacity • Assigning coordinators for quality improvement • Establishing unified cancer screening services • Establishing a quality assurance department for continuous monitoring • Inclusion of cancer screening and early detection in mandatory insurance coverage • Monitoring the outcome of screening on cancer mortality

Table 5.2 (continued)

Cause/factor	Root causes	Reformatory actions
Community awareness regarding the importance of early detection and cancer screening	• Deficiency in awareness campaigns • Poor awareness of cancer etiology, management, and outcomes • Improper integration of cancer-related topics into the educational curriculum • Lack of utilization of smart applications for cancer awareness • Lack of public knowledge or awareness about early detection programs • Lack of publicity for global awareness days, months, or events for different types of cancer	• Conducting community cancer awareness measurement questionnaire studies • Highlighting the importance of early detection through awareness campaigns • Spreading information about the availability and locations of screening services • Involving famous personnel and public figures to disseminate awareness • Emphasize the significance of early cancer detection • Creation of a public information website to facilitate public access to accurate information • Incorporating cancer-related topics into educational curricula

End-user factors influencing the screening and detection service	1. Emotional factors
	    • Fear
	    • Shyness
	    • Anxiety about the outcome
	    • Hospital admission
	2. Practical factors
	    • Absence from work
	    • Transportation
	3. Understanding and education
	    • Lack of knowledge
	    • Misconceptions and stigma
	    • Confidentiality of the information
	    • Awareness of the importance of examination
	4. Cultural differences in society affect concepts of prevention and early detection
		• Universities and school campaigns
		• Include educational authorities' representatives
		• Utilization of ambulatory care services to conduct free campaigns to detect cancer
		• Smart awareness campaigns and applications to help community members make the right decisions and get early detection tests

Capacity building in human resources and logistic support	• Lack of competencies • Inadequate distribution of human resources • Lack of logistical resources and modern equipment	• Increasing the budget allocation for early detection programs • Increasing the logistical and human resources to increase the coverage rate through training and manpower • Linking the annual performance of general practitioners in primary healthcare centers • Hosting independent international experts regularly to monitor and evaluate the staff and program • Implementing continuous training and education to raise the level of efficiency of employees

5.4   Diagnostic Tools and Services

Achieving a clear and definitive diagnosis is crucial for prompt and successful treatment. However, there is a lack of well-established and efficient pathways that ensure the timely referral of patients to the appropriate physicians and centers for diagnosis. During this crucial period, patients should undergo a comprehensive clinical and physical assessment, appropriate radiologic tests, and a biopsy with histopathology for tissue diagnosis. The plan for further management, including assessment and treatment, should be discussed and agreed upon at multidisciplinary team (MDT) tumor board meetings involving various clinical specialties. Unfortunately, not all diagnostic services are universally accessible or provided with the desired equity. Therefore, there is a pressing need for tumor-site-specific diagnostic clinics that offer access to supportive genetic testing. Quality certifications in line with international standards and regulations are essential, often requiring accreditation from recognized international organizations and societies. This process should align with the development and implementation of regulations that adhere to internationally established diagnostic standards [5, 19–23].

5.5   Cancer Management

The establishment, adaptation, integration, and oversight of various cancer-related services are of utmost importance. These services encompass medical oncology, surgical interventions, radiotherapy, pediatric cancer care, and palliative care. In addition, emerging subspecialties such as gyne-oncology, uro-oncology, ortho-oncology, and neuro-oncology require dedicated development to meet the increasing demand and need for specialized services. Currently, these services do not meet the proposed standards. There is a notable deficiency in comprehensive supportive services accessible to all patients, including clinical psychologists, certified dietitians, social and community workers, and nursing care at all healthcare levels. We still face challenges in establishing and implementing universally accepted practice guidelines for physicians, radiologists, and pathologists in the treatment of each specific type of cancer. These guidelines should be customized based on local data, planned objectives, specific needs, disease biology within the local context, and available resources. The role of multidisciplinary tumor boards in cancer institutions is crucial, as they review and discuss new and complex cases to formulate well-documented care plans for each patient involved [19–23].

5.6   The Cancer Control Plan of the UAE

In 2011 [11, 16, 24, 25], the World Health Organization (WHO) emphasized the significant importance of establishing a standardized framework to track trends and evaluate the efforts of individual countries in addressing the worldwide cancer crisis. A comprehensive cancer control plan necessitates the collection of accurate data, the establishment of a reliable and evolving cancer registry, and the monitoring and evaluation of initiatives to ensure effective prioritization and quality implementation [26]. The World Health Organization (WHO) recommended that health regulatory bodies integrate their health information systems with reliable and evidence-based research indicators that have been well-established.

This plan embraces the principle of establishing a collaborative partnership with relevant sectors, both governmental and nongovernmental organizations, to effectively implement its goals. A unified policy has been developed to decrease cancer mortality rates, and numerous initiatives have been undertaken by participating governmental and nongovernmental agencies. The implementation of this plan signifies a significant milestone in advancing policy execution, health initiatives, and services aimed at reducing cancer prevalence and mortality, as well as controlling risk factors in the UAE [2, 17].

The Ministry of Health and Prevention, along with its partners in the UAE, has developed a national comprehensive plan to prevent and control cancer. We propose implementing this plan in line with specific objectives and in alignment with the National UAE Vision 2021. The national plan for cancer prevention and control (2022–2026) has been carefully crafted, taking into consideration the global action plan for cancer prevention and control by the World Health Organization (WHO), as well as the regional executive framework for cancer prevention and control in the Eastern Mediterranean Region Office (EMRO). The structure of this plan reflects the national strategy to reduce cancer incidence, improve the cancer patient's journey, decrease mortality by at least 25%, enhance quality of life, and alleviate the suffering of both cancer patients and caregivers. This approach is based on an integrated model of comprehensive cancer care services. The plan is based on nine strategic axes centered across the healthcare system:

1. Healthcare services
2. Cancer prevention
3. Sustainability
4. Innovation
5. Quality and patient safety
6. Health workforce
7. Research and development
8. Regulation and legislation
9. Strategic partnerships and collaborations

The plan encompasses various domains, such as governance, disease prevention, early detection, treatment, palliative care, surveillance, and research [2, 6, 17].

In collaboration with relevant sectors and stakeholders, including both public and private entities, a unified national plan for cancer prevention and control has been established. The objective of this plan is to strengthen the commitment to global and regional initiatives in combating and managing cancer. The strategic plan provides a comprehensive framework for indicators and controls that guide the development of policies, the formulation of regulations, and the evaluation of material resources to support disease prevention, cancer care, and quality management programs within the country. It also outlines appropriate measures for plan implementation, ongoing assessment, and a follow-up system. Key performance indicators (KPIs) have been identified to assess the effectiveness of proposed and implemented activities. This plan serves as a roadmap for the implementation of a targeted strategy to prevent and control cancer. It is an integral part of the National UAE Agenda for 2021, aiming to reduce cancer mortality rates and contribute to the achievement of the agenda's objectives, as outlined in the "one thousand people" initiative in Table 5.3 [11, 16, 23, 25, 26].

5.6.1 The Principles and Strategic Directions of the National Plan as per WHO and EMRO Guidelines

The national plan for cancer prevention and control draws inspiration from the strategic directives outlined by the World Health Organization for the Eastern Mediterranean Region. It aligns with the executive framework established for the action plan. The recommendations provided by both the Eastern Mediterranean Region Office (EMRO) and the World Health Organization (WHO) are condensed in Table 5.4 [4, 11, 17, 26].

Table 5.3 The vision, objectives, and strategy of the UAE cancer care plan [18]

Vision	To reduce cancer mortality and morbidity and improve survival rate in the UAE
Message	Saving lives and reliving suffering across the UAE population through cancer prevention, early detection, and best curative and palliative care
Objectives of the strategy	1. To strengthen the implementation and planning at the national level aiming to combat cancer in the United Arab Emirates 2. To reduce preventable and early cancer deaths and the risk of developing cancer by 30% by the year 2030 3. To optimize cancer prevention in the UAE society 4. To strengthen cancer prevention, early detection, and treatment 5. To ensure a sustainable and continuous development of cancer prevention, control, and treatment plans and strategies 6. To improve the cancer patients' quality of life 7. To ensure continuity of care through well-defined transition points in the healthcare system 8. To develop a framework to enhance, integrate, and coordinate initiatives to combat cancer and outline principles and regulations to supervise the organization 9. To ensure consistency and standardization in practices and help unify efforts in the fight against cancer. 10. To consolidate the efforts by providing a legal framework of applicable governmental regulations and policies

Table 5.4   Recommendations by EMRO and WHO [4, 11, 17, 26]

1.	**Governance:** Focus on developing a strategy and establishing a multisectoral committee for cancer prevention and control, while ensuring an available and sustained budget, adequate, and well-identified national cancer rates, establishing unified and reasonable costs for cancer care and management packages, and determining a mechanism to ensure treatment expense coverage with equity
2.	**Prevention:** Focus on implementing healthy lifestyle measures by combating smoking and encouraging physical activity and healthy food habits in line with the noncommunicable disease control framework and plan. This focus should also include vaccination strategies against hepatitis and HPV infections
3.	**Early detection:** The directions in this area focus on four main titles: raising public awareness about the importance of early warning signs and symptoms of cancer, mass education, and ongoing focused education for healthcare professionals on the early signs and symptoms of common cancers, easy and accessible diagnosis and referral for patients, effective screening programs, and continuous evaluation and monitoring of these programs. The focus should also be on enhancing the accessibility and affordability of diagnostic tools for suspected patients
4.	**Treatment:** Focus on the development and implementation of protocols and clinical practices based on evidence-based guidelines. Assess human resource availability and focus on cancer care services that are accessible to all with affordable treatment pathways. This also includes the development of an integrated, coordinated, and prompt referral system to avoid delays in diagnosis and treatment
5.	**Palliative care:** There is an unmet need to develop and integrate multidisciplinary palliative care services, including but not limited to pain management and psychological support, available in hospitals and primary healthcare centers. Developing and implementing these standards for best evidence-based practice and comprehensive care, and a smooth and early transition. Palliative care should be highlighted in medical educational programs.
6.	**Research and surveillance:** Developing a national cancer registry and hospital registries. With continuous monitoring of these registries through an accredited quality insurance program. The area includes focusing on the development and utilization of an integrated plan for research according to the priorities of the country

5.6.2   Definitions of the Strategic Axes and Executive Framework of the UAE Cancer Plan

The national cancer control plan of the UAE should encompass three main areas of focus, which are elaborated upon through nine strategic axes (Table 5.5). The first area of emphasis is cancer prevention, which encompasses education, understanding of the disease, prevention strategies, and early detection through prompt and efficient diagnosis. The second area centers around healthcare services, covering continuous care and comprehensive cancer treatment. The third area pertains to sustainability and innovation, which involves performance measurement, human resources, and research. These aspects are further outlined and discussed in detail in Table 5.6.

Table 5.5 Strategic axes of the UAE cancer plan [18]

First area
1. Education and understanding
Enhancing health awareness about the knowledge of cancer, risk factors leading to cancer, and correction of misconceptions about the disease
2. Prevention
Launching awareness campaigns and prevention programs against cancer and known causes
3. Early detection
Detecting cancer in the early stages increases the patients' survival and outcome. It involves periodic clinical assessments and reduces the delays in appropriate treatment referrals to receive treatment promptly
4. Rapid diagnosis
A healthcare center should assess the condition of the patient promptly in a systematic integrated way and take appropriate medical decisions based on a developed pathway
Second area
5. Treatment
Provide appropriately validated clinical practices in line with international guidelines for treating cancer according to disease stage to improve outcomes
6. Ongoing care
Provide timely transition to post-treatment and palliative care services for cancer patients and educate them about the appropriate ways to live with the disease and directions to avoid disease recurrence
Third area
7. Performance measurements
Establishing national records including all data sources in a central place and assembling comprehensive data of high quality, accuracy, and to record information about a disease
8. Workforce capacity building
Providing a qualified and appropriately trained team to deliver prevention, treatment, and continuity of care for patients; and provision of suitable training facilities for the workforce
9. Research
Cancer research improves the diagnosis, treatment, and outcome, and enhances quality of life by translating quality research and clinical trials for improvements in personalized care

Table 5.6 Detailed strategic and executive framework of the UAE cancer plan [18]

Strategic axis	Education and understanding			
Main objectives	Application mechanisms	Measurement indicators	The executing agency	Follow-up
Raising health awareness about cancer and associated risk factors and correcting the misconceptions	Conducting a national survey on awareness in society assessing knowledge of risk factors and opinions about access to services and early examination	Survey completion rate	Ministry of Health and Prevention—Noncommunicable Disease and Mental Health Section	Ministry of Health and Prevention—Noncommunicable Disease and Mental Health Section
	Raising health awareness about cancer risk factors • Inclusion of cancer in scientific curricula • The initiative of the researcher/young intellectual which aims and implements cancer awareness campaigns in school and university	Number of awareness campaigns	Stakeholders	
	Awareness campaigns synchronized with designated international days for each cancer type	Number of awareness campaigns	Stakeholders	

(continued)

Table 5.6 (continued)

Strategic axis				
	II. Prevention			
Main objectives	Application mechanisms	Measurement indicators	The executing agency	Follow-up
Monitoring of risk factors between different groups in society and encouragement to adopt healthy lifestyles	Physical activity • Inclusion and intensification of physical activity compulsory in schools • Campaigns to encourage exercise and walk in the community • Creation of more tracks for walking and parks within reach of people	The number of awareness programs	Ministry of Health and Prevention—Noncommunicable Disease and Mental Health Section	Ministry of Health and Prevention
	Healthy foods • Develop educational programs on a healthy diet	The number of awareness programs		
	Assessment of the presence of carcinogenic factors in the environment and highlighting the environmental pollution and exposure to radiation	The number of awareness programs		
	Awareness campaigns about the harms of smoking and shisha in young people	The number of awareness programs Monitor smoking rates		
Providing preventive vaccinations	Hepatitis B vaccination for prevention of liver cancer for high-risk population	Hepatitis C vaccination coverage rate Children and among those who have major risk	Ministry of Health and Prevention/ All health authorities	
	HPV vaccination in schools and society for girls aged 13–26 years to prevent cervical cancer	Coverage rate of the targeted category		

Strategic axis	III. Early detection			
Main objectives	Application mechanisms	Measurement indicators	The executing agency	Follow-up
Create a national program for early detection of cancer	Development of a central public electronic recall system for early detection services and identification by e-mail Create a national platform or program for the registration of cases that underwent early examination for cancer Increase the capacity of logistical and human resources to increase population coverage	• Completion rate • Population coverage rate in target groups	Ministry of Health and Prevention/Statistics Department and Research	Ministry of Health and Prevention The society
Health insurance and financial coverage for early disclosure	• Insurance coverage for early diagnostic examinations • Discounted packages for early detection and uniform prices in the private and public sectors	Completion rate		

(continued)

Table 5.6 (continued)

Strategic axis					
III. Early detection					
Main objectives	Application mechanisms	Measurement indicators	The executing agency	Follow-up
Increase awareness about the importance of early detection	• Awareness campaigns on the importance of early screening • Facilitate visitors to healthcare centers • Target age and annual performance linked for health workers in early examination centers • The rate of turnout to early examination • Appointing clinical nurse specialists to educate and support the team	• Number of awareness campaign • Percentage of cases from categories of target age transferred for early examination • Number of specialized workforces that were redundant	Ministry of Health and Prevention (community/participants)	Ministry of Health and Prevention
Establish a framework and governance policy for quality assurance and early screening services in the health regions	• Develop a framework for standardization and best practices • Clinical pathways for early cancer examination • For early screening of breast, cervical, and colon cancer among the target age groups in the population	Percentage of policy completion and frameworks	Ministry of Health and Prevention—Noncommunicable Disease and Mental Health Section	
Raising awareness of common symptoms of cancer in society	Awareness campaigns about the symptoms of the most common cancers in the community	Number of awareness campaigns	All participating parties	

Strategic axis			
Main objectives	Application mechanisms	Measurement indicators	The executing agency	Follow-up
IV. Rapid diagnosis				
Establish an effective referral system between different levels of care for cancer patients	1. Implementing the service access policy so that access time is reduced to diagnostic and therapeutic services every year	1. Average waiting time from the time of onset to GP referred to the specialist 2. Average waiting time from doctor's appointment till the diagnosis 3. Average waiting time from diagnosis time until the start of the treatment 4. Average waiting time from GP appointment to time received treatment	Service providers, early examination, and therapeutic services	Ministry of Health and Prevention Participating parties
Rapid diagnostic initiative for lung cancer	Launch of rapid mobile investigation clinics for early detection of lung cancer using X-ray, CT scan, and breath examination for people susceptible to lung cancer	Number of beneficiaries		

(continued)

Table 5.6 (continued)

Strategic axis	V. Treatment			
Main objectives	Application mechanisms	Measurement indicators	The executing agency	Follow-up
Covering the cost of cancer treatment	Adopt a model pay for performance (personalized reimbursement model). Dubai Health Authority is providing therapeutic services where the treatment is covered by insurance and pharmaceutical companies. The results are then evaluated on treatment response/efficacy	Number of beneficiaries	Healthcare service providers	Ministry of Health and Prevention Participating parties
Accreditation of centers of excellence for cancer treatment	Preparing for specialized centers of excellence (third level) in cancer treatment and its complications and rehabilitation centers	Number of accredited centers of excellence	Stakeholders	

Strategic axis	VI. Ongoing care			
Main objectives	Application mechanisms	Measurement indicators	The executing agency	Follow-up
Development of palliative care services	• Studying the work on adding palliative service in health centers with easy access to services and developing a guide for implementing a palliative care program	• Percentage of completion of the guide	Ministry of Health and Prevention Stakeholders	Ministry of Health and Prevention Participating parties
	• Creation of teams to support cancer patients. The team consists of patients who have recovered or are under treatment • The team meets periodically for psychology support among patients	• Team achievement percentage • Number of beneficiaries	Stakeholders	

Strategic axis	VII. Performance measurements			
Main objectives	Application mechanisms	Measurement indicators	The executing agency	Follow-up
Annual evaluation for anticancer performance indicators	• Establishing a registry for early detection of cancer • Preparing and developing a registry for early detection and development • Early detection data collection	The percentage of completion of the national registry, population coverage for early cancer screening among the targeted age groups	Ministry of Health and Prevention Department of Research and Statistics Participating parties	Ministry of Health and Prevention Department of Research and Statistics
	• Create a unified national electronic registry for cancer	The percentage of completion of the National Cancer Registry	Ministry of Health and Prevention Society/Statistics Department of Research and Statistics Participating parties	Ministry of Health and Prevention Department of Research and Statistics
	Measuring indications: • The rate of detection of positive cases by early cancer examination • Average number of cases detected in late stages of cancer • Waiting period, since the case was referred before to the general practitioner to complete the early examinations	Report completion percentage rate of time commitment to access the service	Ministry of Health and Prevention Participating parties	Ministry of Health and Prevention Community/care management Specialty
	• Measurement of the KAP index (knowledge, attitude, and practice—to assess acceptance) for the community's understanding of cancer screening	Report completion rate	Ministry of Health and Prevention Participating parties	Ministry of Health and Prevention Community/care management Specialty

(continued)

Table 5.6 (continued)

Strategic axis	Main objectives	Application mechanisms	Measurement indicators	The executing agency	Follow-up
VII. Research	The priority for epidemiological and clinical research of cancer	Research work to discover concepts, knowledge, and opinions about cancer, risk factors, and screening cancer in the context of encouraging research related to cancer	Research completion rate	Ministry of Health and Prevention Participating parties	Ministry of Health and Prevention Community/care management Specialization/management Statistics and research
		Develop a research agenda for the three most common cancers (breast, colon, and thyroid)	Agenda completion rate	Ministry of Health and Prevention Participating parties	
Strategic axis	Main objectives	Application mechanisms	Measurement indicators	The executing agency	Follow-up
IX. Workforce	Providing qualified human resources in the developing field of cancer	Complete medical team specialized in treating cancer in secondary and specialty care • Specialized doctors • Clinical nurse specialist • X-ray technicians	The percentage of increase in the workforce	Ministry of Health and Prevention Participating parties	Ministry of Health and Prevention Community/care management Specialty
	Raising the efficiency of employees, healthcare professionals	Creation of training programs for healthcare workers in the field of cancer and assess the risk factors such as: • Awareness and health education • Early detection • Palliative care	The number of training programs	Ministry of Health and Prevention Community/training center and development Participating parties	

5.7   Vistas in Cancer Care Plan

5.7.1   Expected Cancer Burden

The documented age-adjusted cancer incidence in the UAE population is lower compared to the reported incidence in countries such as the United States and other similar developed nations. However, it is anticipated that there will be an actual increase of approximately 10–15% in the number of cancer cases annually, aligning with the current trends in cancer occurrence [2, 3, 5, 7, 8, 16].

5.7.2   Cancer Control Program

It is crucial to establish a comprehensive National Cancer Control Program that encompasses strategy development, the implementation of preventive measures, and the execution of a comprehensive cancer management plan. This process requires collaboration among various stakeholders, including government agencies, nongovernmental organizations (NGOs), and the community. An effective and well-rounded plan should prioritize risk reduction, early detection of cancer, and improved management to enhance survivorship and outcomes and ultimately reduce the overall burden of cancer.

The cancer control program should not only simply aim to maintain the current incidence rate of cancer but also strive to reduce it. A thorough national cancer program assesses different approaches to cancer control and implements the most modern and cost-effective methods for the entire population. It emphasizes the significance of cancer prevention and early detection at a treatable stage while also aiming to provide optimal support to patients and their caregivers during the journey of advanced disease. A well-established and effectively implemented cancer control program is expected to lead to substantial risk reduction, improved early detection, enhanced treatment approaches, and ultimately higher rates of survivorship.

In 2022, the National Cancer Committee underwent revisions and welcomed experts from various healthcare sectors. It is crucial to adhere to the WHO Cancer Control Strategy to establish a uniform, equitable, and consistent approach. Strengthening collaboration with other ministries, including the Ministry of Education, the Ministry of Religious Affairs, and the media, is strongly encouraged. A well-defcO strategy to local requirements and existing data [8, 11, 24–26].

5.7.3   Cancer Registry

The cancer registry serves as a primary tool for the Cancer Control Program, providing essential functions. It plays a vital role in determining the extent of the cancer burden, identifying prevalent risk factors, and evaluating the effectiveness of the cancer control program. By collecting and analyzing comprehensive epidemiological data, such as cancer incidence, mortality, prevalence, stage at diagnosis, and patient survival outcomes, the cancer registry enables the development and organization of standardized control plans. It provides valuable insights into cancer patterns and trends over time, aiding in strategic planning and the implementation of effective control measures.

In recent times, registry managers from 19 countries in the Middle East and North Africa (MENA) region have reported the existence of 97 population-based registries, 48 hospital-based registries, and 24 pathology-based registries. The majority of population-based registries were either well-developed or partially developed. However, significant challenges were identified, including the lack of accurate death records, incomplete and unclear medical records, limited communication between various stakeholders, and a shortage of trained personnel. These challenges were particularly pronounced in active conflict zones and neighboring regions. Cancer registration faced additional obstacles, including inadequate health infrastructure, the absence of regulations mandating cancer registration, and disruptions caused by ongoing wars, conflicts, and financial constraints [27].

A fully developed national cancer registry consistently publishes an annual report that provides details on cancer incidence and prevalence. It is essential for the registry to collect extensive information, including disease staging, mortality rates, disease demographics, and data on risk factors. However, the registry often faces challenges such as a limited workforce, inadequate data on patients diagnosed and treated overseas, difficulties in obtaining cooperation from nongovernmental institutions and medical personnel, and the mobility of the population [5, 27–29]. To ensure accurate, consistent, reliable, comprehensive, and valuable data in the cancer registry, the following actions need to be taken:

1. Establish the Cancer Registry Advisory Committee, which will be responsible for planning cancer registry activities, providing feedback on data and reports, and contributing to data dissemination and advocacy efforts.
2. Implement data quality indicators to monitor and improve the accuracy and reliability of the collected data.
3. Collect complete data on cancer stage, mortality, risk factors, and prevalence to enhance the comprehensiveness of the registry.
4. Increase the number of trained personnel in the registry by offering training to new staff members and providing ongoing technical and financial support. This will ensure the completeness of the cancer registration team and the sustainability of the cancer registry.
5. Collect more comprehensive data on different types of cancers, including non-morphological data, to capture a broader range of information.
6. Establish mandatory reporting requirements to ensure that healthcare providers are legally obligated to notify cancer cases.
7. Make adjustments to the cancer notification form and electronic notification forms to improve the efficiency and effectiveness of the reporting process.
8. Foster greater collaboration from healthcare providers to encourage their active participation and contribution to the cancer registry efforts.

5.7.4   Indicators for Monitoring

There are several widely recognized indicators used globally to monitor and assess progress within a national cancer framework. Key performance indicators (KPIs) are utilized, some of which are well-established, while others are specifically developed to suit local circumstances. Examples of these indicators include raising cancer awareness, reducing the number of active and passive smokers, increasing the proportion of early-stage (1 and 2) colon and breast cancer cases, boosting the number of patients diagnosed through screening and/or urgent referral pathways, comparing 1-year and 5-year cancer survival rates to international benchmarks, increasing the percentage of patients with comprehensive treatment plans across the entire care pathway, and decreasing the number of patients with incomplete comprehensive treatment plans [3–5, 10, 16, 17, 23, 26].

5.7.5   Continuing on the Path to Progress and Excellence

While considerable advancements have been achieved, there is still a considerable amount of work that lies ahead as we strive to realize our vision for the future of cancer care, which remains a significant challenge. The UAE is committed to pursuing excellence in health care, particularly in the field of cancer care, by enhancing specialized services and implementing innovative treatment approaches. It is crucial to continuously update and incorporate evidence-based, approved treatment modalities and technologies that have demonstrated positive impacts on patient care into our standardized management strategies. Moreover, there is a need for improvements in the patient experience, with a particular emphasis on ensuring seamless care pathways that align with the principles of the WHO Cancer Care Continuum [24, 26, 27, 29].

5.7.6   Public Education and Understanding

It is imperative to invest in knowledge dissemination and educate the public about the significance of prevention, enhance early detection initiatives, and facilitate accurate cancer diagnoses. Efforts should be made to improve awareness campaigns and debunk misconceptions surrounding cancer. These misconceptions include the belief that cancer only manifests in advanced stages or specific locations, that it is more prevalent in low-income countries, that it is invariably fatal, and that surgery accelerates its spread. In addition, cancer carries numerous stigmas, leading to discomfort and avoidance when discussing the topic. People may refrain from using terms such as "cancer" or "tumor." A diagnosis is often associated with vulnerability, concerns about job security, and apprehension about the impact on sexual activity [2, 4, 28, 29].

5.7.7 Partnership Within the Cancer Community

There is a significant requirement to establish strong connections within the cancer patient community, with the objective of promoting cancer awareness and prevention; fostering public knowledge and comprehension of cancer; facilitating the sharing of experiences; offering financial assistance to eligible cancer patients; promoting collaboration with various stakeholders to organize gatherings, workshops, educational programs, and conferences; and supporting research endeavors that raise awareness about the disease and its prevention methods [27, 29].

5.8 Prevention

Nearly one-third of cancer cases can be attributed to lifestyle and environmental factors. The incidence of these preventable cancers will depend on the success of preventive service initiatives in promoting healthy diets, physical activity, and environmental well-being. By prioritizing public health through primary healthcare and prevention services, we can reduce these risk factors and effectively lower the incidence of cancer. A small proportion of cancer cases (10%) are associated with a significant family history of cancer, potentially resulting from identifiable gene mutations. Enhancing the capacity for genetic testing is necessary to identify these cancer-linked genes. To address this, a comprehensive program for cancer prevention and education should be developed, specifically targeting academic institutions, and regular public education events should be scheduled [1].

5.8.1 Healthy Lifestyle Strategies

Adopting a healthy lifestyle, maintaining a nutritious diet, and engaging in sufficient physical activity play a crucial role in reducing the risk of cancer. It is important to enhance awareness regarding the positive impact of nutrition and regular physical activity. To achieve this, it is necessary to establish and continuously monitor culturally appropriate and sustainable policies and regulations that aim to promote food diversity, encourage healthy eating habits, and facilitate physical activity across various settings, such as the general population, schools, universities, workplaces, and targeted communities. Primary healthcare teams have a vital role in providing education, preventive guidance, and wellness services. The presence of health coaches in primary healthcare centers is essential. Cancer prevention and education messages should be disseminated through family physicians, primary care providers, and community pharmacists, whenever possible, to support community education. These messages should be tailored to local needs and culture, such as enhancing private exercise and recreation areas for women, providing access to exercise facilities in the workplace, and regulating potentially harmful substances used in food and its storage [10, 12, 13].

5.8.2   Access to Genetic Testing

The breast and ovarian cancer screening clinic for individuals at high risk should incorporate genetic testing and counseling services. Detecting individuals with a heightened risk of hereditary cancer allows for prevention and early detection, leading to the potential reduction of cancer cases and related deaths. Consideration should be given to making premarital screening mandatory, and the scope of services can be expanded to encompass other hereditary diseases linked to cancer, including gastrointestinal conditions. It is crucial to foster collaboration and establish an infrastructure for a national program focused on genetic testing, sequencing, and research [26, 28, 29].

5.8.3   Vaccination

Human papillomavirus (HPV) infection is widely recognized as a major cause of cervical cancer. Effective vaccination against the specific HPV strains that can lead to cancer is possible. Many countries with high rates of cervical cancer have implemented vaccination programs targeting preadolescent females. While the incidence of cervical cancer is currently low in the UAE, further reductions can be achieved through vaccination. Therefore, it is crucial to ensure that the HPV vaccine remains accessible to families who choose to immunize their children. In addition, hepatitis vaccination already plays a well-established role in preventing hepatocellular carcinoma and should continue to be included in all national immunization programs [27, 29].

5.8.4   Smoking

It is imperative to maintain our commitment to implementing a public health strategy aimed at eliminating all forms of smoking. This strategy encompasses various components, such as tobacco surveillance systems, warnings on cigarette packaging, a regulatory framework for enforcement, tobacco cessation services, including a national quit helpline and website, access to nicotine replacement therapy, support through primary healthcare services, and a well-designed comprehensive tobacco taxation model encompassing customs and excise taxes on all tobacco products [14, 26, 28, 29].

5.9   Early Detection

Evidence has shown that early detection of cancer significantly improves the likelihood of successful treatment and better outcomes. For instance, in cases of colon cancer diagnosed at the earliest stage, over 90% of individuals survive for at least

10 years. On the other hand, if the cancer is diagnosed at an advanced stage, the 10-year survival rate drops to below 5%.

5.9.1   Increase in Awareness

Ongoing efforts to raise awareness about the accessibility and significance of screening and early detection should persist and be enhanced throughout the entire healthcare system, encompassing public and private primary, secondary, and tertiary healthcare providers. Engaging public figures as advocates for promoting screening as a means of early disease detection can have a substantial impact. It is essential for each country to determine the appropriate age and frequency of screening based on its own national data [29].

5.9.2   Enhancing Cancer Detection Services

Despite having a relatively low incidence of cancer, our priority remains enhancing the treatment outcomes for cancer patients. Early detection plays a critical role in improving cancer survival rates. Given that breast cancer tends to occur at a younger age in the United States compared to other developed nations, it is essential to consider initiating breast screening for women at a younger age [30]. Continuing the implementation of comprehensive national screening programs and establishing effective systems for screening recall are crucial to screening services. By thoroughly examining baseline data, we can assess the advantages and challenges of lowering the age for initial screening and implementing screening for the young adult population. Cervical screening operates on an opportunistic basis due to the relatively low prevalence. A situational analysis is needed to assess the evidence for the adoption of a national population-based cervical cancer screening program. Early detection of lung cancer can significantly improve outcomes, and it is worth considering the evaluation of low-dose computed tomography scanning for lung cancer in older males and smokers. Moreover, given the increasing incidence of thyroid cases, population-based thyroid screening should also be taken into consideration. It is crucial that the guidelines are fully embraced and uniformly implemented by all healthcare providers. These guidelines should be formally adopted and established as national policy while also being utilized for reimbursement by the National Health Insurance. To stay up-to-date with evolving practices and technological advancements in diagnostics, it is necessary to periodically review and update clinical management guidelines and screening protocols through peer review. Implementation of continuous improvement initiatives and the use of performance indicators are essential for effective monitoring and surveillance. Operational standards for all cancer screening programs need to be updated on a regular basis [6, 19–22, 27–29].

5.10   Diagnosis

Ensuring a prompt and accurate diagnosis is crucial when cancer is suspected to enable timely treatment. It is necessary to establish efficient referral pathways to ensure that patients referred to specialists for suspected cancer undergo diagnostic procedures within a predetermined timeframe. During this period, a comprehensive evaluation, including physical examinations, imaging, and pathology testing, should be conducted to establish a diagnosis, which can then be discussed and approved in multidisciplinary team (MDT) meetings. The investment in advanced imaging systems has yielded positive results, enabling clinicians to plan targeted therapies and monitor treatment responses. It is important to have a connected system where diagnostic images can be accessed by all healthcare providers. Centralized pathology laboratories, PET scans, nuclear medicine units, and other support services can enhance diagnostic capabilities. Streamlining the patient experience along the entire care pathway is crucial, with breast clinics following international best practices to offer mammography, ultrasound, core biopsy, and clinical examination in a single visit. In the coming years, site-specific diagnostic clinics will be established, equipped to provide site-specific genetic testing. While considering the pros and cons of centralization, the principle of "localize where possible, centralize where necessary" should guide the planning and development of cancer diagnostic services [3, 8, 19, 20, 26].

We need to ensure our dedication to meeting the highest international standards and adhering to quality control regulations, which often necessitate accreditation from internationally recognized organizations. It is important to establish a steering committee that will be responsible for overseeing the regulation, development, and implementation of diagnostic standards.

5.10.1   Monitoring and Measuring Success

Regular and periodic monitoring of diagnostic pathways is essential to ensure sustained progress. The number of cancer patients seeking healthcare services worldwide is steadily increasing, leading to greater demands on diagnostic services such as pathology and imaging. Meeting targets becomes challenging due to these increasing demands. The implementation of electronic medical records systems in public healthcare providers has improved data exchange capabilities within the system. Prioritizing the recruitment of specialized personnel in pathology and imaging, the continuous development of patient pathways, the adoption of precision diagnostic techniques, the incorporation of precision medicine, and the integration of molecular pathology and molecular genetics are crucial. Gradual decentralization of diagnostic tests and training in cancer diagnosis can help meet timelines for cancer treatment pathways and increase the rate of patient referrals in a cost-effective manner. Providing training for primary care physicians in procedures like colonoscopy, ultrasound, and radiology can support the diagnostic pathway. It is necessary to initiate a communication skills training program as well [28, 31].

Regularly gathering feedback from patients and primary care physicians is important for the ongoing review, development, and enhancement of the training program. This feedback should be consistently communicated across all levels of the healthcare system to ensure uniformity and enable patients to make well-informed decisions about their treatment. It is necessary to create tools that facilitate shared decision-making. Key performance indicators can be established in collaboration with healthcare professionals and the broader public to monitor progress and effectiveness [27, 29].

5.11 Treatment

Following the establishment of up-to-date diagnostic services, it is essential to develop and maintain comprehensive treatment services, which encompass surgical oncology, medical oncology, radiation oncology, and palliative care. Surgical oncology has undergone continuous advancement, and a proficient surgical oncologist collaborates closely with other subspecialties in oncology to provide improved cancer treatment. This collaboration helps reduce the occurrence of unnecessary surgeries, such as those with involved margins, inadequate surgeries, or unplanned procedures. Furthermore, surgical oncologists are skilled in reconstruction surgeries, aiming to enhance the quality of life for patients. Integrating surgical oncology within oncology services and ensuring structured training and fellowship programs for surgeons are crucial. In addition, the introduction of new supportive services like clinical psychologists, clinical dietitians, social workers, and community nursing services is necessary [7, 28].

We should establish and enforce guidelines for referring physicians and radiologists to ensure standardized practices for each type of cancer. Strengthening the tumor boards in cancer institutes is crucial, and all new and complex cases should be reviewed and discussed by a multidisciplinary team (MDT). This collaborative approach will result in a well-documented care plan tailored to each patient's needs. In addition, there should be a mechanism in place to evaluate and incorporate new treatment options based on randomized controlled trials, international approvals, and guidance, ensuring that the most up-to-date and effective treatments are available [23, 27, 31].

5.12 Palliative Care

Palliative care is a crucial and specialized aspect of cancer care services that centers around alleviating symptoms, preventing patient distress, and enhancing overall quality of life [8, 17, 29]. Palliative care encompasses all stages of illness, including those receiving treatment for curable conditions, those living with the disease, and those nearing the end of life. It can be provided in various settings, such as hospitals, homes, or hospices. Palliative care adopts a multidisciplinary approach, involving a team of healthcare professionals including physicians, pharmacists, specialized and community nurses, chaplains, social workers, psychologists, dietitians, and other allied health professionals. Together, they collaborate to create a comprehensive care plan that addresses the physical, emotional, spiritual, financial, and social needs of the patient and their family, aiming to alleviate suffering in all aspects of their lives [24, 28]. We need to identify and address gaps and needs in palliative care and utilize all available resources effectively. Our focus should be on promoting the establishment of palliative care services across different levels of health care, including the community. This entails engaging in advocacy, providing education and training, and raising awareness about palliative care. We should also advocate for the availability and accessibility of essential opioids and other medications used in palliative care for all cancer patients. In addition, we should explore potential legislative and regulatory changes to support and improve palliative care services.

5.13 Services

There will be a growing need to expand services and optimize medical oncology services in terms of human resources and infrastructure to ensure the effective delivery of systemic anticancer therapy. Early diagnosis plays a crucial role in improving cure rates and is both simple and cost-effective. Achieving early diagnosis relies on raising awareness among the general public and healthcare professionals about the early signs and symptoms of cancer. This includes recognizing potential warning signs and taking prompt action. It is important to disseminate knowledge to the public to enhance cancer awareness and provide training to healthcare professionals to improve their understanding and skills in identifying early signs and symptoms of common cancers. Furthermore, it emphasizes the importance of accessible, affordable, and timely access to diagnostic tests, staging investigations, treatment services, and follow-up care within public healthcare services. Screening is a method of identifying individuals who are apparently healthy and asymptomatic but at higher risk of having early-stage disease that may not be detectable clinically. Screening tests can be offered to the general population at regular intervals (population-based screening) or recommended to asymptomatic individuals by healthcare providers during routine healthcare visits (opportunistic or spontaneous screening).

Cancer control and prevention can be accomplished through several crucial measures. Implementing an effective screening program can lead to a decrease in the incidence of new cancer cases, improved outcomes, and reduced mortality rates. Globally recommended screening programs for breast, cervical, and colorectal cancers aim to prevent these diseases by identifying and treating pre-cancerous lesions that may increase the risk of developing malignancies. These screening initiatives play a vital role in detecting and managing these conditions in an effective and timely manner [19–22]. Furthermore, early detection of these diseases, often at a stage where treatment is possible, is facilitated. To effectively plan and address the situation, it is necessary to conduct a comprehensive analysis, evaluating the current status and identifying the need for capacity building. This includes developing training programs for healthcare providers and ensuring readiness and accessibility to timely diagnostic investigations, appropriate treatment options, and follow-up care. Collaboration with the community and the utilization of mass media platforms can help disseminate accurate information to the public. Standardized plans for early cancer diagnosis and screening in primary care facilities should be developed, establishing reliable screening infrastructure tailored to specific cancers and adhering to unified selection criteria for cancer screening [17].

We should provide training to volunteers from nongovernmental organizations (NGOs) to enhance their ability to effectively communicate scientifically based information to the public. In addition, efforts should be made to establish a well-structured screening registry to ensure organized and systematic screening processes.

It is important to initiate women's health programs at the age of 40, focusing on regular clinical breast examinations, educating women about breast self-examination, and providing mammograms as necessary and upon request. We should also prioritize adherence to the Extended Program of Immunization and the National Tobacco Control Program.

In numerous developed nations, well-organized medical oncology services exist, accompanied by comprehensive guidelines for the safe and efficient administration of chemotherapy medications. These services are conveniently located near patients' residences, ensuring easier access to effective treatment and enhancing patient adherence and outcomes. Therefore, there is a continuous requirement to establish models that enable simplified and secure delivery of systemic therapy, including chemotherapy, to eligible patients in the UAE, integrating it as an essential element of a National Cancer Control Program [3, 9, 16, 26, 29].

The implementation of a comprehensive plan to enhance oncology services at the secondary and tertiary healthcare levels will contribute to ensuring equal access to standardized treatment, and delivering safe and optimal therapy in proximity to patients' residences. This involves establishing guidelines for the procurement, storage, and administration of drugs, monitoring treatment outcomes, establishing referral pathways, and creating a network of oncology services within the public healthcare system to facilitate training. To meet the growing demand for cancer care, it is essential to establish satellite or secondary oncology centers that can provide services in close proximity to patients, alleviating the need for extensive travel. These centers should operate under the supervision of the main oncology services, maintaining affiliation and collaboration.

5.14   Resource Allocation and Cost-Effectiveness

Proper allocation of resources is crucial for the development and success of any cancer control plan. It requires strong commitment, dedication, and vision from decision-makers. The rising costs of cancer treatment, driven by advancements in diagnostics and therapies, have made it increasingly unaffordable. In the United States, private insurers rely on the Institute for Clinical and Economic Review (ICER) to assess costs and determine reimbursement decisions. Meanwhile, in

England, the National Institute for Health and Care Excellence (NICE) makes coverage decisions for the publicly funded National Health Service. Many newly developed cancer drugs are found to be lacking cost-effectiveness in evaluations conducted by NICE. However, NICE's ability to negotiate price discounts and implement patient access schemes has helped reduce costs significantly. The challenges in proving the clinical effectiveness of new cancer drugs, including reliance on inadequately validated surrogate measures, contribute to variations in cost-effectiveness. The adoption of ICER assessments by private insurers in the US has provided a standard for measuring the comparative value of new cancer treatments. NICE employs various policy tools, such as value-based pricing, direct price negotiations, and patient access schemes, to inform its recommendations, considering a range of cost-effectiveness values. Spending on expensive new cancer therapeutics is increasing for payers and health systems. Both ICER and NICE have indicated that most new cancer drugs do not offer good value for money at their current prices [28, 31–33]. Hence, it is essential to establish an evaluation system for all treatments, engage in negotiations with suppliers, and develop patient support systems. We need to allocate our existing resources in a fair and reasonable manner to provide patients with equitable access, affordability, and proven therapeutic benefits.

5.14.1 Workforce

There will be a requirement to invest in capacity development by recruiting and nurturing a skilled workforce, providing training opportunities, and establishing local institutions. It will be necessary to enhance the potential and capabilities of the workforce through ongoing training, education, and the implementation of personal development programs for each individual.

We should establish training programs and fellowships in specialized areas of cancer care to strengthen our cancer services by training and incorporating a local workforce. It is essential to ensure quality by implementing accreditation, certification, audits, and regular evaluations [2, 4, 9, 11, 19, 20].

5.15   Research

In 2011, the United Nations highlighted the importance of research to guide action against NCDs at regional and global levels [24, 26, 29] as NCDs, including cancer, are now considered global crises. Every country, culture, and society are different in terms of cancer incidence, cancer types, cancer biology, economy, access, and affordability. Differences in cultural beliefs, customs, and misconceptions about cancer contribute to variations in the acceptance of certain treatments across different societies. Therefore, there is a growing need to develop guidelines, pathways, and practices that are tailored to local factors and take into account the specific cultural contexts.

5.16   Conclusion

Countries in the WHO Eastern Mediterranean Region must prioritize national cancer control planning, as it is crucial to address the rising incidence of cancer in this region. The increasing rates of cancer pose significant challenges, including a high disease burden, premature mortality, and escalating healthcare expenses [4, 17]. Cancer control planning and implementation exhibit significant disparities both between countries and within countries. Over half of the countries (12 out of 22) have individual comprehensive national cancer control plans, while six (27%) have noncommunicable disease plans that encompass cancer [27, 29]. The effective execution of cancer plans has faced obstacles due to inadequate governance structures, limited coordination mechanisms, and insufficient human and financial resources. In the majority of countries (20 out of 22, or 91%), the plan receives either full or partial funding, yet this is often hindered by political instability and conflicts, greatly impacting the planning and implementation of cancer control measures [32].

The UAE, as a growing economy, is facing a significant burden of cancer that is expected to increase, leading to higher rates of illness and death. However, the current efforts in screening and early detection are falling short of reaching the desired coverage among the target population. To effectively combat cancer, it is crucial to implement a comprehensive, well-structured, and efficient national cancer control plan. This plan should involve accurate data collection, the establishment of an organized and efficient cancer registry, and regular monitoring and evaluation of the plan's activities. The UAE's cancer control plan aligns with the cancer control initiatives and frameworks set by the WHO and EMRO, incorporating well-defined objectives and clear targets. These objectives aim to combat cancer, reduce its incidence, minimize cancer-related morbidity and mortality, improve patient outcomes, and enhance the quality of life for individuals affected by cancer [2, 8, 9, 17].

Continual commitment to progress and excellence is essential to our journey. Expanding the cancer registry and providing it with the necessary legal framework is imperative. We must prioritize preventive oncology by integrating the latest knowledge, advancements in technology, and approved medications based on solid evidence from international data and guidelines. Developing clinical pathways and guidelines and implementing them, along with using key performance indicators (KPIs), will enable us to monitor our cancer services effectively. Collaboration with all stakeholders is crucial in enhancing cancer care initiatives and ensuring equitable and affordable services while considering cost-effectiveness. Retaining a skilled workforce and continuously enhancing their expertise through training and ongoing education is vital. Regular monitoring and evaluation of their performance is essential. Seeking international accreditations from reputable organizations will ensure that we remain on the path of progress and excellence [34].

Cancer care services should be easily accessible, consistent, and affordable for everyone. By optimizing resource utilization, we can improve the delivery of cancer care, ensuring that it reaches patients conveniently. It is important to establish connections and collaborations between primary health care, secondary care hospitals, tertiary care centers, and private cancer care facilities equipped with advanced technologies. A significant emphasis should be placed on investing in preventive oncology through

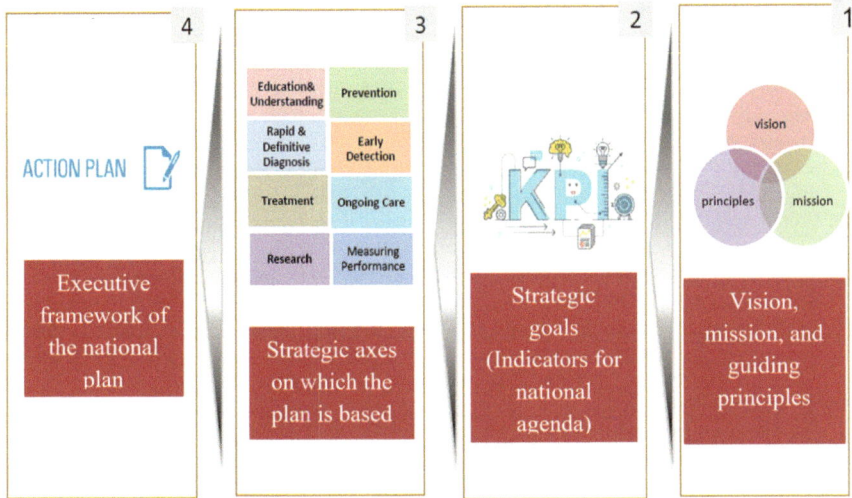

Fig. 5.1 General structure of the plan

enhanced education, screening, and early detection approaches. The patient's path from identifying symptoms to receiving definitive treatment should be well-organized, following both local regulations and international guidelines and protocols.

The primary emphasis should be on enhancing public awareness, enhancing coordination of prevention efforts, expanding initiatives for early detection, ensuring timely diagnosis, expediting treatment processes, and ensuring seamless continuity of care. These collective and integrated efforts should undergo regular evaluation based on predetermined targets and key performance indicators (KPIs), with a strong emphasis on workforce training and research. By addressing all aspects of service and care, we can enhance the delivery of high-quality care and improve patient outcomes (Fig. 5.1).

Conflict of Interest The authors have no conflict of interest to declare.

References

1. National cancer control programs: policies and managerial guidelines. 2nd ed. Geneva: World Health Organization; 2002. http://www.who.int/cancer/media/en/408.pdf.
2. Towards a strategy for cancer control in the Eastern Mediterranean Region, WHO Regional Office for the eastern Mediterranean WHO EMRO. 2009. https://applications.emro.who.int/docs/EMROPUB_2019_NCD_EN_23548.
3. Radwan H, Hasan H, Ballout RA, Rizk R. The epidemiology of cancer in the United Arab Emirates—a systematic review. Medicine (Baltimore). 2018;97(50):e13618. https://doi.org/10.1097/MD.0000000000013618.
4. Regional framework for action on cancer prevention and control Executive summary. WHO Regional Committee for the Eastern Mediterranean sixty-fourth session provisional agenda item 3(a). EM/RC64/3. 2017. https://applications.emro.who.int/docs/RC_technical_papers_2017_3_20037_en.

5. Cancer incidence in United Arab Emirates-Annual Report of the UAE-National Cancer Registry. Statistics and Research Center, Ministry of Health and Prevention.
6. National Cancer Registry (UAE). The Statistics and Research Department—National Disease Registry Section. Year List. 2019, 2017, 2016, 2015, 2014, 2013, 2012, 2011. Report. https:// smartapps.moh.gov.ae/ords.
7. Country Profile United Arab Emirates. https://www.who.int/nmh/countries/are_en.pdf.
8. Shihab M. Economic development in the UAE. In: Abed IA, Hellyer P, editors. United Arab Emirates: a new perspective. London: Trident Press; 2001. p. 249–60.
9. United Arab Emirates National Bureau of Statistics. Population estimates 2006–2010. http:// www.fcsa.gov.ae/.
10. Henry CJK, Lightowler HJ, Al-Hourani HM. Physical activity and levels of inactivity in adolescent females ages 11-16 years in the United Arab Emirates. Am J Hum Biol. 2004;16:346–53.
11. World Health Organization. Regional Office for the Eastern Mediterranean. Country cooperation strategy for WHO and United Arab Emirates: 2012–2017. 2012. http://apps.who.int/iris/ bitstream/10665/113226/1/CCS_UAE_2012_EN_14947.pdf.
12. Ng SW, Zaghloul S, Ali H, et al. Nutrition transition in the United Arab Emirates. Eur J Clin Nutr. 2011;65:1328–37.
13. Ng M, Fleming T, Robinson M, et al. Global, regional, and national prevalence of overweight and obesity in children and adults during 1980–2013: a systematic analysis for the Global Burden of Disease Study. Lancet. 2014;384:766–81.
14. Maziak W, Taleb ZB, Bahelah R, et al. The global epidemiology of waterpipe smoking. Tob Control. 2015;24(suppl 1):i3–i12.
15. Brauer M, Freedman G, Frostad J, et al. Ambient air pollution exposure estimation for the global burden of disease. Environ Sci Technol. 2016;50:79–88.
16. World Health Organization. Regional Office for the Eastern Mediterranean. United Arab Emirates. Noncommunicable Diseases (NCD) Country Profiles. 2014. http://www.who.int/ nmh/countries/are_en.pdf?ua=1.
17. Al-Othman S, Haoudi A, Alhomoud S, Alkhenizan A, Khoja T, Al-Zahrani A. Tackling cancer control in the Gulf Cooperation Council countries. Lancet Oncol. 2015;16(5):E246–57. https://doi.org/10.1016/S1470-2045(15)70034-3.
18. Al-Shamsi HO, Abyad AM, Rafii S. A Proposal for a National Cancer Control Plan for the UAE: 2022-2026. Clin Pract. 2022 Feb 17;12(1):118-132. https://doi.org/10.3390/clinpract12010016. PMID: 35200267; PMCID: PMC8870909.
19. The national guidelines for breast cancer screening and diagnosis. Ministry of Health and Prevention. United Arab Emirates; 2014. www.mohap.gov.ae.
20. Sabih W, Taher J, El Jabari C, Hajat C, et al. Barriers to breast cancer screening and treatment among women in Emirate of Abu Dhabi. Ethn Dis. 2012;22(2):148–54.
21. The national guidelines for colorectal cancer screening and diagnosis. Ministry of Health and Prevention. United Arab Emirates; 2014. www.mohap.gov.ae.
22. The national guidelines for cervical cancer screening and diagnosis. Ministry of Health and Prevention. United Arab Emirates; 2014. www.mohap.gov.ae.
23. Loney T, Aw TC, Handysides DG, et al. An analysis of the health status of the United Arab Emirates: the "Big 4" public health issues. Glob Health Action. 2013;6:20100.
24. United Nations General Assembly. Political declaration of the high-level meeting of the general assembly on the prevention and control of noncommunicable diseases. United Nations. Imprint. New York: UN; 2011.
25. Beaglehole R, Bonita R, Alleyne G, et al. UN high-level meeting on noncommunicable diseases: addressing four questions. Lancet. 2011;378:449–55.
26. World Health Organization. National Cancer Control Programmes (NCCP). 2018. http://www. who.int/cancer/nccp/en/.
27. Abdul-Sater Z, Shamseddine A, Taher A, Fouad F, Abu-Sitta G, Fadhil I, et al. Cancer registration in the Middle East, North Africa, and Turkey: scope and challenges. JCO Glob Oncol. 2021;7:1101–9. https://doi.org/10.1200/GO.21.00065.
28. Fidler MM, Gupta S, Soerjomataram I, Ferlay J, Steliarova-Foucher E, Bray F. Cancer incidence and mortality among young adults aged 20–39 years worldwide in 2012: a population-based study. Lancet Oncol. 2017;18(12):1579–89.

29. Fadhil I, Alkhalawi E, Nasr R, Fouad H, Basu P, Camacho R, Alsaadoon H. National cancer control plans across the eastern Mediterranean region: challenges and opportunities to scale-up. Lancet Oncol. 2021;22(11):e517–29.
30. Al-Shamsi HO, Alrawi S. Breast cancer screening in the United Arab Emirates: is it time to call for a screening at an earlier age? J Cancer Prev Curr Res. 2018;9(3):123–6. https://doi.org/10.15406/jcpcr.2018.09.00334.
31. Gulia S, Sengar M, Badwe R, Gupta S. National Cancer Control Programme in India: proposal for organization of chemotherapy and systemic therapy services. J Glob Oncol. 2017;3(3):271–4.
32. Cherlaa A, Renwicka M, Jhab A, Mossialosa E. Cost-effectiveness of cancer drugs: comparative analysis of the United States and England. EClinicalMedicine. 2020;29–30:100625. https://doi.org/10.1016/j.eclinm.2020.100625.
33. NHS England: Cancer chemotherapy. www.england.nhs.uk.
34. Al-Shamsi HO. The state of cancer care in the United Arab Emirates in 2022. Clin Pract. 2022;12(6):955–85. https://doi.org/10.3390/clinpract12060101. PMID: 36547109; PMCID: PMC9777273.

Prof. Humaid Obaid Al-Shamsi is the Chief Executive Officer of Burjeel Cancer Institute in Abu Dhabi, UAE, President of the Emirates Oncology Society, Lead of the Gulf Cancer Society, Full Professor of Oncology at the Ras Al Khaimah Medical and Health Sciences University, Ras Al Khaimah, UAE, and an Adjunct Professor of Oncology at the College of Medicine, University of Sharjah. He is the first Emirati to be promoted as a professor in oncology in the UAE. He is also the Chairman for Colorectal Cancer in the MENA region, appointed by the prestigious National Comprehensive Cancer Network®. He is also the only member of Lung Cancer Policy Network in the MENA region that aims to advance lung cancer research and screening globally. He is the Chairman of the Oncology and Hematology Fellowship Training Program for the National Institute for Health Specialties in the United Arab Emirates. He is the only member in GCC in the WIN Consortium which is comprised of organizations representing all stakeholders in personalized cancer medicine globally.

He is board-certified in both internal medicine and oncology from the UK, USA (ABIM), and Canada (FRCPC). He has also been awarded the FRCP (London) in 2023 and FRCP (Glasgow) in 2024. He is the only physician in the UAE with a subspecialty fellowship certification and training in gastrointestinal oncology and the first Emirati to train and complete a clinical post-doctoral fellowship in palliative care. He was an assistant professor at the University of Texas MD Anderson Cancer Center between 2014 and 2017. He has published more than 140 peer-reviewed articles in JAMA Oncology, Lancet Oncology, The Oncologist, BMC Cancer, and many others. His area of expertise includes precision oncology and cancer care in the UAE. In 2016, he published with his group from MD Anderson the JCO paper describing a new distinct subgroup of CRC, NON V600 BRAF-mutated CRC. In 2022, he published the first book about cancer research in the UAE and also the first book about cancer in the Arab world, both of which were launched at Dubai Expo 2020. *Cancer in the Arab World* has been downloaded more than 450,000 times in its first 18 months of publication and is the ultimate source of cancer data in the Arab region. He also published the first comprehensive book about cancer care in the UAE which is the first book in UAE history to document the cancer care in the UAE with many topics addressed for the first time, e.g., neuroendocrine tumors in the UAE. He is passionate about advancing cancer care in the UAE and the GCC and has made significant contributions to cancer awareness and early detection for the public using social media platforms. He is considered as the most followed oncologist in the world with over 300,000 subscribers across his social media platforms (Instagram, Twitter, LinkedIn, and TikTok). In 2022, he was awarded the prestigious Feigenbaum Leadership Excellence Award from Sheikh Hamdan Smart University for his exceptional leadership and research and the Sharjah Award for Volunteering. He was also named the Researcher of the Year in the UAE in 2020 and 2021 by the Emirates Oncology Society.

In May 2024, HH Sheikh Mansour bin Zayed Al Nahyan, Vice President of the United Arab Emirates, awarded him the first place in UAE Nafis program for outstanding leadership in private sector across all business and medical disciplines. Beside his clinical and administrative duties, he is engaged in education and various levels of research training for medical trainees to enhance their clinical and research skills. His mission is to advance cancer care in the UAE and the MENA region and make cancer care accessible to everyone in need around the globe.

Dr. Amin M. Abyad earned his medical degree, Bachelor in Medicine and Surgery (MBChB), from Beirut Arab University. After completing his internship, he joined the Internal Medicine Residency at Makassed Hospital in Beirut, Lebanon, which is affiliated with the American University of Beirut Medical Center (AUBMC). Dr. Amin was appointed as Chief Resident of Internal Medicine (2017–2018). Then, Dr. Amin started his fellowship in hematology and medical oncology at Makassed Hospital, where he received intensive training in hematology and medical oncology. Dr. Amin joined Burjeel Medical City in July 2021. Dr. Amin is highly interested in malignant hematology and solid malignancies. He has been highly involved in clinical research, being involved in multiple research projects and publishing in multiple peer-reviewed journals.

Dr. Abyad believes in patient-centered care, trying to enhance patient outcomes through the application of the latest evidence-based practice and personalized medicine.

Comprehensive Cancer Centers in the UAE

6

Humaid O. Al-Shamsi ⓘ and Amin M. Abyad

6.1 Introduction

The UAE is home to more than 30 oncology centers and clinics. In previous discussions, we have provided an overview of the historical context surrounding the prominent cancer centers in the UAE [1]. In the UAE, a specific definition for a comprehensive cancer center (CCC) does not exist officially. The Department of Health (DOH) has issued general criteria for centers of excellence in Abu Dhabi, which are not tailored specifically to oncology [2].

H. O. Al-Shamsi (✉)
Burjeel Cancer Institute, Burjeel Medical City, Burjeel Holdings, Abu Dhabi,
United Arab Emirates

Ras Al Khaimah Medical and Health Sciences University, Ras Al Khaimah, United Arab Emirates

Gulf Medical University, Ajman, United Arab Emirates

Emirates Oncology Society, Emirates Medical Association, Dubai, United Arab Emirates

College of Medicine, University of Sharjah, Sharjah, United Arab Emirates

Gulf Cancer Society, Alsafa, Kuwait
e-mail: alshamsi@burjeel.com; humaid.al-shamsi@medportal.ca

A. M. Abyad
Emirates Oncology Society, Emirates Medical Association, Dubai, United Arab Emirates

Burjeel Medical City, Abu Dhabi, United Arab Emirates

© The Author(s) 2024

127

H. O. Al-Shamsi (ed.), *Cancer Care in the United Arab Emirates*,
https://doi.org/10.1007/978-981-99-6794-0_6

6.2   Comprehensive Cancer Centers in the UAE

The comprehensive cancer center (CCC) should serve as a single destination catering to all requirements related to cancer treatment [3]. According to our perspective, for a facility to be recognized as a comprehensive cancer center (CCC), it must offer the following services: medical oncology and hematology for both adults and children, surgical oncology, radiation oncology, nuclear medicine, and palliative care [3]. At present, there are four centers that meet the criteria for being classified as comprehensive cancer centers (CCCs) (Table 6.1). The cancer centers and hospitals in the UAE are listed in Table 6.2 in alphabetical order.

The Al Jalila Foundation, a member of Sheikh Mohammed Bin Rashid Al Maktoum Global Initiatives, has plans to establish the inaugural comprehensive cancer charity hospital in Dubai, known as the "Hamdan Bin Rashid Cancer Hospital." This initiative aims to bring together renowned experts in order to provide integrated services for cancer prevention, diagnosis, and treatment within a single facility. The hospital, named after the late Sheikh Hamdan Bin Rashid Al Maktoum, will offer a range of services, including outpatient, ambulatory, and diagnostic services, as well as inpatient and surgical care, all delivered in a nurturing environment that prioritizes individualized patient attention. Patients from all parts of the UAE will be accepted, and the medical services provided will be either free or significantly subsidized to alleviate the financial burden on individuals who are unable to afford high-quality healthcare. The Al Jalila Foundation is investing AED 1.2 billion in constructing the region's first fully modular-built hospital. Opening its doors in 2026, the hospital aims to serve as a comprehensive hub for cancer care, offering prevention, diagnosis, and treatment services with a capacity to accommodate 30,000 patients annually [4].

Table 6.1 Established comprehensive cancer centers in the UAE[a] (alphabetical orders)

Hospital	Location	Established	Oncology International Accreditation	Unique services	Services not offered
American Hospital Dubai	Dubai	2010	• N/A	• Only acute hematology in private sector and BMT unit in Dubai • Palliative care unit • Pediatric oncology	Research Unit/ Publication, Genetic Counseling
Burjeel Medical City	Abu Dhabi	2020	• The European Society for Medical Oncology (ESMO) Designated Centres of Oncology and Palliative Care, the only center accredited by ESMO in the UAE	• BMT unit and only pediatric BMT in the UAE • Only palliative care service in Abu Dhabi city • Only acute hematology in private sector in Abu Dhabi • Cancer research unit • Only Brain lab in the UAE • Pediatric oncology	Genetic counseling
Mediclinic City Hospital	Dubai	2016	• JCI accredited breast cancer unit	• Brachytherapy • Palliative care unit • Pediatric oncology	Acute hematology and BMT unit, genetic counseling
Tawam Hospital[b]	Alain	1979	• JCI-accredited breast cancer unit • JCI Clinical Care Program Certification in 2017 • National Accreditation Program for Breast Centers (NAPBC) in 2015	• Palliative service • Genetic counseling • Pediatric oncology	• Hepatobiliary surgery • BMT unit • Research and publications

[a] The following services must be available at the facility to be considered as a comprehensive cancer center: medical adult and pediatric oncology and hematology; surgical oncology; radiation oncology; nuclear medicine and palliative care
[b] The positron emission tomography (PET) scanner which is located near the main building of Tawam hospital is affiliated with Cleveland Clinic Abu Dhabi (CCAD)

Table 6.2 Cancer centers and hospitals in the UAE in alphabetical order

Cancer center	Location	Facility available								
		Medical oncology and infusion unit	Acute hematology service	Bone marrow transplantation unit	Pediatric oncology	Radiation oncology	Surgical oncology	Nuclear medicine/PET imaging	Palliative care unit	Research unit
Abu Dhabi Stem Cells Center	Abu Dhabi	No	No	Yes	No	No	No	No	No	No
Advanced Care Oncology Center	Dubai	Yes	No	No	No	No	No	No	No	No
American Hospital Dubai	Dubai	Yes	Yes	Yes	Yes	Yes	Yes	Yes	Yes	No
Al Zahra Hospital Dubai	Dubai	Yes	No	No	No	No	Yes	No	No	No
Aster Hospital	Dubai	Yes	No	No	No	No	Yes	No	No	No
Burjeel Hospital	Abu Dhabi	Yes	No	No	No	No	Yes	No	No	No
Burjeel Hospital for Advanced Surgery Dubai	Dubai	Yes	No	No	No	No	Yes	No	No	No
Burjeel Medical City	Abu Dhabi	Yes	Yes	Yes	Yes	Yes	Yes	Yes	Yes	Yes
Burjeel Day Surgery Center, Al Reem Island	Abu Dhabi	Yes	No	No	No	No	No	No	No	No
Burjeel Royal Hospital	Alain	Yes	No	No	No	No	Yes	No	No	No
Burjeel Specialty Hospital	Sharjah	Yes	No	No	No	No	Yes	No	No	No
Canadian Hospital	Dubai	Yes	No	No	No	No	Yes	No	No	No

Name	Location									
Cleveland Clinic Abu Dhabi	Abu Dhabi	Yes	No	No	No	No	Yes	Yes^a (off-site)	No	No
Clemenceau Medical Center	Dubai	Yes	No	No	Yes	No	Yes	Yes	No	Yes
Dubai Hospital	Dubai	Yes	Yes	No	Yes	No	Yes	Yes	No	Yes
Gulf International Cancer Center	Abu Dhabi	Yes	No	No	No	Yes	No	Yes	No	No
Lifeline Hospital VPS	Abu Dhabi	Yes	No	No	No	No	No	No	No	No
King's College Hospital London, Dubai	Dubai	Yes	Yes	No	No	No	Yes	No	No	No
Medcare Hospital	Sharjah	Yes	No	No	No	No	Yes	No	No	No
Mediclinic Airport Road	Abu Dhabi	Yes	Yes	No	No	Yes	Yes	No	No	No
Mediclinic City Hospital	Dubai	Yes	Yes	No	Yes	Yes	Yes	Yes	Yes	Yes
Neuro Spinal Hospital	Dubai	Yes	No	No	No	Yes	Yes	No	No	No
NMC Royal Hospital Sharjah	Sharjah	Yes	No	No	No	No	Yes	No	No	No
NMC Specialty Hospital	Abu Dhabi	Yes	No	No	Yes	No	Yes	No	No	No
Saudi German Hospital Dubai	Dubai	Yes	No	No	No	Yes	Yes	No	No	No
Saudi German Hospital Ajman	Ajman	Yes	No	No	No	No	Yes	No	No	No
Sharjah University Hospital	Sharjah	Yes	No	No	No	No	Yes	No	No	Yes

(continued)

Table 6.2 (continued)

Cancer center	Location	Facility available								
		Medical oncology and infusion unit	Acute hematology service	Bone marrow transplantation unit	Pediatric oncology	Radiation oncology	Surgical oncology	Nuclear medicine/PET imaging	Palliative care unit	Research unit
Sheikh Khalifa Specialty Hospital	Rasal-Khaimah	*Yes*	No	No	No	*Yes*	*Yes*	*Yes*	No	No
Sheikh Shakhbout Medical City	Abu Dhabi	*Yes*	*Yes*	No	No	No	*Yes*	No	No	No
Tawam Hospital	Alain	*Yes*	*Yes*	No	*Yes*	*Yes*	*Yes*	No[a]	*Yes*	*Yes*
Yas Clinic	Abu Dhabi	*Yes*	No	No	No	No	No	No	No	No
Zulekha Hospital Sharjah	Sharjah	*Yes*	No	No	No	No	*Yes*	No	No	No

[a] Cleveland Clinic Abu Dhabi's (CCAD) positron emission tomography (PET) scanner is located in Al Ain city (150 km away from CCAD main campus) near the main building of Tawam Hospital

6.3   Conclusion

In the UAE, there are over 30 cancer centers, with four of them meeting the criteria to be classified as comprehensive cancer centers. These comprehensive centers offer a range of services, including medical oncology and hematology for both adults and children, surgical oncology, radiation oncology, nuclear medicine, and palliative care. While there are several other cancer centers in the UAE providing various aspects of cancer care, they may lack certain essential modalities such as radiation and palliative care. However, there are ongoing efforts by additional centers to establish themselves as comprehensive cancer centers in the UAE. It is crucial to prioritize quality to ensure that cancer patients in the UAE receive high-quality care.

Conflict of Interest   The authors have no conflict of interest to declare.

References

1. Al-Shamsi H, Darr H, Abu-Gheida I, et al. The state of cancer care in the United Arab Emirates in 2020: challenges and recommendations, a report by the United Arab Emirates oncology task force. Gulf J Oncolog. 2020;1:71–87.
2. https://www.doh.gov.ae/-/media/F27DED3FBBA340A58EA4FD891EE39215.ashx. Accessed 9 Aug 2022.
3. Grosso D, Aljurf M, Gergis U. Building a comprehensive cancer center: overall structure. In: Aljurf M, Majhail NS, Koh MBC, et al., editors. The comprehensive cancer center: development, integration, and implementation. Cham: Springer International Publishing; 2022. p. 3–13.
4. https://www.aljalilafoundation.ae/what-we-do/treatment/hamdan-bin-rashid-cancer-hospital/. Accessed on 7 Jun 2024.

Prof. Humaid Obaid Al-Shamsi   is the Chief Executive Officer of Burjeel Cancer Institute in Abu Dhabi, UAE, President of the Emirates Oncology Society, Lead of the Gulf Cancer Society, Full Professor of Oncology at the Ras Al Khaimah Medical and Health Sciences University, Ras Al Khaimah, UAE, and an Adjunct Professor of Oncology at the College of Medicine, University of Sharjah. He is the first Emirati to be promoted as a professor in oncology in the UAE. He is also the Chairman for Colorectal Cancer in the MENA region, appointed by the prestigious National Comprehensive Cancer Network®. He is also the only member of Lung Cancer Policy Network in the MENA region that aims to advance lung cancer research and screening globally. He is the Chairman of the Oncology and Hematology Fellowship Training Program for the National Institute for Health Specialties in the United Arab Emirates. He is the only member in GCC in the WIN Consortium which is comprised of organizations representing all stakeholders in personalized cancer medicine globally.

He is board-certified in both internal medicine and oncology from the UK, USA (ABIM), and Canada (FRCPC). He has also been awarded the FRCP (London) in 2023 and FRCP (Glasgow) in 2024. He is the only physician in the UAE with a subspecialty fellowship certification and training in gastrointestinal oncology and the first Emirati to train and complete a clinical post-doctoral fellowship in palliative care. He was an assistant professor at the University of Texas MD Anderson Cancer Center between 2014 and 2017. He has published more than 140 peer-reviewed articles in JAMA Oncology, Lancet Oncology, The Oncologist, BMC Cancer, and many others. His area of expertise includes precision oncology and cancer care in the UAE. In 2016, he published with his group from MD Anderson the JCO paper describing a new distinct subgroup of CRC, NON V600 BRAF-mutated CRC. In 2022, he published the first book about cancer research in the UAE and also the first book about cancer in the Arab world, both of which were launched at Dubai Expo 2020. *Cancer in the Arab World* has been downloaded more than 450,000 times in its first 18 months of publication and is the ultimate source of cancer data in the Arab region. He also published the first comprehensive book about cancer care in the UAE which is the first book in UAE history to document the cancer care in the UAE with many topics addressed for the first time, e.g., neuroendocrine tumors in the UAE. He is passionate about advancing cancer care in the UAE and the GCC and has made significant contributions to cancer awareness and early detection for the public using social media platforms. He is considered as the most followed oncologist in the world with over 300,000 subscribers across his social media platforms (Instagram, Twitter, LinkedIn, and TikTok). In 2022, he was awarded the prestigious Feigenbaum Leadership Excellence Award from Sheikh Hamdan Smart University for his exceptional leadership and research and the Sharjah Award for Volunteering. He was also named the Researcher of the Year in the UAE in 2020 and 2021 by the Emirates Oncology Society.

In May 2024, HH Sheikh Mansour bin Zayed Al Nahyan, Vice President of the United Arab Emirates, awarded him the first place in UAE Nafis program for outstanding leadership in private sector across all business and medical disciplines. Beside his clinical and administrative duties, he is engaged in education and various levels of research training for medical trainees to enhance their clinical and research skills. His mission is to advance cancer care in the UAE and the MENA region and make cancer care accessible to everyone in need around the globe.

Dr. Amin M. Abyad earned his medical degree, Bachelor in Medicine and Surgery (MBChB), from Beirut Arab University. After completing his internship, he joined the Internal Medicine Residency at Makassed Hospital in Beirut, Lebanon, which is affiliated with the American University of Beirut Medical Center (AUBMC). Dr. Amin was appointed as Chief Resident of Internal Medicine (2017–2018). Then Dr. Amin started his fellowship in hematology and medical oncology at Makassed Hospital, where he received intensive training in hematology and medical oncology. Dr. Amin joined Burjeel Medical City in July 2021. Dr. Amin is highly interested in malignant hematology and solid malignancies. He has been highly involved in clinical research, being involved in multiple research projects and publishing in multiple peer-reviewed journals.

Dr. Abyad believes in patient-centered care, trying to enhance patient outcomes through the application of the latest evidence-based practice and personalized medicine.

Emirates Oncology Society

7

Humaid O. Al-Shamsi ⓘ and Amin M. Abyad

7.1 Emirates Medical Association

Emirates Medical Association (EMA) is the mother organization of the Emirates Oncology Society. EMA was established in 1981 as a nonprofit organization under the United Arab Emirates' Ministry of Community Development by a ministerial decree number (24) dated May 9, 1981. The EMA is composed of healthcare providers that are members of the EMA as defined by its established bylaws. EMA is responsible for supervising and establishing subsocieties, scientific training, conferences, and events, as well as collaborating with healthcare organizations. EMA launched its work through its competent executive, supervisory, and specialist elected boards and committee members. The election happens every three years, and all members are eligible to nominate themselves. The last election was in 2021, and the next election will be in 2024. EMA is a scientific organization that has legal personality, and its headquarters are in Dubai.

H. O. Al-Shamsi (✉)
Burjeel Cancer Institute, Burjeel Medical City, Burjeel Holdings, Abu Dhabi, United Arab Emirates

Ras Al Khaimah Medical and Health Sciences University, Ras Al Khaimah, United Arab Emirates

Gulf Medical University, Ajman, United Arab Emirates

Emirates Oncology Society, Emirates Medical Association, Dubai, United Arab Emirates

College of Medicine, University of Sharjah, Sharjah, United Arab Emirates

Gulf Cancer Society, Alsafa, Kuwait
e-mail: alshamsi@burjeel.com; humaid.al-shamsi@medportal.ca

A. M. Abyad
Emirates Oncology Society, Emirates Medical Association, Dubai, United Arab Emirates

Burjeel Medical City, Abu Dhabi, United Arab Emirates

© The Author(s) 2024
H. O. Al-Shamsi (ed.), *Cancer Care in the United Arab Emirates*,
https://doi.org/10.1007/978-981-99-6794-0_7

EMA aims to improve and enhance professional performance, develop, advance, and encourage skills, and enrich scientific publications and practice in all health fields, by:

1. Developing and advancing professional health-related programs and developing sustained medical education programs for healthcare specialties, in line with the general educational framework.
2. Coordinating and planning with other professional health entities, associations, and authorities, inside and outside the United Arab Emirates (UAE).
3. Encouraging and supporting scientific research, publishing scientific publications, and issuing its own journal, Emirates Medical Journal (EMJ), established in 2019.
4. Recommending and planning general plans to develop and qualify human resources in health fields, e.g., physicians, nurses, and technicians.
5. Organizing scientific meetings, seminars, workshops, campaigns, and conferences to discuss national health challenges and provide solutions and recommendations to regulators and decision-makers.
6. Proposing research topics and encouraging health-related scientific research.
7. Approving the establishment of scientific societies or clubs for health specialties under the umbrella of the EMA. EMA has 48 societies and 30 scientific clubs as of September 2022.
8. The establishment of Emirates Medical Day was celebrated on May 9th annually. The day is dedicated to all healthcare workers in the UAE.

7.2   EMA Vision, Mission, and Core Values

The EMA vision is "to play a vital role in the continuous improvement in the quality of health care in the UAE by being an integral part of the professional life of every physician" [1].

To keep the EMA members up-to-date with the global changes and advancements in practice within the medical field.

- Leadership
- Commitment
- Quality
- Integrity
- Ethics

7.3   Emirates Oncology Society Mission and Logo

The Emirates Oncology Society (EOS) was established in 2016. It is dedicated to promoting and enhancing the comprehensive care of cancer patients across the UAE by bringing together all of the practicing cancer specialists and healthcare workers from different backgrounds in the field of oncology.

To help our cancer patients through their treatment journey against different challenges until recovery, this requires their treating team, including medical oncologists, surgeons, radiation oncologists, oncology nurses, pharmacists, cancer researchers, social workers, and dietitians, to gain easy access to the latest scientific knowledge, technology, and treatment protocols at hand. The EOS aims to provide the members with state-of-the-art evidence-based information in order to unify and enhance their practice and achieve this goal for all of their patients. The EOS is also dedicated to cancer research and scientific publications.

The EOS logo highlights the core values and mission: Innovation, Research, and Education.

7.4 Board of Directors

The EOS board of directors is elected every three years; the last election was in November 2021. The following were elected (Fig. 7.1):

President (second term): Professor Humaid Al Shamsi
Vice President (second term): Dr. Falah Al khatib
General Secretary: Dr. Saeed Rafii
Scientific committee Chairperson: Dr. Aydah Alawadi
Cultural committee Chairperson: Dr. Syed Hammad Tirmazy

Prof. Humaid Al-Shamsi
President

Dr. Falah Abdul-Hamid Al Khatib
Vice-President

Dr. Saeed Rafii
General Secretary

Dr. Aydah AlAwadi
Scientific Committee
Chairperson

Dr. Syed Hammad Tirmazy
Cultural Committee
Chairperson

Fig. 7.1 The EOS board of directors (2022)

7.5 EOS Membership

For a candidate to be eligible for membership, he or she must be a physician licensed to practice oncology or hematology in the UAE. There are 80 active members in EOS as of September 2022.

7.6 Continuing Medical Education (CME) Activities and Conferences

The EOS is dedicated to advancing cancer care in the UAE by updating all healthcare providers in the cancer field with the latest advances in cancer. The EOS holds the largest number of accredited conferences and CMEs in the UAE. The EOS delivered more than 200 accredited CME hours in 2020 and 2021. The EOS holds many events and activities as stand-alone activities or in collaboration with international cancer societies and organizations like the European Society of Medical Oncology (ESMO), the International Association for the Study of Lung Cancer (IASLC), and the Union for International Cancer Control (UICC) [2].

The EOS annual conference is held every year in September. The last annual conference was held in September 2022, with over 2000 attendees, both in person and virtually from across the globe. The faculty were highly selected from the UAE, Gulf Cooperation Council (GCC) countries, the Middle East and North Africa (MENA) region, and also internationally, e.g., the USA, Spain, the UK, and Canada.

The EOS is also the official organizer of Best of ESMO© in the UAE and Best of WCLC© in the UAE. There are also many local oncology CME activities organized locally and across the GCC.

Besides the scientific activities, the EOS is dedicated to increasing community awareness about cancer. It publishes and organizes awareness campaigns throughout the year. It also posts articles in Arabic and English about cancer dedicated to the public [3].

7.7 Cancer Care Quality Improvement

The ESO works closely with the regulators in the UAE to identify gaps and improve the quality of cancer care by implementing quality checks and KPIs. The EOS is part of the UAE national committee for cancer care under the Ministry of Health and Prevention (MOHAP). The EOS has also worked closely with the Dubai Health Authority (DHA) as an advisor on a cancer quality improvement project in Dubai. The EOS provides professional consultations to national, regional, and international bodies interested in learning about cancer care in the UAE and quality improvement measures.

7.8 Research and Publications

In 2021, the EOS was named by the EMA as the most scientific publishing society among all the 48 members of the EMA. The EOS has published more than 25 publications in 2021.

7.8.1 Books

1. Cancer in the Arab World
   Al-Shamsi, Humaid O., *Ibrahim H. Abu-Gheida, Faryal Iqbal, and Aydah Al-Awadhi.* [4]
   This book contains up-to-date data on cancer rates across the Arab region, discussing the present state of cancer treatment and potential future developments. It features dedicated sections on breast cancer, colon cancer, and pallia-

tive care specific to the Arab world. Expert authors from various Arab countries contributed to each chapter.

The book has been downloaded more than 300,000 times within 1 year of its launch, making it the most downloaded medical book in the MENA region in 2022.

7.8.2   Book Chapters

1. *Abu-Gheida, I.H., Nijhawan, N., Al-Awadhi, A.,* **Al-Shamsi, H.O.** *(2022). General Oncology Care in the UAE. In: Al-Shamsi, H.O., Abu-Gheida, I.H., Iqbal, F., Al-Awadhi, A. (eds) Cancer in the Arab World. Springer, Singapore. https://doi.org/10.1007/978-981-16-7945-2_19*
2. **Al-Shamsi, H.O.,** *Iqbal, F. (2022). General Oncology Care in Qatar, Comoros, and Djibouti. In: Al-Shamsi, H.O., Abu-Gheida, I.H., Iqbal, F., Al-Awadhi, A. (eds) Cancer in the Arab World. Springer, Singapore. https://doi. org/10.1007/978-981-16-7945-2_21*
3. **Al-Shamsi, H.O.,** *Iqbal, F., Abu-Gheida, I.H. (2022). Introduction. In: Al-Shamsi, H.O., Abu-Gheida, I.H., Iqbal, F., Al-Awadhi, A. (eds) Cancer in the Arab World. Springer, Singapore. https://doi.org/10.1007/978-981-16-7945-2_1*
4. **Humaid Al-Shamsi,** *Siefker-Radtke A, Czerniak B, Dinney C, Millikan RE. Uncommon Cancers of the Bladder. In: Textbook of Uncommon Cancer, 5th ed. Ed(s) Raghavan D, Brecher ML, Johnson DH, Meropol NJ, Moots PL, Rose PG, Mayer IA. John Wiley and Son, 2016.*
5. *Nijhawan N.A.,* **Al-Shamsi H.O.** *(2021) Palliative Care in the United Arab Emirates (UAE). In: Laher I. (eds) Handbook of Healthcare in the Arab World. Springer, Cham. https://doi.org/10.1007/978-3-319-74365-3_102-1*
6. *Abu-Gheida I., Nijwahan N.,* **Al-Shamsi H.O.** *(2020) Oncology Care in the UAE. In: Laher I. (eds) Handbook of Healthcare in the Arab World. Springer, Cham. https://doi.org/10.1007/978-3-319-74365-3_183-1*

7.8.3   Peer-Reviewed Publications

1. *Abboud K, Umoru G, Esmail A, Abudayyeh A, Murakami N,* **Al-Shamsi HO,** *Javle M, Saharia A, Connor AA, Kodali S, Ghobrial RM. Immune Checkpoint Inhibitors for Solid Tumors in the Adjuvant Setting: Current Progress, Future Directions, and Role in Transplant Oncology. Cancers. 2023 Feb 23;15(5):1433.*
2. *Park, S. H., Hong, S. H., Kim, K., Lee, S. W., Yon, D. K., Jung, S. J., ... & Smith, L. (2023). Nonpharmaceutical interventions reduce the incidence and mortality of COVID-19: A study based on the survey from the International COVID-19 Research Network (ICRN). Journal of medical virology, 95(2), e28354.*
3. *Bernstein E, Lev-Ari S, Shapira S, Leshno A, Sommer U,* **Al-Shamsi H,** *Shaked M, Segal O, Galazan L, Hay-Levy M, Sror M. Data From a One-Stop-Shop Comprehensive Cancer Screening Center. Journal of Clinical Oncology. 2023 Jan:JCO-22.*

4. **Al-Shamsi HO**, Abdelwahed N, Al-Awadhi A, Albashir M, Abyad AM, Rafii S, Afrit M, Al Lababidi B, Abu-Gheida I, Sonawane YP, Nijhawan NA. *Breast Cancer in the United Arab Emirates. JCO Global Oncology.* 2023 Jan;9:e2200247.

5. **Al-Shamsi, Humaid O.**, Ibrahim H. Abu-Gheida, Faryal Iqbal, and Aydah Al-Awadhi. *"Cancer in the Arab World."* https://link.springer.com/book/10.1007/978-981-16-7945-2

6. AlZaabi A, AlHarrasi A, AlMusalami A, AlMahyijari N, Al Hinai K, ALAdawi H, **Al-Shamsi HO**. *Early onset colorectal cancer: Challenges across the cancer care continuum. Annals of Medicine and Surgery.* 2022 Aug 22:104453.

7. Nijhawan NA, **Al-Shamsi HO**. *Experiences and challenges of a new palliative care service in the United Arab Emirates. Palliat Med Hosp Care Open J.* 2022; 8(2):30–34. doi: 10.17140/PMHCOJ-8-150.

8. **Al-Shamsi HO**. *The State of Cancer Care in the United Arab Emirates in 2022. Clinics and Practice.* 2022 Nov 23;12(6):955–85.

9. **Al-Shamsi HO**. *Barriers and Facilitators to Conducting Oncology Clinical Trials in the UAE. Clinics and Practice.* 2022 Nov 7;12(6):885–96.

10. Ennab, F., Tsagkaris, C., Babar, M. S., Tazyeen, S., Kokash, D., Nawaz, F. A., & **Al-Shamsi, H. O.** (2022). *A potential rise of breast cancer risk in the UAE post-COVID-19 lockdown: A call for action. Annals of Medicine and Surgery,* 103976.

11. Aoude M; Mousallem M; Abdo M; Youssef B; Kourie HR; **Al-Shamsi HO**. *Gastric cancer in the Arab World: a systematic review. East Mediterr Health J.* 2022;28(7):521–531. https://doi.org/10.26719/emhj.22.051

12. Abou Ghayda, R., Lee, K. H., Han, Y. J., Ryu, S., Hong, S. H., Yoon, S., ... & Shin, J. I. (2022). *The global case fatality rate of coronavirus disease 2019 by continents and national income: A meta-analysis. Journal of Medical Virology,* 94(6), 2402–2413.

13. Al Ashour, B. H., Azam, F., Ibnshamsah, F., Alrowais, F., Al-Garni, A., **Al-Shamsi, H. O.**, & Bukhari, N. (2022). *Metastatic Type II Papillary Renal Cell Carcinoma With Recurrent Complete Responses to Sunitinib: A Case Report With a Literature Review. Cureus,* 14(5).

14. Mula-Hussain, Layth, Hala Mahdi, Zhian Salah Ramzi, Marwan Tolba, Mohamad Saad Zaghloul, Zineb Benbrahim, Atlal Abusanad, **Humaid Al-Shamsi**, Adda Bounedjar, and Abdul-Rahman Jazieh. *"Cancer Burden Among Arab World Males in 2020: The Need for a Better Approach to Improve Outcome." JCO Global Oncology* 8 (2022): e2100407.

15. Abu-Gheida, I.H., Nijhawan, N., Al-Awadhi, A., **Al-Shamsi, H.O.** (2022). *General Oncology Care in the UAE. In: Al-Shamsi, H.O., Abu-Gheida, I.H., Iqbal, F., Al-Awadhi, A. (eds) Cancer in the Arab World. Springer, Singapore.* https://doi.org/10.1007/978-981-16-7945-2_19

16. **Al-Shamsi, H.O.**, Iqbal, F. (2022). *General Oncology Care in Qatar, Comoros, and Djibouti. In: Al-Shamsi, H.O., Abu-Gheida, I.H., Iqbal, F., Al-Awadhi, A. (eds) Cancer in the Arab World. Springer, Singapore.* https://doi.org/10.1007/978-981-16-7945-2_21

17. *Al-Shamsi, H.O., Iqbal, F., Abu-Gheida, I.H. (2022). Introduction. In: Al-Shamsi, H.O., Abu-Gheida, I.H., Iqbal, F., Al-Awadhi, A. (eds) Cancer in the Arab World. Springer, Singapore. https://doi.org/10.1007/978-981-16-7945-2_1*
18. *Mula-Hussain L, Mahdi H, Ramzi ZS, Tolba M, Zaghloul MS, Benbrahim Z, Abusanad A, Al-Shamsi H, Bounedjar A, Jazieh AR. Cancer Burden Among Arab World Males in 2020: The Need for a Better Approach to Improve Outcome. JCO Global Oncology. 2022 Mar;8:e2100407.*
19. *Allehebi A, Kattan KA, Rujaib MA, Dayel FA, Black E, Mahrous M, AlNassar M, Hussaini HA, Twairgi AA, Abdelhafeiz N, Omair AA, Shehri SA, Al-Shamsi HO, Jazieh AR. Management of Early-Stage Resected Non-Small Cell Lung Cancer: Consensus Statement of the Lung cancer Consortium. Cancer Treat Res Commun. 2022 Feb 22;31:100538. doi: 10.1016/j.ctarc.2022.100538. Epub ahead of print. PMID: 35220069.*
20. *Al-Shamsi, H.O., Abyad, A.M. and Rafii, S., 2022. A Proposal for a National Cancer Control Plan for the UAE: 2022–2026. Clinics and Practice, 12(1), pp.118–132.*
21. *Rafii, S., Tashkandi, E., Bukhari, N. and Al-Shamsi, H.O., 2022. Current Status of CRISPR/Cas9 Application in Clinical Cancer Research: Opportunities and Challenges. Cancers, 14(4), p.947.*
22. *Al-Shamsi, H. O., Abyad, A., Kaloyannidis, P., El-Saddik, A., Alrustamani, A., Abu Gheida, I., ... & Mheidly, K. (2022). Establishment of the First Comprehensive Adult and Pediatric Hematopoietic Stem Cell Transplant Unit in the United Arab Emirates: Rising to the Challenge. Clinics and Practice, 12(1), 84–90.*
23. *Ramia P, Bodgi L, Mahmoud D, Mohammad MA, Youssef B, Kopek N, Al-Shamsi H, Dagher M, Abu-Gheida I. Radiation-Induced Fibrosis in Patients with Head and Neck Cancer: A Review of Pathogenesis and Clinical Outcomes. Clinical Medicine Insights: Oncology. 2022 Jan;16:11795549211036898.*
24. *Elsamany, Shereef, Mohamed Elbaiomy, Ahmed Zeeneldin, Emad Tashkandi, Fayza Hassanin, Nafisa Abdelhafeez, Humaid O. Al-Shamsi, Nedal Bukhari, and Omima Elemam. "Suggested Modifications to the Management of Patients With Breast Cancer During the COVID-19 Pandemic: Web-Based Survey Study." JMIR cancer 7, no. 4 (2021): e27073.*
25. *M.H. Hodroj, G. El Hasbani, Humaid O. Al Shamsi, et al., clinical burden of hemophilia in older adults: Beyond bleeding risk, Blood Reviews (2021) https://doi.org/10.1016/j.blre.2021.100912*
26. *Al-Shamsi, H.O., Jaffar, H., Mahboub, B., Khan, F., Albastaki, U., Hammad, S. and Zaabi, A.A., 2021. Early Diagnosis of Lung Cancer in the United Arab Emirates: Challenges and Strategic Recommendations. Clinics and Practice, 11(3), pp.671–678.*
27. *Al-Shamsi, H.O., Abu-Gheida, I., Abdulsamad, A.S., AlAwadhi, A., Alrawi, S., Musallam, K.M., Arun, B. and Ibrahim, N.K., 2021. Molecular Spectra and Frequency Patterns of Somatic Mutations in Arab Women with Breast Cancer. The oncologist.*

28. **Humaid, O. Al-Shamsi,** *Eric A Coomes (2021) Successful Vaccination of Patients with History of Severe Anaphylactic Reaction with Pfizer-Biotech COVID-19 Vaccine. Journal of Oncology Research Review & Reports. SRC/ JONRR-125. occur up to, 20.*

29. **Al-Shamsi, H. O.,** *Alzaabi, A. A., Afrit, M., Abu-Gheida, I., & Musallam, K. M. (2021). Clinicopathological Features of Gastric Cancer in a Cohort of Gulf Council Countries' Patients: A Cross-Sectional Study of 96 Cases. Journal of Oncology Research Review & Reports. SRC/JONRR-133, 3.*

30. **Al-Shamsi, H.,** *Azribi, F., Abu-Gheida, I., Jaafar, H., & Razek, A. A. (2021). The Emirates Oncology Task Force Clinical Practice Guideline on Screening for SARS-CoV-2 in Asymptomatic Adult Cancer Patients Prior to Anti-Cancer Therapy. Journal of Oncology Research Review & Reports. SRC/JONRR-137, 3.*

31. **Humaid O Al-Shamsi,** *Faryal Iqbal, Mohd Subhi Al Saad, Priyanka Ashish Dhemre, Fady Georges Hachem,* et al *(2021) The Burden of Gynecologic Cancers in the UAE. Journal of Oncology Research Review & Reports. SRC/ JONRR-147.*

32. Bukhari N, Alshangiti A, Tashkandi E, Algarni M, **Al-Shamsi HO**, Al-Khallaf H. *Fluoropyrimidine-Induced Severe Toxicities Associated with Rare DPYD Polymorphisms: Case Series from Saudi Arabia and a Review of the Literature.* Clinics and Practice. 2021; 11(3):467–471. https://doi.org/10.3390/ clinpract11030062

33. Azhar, Malik, Faisal Aziz, Salama Almuhairi, Mohammad Alfelasi, Ali Elhouni, Rizwan Syed, **Humaid O. Al-Shamsi,** and Khaled M. Aldahmani. "Decline in radioiodine use but not total thyroidectomy in thyroid cancer patients treated in the United Arab Emirates-A retrospective study." Annals of Medicine and Surgery 64 (2021): 102203.

34. **Humaid, O. Al-Shamsi,** *Eric A Coomes (2021) Higher and Increasing Incidence of Cancer between the Age of 20–49 Years in the UAE Population; A focus Analysis of the UAE National Cancer Registry Data 2015–2017. Journal of Oncology Research Review & Reports. SRC/JONRR-127, 3.*

35. **Humaid, O. Al-Shamsi,** *Sadir Alrawi, Ahmed S. Abdulsamad, Nuhad K. Ibrahim (2021) AStrong Beliefs and Soft Evidence Underlying Mammography Surveillance Recommendations in Breast Cancer Survivors: A Study in Reflexive Science and Decision-Making. Journal of Oncology Research Review & Reports. SRC/JONRR-131, 3.*

36. Bernstein, Ezra, Shiran Shapira, Shahar Lev-Ari, Ari Leshno, Udi A. Sommer, Lior Galazan, **Humaid O. Al-Shamsi** et al. "One-stop-shop for cancer screening: A model for the future." (2021): 10554–10554.

37. Latif, M.F., Azam, F., Tirmazy, S.H., Bashir, S., Ibnshamsah, F., Al Selwi, W.M., Bukhari, N., Alwbari, A., Alshangiti, A., Gabsi, A. and **Al-Shamsi, H.O.,** 2021. *Impact of COVID 19 pandemic on psychological wellbeing of oncology clinicians in MENA (Middle East and North Africa) region.*

38. **Al-Shamsi, H.**, Darr, H., Abu-Gheida, I., Ansari, J., McManus, M. C., Jaafar, H., ... & Al-Khatib, F. (2020). The State of Cancer Care in the United Arab Emirates in 2020: Challenges and Recommendations, A report by the United Arab Emirates Oncology Task Force. The Gulf journal of oncology, 1(32), 71–87.

39. **Al-Shamsi, H. O.**, Coomes, E. A., & Alrawi, S. (2020). Screening for COVID-19 in asymptomatic patients with cancer in a hospital in the United Arab Emirates. JAMA oncology, 6(10), 1627–1628.

40. **Al-Shamsi, H. O.**, Coomes, E. A., Aldhaheri, K., & Alrawi, S. (2020). Serial Screening for COVID-19 in Asymptomatic Patients Receiving Anticancer Therapy in the United Arab Emirates. JAMA oncology. doi:10.1001/jamaoncol.2020.5745.

41. **Humaid O. Al-Shamsi** (2020) Rethinking Cancer Screening and Diagnosis During the Covid-19 Pandemic. Journal of Oncology Research Review & Reports. SRC/JORRR/110.

42. **Humaid O. Al-Shamsi.**, et al. "Early Onset Colorectal Cancer in the United Arab Emirates, Where do we Stand?". Acta Scientific Cancer Biology 4.11 (2020): 24–27.

43. Abdel-Wahab, R., Hassan, M.M., George, B., Pestana, R.C., Xiao, L., Lacin, S., Yalcin, S., Shalaby, A.S., **Al-Shamsi, H.O.**, Raghav, K. and Wolff, R.A., 2020. Impact of Integrating Insulin-Like Growth Factor 1 Levels into Model for End-Stage Liver Disease Score for Survival Prediction in Hepatocellular Carcinoma Patients. Oncology, 98(12), pp.836–846.

44. **Al-Shamsi, H. O.**, Abu-Gheida, I., Rana, S. K., Nijhawan, N., Abdulsamad, A. S., Alrawi, S., ... & McManus, M. C. (2020). Challenges for cancer patients returning home during SARS-COV-19 pandemic after medical tourism-a consensus report by the emirates oncology task force. BMC cancer, 20(1), 1–10.

45. **Al-Shamsi HO**, Alhazzani W, Alhuraiji A, Coomes EA, Chemaly RF, Almuhanna M, Wolff RA, Ibrahim NK, Chua MLK, Hotte SJ, Meyers BM, Elfiki T, Curigliano G, Eng C, Grothey A, Xie C. A Practical Approach to the Management of Cancer Patients During the Novel Coronavirus Disease 2019 (COVID-19) Pandemic: An International Collaborative Group. Oncologist. 2020 Jun;25(6):e936-e945. doi: 10.1634/theoncologist.2020-0213. Epub 2020 Apr 27. PMID: 32243668; PMCID: PMC7288661.

46. **Al Shamsi, H.**, R. Iskanderian, R., Karmstaji, A., Kamal Mohamed, B., Alahmed, S., H. Masri, M., ..., R. Grobmyer, S. (2020). Outcomes and Impact of a Universal COVID-19 Screening Protocol for Asymptomatic Oncology Patients. The Gulf Journal of Oncology, (34), 162–167.

47. Coomes, E. A., **Al-Shamsi, H. O.**, Meyers, B. M., Alhazzani, W., Alhuraiji, A., Chemaly, R. F., ... & Xie, C. (2020). Evolution of Cancer Care in Response to the COVID-19 Pandemic. The oncologist, 25(9), e1426–e1427.

48. **Al-Shamsi HO** et al, Authors Reply: Oncological surgery during COVID-19 pandemic: The need for deep and lasting measures. https://doi.org/10.1634/theoncologist.2020-0451

49. Jain, A., Borad, M.J., Kelley, R.K., Wang, Y., Abdel-Wahab, R., Meric-Bernstam, F., Baggerly, K.A., Kaseb, A.O., *Al-Shamsi, H.O.*, Ahn, D.H. and DeLeon, T., 2018. Cholangiocarcinoma with FGFR genetic aberrations: a unique clinical phenotype. JCO Precision Oncology, 2, pp.1–12.

50. Ehab Abdou, Ravi M Pedapenki, Mohamed Abouagour, Abdul R Zar, Emad Anwar, Dalia Elshourbagy, *Humaid Al-Shamsi* & Enrique Grande (2020) Patient selection and risk factors in the changing treatment landscape of metastatic renal cell carcinoma, Expert Review of Anticancer Therapy, 2020. https://doi.org/10.1080/14737140.2020.1810572.

51. *Al-Shamsi HO* et al, Authors Reply: How to manage febrile neutropenia during the COVID pandemic? In response to, "A Practical Approach to the Management of Cancer Patients During the Novel Coronavirus Disease 2019 (COVID-19) Pandemic." 2020 https://doi.org/10.1634/theoncologist.2020-0329

52. Tashkandi, E., Zeeneldin, A., AlAbdulwahab, A., Elemam, O., Elsamany, S., Jastaniah, W., ... & *Al-Shamsi, H.* (2020). Virtual management of cancer patients in the era of COVID-19 pandemic. J Med Internet Res [Internet].

53. *Al-Shamsi, H. O.* (2020). Mammography screening for breast cancer—the UK Age trial. The Lancet Oncology, 21(11), e505.

54. *Humaid O. Al-Shamsi* (2020) COVID-19 Vaccination for Cancer Patients, What Oncologists and Cancer Patients Need to Know? Journal of Oncology. Research Review & Reports. SRC/JONRR-114.

55. *Humaid O. Al-Shamsi* (2020) Anterior Mediastinal Mass: A Rare Presentation of Thyroid Mass Compressing Mediastinal Structures. Journal of Oncology. Research Review & Reports. SRC/JORRR-113.

56. Atlal M. Abusanad and *Humaid O. Al-Shamsi**, "Tele-Oncology: An Emerging Technology in Developing Countries during the COVID-19 Pandemic," New Emirates Medical Journal (2020) 1: 1. https://doi.org/10.217 4/02506882019992011109160857

57. Benbrahim, Z., Al Asiri, M., Al Bahrani, B., AlNassar, M. A. M. A., *Al-Shamsi, H. O.*, Bounedjar, A., ... & Labidi, S. (2020). 1737P National approaches to managing cancer care: Responses of countries in the MENA region to COVID-19 pandemic. Annals of Oncology, 31, S1016.

58. Abdulsamad, A. S., Inam, A., Oner, M., Darr, H., Madi, T., Alrawi, S. J., & *Al-shamsi, H. O.* (2019). Malignant Peritoneal Mesothelioma Following Wilms' Tumor in a Horse-Shoe Kidney, A Case Report and Review of Literature. Cancer Therapy & Oncology International Journal, 13(2), 66–71.

7.9 EOS Annual Awards

The EOS annual awards are dedicated to researchers, clinicians, healthcare workers, nurses, and other healthcare providers who have made a significant impact on cancer care in the UAE, GCC, MENA region, and globally. The list of award recipients in 2022 is shown in Table 7.1.

Table 7.1 The list of EOS award recipients in 2022

Award title	Award recipients
EOS Lifetime Achievement Global Oncology	Prof. Toni Choueiri
EOS Lifetime Achievement MENA Oncology	Prof. Nagi El Saghir
EOS Lifetime Achievement GCC Oncology	Prof. Bassim Al Bahrani
EOS Lifetime Achievement UAE Oncology	Dr. Hassan Jaafar
EOS Women in Oncology Award MENA	Dr. Omalkhair Abulkhair
EOS Women in Oncology Award UAE	Dr. Aydah Al-Awadhi
EOS Researcher of the Year	Dr. Deborah Mukherji
EOS Publication of the Year	Dr. Ibrahim Abu-Gheida
EOS Industry Researcher of the Year	Amgen Inc
EOS Member of the Year	Dr. Hassan Ghazal
EOS Outstanding Oncology Nursing Award	Mr. Khaled Al Qawasmeh
EOS Cancer Awareness Advocate of the Year	Abu Dhabi Public Health Center
EOS Cancer Awareness Advocate of the Year	Ministry of health and prevention
EOS Cancer Awareness Partner of the Year	MSD Inc
EOS Cancer Awareness Partner of the Year	Ipsen Inc
EOS Patient Advocate of the Year	Emirates Cancer Society
EOS Media Cancer Awareness Award of the Year	Ms. Mona Alhmoudi/Etehad
EOS Media Cancer Awareness Award of the Year	Mr. Wagih El Sebaei/Emarate
EOS Media Cancer Awareness Award of the Year	Ms. Jameelah Ismail/Albayan
EOS Hope Award	Ms. Sumaia Saif Alkaabi
EOS Hope Award	Dr. Hamad Alghamdi/KSA
Appreciation Award	Dr. Mouza AlSharhan/EMA
EOS Special Recognition Award	Prof. Nuhad Ibrahim
EOS Special Recognition Award	Prof. Robert Wolff

7.10   EOS' Achievements

1. **Cancer in the Arab World was named the most downloaded medical book in the MENA region in 2021** [4]. The book has been downloaded more than 190,000 times within the 4 months of publication as of September 2022.

   The book has been downloaded more than 300,000 times within 1 year of its launch, making it the most downloaded medical book in the MENA region in 2022.

2. **The publication of the year 2020**—July 2020, by the Oncologist Journal "Editor-in-Chief Bruce A. Chabner Massachusetts General Hospital Harvard Medical School Boston, MA" for the publication of "A Practical Approach to the Management of Cancer Patients During the Novel Coronavirus Disease 2019 (COVID-19) Pandemic: An International Collaborative Group" [5].

3. **Guinness World record** for the largest cancer awareness ribbon, on Nov 9, 2021, on neuroendocrine tumor day [6].

4. **Guinness World record** for the largest number of cancer awareness ribbons, Guinness World record on February 4, 2022 [7].

The most publishing scientific society in the UAE among all 48 EMA societies with over 25 publications in 2021.

7.11 Future Outlook

The EOS continues to work closely with the regulators in the UAE as an advisor to advance cancer care quality in the UAE. The EOS is participating in the Ministry of Health and Prevention Committee to reduce the cancer mortality rate in the UAE, according to the UAE government's agenda.

Upcoming book projects from the EOS include this book, "Cancer Care in the UAE," and "Healthcare in the UAE," both of which are planned for publication by Springer in the first quarter of 2023. The EOS is also planning to bid on international oncology conferences to be held in the UAE.

The EOS is also supporting the initiative to establish accredited oncology and hematology fellowship training programs.

7.12 Conclusion

The Emirates Oncology Society's mission is to promote and improve the complete care of cancer patients throughout the United Arab Emirates (UAE) by bringing together all practicing oncologists and healthcare workers from various backgrounds. It is the professional organization in the UAE that represents medical oncologists, radiation specialists, and palliative care physicians. The EOS is also working with the UAE to establish accredited oncology and hematology fellowship training programs. The EOS collaborates closely with UAE regulators to improve cancer care quality through the application of quality measures.

Conflict of Interest The authors have no conflict of interest to declare.

References

1. https://www.ema.ae/. Accessed 8 Sept 2022.
2. https://www.uicc.org/membership/emirates-oncology-society. Accessed 8 Sept 2022.
3. https://www.emaratalyoum.com/local-section/health/2022-06-03-1.1637244.
4. https://link.springer.com/book/10.1007/978-981-16-7945-2. Accessed 9 Sept 2022.
5. https://theoncologist.onlinelibrary.wiley.com/journal/1549490x/homepage/anniversaryretro-spective. Accessed 8 Sept 2022.
6. https://www.zawya.com/en/press-release/emirates-oncology-society-and-ipsen-break-guinness-world-record-for-largest-awareness-ribbon-grtn1tcx.
7. https://web-release.com/msd-gulf-honored-by-emirates-oncology-society-for-its-outstanding-support-towards-cancer-patients-in-the-uae/#:~:text=On%20World%20Cancer%20Day%20on,World%20Cancer%20Day%20in%202021.

Prof. Humaid Obaid Al-Shamsi is the Chief Executive Officer of Burjeel Cancer Institute in Abu Dhabi, UAE, President of the Emirates Oncology Society, Lead of the Gulf Cancer Society, Full Professor of Oncology at the Ras Al Khaimah Medical and Health Sciences University, Ras Al Khaimah, UAE, and an Adjunct Professor of Oncology at the College of Medicine, University of Sharjah. He is the first Emirati to be promoted as a professor in oncology in the UAE. He is also the Chairman for Colorectal Cancer in the MENA region, appointed by the prestigious National Comprehensive Cancer Network®. He is also the only member of Lung Cancer Policy Network in the MENA region that aims to advance lung cancer research and screening globally. He is the Chairman of the Oncology and Hematology Fellowship Training Program for the National Institute for Health Specialties in the United Arab Emirates. He is the only member in GCC in the WIN Consortium which is comprised of organizations representing all stakeholders in personalized cancer medicine globally.

He is board-certified in both internal medicine and oncology from the UK, USA (ABIM), and Canada (FRCPC). He has also been awarded the FRCP (London) in 2023 and FRCP (Glasgow) in 2024. He is the only physician in the UAE with a subspecialty fellowship certification and training in gastrointestinal oncology and the first Emirati to train and complete a clinical post-doctoral fellowship in palliative care. He was an assistant professor at the University of Texas MD Anderson Cancer Center between 2014 and 2017. He has published more than 140 peer-reviewed articles in JAMA Oncology, Lancet Oncology, The Oncologist, BMC Cancer, and many others. His area of expertise includes precision oncology and cancer care in the UAE. In 2016, he published with his group from MD Anderson the JCO paper describing a new distinct subgroup of CRC, NON V600 BRAF-mutated CRC. In 2022, he published the first book about cancer research in the UAE and also the first book about cancer in the Arab world, both of which were launched at Dubai Expo 2020. *Cancer in the Arab World* has been downloaded more than 450,000 times in its first 18 months of publication and is the ultimate source of cancer data in the Arab region. He also published the first comprehensive book about cancer care in the UAE which is the first book in UAE history to document the cancer care in the UAE with many topics addressed for the first time, e.g., neuroendocrine tumors in the UAE. He is passionate about advancing cancer care in the UAE and the GCC and has made significant contributions to cancer awareness and early detection for the public using social media platforms. He is considered as the most followed oncologist in the world with over 300,000 subscribers across his social media platforms (Instagram, Twitter, LinkedIn, and TikTok). In 2022, he was awarded the prestigious Feigenbaum Leadership Excellence Award from Sheikh Hamdan Smart University for his exceptional leadership and research and the Sharjah Award for Volunteering. He was also named the Researcher of the Year in the UAE in 2020 and 2021 by the Emirates Oncology Society.

In May 2024, HH Sheikh Mansour bin Zayed Al Nahyan, Vice President of the United Arab Emirates, awarded him the first place in UAE Nafis program for outstanding leadership in private sector across all business and medical disciplines. Beside his clinical and administrative duties, he is engaged in education and various levels of research training for medical trainees to enhance their clinical and research skills. His mission is to advance cancer care in the UAE and the MENA region and make cancer care accessible to everyone in need around the globe.

Dr. Amin M. Abyad earned his medical degree, Bachelor in Medicine and Surgery (MBChB), from Beirut Arab University. After completing his internship, he joined the Internal Medicine Residency at Makassed Hospital in Beirut, Lebanon, which is affiliated with the American University of Beirut Medical Center (AUBMC). Dr. Amin was appointed as Chief Resident of Internal Medicine (2017–2018). Then Dr. Amin started his fellowship in hematology and medical oncology at Makassed Hospital, where he received intensive training in hematology and medical oncology. Dr. Amin joined Burjeel Medical City in July 2021. Dr. Amin is highly interested in malignant hematology and solid malignancies. He has been highly involved in clinical research, being involved in multiple research projects and publishing in multiple peer-reviewed journals. Dr. Abyad believes in patient-centered care, trying to enhance patient outcomes through the application of the latest evidence-based practice and personalized medicine.

Factors Influencing Seeking Cancer Care Abroad for UAE Citizens

Humaid O. Al-Shamsi [ID]

8.1 Introduction

Despite notable progress in healthcare and oncology in the United Arab Emirates (UAE), a considerable proportion of cancer patients still opt for treatment in other countries. According to a report, in 2013, the UAE allocated approximately $163 million US dollars toward government-funded cancer care overseas and medical tourism beyond its borders [1]. There is currently no publicly available official data regarding the specific types and stages of cancer cases treated outside the United Arab Emirates (UAE). However, the most popular destinations for cancer medical tourism from the UAE are the United States of America, Germany, Singapore, South Korea, and Thailand [2, 3]. A study conducted using administrative data obtained from the Dubai Health Authorities focused on UAE nationals who received medical treatment abroad from 2009 to 2016. The study analyzed information from a total of 6557 UAE nationals. The primary destinations for treatment were Germany (46%), the United Kingdom (UK) (19%), and Thailand (14%). The most prevalent medical specialties sought were oncology (13%), orthopedic surgery (13%), and neurosurgery (10%). After accounting for various factors, the study found that oncology had the highest anticipated number of trips, with an incidence rate ratio (IRR) of 1.34 (95% CI: 1.24–1.44) [4, 5].

H. O. Al-Shamsi (✉)
Burjeel Cancer Institute, Burjeel Medical City, Burjeel Holdings, Abu Dhabi, United Arab Emirates

Ras Al Khaimah Medical and Health Sciences University, Ras Al Khaimah, United Arab Emirates

Gulf Medical University, Ajman, United Arab Emirates

Emirates Oncology Society, Emirates Medical Association, Dubai, United Arab Emirates

College of Medicine, University of Sharjah, Sharjah, United Arab Emirates

Gulf Cancer Society, Alsafa, Kuwait
e-mail: alshamsi@burjeel.com; humaid.al-shamsi@medportal.ca

In the UAE, cancer care abroad is supported by various distinct sponsoring agencies. These include presidential affairs offices, the armed forces, the police, all health authorities (such as the Department of Health, the Dubai Health Authority, and the Ministry of Health and Prevention), as well as charitable organizations that cover the costs themselves [2]. The sponsoring requirements and procedures differ among the various sponsoring agencies in the UAE, and this is an important criterion alongside being a UAE citizen. However, exceptions are occasionally made for non-UAE citizens if they can provide evidence that the necessary treatment is not available within the UAE. Nonetheless, despite the existence of cancer treatment options in the UAE, a significant number of patients are granted exemptions to seek treatment overseas. These entities and agencies lack standardized guidelines or criteria for selecting patients to receive treatment abroad [2, 5].

During an internal assessment conducted at a tertiary referral oncology center in the UAE, a review of 273 patients who sought permission to travel abroad between January and September 2017 revealed that 86% of the referrals were deemed unnecessary from a clinical standpoint. This assessment was based on the fact that the required oncology services were already available in the UAE [1]. This assessment was conducted prior to the introduction of bone marrow transplantation services, and a significant number of these cases involved such treatments [6]. Based on our expertise, we estimate that over 95% of cancer cases can now be effectively treated within the UAE [6].

The UAE has very well-established cancer care centers, with over 30 cancer centers and at least five comprehensive cancer centers, yet there are various factors for seeking treatment abroad [5] (Fig. 8.1).

Fig. 8.1 Factors causing Emirati patients to travel abroad for cancer care

8.2   Factors Leading Emirati Patients to Travel Abroad for Cancer Care

8.2.1   Family and Societal Pressure

Family and societal pressure emerged as one of the primary driving factors behind Emirati cancer patients' decision to seek cancer care outside their country [7, 8]. This can be attributed to the familial setup prevalent in the UAE, characterized by extensive and interconnected family units. In such structures, family members play a significant role not only in everyday decision-making but also in matters concerning an individual's health, particularly when it comes to cancer. A study conducted on a diverse patient population in critical care settings revealed that families in the UAE perceived it as their duty to ensure that the patient received appropriate medical care. Therefore, they sought second opinions to confirm whether the provided treatment was suitable or not [8]. In the Australian study conducted by Philip, it was revealed that 70% of the patients received encouragement from their family and friends to seek a second opinion. This was one of several reasons identified in the study [7].

8.2.2   Perception of Patients with Cancer in Cancer Care in the UAE

During the early days of the UAE after the establishment of the UAE as an independent country, as expected, the UAE health system was in its infancy, many healthcare services were not available, and it was depending on some countries like Kuwait to support the health system [9]. During that time, the concept of traveling abroad for medical care started to evolve, and with the generous support of Sheikh Zayed, the founder of the UAE, the sponsorship programs for UAE nationals to seek medical care abroad started the culture of treatment abroad for various medical conditions, from simple to very complex cases. Again, due to the limited healthcare resources and healthcare providers' manpower, the sponsorship programs supported hundreds of thousands of Emirati patients over the last five decades to seek medical treatment in Europe, North America, and Asia. The culture of traveling abroad for medical care continued to evolve and persist despite the advances in the UAE's health system. This culture caused distrust in the health system, and the oncology sector in the UAE health system may be affected the most, in our opinion, as oncology care lagged behind in the development of the UAE health system for various reasons, including the lack of specialized cancer care until around 1979, when Tawam Hospital opened its doors, which was 8 years after the establishment of the UAE in 1971.

From our ongoing discussion with patients during clinical care for cancer patients who expressed their wishes to travel abroad after being diagnosed with a malignancy in the UAE, common themes for reasons why they decline treatment locally include: lack of trust in oncologists and other healthcare providers in the UAE; lack of trust in accurate pathological diagnosis; rumors of errors in diagnosis in the UAE among the members of the UAE community; social media false information and stories about errors in the diagnosis and management of cancer in the UAE; and the wrong belief that patients treated locally tend to die and ones treated abroad tend to survive.

These common misconceptions and the need to gain trust in the health system should be addressed by highlighting the success stories in both traditional and social media and by launching awareness campaigns to highlight the expertise and facilities that are specialized in cancer care in the UAE. Accredited comprehensive cancer centers by the regulators to guide the public to the center of excellence in cancer care. Utilization of virtual consultation with major cancer centers abroad with confirmation of the locally proposed treatment plan may help in establishing trust and acceptance of cancer care locally among Emirati patients. This will also reduce the financial burden on the government from the cost of cancer care abroad.

8.2.3   False Belief in Better Cancer Treatment Abroad

While better cancer care and outcomes would have certainly been true abroad in more advanced health systems between the 1970s and early 2000s in the UAE, this does not hold true nowadays. While survival and outcome data are generally lacking except for a few reports [10], in our experience and that of other colleagues who trained and practiced in the USA, Canada, and the UK, treatment modalities and outcomes are similar for major cancer centers in the UAE. Certainly, patients being treated in nonspecialized cancer centers, either in the UAE or anywhere else, have a lower cure and survival rate [11].

Another common belief among cancer patients and their families is that more effective and reliable anticancer therapies and technologies are available abroad than in the UAE. Again, historically, this was true, but with fast advances in the oncology landscape in the UAE [1, 12], fast-track approvals for anticancer therapies in the UAE, the availability of state-of-the-art centers and technologies, including the latest radiation machines, AI-powered applications used in cancer screening and chemotherapy preparations, and advanced robotic surgical equipment, the cancer treatment technologies in the UAE are very advanced [13].

As mentioned earlier, these advances must be highlighted to the public to change these common misconceptions [14, 15].

There are very generous support programs for Emirati patients to travel to world-class facilities in Europe, Asia, and North America, with the cost of medical treatment completely covered by the UAE government. There are many sponsoring governmental entities that support Emirati patients traveling abroad for medical care. There are various criteria for patients to travel abroad, but all sponsoring entities request medical reports from healthcare providers to confirm the patient's medical condition [1]. All treatment costs are completely covered, with no co-payment by the patients. The support is part of the government's support of UAE nationals, in addition to many other initiatives like free housing schemes, financial support for youth, the "Marriage Grant," and many others [14, 15].

There have been no official reports on the cost of medical treatment abroad over the years, but with the increasing cost of cancer care worldwide, the cost of treatment abroad must be significant.

Another important yet overlooked factor for patients to consider when traveling abroad for cancer care is the travel distance for patients with cancer and their families while being treated in the UAE, as the cost of travel and accommodation can be significant, and this is not covered by the healthcare providers or any governmental organizations. This cost is completely covered for patients and their caregivers while being treated abroad. This is a major issue for many patients with limited budgets, and the choice between staying in the UAE and paying out of pocket or traveling abroad and getting all costs completely covered makes the latter choice more appealing and attractive.

Supporting patients' travel and lodging costs who live far from cancer centers will increase their acceptance of being treated locally rather than abroad.

8.2.4 Extended Sick Leave for Cancer Patients and Their Companions

Patients and one or two family members get extended paid sick leave and companion sick leave. This leave could be for months or years, as long as the medical condition is approved to continue treatment abroad. These sick leaves are not guaranteed, either for patients with cancer or their family members who have to take sick leave, which can be exhausted in a matter of a couple of weeks.

In order to encourage Emirati patients with cancer to get treatment locally, new rules and regulations grant a similar advantage of extended work hours for patients and their caregivers. Studies have shown that patients with cancer have a significant negative impact on their work and finances [16].

Emirati patients' and caregivers' jobs and financial security will be critical in changing the mindset about traveling abroad for cancer treatment.

8.2.5 International Institutions Promotion for Cancer Care Abroad

Many major cancer centers in the USA, for example, have dedicated offices in their centers to facilitate and attract cancer patients from Gulf countries. The usual approach is direct communication with the sponsoring agencies in the UAE, holding scientific meetings to attract oncologists and create referral pathways from the local oncologists to the cancer centers abroad, and having dedicated Arabic websites to attract patients for treatment at their centers [17–19].

8.2.6 Government Financial Incentive for Patients and Their Companions While Abroad

The UAE government has a very generous program to cover the expenses for patients and their families while being treated abroad; this covers the accommodation, food, pocket money, etc. Depending on where the patient and their companion are being treated, the daily allowance can reach a few hundred dollars.

8.2.7 Physicians' Attitude Influencing Patients' Decisions for Traveling Abroad

One of the major setbacks for cancer care in the UAE is the attitude of some oncologists and hematologists, who encourage patients to travel abroad, especially in more complex cases, in order to avoid clinical work and complications and potential medicolegal complaints, which is not an uncommon practice by patients and their caregivers. The direct recommendations in the medical reports to travel abroad, e.g., for advanced, noncurable stage malignancies with limited survival and the availability of the same treatment modalities in the UAE, are not uncommon for the abovementioned reasons. Auditing physicians with a high rate of recommendations adjusted to their volume of patients is a potential solution to address this issue.

8.3 Negative Implications for Cancer Care Abroad

- Treatment postponement as a result of arranging travel and scheduling appointments.
- Language barriers may negatively affect communication and adherence to treatment.
- Patients receiving treatment abroad typically have limited time with their doctors, leading to a lack of follow-up care and a potential failure to identify complications or treatment side effects. It is crucial for patients to be knowledgeable about specialized centers that excel in treating specific types of cancer in order to obtain accurate opinions. Without proper guidance, patients may select a location for a second opinion that lacks the necessary facilities and expertise, which can have adverse effects on their health.
- Significant cost to the health system, which can be utilized to support and improve the UAE health system.
- Abuse of the traveling abroad program by cancer centers abroad: for example, some centers abroad charge the UAE government for clinical trial participation, despite the fact that these trials are already sponsored by pharmaceutical companies.

8.4 Conclusion

The long-term sustainability and exorbitant costs associated with traveling abroad for cancer treatment make it an impractical option. It is recommended that treatment overseas be limited to complex cancer cases requiring specialized care that is not available within the UAE. This decision should be made after a thorough evaluation by an accredited comprehensive cancer center, following a consensus review. To establish public confidence in cancer care within the UAE, it is essential to prioritize outreach programs at the national level, involving regulatory bodies and sponsoring agencies responsible for facilitating treatment abroad. Focused research efforts are necessary to understand the factors motivating Emirati patients and their families to seek treatment overseas as well as the barriers preventing them from receiving treatment locally in the UAE. Such research will ultimately help reduce the demand for and dependence on foreign cancer care.

Conflict of Interest The author has no conflict of interest to declare.

References

1. Al-Shamsi H, Darr H, Abu-Gheida I, Ansari J, McManus MC, Jaafar H, Tirmazy SH, Elkhoury M, Azribi F, Jelovac D, et al. The state of cancer Care in the United Arab Emirates in 2020: challenges and recommendations, a report by the United Arab Emirates Oncology Task Force. Gulf J Oncolog. 2020;1(32):71–87.
2. Sahoo S. Rising numbers in outbound medical tourism. http://www.thenational.ae/business/industry-insights/tourism/rising-numbers-in-outbound-medical-tourism. Accessed 12 Dec 2022.
3. Alnakhi WK, Iqbal F, Nadabi WA, Balushi AA. Challenges associated with medical travel for cancer patients in the arab world: a systematic review. In: Al-Shamsi HO, Abu-Gheida IH, Iqbal F, Al-Awadhi A. Cancer in the Arab world. Singapore: Springer Singapore; 2022: 427–444.
4. Alnakhi WK, Segal JB, Frick KD, Hussin A, Ahmed S, Morlock L. Treatment destinations and visit frequencies for patients seeking medical treatment overseas from the United Arab Emirates: results from Dubai health authority reporting during 2009-2016. Trop Dis Travel Med Vaccines. 2019;5:10.
5. Al-Shamsi HO. The state of cancer care in the United Arab Emirates in 2022. Clin Pract. 2022;12:955–85. https://doi.org/10.3390/clinpract12060101.
6. 95% of global cancer treatments are available in the UAE. https://www.alittihad.ae/news/%D8%A7%D9%84%D8%A5%D9%85%D8%A7%D8%B1%D8%A7%D8%AA/4236923/-95-%D9%85%D9%86-%D8%A7%D9%84%D8%B9%D9%84%D8%A7%D8%AC%D8%A7%D8%AA-%D8%A7%D9%84%D8%B9%D8%A7%D9%84%D9%85%D9%8A%D8%A9-%D9%84%D9%84%D8%B3%D8%B1%D8%B7%D8%A7%D9%86-%D9%85%D9%88%D8%AC%D9%88%D8%AF%D8%A9-%D8%A8%D8%A7%D9%84%D8%AF%D9%88%D9%84%D8%A9. Accessed 12 Dec 2022.
7. Philip J, Gold M, Schwarz M, Komesaroff P. Second medical opinions: the views of oncology patients and their physicians. Support Care Cancer. 2010;18(9):1199–205.

8. Al Shaaibi R, Burney IA. Motives for medical tourism amongst cancer patients in Oman: a perspective from patient's point of view. Columbia Univ J Glob Health. 2019;9(1). https://doi.org/10.7916/thejgh.v9i1.4932.

9. UAE and Kuwait. "6" decades of brotherly relations based on unity of purpose and destin. http://wam.ae/ar/details/1395302599562. Accessed 12 Dec 2022.

10. Elobaid Y, Aamir M, Grivna M, Suliman A, Attoub S, Mousa H, Ahmed LA, Oulhaj A. Breast cancer survival and its prognostic factors in the United Arab Emirates: a retrospective study. PLoS One. 2021;16(5):e0251118.

11. Wolfson JA, Sun CL, Wyatt LP, Hurria A, Bhatia S. Impact of care at comprehensive cancer centers on outcome: results from a population-based study. Cancer. 2015;121(21):3885–93.

12. Abu-Gheida IH, Nijhawan N, Al-Awadhi A, Al-Shamsi HO. General oncology care in the UAE. In: Al-Shamsi HO, Abu-Gheida IH, Iqbal F, Al-Awadhi A, editors. Cancer in the Arab world. Singapore: Springer Singapore; 2022. p. 301–19.

13. Al-Shamsi HO, Abyad AM, Rafii S. A proposal for a National Cancer Control Plan for the UAE: 2022–2026. Clin Pract. 2022;12(1):118–32.

14. Housing and land benefits for UAE nationals. https://u.ae/en/information-and-services/housing/housing-authorities-and-programmes/for-uae-nationals. Accessed 12 Dec 2022.

15. Marriage Grant. https://www.mocd.gov.ae/en/services/request-for-marriage-grant.aspx. Accessed 12 Dec 2022.

16. Blinder VS, Gany FM. Impact of cancer on employment. J Clin Oncol. 2020;38(4):302–9.

17. International patients services. https://www.mskcc.org/ar/experience/become-patient/international-patients. Accessed 13 Dec 2022.

18. Why choosing MD Anderson? https://www.mdanderson.org/ar/why-choose-md-anderson.html. Accessed 13 Dec 2022.

19. Reasons for choosing the children's hospital of Philadelphia Cancer Center. https://gps.chop.edu/ar/%D8%A7%D9%84%D8%A8%D8%B1%D8%A7%D9%85%D8%AC%20%D8%A7%D9%84%D8%B7%D8%A8%D9%8A%D8%A9/%D9%85%D8%B1%D9%83%D8%B2-%D8%A7%D9%84%D8%B3%D8%B1%D8%B7%D8%A7%D9%86. Accessed 13 Dec 2022.

Prof. Humaid Obaid Al-Shamsi is the Chief Executive Officer of Burjeel Cancer Institute in Abu Dhabi, UAE, President of the Emirates Oncology Society, Lead of the Gulf Cancer Society, Full Professor of Oncology at the Ras Al Khaimah Medical and Health Sciences University, Ras Al Khaimah, UAE, and an Adjunct Professor of Oncology at the College of Medicine, University of Sharjah. He is the first Emirati to be promoted as a professor in oncology in the UAE. He is also the Chairman for Colorectal Cancer in the MENA region, appointed by the prestigious National Comprehensive Cancer Network®. He is also the only member of Lung Cancer Policy Network in the MENA region that aims to advance lung cancer research and screening globally. He is the Chairman of the Oncology and Hematology Fellowship Training Program for the National Institute for Health Specialties in the United Arab Emirates. He is the only member in GCC in the WIN Consortium which is comprised of organizations representing all stakeholders in personalized cancer medicine globally.

He is board-certified in both internal medicine and oncology from the UK, USA (ABIM), and Canada (FRCPC). He has also been awarded the FRCP (London) in 2023 and FRCP (Glasgow) in

2024. He is the only physician in the UAE with a subspecialty fellowship certification and training in gastrointestinal oncology and the first Emirati to train and complete a clinical post-doctoral fellowship in palliative care. He was an assistant professor at the University of Texas MD Anderson Cancer Center between 2014 and 2017. He has published more than 140 peer-reviewed articles in JAMA Oncology, Lancet Oncology, The Oncologist, BMC Cancer, and many others. His area of expertise includes precision oncology and cancer care in the UAE. In 2016, he published with his group from MD Anderson the JCO paper describing a new distinct subgroup of CRC, NON V600 BRAF-mutated CRC. In 2022, he published the first book about cancer research in the UAE and also the first book about cancer in the Arab world, both of which were launched at Dubai Expo 2020. *Cancer in the Arab World* has been downloaded more than 450,000 times in its first 18 months of publication and is the ultimate source of cancer data in the Arab region. He also published the first comprehensive book about cancer care in the UAE which is the first book in UAE history to document the cancer care in the UAE with many topics addressed for the first time, e.g., neuroendocrine tumors in the UAE. He is passionate about advancing cancer care in the UAE and the GCC and has made significant contributions to cancer awareness and early detection for the public using social media platforms. He is considered as the most followed oncologist in the world with over 300,000 subscribers across his social media platforms (Instagram, Twitter, LinkedIn, and TikTok). In 2022, he was awarded the prestigious Feigenbaum Leadership Excellence Award from Sheikh Hamdan Smart University for his exceptional leadership and research and the Sharjah Award for Volunteering. He was also named the Researcher of the Year in the UAE in 2020 and 2021 by the Emirates Oncology Society.

In May 2024, HH Sheikh Mansour bin Zayed Al Nahyan, Vice President of the United Arab Emirates, awarded him the first place in UAE Nafis program for outstanding leadership in private sector across all business and medical disciplines. Beside his clinical and administrative duties, he is engaged in education and various levels of research training for medical trainees to enhance their clinical and research skills. His mission is to advance cancer care in the UAE and the MENA region and make cancer care accessible to everyone in need around the globe.

Nongovernmental Organizations' (NGO) Role in Cancer Care in the UAE: Friends of Cancer Patients as an Example

9

Aisha Al Mulla and Majed Mohamed

9.1 Cancer Care and the Role of FOCP in the UAE

Cancer care receives vital attention in the United Arab Emirates (UAE), with Friends of Cancer Patients (FOCP) leading this frontier as an organization exclusively focused on cancer patients and awareness. The nongovernmental organization was founded in 1999 under the patronage of Her Highness Sheikha Jawaher Bint Mohammed Al Qasimi, Wife of the Ruler of Sharjah, and has the vision of a world where cancer no longer holds power over our lives. With the objective of providing financial and emotional support to cancer patients and their families in the UAE, the organization was also established to raise awareness, particularly about the cancers with the highest rates of early detection.

For more than two decades, FOCP has been successful in offering hope and support to cancer patients during their journey toward recovery, as well as assisting families in coming together again. The organization supports both UAE citizens as well as residents and provides crucial guidance on the latest advancements in cancer treatment. Fundraising efforts, either in the physical or digital realm, contribute largely to the financial aid targeting the medical treatment of cancer patients. It also provides donors with the opportunity to make a lasting impact and contribute to the well-being of society through its focus on cancer care. The firm engages with the community through initiatives like cancer walks, awareness campaigns, and involvement in seminars and conferences.

FOCP emphasizes early screening and detection methods in addition to focusing on pediatric cancers and cancers with high incidence rates in the UAE, such as breast, colorectal, lung, thyroid, skin, and leukemia. Cancer screening projects are in place for the early diagnosis of high-risk cancers like breast and cervical cancer. The organization is also involved in discussions and advocacy efforts related to

A. Al Mulla · M. Mohamed (✉)
Friends of Cancer Patients, Sharjah, United Arab Emirates
e-mail: Aisha@focp.ae; Majed@focp.ae

© The Author(s) 2024
H. O. Al-Shamsi (ed.), *Cancer Care in the United Arab Emirates*,
https://doi.org/10.1007/978-981-99-6794-0_9

163

policy change, promotion of healthy lifestyles, tobacco control, and vaccination in order to create a healthier society in the UAE.

In addition to collaborating with international and regional organizations such as the Union for International Cancer Control (UICC), the American Cancer Society, the NCD Alliance, and the Gulf Federation for Cancer Control, FOCP has also partnered with local health bodies such as the UAE Ministry of Health, the Abu Dhabi Health Authority, the Sharjah Health Authority, and the Dubai Health Authority. The organization has received widespread recognition and awards from leading institutions, both locally and internationally, for its efforts in supporting cancer patients and increasing awareness about the disease.

9.2   Beneficiary Support in the UAE

FOCP, the UAE-based cancer-focused society, provides services to all residents of the country without discrimination based on nationality, gender, age, or religion. Since its establishment in 1999, a total of 6217 patients have received various levels of support from the FOCP. The NGO provides annual care to hundreds of cancer patients, with the number reflected in Tables 9.1 and 9.2 based on months of diagnosis and gender for 2020 and 2021, respectively.

For the current year of 2022, the data in the table include analysis based on gender and nationality until November (Table 9.3).

The FOCP is dedicated to providing comprehensive care to cancer patients in the UAE, including early detection and screening, financial and moral support, and awareness campaigns. The organization works with both nonprofits and private sector companies to ensure that patients have access to the resources they need to combat their illness and recover. To receive assistance from the FOCP, cancer patients must provide documents, including data on their social and economic status. In

Table 9.1 Number of FOCP beneficiaries (cancer patients) in 2020

	Male	Female	Children	Total number of patients per month
January	9	25	2	36
February	9	21	2	32
March	11	24	1	36
April	8	26	1	35
May	7	21	3	31
June	7	22	2	31
July	8	26	1	35
August	6	24	2	32
September	7	24	2	33
October	11	22	2	35
November	9	21	1	31
December	9	20	3	32
Total	**101**	**276**	**22**	**399**

Table 9.2 Number of FOCP beneficiaries (cancer patients) in 2021

	Male	Female	Children	Total number of patients per month
January	7	20	1	28
February	5	24	1	30
March	7	20	1	28
April	3	27	1	31
May	6	19	4	29
June	4	24	1	29
July	3	24	1	28
August	5	24	1	30
September	5	23	1	29
October	5	21	1	27
November	5	23	1	29
December	7	28	2	37
Total	**62**	**277**	**16**	**355**

Table 9.3 Number of FOCP beneficiaries (cancer patients) in 2022

	Male	Female	Children	Total number of patients per month
January	2	16	1	19
February	5	17	3	25
March	2	18	1	21
April	2	15	1	18
May	6	13	2	21
June	3	18	1	22
July	3	16	2	21
August	1	20	1	22
September	3	17	1	21
October	3	18	1	22
November	2	19	1	22
December	2	21	1	24
Total	**34**	**208**	**16**	**258**

response to the COVID-19 pandemic, the FOCP created an online portal for patients to submit their applications and continue receiving support.

To ensure the sustainability of patient access to care, the FOCP works with non-profit organizations and private sector companies to provide access programs with free additional packages, including long-term medication support during the recovery process. Over the past 3 years, 3392 patients have received psychological and moral support from the FOCP. On the other hand, the Locks of Hope campaign, which aims to help restore their self-esteem and their confidence, enabling them to face the world with a positive outlook by providing wigs to cancer patients, has received 3244 hair donations, and the Ramadan campaign, which has supported 2047 cancer patients since 2013, are two of the most effective community-focused campaigns run by the FOCP. These campaigns aim to raise awareness about cancer and its treatment, as well as provide support for those affected by the disease.

9.3   Global Sustainable Aid Initiatives

The Ameera Fund is an initiative of the FOCP that aims to address cancer care on a global scale through collaborative projects in the fields of cancer research, capacity building, prevention, and treatment. Established in partnership with The Big Heart Foundation, the Ameera Fund has a vision of a world where everyone has access to cancer care and a mission to improve understanding of cancer through various initiatives and partnerships.

The Ameera Fund is committed to its vision of a world where everyone has access to cancer care and its mission to improve understanding of cancer through a range of initiatives and partnerships. By increasing access to cancer care, strengthening cancer monitoring systems, enhancing capacity building, and supporting cancer research and treatment, the Fund is working toward its goals of improving the lives of cancer patients and their families around the world.

Since its establishment in 2018, the Ameera Fund has spent a total of USD 5,186,421 on cancer care and capacity building, with a total spending of USD 2,863,447 on 222,499 beneficiaries globally (Table 9.4).

One of the main objectives of the Ameera Fund is to increase access to cancer care, particularly for underserved and disadvantaged communities. To achieve this goal, the fund has initiated a number of projects in collaboration with various organizations around the world. For example, in partnership with Tumiani La Maisha in Tanzania, the Ameera Fund helped to construct a pediatric intensive care unit and a neonatal intensive care unit at Muhimbil National Hospital, the largest government hospital in the country. This project aimed to give young patients greater access to cancer care, allowing them to receive treatment closer to home and in a more supportive environment.

In addition to increasing access to cancer care, the Ameera Fund also focuses on capacity building and the strengthening of cancer surveillance systems. To this end, the Fund has collaborated with the UICC in Geneva, Switzerland, on a project aimed at strengthening cancer monitoring through cancer registries and supporting the "Treatment for All" campaign, which advocates for universal access to cancer treatment. The Ameera Fund has also joined forces with Access to Child Cancer Essentials (ACCESS) in Eastern Africa to amplify cancer care facilities and conduct research into the barriers to the availability and accessibility of cancer care resources in the region.

Table 9.4 Ameera fund global funding support (2018–2022)

Year	Total funding
2018	22,95,155
2019	18,67,000
2020	2,40,892
2021–2022	7,83,374

Working on increasing access to cancer care and strengthening cancer monitoring systems, the Ameera Fund also supports cancer research and treatment in various ways. One notable highlight is the collaboration with the King Hussein Cancer Foundation in Jordan, where the Fund supported the treatment and medical costs of refugee cancer patients. Additionally, the Ameera Fund has partnered with BIO Ventures for Global Health in Côte d'Ivoire-Abidjan to ensure access to oncology medicines and build research and healthcare capacity.

The Ameera Fund has also responded to the needs of cancer patients affected by natural disasters and other crises. For instance, in partnership with the Cancer Warriors Foundation—Philippines Childhood Cancer International, the Fund launched a project to support children with cancer who were affected by the Typhoon Goni Hurricane in the Philippines, providing them with treatment, maintenance medications, and food aid for themselves and their families. The Ameera Fund has also provided support to cancer patients in Lebanon affected by the catastrophic explosion in Beirut [1, 2].

The Fund, in partnership with the Bangladesh Ministry of Health and Family Welfare, also organized a conference in Bangladesh, where cancer is the leading cause of death among women of reproductive age. The conference brought together various stakeholders to share evidence-based best practices for noncommunicable diseases (NCDs) and advocate for policy changes to increase investment in prevention, care, and treatment for disadvantaged communities. Additionally, the conference provided an opportunity to establish partnerships for research and community-level model development.

Taking up various initiatives and partnerships, the Ameera Fund has also established projects not only globally but also provided support to cancer treatment and facilities within the UAE. The Fund has joined forces with the American University of Sharjah to improve the effectiveness and efficiency of chemotherapy treatment protocols in the country.

9.4   Advocacy for International Collaboration

The FOCP engages in international collaboration with similar organizations to share knowledge and expertise in cancer screening, diagnosis, treatment, and research. These collaborations take the form of forums, conferences, and roundtables, some of which are pioneering events in the region. The FOCP seeks to ensure that individuals in the region have access to the most advanced cancer treatments in order to improve survival rates and reduce the incidence of the disease.

One example of this collaboration is the Sharjah PORTAGE (Pediatric Oncology Roundtable to Transform Access to Global Essentials), which works to combat pediatric cancer and support efforts to eliminate the disease. The forum has also helped to identify funding sources and partners to implement priority initiatives and develop strategies for effective communication between medical institutions and relevant authorities in order to coordinate efforts to fight pediatric cancer.

The First Childhood Noncommunicable Diseases (NCD) Forum, held in partnership with the Global NCD Alliance, also focused on global efforts to address NCDs, including cancer. The forum featured 110 delegates and 25 speakers from over 20 countries.

The Global NCD Alliance Forum, first held in 2015, provides a platform to advocate for action to implement global pledges and meet the globally agreed NCD targets by 2025. The 2020 forum attracted more than 400 delegates from 80 countries, with 53 national and regional alliances and over 20 youth leaders in attendance. In addition to 20 individuals sharing their experiences living with NCDs, the event featured 140 speakers from 32 countries, as well as five plenary sessions and 18 workshops.

The Cervical Cancer Forum, held in 2019 and 2021, has provided programmatic guidance to the Ministry of Health and other agencies for the development and revision of cervical cancer prevention and control programs. The forum featured best practices from the UAE and introduced the "Sharjah Declaration on Cervical Cancer," developed in partnership with the United Nations Population Fund (UNFPA) and the MENA Coalition for Human Papillomavirus (HPV) Elimination. The declaration outlines the actions and collaborations needed to address cervical cancer and save lives in the region.

The forum advocates for regional and national cervical cancer strategies that align with global initiatives, taking into account national capacities in order to ensure their implementation in each country. It also aims to enhance cooperation in order to improve countries' and organizations' capacity for cervical cancer control and nationwide HPV vaccine program implementation, particularly in the context of current and future health emergencies. The forum also seeks to improve data collection, analysis, and utilization for evidence-based decision-making and to address inequities in access to the HPV vaccine and cervical cancer detection and treatment. The conference has also led to the establishment of a framework for monitoring the progress of HPV vaccine implementation and cervical cancer elimination.

The fifth annual Combined Gulf Cancer Conference in 2022, hosted by the FOCP in Sharjah in collaboration with the Gulf Federation for Cancer Control and the Gulf Center for Cancer Control and Prevention, brought together over 50 international speakers and 500 delegates, including scientists, researchers, cancer prevention advocates, and representatives of leading cancer nonprofit organizations in the region. The conference focused on analyzing the current state of cancer care in the region and discussing best practices to improve the standard of care across the cancer control continuum. It also addressed the potential for creating medical frameworks to improve access to healthcare and facilities for cancer patients during treatment and beyond. Apart from being a networking platform for global cancer experts, the conference also featured interactive panel discussions and presentations by young scientists who presented their research abstracts and hypotheses.

The conference featured six tracks with a variety of panel sessions, dialogues, and keynote speeches. The discussions addressed long-term issues in the sector, such as cultural and socioeconomic barriers to cancer awareness and prevention, as well as the importance of early detection and screening in cancer care and control.

The conference also focused on current issues such as the use of social media and mobile health technology for health promotion and behavior change, the influence of media on behavior, how to build better by learning from the COVID-19 pandemic, and the impact of advanced technology and innovation on cancer screening programs in the region.

9.5 Cancer Awareness Initiatives

The FOCP places a significant emphasis on raising public awareness about the importance of early detection in the fight against cancer. This focus is in line with findings from the World Health Organization, which suggest that approximately 30–50% of cancers can be prevented through the adoption of healthy behaviors, such as maintaining a healthy diet and weight and engaging in regular physical activity [3]. Additionally, the World Health Organization (WHO) reports that another 40% of cancers are curable if diagnosed in their early stages and treated promptly [4].

Conversely, a large percentage of cancer cases that are diagnosed at later stages are difficult to cure. Late detection is often attributed to a lack of knowledge about cancer symptoms and the failure to undergo regular cancer screenings.

To address these issues, the FOCP works with local government organizations to provide educational lectures and workshops, free medical examinations, and support programs for cancer patients and their families. One of the organization's most successful campaigns, the "Pink Caravan," is a UAE-wide initiative focused on raising awareness about breast cancer and promoting early detection and screening methods [1, 2] (Table 9.5).

Cancer awareness is at the heart of FOCP's mission, which involves creating awareness around the six early detectable cancers: breast cancer, cervical cancer, prostate cancer, testicular cancer, colorectal cancer, and skin cancer under the

Table 9.5 Number of detected cancer cases since the launch of the Pink Caravan initiative (2011–2022)

Category	الفئة	2011	2012	2013	2014	2015	2016	2017	2018	2019	2020	2021	2022	Total
Examined	المفحوصين	10,024	6,751	6,102	6,072	2,383	5,059	7,481	7,932	7,206	11,077	2,197	2,702	79,988
Referrals	المحولين													
Mammogram	ماموجرام	3,454	2,152	1,197	1,546	1,722	1,665	2,542	2,226	2,126	2,152	769	891	22,456
Ultrasound	اشعة الصوتية	186	325	189	286	176	460	-	436	707	639	58	8	3,650
Male & Female	الرجال و النساء													
Females Examined	النساء المفحوصات	6,079	6,275	4,980	5,084	6,041	4,406	6,366	7,314	6,570	8,316	2,197	2,702	66,330
Males Examined	الرجال المفحوصين	3,945	476	1,122	988	1,342	653	1,117	618	636	2,761	-	-	13,658
Positive Breast cancer cases detected (Malignant)	عدد الحالات المكتشفة بسرطان الثدي													
Females	نساء	8	5	3	-	6	4	15	17	11	5	-	-	79
Males	رجال	-	-	1	-	-	-	-	-	-	-	-	-	1

Pink Caravan in action since 2011 — مخرجات القافلة الوردية منذ عام 2011

* Detected cancer cases are currently under going treatment adopted by Friends of Cancer Patients

"KASHF" umbrella initiative for early detection. In this regard, FOCP has launched numerous cancer-focused awareness initiatives with a regional and international target scope, including the Pink Caravan for breast cancer, "Ana" for childhood cancer, "Shanab" for men's cancer, and "Mole Talk" for skin cancer.

9.6   Challenges in the UAE for Cancer Care

There are several challenges that cancer awareness initiatives in the UAE must overcome, including misdiagnoses and expensive treatment costs. Much like the global health diagnosis scenario, especially when it comes to cancer. Some patients' tumors may not be identified correctly, and they do not receive the right treatment at the right time, leading to undesired treatment complications or even death.

The cost of treatment can also pose a significant burden for patients, even if they have health insurance or their employer covers some medical expenses. Certain insurance companies may not cover treatments for previously diagnosed cases or for those who have lived with the disease for an extended period without diagnosis or treatment. Also, some insurance providers may not cover treatment costs for individuals over a certain age.

Access to comprehensive cancer care facilities is also a challenge in the UAE, as patients may need to visit multiple hospitals to complete their treatment journey. Specialized cancer facilities such as Tawam Hospital may also be difficult to reach for those living in other emirates.

The societal stigma surrounding cancer can also negatively impact the psychological well-being of patients. Cancer patients may be reluctant to share their experiences for fear of rejection, preventing them from receiving necessary psychological and emotional support during the early stages of diagnosis. A cancer diagnosis may also have economic consequences, as individuals may lose their jobs.

9.7   Strategies to Tackle Cancer Care Challenges

Cancer care is a complex and multifaceted issue that requires the implementation of strategies that address the needs of patients and caregivers, as well as the training and support of healthcare professionals. One key aspect of cancer care is ensuring access to accurate assessment and diagnostic services, as well as reliable information about the condition and available treatment options. This is crucial for empowering patients to make informed decisions about their care and for helping them maintain as normal a lifestyle as possible while minimizing the risk of further complications.

By providing access to information and services, NGOs can support patients in understanding the potential long-term effects of treatment and the possibility of relapse. This can be achieved through the provision of information about new research and the availability of resources for managing the disease.

To further improve cancer care, it is necessary to train healthcare professionals in the delivery of high-quality care and to provide them with access to the latest technologies. This includes the development of frameworks that are tailored to the needs of individual patients and the implementation of strategies to optimize the use of time and resources.

To increase cancer survival rates and bring care up to par with countries with the most advanced cancer care systems, it is necessary for health policymakers to implement strategies that address the unique challenges faced by cancer patients in different regions. This may include the creation of a registry that tracks the experiences and outcomes of cancer patients, allowing for the exchange of best practices and the identification of areas for improvement.

Enhancing cancer care requires the implementation of strategies that address the needs of patients and caregivers, as well as the training and support of healthcare professionals. By working toward a more inclusive and comprehensive approach to cancer prevention, control, and care, we can create a brighter future for those affected by the disease.

9.8   Conclusion

In summary, organizations like the Sharjah-based FOCP play a crucial role in providing aid, raising awareness about cancer, and enhancing cancer care facilities. FOCP focuses on early screening and detection methods as well as pediatric cancers and cancers with high incidence rates in the UAE. The organization also aims to advocate for health, call for better policy change efforts, and promote healthy lifestyles.

FOCP has been successful in offering hope and support to cancer patients and their families, as well as collaborating with international and local organizations to increase awareness about the disease. The organization provides services to all residents of the UAE without discrimination and has helped over 6000 patients since its establishment in 1999. In 2020 and 2021 alone, FOCP provided annual care to hundreds of cancer patients, with the majority being female and receiving support within the first 6 months of their diagnosis.

Apart from providing medical financial aid and social support, FOCP also conducts fundraising events and awareness workshops to increase public knowledge and address misunderstandings about cancer care. By raising awareness and providing support to cancer patients, FOCP is making a positive impact on the lives of individuals affected by the disease and the overall health of UAE society. It is important for organizations like FOCP to continue their efforts in cancer care and for individuals to prioritize their health and seek out early screening and detection methods.

Conflicts of Interest The authors have no conflict of interest to declare.

References

1. FOCP. Ameera fund. Advocacy. Our programs. Friends of Cancer Patients. 2022. https://www.focp.ae/our-programs/ameera-fund/.
2. FOCP. Pink Caravan. Women's health. Our programs. Friends of Cancer Patients. 2022. https://www.focp.ae/our-programs/womens-health/.
3. WHO. Preventing cancer. Activities. World Health Organization. 2022. https://www.who.int/activities/preventing-cancer.
4. UN News. New WHO platform promotes global cancer prevention. Health. United Nations. 2022. https://news.un.org/en/story/2022/02/1111312. Accessed 10 Feb 2023.

Aisha Al Mulla is a leader who possesses a strong commitment to creating a positive impact within her community. Her academic achievements demonstrate her aptitude for finance and her drive for self-improvement.

With a wealth of experience garnered from various roles and organizations, Aisha began her career as a Senior Finance Coordinator at Salam Ya Seghar, where she effectively managed projects for Palestine and Syrian refugees and oversaw the department's yearly and monthly budget and expenses. In September 2015, she assumed the position of Head of Programs at The Big Heart Foundation, where she expertly coordinated with existing and potential partners, and implemented projects for all the initiatives under the foundation. In July 2022, Aisha was promoted to the role of Deputy Director of Friends of Cancer Patients before becoming Director in October 2022.

Throughout her career, Aisha has achieved numerous notable accomplishments. In 2020, she received a Diploma in Humanitarian Assistance from the Ministry of Foreign Affairs x UAE University. In 2019, she completed the Sharjah Leadership Program from Sharjah Capability Development. In 2015, she received a certificate in the Advanced Management Development Program as well as a certificate in Business Etiquette from a course given by Mohamed AlMarzouqi.

Majed Mohamed is the Advocacy and Scientific Affairs Manager at the Friends of Cancer Patients (FOCP) organization in the UAE. Educational background includes a Master of Science in Public Health from Hamdan Bin Mohammed Smart University and a Bachelor of Science in Medical Sciences from the United Arab Emirates University.

Focused on reducing the impact of cancer, Majed leads local and international advocacy efforts, amplifying the voices of cancer patients for impactful change. With an extensive local, regional, and global network, Majed actively contributes to esteemed committees, including the WHO NCD Labs Steering Committee, NCD Child Young Leader, CCI Asia Regional Committee (ASRC), and MOHAP National Policy Healthy Lifestyle Committee.

He is recognized for his efforts in advancing cancer awareness, prevention, and treatment through local and international advocacy efforts to ensure that cancer patients' voices are heard and valued to drive better change.

Clinical Cancer Research in the UAE

10

Subhashini Ganesan ⓘ, Humaid O. Al-Shamsi ⓘ,
Mohamed Mostafa, and Walid Abbas Zaher ⓘ

10.1 Need for Cancer Research in the UAE

The worldwide cancer incidence is on the rise, and according to the World Health Organization (WHO), cancer is the first or second leading cause of death in about 112 countries [1]. In the United Arab Emirates (UAE), according to the UAE

S. Ganesan
G42 Healthcare, Abu Dhabi, United Arab Emirates

IROS (Insights Research Organization and Solutions), Abu Dhabi, United Arab Emirates
e-mail: subhashini.g@iros.ai

H. O. Al-Shamsi
Burjeel Cancer Institute, Burjeel Medical City, Burjeel Holdings, Abu Dhabi,
United Arab Emirates

Ras Al Khaimah Medical and Health Sciences University, Ras Al Khaimah, United Arab Emirates

Gulf Medical University, Ajman, United Arab Emirates

Emirates Oncology Society, Emirates Medical Association, Dubai, United Arab Emirates

College of Medicine, University of Sharjah, Sharjah, United Arab Emirates

Gulf Cancer Society, Alsafa, Kuwait
e-mail: alshamsi@burjeel.com; humaid.al-shamsi@medportal.ca

M. Mostafa
Science Park, Dubai, United Arab Emirates
e-mail: mohamed.mostafa@pdc-cro.com

W. A. Zaher (✉)
Science Park, Dubai, United Arab Emirates

College of Medicine and Health Sciences, Khalifa University,
Abu Dhabi, United Arab Emirates

College of Medicine and Health Sciences, United Arab Emirates University,
Abu Dhabi, United Arab Emirates
e-mail: walid.zaher@carexso.com

© The Author(s) 2024
H. O. Al-Shamsi (ed.), *Cancer Care in the United Arab Emirates*,
https://doi.org/10.1007/978-981-99-6794-0_10

National Cancer Registry annual report 2021, cancer is the fifth leading cause of death in the United Arab Emirates. The report showed that breast, thyroid, colorectal, leukemia, and skin (carcinoma) were the top-ranked cancers among all new cancer cases in both genders. Colorectal, prostate, leukemia, and thyroid were the top-ranked cancers among the males. Among females, breast, thyroid, colorectal, uterus, and cervix uteri were the top-ranked cancers. The overall age-standardized incidence rate was 107.8 per 100,000 and an overall crude incidence rate of 60.5/100,000 for both genders. Among females, the crude incidence rate was reported to be 108.7 per 100,000 females and 39.5/100,000 for males [2]. These numbers distinctly suggest that the UAE population differs from the west in their ethnicity and genetic makeup, which, to a large extent, can affect the diagnosis, treatment, and prognosis of this population. There is a difference in genomic structure between ethnic groups, and they differ in the strength of their association with cancer risk. As a result, some research findings may not be validated in ethnic groups other than the study cohort, affecting the study's reproducibility in another ethnic group [3]. Moreover, the sociocultural aspects of ethnicity and race, like food habits, lifestyle, and environment, contribute to cancer risk and make a significant contribution to cancer research [4]. As a result, robust cancer research among the Emirati population is required to improve our understanding of such differences and their impact on diagnostic and treatment strategies.

10.2 Clinical Trials in Oncology

The first clinical trial for the evaluation of cancer treatments was conducted in the mid-1950s by the National Cancer Institute of NIH [5]. Since then, hundreds of studies have been conducted and have provided substantial data, improving the knowledge base and thereby providing supporting evidence for newer cancer treatments or changes in regimens. In the hierarchy of evidence, randomized control trials (RCTs) are the most robust designs, and clinical trials produce evidence that can inform treatment policies and support changes [6]. Clinical trials have grown and evolved significantly over time, both in terms of their scientific impact and the regulations that govern them. However, clinical trials have their critics regarding the external validity of the trials as they are conducted in a highly controlled environment and the challenges such as globalization of trials, operational complexities like long time periods, regulatory and ethical approvals, recruiting and retaining the participants in the study, and the increasing cost of conducting clinical trials [7, 8].

In this chapter, we will discuss the clinical trials and the randomized controlled trials conducted in the UAE specific to oncology and their characteristics.

10.3 Clinical Trials in Oncology in the UAE

10.3.1 Search Criteria

A PubMed search was performed to find all cancer-related articles published in the United Arab Emirates (UAE) in the last ten years, from 2012 to 2022. The database was searched using the key words with Boolean operators as "OR" and "AND." The keywords used were "cancer," "oncology," "tumor," "tumor," "United Arab Emirates," and "UAE." The search was filtered by article type, which included only clinical trials and randomized controlled trials, and an additional filter on publication date, focusing on the last 10 years, was used to refine the search.

The search strategy included the following keywords: (Cancer) OR (TUMOR) OR (TUMOUR) AND (United Arab Emirates), which resulted in 28 studies. The abstracts of all 28 studies were screened by two coauthors, and studies that did not specifically relate to cancer were excluded. Following this exclusion, 11 studies were considered that were either conducted among the UAE population or had one of the authors affiliated with a UAE-based institute (Fig. 10.1).

Among these 11 studies, 3 (27.3%) focused on breast cancer, followed by 2 (18%) on leukemia. Ten of these 11 studies were based on therapy for cancer, and one was based on cancer screening. Out of the 11 studies, 6 (55%) were interventional trials, 2 (18%) were observational, 2 (18%) were phase III trials, and 1 (9%) was a phase II trial.

About 6 (55%) of these 11 studies were published in journals with an impact factor of 3.0 or higher, while 2 (18%) were published in journals with an impact factor of 10 or higher. About 7 out of 11 (64%) studies have gotten more than or equal to 15 citations (Table 10.1).

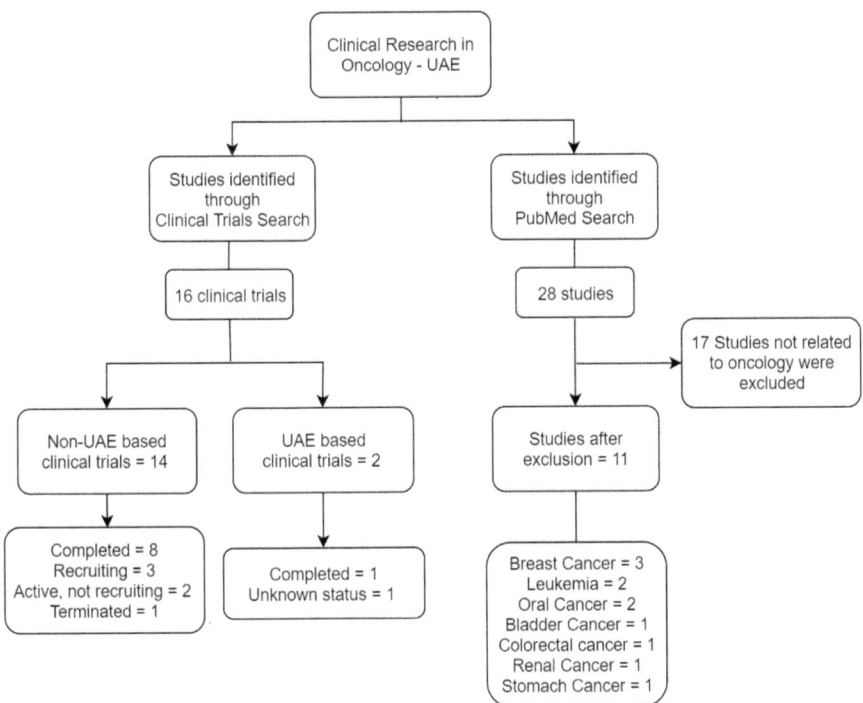

Fig. 10.1   Clinical research in oncology in the UAE

Table 10.1 Characteristics of clinical trials registered in oncology in the UAE

	Title	NCT	Type of cancer	Study type	Year	UAE-based	Status	Study results
1.	Epidemiological Study to Describe Non-small Cell Lung Cancer Clinical Management Patterns in MENA. Lung-EPICLIN/Gulf	NCT01562665	Lung cancer	Observational	2012	No	Completed	No results
2.	A Study of Trastuzumab Emtansine in Participants with Human Epidermal Growth Factor Receptor 2 (HER2)-Positive Breast Cancer Who Have Received Prior Anti-HER2 And Chemotherapy-based Treatment	NCT01702571	Breast cancer	Interventional	2012	No	Completed	Has results
3.	A Study of Pertuzumab in Combination With Trastuzumab (Herceptin) and a Taxane in First-Line Treatment in Participants With Human Epidermal Growth Factor 2 (HER2)-Positive Advanced Breast Cancer	NCT01572038	Breast cancer	Interventional	2012	No	Completed	Has results
4.	A Safety and Tolerability Study of Assisted and Self-Administered Subcutaneous (SC) Herceptin (Trastuzumab) as Adjuvant Therapy in Early Human Epidermal Growth Factor Receptor 2 (HER2)-Positive Breast Cancer	NCT01566721	Breast cancer	Interventional	2012	No	Completed	Has results
5.	Study of Efficacy and Safety in Premenopausal Women With Hormone Receptor-Positive, HER2-negative Advanced Breast Cancer	NCT02278120	Breast cancer	Interventional	2014	No	Active, not recruiting	Has results
6.	Retrospective Epidemiology Study Of ALK Rearrangement In Non-Small Cell Lung Cancer Patients In The Middle East & North Africa	NCT02304406	Lung cancer	Observational	2015	No	Completed	Has results

7.	Multicenter Registry of Treatments and Outcomes in Patients With Chronic Lymphocytic Leukemia (CLL) Or Indolent Non-Hodgkin's Lymphoma (iNHL)	NCT02273856	Lymphoma leukemia	Observational	2015	No	Terminated	No results
8.	Prevalence of BRCA1 and BRCA2 Mutations in Ovarian Cancer Patients in the Gulf Region	NCT03082976	Ovarian cancer	Observational	2017	No	Completed	No results
9.	Fulvestrant Versus Fulvestrant Plus Palbociclib in Operable Breast Cancer Responding to Fulvestrant	NCT03447132	Breast cancer	Interventional	2017	Yes	Completed	No results
10.	A Study of Atezolizumab (Tecentriq) to Investigate Long-term Safety and Efficacy in Previously treated Participants With Locally Advanced or Metastatic Non-Small Cell Lung Cancer (NSCLC)	NCT03285763	Lung cancer	Interventional	2017	No	Completed	No results
11.	Pembrolizumab And Tamoxifen Among Women With Advanced Hormone Receptor-Positive Breast Cancer And Esr1 Mutation	NCT03879174	Breast cancer	Interventional	2019	Yes	Unknown status	No results
12.	A Study Investigating the Outcomes and Safety of Atezolizumab Under Real-World Conditions in Patients Treated in Routine Clinical Practice	NCT03782207	Urothelial carcinoma/ non-small cell lung cancer/small cell lung cancer/ hepatocellular carcinoma	Observational	2019	No	Recruiting	No results
13.	Prevention of Colorectal Cancer Through Multiomics Blood Testing	NCT04369053	Colorectal cancer	Observational	2020	No	Active, not recruiting	No results

(continued)

Table 10.1 (continued)

	Title	NCT	Type of cancer	Study type	Year	UAE-based	Status	Study results
14.	Retrospective Study to Describe the Real-world Treatment Patterns and Associated Clinical Outcomes in Patients With Metastatic Castration-resistant Prostate Cancer	NCT04801186	Prostate cancer	Observational	2021	No	Recruiting	No results
15.	Study to Determine the Prevalence of Homologous Recombination Deficiency Among Women With Newly Diagnosed, High-grade, Serous or Endometrioid Ovarian, Primary Peritoneal, and/or Fallopian Tube Cancer	NCT04991051	Fallopian tube cancer	Observational	2021	No	Completed	No results
16.	An Observational Study to Evaluate the Real-World Clinical Management and Outcomes of ALK-Positive Advanced NSCLC Participants Treated With Alectinib	NCT04764188	Lung cancer	Observational	2021	No	Recruiting	No results

10.3.2 General Characteristics of the Clinical Trials

The clinical trials on cancer conducted or ongoing in the UAE were gathered from the clinicaltrials.gov site, and the search keywords were "tumor," "cancer," and "United Arab Emirates." The search resulted in 25 studies. When restricting the search to studies initiated in the last 10 years, only 16 clinical trials were identified. Only 2 of the 16 trials were initiated and sponsored by an institute or hospital in the UAE. The rest were sponsored by non-UAE-based organizations, of which the UAE was one of the study sites. Of these 16 trials, 5 were active or recruiting, 9 were completed, 1 study was terminated, and the status of one of the trials was unknown. Among the completed trials, results were available for only four studies (Fig. 10.1; Table 10.2).

About 6 (38%) of the cancer therapeutic trials were on breast cancer, 5 (31%) were on non-small cell lung cancer (NSCLC), and the others were on leukemia, colorectal, ovarian, prostate, and fallopian tube cancer. Regarding the study types, 7 (44%) were interventional trials, and the other 9 (56%) studies were observational. About 15 of 16 studies (94%) were solely sponsored by the pharmaceutical industry, and one was sponsored by Mediclinic in collaboration with AstraZeneca, Pfizer, and Genomic Health. The main pharmaceutical industry sponsors were Hoffmann-La Roche, which sponsored 6 (38%) of these trials, followed by AstraZeneca, which sponsored 4 (25%) of the trials; the other sponsors were Astellas Pharma International, Novartis, Pfizer, and Freenome Holdings (a biotechnology company). Ten of these 16 (63%) trials were registered between 2012 and 2017, and 6 (37%) were registered between 2018 and 2022 (Fig. 10.2).

Table 10.2 Characteristics of studies published in PubMed on cancer in the UAE

	Title	Cancer type	Study type	Year	Journal	Impact factor
1.	Reinduction chemotherapy using FLAG-mitoxantrone for adult patients with relapsed acute leukemia: a single-center experience from United Arab Emirates	Leukemia	Interventional clinical trial	2018	International Journal of Hematology	2.31
2.	Response to Induction Neoadjuvant Hormonal Therapy Using Upfront 21-Gene Breast Recurrence Score Assay-Results From the SAFIA Phase III Trial	Breast cancer	Phase III	2021	JCO Global Oncology	4.33
3.	Parts greater than their sum: randomized controlled trial testing partitioned incentives to increase cancer screening	Colorectal cancer	Interventional clinical trial	2019	Annals of the New York Academy of Sciences	6.49
4.	Safety and tolerability of subcutaneous trastuzumab for the adjuvant treatment of human epidermal growth factor receptor 2-positive early breast cancer: SafeHer phase III study's primary analysis of 2573 patients	Breast cancer	Phase III	2017	European Journal of Cancer	10.00
5.	Autologous Bone Marrow Concentrates and Concentrated Growth Factors Accelerate Bone Regeneration After Enucleation of Mandibular Pathologic Lesions	Oral cancer	Interventional clinical trial	2018	Journal of Craniofacial Surgery	0.92
6.	Standard triple therapy versus sequential therapy for eradication of Helicobacter pylori in treatment naïve and retreat patients	Stomach cancer	Interventional clinical trial	2016	Arab Journal of Gastroenterology	1.80
7.	Outcome of pregnancy in chronic myeloid leukemia patients treated with tyrosine kinase inhibitors: short report from a single center	Leukemia	Observational clinical trial	2015	Leukemia Research	3.71
8.	Association of AXL and PD-L1 Expression with Clinical Outcomes in Patients with Advanced Renal Cell Carcinoma Treated with PD-1 Blockade	Renal cancer	Observational clinical trial	2021	Clinical Cancer Research	13.80

9.	A multicenter prospective phase II trial of neoadjuvant epirubicin, cyclophosphamide, and 5-fluorouracil (FEC100) followed by cisplatin-docetaxel with or without trastuzumab in locally advanced breast cancer	Breast cancer	Phase II	2015	Cancer Chemotherapy and Pharmacology	3.28
10.	Monopolar vs. bipolar transurethral resection for non-muscle invasive bladder carcinoma: A post-hoc analysis from a randomized controlled trial	Bladder cancer	Interventional clinical trial	2018	Urologic Oncology: Seminars and Original Investigations	2.95
11.	Efficacy of Spirulina 500 mg vs. Triamcinolone Acetonide 0.1% for the Treatment of Oral Lichen Planus: A Randomized Clinical Trial	Oral cancer	Interventional clinical trial	2022	The Journal of Contemporary Dental Practice	1.01

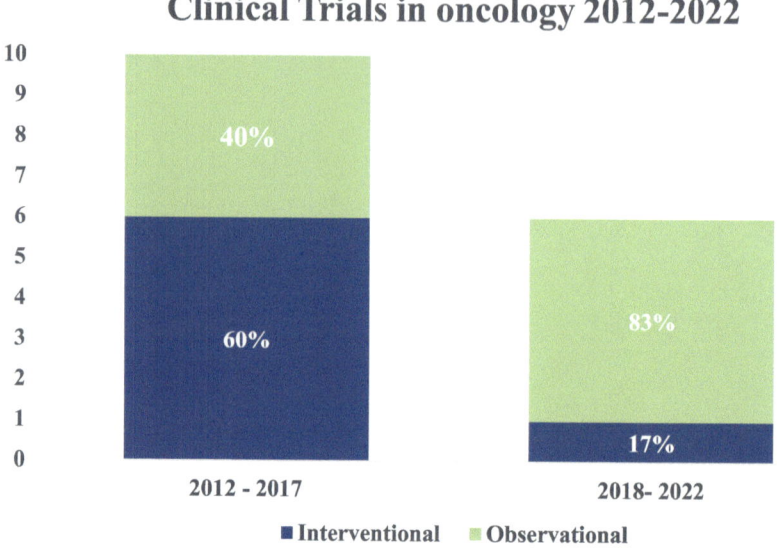

Fig. 10.2 Clinical trials in oncology in the UAE

10.4 Data Utilization

Clinical trial data that were analyzed included the trial start date, study type, study participants (including inclusion and exclusion criteria), intervention or treatment details, type of cancer studied, research site, research institute, phase of the trial, outcome measures, and availability of results.

10.4.1 Characteristics of Inclusion Criteria

The common inclusion criteria across all interventional trials included the ability to give signed informed consent, being over 18 years old, being willing to be randomly assigned to any treatment arms, and agreeing not to participate in any other trial until the completion of the follow-up period. Other criteria included histologically confirmed cancers by tests like immunohistochemistry (IHC), fluorescent in situ hybridization (FISH), next-generation sequencing (NGS), or other nonspecified sequencing methods depending on the type of cancer. Most trials limited the inclusion criteria to subjects who were newly diagnosed or who were in the early stages of cancer. However, few studies included patients with severe or advanced cancer stages. To avoid pregnancy during the study period, most studies required the use of highly effective contraception as defined by the protocol.

10.4.2 Characteristics of Exclusion Criteria

The most common exclusion criteria across most studies were patients diagnosed with any severe acute or chronic medical or psychiatric conditions that might increase the risk associated with study participation or interfere with the interpretation of the study result. Additionally, history of any investigational drug or device use within 4 weeks of recruitment and chronic medical conditions like uncontrolled diabetes, progressive neurological disorders, and any other medical condition that, in the opinion of the investigator, should preclude enrollment in the study were considered as exclusion criteria.

10.4.3 Study Type

Among the 16 studies registered, 9 (56%) were observational and 7 (44%) were interventional, and the main outcomes were to study the safety and efficacy of the treatment drug. Among the interventional studies, Pembrolizumab, Tamoxifen, Trastuzumab Emtansine, Letrozole, Anastrozole, Goserelin, LEE011, Fulvestrant, Palbociclib, Goserelin, Docetaxel, Nab-paclitaxel, Paclitaxel, Pertuzumab, Trastuzumab, and Herceptin were the drugs studied among the breast cancer disease patients. Alectinib and Atezolizumab were studied among non-small cell lung cancer (NSCLC) patients. The main routes of administration of the drugs included oral, intravenous, and subcutaneous injections. One of the studies was to validate a diagnostic test (the Freenome test), a blood-based test for the early detection of colorectal cancer.

Seven out of 16 (44%) studies focused on breast cancer, 5 (31%) focused on NSCLC, and the remaining included colorectal cancer, ovarian, prostate, fallopian tube cancer, and non-Hodgkin's lymphoma (NHL). Among the 7 interventional studies, 6 (86%) focused on breast cancer, and one was on NSCLC. In four of these trials, a combination of drugs was used.

10.4.4 Outcomes Measured

The outcomes of the trials included the progression of the disease, overall survival, and adverse effects. The parameters used for assessing the disease progression were radiological findings, grading of the tumor, duration of response, clinical evaluation of the tumor, and molecular responses. The follow-up period for interventional studies ranged from 1 year to 7 years, depending on the nature of the study. In real-world safety assessment studies, observational studies had follow-up periods ranging from 3 months to 6 years.

10.5   Discussion

Scientific studies producing high-quality evidence are essential for advancement in medical practices and thereby influencing policy changes and guidelines for the treatment and management of diseases. Cancer epidemiology varies among different ethnic groups, and a variety of sociocultural factors play a significant role in cancer detection and treatment; therefore, it is necessary to conduct studies and gather evidence among the local population.

The PubMed search on clinical trials or randomized control trials in oncology conducted in the UAE showed that there were only 11 studies published in the past 10 years. The majority of them were published in the last 5 years. However, out of these 11 studies, three of them focused on breast cancer, which is the most common cancer among women in the UAE, followed by two studies on leukemia. The majority (38%) of clinical trials registered on clinicaltrials.gov focused on breast cancer, followed by 31% on non-small cell lung cancers. This indicates that a higher number of oncology trials were conducted on conditions like breast cancer and lung cancer, which are the most common cancers documented among women and men in the Emirati population, respectively. However, no trials were registered on other common cancers among this population, like cervical, thyroid, and gastric cancers. This results in a lack of sufficient evidence for the other common cancers in this population. It is a well-established fact that the external validity of clinical trials is often questionable as they do not adequately represent real-world patient populations, where a variety of patient characteristics can interfere [7]. The validity of these study results is challenged when the treatment protocols are based on research conducted in other ethnic groups. Therefore, studying these cancer characteristics in the local population is highly critical, as the application of trial findings conducted on other populations might not help us completely understand the disease characteristics and the treatment effects specific to this ethnic group.

Only 2 of the 16 clinical trials were based in the UAE, and the rest had the UAE as one of the trial sites where the study was conducted. This is primarily due to the increasing trend toward conducting trials outside of western countries like the United States. Most pharmaceutical companies from the west now conduct trials outside the country, as this costs less than running the trial in their country [9]. The cost of conducting a trial is huge, depending on the number of participants needed, the number of research sites, the country in which the site is located, the complexity of the trial protocol, and the incentives provided to investigators and reimbursements for patients. This globalization of clinical trials has also increased the efficiency of clinical drug development, which can support the marketing authorization of new drugs globally [7].

Exploration of the inclusion and exclusion criteria of the clinical trials showed that all studies except for one epidemiological study on NSCLC were conducted among the adult population. Further, most of the studies excluded participants with comorbid conditions like diabetes, hypertension, severe kidney diseases, and pregnancy, which led to a lack of evidence in such groups. In the UAE, where diabetes and hypertension are major noncommunicable diseases, excluding people with these comorbid conditions can result in understudying of a significant percentage of the population and a lack of supportive evidence in this most vulnerable group of population [10–12].

Further, the majority of the interventional trials were in phase III, which was mainly to test the safety and effectiveness of the new treatment against the current standard treatment. Phase I and II studies were very limited; they were primarily to study the pharmacokinetics and pharmacodynamics and to test whether a new treatment worked for a certain type of cancer [13]. This indicates that we need more resources and funding for the development of new molecules for cancer, which are concerning and more specific to the Emirati population, to improve cancer treatment and care.

Among the nine completed trials, results were available for only four trials on the clinicaltrials.gov site; one more study result, though not available at this site, was published in PubMed with a ClinicalTrials.gov identifier (NCT number). This shows that only about 50% of the trial results are published.

The number of registered clinical trials in oncology has decreased in the last 5 years when compared to the 2012–2017 time period; this could be because the recent pandemic has diverted grants and research capabilities into COVID-19 research, decelerating research and development in other critical diseases [14, 15].

In the context of oncology-based clinical trials, there are several barriers and facilitators to conducting clinical trials. Healthcare providers face barriers such as a lack of time and research experience; institutional barriers such as a lack of research and clinical trial units, and so on; and patient and community barriers such as a lack of understanding, unawareness, mistrust, and fear of clinical trials. Apart from these, there are challenges from the global community, which include a lack of trust in research capabilities in developing countries, the exclusion of researchers from developing countries from research networks, and a lack of support for the publication of research from non-Western countries. Finally, the barriers from the pharmaceutical industry include concerns about research capabilities and regulatory delays, a lack of accredited centers of excellence, and gaps in global and local pharmaceutical funding [16].

10.6   Way Forward for Clinical Trials in Oncology in the UAE

Overcoming the barriers of clinical research in oncology is of paramount importance and a critical step in improving cancer care in the UAE. To overcome these barriers in clinical research, there is a need for all stakeholders and policymakers to come together to improve local research capabilities. There is a need to establish a research consortium specially dedicated to oncology research, which will work to expand research infrastructure. Recently, the Emirates Oncology Society established an oncology research program to advance cancer research in the UAE [17]. These initiatives should focus on training physicians in conducting clinical trials, incentives for physicians, and awareness and training of investigators in clinical trial protocols and procedures, along with regulatory aspects that can improve participation in clinical trials.

Implementing a successful clinical trial requires a team of multidisciplinary support staff that includes skilled technicians, epidemiologists, biostatisticians, administrative support staff, hospital staff to handle patients, a data monitoring committee, etc. This again reiterates the fact that training is not only needed for investigators but for the whole team, which needs to be knowledgeable as well as trained in all aspects of clinical trials.

Apart from training physicians and their teams, training patients also becomes essential, as patient participation is key to clinical trials. The patient must be made aware of their rights, their autonomy, and the opportunities to participate in clinical trials. This could help to dispel widespread skepticism about industry-sponsored trials among the general public [18].

Organizing research grants specific to research on common cancers among the Emirati population can improve scientific output on these cancers. Collaboration with international organizations and partnerships with established research organizations would go a long way toward including the UAE in clinical trials and provide an exceptional opportunity for knowledge sharing and transfer among collaborators.

10.7   Limitations

The limitations of this study are that some studies might have been missed due to the restriction of searches to PubMed and clinicaltrial.gov and limiting the search to only English-language publications.

10.8   Conclusion

There are a limited number of clinical trials in oncology, where cancer is the third leading cause of death among people in the UAE. More so, UAE-based trials are fewer in number, which indicates the necessity to initiate specific cancer-based research in the UAE. The country needs to allocate adequate funds and establish a research unit pertaining to cancer research to increase the evidence bank for cancer diagnosis and treatment that is precise for the Emirati population.

Conflict of Interest   The authors have no conflict of interest to declare.

References

1. World Health Organnization (WHO). Global health estimates: leading causes of death. 2019 [cited 2022 21 Nov]. https://www.who.int/data/gho/data/themes/mortality-and-global-health-estimates/ghe-leading-causes-of-death.2.
2. Cancer incidence in United Arab Emirates, Annual report of the UAE - National Cancer Registry, 2021. Statistics and Research Center, Ministry of Health and Prevention. (Accessed on 08 Apr 2024).

3. Jing L, Su L, Ring BZ. Ethnic background and genetic variation in the evaluation of cancer risk: a systematic review. PLoS One. 2014;9(6):e97522.

4. Newman LA, Carpten J. Integrating the genetics of race and ethnicity into cancer research: trailing Jane and John Q. Public. JAMA Surg. 2018;153(4):299–300.

5. Frei E III, et al. A comparative study of two regimens of combination chemotherapy in acute leukemia. Blood. 1958;13(12):1126–48.

6. Concato J, Shah N, Horwitz RI. Randomized, controlled trials, observational studies, and the hierarchy of research designs. N Engl J Med. 2000;342(25):1887–92.

7. Eichler H-G, Sweeney F. The evolution of clinical trials: can we address the challenges of the future? Clinical Trials. 2018;15(1_suppl):27–32.

8. English R, Lebovitz Y, Giffin R. Transforming clinical research in the United States: challenges and opportunities. In: Workshop summary. 2010.

9. Lahey T. The ethics of clinical research in low-and middle-income countries. Handb Clin Neurol. 2013;118:301–13.

10. Mamdouh H, et al. Prevalence and associated risk factors of hypertension and pre-hypertension among the adult population: findings from the Dubai Household Survey, 2019. BMC Cardiovasc Disord. 2022;22(1):1–9.

11. Shehab A, et al. Prevalence of cardiovascular risk factors and 10-years risk for coronary heart disease in the United Arab Emirates. Curr Diabetes Rev. 2022;19:e210422203892.

12. Hamoudi R, et al. Prediabetes and diabetes prevalence and risk factors comparison between ethnic groups in the United Arab Emirates. Sci Rep. 2019;9(1):1–7.

13. Cancer Research UK. Phases of clinical trilas [cited 2022 5 Dec]. https://www.cancerresearchuk.org/about-cancer/find-a-clinical-trial/what-clinical-trials-are/phases-of-clinical-trials.17.

14. Harper L, et al. The impact of COVID-19 on research. J Pediatr Urol. 2020;16(5):715.

15. Riccaboni M, Verginer L. The impact of the COVID-19 pandemic on scientific research in the life sciences. PLoS One. 2022;17(2):e0263001.

16. Al-Shamsi HO. Barriers and facilitators to conducting oncology clinical trials in the UAE. Clin Pract. 2022;12(6):885–96.

17. Emirates Oncology Society. [cited 2022 5 Dec]. https://eos-uae.com/.

18. Breault JL, Knafl E. Pitfalls and safeguards in industry-funded research. Ochsner J. 2020;20(1):104–10.

Dr. Subhashini Ganesan is a researcher at G42 Healthcare. She has contributed efficiently to research on various aspects of the COVID-19 pandemic and published her findings in well-reputed journals, supporting the company's efforts in fighting the COVID-19 pandemic. She also works as a Senior Researcher at IROS, a clinical research organization, where she consults on research studies and clinical trials. Before joining G42 Healthcare, Dr. Subhashini was working as an assistant professor at the PSG Institute of Medical Sciences and Research, India, and has been an active member of the Research and Ethics committee. She has more than 25 scientific publications in peer-reviewed journals, including Nature. She has worked as a consultant on various community-based research projects in India. She is trained at the World Health Organization (WHO), Geneva, in adolescent and reproductive health research. She graduated with a Bachelor of Medicine and Surgery (MBBS) and an MD in Public Health from the PSG Institute of Medical Sciences and Research, India. She also has a PG Diploma in Research and Bioethics.

Prof. Humaid Obaid Al-Shamsi is the Chief Executive Officer of Burjeel Cancer Institute in Abu Dhabi, UAE, President of the Emirates Oncology Society, Lead of the Gulf Cancer Society, Full Professor of Oncology at the Ras Al Khaimah Medical and Health Sciences University, Ras Al Khaimah, UAE, and an Adjunct Professor of Oncology at the College of Medicine, University of Sharjah. He is the first Emirati to be promoted as a professor in oncology in the UAE. He is also the Chairman for Colorectal Cancer in the MENA region, appointed by the prestigious National Comprehensive Cancer Network®. He is also the only member of Lung Cancer Policy Network in the MENA region that aims to advance lung cancer research and screening globally. He is the Chairman of the Oncology and Hematology Fellowship Training Program for the National Institute for Health Specialties in the United Arab Emirates. He is the only member in GCC in the WIN Consortium which is comprised of organizations representing all stakeholders in personalized cancer medicine globally.

He is board-certified in both internal medicine and oncology from the UK, USA (ABIM), and Canada (FRCPC). He has also been awarded the FRCP (London) in 2023 and FRCP (Glasgow) in 2024. He is the only physician in the UAE with a subspecialty fellowship certification and training in gastrointestinal oncology and the first Emirati to train and complete a clinical post-doctoral fellowship in palliative care. He was an assistant professor at the University of Texas MD Anderson Cancer Center between 2014 and 2017. He has published more than 140 peer-reviewed articles in JAMA Oncology, Lancet Oncology, The Oncologist, BMC Cancer, and many others. His area of expertise includes precision oncology and cancer care in the UAE. In 2016, he published with his group from MD Anderson the JCO paper describing a new distinct subgroup of CRC, NON V600 BRAF-mutated CRC. In 2022, he published the first book about cancer research in the UAE and also the first book about cancer in the Arab world, both of which were launched at Dubai Expo 2020. *Cancer in the Arab World* has been downloaded more than 450,000 times in its first 18 months of publication and is the ultimate source of cancer data in the Arab region. He also published the first comprehensive book about cancer care in the UAE which is the first book in UAE history to document the cancer care in the UAE with many topics addressed for the first time, e.g., neuroendocrine tumors in the UAE. He is passionate about advancing cancer care in the UAE and the GCC and has made significant contributions to cancer awareness and early detection for the public using social media platforms. He is considered as the most followed oncologist in the world with over 300,000 subscribers across his social media platforms (Instagram, Twitter, LinkedIn, and TikTok). In 2022, he was awarded the prestigious Feigenbaum Leadership Excellence Award from Sheikh Hamdan Smart University for his exceptional leadership and research and the Sharjah Award for Volunteering. He was also named the Researcher of the Year in the UAE in 2020 and 2021 by the Emirates Oncology Society.

In May 2024, HH Sheikh Mansour bin Zayed Al Nahyan, Vice President of the United Arab Emirates, awarded him the first place in UAE Nafis program for outstanding leadership in private sector across all business and medical disciplines. Beside his clinical and administrative duties, he is engaged in education and various levels of research training for medical trainees to enhance their clinical and research skills. His mission is to advance cancer care in the UAE and the MENA region and make cancer care accessible to everyone in need around the globe.

Mohamed Mostafa is Chief Executive Officer at PDC CRO from 2020; as such, he is responsible for overall management and strategic initiatives within the organization. Mohamed holds a bachelor's degree in Pharmaceutical Science with more than 15 years' experience in the Pharma/CRO industry in the MEA region. Mohamed was responsible for various scientific initiatives with local and international partners and worked closely with regulatory agencies across the region on various guidelines and strategic plans. Mohamed has a solid understanding of the MEA region's pharma, biotech, and healthcare markets and continues to work closely with colleagues to further develop the clinical research capabilities within the MEA region.

Dr. Walid Abbas Zaher is a highly awarded scientist, medical doctor, and businessman from the Kingdom of Saudi Arabia. He is the current CEO of Carexso, the MENA region's first site management organization for biotech and medical research. He previously founded and was CEO of the UAE's first contract research organization, as well as CRO of G42 Healthcare, and, before that, was Corporate Group R&D Director of SEHA in Abu Dhabi. He was instrumental in revamping the clinical and R&D ecosystem in Abu Dhabi, as exemplified by his leadership of 4Humanity, the Middle East's largest clinical trial, the empowerment of COVID-19 vaccine manufacturing, and spearheaded the Emirati Genome Program, one of the most ambitious population genomics and precision medicine programs to date. He also helped achieve over 100x company growth in 2 years. He drove innovative health technology platforms, regenerative medicine, and longevity initiatives. He has over 78 publications in peer-reviewed journals and book chapters, including papers in Nature and JAMA. Dr. Walid Zaher has a degree in medicine from King Saud University with a residency in Obstetrics & Gynecology, followed by two MScs and a PhD in Regenerative medicine from Odense University Hospital in Denmark, with a visiting research period at Harvard Medical School. He is known for his work in transformative genomics, pioneering health and longevity, and next-generation implementation of research and innovation.

Basic Cancer Research in the UAE 11

Ibrahim Yaseen Hachim ⓘ, Saba Al Heialy ⓘ,
and Mahmood Yaseen Hachim ⓘ

11.1 Introduction

In the last few decades, there has been a significant improvement in our understanding of cancer's etiology, pathogenesis, and progression [1, 2]. Basic cancer research breakthroughs were critical in determining the genetic, molecular, and clinical heterogeneity of cancer's various cancers and their interactions with the tumor microenvironment [3]. Indeed, such discoveries were essential for the personalized cancer medicine concept, which focuses mainly on tailoring drugs to specifically target the driver mutations in cancer patients according to their genetic and molecular fingerprint [4]. Therefore, better allocation of resources for basic cancer research that might help improve patient stratification will be essential for accelerating discoveries in cancer management and care to improve patient outcomes and reduce side effects. This might be achieved through collaboration between laboratories and hospitals to incorporate their discoveries into routine practice [5].

I. Y. Hachim
Sharjah Institute for Medical Research, University of Sharjah, Sharjah, United Arab Emirates

Clinical Sciences Department, College of Medicine, University of Sharjah,
Sharjah, United Arab Emirates
e-mail: ibrahim.hachim@sharjah.ac.ae

S. Al Heialy
Immunology, College of Medicine, Mohammed bin Rashid University of Medicine and
Health Sciences, Dubai, United Arab Emirates

Division of Respiratory Diseases, Department of Medicine, McGill University,
Montreal, QC, Canada
e-mail: saba.alheialy@mbru.ac.ae

M. Y. Hachim (✉)
Molecular Medicine, College of Medicine, Mohammed bin Rashid University of Medicine
and Health Sciences, Dubai, United Arab Emirates
e-mail: mahmood.almashhadani@mbru.ac.ae

© The Author(s) 2024   193
H. O. Al-Shamsi (ed.), *Cancer Care in the United Arab Emirates*,
https://doi.org/10.1007/978-981-99-6794-0_11

11.2   Why Basic Cancer Research Is Needed in the United Arab Emirates (UAE)

In the last few decades, there has been rapid and significant progress in understanding cancer etiology and tumorigenesis. This was achieved through advancements in basic cancer research that implement new technologies that allow high-resolution genetic mapping of thousands of cancers and tumors [2, 6]. This has led to the discovery of specific mutations and aberrations that can be specifically targeted by novel drugs [2]. Indeed, identifying patients with those mutations helps tailor patients' care to be more personalized. For that reason, basic cancer research in the UAE is essential not only for the validation of results obtained from the Western population but also for the discovery of distinct UAE population-specific genetic, molecular, and biological variations that might lead to the discovery of novel molecular markers and targets that are associated with cancer risk and response to therapy unique to the UAE population [7, 8]. For example, researchers discovered a distinct molecular profile of breast cancer in women from Arab countries, including the UAE, compared to the Western population. This includes a high tumor grade, fewer luminal subtypes of tumor, and a higher rate of HER-2 positivity and triple-negative breast cancer (TNBC) tumors in breast cancer patients [9–12]. A study done by Al-Shamsi et al. revealed that evaluation of the molecular characteristics of colorectal cancer in patients from the Arab Gulf region showed a similarity in the frequency of *KRAS, BRAF, NRAS, TP53, and APC* as well as *PIK3CA* mutations between Arab and Western populations; however, the *SMAD4* and *FBXW7* mutation frequencies were distinct [13].

11.3   State of Cancer Research in the United Arab Emirates

In the past decades, there has been a significant advancement in cancer research in the United Arab Emirates. Therefore, an in-depth analysis of the cancer research activity might be essential not only to understand the trends of cancer research but also to highlight the advances, achievements, gaps, and obstacles in this field and provide information for scientific, funding, and governmental institutions that might help in strengthening the research productivity [14].

One of the significant sources to investigate the state of cancer research activity in the UAE is through an analysis of the number of cancer-related articles in the UAE using the National Library of Medicine (NLM) database MEDLINE and its search engine PubMed (https://www.ncbi.nlm.nih.gov/pubmed), which includes more than 34 million citations for biomedical literature [15]. Globally, the proportion of cancer-related entries per year significantly increased from 6% in 1950 to around 16% in 2016 [15]. Similarly, a recent study showed that 26,656 cancer-related studies published in the Arab world represent around 13.4% of the entire Arab world's biomedical research manuscripts between 2005 and 2019 [16].

Using a similar methodology [16], we investigated the number of cancer-related publications in the UAE from 1987 to 2021 using the Boolean operator in the

PubMed search engine (Fig. 11.1a). In the advanced article search, we used the following MeSH terms and formula: cancer, malignant, oncology, tumor, tumour, neoplasm, carcinoma, adenocarcinoma, sarcoma, leukemia, lymphoma, metastasis, oncogene, chemotherapy. The affiliation, country, or territory should be the United Arab Emirates.

Our results showed an exponential increment in cancer-related publications in the last decade, from only 66 publications in 2011 to around 865 in 2021. Using the

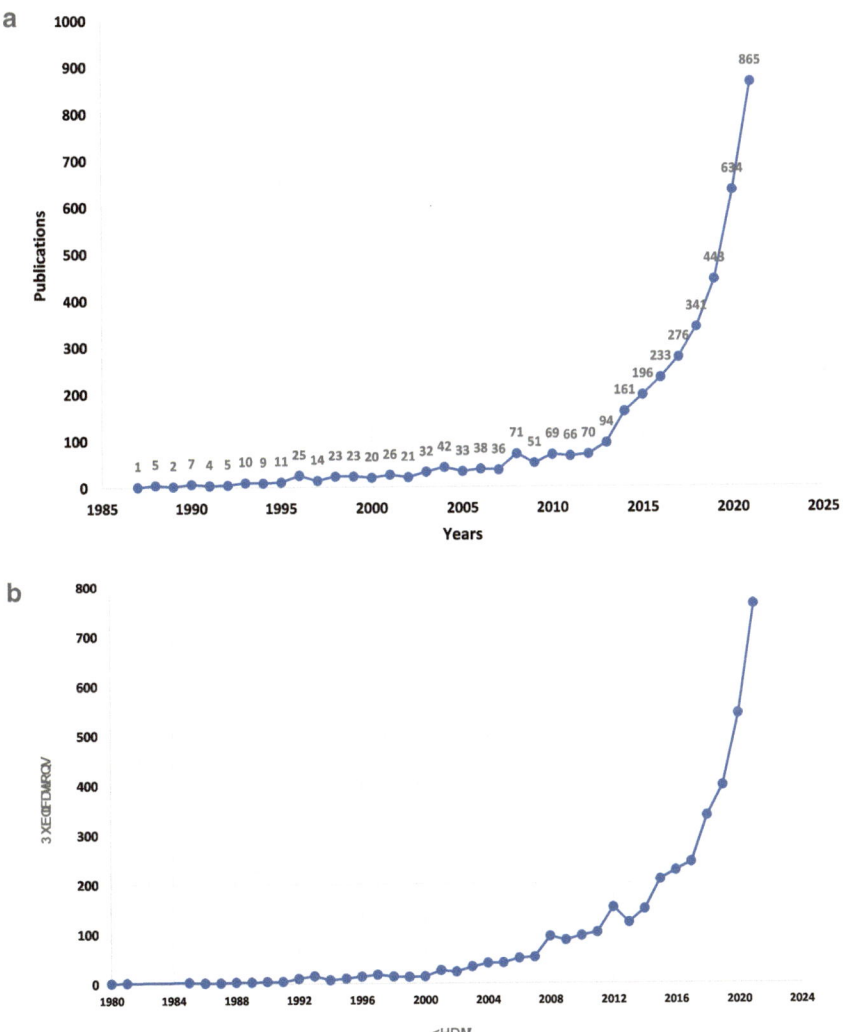

Fig. 11.1 (a) Number of cancer-related publications in the United Arab Emirates generated from PubMed between 1987 and 2021. (b) Number of cancer-related publications in the United Arab Emirates generated using Scopus database

same MeSH terms used above, we also investigated the number of cancer-related publications in the UAE during the same period using the Scopus database (Fig. 11.1b). Our results revealed 746 cancer-related documents in 2021, compared to only one publication in 1980. This goes with a recent report that showed a 16-fold increase in research publications in the last two decades in the UAE [17]. An interesting finding is that in 1998, around 50% of the UAE-based publications were in medicine and life sciences [17]. Since then, the research output growth has become more diverse, and the contribution of other disciplines has started to increase. An analysis of the research publications according to significant research disciplines in the UAE revealed that in 2017, medicine and life sciences represented around 20% of the research productivity, preceded by engineering, energy, and environmental sciences [17].

11.4   Research Infrastructure in the United Arab Emirates

Previously, the lack of research infrastructure and the limited number of researchers and funding opportunities represented a significant challenge for conducting research in Middle Eastern and North African countries [18]. However, some countries, like the United Arab Emirates, showed a ground-breaking improvement in their biomedical research productivity due to their investment in educational institutes, foundations, and multidisciplinary medical research centers.

Moreover, several actions were taken as part of the country's efforts to strengthen the research capacity within the UAE. This includes the investment in state-of-the-art research infrastructure and facilities, the recruitment of experienced researchers, the establishment of various training programs, and the encouragement of international collaboration. This was reflected by the exponential increase in higher education and research institutions [17]. For example, a report published in 2007 showed that while in 1998 there were only 29 institutions with biomedical publications, this number increased to 103 institutions (including hospitals, universities, and research centers) after 8 years [19].

11.5   Cancer Research Centers and Institutes in the United Arab Emirates

In recent years, there have been significant initiatives for the establishment of several multidisciplinary medical research centers that aim to find solutions to several health challenges and diseases, including cancer, through innovation, providing an interactive environment for scientists and clinicians to work together, and encouraging local, regional, as well as international scientists' collaboration.

11.5.1 Zayed Bin Sultan Al Nahyan Center for Health Sciences (ZCHS)

This center was established as part of UAE University's efforts to become a world-class center for applied health research. This center supports researchers from a wide range of backgrounds, including medicine, biology, engineering, chemistry, information technology, and nutrition, to advance innovation, discovery, and improve health practices. ZCHS has established more than 15 programs in various fields, including cancer, molecular genetics and genomic medicine, immunoregulation and infection, artificial intelligence, robotics applications in health, and nanotechnology.

11.5.2 Research Institute of Medical and Health Sciences (RIMHS) at the University of Sharjah

Since its establishment in 2015, this institute has been one of the top research and innovation centers in the UAE and the region. This was achieved by supporting more than 170 distinguished faculties and young researchers and providing them with a supportive research environment, including state-of-the-art lab facilities and top-notch research equipment. The institute includes 27 research groups led by distinguished scientists from different disciplines and colleges, including medicine, dentistry, pharmacy, health sciences, and arts and sciences. These focus groups are directed toward providing solutions for health problems, including cancer, inflammatory diseases, immunological disorders, and genetic disorders, in addition to drug discovery.

11.5.3 Mohammed Bin Rashid Medical Research Institute

This world-class biomedical research institute is an initiative of the Al Jalila Foundation. The center aims to encourage local and international scientists to collaborate to provide solutions for many human diseases, including cancer. The Al Jalila Foundation fully funds this life sciences research center, and it is equipped with the latest research technology. Furthermore, it recruits scientists and postdoctoral fellows who graduated from leading institutions across the globe. The collaboration between this center, the Mohammed Bin Rashid University of Medicine and Health Sciences (MBRU), and the Al Jalila Foundation resulted in significant scientific discoveries translated into landmark publications in well-known and highly reputed scientific journals.

11.6 Top Institutional Affiliation and Internal Funding Sources in the United Arab Emirates

To improve our understanding of the currently available cancer research institutes and infrastructures, we investigated the affiliations and funding bodies of cancer-related manuscripts within the UAE between 1978 and 2021. Our analysis revealed that United Arab Emirates University is the first UAE-affiliated research institute, followed by the University of Sharjah, Tawam Hospital, Dubai Hospital, Khalifa University of Science and Technology, Cleveland Clinic Abu Dhabi, Gulf Medical University, New York University (NYU) Abu Dhabi, American University of Sharjah, Ajman University, Rashid Hospital, and Mohammed Bin Rashid University of Medicine and Health Sciences (MBRU) (Fig. 11.2a). Most institutions include state-of-the-art specialized research centers and research groups that aim to enrich the research environment and facilitate integration and collaboration between colleges, hospitals, and healthcare providers.

Analysis of the top funding local sponsors of the cancer research publication, which is one of the main outputs of funding in the UAE, showed United Arab Emirates University to have the largest number of funded research publications, followed by the University of Sharjah, Al Jalila Foundation, American University of Sharjah, NYU Abu Dhabi, Sheikh Hamdan Bin Rashid Al Maktoum Award for Medical Sciences, and Khalifa University of Science and Technology (Fig. 11.2b). The presence of this large number and various types of internal funding institutions reflects the UAE strategy to support cancer research as a step to provide novel solutions to improve cancer patients' management and care.

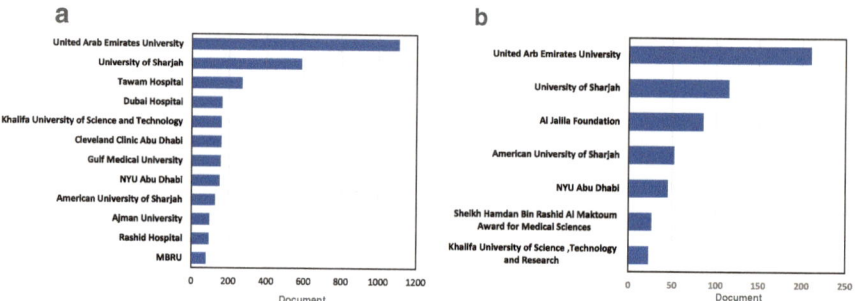

Fig. 11.2 Bibliometric analysis was extracted from the Scopus database to investigate the affiliation, funding, sponsors, and output of cancer-related publications in the United Arab Emirates. (**a**) The top affiliation of cancer-related publications in the United Arab Emirates using the Scopus database between 1978 and 2021. (**b**) The top internal funding institutions of cancer-related publications in the United Arab Emirates using the Scopus database between 1978 and 2021. (**c**) The top external funding institutions of cancer-related publications in the United Arab Emirates using the Scopus database between 1978 and 2021. (**d**) Publication types of cancer-related publications in the United Arab Emirates using the Scopus database between 1978 and 2021. (**e**) Subject-wise distribution of cancer-related publications in the United Arab Emirates using the Scopus database between 1978 and 2021

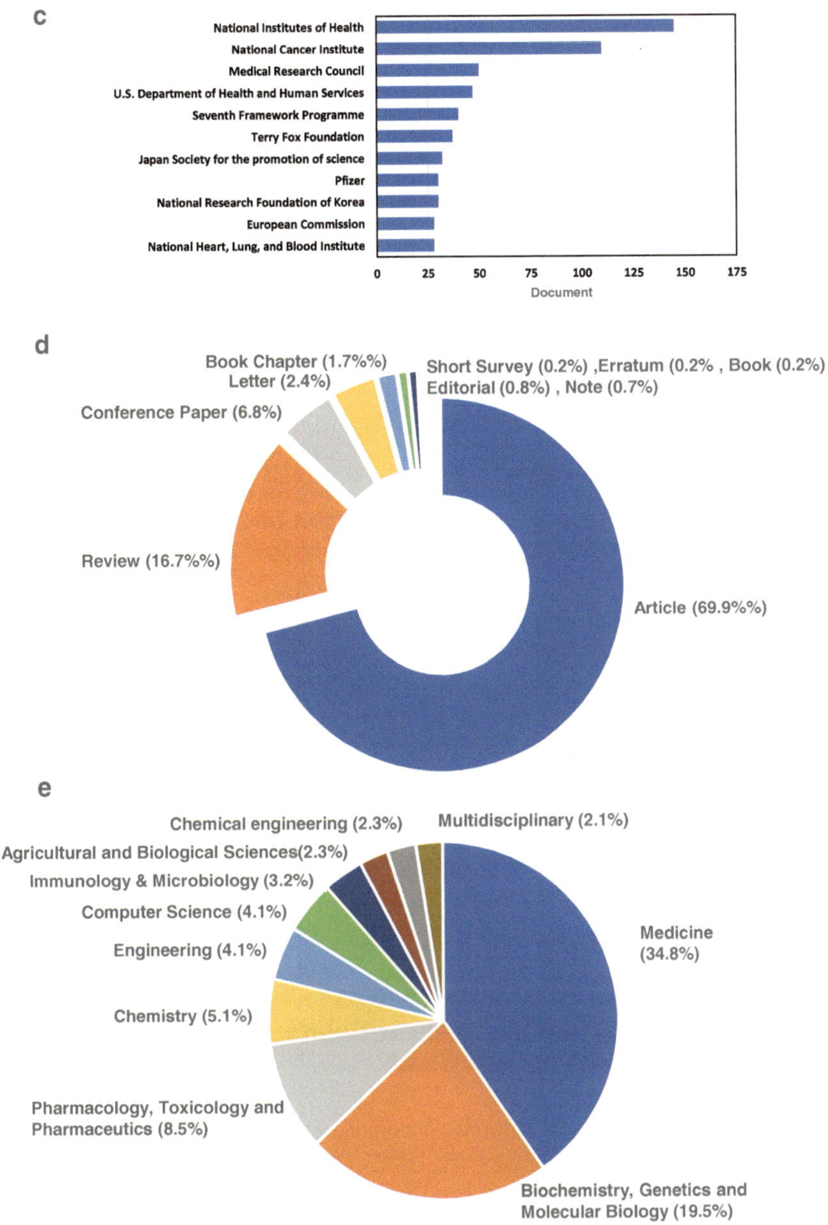

c
National Institutes of Health
National Cancer Institute
Medical Research Council
U.S. Department of Health and Human Services
Seventh Framework Programme
Terry Fox Foundation
Japan Society for the promotion of science
Pfizer
National Research Foundation of Korea
European Commission
National Heart, Lung, and Blood Institute
0 25 50 75 100 125 150 175
Document
d
Book Chapter (1.7%%)
Letter (2.4%)
Short Survey (0.2%) ,Erratum (0.2% , Book (0.2%)
Editorial (0.8%) , Note (0.7%)
Conference Paper (6.8%)
Review (16.7%%)
Article (69.9%%)
e
Chemical engineering (2.3%)
Multidisciplinary (2.1%)
Agricultural and Biological Sciences(2.3%)
Immunology & Microbiology (3.2%)
Computer Science (4.1%)
Engineering (4.1%)
Medicine (34.8%)
Chemistry (5.1%)
Pharmacology, Toxicology and Pharmaceutics (8.5%)
Biochemistry, Genetics and Molecular Biology (19.5%)

Fig. 11.2 (continued)

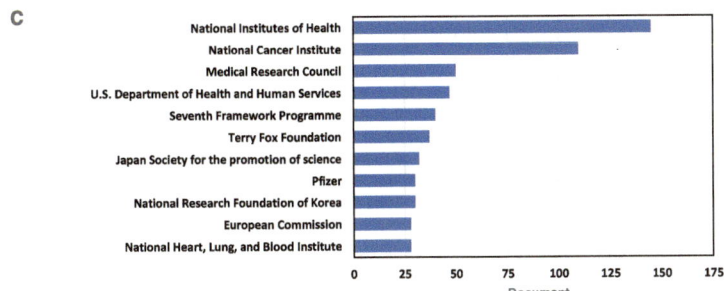

11.7 External Collaboration and External Funds for Cancer-Related Research in the United Arab Emirates

The number of cancer-related research projects in the UAE that include international collaboration and external funding has significantly increased in the last decade. Our analysis revealed a wide spectrum of highly recognized funding bodies involved in cancer-related research projects, including UAE-based scientists. This includes the National Institutes of Health, the National Cancer Institute, the Medical Research Council (MRC), the US Department of Health and Human Services, the Seventh Framework Program, the Terry Fox Foundation, and others. This indicates the trust of well-known international funding bodies and collaborators in UAE-based cancer research and the attempt of UAE scientists and institutes to produce high-quality scientific research projects (Fig. 11.2c).

11.8 Type and Subject-Wise Distribution of Cancer-Related Research in the United Arab Emirates

Our results showed that most of the cancer-related publications in the UAE were research articles (69.9%), followed by reviews (16.7%), conference papers (6.8%), letters (2.4%), and book chapters (1.7%). Most of the published manuscripts in the cancer-related field indicate that UAE scientists and institutes focus mainly on innovative primary research projects to improve our understanding of cancer and discover new solutions to treat this disease (Fig. 11.2d).

Subject-wise distribution of cancer-related manuscripts revealed that 34.8% of manuscripts fall into the medicine subject area, followed by biochemistry, genetics, and molecular biology (19.5%), pharmacology, toxicology, and pharmaceutics (8.5%), chemistry (5.1%), computer sciences (4.1%), engineering (4.1%), immunology and microbiology (3.2%), and agricultural and biological sciences (2.3%) (Fig. 11.2e).

11.9 Top Journals in Which Cancer-Related Manuscript from the UAE Was Published

As seen in Fig. 11.3, the local Emirates Medical Journal was the dominant journal for cancer-related manuscripts between 1980 and 2006. Since then, the diversity and ranking of UAE-based cancer-related manuscript publishing journals have significantly improved (Table 11.1). This includes PLOS One (CiteScore = 5.6), Scientific Reports (CiteScore = 6.9), Molecules (CiteScore = 5.9), Asian Pacific Journal of Cancer Prevention (CiteScore = 3.1), International Journal of Molecular Sciences (CiteScore = 6.9), Annals of the New York Academy of Sciences (CiteScore = 10), and Lancet (CiteScore = 115). This clearly indicates that the increase in cancer research productivity was coupled with an improvement in the quality of research output.

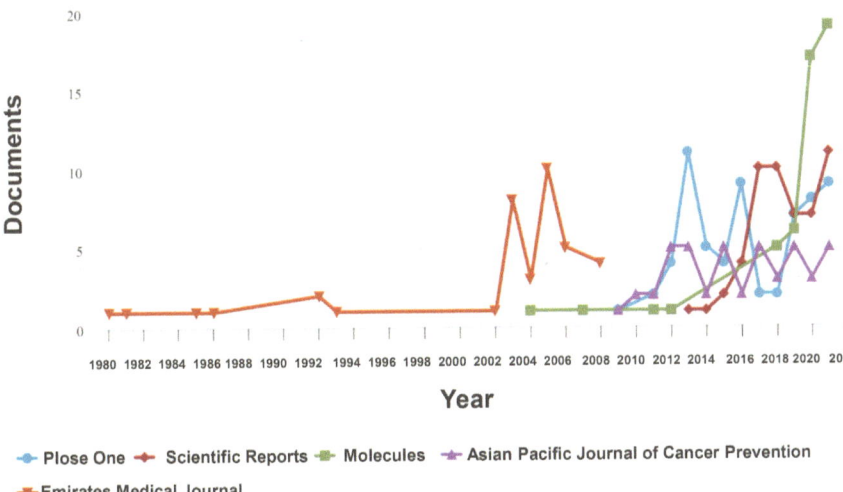

Fig. 11.3 The top publishing journals of cancer-related publications in the United Arab Emirates using the Scopus database in the period between 1978 and 2021

Table 11.1 The top 15 journals that published cancer-related publications in the UAE

Journal	No. of publications
PLOS One	64
Scientific Reports	53
Molecules	51
Asian Pacific Journal of Cancer Prevention	45
Emirates Medical Journal	38
International Journal of Molecular Sciences	34
Annals of the New York Academy of Sciences	25
Lancet	24
Frontiers in Immunology	23
BMJ Case Reports	22
Cancers	20
Saudi Medical Journal	20
Cellular Physiology and Biochemistry	19
European Journal of Medicinal Chemistry	19
Frontiers in Pharmacology	18

Table 11.2 Classification of cancer-related manuscripts according to cancer type

Cancer type/keywords	Number of publications
Breast, mammary	370
Liver, hepatocellular carcinoma	181
Colorectal, colon, rectum	137
Gastrointestine, stomach	98
Lung	98
Renal, kidney	78
Skin, melanoma	71
Prostate	67
Cervix, cervical	67
Leukemia	45
Lymphoma	49
Brain, CNS	49
Thyroid	44
Bladder	31
Ovary	24
Uterine, endometrial	17
Total	**1426**

11.10 Cancer-Related Manuscript Distribution According to Cancer Type in the UAE

We also tracked the distribution of cancer-related publications based on their cancer type. To achieve this, we used the anatomical site and common keywords for each cancer type (Table 11.2). Breast cancer research output was the highest investigated cancer type with 370 publications, followed by liver (181), colorectal (137), gastrointestinal (98), lung (98), kidney (78), skin (71), prostate (67), and cervix (67). This distribution of cancer-related manuscripts according to cancer types showed similarity with the incidence of various cancers in the UAE population. For example, a recent study investigating the epidemiology of cancers in the UAE population found breast, cervical, and thyroid cancer to be the most common cancers in Emirati females, compared to lung, gastric, and prostate cancers in male patients [20, 21].

11.11 Common Scientific Outline (CSO) of Cancer Research in the UAE

To understand the focus of cancer research groups in the UAE, we stratified cancer-related articles according to the Common Scientific Outline (CSO) implemented by the International Cancer Research Partnership (ICRP) coding system [14, 22]. This classification subdivides cancer research projects into six broad areas including:

1. Cancer biology
2. Cancer etiology

3. Prevention
4. Early detection, diagnosis, and prognosis
5. Treatment
6. Cancer control survivorship and outcomes

11.11.1 Cancer Biology and Etiology Research in the UAE

To extract the research articles related to cancer biology and etiology, we used further filtration of our cancer-related articles using the keywords (biology OR initiation OR progression OR metastasis) as additional MeSH terms. Our results showed 538 documents that fulfilled those criteria. The top active institutes in this category were the United Arab Emirates University and the University of Sharjah. International Journal of Molecular Sciences, Scientific Reports, Seminars in Cancer Biology, PLOS One, and Frontiers in Oncology and Cancers were the top journals that publish manuscripts in this category. Research projects in this category were comprehensive and included different aspects of cancer biology, including genetic [23–25] and epigenetic studies [26, 27]. They investigate factors involved in different stages of cancer development, including tumor initiation, progression, and metastasis [28–32]. Many of the research projects include the use of new advanced and state-of-the-art techniques and technologies like next-generation sequencing, which allows whole genome, exome, and transcriptome sequencing [13, 33]. Many projects also introduced multiomics approaches to investigate tumor heterogeneity, complex biological derangements, and tumor subtyping [34, 35].

11.11.2 Cancer Prevention Research in the UAE

A better understanding of the risk factors involved in cancer development is essential to cancer prevention. For that reason, many research projects focused on understanding cancer risk factors and raising awareness among community members and patients of those factors to reduce cancer risk by avoiding exposure to carcinogens and adopting a healthier lifestyle. Our analysis revealed that more than 300 cancer-related manuscripts were related to risk factor assessment or cancer prevention. Some of those manuscripts include collaborative work with other international and regional collaborators to evaluate risk factors and the status of cancer prevention [36, 37]. Other manuscripts try to investigate the awareness and knowledge of some community groups on the role of lifestyle and diet in preventing different cancer types [38–40].

11.11.3 Early Detection, Diagnosis, Prognosis, and Outcome Prediction-Related Cancer Research in the UAE

Early detection, diagnosis, and prognosis represent a significant research area in the UAE; this was evident by the number of cancer-related articles related to this category. Our results showed more than 200 cancer-related manuscripts related to early detection, diagnosis, and prognosis for various cancers. This includes using genetic, molecular, clinical, and novel algorithms to improve the early detection, diagnosis, and prognostication of cancers [41–43]. In addition, artificial intelligence, machine learning, bioinformatics, and computer-aided techniques were also introduced to enhance and refine the diagnostic, predictive, and prognostic value of various markers and techniques in different cancers [34, 42, 44–46].

11.11.4 Cancer Treatment and Anticancer Drug Discoveries in the UAE

In recent years, there has been a significant shift in cancer research trends from fundamental basic cancer research into more translational research. The first step in the translational cancer research paradigm and new drug creation is the discovery of molecules, components, and compounds that specifically target specific biological processes that are deranged in malignant cells compared to healthy cells [47]. Implementing such an approach is essential to increasing the efficacy of anticancer drugs and minimizing the toxicity usually observed in conventional anticancer drugs [47].

In the last decades, there has been an exponential increase in the number of research projects aiming to discover novel targets, molecules, and compounds that might have anticancer activity. We performed a bibliometric analysis of the Scopus database to investigate the drug discovery-related articles among the cancer-related manuscripts within the UAE. In addition to the MeSH terms we used for investigating cancer-related articles in the UAE, we further filtrated our data to be limited to articles that contain (drug OR compound OR anticancer) in their title or abstract.

Our analysis revealed that 723 cancer research articles (excluding reviews, conference papers, and book chapters) investigated molecules, compounds, and drugs with anticancer activity, including around 139 articles in 2021 alone. The University of Sharjah was the top UAE institute in compounds and anticancer drug discovery-related articles, followed by the United Arab Emirates University, the American University of Sharjah, Ajman University, and NYU Abu Dhabi. Some of those reports include nanoparticles used as a drug delivery system in cancer therapeutics [48–53].

Interestingly, our analysis revealed the presence of 42 patents from UAE-based researchers related to the discovery of molecules, compounds, and drugs with potential use in the treatment of various cancers. This includes 33 patents from the United States Patent and Trademark Office, 5 from the Japan Patent Office, 2 from the World Intellectual Property Organization, 1 from the European Patent Office, and another from the United Kingdom Intellectual Property Office.

11.12 Conclusion

In conclusion, our analysis revealed a significant improvement in the quantity and quality of cancer-related projects and manuscripts. This was also associated with more diverse cancer research and an exponential increase in cancer research productivity, coupled with improvements in the quality and impact of those research activities. The cancer research activities were distributed among a wide range of cancer types, subjects, and scientific outlines covering all aspects of cancer biology. This was achieved through investment in research infrastructure, recruitment of experienced researchers, and the establishment of various training programs.

Conflict of Interest The authors have no conflict of interest to declare.

References

1. Lowy DR, Collins FS. Aiming high—changing the trajectory for cancer. N Engl J Med. 2016;374(20):1901–4.
2. Yarden Y, Caldes C. Basic cancer research: why it is essential for the future of cancer therapy. Bull Cancer. 2014;101(9):E25–6.
3. Elmore LW, et al. Blueprint for cancer research: critical gaps and opportunities. CA Cancer J Clin. 2021;71(2):107–39.
4. Krzyszczyk P, et al. The growing role of precision and personalized medicine for cancer treatment. Technology (Singap World Sci). 2018;6(3–4):79–100.
5. Jackson SE, Chester JD. Personalised cancer medicine. Int J Cancer. 2015;137(2):262–6.
6. Yarden Y, Caldas C, European Association for Cancer. Basic cancer research is essential for the success of personalised medicine. Eur J Cancer. 2013;49(12):2619–20.
7. Savage SA, Chanock SJ. Genetic association studies in cancer: good, bad or no longer ugly? Hum Genomics. 2006;2(6):415–21.
8. Al-Shamsi HO, et al. Cancer in the Arab World. Singapore: Springer; 2022.
9. Sami MM, et al. Breast cancer profile in Ras Al Khaimah, United Arab Emirates—a histopathological and immunohistochemical study. Hamdan Med J. 2014;7(1):70–92.
10. Chouchane L, Boussen H, Sastry KS. Breast cancer in Arab populations: molecular characteristics and disease management implications. Lancet Oncol. 2013;14(10):e417–24.
11. Al-Shamsi HO, et al. Molecular spectra and frequency patterns of somatic mutations in Arab women with breast cancer. Oncologist. 2021;26(11):e2086–9.
12. Alkhayyal N, et al. Correlation of insulin-like growth factor 1 receptor expression with different molecular subtypes of breast cancer in the UAE. Anticancer Res. 2020;40(3):1555–61.
13. Al-Shamsi HO, et al. Molecular spectrum of KRAS, NRAS, BRAF, PIK3CA, TP53, and APC somatic gene mutations in Arab patients with colorectal cancer: determination of frequency and distribution pattern. J Gastrointest Oncol. 2016;7(6):882–902.
14. Moodley J, et al. A bibliometric analysis of cancer research in South Africa: study protocol. BMJ Open. 2015;5(2):e006913.
15. Reyes-Aldasoro CC. The proportion of cancer-related entries in PubMed has increased considerably; is cancer truly "The Emperor of All Maladies"? PLoS One. 2017;12(3):e0173671.
16. Machaalani M, Masri JE, El Ayoubi LM. Cancer research activity in the Arab world: a 15-year bibliometric analysis. J Egypt Public Health Assoc. 2021;97(1):26.
17. Al Marzouqi AHH, et al. Research productivity in the United Arab Emirates: a 20-year bibliometric analysis. Heliyon. 2019;5(12):e02819.
18. Copur MS. State of cancer research around the globe. Oncology (Williston Park). 2019;33(5):181–5.
19. Neves K, Lammers WJ. Growth in biomedical publications and scientific institutions in the Emirates (1998–2004): an Arabian renaissance? Health Inf Libr J. 2007;24(1):41–9.
20. Radwan H, et al. The epidemiology of cancer in the United Arab Emirates: a systematic review. Medicine (Baltimore). 2018;97(50):e13618.
21. Al-Shamsi H, et al. The state of cancer care in the United Arab Emirates in 2020: challenges and recommendations, a report by the United Arab Emirates oncology task force. Gulf J Oncol. 2020;1(32):71–87.

22. ICRP, ICRP International Cancer Research Partnership (ICRP). Common Scientific Outline 2022. https://www.icrpartnership.org/cso.
23. Lopez-Ozuna VM, et al. Identification of predictive biomarkers for lymph node involvement in obese women with endometrial cancer. Front Oncol. 2021;11:695404.
24. Muhammad JS, et al. Estrogen-induced hypomethylation and overexpression of YAP1 facilitate breast cancer cell growth and survival. Neoplasia. 2021;23(1):68–79.
25. Venit T, et al. Nuclear myosin 1 activates p21 gene transcription in response to DNA damage through a chromatin-based mechanism. Commun Biol. 2020;3(1):115.
26. Kedhari Sundaram M, et al. Epigallocatechin gallate inhibits HeLa cells by modulation of epigenetics and signaling pathways. 3 Biotech. 2020;10(11):484.
27. Sasidharan Nair V, et al. DNA methylation and repressive H3K9 and H3K27 trimethylation in the promoter regions of PD-1, CTLA-4, TIM-3, LAG-3, TIGIT, and PD-L1 genes in human primary breast cancer. Clin Epigenetics. 2018;10:78.
28. Cheratta AR, et al. Caspase cleavage and nuclear retention of the energy sensor AMPK-alpha1 during apoptosis. Cell Rep. 2022;39(5):110761.
29. Benhalilou N, et al. Origanum majorana ethanolic extract promotes colorectal cancer cell death by triggering abortive autophagy and activation of the extrinsic apoptotic pathway. Front Oncol. 2019;9:795.
30. Khan F, et al. Rutin mediated apoptotic cell death in Caski cervical cancer cells via Notch-1 and Hes-1 downregulation. Life (Basel). 2021;11(8):761.
31. Alsamri H, et al. Carnosol, a natural polyphenol, inhibits migration, metastasis, and tumor growth of breast cancer via a ROS-dependent proteasome degradation of STAT3. Front Oncol. 2019;9:743.
32. Bajbouj K, et al. Vitamin D exerts significant antitumor effects by suppressing vasculogenic mimicry in breast cancer cells. Front Oncol. 2022;12:918340.
33. Azribi F, et al. Prevalence of BRCA1 and BRCA2 pathogenic sequence variants in ovarian cancer patients in the Gulf region: the PREDICT study. BMC Cancer. 2021;21(1):1350.
34. Hachim MY, et al. M1 polarization markers are upregulated in basal-like breast cancer molecular subtype and associated with favorable patient outcome. Front Immunol. 2020;11:560074.
35. Nelson DR, et al. Molecular mechanisms behind Safranal's toxicity to HepG2 cells from dual omics. Antioxidants (Basel). 2022;11(6):1125.
36. Elmusharaf K, et al. The case for investing in the prevention and control of non-communicable diseases in the six countries of the Gulf Cooperation Council: an economic evaluation. BMJ Glob Health. 2022;7(6):e008670.
37. Ong JS, et al. A comprehensive re-assessment of the association between vitamin D and cancer susceptibility using Mendelian randomization. Nat Commun. 2021;12(1):246.
38. Hashim M, et al. Knowledge, awareness, and practices of university students toward the role of dietary and lifestyle behaviors in colorectal cancer: a cross-sectional study from Sharjah/UAE. Asian Pac J Cancer Prev. 2022;23(3):815–22.
39. Ahmed SBM, et al. Assessing the knowledge of environmental risk factors for cancer among the UAE population: a pilot study. Int J Environ Res Public Health. 2020;17(9):2984.
40. Al Abdouli L, et al. Colorectal cancer risk awareness and screening uptake among adults in the United Arab Emirates. Asian Pac J Cancer Prev. 2018;19(8):2343–9.
41. Idris NF, Ismail MA. Breast cancer disease classification using fuzzy-ID3 algorithm with FUZZYDBD method: automatic fuzzy database definition. PeerJ Comput Sci. 2021;7:e427.
42. Mustapha MT, et al. Breast cancer screening based on supervised learning and multi-criteria decision-making. Diagnostics (Basel). 2022;12(6):1326.
43. Umapathy VR, et al. Emerging biosensors for oral cancer detection and diagnosis—a review unravelling their role in past and present advancements in the field of early diagnosis. Biosensors (Basel). 2022;12(7):498.
44. Carreras J, et al. Artificial neural networks predicted the overall survival and molecular subtypes of diffuse large B-cell lymphoma using a pancancer immune-oncology panel. Cancers (Basel). 2021;13(24):6384.

45. Shen Y, et al. Artificial intelligence system reduces false-positive findings in the interpretation of breast ultrasound exams. Nat Commun. 2021;12(1):5645.
46. Wallace MB, et al. Impact of artificial intelligence on miss rate of colorectal neoplasia. Gastroenterology. 2022;163(1):295–304 e5.
47. Goldblatt EM, Lee WH. From bench to bedside: the growing use of translational research in cancer medicine. Am J Transl Res. 2010;2(1):1–18.
48. Al-Humaidi RB, et al. Optimum inhibition of MCF-7 breast cancer cells by efficient targeting of the macropinocytosis using optimized paclitaxel-loaded nanoparticles. Life Sci. 2022;305:120778.
49. Benyettou F, et al. Synthesis of silver nanoparticles for the dual delivery of doxorubicin and alendronate to cancer cells. J Mater Chem B. 2015;3(36):7237–45.
50. Palanikumar L, et al. pH-responsive high stability polymeric nanoparticles for targeted delivery of anticancer therapeutics. Commun Biol. 2020;3(1):95.
51. Soares NC, et al. Unveiling the mechanism of action of nature-inspired anti-cancer compounds using a multi-omics approach. J Proteome. 2022;265:104660.
52. Palanikumar L, et al. Protein mimetic amyloid inhibitor potently abrogates cancer-associated mutant p53 aggregation and restores tumor suppressor function. Nat Commun. 2021;12(1):3962.
53. Agrawal YO, et al. Methotrexate-loaded nanostructured lipid carrier gel alleviates imiquimod-induced psoriasis by moderating inflammation: formulation, optimization, characterization, in-vitro and in-vivo studies. Int J Nanomed. 2020;15:4763–78.

Dr. Ibrahim Yaseen Hachim is an Assistant Professor in Pathology, the Department of Clinical Sciences, College of Medicine, University of Sharjah. Dr. Ibrahim is a medical doctor with an M.Sc. in histopathology. He obtained his Ph.D. from the Cancer Research Program, College of Medicine, McGill University, Montreal, Canada.

Dr. Hachim is known for his research work in the fields of molecular pathology and cancer biology. His research interests include molecular pathology, cancer biology, molecular subtyping of cancers, biomarkers, cancer progression, and metastasis.

Dr. Ibrahim had more than 60 publications in different prestigious journals. During his career, Dr. Ibrahim received several awards, including the George G. Harris Fellowship in Cancer, McGill University, Montreal, Canada, and the Pauline Blinder Krupp Award for the most promising clinical researchers in Montreal, Canada.

Dr. Saba Al Heialy is an Associate Professor of Immunology at Mohammed bin Rashid University of Medicine, Dubai, UAE and Health Sciences, and an Adjunct Professor at McGill University, Montreal, Canada. Her research revolves around immune dysregulation in various respiratory diseases. She has a particular interest in asthma and its association with other diseases such as obesity and lung cancer. Asthma is a heterogeneous disease of the airways characterized by airway hyperresponsiveness and inflammation. Therefore, she has been interested in exploring its association with lung cancer, which has shown conflicting data in the past. Through funding from the Al Jalila Foundation, her group has been able to investigate this association in the UAE and identify candidate biomarkers that may guide us in better understanding the progression of lung cancer.

Dr. Mahmood Hachim 's multidisciplinary background shaped his medical degree and research career. With two Ph.D.s in molecular medicine and translation research from the University of Sharjah and Lubeck University in Germany, a master of research in cancer biology from the University of Dundee, and a master of science in medical microbiology and immunology from Al Nahrain University in Iraq, Dr. Mahmood made a research track and experience in systems biology to apply its advances to decipher complex diseases like breast cancer, diabetes, and asthma, as well as an understanding of the human microbiome by applying novel bioinformatics approaches and a state-of-the-art omics approach. His goal is to understand the molecular basis of such diseases to identify novel diagnostic and predictive clinically proven biomarkers. Specifically, to understand the exact role of host immune response and microbiome interaction in susceptibility and the development of clinical heterogeneity in such chronic diseases, Dr. Mahmood had more than 75 Scopus-indexed publications.

Humaid O. Al-Shamsi

12.1 Introduction

The United Arab Emirates (UAE) lacks systematized and well-structured advanced fellowship training programs in hematology and oncology. The medical oncology fellowship training program at Tawam Hospital, which was launched in August 2019, is the UAE's sole approved program by the Accreditation Council for Graduate Medical Education-International (ACGME-I). The program consists of 3 years of medical oncology fellowship training (with no hematology training). Three fellows were admitted at that time; however, the first graduate of the medical oncology program was Dr. Ali Yousif, a Sudanese doctor who completed his training in December 2022 and obtained the Jordanian Board of Medical Oncology Certification in August 2022. The UAE has only one hematology fellowship training program, which started in the Emirate of Dubai in 2020 [1].

12.2 Current Status

The establishment of the National Institute for Health Specialties (NIHS) was authorized through Cabinet Decree No. 28 of 2014. Its primary objective is to lead,

H. O. Al-Shamsi (✉)
Burjeel Cancer Institute, Burjeel Medical City, Burjeel Holdings, Abu Dhabi, United Arab Emirates

Ras Al Khaimah Medical and Health Sciences University, Ras Al Khaimah, United Arab Emirates

Gulf Medical University, Ajman, United Arab Emirates

Emirates Oncology Society, Emirates Medical Association, Dubai, United Arab Emirates

College of Medicine, University of Sharjah, Sharjah, United Arab Emirates

Gulf Cancer Society, Alsafa, Kuwait
e-mail: alshamsi@burjeel.com; humaid.al-shamsi@medportal.ca

regulate, and coordinate the professional development of the healthcare workforce, with a specific focus on specialized training. Affiliated with the United Arab Emirates University, the country's leading institution of higher education, the NIHS operates under the guidance of a prominent Board of Directors led by H.E., the Minister of Education. In May 2022, the NIHS formed a fellowship committee dedicated to oncology and hematology for the Emirati Board of Medical Oncology and Hematology. Professor Humaid Al-Shamsi was appointed as the committee's chairman [2]. According to the NIHS, the Emirati Board in Oncology is responsible for defining its program objectives in alignment with the sponsoring institution's overall mission, the community's needs, the intended beneficiaries of its graduates, and the unique skills required of physicians. The program is expected to demonstrate substantial adherence to both general and specialty-specific program requirements. The committee comprises specialists from major cancer centers across the UAE, encompassing medical oncology for adults, hematology for adults, and pediatric hematology and oncology. Together, they have established four distinct pathways for fellowship training:

1. Adult medical oncology for 3 years.
2. Adult hematology for 2 years.
3. Adult combined medical oncology and hematology for 4 years.
4. Pediatric hematology and oncology for 3 years.

According to the NIHS, the eligibility requirements for the fellowship training are [3]:

- Completion of Clinical Education: To qualify for enrollment in an NIHS-accredited fellowship program, applicants must have successfully completed all clinical education requirements within an NIHS-accredited residency program, a program with ACGME-International (ACGME-I) Advanced Specialty Accreditation, or any other structured residency program that has been approved by the Central Accreditation Committee.
- Completion of Internal Medicine Program: Prior to joining the fellowship, candidates should have fulfilled the requirements of an internal medicine program.
- Compliance with NIHS Criteria: It is essential to refer to the NIHS criteria outlined in the Training Bylaw for additional details and specific requirements.
- Fellow Eligibility Exception: Certain exceptions to the eligibility requirements may be considered for prospective fellows, subject to further evaluation and assessment.

The Central Accreditation Committee for Internal Medicine will allow the following exception to the fellowship eligibility requirements:

*In exceptional cases, an NIHS-accredited fellowship program has the option to consider an international graduate applicant who does not meet the standard eligibility requirements. However, the applicant must fulfill the following additional qualifications and conditions:

1. Eligibility for Specialist License: The applicant should be eligible for a specialist license in internal medicine as per the Professional Qualification Requirement (PQR) set by the UAE Health Authority.
2. Evaluation by Program Director and Selection Committee: The program director and fellowship selection committee will assess the applicant's prior training and review summative evaluations of their training in the core specialty.
3. Approval by the Graduate Medical Education Committee (GMEC): The exceptional qualifications of the applicant will be thoroughly reviewed and approved by the Graduate Medical Education Committee (GMEC).

It is important to note that this provision is reserved for highly qualified international applicants who possess exceptional qualifications and meet the specified conditions. The final decision to accept such applicants rests with the NIHS-accredited fellowship program.

*Applicants accepted through this exception must have an evaluation of their performance by the Clinical Competency Committee within 12 weeks of matriculation [3].

- All prerequisite postgraduate clinical education necessary for admission or transfer into NIHS-accredited fellowship programs must be completed within a NIHS-accredited program authorized by the NIHS.
- Prior to joining the program, fellows must satisfy the eligibility criteria established by the NIHS.
- Collaborations with prominent international cancer centers offer UAE fellowship trainees valuable opportunities to gain specialized experience in oncology and hematology rotations, including advanced treatments like CAR-T cell therapy that are currently unavailable within the UAE.
- The interest in medical oncology and hematology as subspecialties is relatively uncommon, resulting in a limited number of UAE trainees pursuing these areas. It is crucial to enhance awareness and attraction toward these subspecialties. This can be accomplished by incorporating mandatory oncology rotations during medical school and internal medicine training, as well as conducting informative lectures for trainees to emphasize the benefits of these subspecialties.
- Presently, there are no fellowship training programs in surgical oncology, palliative care, or radiation therapy available within the UAE. Establishing fellowship training in these subspecialties is essential to meeting the healthcare needs in the field of oncology.
- The UAE currently lacks residency training programs in oncology for nurses. It is imperative to develop robust oncology training programs that encompass both physicians and nurses, particularly considering the global shortage of oncology-trained nurses [1, 4].

12.3 Conclusion

Advanced oncology and hematology fellowship training in the UAE is still evolving. In Tawam Hospital, there is one oncology fellowship, and there is one hematology fellowship program in Dubai. The medical oncology fellowship training program at Tawam Hospital, which was launched in August 2019, is the UAE's sole approved program by the Accreditation Council for Graduate Medical Education-International (ACGME-I) and has graduated only one fellow as of December 2022. In May 2022, the NIHS established an oncology and hematology fellowship committee for the Emirati Board of Medical Oncology and Hematology, and Prof. Humaid Al-Shamsi was named the chairman of this committee. The hematology fellowship and medical oncology fellowship were both approved by the NIHS in November 2022 and December 2022, respectively, and as of December 2022, no program has been accredited by the NIHS in these two programs, yet it is expected that multiple hospitals will apply for accreditation in 2023. The interest in medical oncology and hematology as subspecialties is not common, and only a limited number of UAE trainees are joining these subspecialties. There is a need to increase awareness of and attraction to this subspecialty.

Conflict of Interest The author has no conflict of interest to declare.

References

1. Al-Shamsi HO. The state of cancer care in the United Arab Emirates in 2022. Clin Pract. 2022;12(6):955–85.
2. Adopting criteria for evaluating national programs for training physicians in the field of cancer and oncology in the country. https://www.wam.ae/ar/details/1395303106331. Accessed 6 Dec 2022.
3. https://nihs.uaeu.ac.ae/en/docs/hematology.pdf (Accessed on 08 Apr 2024).
4. Young A, Samadi M. The global power of oncology nurses in low- and middle-income countries. Asia-Pac J Oncol Nurs. 2022;9:131–2.

Prof. Humaid Obaid Al-Shamsi is the Chief Executive Officer of Burjeel Cancer Institute in Abu Dhabi, UAE, President of the Emirates Oncology Society, Lead of the Gulf Cancer Society, Full Professor of Oncology at the Ras Al Khaimah Medical and Health Sciences University, Ras Al Khaimah, UAE, and an Adjunct Professor of Oncology at the College of Medicine, University of Sharjah. He is the first Emirati to be promoted as a professor in oncology in the UAE. He is also the Chairman for Colorectal Cancer in the MENA region, appointed by the prestigious National Comprehensive Cancer Network®. He is also the only member of Lung Cancer Policy Network in the MENA region that aims to advance lung cancer research and screening globally. He is the Chairman of the Oncology and Hematology Fellowship Training Program for the National Institute for Health Specialties in the United Arab Emirates. He is the only member in GCC in the WIN Consortium which is comprised of organizations representing all stakeholders in personalized cancer medicine globally.

He is board-certified in both internal medicine and oncology from the UK, USA (ABIM), and Canada (FRCPC). He has also been awarded the FRCP (London) in 2023 and FRCP (Glasgow) in 2024. He is the only physician in the UAE with a subspecialty fellowship certification and training in gastrointestinal oncology and the first Emirati to train and complete a clinical post-doctoral fellowship in palliative care. He was an assistant professor at the University of Texas MD Anderson Cancer Center between 2014 and 2017. He has published more than 140 peer-reviewed articles in JAMA Oncology, Lancet Oncology, The Oncologist, BMC Cancer, and many others. His area of expertise includes precision oncology and cancer care in the UAE. In 2016, he published with his group from MD Anderson the JCO paper describing a new distinct subgroup of CRC, NON V600 BRAF-mutated CRC. In 2022, he published the first book about cancer research in the UAE and also the first book about cancer in the Arab world, both of which were launched at Dubai Expo 2020. *Cancer in the Arab World* has been downloaded more than 450,000 times in its first 18 months of publication and is the ultimate source of cancer data in the Arab region. He also published the first comprehensive book about cancer care in the UAE which is the first book in UAE history to document the cancer care in the UAE with many topics addressed for the first time, e.g., neuroendocrine tumors in the UAE. He is passionate about advancing cancer care in the UAE and the GCC and has made significant contributions to cancer awareness and early detection for the public using social media platforms. He is considered as the most followed oncologist in the world with over 300,000 subscribers across his social media platforms (Instagram, Twitter, LinkedIn, and TikTok). In 2022, he was awarded the prestigious Feigenbaum Leadership Excellence Award from Sheikh Hamdan Smart University for his exceptional leadership and research and the Sharjah Award for Volunteering. He was also named the Researcher of the Year in the UAE in 2020 and 2021 by the Emirates Oncology Society.

In May 2024, HH Sheikh Mansour bin Zayed Al Nahyan, Vice President of the United Arab Emirates, awarded him the first place in UAE Nafis program for outstanding leadership in private sector across all business and medical disciplines. Beside his clinical and administrative duties, he is engaged in education and various levels of research training for medical trainees to enhance their clinical and research skills. His mission is to advance cancer care in the UAE and the MENA region and make cancer care accessible to everyone in need around the globe.

Oncology Nursing in the UAE 13

Lois Nyakotyo

13.1 Introduction

13.1.1 History of Oncology Nursing

The cancer care continuum spans prevention, early detection and screening, diagnosis and treatment, living with and beyond cancer, palliative care, and the end of life. A cancer or oncology nurse is a qualified nurse who has the knowledge, skills, and complete responsibility to provide critical nursing care to cancer patients and their families based on evidence-based, specialized, ethical, knowledge and skills [1].

Globally, the development of medical oncology as a medical specialty created the need for the development of nursing through academic education and the practice of oncology with a foundation in research-based evidence. Before oncology nursing existed as a specialty; cancer patients received care from general nurses who provided bedside care without specialist knowledge, skills, or practice [2]. However, the introduction of chemotherapy in the early 1970s and 1980s created the need for nurses to be educated in the administration and management of the side effects it posed for patients and the hazards it posed for staff. Thus, more focus emerged on cancer as a specialty, which further developed into the delivery of chemotherapy in specialized oncology centers [3].

Since then, oncology nursing has evolved significantly from physician-led inpatient care to oncology nurse practitioner-led outpatient care. Oncology nursing professional organizations have emerged, bringing with them the development of standards and guidelines for practice that outline responsibilities and expectations for the specialty role of nurses in cancer care, illustrating a complete transformation from general, basic nursing care to advanced practice [4]. Oncology nurses are responsible for everything from performing invasive procedures to diagnostic

L. Nyakotyo (✉)
Mediclinic Middle East, Dubai, United Arab Emirates
e-mail: Lois.nyakotyo@mediclinic.ae

© The Author(s) 2024
H. O. Al-Shamsi (ed.), *Cancer Care in the United Arab Emirates*,
https://doi.org/10.1007/978-981-99-6794-0_13

interpretation and screening for cancer prevention. In contrast to the preceding illustration of the oncology nursing role, oncology nursing in the United Arab Emirates (UAE) is in its infancy, faced with the challenges of a transient workforce, a lack of a defined oncology nurse role, a lack of licensure as a specialist in oncology, and wage-related benefits in recognition of specialist skills [5]. This chapter explores oncology nursing in the UAE, with a particular focus on the role of the oncology nurse and the training required to develop specialist oncology nursing practice in the UAE.

13.2　Impact of the Healthcare System on the Development of Oncology Nursing in the UAE

The UAE has an outstanding government-funded health service that delivers timely, effective, and safe care for Emirati nationals and expats in emergency cases as needed. So, the private healthcare sector is rapidly expanding [6]. In the UAE, there are 157 hospitals and 5369 health centers in private care that continuously develop services in response to the economy, health policy, and advancements by competitors [7]. Health care in the UAE's public and private sectors delivers efficient, effective, and safe care that is closely monitored and regulated by government agencies. However, there is a lack of government agency-led drive for collaboration in training and development or sharing best practices among oncology nurses.

Nurses are critical to achieving the goal of safe health care for all globally, and the current staffing shortages are echoed in the oncology nursing setting [8]. The pandemic has magnified and exacerbated the global nursing shortage issues, which has led to changes in immigration laws in the United Kingdom (UK), Australia, and the USA regarding nurses [9]. In the UAE, it has manifested in a significant number of experienced expat oncology nurses of Indian and Filipino descent migrating from the UAE to the UK, the USA, or Australia, where there is recognition of specialist skills, increased wages, and the prospects of citizenship over a length of time [10]. To attract more nurses, the UAE government has removed the pre-requisite of 2 years of experience before registering as a nurse in the UAE. Therefore, no experience is required for nurses to get a license in the UAE if they have a current active registration in Canada, the USA, the UK, Ireland, South Africa, New Zealand, or Australia [11].

Training and further professional development opportunities for oncology nurses in the UAE vary between employers, making it difficult to benchmark practice. In addition, the competitive private care setting in the UAE does not promote sharing best nursing practices across healthcare facilities through an agreed-upon and standardized framework for training and developing oncology nurses. As a result, newly qualified nurses (NQN) who join the nursing workforce in the UAE will receive a non-standardized orientation from their employee, which does not necessarily ensure appropriate experience and exposure to develop the good communication and practice skills required for cancer care. This lack of recognized standards of practice creates a barrier to developing experienced and well-trained specialist oncology nurses, as there is a pool of varying abilities and levels of training and development across the private healthcare sector's oncology services [12].

Nursing training worldwide aims to produce qualified nurses who employ evidence-based practice, work as part of a multidisciplinary team, continuously apply quality improvement principles, and utilize informatics to inform innovation in practice [12]. Therefore, there must be an agreed-upon framework for NQN oncology orientation, development, and training to ensure that the UAE maintains a well-defined, high level of oncology nursing practice. NQN in the UK joins the workforce in an oncology nursing preceptorship capacity. The goal of preceptorship is for the NQN to develop their confidence and autonomy. In organizations with well-established preceptorship as part of the culture, there are significant benefits for newly registered nurses, patients, and wider teams. Particularly in terms of retention, recruitment, and staff engagement [7].

National guidance from the National Health Service (NHS) Executive and Innovation recommends 2 weeks or 75 h of supernumerary time, not including induction or orientation time. After this supernumerary period, the newly qualified nurse undergoes a preceptorship period [12]. Only once one has completed a year of preceptorship can they consider starting a career path as a junior oncology nurse, focusing on outpatient chemotherapy, hematology inpatients, oncology inpatients, oncology home care, or palliative care. Newly registered nurses become accountable as soon as they are registered, and this transition from student to responsible practitioner is known to be challenging [13]. The purpose of the preceptorship is to provide support during this transition. Preceptorship programs may include classroom teaching and the attainment of role-specific competencies. However, the essential element is the individualized support provided in practice by the preceptor [14].

13.3 Role of Oncology Nurses

Cancer is a life-threatening disease that is reported to cause 1 in 6 deaths per year across the world [15]. Research shows that cancer patients interact with nurses more than other members of the multidisciplinary team. Cancer patients have frequent hospital visits due to the nature of cancer treatment management plans, and such patients experience most of their time in the healthcare setting under the care and guidance of a cancer nurse [16]. The cancer nurse's role is pivotal in delivering cancer care. A cancer nurse's role is to coordinate, facilitate, and provide care at all critical points in a patient's cancer journey [2]. Cancer care aims to effectively advocate for prevention, early diagnosis, effective treatment, survivorship, and providing end-of-life care that incorporates the patients' priorities of care. This approach to cancer care promotes quality of life throughout the cancer journey.

To provide the best evidence-based cancer care investment in training and developing oncology nurses [13], specific job descriptions for oncology nurses differ between healthcare organizations in the UAE. Still, it is widely reflected in the job descriptions that oncology nurses provide nursing care for cancer patients, but the level of responsibility varies depending on the employer's vision and priorities for cancer care. A well-established oncology service, as shown in the UK guidelines 2020 [13], should have well-defined oncology nursing roles, and the nurses should perform the following duties:

- History taking.
- Physical and psychological assessment.
- Monitoring and reviewing test results.
- Administration of systemic anti-cancer treatments.
- Key point of contact for the patient and a key member of the multidisciplinary team.
- Facilitating patient education across the clinical pathway.
- Patient advocate.
- Promoting self-management.

The UAE healthcare system is essentially physician-led; insurance charges are largely related to physician activity; and service users value the input of a specialist physician rather than a nurse. Furthermore, insurance approval is required before delivering oncology services, except in an emergency. This results in oncology services being facilitated with the involvement of nurses carrying out extensive administrative tasks, including liaising with the internal insurance department to ensure that the proper clinical updates are given to meet requests from insurance companies. This further influences the perception patients have of the role of an oncology nurse, which limits the engagement that patients have with nurses as they perceive the physician to be the critical healthcare professional throughout their cancer journey and do not get to fully experience the full extent of care that a skilled oncology nurse can deliver through assessment, management of side effects, care planning, education, and psychological support. In addition, the revenue margins for healthcare organizations are directly linked to physical activity, and as such, organizations invest in physicians rather than nurses. Therefore, there is a need for a culture shift in practice; this involves educating healthcare leaders on the value of oncology nurses and patients on the role of an oncology nurse and the benefits for patients.

13.4 Oncology Nurse Training and Development

Nurses who provide specialist cancer nursing are encouraged to undertake postgraduate courses that enhance their research and evidence-based practice. However, there are different levels of academic qualifications that range across certificate, master's, and doctorate levels. This enables nurses to practice as oncology-certified nurses, specialist nurses, or advanced nurse practitioners or nurse consultants. To this end, several organizations have established well-defined competencies that are required for a cancer nurse to practice effectively, including EONS and the Oncology Nursing Society (ONS) [17].

Cancer nursing in the UAE is a relatively new concept supported by the Emirates Oncology Nursing Society, founded in 2017. The Emirates Oncology Nursing Society (EOHNS) focuses on promoting oncology nursing and sharing advancements in nursing practice among different oncology nurses within the UAE. EOHNS is in its infancy; there is a need for solid nursing leadership across the UAE to form a working group in pursuit of developing well-defined oncology nurse specialist roles and pre-requisite education, training, and development frameworks with support from a government healthcare agency to address the barriers to collaboration across healthcare organizations.

NHS England 2017 [12] illustrates the four components of advanced specialist nursing:

1. Providing specialist clinical practice
2. Clinical leadership within a multidisciplinary setting
3. Continuous improvement of quality of cancer nursing care through professional development
4. Continuing professional development and sharing best practices to develop others

An educational program for cancer nurses aims to provide evidence-based research that develops good cancer nursing practice using a wide range of skills and knowledge in pediatric or adult cancer services. The level of competency is defined by the varying depth and complexity of training and practice. This separates the novice from the expert in terms of a cancer nurse providing routine ward care, a clinical nurse specialist, or an advanced nurse practitioner. The key principle is to create a program that enables both employers and employees to understand the cancer nursing career pathway. This will ensure that cancer nurses with the right skillset are employed in the relevant cancer nursing positions to ensure the provision of quality cancer care [17]. Research shows that cancer nurses contribute substantially to the quality of cancer services across the clinical pathway, from diagnosis, treatment, survivorship, and palliative care, at all levels of health care, from primary to tertiary care [18].

As educational programs develop, it is essential to simultaneously have a structure that reflects job roles, titles, and the required competence level that is related to academic qualifications that include assessment in practice. This is particularly important in terms of the consistency of the standard of cancer care delivery. Policymakers, regulators, healthcare professionals, and the wider public will have the same expectation of care provided by cancer nurses according to their title, which inherently indicates a well-defined level of competence. In addition, cancer nurses will be more accountable for their practices or omissions. Particularly, as there are currently significant inconsistencies in relation to titles and competence across the UK. For example, the title specialist nurse is used in different settings for nurses with varying skillsets [17]. Therefore, the introduction and establishment of a well-defined education program, clear role definitions, and skillset requirements for every level of practice to address the gap of certified cancer nurses in the UAE are imperative to further develop the quality of cancer nursing care.

13.5 Assessment and Care Plan Development

Cancer nursing practice provides patient care in all healthcare settings. The scope of cancer nursing practice includes screening, diagnosis, all treatment modalities, survivorship, and palliative care. Cancer nurses are an instrumental part of the multidisciplinary team [19]. They coordinate and facilitate the delivery of cancer through effective communication with other healthcare professionals, patients, and their families to ensure the delivery of the best evidence-based cancer care.

Cancer nurses need to have a good understanding of all treatment modalities, management of side effects, psychological and emotional impact of the disease, or treatment on individual patients [19]. This enables them to formulate effective cancer nursing plans that address the patients' health needs effectively with appropriate and timely referrals to members of the wider multidisciplinary team. The aim is to minimize the impact of side effects of treatment and to manage disease-related symptoms effectively. The role of the cancer nurse has become pivotal in cancer care as more complex protocols emerge. The key is to have cancer nurses who provide robust individualized education, cancer care, and monitoring at all points throughout the cancer journey [20].

13.5.1 Components of Assessment and Management

- Management plan for common side effects of treatment and disease-related symptoms
- Strategies for self-management of symptoms such as shortness, pain, or peripheral neuropathy
- Promotes evidence-based supportive care such as acupuncture
- Prioritizes facilitating a seamless pathway to living beyond cancer
- Facilitates follow-up and after care
- Facilitates individualized, high-quality clinical pathways that are affordable and easy to access
- Good knowledge of ongoing clinic research studies and referring patients appropriately [2]

The peer-review process requires healthcare providers to share evidence of the quality and outcomes of their service against agreed-upon measures. Their peers then review these to monitor adherence and effectiveness [17]. However, the UAE does not currently have peer reviews across healthcare organizations for oncology nursing. Peer review would be beneficial in establishing good oncology nursing practice in assessment and care plan development.

13.5.2 Cancer Nursing Development in the UAE

There is a need for further development of cancer nursing in the UAE. There are varied levels of training across the healthcare system. The level of commitment to developing oncology nursing is dependent on each healthcare facility's priorities. The nature of the private healthcare system is to be profitable. In an economy where the nursing workforce is as transient as the patients, healthcare organizations lack the appetite to invest in cancer nursing practice as the return on investment is not guaranteed.

Specialist cancer nursing practice delivers expert cancer with the ability to recognize and manage complications independently according to the level of competence. This approach enables quality improvement in cancer nursing through setting good standards of practice, clinical audits, and supervised practice to ensure quality nursing practice and leadership [16].

Nursing research across the world has made a significant contribution to the cancer care provided to patients and has also improved patient outcomes. Particularly in relation to patient self-management, reducing the time patients spend in hospital beds, and improving the overall patient experience [18]. As healthcare systems continue to adjust to innovations, so too will cancer nursing evolve to meet the changing demands [21]. The healthcare landscape for cancer services is rapidly transforming, and it is key that cancer nursing care be encouraged to keep up with this transformation. The UAE has exceptional technological advances that would support innovation in oncology nursing practice, but this would require support from government agencies and collaboration between private healthcare organizations.

13.6 Conclusion

The UAE healthcare system delivers a high standard of care to oncology patients through a sound, coordinated care pathway facilitated by well-trained and experienced surgeons, radiologists, and pathologists. The vast majority of people in the UAE prefer to be treated by a specialist consultant physician. The key to developing oncology nurse practice at the international level in the well-researched and recognized role of specialist or advanced nurses is engagement, investment, and commitment from the health governing bodies and private and public healthcare organizations. It is a role that will address patients' complex physical, social, and psychological needs and provide additional support for their families. There is a need for nursing leaders at the executive level to drive this forward. This will be a significant challenge, but the benefits of a good oncology nursing framework, training, and development supported by recognized licensure and remuneration for specialist oncology nursing skills are well-researched. This would be the new frontier of oncology services and care delivered to cancer patients across the UAE in an inpatient or outpatient setting and at the patient's home.

Conflict of Interest The author has no conflict of interest to declare.

References

1. Jacobs LA. Standards of oncology nursing education generalist and advanced practice levels. 3rd ed. Pittsburgh: Oncology Nursing Society; 2003.
2. Royal College of Nursing. Career pathway and education framework for cancer nursing.
3. Rosenzweig MQ, Kota K, van Londen G. Interprofessional management of cancer survivorship: new models of care. Semin Oncol Nurs. 2017;33(4):449–58.
4. Ferlay J, Ervik M, Lam F, Colombet M, Mery L, Piñeros M, et al. Global cancer observatory: cancer today. Lyon: International Agency for Research on Cancer; 2020.
5. El-Haddad M. Nursing in the UAE: an historical background. International Nursing Review. 2007. International Council of Nurses.
6. UAE Annual Statistical Report. Statistics research center Ministry of Health and community. UAE Ministry of Health and Prevention; 2020.
7. United Arab Emirates Ministry of Health and Prevention. UAE statistical annual report. 2020. https://mohap.gov.ae/en/open-data/mohap-open-data.
8. International Council of Nurses. The global nursing shortage and nursing retention. 2021.
9. Marć M, Bartosiewicz A, Burzyńska J, Chmiel Z, Januszewicz P. A nursing shortage—a prospect of global and local policies. Int Nurse Rev. 2021;66(1):9–16.
10. England K. Care migration and citizenship: nurse migration. Soc Cult Geogr J. 2022;14(5):558–74.
11. UAE Ministry of Health and Prevention Unified Healthcare professional qualification requirements. 2022.
12. NHS England National Preceptorship Framework for Nursing. 2022. https://www.england.nhs.uk/long-read/national-preceptorship-framework-for-nursing/#:~:text=The%20goal%20of%20preceptorship%20is,newly%20registered%20practitioners%20in%20England.
13. Nursing and Midwifery Council UK. Standards for specialist education and practice. 2015.
14. Higgins G, Spencer RL, Kane R. A systematic review of the experiences and perceptions of the newly qualified nurse in the United Kingdom. Nurse Educ Today. 2010;30:499–508.
15. Torre LA, Bray F, Siegel RL, Ferlay J, Lortet-Tieulent J, Jemal A. Global cancer statistics, 2022. CA Cancer Clin J. 2022;65(2):87–108.
16. Oncology Nursing Society. Oncology nurse generalist competencies. 2016.
17. Institute of Medicine. The future of nursing: leading change, advancing health. Washington, DC: National Academies Press; 2018.
18. Lubejko BG, Wilson BJ. Oncology nursing: scope and standards of practice. 2019.
19. Gaguski ME, George K, Bruce SD, Brucker E, Leija C, LeFebvre KB, Mackey H. Oncology. Nurse generalist competencies: oncology nursing society's initiative to establish best practice. Clin J Oncol Nurse. 2017;21(6):679–87.
20. Coolbrandt M, Wildiers H, Aertgeerts B. Systematic development of chemotherapy support a nursing intervention to support adult patients with cancer in dealing with chemotherapy related symptoms. BMC Nurs J. 2018;17:28.
21. World Health Organization. Cancer—key facts. 2022. https://www.who.int/news-room/fact-sheets/detail/cancer.

Lois Nyakotyo gained her nursing degree from Nottingham University, her postgraduate oncology degree from Anglia Ruskin University, and her MSc in healthcare management from Westminster University, London, United Kingdom. She is the Regional Cancer Programme Manager for Mediclinic Middle East. She is registered with the DHA in the UAE and the Nursing and Midwifery Council, UK. She is a member of the UK Oncology Nursing Society. She has over 20 years of experience in specialist oncology nursing, operational healthcare leadership, and management. She has worked in the National Health Service, Private care in the UK and the UAE. This included roles at Cambridge University Hospitals, Imperial Healthcare College NHS Trust, The Royal Marsden NHS Trust, King's College London Hospital UAE, and Mediclinic Middle East. Her achievements include being the chairperson of the lung cancer nurses' committee for the East of England, member of the advisory panel for Roche and advisory panel for the Roy Castle Foundation. She has a passion for facilitating learning in practice and has previously worked as a link lecturer for Anglia Ruskin University in Cambridge, currently coordinates the oncology rotation for MBRU students at Mediclinic Middle East. She has a particular interest in developing innovative operational efficiency that delivers excellence in all aspects of cancer care.

Genomic Medicine in Cancer Care in the UAE

14

Faraz A. Khan ⓘ and Maroun El Khoury

14.1 Introduction

The importance of genomics in cancer dates back over 100 years. Theodor Boveri, a German zoologist, made the initial proposition about the potential role of alterations in the genetic material and the development of cancer. His publication was later translated and published by Henry Harris, where he made the observation that malignant tumor cells develop from normal tissue and postulated the potential role of abnormal genetic material and chromosomes in tumorigenesis. He partly supported the observations made earlier by *Von Hansemann,* who is regarded as one of the pioneers in human cancer genetics [1].

However, only later in the last century did we witness some key developments, not only in the understanding of the role of genomic alterations in the development of cancer but also in the concept of driver mutations and the development of drugs as a therapeutic strategy.

14.2 Milestones in Cancer Genomics and Therapeutics

In 1959, David Hungerford and Peter Nowell first discovered the Philadelphia chromosome in patients with chronic myeloid leukemia [2]. It took more than 30 years, and only in the 1990s did Brian Drucker and his colleagues start working on potential drugs that could block the BCR-ABL pathway, as they hypothesized that blocking the BCR-ABL pathway may halt the progression of the leukemia, considering that the mutation is not present in normal cells. In 1998, 40 years after the discovery of the Philadelphia chromosome, the data of the first phase 1 clinical trial utilizing STI-571, a compound known as imatinib, was reported, which changed the

F. A. Khan (✉) · M. El Khoury
American Hospital, Dubai, United Arab Emirates
e-mail: fkhan@ahdubai.com; mkhoury@ahdubai.com

© The Author(s) 2024
H. O. Al-Shamsi (ed.), *Cancer Care in the United Arab Emirates*,
https://doi.org/10.1007/978-981-99-6794-0_14

prognosis and dynamics of a disease that, in the absence of a bone marrow transplant, carried a poor prognosis. The Food and Drug Administration (FDA) approved imatinib for the treatment of chronic myeloid leukemia in 2001 [3].

In 1984, the oncogene "neu" was discovered in rat cells, and the human "neu" oncogene coding for the epidermal growth factor receptor was reported in 1985 [4]. More than a decade later, after confirming its role in the biological behavior of breast cancer, trastuzumab was first approved by the FDA for the treatment of metastatic breast cancer in 1998 [5].

In 1994 and 1995, the *BRCA1* and *BRCA2* tumor suppressor genes were identified, and their association with familial breast and ovarian cancer was established. The discovery led to additional research, and using the mutation as a predictor to guide therapy in patients with ovarian cancer harboring the BRCA mutation with a PARP inhibitor, olaparib, came only in 2014 [6, 7].

Similar is the story of the identification of epidermal growth factor receptor (EGFR) and anti-EGFR inhibitors like gefitinib in non-small cell lung cancer [8].

However, what really transformed the landscape was the start of a landmark program in 2006 known as the Human Cancer Genome Atlas program, a joint effort between the National Cancer Institute and the National Human Genome Research Institute. The collaborative work of researchers from various disciplines and institutions was successful in generating an enormous amount of genomic and molecular data, which has played a pivotal role in the understanding of cancer genetics and paved the way for the development of new drugs. The analysis of over 11,000 tumors with over 33 different subtypes of cancer created a vital resource that has become the key to the development of new treatments based on genomic data [9].

14.3 The Era of Precision Medicine

In the last two decades, we have witnessed revolutionary advancements in biotechnology utilizing sequencing platforms and bioinformatics. In the field of medicine, the most significant impact of progress in genomics and proteomics is seen in cancer medicine. The identification of new molecular targets and their significance in tumorigenesis has led to the approval of many new classes of drugs with unique mechanisms of action. However, this required changes at the regulatory level to allow the incorporation and approval of new clinical trial designs like master protocols and basket trials. The passage of the twenty-first Century Cures Act by the US Congress in 2016 enabled the FDA to approve such modern clinical trial designs that have facilitated rapid and fast-track drug development and approval [10].

In 2017, researchers from Memorial Sloan Kettering (MSK) published their data from a next-generation sequencing test called MSK-IMPACT using over 10,000 patient tumor specimens and identified that 37% had relevant mutations and alterations, which allowed 11% of the first 5009 patients to enroll in a relevant clinical trial, which also led to the approval of the test as the first comprehensive next-generation sequencing panel testing 468 genes [11]. A few days later, the FDA approved the FoundationOne CDx panel as well as the 324-gene next-generation sequencing assay [12].

In the last decade, at least 52 drugs for hematologic malignancies were approved, including small molecules like Bruton's tyrosine kinase (BTK) inhibitors, monoclonal antibodies, bispecific T-cell engager antibodies, antibody-drug conjugates, and CAR T-cell products. All this is possible only due to an improved understanding of cancer genomics and tumorogenesis [13]. In 2020 alone, half of the 53 new small molecules approved were for oncology [14]. Every year, we see hundreds of new drugs with novel mechanisms of action based on genomic data, which has accelerated and transformed drug development, clinical trials, and approval.

14.4 Cancer Care in the UAE

Al-Shamsi et al. published a narrative description of the development of cancer care facilities, programs, manpower resources, and access to modern cancer treatments and technology in the United Arab Emirates (UAE) over the last decade with historical timelines in 2020 [15].

In this chapter, we reflect on the development and progress in cancer genomics and molecular diagnostics, including the utilization of comprehensive tumor profiling in the UAE, with their limitations and challenges.

14.5 Genomic Testing Facilities and Resources

Even though access to commercial genomic tests has been available for several years, only a few centers have developed the facilities and expertise for in-house testing, particularly with reference to the genomic analysis of cancer cases.

Nevertheless, in the UAE, major milestones have been achieved. In 1995, the first of its kind in the region, the Dubai Genetics Center, was inaugurated under the patronage of HH Sheikh Hamdan bin Rashid Al-Maktoum. Since its inception, the center has reported genomic data on various hereditary disorders like thalassemia, sickle cell disease, and other hemoglobinopathies. For cancer genomics, the center has developed facilities offering molecular testing, utilizing fluorescence in situ hybridization (FISH), polymerase chain reaction (PCR), and lately next-generation sequencing (NGS), and is CAP accredited. The center is offering analysis for diagnosis and monitoring response to therapies in various hematologic malignancies, including acute and chronic myeloid and lymphoid leukemias and multiple myeloma [16].

In 2020, Sheikh Khalifa Specialty Hospital in Ras Al Khaimah was the first center to start in-house next-generation sequencing analysis for solid tumors that includes a 52-gene solid tumor panel test; however, testing is limited to patients seen at the hospital [17].

The government in Abu Dhabi has also started an ambitious national project called the Emirati Genome Program with the aim of profiling and completing the

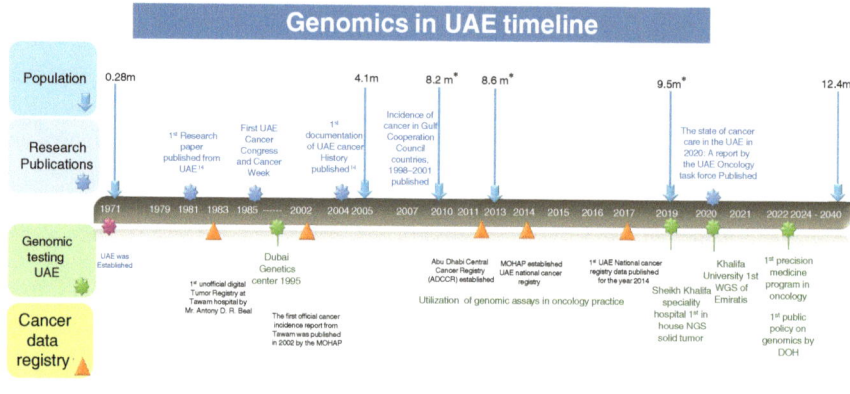

Fig. 14.1 Timeline of genomics in the UAE

gene sequencing of UAE nationals to aid in the prevention and treatment of chronic diseases. In 2019, Al-Safar and her colleagues from Khalifa University in Abu Dhabi reported the whole-genome sequencing data of the first two Emiratis [18]. It is expected that this program will facilitate the identification of local populations at genetic risk for cancer and help develop preventative strategies. Initiatives are expected to further research and the development of more comprehensive programs in cancer genomics within the UAE (Fig. 14.1).

14.6   Utilization of Genomic Assays in Cancer Care and Research

Despite the limited number of centers performing in-house genomic testing, oncologists and hematologists in the UAE have been able to have access to many of the commercial assays, like the Molecular Intelligence Profile by Caris and FoundationOne CDx for solid tumors and FoundationOne Heme for hematologic malignancies offered by Roche. There are a number of other commercial companies that offer both germline (hereditary cancer panels offered by Centogene and Myriad) and somatic mutation testing, or more precise genomic signatures, to determine and guide therapy based on the risk of cancer recurrence, like Oncotype Dx in breast cancer, which is routinely used. There are also a number of commercial companies offering liquid biopsy as an alternative to tissue in cases where tissue quantity is inadequate or follow-up serial monitoring of the mutational landscape is desired (FoundationOne Liquid, Guardant 360). There are also other well-reputed international laboratories, like the Mayo Clinic, that offer cancer-specific gene panel tests for both germline and somatic mutations.

Most of the institutions providing cancer care in the UAE have access to these tests. In many cases, these are covered and reimbursed by some local health authorities and insurance providers. Many drug companies have also chipped in with third-party support programs to cover the cost of limited genomic testing for their partner drugs. However, there are still barriers and challenges with coverage and reimbursement in many cases. Some insurance providers refuse to cover genomic testing and consider it a genetic test for hereditary diseases. There is no one central genomic laboratory in any of the emirates that can offer a more comprehensive scope of testing for solid tumors and hematologic malignancies. Most of the centers have developed institutional relationships with international laboratories and institutions for sample testing. However, in spite of the barriers, utilization of genomic testing in cancer care in the UAE has steadily increased over the last decade, and some have reported the data, which remains mostly retrospective and observational.

In patients with breast cancer, Dawood et al. reported the data of 363 patients tested in the UAE using an NGS-based panel and found that 32 (8.8%) had a pathogenic variant identified in the *BRCA1*, *BRCA2*, or *CHEK2* genes. In 89 patients that were tested using a 33-gene panel, they found additional pathogenic variants in 7 (7.8%) of the *MUTYH*, *RAD51C*, *RAD50*, and *PALB2* genes. The data was used to enroll patients in the pilot phase of the tele-genetics program, where patients were offered genetic counseling using either a telephone or an online Skype platform [19].

Al-Shamsi recently published a study that used an AmpliSeq 50-gene panel to look at the pattern of somatic mutations in Arab women with breast cancer [20]. Earlier, they also reported the frequency of somatic mutations in colorectal cancer using next-generation sequencing and reported a similar prevalence of common mutations like *KRAS*, *NRAS*, and *BRAF* [21].

We have also reported retrospective data from 25 patients with advanced and refractory solid tumors using a combination of next-generation sequencing (NGS), protein expression (IHC), gene amplification (CISH or FISH), and RNA fusion analysis, showing an overall disease control rate of 73% and a median duration of response of 7 months when treated according to the genomic profile data [22].

14.7 Challenges, Limitations, and Recommendations

However, there are many challenges that limit the optimal utilization of genomics in cancer care in the UAE, both in clinical practice and research. Some of the difficulties stem from the diverse configurations of healthcare delivery systems in each emirate, as well as the lack of a broader, unified national policy on healthcare delivery systems.

Over the years, health care has been moving from government-funded to private in many emirates. The private hospitals operate on a commercial basis with limited interest in investment in research and a lack of private (not-for-profit) institutions. There is no national cancer care body, such as the National Cancer Institute (NCI) in the United States, to facilitate regulation, support, and fund clinical research. Hence, most of the effort is individual-driven, and there is an absence of structured utilization of genomics in cancer in the context of clinical trials.

There is currently no single, central facility, either in the public or private sector, that can offer more comprehensive genome sequencing for patients with both hematologic and solid tumors. This limits the availability of a reliable genomic database of the local population, which is quite diverse. This also allows many small or less-known genomic laboratories to commercially operate and market their tests with no real oversight on quality, reproducibility, or reliability.

In 2018, the European Society of Medical Oncology provided the framework for the optimal utilization and categorization of alterations and mutations, paving the way for recommendations on the utilization of genomic tumor profiling both in clinical practice and research [23]. There is no centralized body in the UAE that is able to oversee optimal utilization, particularly in clinical practice. More recently, the Department of Health (DOH) in Abu Dhabi, after the initiative of the National Genome Project, has published a policy on genomics. However, currently there is no specific document or policy on genomics in cancer [24].

There is also a need for local consensus guidelines on the utilization and interpretation of these assays.

Even though tumor profiling is increasingly used in clinical practice, there is a lack of precision medicine clinics with a trained workforce. Considering that the field is relatively new, most practicing oncologists lack training in molecular diagnostics and genomics and rely on the interpretation of the genomic report to make therapeutic decisions, which at times can be misleading. Most of the time, molecular data is discussed in multidisciplinary tumor boards (MDTs) and therapeutic decisions are made, but most MDTs may lack experts in the genomic field, which can limit the strength of recommendations [25].

There are individual efforts to develop molecular tumor boards, but most lack structure and expertise. International and institutional collaboration within the region and abroad can address the issue in the short term by using virtual platforms and seeking input from tertiary centers and precision clinics. We and others have shown that utilizing virtual platforms and MDTs with international collaboration can improve patient care and decision-making, and this approach can be extended to genomic medicine in the form of well-structured molecular tumor boards [26, 27]. Long term, there is a need for dedicated training and education for trainees and practicing oncologists. Khalifa University in Abu Dhabi and Mohammed Bin Rashid University in Dubai (MBRU) have started to offer similar programs, but training the treating oncologist and developing precision medicine clinics manned by trained professionals are important.

Recently, DOH Abu Dhabi announced the launch of the first Precision Medicine Program for Oncology in the region, in collaboration with Mubadala Health, Cleveland Clinic Abu Dhabi, NYU Abu Dhabi, Mohamed bin Zayed University of Artificial Intelligence, and G42 Healthcare [28].

However, to further the progress of genomic medicine in cancer in the UAE, it will require setting up a clinical trial network where, based on their genomic profiles, patients could enroll in clinical trials.

Table 14.1 Recommendations to advance genomics in cancer care in the UAE

Recommendations to advance genomics in cancer care in UAE
Formulation of a national policy on genomics in cancer care
Development of a national cancer institute to enhance collaboration and coordination among different emirates for cancer care and research
Regulation and legislation to develop national tumor banks
Development of clinical trial networks
Local consensus guidelines on the utilization of genomic assays in cancer care
Education and training to incorporate molecular tumor boards and precision medicine programs as part of comprehensive cancer care in the UAE
Regulation for payers to ensure uniform access to genomic testing, both in the private and public sectors

There is a need to create tumor banks at the regional or national level, which will be essential for any meaningful research in genomics and cancer in the UAE and the region in the future [29, 30].

Perhaps the formation of a national subcommittee of genomic medicine in cancer care, operating under the auspices of the Ministry of Health and other health regulatory bodies such as the Department of Health (DOH), could develop a framework for advancing standards of practice and research in the field (Table 14.1).

In summary, even though there has been an increase in the utilization of genomics in clinical practice in cancer care over the last several years, utilization remains sporadic, and the challenges outlined above remain barriers to the optimal utilization of genomics in cancer care in the UAE.

With recent initiatives such as the development of the first genomic policy and the launch of a precision medicine program, there is hope that the era of genomic and precision medicine in cancer care in the UAE has begun. However, the success of these initiatives will depend on the development of robust research and clinical trial networks.

14.8 Conclusion

Genomics is reshaping the treatment paradigm in cancer medicine in the modern era. The outcomes of various cancers continue to improve with the utilization of genomic medicine in cancer care. The UAE has also joined the evolving healthcare system in the utilization of genomics in cancer treatment. It is hoped that new initiatives and regulatory frameworks will transform the practice from utilization to research and clinical trial development.

Conflict of Interest The authors have no conflict of interest to declare.

References

1. Boveri T. Concerning the origin of malignant tumours. Translated and annotated by Henry Harris. J Cell Sci. 2008;121(Supplement_1):1–84.
2. DeWeerdt S. Genetics: written in blood. Nature. 2013;498:S4–6.
3. Drucker BJ, et al. Efficacy and safety of his specific inhibitor of BCR–ABL tyrosine kinase in chronic myeloid leukemia. N Engl J Med. 2001;344(14):1084.
4. Bargmann CI, Hung MC, Weinberg RA. The neu oncogene encodes an epidermal growth factor receptor-related protein. Nature. 1986;319(6050):226–30.
5. Slamon DJ, Leyland-Jones B, Shak S, Fuchs H, Paton V, Bajamonde A, Fleming T, Eiermann W, Wolter J, Pegram M, Baselga J, Norton L. Use of chemotherapy plus a monoclonal antibody against HER2 for metastatic breast cancer that overexpresses HER2. N Engl J Med. 2001;344(11):783–92.
6. Hall JM, Lee MK, Newman B, Morrow JE, Anderson LA, Huey B, King MC. Linkage of early-onset familial breast cancer to chromosome 17q21. Science. 1990;250(4988):1684–9.
7. Ledermann J, Harter P, Gourley C, Friedlander M, Vergote I, Rustin G, Scott C, Meier W, Shapira-Frommer R, Safra T, Matei D, Macpherson E, Watkins C, Carmichael J, Matulonis UN. Olaparib maintenance therapy in platinum-sensitive relapsed ovarian cancer. Engl J Med. 2012;366(15):1382–92.
8. Fukuoka M, Yano S, Giaccone G, Tamura T, Nakagawa K, Douillard J-Y, Nishiwaki Y, Vansteenkiste J, Kudoh S, Rischin D, Eek R, Horai T, Noda K, Takata I, Smit E, Averbuch S, Macleod A, Feyereislova A, Dong R-P, Baselga J. Multi-institutional randomized phase II trial of gefitinib for previously treated patients with advanced non-small-cell lung cancer. J Clin Oncol. 2003;21:2237–46.
9. The Cancer Genome Atlas Research Network, Weinstein J, Collisson E, et al. The Cancer Genome Atlas pan-cancer analysis project. Nat Genet. 2013;45:1113–20. https://doi.org/10.1038/ng.2764.
10. Janiaud P, Serghiou S, Ioannidis JPA. New clinical trial designs in the era of precision medicine: an overview of definitions, strengths, weaknesses, and current use in oncology. Cancer Treat Rev. 2019;73:20–30. https://doi.org/10.1016/j.ctrv.2018.12.003. Epub 2018 Dec 11.
11. Zehir A, Benayed R, Shah RH, Syed A, Middha S, Kim HR, Srinivasan P, Gao J, Chakravarty D, Devlin SM, Hellmann MD, Barron DA, Schram AM, Hameed M, Dogan S, Ross DS, Hechtman JF, DeLair DF, Yao J, Mandelker DL, Cheng DT, Chandramohan R, Mohanty AS, Ptashkin RN, Jayakumaran G, Prasad M, Syed MH, Rema AB, Liu ZY, Nafa K, Borsu L, Sadowska J, Casanova J, Bacares R, Kiecka IJ, Razumova A, Son JB, Stewart L, Baldi T, Mullaney KA, Al-Ahmadie H, Vakiani E, Abeshouse AA, Penson AV, Jonsson P, Camacho N, Chang MT, Won HH, Gross BE, Kundra R, Heins ZJ, Chen HW, Phillips S, Zhang H, Wang J, Ochoa A, Wills J, Eubank M, Thomas SB, Gardos SM, Reales DN, Galle J, Durany R, Cambria R, Abida W, Cercek A, Feldman DR, Gounder MM, Hakimi AA, Harding JJ, Iyer G, Janjigian YY, Jordan EJ, Kelly CM, Lowery MA, Morris LGT, Omuro AM, Raj N, Razavi P, Shoushtari AN, Shukla N, Soumerai TE, Varghese AM, Yaeger R, Coleman J, Bochner B, Riely GJ, Saltz LB, Scher HI, Sabbatini PJ, Robson ME, Klimstra DS, Taylor BS, Baselga J, Schultz N, Hyman DM, Arcila ME, Solit DB, Ladanyi M, Berger MF. Mutational landscape of metastatic cancer revealed from prospective clinical sequencing of 10,000 patients [Erratum in: Nat Med. 2017 Aug 4;23(8):1004. PMID: 28481359; PMCID: PMC5461196]. Nat Med. 2017;23(6):703–13. https://doi.org/10.1038/nm.4333. Epub 2017 May 8.
12. https://www.fda.gov/downloads/medicaldevices/productsandmedicalprocedures/invitrodiagnostics/ucm584603.pdf.
13. Sochacka-Ćwikła A, Mączyński M, Regiec A. FDA-approved drugs for hematological malignancies-the last decade review. Cancers (Basel). 2021;14(1):87. https://doi.org/10.3390/cancers14010087.
14. Cristina Mendonça Nogueira T, Nora V, de Souza M. New FDA oncology small molecule drugs approvals in 2020: mechanism of action and clinical applications. Bioorg Med Chem. 2021;46:116340. https://doi.org/10.1016/j.bmc.2021.116340. Epub 2021 Aug 9.

15. Al-Shamsi H, Darr H, Abu-Gheida I, Ansari J, McManus MC, Jaafar H, Tirmazy SH, Elkhoury M, Azribi F, Jelovac D, Doufan TA, Labban AR, Basha AA, Samir A, Maarraoui A, Al Dameh A, Al-Awadhi A, Al Haj Ali B, Aboud B, Elshorbagy D, Trad D, Abdul Jabbar D, Hamza D, Ashtar E, Dawoud E, Aleassa EM, Khan F, Iqbal F, Abdellatif H, Afrit M, Masri MH, Abuhaleeqa M, Alfalasi M, Omara M, Diab M, Latif MF, Oner M, Dreier N, Almarzouqi O, Singarachari RA, Bendardaf R, Alrawi S, Aldeen SS, Rana S, Talima S, Abdelgawad T, Ahluwalia A, Alkasab T, Madi T, Alkhouri T, Ul Haq U, Alabed YZ, Azzam M, Ali Z, Abbas MA, Razek AA, Al-Khatib F. The state of cancer care in the United Arab Emirates in 2020: challenges and recommendations, a report by the United Arab Emirates oncology task force. Gulf J Oncol. 2020;1(32):71–87.
16. https://dha.gov.ae/en/facilities/speciality-centers/16.
17. https://www.sksh.ae/sksh-achieves-a-breakthrough-in-genetic-testing-for-solid-tumors/.
18. AlSafar HS, Al-Ali M, Elbait GD, et al. Introducing the first whole genomes of nationals from the United Arab Emirates. Sci Rep. 2019;9:14725. https://doi.org/10.1038/s41598-019-50876-9.
19. Dawood S, Khoury M, Khan F, Al Khatib F, Bello M, Hamadi A, Kazim H, Dhar A, Nasioulas G, Khoury R. Ann Oncol. 2018;29(suppl_9):ix113–20. https://doi.org/10.1093/annonc/mdy441.
20. Al-Shamsi H, Abu-Gheida I, Abdulsmad SA, AlAwadhi A, Alrawi S, Musallam KM, Arun B, Ibrahim NK. Molecular spectra and frequency patterns of somatic mutations in Arab women with breast cancer. Oncologist. 2021;26(11):e2086–98. https://doi.org/10.1002/onco.13916. Epub 2021 Aug 14.
21. Al Shamsi H, Jones J, Fahmawi Y, Dahbour I, Tabash I, Abdel-Wahab R, Abousamra AO, Shaw KR, Xiao L, Hassan MM, Kipp BR, Kopetz S, Soliman AS, McWilliams RR, Wolff RA. Molecular spectrum of KRAS, NRAS, BRAF, PIK3CA, TP53 and APC somatic gene mutations in Arab patients with colorectal cancer: determination of frequency and distribution pattern. J Gastrointest Oncol. 2016;7(6):882–902. https://doi.org/10.21037/jgo.2016.11.02.
22. Khan FA, El Khoury M, Safar T, Tareen M, Sayed K. J Clin Oncol. 37(15_suppl):e14730. https://doi.org/10.1200/JCO.2019.37.15_suppl.e14730.
23. Mateo J, Chakravarty D, Dienstmann R, Jezdic S, Gonzalez-Perez A, Lopez-Bigas N, Ng CKY, Bedard PL, Tortora G, Douillard JY, Van Allen EM, Schultz N, Swanton C, André F, Pusztai L. A framework to rank genomic alterations as targets for cancer precision medicine: the ESMO Scale for Clinical Actionability of molecular targets (ESCAT). Ann Oncol. 2018;29(9):1895–902. https://doi.org/10.1093/annonc/mdy263.
24. https://www.doh.gov.ae/-/media/1C68EF54799945839198F0FF27C04016.ashx.
25. Specchia ML, Frisicale EM, Carini E, Di Pilla A, Cappa D, Barbara A, Ricciardi W, Damiani G. The impact of tumor board on cancer care: evidence from an umbrella review. BMC Health Serv Res. 2020;20(1):73.
26. Khan F, Maqbool S, El Khoury M, Habermann TM, Dufan T, Siddiqi AH, Han A, Sayed K, Ahmed Y, Kantawala K, Hamza M, Sonal A, Ahmed S, Yang C, Abdelrahman D, AHD-MDT members, Mayo eTumor Board members; American Hospital Dubai, Dubai, United Arab Emirates; American Hospital, Dubai, United Arab Emirates; Mayo Clinic, Rochester, MN; American hospital, Dubai, United Arab Emirates. Impact of international collaboration utilizing E-consult and E-tumor board in the multidisciplinary management of cancer patients: a study from Mayo Clinic Care Network (MCCN) members. J Clin Oncol. 2019;37:abstr e18261.
27. Habermann TM, Khurana A, Lentz R, Schmitz JJ, Bormann AG, Young JR, Hunt CH, Christofferson SN, Nowakowski GS, McCullough KB, Horna P, Wood AJ, Macon W, Kurtin PJ, Lester SC, Stafford SL, Chamarthy U, Khan F, Ansell S, King RL. Analysis and impact of a multidisciplinary lymphoma virtual tumor board. Leuk Lymphoma. 2020;61(14):3351–9.
28. https://www.doh.gov.ae/en/news/doh-launches-the-first-personalised-precision-medicine-programme-for-oncology-in-the-region.
29. Oosterhuis JW, Coebergh JW, van Veen EB. Tumour banks: well-guarded treasures in the interest of patients. Nat Rev Cancer. 2003;3(1):73–7. https://doi.org/10.1038/nrc973.
30. Qualman S, France M, Grizzle W, et al. Establishing a tumour bank: banking, informatics and ethics. Br J Cancer. 2004;90:1115–9. https://doi.org/10.1038/sj.bjc.6601678.

Dr. Faraz Khan is US board-certified in internal medicine, hematology, and medical oncology. He is a consultant medical oncologist and hematologist, head of the hematology program at the American Hospital Dubai, and an assistant professor at Sharjah Medical University. He is currently also serving as secretary general of the Emirates Society of Haematology.

Dr. Maroun El Khoury is US board certified in internal medicine, hematology, and medical oncology. He is a consultant medical oncologist and hematologist, director of the cancer program at the American Hospital Dubai, and an assistant professor at Sharjah Medical University. He is also a member of the Emirates Oncology Society Research Council.

Dr. Maroun El Khoury is US board-certified in internal medicine, hematology, and medical oncology. He is a consultant medical oncologist and hematologist, director of the cancer program at the American Hospital Dubai, and an assistant professor at Sharjah Medical University. He is also a member of the Emirates Oncology Society Research Council.

Genetic Testing for Cancer Risk in the UAE

15

15

Rita A. Sakr ⓘ and Hassan Ghazal ⓘ

15.1 Introduction

Hereditary cancers account for around 10% of all cancers. Of all cancer patients, 15–20% are estimated to have a positive family history, and mutations in highly penetrant genes were identified in 20% of high-risk families [1]. For example, hereditary breast and ovarian cancers originate from specific gene mutations called *BRCA1* and/or *BRCA2* (Table 15.1). Multiple studies into hereditary cancer genes were performed in Asian and European populations, but very few were completed in Arab countries [2]. Furthermore, consanguineous marriages

Table 15.1 Clinical features of hereditary breast cancer genes

Gene	Syndrome	Breast cancer risk	Other
BRCA1 BRCA2	Hereditary breast and ovarian cancers	BRCA1 60% by age 70 BRCA2 55% by age 70	Ovarian cancer Pancreatic cancer Prostate cancer
TP53	Li-Fraumeni syndrome	High	Sarcoma Brain tumors Adrenocortical carcinomas
PTEN	Cowden syndrome	77% by age 70	Thyroid cancer Endometrial cancer Gastrointestinal cancer Renal cell cancer
CDH1	Hereditary diffuse gastric cancer	39% by age 80	Gastric cancer

R. A. Sakr (✉) · H. Ghazal
Emirates Oncology Society, Dubai, United Arab Emirates

King's College Hospital London, Dubai, United Arab Emirates

© The Author(s) 2024
H. O. Al-Shamsi (ed.), *Cancer Care in the United Arab Emirates*,
https://doi.org/10.1007/978-981-99-6794-0_15

can increase the risk of inheriting a gene mutation, with, for example, 21–28% of all Emirati marriages happening between cousins, according to current research by the Centre for Arab Genomic Studies based in Dubai [3]. Therefore, it can be of great value to better understand targetable familial cancer genes in the United Arab Emirates (UAE) and the Arab region.

15.2 Oncogenesis

Cancer can develop after multiple damages to cellular DNA. By consequence, the cell will not follow the normal process for cell proliferation, and it will engage in dysregulated cell division. In oncogenesis, the tumor will initiate with damage to DNA, which will cause a mutation in a critical gene. Following that, the tumor cell is exposed to stimuli that will promote clonal proliferation and genetic instability. New mutations will provide a selective advantage for tumor cells to grow. Somatic mutations can happen in cells throughout their lifetime, with a number of them being cancer-causing mutations, whereas in hereditary cases, cancer occurs due to a constitutional mutation in a gene [4, 5] (Figs. 15.1 and 15.2).

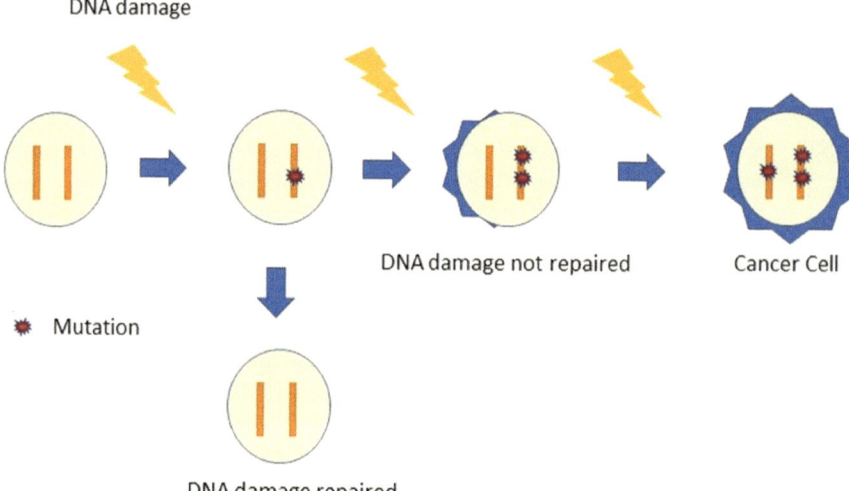

Fig. 15.1 The cell accumulates somatic mutations. DNA damage can be repaired by a cell. In cases of deficient DNA damage repair, the mutations will accumulate and the cell will become malignant

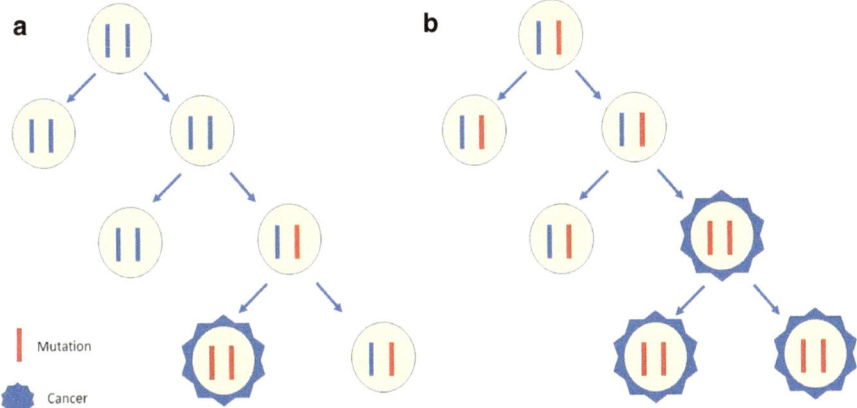

Fig. 15.2 (**a**) When cancer is non-hereditary, the somatic mutations will occur later, and the cancer is happening at a later age. (**b**) When cancer is hereditary, the mutations are already present in the cells. The tumor will initiate with only one somatic mutation, and thus, the cancer is happening at an earlier age

15.3   Genetic Counseling

In cases of a risk of hereditary cancer predisposition syndrome, we should have the risk assessed through a three-generation family. It should be done as part of the regular doctor's family history assessment, followed by the assessment with a clinical genetic specialist. The guidelines for whom to refer to clinical genetic services in cases of suspected hereditary cancer are publicly available (National Institute for Health and Care Excellence, National Comprehensive Cancer Network) [6, 7]. Many prediction scoring tools and many computerized models were also identified as being helpful in the clinical setting (e.g., the MSS scoring system, Gail model, Claus model, Tyrer-Cuzick model, or IBIS model in breast cancer) [8, 9] in order to identify the necessity for a genetic test. However, they tend to be less accessible because of the need to collect information, which can be time-consuming and less accurate [10]. Genetic testing can be either diagnostic or predictive. The diagnostic testing is performed on an individual with cancer to help guide further treatment options. The predictive testing is performed on any individual with a high familial risk of carrying a gene mutation. As genetic testing technology expands, it is becoming easier to test a large panel of genes involved in a higher risk for cancer. In breast cancer, for example, it allows for the identification of patients who have a specific constitutional gene mutation, thus enabling more personalized management. As access to technology is widely increasing, the tests are likely to be provided by more specialties outside of clinical genetics with respect to the guidelines and recommendations.

15.4   BRCA1/2 Mutation

BRCA gene mutations can increase an individual's risk of developing cancers like breast cancer (up to 87% lifetime risk), ovarian cancer (45%), prostate cancer (20%), pancreatic cancer (7%), and male breast cancer (8% lifetime risk). There are two genes, *BRCA1* and *BRCA2*, that make proteins to help repair damaged DNA. Every one of us has two copies of each of these genes, called tumor suppressor genes, and any defect in them will usually lead to cancer development at a younger age. In contrast to acquired or somatic mutations, germline mutations are inherited and carried by one of the parents at a 50% risk and have been present since birth.

So, in breast cancer, about 13% of women will develop it, and that risk goes up to 55–72% in *BRCA1* carriers and 45–69% in *BRCA2* carriers. Furthermore, contralateral breast cancer may develop in 25% of women after 10 years and up to 40–50% by year 20. Ovarian cancer risk is up to 45% in *BRCA1* carriers and 12–18% in *BRCA2* carriers. Prostate cancer may occur in *BRCA2* carriers mostly, while pancreas cancer can occur in both *BRCA1* and *BRCA2* carriers.

Other mutations that can increase one's risk for cancer are: *PALB2*, *ATM*, the *CHECK2* mutation, and *TP53*.

Ways to reduce this high lifetime risk include:

– Surgery, like risk-reducing mastectomies and bilateral salpingo-oophorectomies.
– Chemoprevention by taking drugs like tamoxifen can reduce breast cancer incidence by nearly 50%.
– Intense screening with breast MRI, ultrasound of both breasts and ovaries, and still mammograms.

In the UAE, the most common gene causing breast cancer is *BRCA2* (19%), followed by *BRCA1* (17%), as per a Tawam Hospital study [11, 12], but it can vary from one Arab country to the next: in Lebanon, up to 6%; in Oman, about 7%; and probably a little higher in the UAE [13–16].

15.5   Lynch Syndrome

Another important inherited mutation is along the Lynch syndrome family of genes, like *MLH1*, *MSH2*, *MSH6*, *PMS2*, and *EPCAM*. These could be due to a germline mutation, an inherited form, or a somatically acquired one. Patients with those mutations are more at risk of developing colorectal cancer, endometrial cancer, breast cancer, stomach cancer, and even skin cancer and urinary malignancies at a young age.

So, taking family history into account is critical in this regard. Among colon cancer patients, only about 4–5% can be attributed to Lynch syndrome, and some strategies to decrease that risk include:

– Surgical intervention, to remove the whole colon or uterus, or to do frequent surveillance like a colonoscopy annually, as well as endometrial surveillance with ultrasound and biopsies.

Testing for this can be done in two ways:

1. Immunohistochemistry, or IHC, checks for the above proteins, like *MLH1*, *MSH2*, and *MSH6*, to see if a person is proficient in or deficient in MMR, or mismatch repair.
2. MSI, or microsatellite instability DNA testing, so if the result is MSI-H high, that means that person has Lynch syndrome [17].

15.6   Available Genetic Testing in the UAE

Technologies for gene sequencing have advanced dramatically. The diagnostic testing is a complete gene screen performed on a cancer patient. It can help guide future therapeutic options and identify family members who are at risk of developing cancer. The predictive testing is typically performed on an individual without cancer but with a high family risk of carrying a gene mutation that was previously identified in a family member. It can help identify high-risk people before they get cancer, therefore allowing proper surveillance and prophylactic treatment before cancer can occur. However, cancer will not happen to every individual who is a mutation carrier. Therefore, pre-test counseling is recommended in order to identify the candidates for genetic testing and to reduce unnecessary adverse psychological effects on the candidate and his family. Gene testing was highly revolutionized by the introduction of next-generation sequencing (NGS), but still, good, detailed clinical information is highly required for a good interpretation of the results and the proper advice of the candidate [18]. The more genetic testing technology expands, the easier it becomes for clinicians to test much larger panels of genes [19]. As a consequence, this large gene panel testing will probably not only increase the detection rate of pathogenic mutations but also the detection rate of variants of uncertain significance (VUS). This detection of VUS can induce confusion among families and patients; therefore, the NCCN highlighted that proper evaluation for clinical suspicion of hereditary cancer susceptibility genes should be considered before the multigene testing (Table 15.2). In the UAE, a multitude of panels testing for mutations in cancer genes are available. They are employed to confirm large deletions or duplications in the number of genes and are thus able to report pathogenic variants (Table 15.3). By consequence, it allows the clinician to provide proper counseling, surveillance management, and prophylactic management (medical or surgical).

Table 15.2 Representation of guidelines for genetic testing in breast cancer

Personal history of breast cancer
Age at diagnosis <45
Age at diagnosis 46–50
– Another primary breast cancer
– One or more close relative(s) with breast, pancreatic, or prostate cancer
– Unknown family history
Age at diagnosis <60
– Triple-negative breast cancer
Diagnoses at any age
– One or more close relative(s)—breast cancer diagnosed at <50
– Two or more close relatives—breast cancer at any age
– One or more close relative(s)—invasive ovarian cancer
– Two or more relatives—pancreatic cancer
– Male relative—breast cancer
– Ashkenazi Jewish ancestry
Personal history of invasive ovarian cancer
Personal history of male breast cancer
Personal history of high-grade prostate cancer—one or more close relative(s) with breast, ovarian, pancreatic, or prostate cancer

Table 15.3 List of mostly analyzed genes

APC	CTNNA1	NBN	SDHA
ATM	DICER1	NF1	SDHB
AXIN2	EPCAM	NTHL1	SDHC
BARD1	GREM1	PALB2	SDHD
BMPR1A	HOXB13	PDGFRA	SMAD4
BRCA1	KIT	PMS2	SMARCA4
BRCA2	MEN1	POLD1	STK11
BRIP1	MLH1	POLE	TP53
CDH1	MSH2	PTEN	TSC1
CDK4	MSH3	RAD50	TSC2
CDKN2A	MSH6	RAD51C	VHL
CHEK2	MUTYH	RAD51D	

15.7 Challenges in the UAE

So far, when required, genetic testing has been performed by oncologists on patients with cancer. Extending genetic testing to individuals with a strong family history of cancer is still limited, even if they do fall into the category of recommended genetic testing as per the international guidelines. Challenging reasons are multiple. Doctors need to be encouraged to ask their patients about their cancer family history, and patients need to be encouraged to mention their cancer family history to the doctor. Thus, individuals with a family history of cancer can be identified and referred for genetic clinical counseling, if available, or to the specialized doctor in the team who is able to deliver proper genetic counseling. Another significant challenge is the lack of insurance coverage for genetic testing, even for people who have a strong family history of cancer or those with a family member already carrying the cancerogenic mutation. Many of whom are at high risk for cancer and are potential carriers of carcinogenic mutations will find themselves postponing the test that could allow them to decide on prophylactic measures to avoid cancer.

15.8 Conclusion

The care of familial breast cancer was revolutionized in the 1990s by the genes *BRCA1* and *BRCA2*. However, the major challenge is how to translate the advances in our understanding of genetic susceptibility into improving patient outcomes in cancer care. With increasing access to technology, the threshold for cancer susceptibility testing falls. As a consequence, it allows for the identification of patients with a specific gene mutation earlier, thus enabling personalized management. The actual model of delivered genetic testing will surely need to follow changing needs, and clinical genetics might not be the only specialty providing the genetic testing. The development of those services would certainly require close communication between all involved clinicians, who need to appreciate the challenges in the interpretation of genetic test results because of the serious potential psychosocial consequences for both individuals and families.

Conflict of Interest The authors have no conflict of interest to declare.

References

1. Shiovitz S, Korde LA. Genetics of breast cancer: a topic in evolution. Ann Oncol. 2015;26(7):1291–9.
2. Elobaid Y, Aw TC, Lim JNW, et al. Breast cancer presentation delays among Arab and national women in the UAE: a qualitative study. SSM Popul Health. 2016;2:155–63.
3. Catalogue for Transmission Genetics in Arabs. Centre for Arab Genomics Studies, Dubai, UAE. http://www.cags.org.ae/.
4. Hanahan D, Weinberg RA. Hallmarks of cancer: the next generation. Cell. 2011;144:646–74.
5. Hyndman IJ. Review: the contribution of both nature and nurture to carcinogenesis and progression in solid tumours. Cancer Microenviron. 2016;9:63–9.
6. National Institute for Health and Care Excellence. Familial breast cancer and related risks and people with a family history of breast cancer. 2018.
7. Hereditary cancer risk assessment and referral guidelines for clinicians. www.ubqo.com/cancergenetics. Accessed 15 Oct 2016.
8. Gail M, Brinton L, Byar D, et al. Projecting individualized probabilities of developing breast cancer for white females who are being examined annually. J Natl Cancer Inst. 1989;81:1879–86.
9. Claus EB, Risch N, Thompson WD. The calculation of breast cancer risk for women with a first degree family history of ovarian cancer. Breast Cancer Res Treat. 1993;28:115–20.
10. Amir E, Freedman OC, Seruga B, Evans G. Assessing women at high risk of breast cancer: a review of risk assessment models. J Natl Cancer Inst. 2010;102:680–91.

11. Altino A, Abdel-Aziz A, Al Ameri M. Genetic predisposition of breast cancer in the United Arab Emirates. Arch Cancer Biol Ther. 2021;2(2):35–6.
12. Altinoz A, Al Ameri M, Quresi W, et al. Clinicopathological characteristics of gene-positive breast cancer in the United Arab Emirates. Breast. 2020;53:119–24.
13. El Saghir NS, Zgheib NK, Assi HA, et al. BRCA1 and BRCA2 mutations in ethnic Lebanese Arab women with high hereditary risk breast cancer. Oncologist. 2015;20(4):357–64.
14. Rahman S, Zayed H. Breast cancer in the GCC countries: a focus on BRCA1/2 and non-BRCA1/2 gens. Gene. 2018;668:73–6.
15. Azribi F, Abdou E, Dawoud E, et al. Prevalence of BRCA1 and BRCA2 pathogenic sequence variants in ovarian cancer patients in the Gulf region: the PREDICT study. BMC Cancer. 2021;21:1350.
16. Dawood SS, Apessos A, Elkhoury M, et al. Analysis of hereditary cancer syndromes in patients from The United Arab Emirates. J Clin Oncol. 2018;36:e13509.
17. Lynch Syndrome - GeneReviews® - NCBI Bookshelf. https://www.ncbi.nlm.nih.gov/books/NBK1211/.
18. Behjati S, Tarpey PS. What is next generation sequencing? Arch Dis Child Educ Pract Ed. 2013;98:236–8.
19. Easton DF, Eeles RA. Genome-wide association studies in cancer. Hum Mol Genet. 2008;17:109–15.

Dr. Rita A. Sakr (MD, PhD, associate professor) is a French/European board-certified, American fellow, consultant breast oncoplastic surgeon, and obstetrician-gynecologist with over 20 years of surgical, clinical, academic, and research experience in the field. After training in both France and the USA, Dr. Rita Sakr was appointed consultant breast oncoplastic surgeon and obstetrician-gynecologist in several institutions, including the Institut Curie, Paris, the Memorial Sloan Kettering Cancer Centre, New York, and Assistance Publique—Hôpitaux de Paris. She also held the position of associate professor and co-head of the Breast and Gynecology Care Unit for high-risk patients at Pierre & Marie Curie/Sorbonne University, France. She later relocated to Dubai as a consultant breast surgeon and obstetrician-gynecologist at the Dr. Sulaiman Al Habib Hospital and then the American Hospital.

Dr. Rita Sakr is a member of various clinical societies such as the American Society of Clinical Oncology, the American Association for Cancer Research, the Lebanese Society of Obstetrics and Gynecology, the Collège National des Gynécologues et Obstétriciens Français, Société Française de Sénologie et Pathologie Mammaire, the New York Academy of Sciences, the European Society of Surgical Oncology, the Emirates Oncology Society, and the Europa Donna. Dr. Rita Sakr has more than 100 publications and abstracts in peer-reviewed journals and international conferences, and has been awarded many international recognitions, such as the American Association for Cancer Research Award and the San Antonio Breast Cancer Symposium Award.

In addition to her wide surgical, clinical, and academic experience, Dr. Rita Sakr has a passion for patient advocacy and has been actively involved in multiple associations for breast cancer survivorship in Europe, the USA, and the UAE.

Dr. Hassan Ghazal is an American board-certified consultant medical oncologist and a consultant clinical hematologist, offering more than three decades of clinical experience.

Dr. Hassan Ghazal, after completing his basic medical education at the American University of Beirut, moved to the USA for his residency program in internal medicine at the Sinai Hospital of Baltimore, MD. In 1992, after completing the residency program, Dr. Hassan Ghazal joined the George Washington University Medical Center for his fellowship in medical oncology and completed the same in 1995. He stayed at George Washington University Medical Center for another 2 years, earning a senior fellowship in bone marrow transplantation in 1997. He got his US board certification in internal medicine, medical oncology, and hematology in 1992, 1995, and 1996, respectively.

Except for his association with CMC for the preceding 3 years, all his clinical practices were in the USA in various capacities, including director, Kentucky Cancer Clinic, and staff oncologist/hematologist at Hazard Appalachian Regional Medical Center.

Dr. Hassan Ghazal is a member of various clinical societies, such as the American College of Physicians, the American Society of Haematology, the American Society of Clinical Oncology, and the European Society of Medical Oncology. Dr. Hassan Ghazal has many publications and abstracts to his credit in peer-reviewed journals, and there are many international recognitions and accolades to his name.

He has been involved in numerous clinical trials—nearly 100 in the USA—serving as the principal investigator for over 25 years and enrolling hundreds of patients in these trials.

Fertility Preservation and Oncofertility in the UAE

16

Nahla Kazim

16.1 Introduction

Cancer burden has risen worldwide by 26.3% in the past decade, concurrently with the rise in the number of new cancer cases that have increased globally from 18.7 million in 2010 to 23.6 million in 2019. Global age-standardized incidence remained at a similar rate of -1.1% (95% UI, -5.8% to 3.5%), while mortality rates decreased by -5.9% (95% UI, -11.0% to -0.9%) [1]. Higher numbers of cases and deaths occurred in the low to middle sociodemographic index (SDI) groups (Tables 16.1 and 16.2) [1].

Table 16.1 Cancer type distribution (United Arab Emirates (UAE) and global) in females [1, 2]

Rank in the UAE	Global rank, 2019	Female cancer in the UAE, 2021
1	1	Breast
2	21	Thyroid
3	3	Colorectal
4	13	Uterine
5	4	Cervical
6	24 (non-melanoma)	Skin
7	6	Ovarian
8	8	Leukemia
9	12	Non-Hodgkin lymphoma
10	2	Tracheal, bronchus, and lung (TBL)

Source: Ministry of Health and Prevention, Statistics and Research Center, National Disease Registry—UAE National Cancer Registry Report, 2021

N. Kazim (✉)
Bourn Hall Fertility Center, Mediclinic Hospital, Al Ain, United Arab Emirates

Department of Obstetrics and Gynecology, College of Medicine and Health Sciences, UAE University, Al Ain, United Arab Emirates

© The Author(s) 2024   245
H. O. Al-Shamsi (ed.), *Cancer Care in the United Arab Emirates*,
https://doi.org/10.1007/978-981-99-6794-0_16

Table 16.2 Cancer type distribution (UAE and global) in males [1, 2]

Rank in the UAE	Global rank, 2019	Male cancer in the UAE, 2021
1	3	Colorectal
2	6	Prostate
3	7	Leukemia
4	23	Thyroid
5	20 (non-melanoma)	Skin
6	1	Bronchus and lung
7	10	Non-Hodgkin lymphoma
8	11	Lip, oral cavity, and pharynx
9	14	Kidney and renal pelvis
10	12	Bladder

Source: Ministry of Health and Prevention, Statistics and Research Center, National Disease Registry—UAE National Cancer Registry Report, 2021

In 2021, the UAE National Cancer Registry (UAE-NCR) (MOHAP, 2021) reported 5830 newly diagnosed cancer cases, of which 96% were malignant and 4% were in situ cases. Comparatively, cancer was more common among women (55.1%) than men (44.9%), while the incidence of cancer increases with advancing age. There were 154 pediatric malignancies reported, with leukemia being the most common (42.9%). Notably, boys had a higher cancer frequency (55%) than girls (45%), and 27.9% of cases were seen in children aged 10–14 years, making this group unique because it represents children in a transitional phase between childhood and adulthood and may not fit perfectly into either category depending on pubertal status (Table 16.1) [2].

The differences in cancer burden worldwide may reflect several factors, including population demography, shifting population age structures, the infrastructure of the healthcare system, improved capacity for diagnosis and registration of cancer cases and deaths, and other influences modifying cancer risk such as metabolic, diet, lifestyle, environmental, and occupational exposures.

Modern anticancer treatments have resulted in higher survival rates (up to 90% in young women when diagnosed early), but at the expense of fertility damage due to unwanted side effects (Dolmans et al. 2019; Oktay et al. 2018) [3, 4]. Irreversible damage occurs when reproductive organs are subjected to aggressive chemotherapy and radiotherapy, resulting in a reduced or permanent loss of fertility. The likelihood of conception is reduced by 30–50% in female cancer survivors following radiotherapy or chemotherapy, with an apparent increase in the risk of obstetric and neonatal complications [3, 4]. Preconception counseling and appropriate obstetric monitoring are recommended for women intending to become pregnant after gonadotoxic treatments. Large reports based on cancer registry data, including the Scottish, Finnish, and North Carolina Central Cancer Registry, concluded that women who have had cancer have higher rates of postpartum hemorrhage, operative or assisted delivery, preterm labor, and a variety of other maternal and fetal complications if they conceive within a year of starting treatment. Similar adverse maternal and fetal outcomes are also found post-radiotherapy [5].

Cancer treatments, including chemotherapy using alkylating agents such as cyclophosphamide, ifosfamide, and busulfan, in addition to ionizing radiotherapy to the abdomen and pelvis or total body irradiation, are all gonadotoxic, causing dose-dependent impairment to the ovaries and uterus [3, 4]. Aside from the dose, the site, the fractionation of the chemotherapy and radiotherapy, and the patient's age at the start of treatment, all contribute to the degree of ovarian and uterine damage. Other benign medical conditions that include autoimmune diseases like systemic lupus erythematosus (SLE) and hematological diseases such as sickle cell anemia and thalassemia may require chemotherapy, radiotherapy, or a combination of both, with some cases requiring bone marrow transplantation. All such interventions may result in an early loss of fertility [6]. Severe ovarian damage can increase the likelihood of diminished ovarian reserve and premature ovarian failure due to the exhaustion of primordial follicles and oocytes. Similarly, severe uterine damage can lead to complications such as pelvic synechia, recurrent miscarriage, pregnancy loss, preterm delivery, and low birthweight due to uterine vasculature disruption. Chemotherapeutic agents can directly damage the DNA of the mature follicle, subsequently causing apoptosis and temporary amenorrhea. Ovarian function in terms of folliculogenesis and resumption of menstruation may occur post-completion of cancer treatment if the primordial follicle pool is unaffected. However, the resumption of menses does not provide evidence of a complete recovery of ovarian function or predict reproductive outcomes. While some metals are considered highly gonadotoxic, causing acute ovarian failure, regimens containing cisplatin, carboplatin, and adriamycin are considered to have a moderate risk for gonadotoxicity, while patients receiving bleomycin, actinomycin D, vincristine, methotrexate, or 5-fluorouracil are considered to have a low risk for gonadotoxicity [7].

Similar to chemotherapy, a high risk of gonadotoxicity is noted post-radiotherapy among women receiving pelvic or whole abdominal radiation doses >6 Gy or total body irradiation (prior to stem cell or bone marrow transplantation) and pelvic or whole abdominal radiation doses >10 Gy among post-pubertal girls. Craniospinal irradiation alone, without additional treatment, is not considered to pose a high gonadotoxicity risk.

Risk-based assessment of cancer patients and survivors depending on chemotherapy and radiation allows appropriate patient counseling based on the degree of gonadotoxicity as well as discussion of fertility preservation options, with early referral of patients planned to undergo a gonadotoxic, high-moderate-risk regimen for freezing options and fertility treatment that is most appropriate for their situation.

International guidelines recommend that physicians discuss with all patients of reproductive age their risk of infertility from cancer and its treatment and streamline fertility preservation options by referring them to a fertility specialist at the earliest possible time to discuss their interest in having children after cancer and the available options [5, 8, 9]. Improved quality of life in survivorship remains the collective goal through a multidisciplinary approach.

As recommended by the American Society of Clinical Oncology [8] and the European Society for Medical Oncology [5], sperm cryopreservation for males and embryo/oocyte cryopreservation for females are standard strategies for fertility

Table 16.3 Female fertility preservation options and current status in the UAE [5, 8, 9]

Characteristics	Option 1	Option 2	Option 3	Option 4
Methods	Embryo freezing following ovarian stimulation and egg retrieval by TVS/TRUS/TAS	Egg freezing following ovarian stimulation and egg retrieval by TVS/TRUS/TAS	Ovarian protection techniques GnRH analog, ovarian transposition, pelvic shielding, and chemotherapy and radiotherapy fractionated dose	Ovarian tissue cryopreservation and autotransplantation of frozen-thawed tissue
Status	Established	Established	Debatable	Established
Availability in the UAE	Yes	Yes	Yes	No reports
Contraindication	Single women and unmarried girls	Prepubertal girls	Oophoropexy and pelvic shielding not useful in chemotherapy Hormonal GnRH analog not useful in radiotherapy	Frozen-thawed ovarian tissue autotransplantation in ovarian cancer and malignancies with risk of metastasis into ovaries

preservation. Other strategies, such as pharmacological protection of the gonads and gonadal tissue cryopreservation, are either yet considered experimental or require further investigation and validation (Table 16.3).

16.2 Fertility Preservation Methods in Females

16.2.1 Ovarian Stimulation

Fertility preservation (FP) is an option to store oocytes or embryos prior to cancer therapy; both require controlled ovarian stimulation (COS), commonly using an antagonist protocol [10]. In brief, daily subcutaneous injections of gonadotropins such as recombinant follicular-stimulating hormone (rFSH) (follitropin alpha, Gonal-F, Merck, Germany; follitropin beta, Puregon, Organon, USA; corifollitropin alpha, Elonva, Organon, USA) or urinary human menopausal gonadotropin (HMG) (Ferring Pharmaceuticals, USA) start from the second to the fourth day of the menstrual cycle. When the leading follicle measures more than 12 mm, 0.25 mg of GnRH antagonist cetrorelix (Cetrotide, Merck, Germany) or ganirelix (Orgalutran, Organon, USA) is added daily to prevent premature LH surge. Finally, recombinant hCG (Ovitrelle, Merck, Germany) and/or a GnRH agonist (Ferring Pharmaceuticals, USA) are given subcutaneously for final oocyte maturation when three or more follicles reach 17 mm in diameter and after assessing the risk of ovarian hyperstimulation syndrome (OHSS). Oocyte retrieval under general anesthesia is performed 36 hours later as a day procedure. Routinely, egg collection is done under transvaginal scan guidance, but other routes such as transrectal, transabdominal, and laparoscopic recovery may be considered for unmarried females, adolescents, and children who have reached puberty. An oocyte is cryopreserved by rapid vitrification or fertilized with her husband's sperm to proceed with embryo cryopreservation.

Other non-conventional ovarian stimulation protocols, such as random start, dual stimulation (DUOSTIM), and luteal phase stimulation, have shown similar outcomes to the conventional cycles, potentially optimizing the number of oocytes retrieved within the shortest possible timeframe while providing tailored care management for cancer patients [11].

16.2.2 Special Considerations

The age of a patient undergoing fertility preservation strongly impacts the outcome of freezing in regard to the number of retrieved mature oocytes, the survival rate of oocytes during the thawing process, and the fertilization rate when injected with sperm, with higher chances of pregnancy, including a live birth rate per patient, seen among women aged <35 years. Outcomes of 137 women returning to use their vitrified oocytes for non-oncologic reasons showed cumulative live birth rates (CLBRs) of 15.4% when five oocytes were used, 40.8% when eight eggs were used, and 60.5% when 10 eggs were used among patients below <35 years of age. In women aged >35 years, lower CLBRs were noted, showing 5.1% (5 eggs), 19.9% (8 eggs), and 29.7% (10 eggs), respectively [12].

While several studies have confirmed an association between older age and lower retrieved oocyte yields and pregnancy chances, such data should be interpreted with caution because the majority of these studies are retrospective and report outcomes on healthy women undergoing elective oocyte freezing. Reduced live birth rates (LBR) were observed in women attempting pregnancy who had undergone FP for malignant indications versus women who had FP for benign indications [11, 12].

The aim of gonadotropin injections for fertility preservation is to produce multifollicular development in a short time; this may increase the risk of ovarian hyperstimulation syndrome (OHSS), particularly in young lean patients with high ovarian reserve. OHSS is an iatrogenic complication associated with ovarian gonadotropin stimulation that is followed by the hCG trigger given for final oocyte maturation. The etiopathogenesis of OHSS includes ovarian enlargement, secretion of vasoactive substances, ascites, hypovolemia resulting from an acute extravasation of fluid into the interstitial space, hemoconcentration, hypercoagulation, and electrolyte imbalances, all leading to life-threatening cascades. The incidence of moderate-to-severe OHSS is approximately 1–5% of cycles, and many strategies are used effectively to prevent or reduce OHSS incidence during the early stages, including using an antagonist protocol with a GnRH agonist for triggering ovulation instead of hCG [13].

Other less common complications of ovarian stimulation with oocyte retrieval include vaginal bleeding, pelvic infection or abscess, injury to pelvic structures, and ovarian torsion [14].

An increased frequency of venous thromboembolism (VTE) is noted with assisted reproduction, complicating 0.1% of cycles, although the exact mechanism remains unknown. High estrogen levels following ovarian stimulation may exhibit a procoagulant effect by increasing levels of von Willebrand factor, factor V, factor VIII, and fibrinogen while decreasing levels of the anticoagulants protein S and antithrombin. The clinical implication of these changes in healthy females is unclear since most variables remain within normal limits. Low-molecular-weight heparin (LMWH) prophylaxis may be considered during ovarian stimulation for patients with an increased risk of VTE, including thrombophilia, OHSS, obesity, prior VTE, or a strong family history of venous thrombosis [15].

To prevent venous thromboembolism and estrogenic symptoms in estrogen-sensitive cancers, an antagonist protocol with the addition of letrozole or tamoxifen may also be added during stimulation [16].

Ovarian gonadotrophic stimulation among females with possible steroid-sensitive tumors requires concurrent use of an aromatase inhibitor to minimize the abnormal iatrogenic rise in estradiol that may accompany COS, thereby reducing the potential risk of exacerbating malignant breast disease [17].

While available data in the literature is reassuring, a theoretical link to disease progression or a high risk of cancer recurrence exists due to a lack of publications on the long-term effect of COS on the prognosis of breast cancer [5, 9, 17].

A cohort analysis of 155 cycles evaluating the effect of different hormonal receptor profiles on stimulation cycle parameters among women with breast cancer found that patients with a triple-negative breast cancer profile (TNBC) had a significantly lower number of mature oocytes when compared to the ER + PR+ and non-TNBC groups [18]. Multivariate analysis using a threshold of 10 mature oocytes revealed a high cumulative live birth rate of 50% among 18–38 years of age. The study concluded that the TNBC subtype, with an average number of 7 mature oocytes collected, had a negative effect on fertility preservation outcomes. While such studies are useful in the fertility preservation decision-making process before deciding on stimulation protocol or gonadotropin dose for women with different breast cancer subtypes, further studies should explore other factors in cancer, such as BRCA mutations, that may hinder the ovarian response to stimulation, reducing the probability of having offspring in the future.

There is a lack of data on oocyte quality in women with cancer. A retrospective case-control study using univariate and multivariate analyses among 105 women with breast cancer and 189 healthy women in the control group, both undergoing controlled ovarian stimulation cycles for fertility preservation, found a fourfold increased risk of retrieving dysmorphic and poor-quality oocytes among breast cancer women with no effect on the ovarian reserve or its response to stimulation [19].

A retrospective, observational multicenter study comparing the success of IVF cycles after elective fertility preservation (EFP) for age-related fertility decline and oncofertility preservation (OFP) prior to cancer treatment showed lower live birth rates after oocyte vitrification among cancer patients. However, after controlling for age and ovarian stimulation (COS) regime, there was no statistically significant association between malignant disease and reproductive outcome. The limitation in statistical power to compare IVF outcomes was attributed to fewer women returning to use their oocytes in the Onco-FP group, and the lower implantation rate failed to prove the impact of cancer disease or the type of cancer treatment as a causative factor per se. While encouraging data is available for donor oocyte vitrification, however, the evidence cannot be applied to oocyte vitrification for infertile patients or women with pre-existing medical or oncological conditions, as different oocyte sources from different conditions may vary in their inherent qualities that may affect vitrification outcomes. Nevertheless, there is an ample amount of data in the literature confirming a significantly higher cumulative probability of live birth in patients <35 years of age versus >35 in the EFP, with improved outcomes when more oocytes were available for IVF [11, 20].

Pregnancy is considered safe in women who have survived breast cancer, independent of the estrogen receptor status of the tumor (strong ESHRE recommendation). Women may be advised to stop tamoxifen treatment due to the risk of abnormal fetal development and wait at least 3 months before attempting conception to allow an appropriate washout period from the drug [5].

Knowledge of the overall risks associated with ART cycles is important for cancer patients to avoid further delays in their treatment or adding to their disease's health burden. Gonadotropin cycles should be managed by a clinician with the requisite training and experience and, in the UAE, by licensed reproductive medicine specialists or consultants.

16.3 Embryo Freezing

Embryo cryopreservation is an effective fertility preservation technique that is the first choice for married women whose mature oocytes can be collected after 10–14 days of ovarian stimulation and injected with their husband's sperm in an embryology laboratory. This technique is not appropriate for unmarried females or children who have not reached puberty, according to the UAE's fertility legislation. The proportion of cryopreserved embryo transfer cycles compared with fresh cycles has grown worldwide. In Europe, it was estimated that cryopreserved cycles contributed to 32% of the transfers in 2011, while around 50% of all ART transfers are now FET. Current UAE fertility legislation that allows embryo and egg cryopreservation has resulted in an increase in the proportion of cryopreserved cycles performed, exceeding 80% of ART transfers.

Data available from cohort studies, large RCTs, and meta-analysis show that the newer rapid vitrification and warming technique is superior to the slower freezing and thawing methods that were used earlier in terms of improved clinical outcomes as well as better cryosurvival rates for oocytes, cleavage-stage embryos, and blastocysts. The longest frozen embryos resulting in a live birth on record were twins born in October 2022 in the USA from embryos frozen 30 years ago. Generally, it is difficult to track pregnancies resulting from older frozen embryos, as the US Centers for Disease Control and Prevention (also the Department of Health in the UAE) track data and success rates around reproductive technologies but do not track how long the embryos have been frozen.

Embryo cryopreservation showed excellent success rates and emerging long-term data supporting the safety of the procedure. The post-thaw survival rates of embryos are up to 90%, the implantation rate is 80%, and the cumulative pregnancy rates are over 50% [21].

The majority of fertility centers in the UAE have adopted preimplantation genetic testing for aneuploidies (PGT-A) in their laboratories, often using ICSI as the fertilization method of choice. PGT-A provides some confidence before embryo transfer for choosing the best embryos that are chromosomally normal with high implantation potential but also reduces the time to achieve pregnancy [22].

Women at risk for or confirmed carriers of hereditary cancers can use both PGT-A and PGT-M, allowing identification of embryos that are both euploid and unaffected/non-pathogenic carriers, thereby reducing the risk of transmission in offspring. However, proper patient counseling is warranted, as this increases the possibility of having fewer euploid and non-affected embryos (or none) that are available for transfer than expected [22, 23].

Although embryos and oocyte vitrification procedures are well-established worldwide, as well as in the UAE, long-term follow-up studies of children are mandated. Data in the literature is conflicting, with some suggesting that pregnancies obtained from a cryopreserved oocyte and/or embryo transfer are associated with increased perinatal and obstetrical risks, while larger systematic reviews and cohort studies mostly show reassuring results and lower obstetric and perinatal complications such as antepartum hemorrhage, preterm delivery, low birthweight, and perinatal mortality with frozen embryo transfer, irrespective of their cleavage stage, compared to fresh embryo transfer [24, 25].

Once again, emphasis is placed on the importance of careful counseling of women with cancer desiring offspring in the future, and case-by-case selection of suitable candidates is recommended. Women should be informed of accurate, center-specific ART performance indicators and live birth rates. Women with a partner may be offered a combination of options for embryo and oocyte cryopreservation [5].

16.4 Oocyte Freezing

Oocyte freezing is no longer considered an experimental method by the American Society for Reproductive Medicine and has merged as a successful technique among unmarried females, provided they attained menarche, in countries where embryo freezing is legally restricted or when sperm is unavailable on the day of oocyte retrieval [5, 8, 9].

Post-pubertal patients can delay chemotherapy by 2 weeks, allowing the retrieval of mature oocytes following ovarian stimulation and eventually being vitrified.

Later on, the oocytes may be thawed, injected with the woman's husband's sperm, forming embryos, and then transferred to the uterine cavity for pregnancy.

In addition to infertility and medical and oncological indications, the trend in the UAE in recent years has been elective fertility preservation for women (married and single) who are conscious of the decline in oocyte quality and quantity with advancing maternal age with the intention of postponing pregnancy for a later age.

The combined technique may also be applied, involving ovarian tissue cryopreservation, IVM, and mature oocyte vitrification following controlled ovarian stimulation. This combined technique theoretically yields more than a 50% chance of achieving a live birth [26, 27].

Patients pursuing oocyte cryopreservation need to be informed about the duration of oocyte cryopreservation as per the law and counseled regarding their likelihood of live birth with autologous thawed oocytes after cancer treatment and if the duration of freezing has any impact on the outcome. A study by Whiteley et al. of 530 IVF cycles using autologous vitrified or thawed oocytes from 2010 to 2020 found no impact of the duration of oocyte vitrification on the live birth rate following fresh or frozen embryo transfer [28].

Most studies have shown that the number of live births obtained from oocyte cryopreservation decreases with advancing age due to low ovarian reserve and poor-quality eggs, with a resultantly high rate of aneuploidy. Doyle et al. [29] estimated that to achieve a 70% chance of one live birth in their cohort, i.e., for women aged 30–34 years, 14 oocytes would need to be frozen, 35–37-year-olds would need 15 oocytes, and 38–40-year-olds would need 26 mature oocytes in their cohort of 128 autologous thawed or warmed treatment cycles. Cobo et al. demonstrated cumulative live birth rates of 43% and 70% when 10 and 15 oocytes were vitrified, respectively, in women aged <35 years, confirming that every 15 cryopreserved oocytes can result in a live birth [11]. Women aged 36–40 years would need to freeze 16–25 oocytes to increase their chances of having one live birth, whereas women aged ≥41 years would need to freeze more than 40 eggs. It is extremely challenging to have similar oocyte yields in older women and in cancer patients due to multiple reasons, including poor ovarian reserve and a time constraint before starting gonadotoxic treatment. Women need to be informed that success rates after cryopreservation of oocytes at the time of a cancer diagnosis may be lower than in women without cancer [5].

16.5 IVM

This technique involves harvesting immature oocytes from unstimulated or minimally stimulated ovaries. While oocytes at the diplotene stage of prophase I, or germinal vesicle (GV) stage, have been shown to survive cryopreservation better than those frozen at the mature stage, a lower blastulation rate with lower implantation and pregnancy rates has been reported when compared to conventional IVF due to suboptimal nuclear and cytoplasmic maturation. IVM gained popularity as an alternative treatment protocol in certain cases, such as women who are at risk of

OHSS or women who have a limited time to begin fertility preservation prior to gonadotoxic cancer treatment. It may also reduce patient burden due to shorter stimulation cycles, fewer injections, and associated reduced drug and monitoring costs.

According to ASRM, this technology is no longer considered experimental, and the procedure should always be performed by experts with specific training and accompanied by appropriate counseling about expected outcomes and informed consent [9].

So far, a relatively small number of children are born worldwide with IVM, with no available long-term studies assessing the safety of IVM with regard to fetal malformations and developmental outcomes [30].

In the UAE, any attempted in vitro gamete maturation from immature sperm or eggs followed by cryopreservation has only been reported in the internal communication of the embryology laboratories of some of the private clinics or is used as a marketing strategy on their websites with no published data on its efficiency or success. An observational study was performed at ART Fertility Clinics, Abu Dhabi, UAE, between January 2019 and June 2021, wherein a total of 5454 cumulus-oocyte complexes (COC) were retrieved from 469 ovarian stimulation cycles, showing no difference in the blastocyst euploidy rates in embryos resulting from mature oocytes at the time of retrieval compared to immature eggs undergoing IVM. The humble data identified a group of patients' populations that may benefit from rescue IVM within a routine ART scenario. Patients with ≤59% mature oocytes at retrieval and/or anti-Mullerian hormone (AMH) >2.52 ng/mL have increased chances of obtaining an euploid embryo from immature eggs progressing to matured oocytes without adhering to unnecessary costs and workload [31].

16.6   Ovarian Tissue Cryopreservation (OTC) and Autotransplantation (OTT)

Due to the increasing success rates of ovarian cryopreservation and autologous transplantation, this technique is no longer considered experimental by the ASRM [9]. Nonetheless, a recent 20-year prospective study found no reduction in reproductive lifespan or effect on the natural age of menopause when one-fifth of ovarian tissue cortex was harvested among 48 women undergoing benign non-ovarian OBGYN procedures for the purpose of fertility cryopreservation to delay reproductive aging when compared to controls who had declined an ovarian biopsy but underwent similar benign surgeries. The study highlighted the scope of ovarian tissue cryopreservation in providing a natural form of hormone replacement therapy [32]. The same should be investigated as an option for OFP and improving reproductive outcomes following cancer treatment.

OTC is recommended for patients undergoing moderate- or high-risk gonadotoxic treatment when oocyte or embryo cryopreservation is not possible or at the patient's preference [5].

Over 100 successful live births have been reported following ovarian tissue replacement. So far, no official report of ovarian tissue cryopreservation and autotransplantation in the UAE is available. A multidisciplinary team should carefully select cases with a focused effort to appropriately select the fertility preservation procedure based on the patient's profile, avoiding harm to the patient by delaying cancer treatment, and ensuring that all attempts are made to provide information on access to such services in the UAE or abroad. Patients of advanced age or with poor ovarian reserve with an AMH <0.5 ng/mL and an AFC <5 may not be suitable for OTC as the risks of the procedure may outweigh the benefits, with experts suggesting alternative FP methods [5]. Furthermore, the OTC method is not suitable for all patients, as there is a theoretical risk of transplanting cancer cells into disease-free patients [5, 9, 33].

Slow freezing protocol for OTC is well-established and commonly used, but setbacks include the lack of FDA-approved media for transportation or cryopreservation of ovarian tissue, and there is no formal system for tracking the outcomes, nationally or internationally. It is recommended to limit tissue harvesting to ≤ one-third of the cortex of one ovary to minimize the risk of inducing early menopause. A superficial cortical biopsy is preferred over performing a wedge resection to minimize adhesion risks.

Ovarian tissue replacement, or autotransplantation, is considered a safe procedure that can be done laparoscopically. Orthotopic sites, such as the placement of thawed slices into a peritoneal pocket, are recommended to restore fertility and natural conception, whereas heterotopic transplantation requires ART. The full potential of subcutaneous heterotopic transplantation techniques in restoring ovarian endocrine function remains unclear; however, they may be considered a natural source of autologous HRT in cases where synthetic hormonal replacement therapy is contraindicated [33].

It is highly recommended to assess the risks of cancer recurrence and the safety of pregnancy after a complete remission of the disease. The decision to perform OTT on oncological patients necessitates a multidisciplinary approach once again.

Parents of patients under the age of 18 should be provided with all available information and future prospects regarding ovarian tissue cryopreservation. Candidates should be informed about the uncertainty of benefits, risks, and alternatives and perform the procedure, preferably under an Institutional Review Board (IRB)-approved protocol.

16.7 GnRH Agonist

GnRH agonists are the only medical strategy available for clinical use during chemotherapy as an option for ovarian function protection in premenopausal cancer patients [5, 8, 9]. A pituitary gonadotropin flare-up is observed after an intramuscular or subcutaneous injection of a GnRH agonist, followed by a prolonged down-regulation and ovarian suppression. However, the benefit of this flare-up is unclear, as the primordial follicles are not gonadotropin-sensitive. Some of the proposed mechanisms by which GnRH agonists exert a follicle protective effect are by reducing primordial follicle recruitment and differentiation along with lowering gonadal vascularity, hence reducing levels of gonadotoxic agents in the targeted organs [34].

GnRH agonist depot may be administered at a monthly or 3-monthly interval, with downregulation lasting 1–3 months and about 1–2 weeks beyond the end of the last chemotherapy cycle. The major adverse effects involve climacteric symptoms such as hot flushes, vasomotor symptoms, and sexual problems [5].

Data on the protective effect of GnRH agonists on ovarian reserve is heterogeneous and has an unproven role in future pregnancies. A potential protective effect was demonstrated when GnRH analogs were administered during gonadotoxic chemotherapy with the intention of preserving ovarian function in a large systematic review and meta-analysis of 12 RCTs and 7 cohort studies published between 1987 and 2015, as measured by the resumption of menstruation [35].

While high-quality evidence shows concurrent use of GnRH agonists during chemotherapy significantly reduces the risk of developing chemotherapy-induced POI with no negative impact on survival, concerns exist about possible irreversible loss of bone density if GnRHa is administered for >6 months, cardiovascular health, and overall quality of life, with no large, randomized control data validating such concerns [36].

It is noteworthy to mention that most of the data from randomized trials assessing the use of GnRHa during chemotherapy have been conducted in breast cancer patients, with limited and mostly negative evidence in women with other malignancies such as ovarian cancer and Hodgkin's and non-Hodgkin's lymphoma [37].

GnRH agonists should be applied in addition to other FP interventions, such as oocyte or embryo cryopreservation. It may be used as a single FP option where oocyte or embryo cryopreservation is not feasible [5].

GnRH agonists should not be routinely prescribed in malignancies such as lymphomas or other than breast cancer without first discussing with the patient the unproven benefit of ovarian function protection or post-treatment pregnancy. Furthermore, a small trial showed the potential benefit of GnRHa for the protection of ovarian function in cases of ovarian cancer, adding to uncertainty about its benefit in the absence of large data [5].

16.8   Oophoropexy

Women may be offered surgical ovarian transposition using either lateral or medial transposition approaches with the aim of preventing premature ovarian insufficiency when pelvic radiotherapy without chemotherapy is planned [5]. The ovaries are mobilized and affixed to a location in the abdomen free of radiation, which is carried out shortly before radiotherapy to prevent the ovaries from returning to their original position. The preservation of ovarian function assessed by the absence of amenorrhea is noted in 71% of cases [38].

Radiation therapy is indicated for the treatment of pelvic malignancies, including cervical, endometrial, rectal, and bladder cancers, as well as sarcomas and lymphomas involving the pelvic region.

The recommendation to offer ovarian transposition as a fertility preservation option to women prior to pelvic radiotherapy is in line with recommendations from the American Society of Reproductive Medicine, the American Society of Clinical Oncology, and the National Comprehensive Cancer Network. While well-designed clinical trials on the efficacy and safety of ovarian transposition are lacking, small observational studies show it is an effective method for preserving ovarian function. Reproductive and pregnancy outcomes are rarely reported as a whole cohort. In the context of safety, an overall complication rate of 12.8% has been reported, mainly ovarian cysts, with no additional intervention or treatment necessary [10]. Again, GDG advises against ovarian transposition for women with reduced ovarian reserve, again highlighting the importance of effective fertility preservation methods to protect ovarian function, which is dependent on patient characteristics, including her age [5]. The risks of ovarian transposition may outweigh the benefits for women of advanced age with poor ovarian reserve, as high chances of ovarian failure exist. Also, women who are at risk of having ovarian metastases are inappropriate candidates for ovarian transposition [5].

16.9 Male Fertility Preservation

Male fertility preservation options are less complex and faster than female fertility preservation options, requiring sperm collection via masturbation, electroejaculation, or testicular biopsy, followed by cryopreservation of semen and testis tissue.

16.9.1 Sperm Cryopreservation Through Ejaculation

Sperm cryopreservation has been used worldwide since the 1970s and in the UAE since the early 1990s to treat couples with infertility.

There is no recommended medical treatment to protect spermatogenesis in males undergoing gonadotoxic treatment as opposed to their female counterparts. Studies to date have failed to confirm the significant benefit of hormonal suppression of testicular function and spermatogenesis with the use of GnRH agonists, testosterone, androgenic progestogens, or anti-androgens [39, 40].

Masturbation-based sperm collection is feasible and successful for the majority of adult and post-pubertal male cancer patients.

As per the AUA/ASRM Guideline (2020), males should be counseled about the negative impact of cancer treatment on their reproductive potential, and early referral to a reproductive specialist or urologist is recommended for options of fertility preservation prior to the initiation of gonadotoxic therapies such as chemotherapy or radiotherapy [40]. To ensure successful sperm oncofertility freezing, ejaculated samples should be obtained twice or thrice to yield several vials for cryopreservation [40, 41].

Sperm can be used successfully during fertility therapy, even after 40 years of cryopreservation. The recent birth of a baby boy in October 2022, in the USA, from sperm frozen in 1996 when his father (21 years old at the time) was diagnosed with Hodgkin's lymphoma made headlines while sparking scientific and ethical debate about the legal limit for the allowed number of years of sperm cryostorage versus sperm shelf-life (news). The duration of sperm freezing in clinical practice world-wide varies depending on local legislation, governing authorities, clinical require-ments, and funding. While the Human Fertilisation and Embryology Authority (HFEA) allows patients with persistent infertility to consent for sperm freezing for up to 55 years, National Health Service (NHS) funding for sperm freezing may be limited to 5–10 years per patient, depending on the region of the UK.

In the UAE, males who consent and proceed with sperm cryopreservation are counseled about the quality and quantity of their cryopreserved sample, the number of years of cryostorage allowed under UAE fertility legislation (up to 5 years, extendable further for onco cases and severe male factor infertility), and how to renew their cryostorage consent on a yearly basis with rights for future use within the same fertility center or after transporting it to another center within the UAE.

Semen parameters prior to cryopreservation have been shown to be an accurate predictor of post-thaw sperm motility and viability, and spermatogenesis recovery rate is dependent on cancer type, treatment modality, type of chemotherapy regi-men, radiation dose, and underlying testicular function [42].

As is the case with women, advanced paternal age has been linked to an increas-ing number of defects in sperm DNA integrity and genomes, influencing assisted conception outcomes and offspring health [43].

Despite this, patients with testicular cancer can experience spontaneous sper-matogenesis recovery in up to 50% of cases 2 years after gonadotoxic treatment, with rates of long-term azoospermia (more than 2 years) ranging from 5 to 18% in patients who have undergone orchiectomy and radiotherapy [40]. The rates of long-term azoospermia post-chemotherapy range from 0 to 82% for men with Hodgkin's lymphoma and 19–55% among men with leukemia [40].

Both radiotherapy and chemotherapy affect the differentiating spermatogenic cells, including the spermatocytes and the spermatids, but do not kill the spermato-gonial stem cells in the testis, thus causing a temporary decline or cessation of spermatogenesis followed by a gradual recovery of sperm production after complet-ing gonadotoxic therapy. Furthermore, a man can produce an increased proportion of genetically abnormal spermatozoa for a specific period of time following radia-tion and/or chemotherapy exposure, which can significantly increase the risk of genetic mutations in the offspring if the spouse conceives during this time. As per ample data given in the literature, AUA/ASRM recommends that clinicians inform patients undergoing chemotherapy and/or radiation therapy to avoid pregnancy for a period of at least 12 months after completion of treatment [40]. It is recommended that clinicians advise patients to delay SA for assessing sperm recovery after gonadotoxic therapies, which should be done at least 12 months (and preferably 24 months) after treatment completion [40].

Concerns over the years have been raised about the freeze-thawing technique that may lead to a reduction in the number of normally functional sperm as a result of osmotic and oxidative stress, toxicity from the cryoprotectant, and the formation of intracellular ice crystals [41]. In recent years, cryopreservation has been improved to reduce the risk of viral cross-contamination or sperm damage by incorporating cryoprotectants and antioxidants. Again, caution is needed for the routine use of additives in sperm cryopreservation, which requires further investigation into the outcome of pregnancy and the safety of offspring.

Cancer itself may impair semen parameters prior to the initiation of oncotherapy. Many retrospective studies evaluating sperm samples obtained from the semen of cancer patients who were cryopreserved before chemotherapy and those who underwent semen analysis for infertility investigation found that men undergoing prechemotherapy had lower sperm concentrations and reduced total motility when compared to healthy controls with semen parameters above the World Health Organization (WHO) 2010 reference limit. Reduced semen parameters were seen irrespective of tumor type and were lowest in individuals with germ cell tumors [44].

The effect of malignancy on spermatogenesis may be due to several factors, including disruption of the normal hypothalamic–pituitary–gonadal axis and injury to the germinal epithelium from cytotoxic immune responses to cancer, fever, and malnutrition [44].

Again, proper counseling is empirical, and assessment of health conditions including erectile dysfunction, comorbidities such as diabetes, and neurological conditions requiring long-term use of medications all play vital roles in successful male fertility preservation.

While a comfortable environment for semen collection is desirable for young men, teenagers, or men who are unable to ejaculate, other alternative options may also be considered to obtain ejaculated sperm for cryopreservation.

16.9.1.1 Vibratory Stimulation and Electroejaculation

Penile vibratory stimulation induces ejaculation in males by increasing penile stimulatory input and triggering the ejaculatory reflex among males who are unable to provide semen through conventional masturbation in cases of neurologic injuries or factors negatively impacting the ejaculatory reflex, including psychogenic anejaculation [9]. Alternatively, electroejaculation may be offered for men and peri-pubertal males who are non-responsive to penile vibratory stimulation. The collection of seminal emission following pelvic tissue stimulation, including the prostate and seminal vesicles, with a transrectal probe may be performed under anesthesia.

Pseudoephedrine (alpha-agonists) may be used cautiously to restore antegrade ejaculation in men suffering from retrograde ejaculation due to autonomic or pelvic nerve injury or bladder neck injury.

For men who failed to respond to alpha-agonist therapy or whose use is contraindicated, the isolation of sperm for cryopreservation is done through the collection and processing of the urine after ejaculation, following urinary alkalization with or without the use of sperm wash media into the bladder just prior to ejaculation. For cases of erectile dysfunction (ED), oral agents such as phosphodiesterase type 5 (PDE-5) inhibitors (sildenafil) may be used. The patient should be evaluated and counseled on its contraindications, timing of administration, need for sexual stimulation, and any other side effects prior to initiating any of these agents [9, 40].

16.9.1.2 Surgical Aspiration and Extraction of Sperm

In cases of azoospermia, severe oligoasthenoteratospermia (OAT), or hypospermia, TESA/PESA/testicular biopsy and microscopic testicular sperm extraction (micro-TESE) may be considered. Although surgical aspiration may be done under local anesthesia, most of these procedures are conducted in the operating room as an outpatient day-case procedure and hence require scheduling.

Testicular tissue cryopreservation is the only method of fertility preservation for prepubertal boys who are not yet producing sperm or for pubertal patients who cannot or will not produce a semen sample. The procedure, done under GA, requires the removal of testicular tissue by open biopsy of one testis, cutting the biopsied testicular tissues into small pieces ($1-25$ mm^3), and then freezing at a slow rate or freezing the intact pieces of testicular tissue.

It is important that such procedures are done with ease of access to fertility centers with a cryopreservation facility, whereby sperm found in testicular tissue is processed and cryopreserved immediately after the procedure. Immature testicular tissue can be cryopreserved in case no mature sperm is recovered, as it may be the only hope for future fatherhood. While testicular biopsy in young patients is generally considered safe with no reported long-term impacts on testicular anatomy, growth, or hormonal function, it is still considered investigational by the ASRM and must be performed in the setting of a clinical trial [9]. Autologous testicular tissue transplantation of cryopreserved testicular cells or tissues may not be appropriate for patients with blood-borne cancers or testicular cancers due to the theoretical risk of reseeding tumors. Moreover, combining multiple techniques, such as sperm extraction from the affected testis tissues immediately after orchiectomy, may also be considered, referred to in the literature as "onco-TESE," as this may be the only potential source or instance for finding viable sperm for cryopreservation.

No report for testicular tissue cryopreservation among pubertal or prepubertal boys in the UAE is available. A patient as young as 15 years of age has undergone oncofertility sperm cryopreservation as per Tawam Fertility Center sperm cryostorage records (internal, unpublished report).

16.10 UAE Fertility Preservation Legislations and Timeline

As per Federal Law No. (11) of 2008 [45] Concerning Licensing of Fertilization Centres and based on the proposal of the Minister of Health, with the approval of the Cabinet and the Federal National Council and the ratification of the Supreme Council of the Federation, this law was applied to fertility centers operating in or applying for a license to operate in the UAE.

- The law advocated for the formation of the Fertilization Centres Oversight and Control Committee under the Ministry of Health to include technical, Sharia (Islamic law), and legal members (Article 3).

- Assisted reproductive techniques (ART) included intrauterine insemination (IUI), in vitro fertilization (IVF), intracytoplasmic sperm injection (ICSI) of the ovum, and subsequent embryo/gamete transfer (Article 8).
- ART is provided for married couples only (Article 9).
- It is prohibited to perform ART using donor embryos, donor eggs, or donor sperm (Article 10).
- Embryo freezing was not allowed with surplus fertilized ova that were left without medical attention until they perished naturally. However, preserving unfertilized ova (egg freezing) and sperm for future use was possible (Article 10).
- Transportation of sperm, ovum, and embryos within the UAE was allowed; however, it was prohibited across borders (Article 18).
- It is prohibited to establish or deal with embryo banks (Article 20).
- The punishments and penalties stipulated under the law in cases of offenses or regulation violations (Article 28–34).

Cabinet Decision No. (36) of 2009 came in soon after issuing the Implementing Regulation of Federal Law No. (11) of 2008. Fertility centers that had frozen embryos in their possession prior to the issuance of the law were allowed to dispose of the stored embryos within 6 months from the date that the law was published (Article 25). The consent for egg freezing required the signatures of both husband and wife (Form 4), whereas the consent for sperm freezing could be given by the male alone (married or unmarried) (Form 5).

On December 19, 2019, the late His Highness Sheikh Khalifa Bin Zayed Al Nahyan, President of the United Arab Emirates at the time, announced the new IVF law that came officially into effect on January 1, 2020, repealing the old IVF law [46].

Freezing of human embryos by fertility centers is now permitted for a period of 5 years (extendable upon request). Prior to that, embryo freezing was done for medical reasons after obtaining special approvals from the Ministry of Health on a case-by-case basis.

Unmarried individuals are also allowed to freeze their eggs or sperm for a period of 5 years, which is also extendable upon request. This allows options for fertility preservation for individuals with medical or oncological conditions and treatments affecting fertility. Sperm, egg, and embryo donation are still considered illegal. Frozen samples of eggs, embryos, or sperm may be taken abroad if prepared in the UAE or brought into the UAE if prepared abroad, of course, subject to compliance with certain controls and procedures [46]. While Muslim couples must provide proof of marriage before proceeding with IVF, non-Muslim unmarried couples can now undergo IVF treatment after seeking permission from the health authority to utilise IVF techniques under Article 8, of the new ruling published in the Official Gazette. The unborn child's rights are protected by requiring the parents to register the baby under both their names. This change to UAE's family laws is part of a wider effort to update laws in line with the needs of all those living in the country [47].

While scientific research on gametes can be carried out for the purpose of increasing knowledge or developing treatments for severe cases or diseases, reproductive cloning, altering genetic traits for the purposes of changing the human genetic structure, and commercial purposes are all prohibited.

Under the Daman-Thiqa plan, UAE nationals can receive total coverage for IVF, embryo freezing, and oocyte freezing for medical indications or oncological diagnosis patients for up to 6 cycles of ovarian stimulation per year. ART funding is provided to women between 18 and 47 years of age with cryostorage cost coverage of their embryos or eggs for the duration of 5 years, with a yearly renewal of consent to continue freezing. Cryostorage can extend beyond 5 years with the consent of the MOH and the patient. It is worth noting that there is no age limit for sperm cryopreservation, nor is there a time limit for women to use their eggs or embryos at a later stage in their lives [48].

Women undergoing chemotherapy or radiation treatments do not need to bear the costs of fertility preservation. While this alleviates the psychological burden, it also raises ethical concerns about the risks associated with early parenthood.

Thiqa coverage for assisted reproductive treatment for UAE nationals, including fertility preservation, is available for oncology patients, which also includes women with *BRCA1* or *BRCA2* genetic mutations and young women with borderline ovarian tumors. Usually, only one cycle can be done on an emergency basis due to the time constraint for cancer patients; however, exceptions are made for up to six cycles of oocyte retrieval in a year in cases of borderline tumors, where more stimulations can take place before definitive treatment. The cost of storage is covered for 5 years; beyond that, it is collected directly from patients [48].

The DOH Policy on THIQA Coverage for Assisted Reproductive Treatment and Services requires gonadotropin injections to be administered by DOH-licensed reproductive endocrinologists, IVF specialists, and consultants [48].

Tawam Fertility Center in Al Ain opened in 1990, thanks to the foresight of the late Sheikh Zayed Bin Sultan Al Nahyan, and was quickly followed by Dubai Gynaecology and Fertility Center (DGFC) in 1992, thanks to the generous support of the late Sheikh Hamdan Bin Rashed Al Maktoum. Tawam Fertility Center celebrated the birth of the first child born following IVF in the UAE in 1991 [49], just a decade after the announcement of the first IVF birth in the USA in 1981, and in the same year of the introduction of pioneering techniques such as laser-assisted Zona Pellucida drilling and the use of antagonists for preventing LH surge [50]. Assisted reproductive technology (ART) procedures including intrauterine inseminations (IUI), fresh IVF cycles using the agonist protocol with day 2–3 embryo transfers, laparoscopic gamete and zygote intrafallopian transfer (GIFT), and ZIFT were practiced commonly, with eventual sperm and embryo freezing using the older slow freezing methods (internal email communication with DGFC). The first ICSI using percutaneous epididymal sperm aspiration was done in August 1992 at the DGFC, resulting in a live birth in May 1993 [51]. Egg and embryo vitrification (an efficient newer technique known for higher pregnancy rates) were introduced by DGFC in 2007 and 2008, respectively (internal communication). The center announced the birth of the first baby after injecting a vitrified-thawed egg with her husband's fresh sperm while using the fluorescence in situ hybridization (FISH) technique for

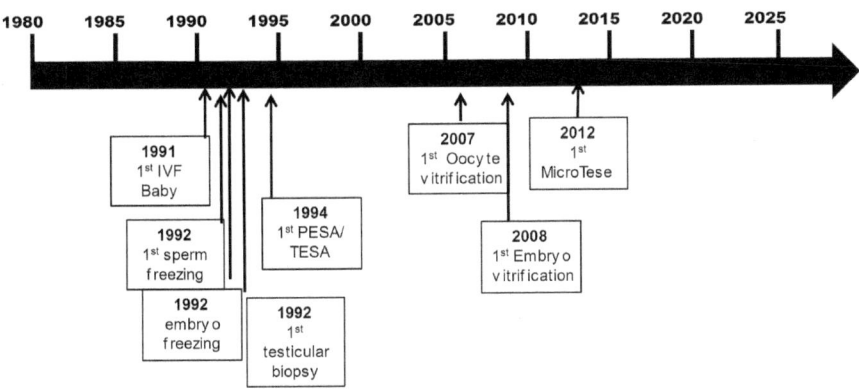

Fig. 16.1 Fertility preservation legislations and timeline in UAE

gender selection in 2008 [52]. Most of the sperm surgical retrieval procedures and andrology services, including testicular biopsies, were done earlier by gynecologists practicing reproductive endocrinology and infertility. While Tawam Fertility Center made a niche for itself by being known as the biggest and longest-running center offering oncofertility sperm cryopreservation, other surgical sperm retrieval methods were also introduced in the UAE, including PESA by DGFC (1994) [51] and the first microscopic testicular sperm extraction (microTESE) (2012) by a visiting doctor at Fakih IVF clinic (internal communication). Over the years, numerous advances have been made in the UAE, catching up with the worldwide pace of fertility treatment protocols and expanding freezing technique services (Fig. 16.1, Table 16.3). Around 25 centers now facilitate state-of-the-art fertility services, including cryopreservation, four of which belong to their respective local governments (i.e., Corniche Fertility Center in Abu Dhabi and Tawam Fertility Center in Al Ain, both under the SEHA umbrella, Dubai Gynecology and Fertility Center, and Sharjah Fertility Center in University Hospital Sharjah). Private fertility clinics such as Fakih IVF, Health Plus, ART Clinic, Bourn Hall Fertility Center, and Al Ain Fertility Center provide services on a larger scale, with multiple branches across the Emirates and some offering in-house genetic testing. Egg cryopreservation for unmarried females is performed laparoscopically and transrectally in a few centers. Recently, in Bourn Hall Mediclinic Al Ain, a 14-year-old girl with Hodgkin's lymphoma underwent oocyte retrieval through a novel transabdominal route, resulting in the freezing of 25 eggs, just 1.6 years after attaining her menarche [53]. She is thought to be the UAE's youngest female cryopreservation patient, expanding the scope of oncofertility cryopreservation services to young adolescents.

16.11 Challenges and Opportunities

The quality of health services in the UAE is on par with international best practices in the fertility market. Furthermore, the UAE has a strong health regulatory framework that ensures top-notch quality health care with cutting-edge infrastructure run by both the public and private sectors and performing 15,000–16,000 cycles per year, which is expected to increase further with population growth. Moreover, current lenient visa rules position the UAE as a leading catalyst for medical tourism, boosting the growth of IVF services [54]. The IVF law changes in recent years are in line with the UAE's National Agenda 2021 and the UAE Centennial 2071 project, which aim to elevate the UAE's position in the global community. It is worth highlighting that the UAE leadership's continuous commitment to providing optimal health and prenatal care to women in the country and prioritizing mother, child, and youth strategies has all increased public confidence in the healthcare system.

Pregnancy rates in the UAE range from 50 to 80%, while the average success rate in European countries, as per ESHRE, is less than 40%, with discrepancies possibly attributed to the reporting methods of cycle outcome. Patients also come for gender selection, as the UAE is one of the few countries where it is permitted. Owing to its high success rates and the quality of treatment offered, the UAE is in the best position to become a reliable hub for ART services, including fertility preservation, both nationally and internationally.

Efforts to track IVF activity and its outcomes in the UAE have been seen since the start of assisted conception treatment services, which initially were reported on a voluntary basis. However, ART providers are now required on a mandated basis to report ART outcomes, including JAWDA KPIs for ART and reporting of serious untoward incidents associated with ART treatments, aiming at continued improvement and evaluation of ART programs among service providers.

Also, public reporting through detailed analysis of data allows for transparency and continued opportunities for evaluation of ART programs, aiding in the continuous improvement of healthcare outcomes.

Since the US Congress passed the Fertility Clinic Success Rate and Certification Act in 1992, clinics have been required to report IVF outcome data to the Centers for Disease Control (CDC) to provide transparency and protect patients from false claims of IVF success.

IVF success rates for all reputable clinics are now available on the web from both the CDC and the Society for Assisted Reproductive Technology (SART), an affiliate of the American Society for Reproductive Medicine. SART is a fantastic resource for both patients and physicians, providing useful information such as detailed guides to various ART protocols and procedures as well as success rates of individual technologies at practices across the country. There is an urgent need for similar resources to be made available in Arabic.

There should also be a multidisciplinary approach for diagnosing and treating neoplastic diseases involving oncologists, surgeons, reproductive endocrinologists, gynecologists, urologists, mental health professionals, and genetic counselors.

The priority of the multi-specialty team is to address all possible available resources and access to oncofertility preservation, including their risks and benefits, prior to starting gonadotoxic treatment. It is important to effectively address patients' psychosocial distress and provide reproductive counseling at the time of diagnosis or soon after the diagnostic therapeutic process.

There is no standardized protocol to streamline oncofertility referrals or model of care for preservation in the UAE. Referring to qualified reproductive specialists, counseling, proper documentation in medical records with informed consent based on what has been determined by legal regulations, and medical ethics provisions wherever necessary have been suggested and considered worldwide (ESHRE FP Guideline Development Group, 2020) [5]. There are no local or regional oncofertility organizations, leaving a niche in the global community of practice for oncofertility, although professionals may have individual membership in societies such as the International Society for Fertility Preservation (ISFP) or the ARSM Fertility Preservation Special Interest Group [55].

16.12 Summary

Before starting cancer therapy, patients and the parent(s) of children and adolescent patients should be educated about the possibility of infertility resulting from cancer therapy.

A multidisciplinary team of medical providers should discuss fertility preservation options and refer patients to appropriate reproductive specialists at the earliest opportunity.

Male fertility preservation options include sperm cryopreservation, and female fertility preservation options include egg and embryo freezing via assisted reproductive technology. While ovarian tissue cryopreservation is no longer considered experimental, there has been no report of cryopreservation using this method in the UAE. Also in the UAE, any attempted in vitro gamete maturation from immature sperm or eggs followed by cryopreservation has only been reported in the internal communications of the embryology laboratories of some of the private clinics or used as a marketing strategy on their websites, with no published data on its efficiency or success. Other techniques, like ovarian suppression and ovarian transposition, are also used but not recommended as the sole option for fertility preservation. Techniques for ovarian protection include oophoropexy, gonadotropin-releasing hormone analogs, pelvic shielding, fractionated doses of chemotherapy, and radiotherapy.

In the literature, several studies have shown a lower success rate of fertility preservation methods in women with advancing age due to a natural decline in ovarian reserve, which may impact the chances of achieving a successful live birth.

Challenges in choosing fertility preservation options include lack of knowledge both from the physician's and the patient's and family's sides, ethical and UAE fertility legislation considerations, time to treatment, and availability of financial resources. To shorten the time frame for egg collection, ovarian stimulation protocols such as random and luteal phase start have shown similar efficiency to the conventional start of stimulation during the early follicular phase of the menstrual cycle days 2–4, allowing for less delay in the cancer treatment plan. Routine use of a GnRH agonist to protect gametes from the gonadotoxic effects of chemotherapy is not only questionable but also should be considered carefully, as it may exacerbate OHSS in PCOS patients if ovarian stimulation is attempted within a short time after downregulation. Also, it may negatively impact the outcome of oncofertility preservation by further causing a poor response to ovarian stimulation through the downregulation of the ovaries in patients with a low pre-existing ovarian reserve. For the sake of collective, efficient, and organizational fertility preservation, there needs to be a balanced tuning of the patient's physical and mental well-being, legal and moral considerations within the available resources, and access to care in accordance with law provisions and the Code of Medical Ethics, working together with multidisciplinary professionals participating in decision-making. The mission of the FP network will extend to enhancing knowledge among healthcare workers and the public on reproductive disorders induced by cancer therapy and other medical treatments through national and international collaboration among reproductive specialists, oncologists, and allied health workers, promoting research and education, and keeping pace with the development of new innovative strategies of fertility preservation customized to UAE demography and legislation.

Conflict of Interest   The author has no conflict of interest to declare.

References

1. Global Burden of Disease 2019 Cancer Collaboration, Kocarnik JM, Compton K, et al. Cancer incidence, mortality, years of life lost, years lived with disability, and disability-adjusted life years for 29 cancer groups from 2010 to 2019: a systematic analysis for the global burden of disease study 2019. JAMA Oncol. 2022;8(3):420–44. https://doi.org/10.1001/jamaoncol.2021.6987.
2. Cancer incidence in United Arab Emirates, Annual Report of the UAE - National Cancer Registry, 2021. Statistics and Research Center, Ministry of Health and Prevention.
3. Dolmans MM, Manavella DD. Recent advances in fertility preservation. J Obstet Gynaecol Res. 2019;45(2):266–79.
4. Oktay K, Harvey BE, Partridge AH, Quinn GP, Reinecke J, Taylor HS, Wallace WH, Wang ET, Loren AW. Fertility preservation in patients with cancer: ASCO clinical practice guideline update. J Clin Oncol. 2018;36(19):1994–2001.
5. ESHRE Guideline Group on Female Fertility Preservation, Anderson RA, Amant F, Braat D, D'Angelo A, Chuva de Sousa Lopes SM, Demeestere I, Dwek S, Frith L, Lambertini M, Maslin C, Moura-Ramos M, Nogueira D, Rodriguez-Wallberg K, Vermeulen N. ESHRE guideline: female fertility preservation. Hum Reprod Open. 2020;2020(4):hoaa052.
6. Sleiman Z, Karaman E, Terzic M, Terzic S, Falzone G, Garzon S. Fertility preservation in benign gynecological diseases: current approaches and future perspectives. J Reprod Infertil. 2019;20(4):201–8.
7. Karavani G, Rottenstreich A, Schachter-Safrai N, Cohen A, Weintraub M, Imbar T, Revel A. Chemotherapy-based gonadotoxicity risk evaluation as a predictor of reproductive outcomes in post-pubertal patients following ovarian tissue cryopreservation. BMC Womens Health. 2021;21(1):201. https://doi.org/10.1186/S12905-021-01343-Z.
8. Oktay K, Harvey BE, Loren AW. Fertility preservation in patients with cancer: ASCO clinical practice guideline update summary. J Oncol Pract. 2018;14(6):381–5. https://doi.org/10.1200/JOP.18.00160.
9. Practice Committee of the American Society for Reproductive Medicine. Fertility preservation in patients undergoing gonadotoxic therapy or gonadectomy: a committee opinion. Fertil Steril. 2019;112(6):1022–33.

10. The ESHRE Guideline Group on Ovarian Stimulation, Bosch E, Broer S, Griesinger G, Grynberg M, Humaidan P, Kolibianakis E, Kunicki M, La Marca A, Lainas G, Le Clef N, Massin N, Mastenbroek S, Polyzos N, Sunkara SK, Timeva T, Töyli M, Urbancsek J, Vermeulen N, Broekmans F. ESHRE guideline: ovarian stimulation for IVF/ICSI†. Hum Reprod Open. 2020;2020(2):hoaa009. https://doi.org/10.1093/hropen/hoaa009.

11. Cobo A, García-Velasco J, Domingo J, Pellicer A, Remohí J. Elective and onco-fertility preservation: factors related to IVF outcomes. Hum Reprod. 2018;33(12):2222–31.

12. Cobo A, García-Velasco JA, Coello A, Domingo J, Pellicer A, Remohí J. Oocyte vitrification as an efficient option for elective fertility preservation. Fertil Steril. 2016;105(3):755–764.e8. https://doi.org/10.1016/j.fertnstert.2015.11.027. Epub 2015 Dec 10.

13. Practice Committee of the American Society for Reproductive Medicine. Prevention and treatment of moderate and severe ovarian hyperstimulation syndrome: a guideline. Fertil Steril. 2016;106(7):1634–47. https://doi.org/10.1016/j.fertnstert.2016.08.048. Epub 2016 Sep 24.

14. ESHRE Working Group on Ultrasound in ART, D'Angelo A, Panayotidis C, Amso N, Marci R, Matorras R, Onofriescu M, Turp AB, Vandekerckhove F, Veleva Z, Vermeulen N, Vlaisavljevic V. Recommendations for good practice in ultrasound: oocyte pick up†. Hum Reprod Open. 2019;2019(4):hoz025. https://doi.org/10.1093/hropen/hoz025. PMID: 31844683; PMCID: PMC6903452.

15. Shannon MB. Anticoagulation and in vitro fertilization and ovarian stimulation. Hematol Am Soc Hematol Educ Prog. 2014;2014(1):379–86. https://doi.org/10.1182/asheducation-2014.1.379.

16. Kim J, Turan V, Oktay K. Long-term safety of letrozole and gonadotropin stimulation for fertility preservation in women with breast cancer. J Clin Endocrinol Metab. 2016;101(4):1364–71.

17. Reddy J, Oktay K. Ovarian stimulation and fertility preservation with the use of aromatase inhibitors in women with breast cancer. Fertil Steril. 2012;98(6):1363–9.

18. Balayla J, Tulandi T, Buckett W, Holzer H, Steiner N, Shrem G, Volodarsky-Perel A. Outcomes of ovarian stimulation and fertility preservation in breast cancer patients with different hormonal receptor profiles. J Assist Reprod Genet. 2020;37(4):913–21. https://doi.org/10.1007/s10815-020-01730-9. Epub 2020 Mar 6. PMID: 32144524; PMCID: PMC7183026.

19. Fabiani C, Guarino A, Meneghini C, Licata E, Paciotti G, Miriello D, Schiavi MC, Spina V, Corno R, Gallo M, Rago R. Oocyte quality assessment in breast cancer: implications for fertility preservation. Cancers (Basel). 2022;14(22):5718. https://doi.org/10.3390/cancers14225718.

20. Rodriguez-Wallberg KA, Marklund A, Lundberg F, Wikander I, Milenkovic M, Anastacio A, et al. A prospective study of women and girls undergoing fertility preservation due to oncologic and non-oncologic indications in Sweden—trends in patients' choices and benefit of the chosen methods after long-term follow up. Acta Obstet Gynecol Scand. 2019;98:604–15. https://doi.org/10.1111/aogs.13559.

21. Vladimirov IK, Tacheva D, Dobrinov V. The present and future of embryo cryopreservation. In: Wu B, Feng HL, editors. Embryology—theory and practice. London: IntechOpen; 2018. https://doi.org/10.5772/intechopen.80587.

22. Practice Committees of the American Society for Reproductive Medicine and the Society for Assisted Reproductive Technology. The use of preimplantation genetic testing for aneuploidy (PGT-A): a committee opinion. Fertil Steril. 2018;109(3):429–36. https://doi.org/10.1016/j.fertnstert.2018.01.002.

23. Chen LM, Blank SV, Burton E, Glass K, Penick E, Woodard T. Reproductive and hormonal considerations in women at increased risk for hereditary gynecologic cancers: Society of Gynecologic Oncology and American Society for Reproductive Medicine Evidence-Based Review [Erratum in: Gynecol Oncol. 2020 Mar;156(3):748.]. Gynecol Oncol. 2019;155(3):508–14. https://doi.org/10.1016/j.ygyno.2019.06.017. Epub 2019 Oct 9.

24. Wennerholm UB, Bergh C. Perinatal outcome in children born after assisted reproductive technologies. Ups J Med Sci. 2020;125(2):158–66. https://doi.org/10.1080/03009734.2020.1726534. Epub 2020 Mar 3. PMID: 32124667; PMCID: PMC7720966.

25. Rienzi L, Gracia C, Maggiulli R, LaBarbera AR, Kaser DJ, Ubaldi FM, Vanderpoel S, Racowsky C. Oocyte, embryo and blastocyst cryopreservation in ART: systematic review and meta-analysis comparing slow-freezing versus vitrification to produce evidence for

the development of global guidance. Hum Reprod Update. 2017;23(2):139–55. https://doi.org/10.1093/humupd/dmw038.

26. von Wolff M, Germeyer A, Liebenthron J, Korell M, Nawroth F. Practical recommendations for fertility preservation in women by the FertiPROTEKT network. Part II: fertility preservation techniques. Arch Gynecol Obstet. 2018;297(1):257–67. https://doi.org/10.1007/s00404-017-4595-2. Epub 2017 Nov 27. PMID: 29181578; PMCID: PMC5762782.

27. Donnez J, Dolmans M-M. Fertility preservation in women. N Engl J Med. 2017;377(17):1657–65.

28. Whiteley GE, Martini AE, Jahandideh S, Devine K, Hill MJ, DeCherney AH, Doyle J, Caleb K. The impact of duration of oocyte cryopreservation on live birth outcomes in IVF cycles using autologous thawed oocytes. Fertil Steril. 2021;116(3):33. https://doi.org/10.1016/j.fertnstert.2021.07.033.

29. Doyle JO, Richter KS, Lim J, Stillman RJ, Graham JR, Tucker MJ. Successful elective and medically indicated oocyte vitrification and warming for autologous in vitro fertilization, with predicted birth probabilities for fertility preservation according to number of cryopreserved oocytes and age at retrieval. Fertil Steril. 2016;105(2):459–66.e2. https://doi.org/10.1016/j.fertnstert.2015.10.026. Epub 2015 Nov 18.

30. Practice Committees of the American Society for Reproductive Medicine, the Society of Reproductive Biologists and Technologists, and the Society for Assisted Reproductive Technology. In vitro maturation: a committee opinion. Fertil Steril. 2021;115(2):298–304. https://doi.org/10.1016/j.fertnstert.2020.11.018. Epub 2020 Dec 24.

31. Elkhatib I, Bayram A, Abdala A, Arnanz A, Melado L, Eldamen A, Lawrenz B, Fatemi H. P-201 Identifying patients benefiting from delayed-matured oocytes insemination. Hum Reprod. 2022;37(1):deac107.194. https://doi.org/10.1093/humrep/deac107.194.

32. Petrikovsky B, Marin L, Oktay KH. Does harvesting ovarian tissue to delay reproductive aging have a negative impact on the natural age of menopause in healthy women? Fertil Steril. 2021;116(3):e9.

33. Oktay KH, Marin L, Petrikovsky B, Terrani M, Babayev SN. Delaying reproductive aging by ovarian tissue cryopreservation and transplantation: is it prime time? Trends Mol Med. 2021;27(8):753–61. https://doi.org/10.1016/j.molmed.2021.01.005. Epub 2021 Feb 4. PMID: 33549473; PMCID: PMC8427891.

34. Hickman LC, Valentine LN, Falcone T. Preservation of gonadal function in women undergoing chemotherapy: a review of the potential role for gonadotropin-releasing hormone agonists. Am J Obstet Gynecol. 2016;215:415–22.

35. Sofiyeva N, Siepmann T, Barlinn K, Seli E, Ata B. Gonadotropin-releasing hormone analogs for gonadal protection during gonadotoxic chemotherapy: a systematic review and meta-analysis. Reprod Sci. 2019;26(7):939–53. https://doi.org/10.1177/1933719118799203.

36. Hickman LC, Llarena NC, Valentine LN, Liu X, Falcone T. Preservation of gonadal function in women undergoing chemotherapy: a systematic review and meta-analysis of the potential role for gonadotropin-releasing hormone agonists. J Assist Reprod Genet. 2018;35(4):571–81. https://doi.org/10.1007/s10815-018-1128-2. Epub 2018 Feb 22. PMID: 29470701; PMCID: PMC5949114.

37. Arecco L, Ruelle T, Martelli V, Boutros A, Latocca MM, Spinaci S, Marrocco C, Massarotti C, Lambertini M. How to protect ovarian function before and during chemotherapy? J Clin Med. 2021;10(18):4192. https://doi.org/10.3390/jcm10184192.

38. Swift BE, Leung E, Vicus D, Covens A. Laparoscopic ovarian transposition prior to pelvic radiation for gynecologic cancer. Gynecol Oncol Rep. 2018;24:78–82. https://doi.org/10.1016/j.gore.2018.04.005.

39. Osterberg EC, Ramasamy R, Masson P, Brannigan RE. Current practices in fertility preservation in male cancer patients. Urol Ann. 2014;6(1):13–7. https://doi.org/10.4103/0974-7796.127008.

40. Schlegel PN, Sigman M, Collura B, De Jonge CJ, Eisenberg ML, Lamb DJ, Mulhall JP, Niederberger C, Sandlow JI, Sokol RZ, Spandorfer SD, Tanrikut C, Treadwell JR, Oristaglio JT, Zini A. Diagnosis and treatment of infertility in men: AUA/ASRM guideline part I. J Urol. 2021;205(1):36–43. https://doi.org/10.1097/JU.0000000000001521. Epub 2020 Dec 9.

41. Rozati H, Handley T, Jayasena CN. Process and pitfalls of sperm cryopreservation. J Clin Med. 2017;6(9):89. https://doi.org/10.3390/jcm6090089.
42. Okada K, Fujisawa M. Recovery of spermatogenesis following cancer treatment with cytotoxic chemotherapy and radiotherapy. World J Mens Health. 2019;37(2):166–74. https://doi.org/10.5534/wjmh.180043. Epub 2018 Nov 27. PMID: 30588779; PMCID: PMC6479085.
43. Halvaei I, Litzky J, Esfandiari N. Advanced paternal age: effects on sperm parameters, assisted reproduction outcomes and offspring health. Reprod Biol Endocrinol. 2020;18:110. https://doi.org/10.1186/s12958-020-00668-y.
44. Shrem G, Azani L, Feferkorn I, Listovsky T, Hussaini S, Farber B, Dahan M, Salmon-Divon M. Effect of malignancy on semen parameters. Life. 2022;12:922. https://doi.org/10.3390/life12060922.
45. Federal Law No. (11) of 2008 Concerning Licensing of Fertilization Centres.
46. Federal Law No. (07) of 2019 Concerning the Medically Assisted Reproduction. Cabinet Decision (No. 64) for 2020 on Issuing the Implementation Regulation of Federal Law No. (09) of 2019 Concerning Medically Assisted Reproduction and its updated versions.
47. The National. https://www.thenationalnews.com/uae/2023/10/26/change-in-law-allows-unmarried-non-muslim-couples-to-undergo-ivf-in-uae/.
48. Standard for Assisted Reproductive Technology Services and Treatment. DOH/SD/ART/1.3/2022. V1.0: Release of DOH Policy on THIQA Coverage for Assisted Reproductive Treatment and Service DOH/STR/CPL/0.9/2022.
49. http://wam.ae/en/details/1395280845362.
50. https://ivf-worldwide.com/ivf-history.html.
51. Shrivastav P, Nadkarni P, Wensvoort S, Craft I. Percutaneous epididymal sperm aspiration for obstructive azoospermia. Hum Reprod. 1994;9(11):2058–61. https://doi.org/10.1093/oxfordjournals.humrep.a138393.
52. http://wam.ae/en/details/1395228472461.
53. Detho S. The first UAE youngest girl fertility preservation. O68 Oral presentation session 3, Middle East Fertility Society, Abu Dhabi. 2022.
54. Colliers Market Research 2021. IVF and fertility in the MENA Region.
55. Ataman LM, Rodrigues JK, Marinho RM, Caetano JP, Chehin MB, Alves da Motta EL, et al. Creating a global community of practice for oncofertility. J Glob Oncol. 2016;2(2):83–96. https://doi.org/10.1200/JGO.2015.000307.

Dr. Nahla Kazim is one of the first UAE physicians to attain sub-specialty degrees of Master's (2001) and Doctorate (2010) in the field of Reproductive and Developmental Sciences from the University of Edinburgh, UK. She is a Consultant Reproductive Medicine and Infertility at Bourn Hall Fertility Clinic, a Mediclinic Middle East company, and is serving the role of Director of Fertility Preservation in the Al Ain Branch. She has been appointed as an Adjunct Assistant Professor in the Department of Obstetrics and Gynecology, College of Medicine and Health Sciences, UAE University (2023). Her past work achievements include serving as a Scientific Director of Fakih IVF Fertility Center L.L.C., Al Ain, 2019–2022. She also worked as a Senior Specialist in Physician–Assisted Conception-Infertility Services, Tawam Fertility Center, Al Ain (2010–September 2019). Dr. Kazim was the Chief Administrator, Tawam Fertility Center, 2018–2019. She worked as a Specialist Obstetrics and Gynecology/Reproductive Health in Mafraq Hospital, Abu Dhabi, UAE (2006–2010), and as a registrar in Ob/Gyn, Mafraq Hospital, starting from 1999.

Dr. Kazim has been nominated and shortlisted for the prestigious Abu Dhabi Medical Distinction Award in the category of medical volunteer twice, in 2012 and 2013. She received the Sheikh Rashid Al Maktoum Award for Educational Excellence in 2002 and 2011. Her special interest is in implementing innovative technologies and procedures, including artificial intelligence and ovarian PRP, into current clinical practice. She is the first in the UAE to explore the uses of virtual and augmented reality (VR and AR) technology in fertility treatment. She was the abstract reviewer for the ASRM Scientific Congress 2018–2022 and the grant reviewer for the ASRM Research Institute 2023. She is a member of the first UAE National Anti-Doping Committee and a member of many known professional societies, including the American Society of Reproductive Medicine, the British Fertility Society, the European Society for Human Reproduction and Embryology, the European Fertility Society, the Society for Reproduction and Fertility, the Society for Reproductive Endocrinology and Infertility, the International Federation of Fertility Societies, and the ASRM Special Interest Groups, including International Membership in Fertility Preservation, Ovarian Insufficiency and Menopause, Reproductive Immunology, and the Women's Council.

Melanie C. Schlatter ⓘ

17.1 Introduction

The field of psycho-oncology originated in the 1970s in America, and it takes into account specific dynamics of an individual undertaking cancer treatment—the psychological impact of cancer on the individual, their family, caregivers, and medical staff, as well as the impact of the individual (behaviors and psychosocial factors) on cancer morbidity and mortality [1]. The field has gradually been embraced in other countries with the development of centralized evidence-based cancer guidelines [2], and proposals have been initiated in the United Arab Emirates (UAE) recently also [3]. Guidelines advocate for a multidisciplinary team (MDT) approach in the management of cancer patients, inclusive of psychology and psychiatry [4, 5], so the UAE has some unique advantages, given that it is recognized as having one of the leading healthcare systems in the world [6].

Unfortunately, there is an increasing prevalence of cancer in the Middle East [7, 8] due to rapid population expansion [6], and although there have been continued advancements in community screening initiatives, awareness campaigns, medical and treatment protocols, and multidisciplinary frameworks to enhance quantity and quality of life [3], it is estimated that, on average, anywhere between 10 and 30% of patients diagnosed with cancer are also at risk of psychological distress (anxiety and depression) and adjustment difficulties at some point during the illness trajectory [9, 10] and more so in inpatient settings [10, 11]. While some distress is to be expected with such a life-changing event, if untreated or unrecognized, these diagnoses can last months or years after diagnosis, even if the prognosis is favorable [12–14]. One study reported the prevalence of depression to be as high as 58% in advanced cancer cases [15]. Another study illustrated that the highest rates of depression were associated with testicular cancer, which actually has one of the best prognoses for cancer

M. C. Schlatter (✉)
American Hospital Dubai, Dubai, United Arab Emirates
e-mail: mschlatter@ahdubai.com

© The Author(s) 2024
H. O. Al-Shamsi (ed.), *Cancer Care in the United Arab Emirates*,
https://doi.org/10.1007/978-981-99-6794-0_17

in general [16]. This indicates a need to be cautious when making assumptions about how an individual might cope post-cancer diagnosis.

17.2 Psycho-Oncology Diagnoses and Risk Factors

Psychological diagnoses are often associated with specific individual characteristics (e.g., age, gender, marital status), social and contextual factors (e.g., level of education, income, and social support), premorbid psychological factors (e.g., personality and prior coping behaviors), the response to illness (e.g., hopelessness, resilience), the characteristics of cancer (e.g., type, stage, and grade), and the type of cancer treatment [17] (Fig. 17.1). The risk factor most likely associated with the development of mental health issues (anxiety, depression, and personality disorders) is treatment involving all three protocols of chemotherapy, radiation, and surgery [16].

Although psychological distress has not been directly linked to an increased incidence of cancer, it has been linked with higher rates of cancer-specific mortality [16] and a greatly reduced quality of life overall [18]. Indeed, in other countries, the risk of self-harm and suicide has been reported to be at its highest within 6-12 months of

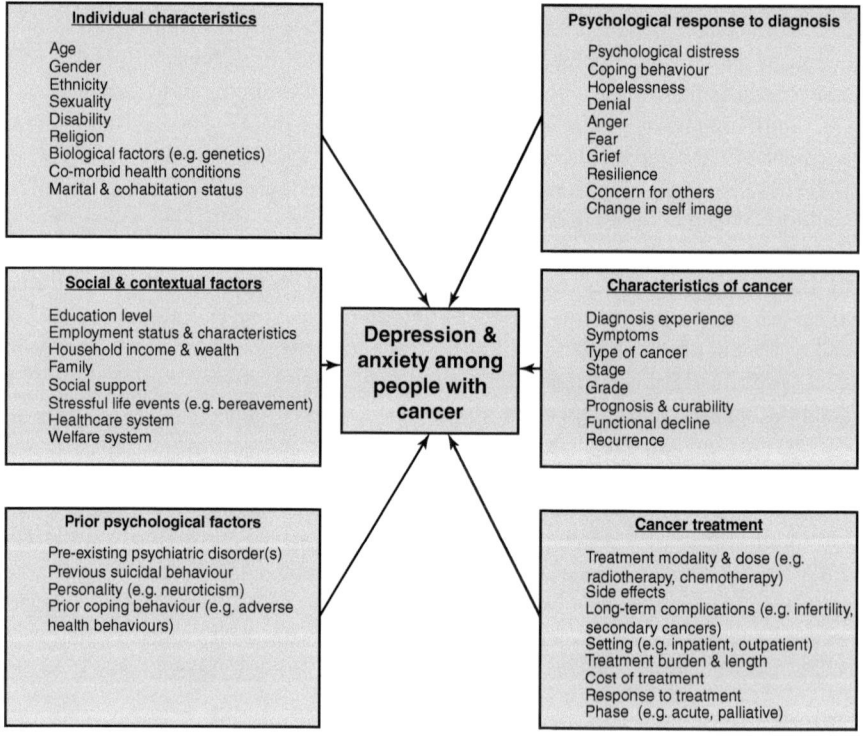

Fig. 17.1 Factors that may contribute to depression and anxiety among people living with and beyond cancer [17]. (Figure used with permission from Dr. Claire Niedzwiedz)

a mental health diagnosis, which suggests that the initial months after a cancer diagnosis necessitate a critical time period for psychological intervention [12, 15, 19].

17.3 Interventions in Psycho-Oncology

Psychological interventions address several targets—physical, emotional, family, and social problems; treatment optimization; improvement of health; and spiritual aspects. Furthermore, the benefits of psychological intervention are numerous: a reduction in disease progression, increased survival rates, enhanced health-related quality of life (HRQoL), a sense of purpose and overall well-being, less hospitalization, and lower medical costs [20–22].

Nonetheless, an issue commonly faced by patients around the world is the lack of appropriate psycho-oncological support services available. Only 38% of patients are thought to be aware of available help [23], and only 5–10% of patients seek professional help voluntarily during these times [24, 25]. These statistics are likely to be even lower in the UAE given various cultural dynamics, beliefs, and language barriers, as well as the lack of specialized support programs and associated research in this area. Indeed, pure medical management of both the condition and treatment-related side effects (the "medical model") typically supersedes recognition and treatment of psychiatric needs, even though the psychological ramifications remain months or even years after an individual has completed treatment [16]. Physicians typically cite a lack of time or knowledge when it comes to engaging with their patients on nonmedical or sensitive issues, and nurses have traditionally taken on the role of emotional caregivers on the ward; however, burnout and compassion fatigue are common among all oncology practitioners [26–29]. Ironically, screening questionnaires for psychological aspects of the cancer trajectory are easy to administer, short, and generally culturally acceptable, and some have been translated into Arabic [30–32], the official language of the UAE. Unfortunately, the Emirates faces a dearth of specialized psychology practitioners embedded directly within oncology units who are trained to regularly administer these tools and report results back to the team in an ethically appropriate and timely manner.

17.4 A Psycho-Oncologist's Approach Toward Patient Care

A psycho-oncologist is a crucial member of the MDT who has the time to provide supportive and educational psychological intervention right from the time of diagnosis and beyond. This may include preoperative and postoperative counseling, learning how to cope with a new diagnosis, waiting for results, understanding multifactorial treatment/treatment side effects such as pain, nausea, and fatigue; changes in relationships (personal, sexual, workplace), family dynamics, changes in body image, loss (of autonomy, physical ability, perceived roles, income, friendships, reproductive ability); fear of recurrence, uncertainty regarding genetic predisposition or risk; and lastly, the management of anxiety (including existential), overwhelm, grief, depression, guilt, and anger. Regardless of the situation, the goal is to compassionately weave in resilience, purpose, a sense of control, meaning, and commitment to a new way of life through education, coping strategies, cognitive techniques [33], supportive values, and compassion-based exercises [34] in either individual or group settings, for both outpatients and inpatients. Liaison with oncology-specific practitioners in the unit is also necessitated for more complex or delicate cases, such as pregnancy-associated breast cancer, anxiety associated with hereditary cancers, lack of treatment adherence, needle phobias, or individuals that need psychiatric management, in which case involvement from gynecological, genetic, nursing, and pharmacological counseling support and psychiatry is important.

For more advanced cases of cancer, where the sanctity of life is often seen as paramount for families due to cultural norms [35], psychological intervention and support can address diminished quality of life as a result of recurrence and/or progression of disease, especially if death is imminent [36], but it is often preferred that psychologists have additional training for end-of-life issues, especially if the patient is a child. Collaboration with palliative services and the use of specific forms of expertise and therapy are frequently required at these times to maintain pride, hope, and continuity of self; reduce perceived burden; and enhance dignity so that individuals can believe, where possible, that their life has stood for something transcendent of death [37] and that they will be able to face whatever they believe is next on their journey. In some cases, though, families may be fearful of broaching information about continued decline or metastatic spread to their loved ones, so these nuances must be carefully navigated and sensitively addressed by all practitioners involved. Ideally, information about support services and psychological support should be accessible as soon as possible to patients, their families, and caregivers throughout the cancer trajectory, including age-appropriate information for children.

17.5   Barriers in Psycho-Oncology in the UAE

Although the evidence thus far is strongly in favor of the provision of psychological support for cancer patients, there are noted barriers to uptake, including older age, less perceived need, a preference for self-reliance, negative beliefs about mental illness and/or "labels," and dissatisfaction from previous help-seeking or healthcare interactions [38, 39]. Barriers may prevent individuals from expressing their true difficulties and from getting their needs met [12], leading to more anxiety, depression, resentment, and a reduced quality of life [40], but research on unmet needs does provide some insight into individuals' experiences of the cancer process [41]. One recent study of attendees to a cancer survivorship clinic in Jordan showed that late-stage diagnosis and quality of life score were significant predictors of need in the physical, psychological, health system, information, and financial domains [42]. Some families decline psychological input because they may prefer either privacy, to rely on their medical practitioner's advice alone, to speak with their families, elders, or selected others, or to seek guidance from pastoral or chaplaincy care instead [43]. Unfortunately, research on the needs of various religious groups, particularly Muslims (with Islam being the official and majority religion in the UAE), is lacking [44, 45].

17.6 Conclusion

Although there is much evidence for the value of psychologists within oncology, particularly with respect to improvements in HRQoL and reductions in psychological and emotional burden for the patient and their loved ones, for many individuals in the UAE, the diagnosis of cancer is still seen as a stigma or issue that must be privately addressed with the immediate or most necessary members of the medical team, family, or trusted others. Many others live alone in the region and are focused on retaining their jobs, so trying to navigate the distressing new world of oncology, even just from a financial perspective, is challenging, let alone from a psychological perspective. The involvement of an individual whose role is to assess how an individual is coping at the psychosocial level can thus be met with skepticism, quiet intrigue, doubt, and confusion during a time when the urgency of the medical situation and treatment predominates. Given the lack of multidisciplinary teams inclusive of both psychology and psychiatry in oncology, as well as the scarcity of time to address psychological concerns by medical teams, it is imperative that the region integrate more specialized psychologists directly into oncology units as part of standard care [46], where practitioners can easily identify those in need and adjust to the flow of an individual's treatment pathway and personal needs, as well as provide relevant information back to the team as permissible and where warranted. The psychologist's role in recognizing one's emotional needs and providing coping strategies as a simple and natural sequel to addressing one's physical or medical needs must be supported by the primary practitioners in charge and, where possible, normalized for patients. Those practitioners should also be very comfortable raising the topic so that patients never feel they have been singled out because they are "not coping."

The UAE is also an ideal place to research and directly provide both structured prehabilitation [47, 48] and survivorship programs to address and manage traditional predictors of poorer outcomes and longer-term impairments in physical, sexual, psychological, cognitive, and social functioning [40]. Psychologists with training in research and clinical methodology would be ideally suited to these programs, especially given advances in technology, which will be a viable source of additional measurement and support in the future. It follows that, given the large number of nationalities in the region, psychology practitioners also need to be sensitive to the widely differing beliefs of individuals under treatment. They need to be open to often strong opinions and perceptions around psychology, as well as the importance of family and community contributions within decision-making strategies throughout care, and they should appreciate the additional time it may take to build a therapeutic relationship. New practitioners to the region cannot utilize a "one-size-fits-all" approach, but they should strive to normalize the expected psychological ramifications of a serious medical diagnosis and provide applicable coping strategies. The integration and acceptance of psycho-oncology into standard oncology care will symbolize a significant step toward comprehensive multidisciplinary support for patients in the UAE.

Conflict of Interest The author has no conflict of interest to declare.

References

1. Akechi T. Psycho-oncology: history, current status, and future directions in Japan. JMA J. 2018;1(1):22–9.
2. Li M, Kennedy EB, Byrne N, Gérin-Lajoie C, Katz MR, Keshavarz H, et al. Management of depression in patients with cancer: a clinical practice guideline. J Oncol Pract. 2016;12(8):747–56.
3. Al-Shamsi HO, Abyad AM, Rafii S. A proposal for a National Cancer Control Plan for the UAE: 2022–2026. Clin Pract. 2022;12(1):118–32.
4. Berardi RMF, Rinaldi S, Torniai M, Mentrasti G, Scortichini L, Giampieri R. Benefits and limitations of a multidisciplinary approach in cancer patient management. Cancer Manag Res. 2020;12:9363–74.
5. Scott B. Multidisciplinary team approach in cancer care: a review of the latest advancements featured at ESMO 2021. Eur Med J Oncol. 2022;10:2–11.
6. Abu-Gheida IH, Nijhawan N, Al-Awadhi A, Al-Shamsi HO. General oncology care in the UAE. In: Al-Shamsi HO, Abu-Gheida IH, Iqbal F, Al-Awadhi A, editors. Cancer in the Arab world. Singapore: Springer Singapore; 2022. p. 301–19.
7. Arafa MA, Farhat KH. Why cancer incidence in the Arab counties is much lower than other parts of the world? J Egypt Natl Canc Inst. 2022;34(1):41.
8. Nair SC, Jaafar H, Jaloudi M, Qawasmeh K, AlMarar A, Ibrahim H. Supportive care needs of multicultural patients with cancer in the United Arab Emirates. Ecancermedicalscience. 2018;12:838.
9. Pitman A, Suleman S, Hyde N, Hodgkiss A. Depression and anxiety in patients with cancer. BMJ. 2018;361:k1415.
10. Naser AY, Hameed AN, Mustafa N, Alwafi H, Dahmash EZ, Alyami HS, et al. Depression and anxiety in patients with cancer: a cross-sectional study. Front Psychol. 2021;12:585534.
11. Clark PG, Rochon E, Brethwaite D, Edmiston KK. Screening for psychological and physical distress in a cancer inpatient treatment setting: a pilot study. Psychooncology. 2011;20(6):664–8.
12. Lang-Rollin I, Berberich G. Psycho-oncology. Dialog Clin Neurosci. 2018;20(1):13–22.
13. Grassi L. Psychiatric and psychosocial implications in cancer care: the agenda of psycho-oncology. Epidemiol Psychiatr Sci. 2020;29:e89.
14. Smith HR. Depression in cancer patients: pathogenesis, implications and treatment (review). Oncol Lett. 2015;9(4):1509–14.
15. Meyer HA, Sinnott C, Seed PT. Depressive symptoms in advanced cancer. Part 2. Depression over time; the role of the palliative care professional. Palliat Med. 2003;17(7):604–7.
16. Chang WH, Lai AG. Cumulative burden of psychiatric disorders and self-harm across 26 adult cancers. Nat Med. 2022;28(4):860–70.
17. Niedzwiedz CL, Knifton L, Robb KA, Katikireddi SV, Smith DJ. Depression and anxiety among people living with and beyond cancer: a growing clinical and research priority. BMC Cancer. 2019;19(1):943.
18. Wang Y-H, Li J-Q, Shi J-F, Que J-Y, Liu J-J, Lappin JM, et al. Depression and anxiety in relation to cancer incidence and mortality: a systematic review and meta-analysis of cohort studies. Mol Psychiatry. 2020;25(7):1487–99.

19. Henson KE, Brock R, Charnock J, Wickramasinghe B, Will O, Pitman A. Risk of suicide after cancer diagnosis in England. JAMA Psychiatry. 2019;76(1):51–60.
20. Pinquart M, Duberstein PR. Depression and cancer mortality: a meta-analysis. Psychol Med. 2010;40(11):1797–810.
21. Mausbach BT, Bos T, Irwin SA. Mental health treatment dose and annual healthcare costs in patients with cancer and major depressive disorder. Health Psychol. 2018;37(11):1035–40.
22. Mausbach BT, Decastro G, Schwab RB, Tiamson-Kassab M, Irwin SA. Healthcare use and costs in adult cancer patients with anxiety and depression. Depress Anxiety. 2020;37(9):908–15.
23. Faller H, Koch U, Brähler E, Härter M, Keller M, Schulz H, et al. Satisfaction with information and unmet information needs in men and women with cancer. J Cancer Surviv. 2016;10(1):62–70.
24. Walker J, Hansen CH, Martin P, Symeonides S, Ramessur R, Murray G, et al. Prevalence, associations, and adequacy of treatment of major depression in patients with cancer: a cross-sectional analysis of routinely collected clinical data. Lancet Psychiatry. 2014;1(5):343–50.
25. Holland JC. IPOS Sutherland memorial lecture: an international perspective on the development of psychosocial oncology: overcoming cultural and attitudinal barriers to improve psychosocial care. Psychooncology. 2004;13(7):445–59.
26. Bui S, Pelosi A, Mazzaschi G, Tommasi C, Rapacchi E, Camisa R, et al. Burnout and oncology: an irreparable paradigm or a manageable condition? Prevention strategies to reduce burnout in oncology health care professionals. Acta Biomed. 2021;92(3):e2021091.
27. Eelen S, Bauwens S, Baillon C, Distelmans W, Jacobs E, Verzelen A. The prevalence of burnout among oncology professionals: oncologists are at risk of developing burnout. Psychooncology. 2014;23(12):1415–22.
28. Kleiner S, Wallace JE. Oncologist burnout and compassion fatigue: investigating time pressure at work as a predictor and the mediating role of work-family conflict. BMC Health Serv Res. 2017;17(1):639.
29. Dreismann L, Goretzki A, Ginger V, Zimmermann T. What if… I asked cancer patients about psychological distress? Barriers in psycho-oncological screening from the perspective of nurses-a qualitative analysis. Front Psych. 2021;12:786691.
30. Alzahrani AS, Demiroz YY, Alabdulwahab AS, Alshareef RA, Badri AS, Alharbi BA, et al. The diagnostic accuracy of the 9-item patient health questionnaire as a depression screening instrument in Arabic-speaking cancer patients. Neurol Psychiatry Brain Res. 2020;37:110–5.
31. Suleiman K, Al Kalaldeh M, AbuSharour L, Yates B, Berger A, Mendoza T, et al. Validation study of the Arabic version of the brief fatigue inventory (BFI-A). East Mediterr Health J. 2019;25(11):784–90.
32. Alosaimi FD, Abdel-Aziz N, Alsaleh K, AlSheikh R, AlSheikh R, Abdel-Warith A. Validity and feasibility of the Arabic version of distress thermometer for Saudi cancer patients. PLoS One. 2018;13(11):e0207364.
33. Getu MA, Chen C, Panpan W, Mboineki JF, Dhakal K, Du R. The effect of cognitive behavioral therapy on the quality of life of breast cancer patients: a systematic review and meta-analysis of randomized controlled trials. Qual Life Res. 2021;30(2):367–84.
34. Zhao C, Lai L, Zhang L, Cai Z, Ren Z, Shi C, et al. The effects of acceptance and commitment therapy on the psychological and physical outcomes among cancer patients: a meta-analysis with trial sequential analysis. J Psychosom Res. 2021;140:110304.
35. Nijhawan N, Al-Shamsi HO. Experiences and challenges of a new palliative care service in the United Arab Emirates. Palliat Med Hospice Care Open J. 2022;8(2):30–4.
36. Diarmuid ÓC, Prizeman G, Korn B, Donnelly S, Hynes G. Dying in acute hospitals: voices of bereaved relatives. BMC Palliat Care. 2019;18(1):91.
37. Chochinov HM, Hack T, Hassard T, Kristjanson LJ, McClement S, Harlos M. Dignity therapy: a novel psychotherapeutic intervention for patients near the end of life. J Clin Oncol. 2005;23(24):5520–5.
38. Shi W, Shen Z, Wang S, Hall BJ. Barriers to professional mental health help-seeking among Chinese adults: a systematic review. Front Psych. 2020;11:442.

39. Mosher CE, Winger JG, Hanna N, Jalal SI, Fakiris AJ, Einhorn LH, et al. Barriers to mental health service use and preferences for addressing emotional concerns among lung cancer patients. Psychooncology. 2014;23(7):812–9.
40. Schmidt ME, Goldschmidt S, Hermann S, Steindorf K. Late effects, long-term problems and unmet needs of cancer survivors. Int J Cancer. 2022;151(8):1280–90.
41. Evans Webb M, Murray E, Younger ZW, Goodfellow H, Ross J. The supportive care needs of cancer patients: a systematic review. J Cancer Educ. 2021;36(5):899–908.
42. Al-Omari A, Al-Rawashdeh N, Damsees R, Ammar K, Alananzeh I, Inserat B, et al. Supportive care needs assessment for cancer survivors at a Comprehensive Cancer Center in the Middle East: mending the gap. Cancers. 2022;14(4):1002.
43. Hyer JM, Paredes AZ, Kelley EP, Tsilimigras D, Meyer B, Newberry H, et al. Characterizing pastoral care utilization by cancer patients. Am J Hosp Palliat Care. 2021;38(7):758–65.
44. Malas E, Chaar B, Krayem G. End-of-life treatment decisions in adult Muslims: a scoping review protocol. JBI Evid Synth. 2020;18(7):1528–36.
45. Gustafson C, Lazenby M. Assessing the unique experiences and needs of Muslim oncology patients receiving palliative and end-of-life care: an integrative review. J Palliat Care. 2019;34(1):52–61.
46. Al-Shamsi H, Darr H, Abu-Gheida I, Ansari J, McManus MC, Jaafar H, et al. The state of cancer Care in the United Arab Emirates in 2020: challenges and recommendations, a report by the United Arab Emirates oncology task force. Gulf J Oncol. 2020;1(32):71–87.
47. Santa Mina D, van Rooijen SJ, Minnella EM, Alibhai SMH, Brahmbhatt P, Dalton SO, et al. Multiphasic prehabilitation across the cancer continuum: a narrative review and conceptual framework. Front Oncol. 2020;10:598425.
48. Giles C, Cummins S. Prehabilitation before cancer treatment. BMJ. 2019;366:l5120.

Dr. Melanie Schlatter (PhD) is a clinical health psychologist of dual nationality (New Zealand and Swiss) who has lived in Dubai since 2006. She presents a unique advantage to the field of oncology, given her qualifications and research in psychoneuroimmunology. After completing an honorary research fellowship, she became the region's first health psychologist, working for 13 years with Dr. Houriya Kazim, the first Emirati female surgeon, while also consulting as a community psychologist for the American Hospital Dubai until she joined them full-time in oncology and psychiatry in 2021.

Melanie is a board member and secretary of Brest Friends, Dubai's first breast cancer support group, and she helped to develop Majlis Al Amal, the UAE's first cancer drop-in center, both of which are Dr. Kazim's original concepts. Melanie has published in the Annals of Behavioral Medicine, the Journal of Psycho-Oncology, Psychosomatic Medicine, the Journal of Vestibular Research, Brain Behavior and Immunity, and the Middle East Journal of Positive Psychology. Her most recent contribution was a chapter on health psychology for a UAE university textbook. She has also taught at universities in New Zealand and Dubai, and she is a reviewer for two scientific journals.

Artificial Intelligence (AI) in Oncology in the UAE

18

Khalid Shaikh and Sreelekshmi Bekal

18.1 Introduction

Ever since its inception decades ago, artificial intelligence (AI) has been a catchphrase for the sustainable future ahead. Like in every other sector, AI is believed to be the potential panacea to radically alter the field of healthcare. Oncology and related fields are examples of such focus areas where AI tools are now widely used. It is critical to embrace this new wave of technological revolution in order to maximize potential and modify future strategies.

Cancer has a significant impact on health worldwide, leading to a high level of sickness and death. While progress has been made in recent decades, there are still challenges in providing individualized care. Artificial intelligence (AI) has emerged as a technology that can enhance cancer care. Its applications in oncology range from optimizing cancer research to improving clinical practices, such as predicting patient outcomes and treatment responses, as well as gaining a better understanding of tumor characteristics [1].

To better understand AI, it is important to grasp its history and key areas, which will help us comprehend its current capabilities and future possibilities more effectively. The concept of using computers to imitate intelligent behavior and critical thinking was first proposed by Alan Turing in 1950. The term "artificial intelligence" (AI) was coined by John McCarthy in 1956, defining it as the science and engineering behind creating intelligent machines [2].

AI initially consisted of simple sets of rules ("if, then" statements) and has evolved over time to include more complex algorithms that can perform tasks resembling human brain functions [2, 3]. It relies on computers following algorithms created by humans or learned through computer-based methods to support decision-making or perform specific tasks [1]. Machine learning, a subfield of AI,

K. Shaikh (✉) · S. Bekal
Prognica Labs, Dubai, United Arab Emirates
e-mail: khalid@prognica.com; sreelekshmi@prognica.com

© The Author(s) 2024 281
H. O. Al-Shamsi (ed.), *Cancer Care in the United Arab Emirates*,
https://doi.org/10.1007/978-981-99-6794-0_18

enables computers to enhance their own performance by continuously incorporating new data into existing iterative models. Deep learning, another subfield of machine learning, utilizes mathematical algorithms implemented through multi-layered computational units that resemble human cognition. This includes various types of neural networks such as recurrent neural networks, convolutional neural networks, and long-term short memory networks [1]. These two concepts, machine learning and deep learning, are central to the AI revolution in managing cancer patients [2].

AI is a rapidly evolving model that encompasses various scientific disciplines, including those focused on managing cancer patients. It can be defined as the capacity of a machine to learn from representative models, recognize patterns and interactions, and utilize this knowledge to enhance decision-making processes in a specific field [2]. In the field of precision oncology, AI is transforming the current landscape by integrating vast amounts of data from multi-omics analyses with advancements in high-performance computing and innovative deep-learning techniques. Notably, the scope of AI applications is expanding and now includes novel approaches for cancer detection, screening, diagnosis, and classification. It also involves the characterization of cancer genomics, analysis of the tumor microenvironment, evaluation of biomarkers for prognostic and predictive purposes, as well as strategies for follow-up care and drug discovery [2].

18.2 Global Cancer Statistics 2020

According to the data from GLOBOCAN 2020, there were 19.3 million new cancer cases diagnosed in 2020, resulting in approximately 10.0 million deaths. Projections from GLOBOCAN suggest that the number of cancer cases will rise to 28.4 million by 2040. Globally, female breast cancer has become the most prevalent type of cancer, accounting for 11.7% of all cases, followed by lung cancer (11.4%), colorectal cancer (10.0%), prostate cancer (7.3%), and stomach cancer (5.6%). In terms of cancer-related deaths, lung cancer is the leading cause, causing 1.8 million deaths (18%), followed by colorectal cancer (9.4%), liver cancer (8.3%), stomach cancer (7.7%), and female breast cancer (6.9%). Among men, the most common cancer types are lung, prostate, and colorectal cancer, while breast, colorectal, and lung cancer are the most common among women. Overall, the top 10 cancers contribute to more than 60% of cancer incidence and 70% of cancer-related mortality [4].

Based on the annual report of the United Arab Emirates (UAE) National Cancer Registry in 2021, malignant neoplasm of colon was the leading cause of cancer-related deaths, accounting for an estimated average of 11.49% of all cancer deaths per year. Trachea, bronchus and lung cancer ranked second as the most common cause of cancer death in both males and females, and breast cancer was the third most common cause of cancer death for both sexes [5].

Emerging technologies have the potential to address the gaps in healthcare and improve the continuum of care for cancer patients. Among these technologies, artificial intelligence (AI) has emerged as a transformative force. AI-guided clinical care has the capacity to significantly reduce health disparities, especially in resource-limited settings. By incorporating AI technology into cancer care, we can enhance the precision and efficiency of diagnosis, assist in clinical decision-making, and ultimately achieve improved health outcomes [6].

18.3   Fundamentals of Artificial Intelligence and Learning Algorithms

An AI system refers to a computer system capable of performing tasks that typically require human intelligence. These systems utilize various learning methods, including machine learning (ML) and deep learning (DL), to operate [7]. ML, which falls under the umbrella of AI, involves algorithms that enable systems to learn automatically. The ML process involves training and testing images, as depicted in Fig. 18.1, allowing the system to enhance its learning experience without the need for complex programming. These algorithms find extensive applications in fields such as medical imaging, computer vision, biometric recognition, object detection, and automation [7, 8]. Within ML, there are three key types: supervised learning, unsupervised learning, and reinforcement learning [7, 8].

Deep learning (DL), a subset of machine learning (ML) and artificial intelligence (AI), is recognized as a fundamental technology in the Fourth Industrial Revolution (4IR) or Industry 4.0. With its ability to learn from data, DL has emerged as a prominent field in computing and finds applications in diverse areas such as healthcare, visual recognition, text analytics, cyber security, and more. DL utilizes multiple layers to abstract and represent data, enabling the construction of computational models [9].

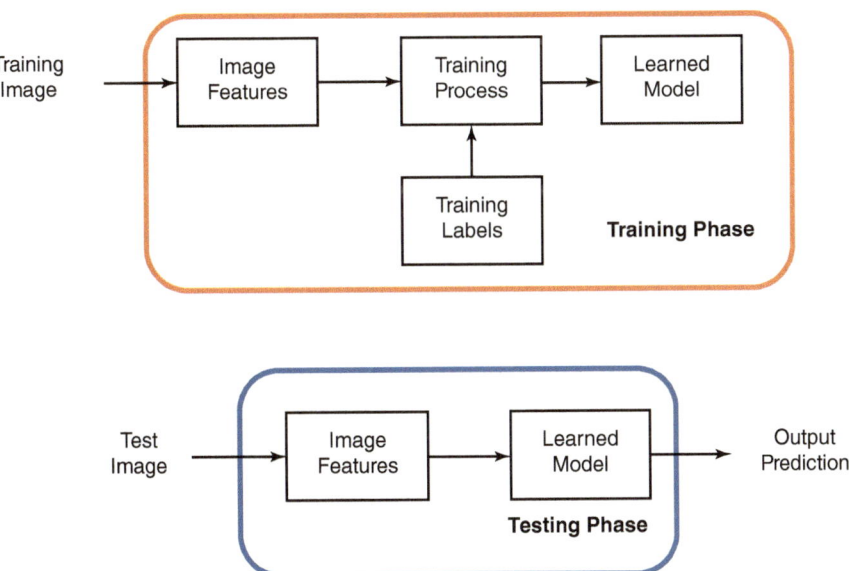

Fig. 18.1   Steps of machine learning algorithm [7]

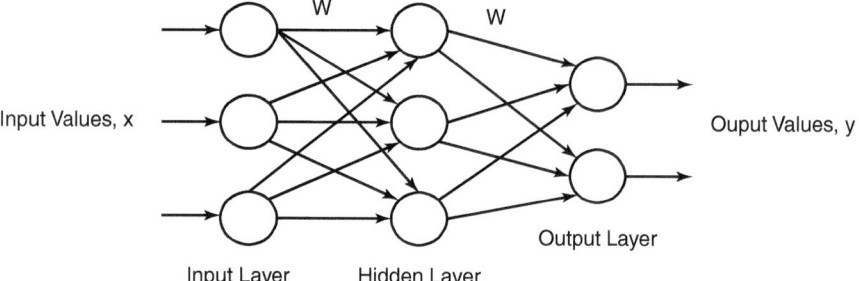

Fig. 18.2 Basic model of artificial neural network (ANN) [7]

DL was introduced by Hinton et al. and is built upon the concept of artificial neural networks (ANNs) [9]. ANNs draw inspiration from the human brain's neurobiology and its capacity to learn complex patterns through cascading, layered combinations of neurons that gradually extract more intricate features [10, 11]. Early ANNs were designed by simulating human neurons in computers. A deep learning algorithm is an extension of the ANN, consisting of an input layer, several hidden layers, and an output layer (see Fig. 18.2). Each layer is connected through nodes, with each hidden layer providing predictions based on the input received from the previous layer. The primary distinction between ANNs and deep learning algorithms lies in the number of hidden layers, with ANNs having a single hidden layer, while deep learning algorithms have two or more hidden layers [7].

18.4   AI in Oncology and Related Fields

Similar to other areas of healthcare, AI is extensively employed in oncology to interpret, analyze, and visualize complex medical data using diverse algorithms [7]. AI has demonstrated promise in enhancing cancer imaging diagnostics, evaluating treatment responses, predicting clinical outcomes, optimizing research, enabling personalized medicine, facilitating drug development, advancing translational oncology, and supporting precision oncology. Unlike traditional healthcare technologies, AI technology collects and processes data before presenting it to end users for informed decision-making. It leverages various machine learning and deep learning algorithms that can analyze patterns in medical data and make autonomous decisions [7].

In the field of cancer, diagnostics serve as a crucial starting point for designing appropriate therapeutic approaches and clinical management, and the refinement of diagnostics through AI represents a significant accomplishment. Furthermore, this highlights the importance of future AI developments in unexplored yet critical areas such as drug discovery, therapy administration, and follow-up strategies [2].

Breast cancer, lung cancer, and prostate cancer are the specific types of cancer that currently benefit the most from AI-based devices in clinical practice. This primarily stems from their higher incidence compared to other tumor types. However,

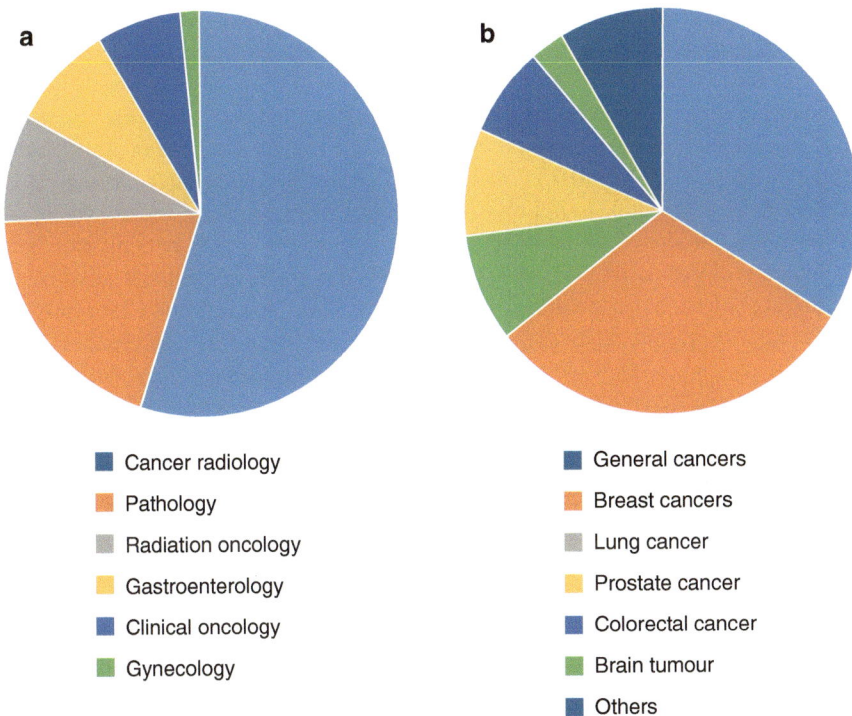

Fig. 18.3 Current status of artificial intelligence in oncology and related fields [2]

in the future, additional tumor types, including rare cancers that lack standardized approaches, should also be considered. Since AI relies on large datasets of cases for analysis, the improvement of treatment for rare neoplasms may be a later achievement [2]. The current status of artificial intelligence in oncology and related fields is depicted in Fig. 18.3.

AI solutions have been created to address diverse challenges associated with cancer. Various stakeholders, including medical institutions, hospital systems, and technology companies, are actively developing AI tools with the goal of enhancing clinical decision-making, expanding access to cancer care, and improving overall efficiency in delivering safe and valuable oncology services. AI applications in the field of oncology have exhibited precise technical performance in tasks such as image analysis, predictive analytics, and the implementation of precision oncology approaches [12].

18.4.1 Artificial Intelligence and Cancer Care in the UAE

The burden of cancer in the United Arab Emirates (UAE) is significant, ranking as the second leading cause of non-communicable disease (NCD)-related mortality in the country. The healthcare system in the UAE has experienced rapid growth and development, resulting in its ranking as the 27th worldwide by the World Health Organization (WHO) [13–15].

The earliest published record of oncology care in the UAE dates back to 1981 when five cases of hepatocellular carcinoma were documented. The first major cancer care facility, Tawam Hospital, was established in Al-Ain, one of Abu Dhabi's main cities, in 1979. By 1983, Tawam Hospital had been designated as the UAE's cancer referral hospital. To improve accessibility, several general oncology care services were introduced throughout the country, enabling cancer patients to receive healthcare closer to their homes. Until 2007, the UAE government covered the entire cost of oncology treatment for all UAE citizens and residents. Presently, the government continues to bear the cost of cancer treatment for UAE citizens, while non-citizens are covered by insurance plans. However, expatriates sometimes face the challenge of having to return to their home countries for ongoing medical treatment due to insurance plans with expiration dates. In response to this issue, the prestigious Emirates Oncology Society (EOS) recently published a collaborative document proposing alternative solutions to adjust cancer insurance packages nationwide [13].

The leadership of the UAE has played a pioneering role in establishing robust cancer screening programs. In 2009, an annual mammography screening program was initiated, advising all UAE national women aged 40 and above to undergo screening. Subsequently, in July 2010, a nationwide screening program for colorectal cancer was launched, and by 2014, screening programs for breast, colorectal, and cervical cancers were established. In 2017, following the release of lung cancer data, a low-dose CT scan-based screening program for lung cancer was implemented. Furthermore, several initiatives have been undertaken to raise awareness and promote cancer screening, such as the annual "Pink Caravan" event, which reaches over 45,000 women across the UAE, focusing on breast cancer awareness and encouraging screening [13]. Alongside screening programs, the healthcare sector in the UAE leads the way in cancer prevention and diagnosis programs supported by extensive research efforts.

The UAE has been at the forefront among regional countries in embracing and fostering cutting-edge technologies like artificial intelligence (AI) in cancer care. Currently, multiple AI platforms are being utilized in the UAE, particularly for assisting in cancer diagnosis through imaging, including breast and lung cancer screening. Table 18.1 presents an overview of the current status of clinical and research initiatives in the UAE [16].

Health-tech companies like Prognica Labs are dedicated to enhancing clinical outcomes in the battle against breast cancer. They employ AI and deep learning to analyze medical images and generate valuable information and data for cancer prediction and diagnosis. Through meticulous research, they generate new knowledge to drive effective innovation, education, and practice. In collaboration with top-tier hospitals and universities in the region, they have launched a project called "Retrospective analysis of breast cancer diagnosis among young and older women in the UAE." These retrospective studies will establish a benchmark to assess the accuracy of diagnosis and ensure the provision of high-quality care.

Table 18.1 The current status of clinical and research initiatives in the UAE [16]

AI technology	Facility	Year	Format	Status
IBM™ Watson Oncology—Pilot	SEHA— Tawam Hospital	2016	Clinical decision in oncology	Suspended
AI enabled Digital mammography system, Lunit INSIGHT MMG Lung Cancer screening— Coreline—Medical AI solutions	International Radiology Centre— Sharjah Commercial	2021	AI-enabled independent reader for breast cancer screening and lung cancer screening	Active
Prognica Labs	Dubai Commercial	2021	Prognica Labs uses artificial intelligence to detect masses in mammography screenings	Active
Mammography Intelligent Assessment (Mia)™	UAE Commercial	2021	First and only AI-enabled independent reader for breast cancer screening to be commercially available in the UAE	Active
The GI Genius™ intelligent endoscopy module	Sheikh Shakhbout Medical City Abu Dhabi	2021	Is the first-to-market, computer-aided polyp detection system powered by AI	Active
Khalifa University researchers	Research Abu Dhabi	2021	To identify cancer in tissue samples, which could speed up diagnosis and improve outcomes in patients with colorectal cancer	Active
DoH—Abu Dhabi	Research Abu Dhabi	2022	First Personalised Precision Medicine for oncology in collaboration with Mubadala Health, Cleveland Clinic Abu Dhabi, NYU Abu Dhabi, Mohamed bin Zayed University of Artificial Intelligence and G42 Healthcare	Active
Mohamed bin Zayed University of Artificial Intelligence team	Research Abu Dhabi	2022	AI tool to better diagnosis and treatment of pancreatic cancer	Active

The gastroenterology team at Sheikh Shakhbout Medical City (SSMC) in Abu Dhabi has introduced an advanced AI system for gastrointestinal intestinal endoscopy. This system significantly improves the detection of precancerous polyps in the colon. While conventional medical examinations detect these growths in about 30–40% of individuals, AI technology increases the detection rate to approximately 50–55%, resulting in a substantial increase in the adenoma detection rate. This benchmark is crucial as every 1% increase in the adenoma detection rate reduces the risk of colon cancer by 3% and the risk of death from colon cancer by 5%. With a

15% increase, there will be a 10% decrease in adenoma detection, leading to a 30% decrease in the risk of colorectal cancer and a 50% decrease in the risk of death from colon cancer [17].

A team of researchers from New York University (NYU) and NYU Abu Dhabi has developed an innovative AI system capable of identifying breast cancer in ultrasound images. With "radiologist-level accuracy," this system serves as a decision-support tool for clinicians [18].

In collaboration with Mubadala Health, Cleveland Clinic Abu Dhabi, NYU Abu Dhabi, the Mohamed bin Zayed University of Artificial Intelligence, and G42 Healthcare, the Department of Health—Abu Dhabi (DoH) has launched the first personalized precision medicine program for oncology in the region. Initially focusing on breast cancer patients, this program aims to treat patients and reduce the risk of disease recurrence [19].

18.4.2 Artificial Intelligence in Cancer-Related Image Analysis

Image analysis has emerged as a highly impactful application of AI, particularly within the field of oncology, due to the abundance of digital imaging data in medicine. The development of convolutional neural networks (CNNs) has revolutionized image analysis by enabling pixel-level examination. CNNs have the advantage of considering pixel orientation, allowing them to recognize lines, curves, and objects within images. Recent studies have demonstrated that CNN-based models are on par with humans in picture classification and object detection [11, 20, 21].

AI has the potential to enhance traditional medical imaging techniques such as computed tomography (CT), magnetic resonance imaging (MRI), and X-rays by offering computational capabilities that enable faster and more accurate image processing at scale [7]. The benefits of AI in medical imaging include higher automation, increased productivity, standardized processes, more accurate diagnosis, computing quantitative data, and assistance for doctors. For example, AI models developed by Google have achieved a 99% accuracy rate in diagnosing breast cancer from medical images, surpassing the performance of some doctors [7].

Computer-aided detection (CADe) and computer-aided diagnosis (CADx) systems are designed to assist doctors in interpreting medical images. Figure 18.4 depicts the general framework of CADe/CADx system. These interdisciplinary systems combine technologies such as AI and computer vision to extract essential information from various imaging techniques like X-ray, MRI, and CT. CADe systems detect conspicuous structures, while CADx systems evaluate these structures. Although CAD systems have been used in clinical environments for the past 50 years, they serve as supportive tools rather than providing a complete solution. Ultimately, doctors are responsible for interpreting medical images. However, CAD systems aim to detect early signs of abnormalities that may go unnoticed by doctors, such as cancerous tumors or glaucoma [7, 22, 23].

Fig. 18.4 General framework of CADe/ CADx system [7]

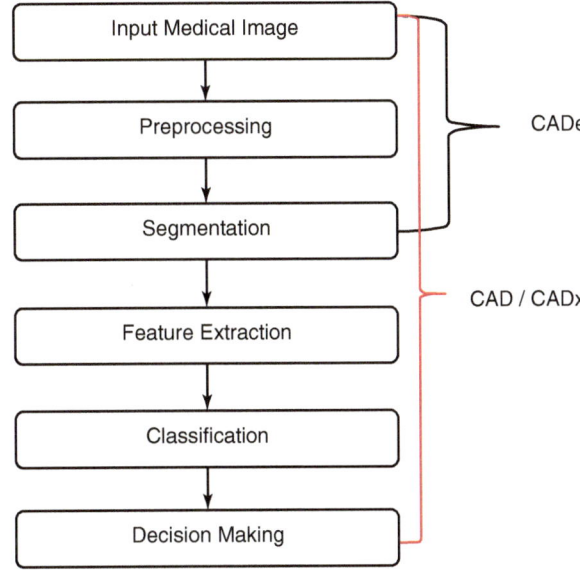

AI has found various applications in the field of medical imaging, encompassing several areas [7]:

- Medical image analysis: AI technology has demonstrated superior capability in identifying anomalies and diseases by analyzing medical images compared to human doctors.
- Neurological condition diagnosis: AI can assist in diagnosing neurological diseases such as amyotrophic lateral sclerosis (ALS), and it has shown potential in predicting Alzheimer's disease years before clinical manifestation.
- Detection of cardiovascular abnormalities: AI algorithms can assess a patient's heart structure and provide insights into their risk of cardiovascular disease or the need for surgical intervention. Automated AI systems can analyze common medical tests like chest X-rays, leading to faster identification of abnormalities and reducing the likelihood of misdiagnosis.
- Cancer screening: Early detection of cancer significantly improves patient outcomes. Recent advancements in AI, specifically utilizing convolutional neural networks (CNNs), have shown remarkable success in accurately identifying various types of cancer. These experiments highlight the potential of AI to reduce detection times and enhance diagnostic rates.

18.4.2.1 Relevant Case Studies of AI in Cancer Imaging

AI is finding applications in oncologic radiographic imaging, specifically in the areas of detection and diagnosis. AI-powered imaging algorithms are being employed in clinical settings to detect and monitor potentially cancerous lesions and provide guidance for patient management [24].

Clinical Photographs

A pioneering study demonstrated the potential of deep learning (DL) in cancer imaging by successfully identifying skin cancer based on skin photographs [11, 25]. The study trained a convolutional neural network (CNN) system on a dataset of 130,000 skin images, achieving higher sensitivity and specificity in classifying malignant lesions compared to a panel of 21 board-certified dermatologists. This breakthrough has led to practical applications in detecting skin pathology using patient-generated imaging data [11, 26]. Another application of CNNs involves the automatic detection of polyps during colonoscopy through digital photography. A study showcased the ability of CNNs not only for image classification but also for identifying regions of clinical significance. By training a CNN on colonoscopic images from 1290 patients, researchers achieved a remarkable 94% sensitivity in polyp detection [11, 27].

Radiographic Imaging

Given the remarkable success of AI techniques in computer vision, there is considerable anticipation within the field of radiology, which deals with a multitude of digitized images. The objectives of AI algorithms in this domain have encompassed assisted diagnosis and outcome prediction [11].

AI algorithms have demonstrated effectiveness in streamlining cancer screening and detection. A significant focus has been on automated lung nodule detection and classification, which was the basis of the 2017 Kaggle Data Science Bowl, an international competition for machine learning scientists [11, 28]. Several CNN-based models, arising from this competition and other research groups, have achieved accuracy ranging from 80 to 95%, showcasing promise for lung cancer screening [11, 29–34]. Additionally, CNNs have exhibited success in segmenting tumor volumes, potentially influencing radiotherapy treatment planning [11, 35]. The enhancement of breast cancer screening through AI has also been an active area of investigation, including dedicated data science competitions [11, 36], leading to the development of a CNN algorithm capable of detecting breast malignancy with a sensitivity of 90% [11, 37, 38].

AI has shown promise in detecting radiographic anatomical features of malignancies that surpass the reliability of human clinicians. For instance, diagnosing extranodal extension (ENE) in head and neck cancer lymph nodes has historically posed challenges, but a CNN-based model achieved an accuracy of over 85% in identifying this feature on diagnostic contrast-enhanced CT scans [11, 39]. Since identifying ENE is crucial for prognosis and management decisions in head and neck cancer patients, this model holds potential as a clinical decision-making tool.

Expanding beyond anatomical characterization, AI has demonstrated promise in the emerging field of radiogenomics, where radiographic image analysis is employed to predict underlying genotypic traits. CNNs applied to brain MRIs of patients with low-grade glioma have successfully predicted both IDH mutation and MGMT methylation status with accuracy rates of 85–95% and 83%, respectively, using raw imaging data alone [11, 40, 41].

DL also holds the potential in predicting treatment response based on imaging findings. A recent CNN model achieved an 80% accuracy in predicting a complete response to neoadjuvant chemoradiation [11, 42]. Furthermore, a radiomics signature utilizing extracted features from CT data and a machine learning algorithm was able to predict underlying CD8 cell tumor infiltration and, notably, response to immunotherapy across various advanced cancers [11, 43].

Digital Pathology

In the realm of digital pathology, the increasing digitization of histopathologic tumor specimen slides provides a robust 2D image suitable for DL analysis. DL CNN algorithms have proven to diagnose breast cancer metastasis in lymph nodes with equivalent performance to a panel of pathologists and in a more time-efficient manner [11, 44]. DL has also shown usefulness in the automated Gleason grading of prostate adenocarcinoma hematoxylin and eosin-stained specimens, achieving a 75% agreement rate between the algorithm and pathologists [11, 45].

DL algorithms have advanced beyond automating pathologic diagnosis and have been utilized to characterize the correlation between genotype and phenotype within tumor specimens. By utilizing digitized tissue from lung cancer biopsies, a CNN was trained to predict six different genetic mutations (STK11, EGFR, FAT1, SETBP1, KRAS, and TP53), demonstrating that histopathologic architectural patterns can provide insight into genotypic information [11, 46]. These methods have the potential to assist pathologists in detecting cancer gene mutations and may offer a more cost-effective alternative to direct mutational analysis. In the realm of endoscopic imaging, AI augmentation has consistently shown improved accuracy in detecting esophageal cancer [24, 47].

AI-based models have become an integral part of breast imaging and are now being used in clinical settings. Several breast imaging detection and diagnosis algorithms have received approval from the U.S. Food and Drug Administration [24, 48]. A significant study published in The Lancet Digital Health directly compared the performance of an AI system in breast cancer screening when operating independently versus when assisting a human expert. Through evaluation using retrospectively collected mammographic images of 4463 screen-detected cancers and 100,055 confirmed normal studies, the study demonstrated the potential application of AI through a decision-referral approach, hybrid triaging approach, and cancer detection approach. Simulating the decision-referral approach revealed substantial improvements in the sensitivity and specificity of individual radiologists compared to the consensus conference when combining the strengths of radiologists and AI. While the standalone use of the AI system on the external test dataset resulted in a statistically significant reduction in radiologist sensitivity by 2.6% points and specificity by 2.0% points, the same models could be employed in collaboration with radiologists within the decision-referral mode. In fact, the AI system's optimal configuration within the decision-referral approach increased radiologist sensitivity by 2.6% points and specificity by 1.0% point while automatically triaging 63.0% of the studies [49] (Fig. 18.5).

Fig. 18.5 Stages of breast cancer detection and diagnosis system [7]

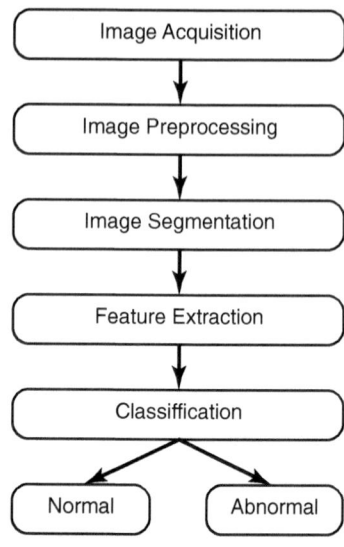

A retrospective analysis carried out by academic hospitals in Korea showcased the advantages of using AI for breast cancer detection through mammography images. The study revealed that the AI system achieved a sensitivity of 88.8% when operating independently, surpassing the sensitivity of radiologists at 75.3%. However, when radiologists received assistance from AI, the accuracy improved by 9.5%, resulting in an overall accuracy of 84.8% [50]. Figure 18.6 illustrates a visual depiction of an AI-powered diagnostic support software.

AI-based imaging models are being utilized for tumor characterization in medical practice. These models can perform tasks such as anatomic segmentation, which involves identifying the boundaries of diseased tissue in relation to normal anatomy, and tumor subtype classification, which uses various features like signal intensity, texture, shape, and other descriptors to make accurate diagnoses. Anatomic segmentation, whether in 2D or volumetric form, is employed in clinical settings for treatment decisions like radiation planning. However, manual tumor segmentation is subject to variability among observers. AI algorithms have the potential to overcome such biases and improve segmentation accuracy [24, 51].

Radiomic analysis is another technique that involves automated extraction of clinically relevant information from radiologic images. It enables the development of radiomic biomarkers by correlating radiomic signatures with genetic, histologic, and other data. This approach holds promise for providing additional insights into tumor pathology and improving diagnosis without the need for invasive sampling, giving rise to the concept of "virtual biopsy" [24, 52].

Moreover, imaging-based machine learning models can be used to predict future outcomes for cancer patients, including locoregional recurrence, distant recurrence, and mortality. For instance, emerging ML models based on imaging data have

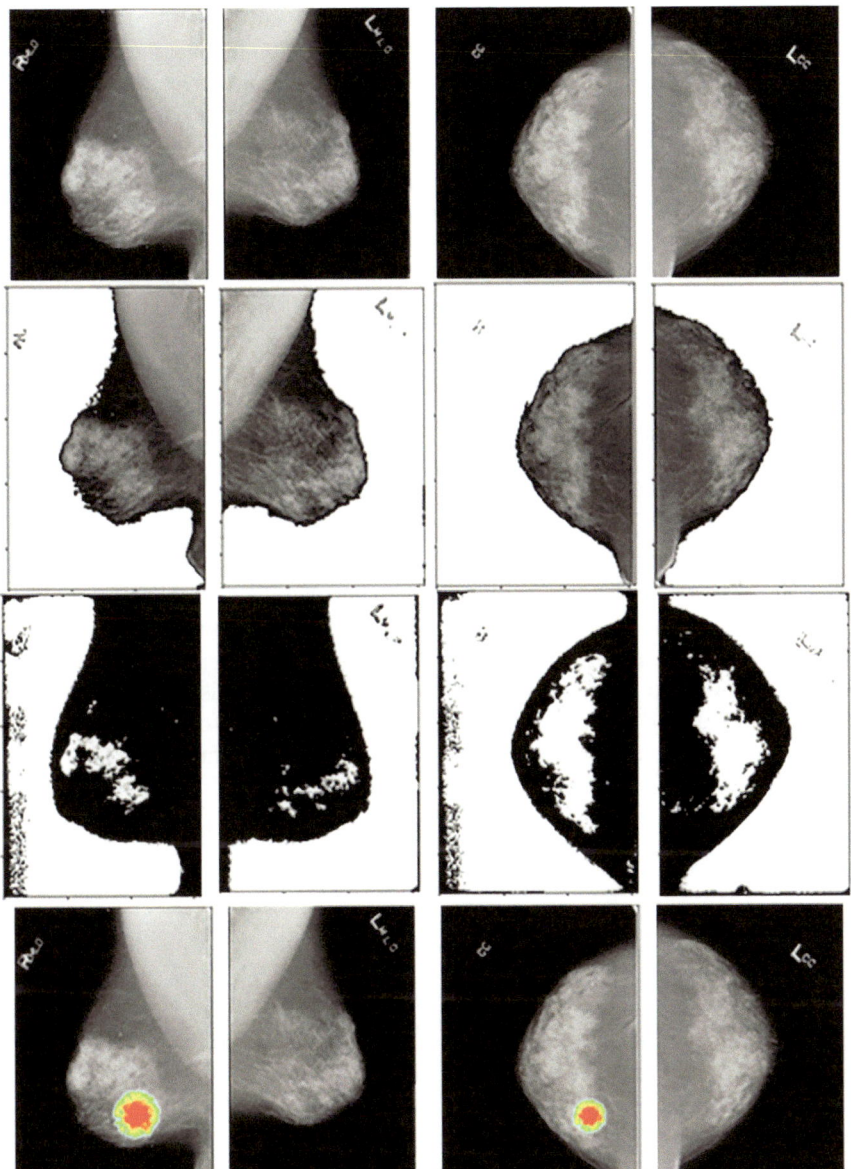

Fig. 18.6 AI-based diagnostic support software. (Source: Adapted from Kim HE et al. [50])

shown predictive capabilities for pancreatic cancer outcomes, such as overall survival and disease-free survival. In the future, this information may guide personalized care for cancer survivors, including surveillance and strategies to prevent recurrence. Radiomic analysis and evolving imaging-based ML models also demonstrate potential in predicting tumor pathology and genomic alterations. This can enable diagnosis and provide biomarker information without the need for invasive sampling, leading to the concept of "virtual biopsy." Notably, noninvasive imaging-based models are being developed for glioblastoma to predict genetic alterations within tumors and influence clinical management [24, 53–55].

The rapid expansion of digital pathology has paved the way for numerous applications of AI in the analysis of pathologic images, offering benefits in diagnosis, grading, and interpreting prognostic biomarkers. A significant focus has been on automating time-consuming tasks, enabling pathologists to enhance their efficiency and allocate more time to complex decision-making tasks associated with disease presentation [24, 56].

Several noteworthy examples highlight the progress in this field. For instance, a DL system has been developed to assess Gleason scores, outperforming general pathologists by analyzing whole-slide images of radical prostatectomy specimens [24, 57]. In addition, a convolutional neural network has been employed to automate the identification of tumor-infiltrating lymphocytes in tissue slide images from The Cancer Genome Atlas. This feature serves as a prognostic indicator for patients with various cancer subtypes [24, 58]. Furthermore, AI has demonstrated success in the classification of dermoscopy images and the annotation of skin lesions, including melanoma, achieving precision levels comparable to those of expert dermatologists [24, 59, 60].

These AI-based algorithms in oncologic imaging analysis are improving diagnostic accuracy and workflow efficiency, with ongoing efforts to translate them from research to clinical practice. As these algorithms continue to advance, the future holds great potential for further AI applications in oncologic imaging, offering opportunities for detection and management that were previously considered unattainable [24].

18.4.3 AI Applications in Precision Oncology and Cancer Genomics

Oncology heavily relies on evidence-based medicine scoring systems for various aspects of cancer management, including risk assessment, diagnosis, prognostic staging, treatment selection, and surveillance monitoring. Over time, these systems have evolved from basic observations using light microscopy to incorporate advanced techniques such as gene expression assays and next-generation sequencing of genomes, both somatic and germline. This evolution has led to an expanding array of prognostic and predictive factors specific to each disease, as exemplified by the growing prevalence of genomic-informed clinical models [24, 61–63].

Precision medicine, also known as personalized medicine, aims to provide tailored therapies based on the unique characteristics of a patient or a specific population group. These characteristics typically involve the patient's genome, transcriptome, and proteome and may include other factors like lifestyle, environment, and socioeconomic status. Sequencing or analyzing the patient's genome, transcriptome, or proteome often plays a central role in this approach. In the context of precision medicine, a "digital twin" refers to a virtual replica of a real-world object. The accuracy and level of detail in a digital twin depend on the precision, detail, and currency of the information describing the object. In the context of precision medicine, this concept can be applied to create a digital twin of an individual patient or a specific population group [64, 65].

Artificial intelligence (AI) encompasses algorithms and computing frameworks designed to perform tasks that typically require human intelligence, such as reasoning, decision-making, speech recognition, language understanding, and visual perception [64, 66]. In simpler terms, AI can be described as software that attempts to mimic human thought processes to accomplish tasks in a manner similar to human experts in the respective field [64, 67]. In precision medicine, the ultimate goal of AI is to identify patterns in data using models and algorithms, enabling predictions. These predictions are initially performed and then refined through machine learning using the software's learning algorithms [64, 68].

The recent advancements in technology have led to the generation of vast amounts of omics data, including genomic, transcriptomic, proteomic (phenotypic), and epigenomic data. This increase is attributed to next-generation sequencing (NGS) for genomic and transcriptomic data and mass spectrometric analysis for proteomic data [64, 66]. To advance precision oncology and provide accurate interpretations of an individual's cancer status, it is crucial for researchers and clinicians to utilize all available information, allowing computational models to capture the complexity of the biological system. AI, supported by high-performance computing and innovative deep learning techniques, offers the only feasible approach to synthesize and understand the magnitude and interdependencies present in multimodal data [24]. The general application of artificial intelligence to genomics data is illustrated in Fig. 18.7.

Precision medicine holds the potential to revolutionize patient care and cancer treatment by tailoring therapies to individual needs. Currently, a patient's ethnicity is often determined based on self-reporting or visual appearance, which may overlook the patient's actual genetic background. To achieve the highest level of accuracy in this regard, the creation of a "digital twin" of the patient becomes necessary. Creating such a digital twin requires capturing and curating extensive data that describes the patient's lifestyle and biology. The management and effective utilization of these large datasets to develop a digital twin for cancer control purposes would be challenging without the assistance of AI. AI plays a vital role in processing and analyzing this data, enabling timely and accurate generation of digital twins. By leveraging AI, it becomes possible to monitor a patient's response to a specific treatment, track their recovery, and predict treatment outcomes. This capability empowers healthcare professionals to fine-tune treatments based on the individual

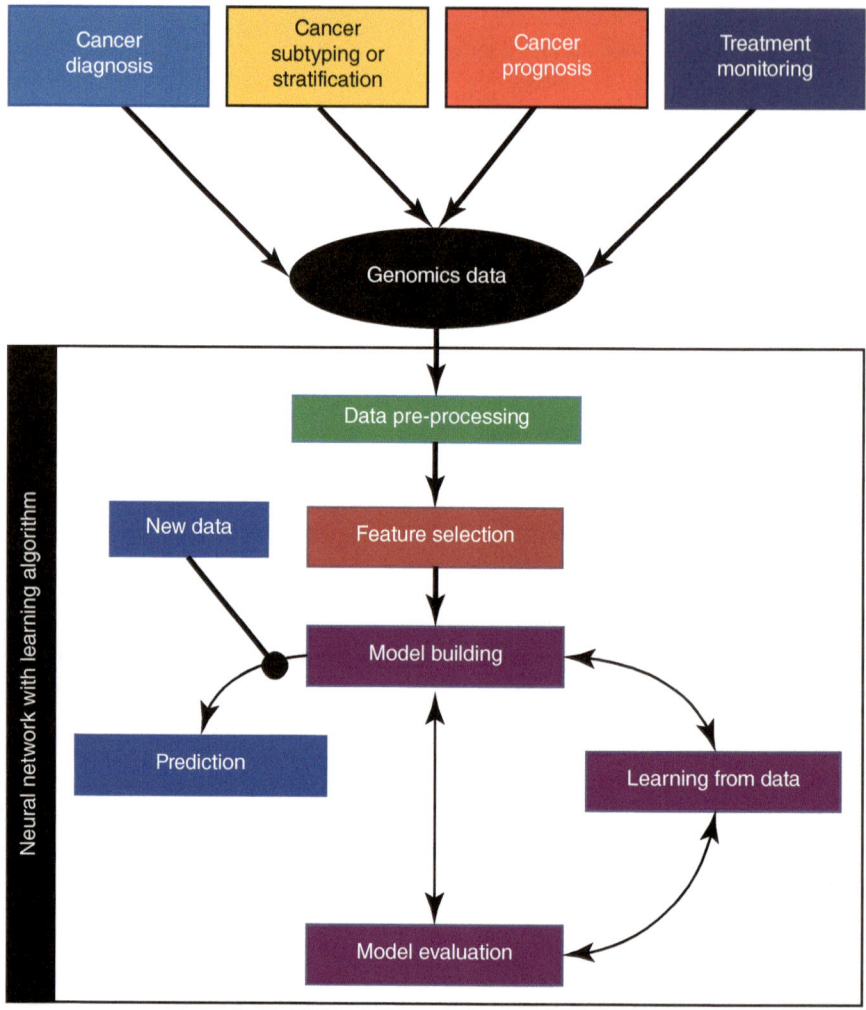

Fig. 18.7 General application of artificial intelligence to genomics data [64]

patient's situation and needs [64]. Figure 18.8 illustrates the application of AI in precision medicine.

The utilization of personalized genomic data obtained from the patient, analyzed through AI, has the potential to enhance cancer screening and diagnosis, thereby enabling the prevention of severe illnesses. Simultaneously, AI-driven analysis of this data can facilitate the development of more precise and targeted treatments, as well as enhance the monitoring of treatment outcomes. These advancements in precision oncology aim to improve patient care and align with the ultimate objective of delivering better healthcare [64].

Fig. 18.8 The application of AI to precision medicine [64]

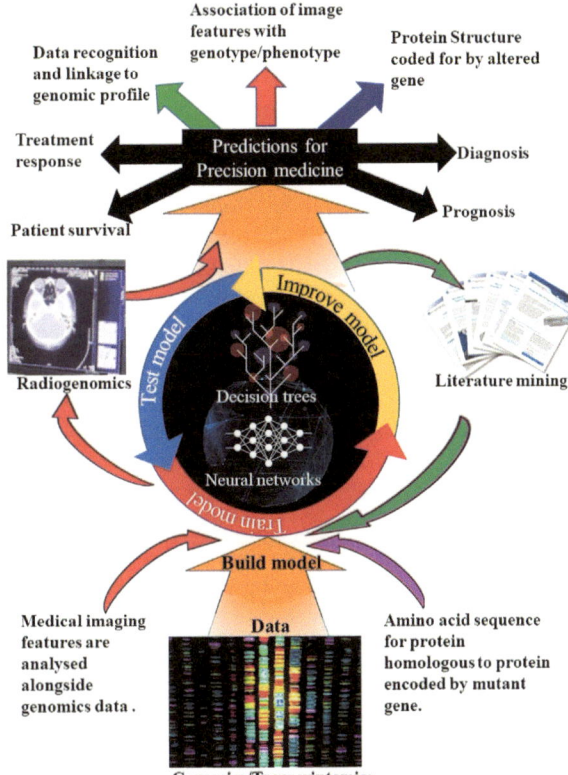

18.4.4 AI in Cancer Research and Clinical Outcomes

In conjunction with the increasing wealth of contemporary biomedical data, artificial intelligence (AI) and specifically deep learning (DL) have achieved notable successes in cancer clinical research. AI-based methods are being increasingly employed across various domains of cancer clinical research to enhance accuracy and efficiency. These applications encompass the utilization of AI in cancer imaging recognition, genomic analysis, mining medical records, drug discovery, and leveraging biomedical literature [69].

Within the field of clinical oncology, AI has been increasingly utilized to harness the potential of electronic health records (EHRs). Particularly, AI-based natural language processing techniques have shown promise in predicting the onset of diseases within extensive healthcare systems. An exemplary instance involves a DL-based AI algorithm developed by researchers at Mount Sinai, which accurately predicted the development of diverse diseases, including prostate, rectal, and liver cancers, with an overall accuracy of 93% [11, 70]. Acquiring clinical data poses challenges for data science experts due to limited opportunities for clinical practice or the requirement of institutional approval. Additionally, the labor-intensive nature of manual data collection has hindered the incorporation of firsthand clinical data into models. Leveraging Electronic Health Record (EHR) data has the potential to enhance the outcomes of biomedical research, as these systems integrate various clinical narratives, laboratory results, procedure and radiology reports, primary care notes, and gastroenterology clinic notes [69].

Recent advancements have showcased several promising outcomes through the application of AI in drug development, drug-target profiling, and drug repurposing and repositioning [71]. In the realm of small-molecule drug design, AI can enhance target specificity, selectivity, and account for pharmacodynamic, pharmacokinetic, and toxicological effects. Various types of data have been utilized in AI-driven cancer-related drug discovery research. Traditional data types include drug chemical structures, physicochemical properties, and molecular targets [72]. Moreover, RNA microarray, single nucleotide polymorphism (SNP) array, RNA sequencing (RNA-Seq), reverse phase protein array, exome sequencing, and DNA methylation status hold promise for identifying biomarkers and generating predictive models for drug sensitivity [73]. Existing resources that facilitate cancer drug discovery include DepMap, Genomics of Drug Sensitivity in Cancer (GDSC), canSAR, Open Targets, TG-GATE, drugBank, and others. These databases and resources enable the correlation of drug sensitivity and provide potential biomarkers for drug response [69].

18.4.4.1   AI in Cancer Clinical Research: Method and Application
The different subareas of cancer clinical research that have benefited from incorporating AI are given below.

Cancer Imaging Recognition
The advancement of computational capabilities and algorithms has facilitated the successful application of artificial intelligence (AI) in radiology, aiding radiologists in disease identification [74]. Prior to being inputted into the model, raw images may undergo basic preprocessing to eliminate irrelevant image portions. This involves extracting regions of interest (ROIs) by segmenting lesions within an image, and then only the information within these ROIs is utilized for model predictions. The annotation of regions can be performed by experts or assigned based on diagnosis labels [74, 75]. However, in contrast to other typical medical image formats, whole slide images (WSIs) are excessively large to be processed in their entirety by DL models [76]. To address this limitation, WSIs are divided into numerous small image patches, which are subsequently combined to generate predictions at the patch level [77]. In some cases, a tumor probability heatmap can be utilized to select geometrical and morphological features, serving as input to the model and enabling the identification and characterization of disease patterns on digitized tissue slides [78]. Traditional image recognition approaches employ manually designed image features, including texture, shape, color, pixel density, and contrast/brightness, to capture tumor or cell morphology [79]. However, these feature-based algorithms possess certain limitations: (1) they rely on a separate feature extraction step [80], and (2) these features may not consistently perform well under different scanning conditions [81]. On the other hand, automatic feature extraction allows for

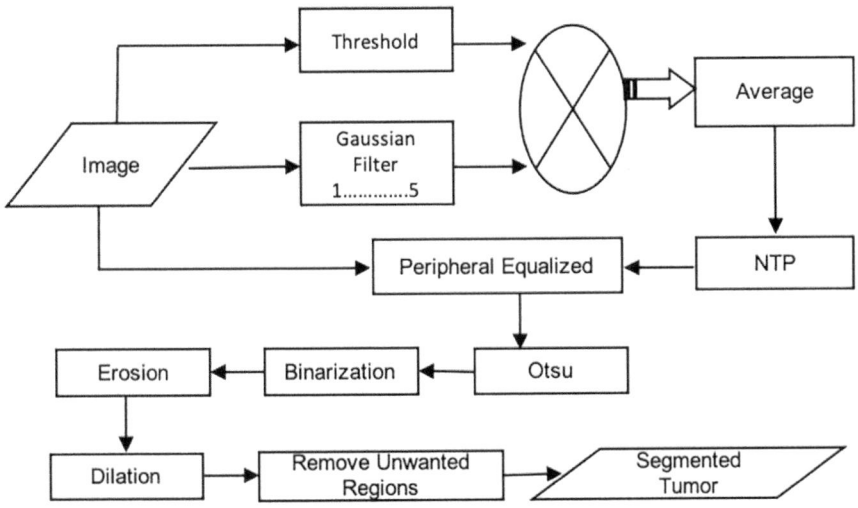

Fig. 18.9 Computational imaging recognition for cancer clinical research. (Source: Adapted from Shao D et al. [69])

the direct use of raw images as input to the model (end-to-end), enabling simultaneous image classification [82]. The process of computational imaging recognition in cancer clinical research is depicted in Fig. 18.9.

Genomic Analysis

In recent times, significant advancements have been made in applying artificial intelligence (AI) to cancer research, particularly in utilizing various genomic data types as input for models. Morais-Rodrigues et al. [83] developed a modified logistic regression approach to analyze microarray gene expression data for breast cancer progression. Similarly, Maros et al. [84] designed machine learning workflows to estimate class probabilities for cancer diagnosis using DNA methylation microarray data. Another noteworthy study by Albaradei et al. [85] introduced a deep learning-based model that differentiated pan-cancer metastasis status by integrating three heterogeneous data layers from TCGA: RNA-Seq, microRNA-Seq, and DNA methylation data. The model employed a convolutional variational autoencoder for feature extraction and a deep neural network for classification. The results demonstrated that the integration of multiple data types improved performance compared to using mRNA data alone.

Additionally, AI models have been employed in cancer-grade prediction. Yamamoto et al. [86] trained a support vector machine (SVM) classifier using morphometric classification of microenvironmental myoepithelial cells to quantitatively diagnose breast tumors. The study involved the quantitative measurement of 11,661 nuclei across four histological types: normal cases, usual ductal hyperplasia, low-grade ductal carcinoma in situ (DCIS), and high-grade DCIS. The model achieved an accuracy of 90.9% in classifying these histological types, with at least three pathologists independently diagnosing and scoring all cases.

Genomic data also enables the identification of disease-related biomarkers. For example, Zeng et al. [87] utilized deep forests in combination with positive-unlabeled learning methods to predict potential disease-related circular RNAs (cir-cRNAs). Furthermore, Radhakrishnan et al. [88] combined fluorescence imaging and deep learning techniques to detect subtle changes in nuclear morphometrics at the single-cell level. This approach opens new avenues for early disease diagnostics and drug discovery.

Electronic Medical Record Mining

Certain AI-based models leverage integrated medical record data, which includes genomic information, unstructured health records, and family history, to enhance the accuracy of cancer prediction [69, 89]. For instance, when predicting the survival outcomes of lung cancer, a dataset incorporating observed cancer-related characteristics of individuals is considered. These characteristics may include lung cancer pathology images, age, gender, smoking status, and stage. Moreover, tumor shape parameters such as area, perimeter, convex area, filled area, major axis length, and minor axis length also contribute to the outcome [69, 75].

Natural language processing (NLP) systems play a significant role in capturing relevant information for cancer research projects. As an example, a preprocessor integrated with an existing NLP system, such as MedLEE, was developed as part of an ongoing clinical research endeavor that focuses on assessing disparities and risks associated with breast cancer development in minority women [90]. NLP algorithms are utilized to identify primary and recurrent cancers by extracting pertinent information from electronic pathology reports [91]. Additionally, NLP can enhance the identification of cancer testing within electronic medical records [92].

Drug Discovery

AI has emerged as a valuable asset in cancer drug research due to the availability of extensive and refined public databases and resources. Choi et al. [69, 93] introduced an innovative deep neural network model that enhances the prediction of drug resistance and the identification of biomarkers associated with drug response. Huang et al. [94] utilized gene expression profiles (RNA-seq or microarray) from individual patient tumors to predict the responses of 175 cancer patients to various standard-of-care chemotherapeutic drugs. Borisov et al. [95] predicted the clinical effectiveness of anti-cancer drugs for individual patients by transferring features obtained from expression-based data derived from cell lines. Another study by Chang et al. [71] presented the Cancer Drug Response Profile Scan (CDRscan), which predicts the responsiveness of anticancer drugs based on large-scale drug screening assay data, encompassing genomic profiles of 787 human cancer cell lines and structural profiles of 244 drugs. Moreover, the application of computational biology approaches to predict and interpret cancer drug response at the single-cell level has demonstrated significant value. Yanagisawa et al. [96] constructed a convolutional neural network (CNN) model to forecast the efficacy of antitumor drugs at the single-cell level.

Numerous computational tools have been proposed for cancer-related drug discovery, employing various AI methodologies. Examples of these applications include DeepChem [97], DeepTox [98], gene2drug [99], STITCH [100], AlphaFold [101], and DeepNeuralNetQSAR [102]. For instance, the DeepTox algorithm employs machine learning to predict the toxic effects of 12,000 environmental chemicals and drugs in specifically designed assays [98]. AlphaFold utilizes deep neural networks to predict the three-dimensional structure of drug target proteins [101]. The development of these tools has contributed to the reduction in the cost of drug discovery [69].

Biomedical Literature Utility
Over the past few decades, significant efforts by large consortiums have led to the development of community-based knowledge bases in the field of cancer clinical research, leveraging extensive collections of published literature. For instance, the National Lung Screening Trial (NLST) [69, 102] serves as a comprehensive data-sharing platform that enables users to search, browse, download, and analyze tumor regions of lung adenocarcinoma (ADC) patients. The National Cancer Institute (NCI) Genomic Data Commons (GDC) [69, 103] acts as a unified knowledge base that integrates genomic and clinical data from various research programs for the cancer research community. In a typical study, a deep convolutional neural network (CNN) model is trained on the systematically studied tumor regions of lung cancer patients from the NLST cohort, enabling automatic recognition of tumor regions. The performance of the model developed from the NLST cohort is then independently validated in the TCGA cohort for prognostic assessment [69, 75].

The Sheikh Khalifa Bin Zayed Al Nahyan Institute for Personalized Cancer Therapy (IPCT) at MD Anderson Cancer Center has developed a knowledge base that provides valuable information on the functions of common genomic alterations and their therapeutic implications, guiding personalized treatments in oncology [69, 104]. A precision oncology decision support (PODS) team comprising oncologists, geneticists, molecular biologists, computational scientists, computer programmers, and bioinformaticians manually reviews the literature for this knowledge base [69, 105]. Additionally, an integrated Precision Medicine Knowledgebase (PreMedKB) has been established to seamlessly interpret the four essential components of precision medicine: diseases, genes, variants, and drugs [69, 106]. These knowledge bases offer a wealth of information and serve as excellent resources and tools for the research community.

18.4.5 AI and Translational Oncology

AI is gradually emerging in the field of translational oncology. In the past decade, there has been a significant expansion of biological quantitative data, commonly known as "-omic" data. Due to the inherent complexities and heterogeneity of this data, deep learning (DL) has become an appealing approach for analysis. DL neural networks have been successfully applied to various tasks in translational oncology. For instance, they have been used to predict protein structure [11, 107], classify cells into specific mitotic stages [11, 108], and even forecast the future lineage of progenitor cells based on microscopy images [11, 109].

DL has also found applications in drug development and repurposing, which have garnered considerable interest. In one study, DL artificial neural networks (ANNs) were trained on transcriptomic response signatures to drugs, achieving high accuracy in predicting the likelihood of failure in clinical trials for over 200 example drugs [11, 110]. Another study utilized an ANN to predict the sensitivity of cancer cells to therapeutics by combining genomic and chemical properties [11, 111]. Convolutional neural networks (CNNs) have also been employed to predict peptide-major histocompatibility complex binding [11, 112], a factor that holds potential implications for the development of oncologic immunotherapy.

Overall, DL has shown promise in translational oncology, enabling valuable insights and contributing to drug development, drug repurposing, and understanding cellular processes in cancer research.

18.5   AI Challenges and Future Directions

In various domains of biomedicine, AI has demonstrated comparable performance to that of human experts. However, despite the availability of certain AI solutions, there are still numerous challenges that need to be addressed for AI to transition from theoretical studies to real-world applications [69].

A significant challenge faced by AI, in general, is the need for large amounts of data. In the context of cancer research, acquiring a sufficiently large and well-annotated dataset is an ongoing requirement for AI algorithms. While the inclusion of images, genomic data, and clinical outcomes in certain open databases has greatly advanced computational clinical research, there is still a need to obtain data of adequate scale, quality, and diversity. For instance, patient histories documented in previous reports contain valuable information relevant to cancer risk and progression, but gathering such data can be time-consuming. To address this challenge, data-sharing agreements play a crucial role. Sharing large datasets with the research community can be facilitated through cloud computing and the continued development of advanced predictive cancer models [69].

Furthermore, the effectiveness of an AI model heavily relies on the availability of high-quality data. Despite the increasing volume and diversity of available data, there is a lack of standardized methods to assess data quality [69].

Challenges in the deployment of AI include the following:

Proving Generalizability and Real-World Applications

Despite the rapid integration of AI into oncology research, there is still work to be done in order to translate these studies into practical and clinically relevant applications. One major challenge involves the external validation and demonstration of the generalizability of deep learning (DL) models. Due to the intricate nature of neural networks and their extensive parameterization (often involving millions of parameters), there is a considerable risk of developing overfitted models that lack the ability to generalize across different populations. Moreover, the presence of significant heterogeneity in medical data across various institutions necessitates the need for multiple external validation sets to establish the performance of an AI application [11, 113].

Data Access and Equity

As mentioned earlier, the problem of overfitting is exacerbated by limitations in data access and quality. Deep learning (DL) neural networks, more than other machine learning algorithms, require large volumes of data, which can be challenging in healthcare settings where diseases with lower prevalence are involved. Additionally, data is often fragmented within individual institutions due to concerns regarding the transmission of protected patient health information. The lack of data-sharing infrastructure, along with heterogeneity and incompleteness in data collection, as well as competition between institutions, contribute to this scarcity of data. However, efforts are being made to address these challenges, with a growing focus on streamlined data capture and the establishment of multi-institutional data-sharing agreements. Guidelines promoting the use of FAIR (findable, accessible, interoperable, and reusable) data have been proposed, and there are now opportunities for research groups to publish their data, which may encourage greater openness in data sharing [11, 114–120].

Interpretability and the Black Box Problem

One of the primary obstacles to the widespread adoption of AI in healthcare is the issue of interpretability. Despite achieving impressive performance, AI models often lack transparency. For example, a deep learning (DL) model might accurately predict that a patient will develop pancreatic cancer based on their 2 years of past data, but the precise reasoning behind this prediction remains unclear. This challenge is commonly referred to as the "black box" problem [11, 121]. In clinical decision-making, understanding the rationale behind each decision has always been crucial. Traditional machine learning (ML) algorithms like linear regression, while limited in modeling complex relationships, offer interpretability. These algorithms provide pre-defined features and corresponding feature weights that indicate their impact. In contrast, DL models utilize unstructured input data, and the majority of knowledge generation occurs within hidden layers, making it difficult to identify which specific characteristics of the input data contribute to the outcome. This lack of interpretability has significant implications for the acceptance of AI-based algorithms in healthcare, both from the perspectives of practitioners and regulatory bodies [11, 122–125].

Realizing the Potential of AI in Oncology: Overcoming Challenges and Maximizing Benefits

The potential applications of AI in medicine and cancer research offer great promise. However, to leverage these opportunities, it is necessary to make increased investments and address several challenges [126]. The National Cancer Institute (NCI) of the USA has put forward the following strategies to advance the field [126]:

1. Establishing an AI cancer research community: Collaboration between the data science, AI, and cancer research communities is crucial to harness the potential of AI in cancer research. The NCI can facilitate this collaboration by providing funding opportunities and data access, fostering connections between cancer researchers and AI experts, and supporting the training and development of professionals with expertise in AI, data science, and cancer. Workshops and initiatives are being organized, building upon the NCI's collaboration with the Department of Education, to encourage a community that pushes the boundaries of computational practices in cancer research and develops new computational technologies.

2. Bridging the gap between research and practice: Currently, the use of AI in cancer research and care is in its early stages. Most research focuses on developing methods rather than implementing them in clinical practice. The NCI can lead the way in integrating AI into cancer care by supporting research to identify effective pathways for clinical integration. This includes understanding uncertainty and validating AI approaches, educating medical professionals about the strengths and limitations of AI technology, and conducting rigorous assessments of its benefits in terms of clinical outcomes, patient experience, and cost-effectiveness.

3. Accessing high-quality cancer data: The scarcity of large, publicly available, and well-annotated cancer datasets has been a significant barrier to AI research and algorithm development. The absence of benchmarking datasets in cancer research hampers reproducibility and validation. To drive AI innovation and facilitate the training and validation of AI models, there is a need for support in annotating, harmonizing, and sharing standardized cancer datasets. As data volumes are expected to increase, it is critical to develop approaches that generate and aggregate new research and clinical data coherently. The NCI aims to refine its policies and practices to enhance and improve data sharing, making cancer data broadly available for all types of research.

By implementing these strategies, the NCI seeks to overcome challenges and promote the effective utilization of AI in oncology, ultimately leading to improved cancer research, diagnosis, and treatment outcomes.

18.6 Conclusion

Despite the hurdles and growing concerns, it is a well-known truism that the integration of AI into the healthcare ecosystem allows for a multitude of benefits. With its myriad of applications, AI is recasting the layout of oncology and the associated sectors by maximizing its potential to facilitate efficient use of healthcare resources. Integration of AI technology in cancer care could improve the accuracy and speed of diagnosis, aid clinical decision-making, and lead to better health outcomes. AI-guided clinical care has the potential to play an important role in reducing health disparities, particularly in low-resource settings. The UAE healthcare system, with its impressive trajectory, has the necessary infrastructure to develop and thrive in a tech-enabled healthcare ecosystem. Given how rapidly the technology is evolving and by recognizing the many potential applications in cancer science, AI will undeniably revolutionize oncology.

Conflict of Interest The authors have no conflict of interest to declare.

Disclosure QuillBot and ChatGPT are used to paraphrase selected segments.

References

1. Farina E, Nabhen JJ, Dacoregio MI, Batalini F, Moraes FY. An overview of artificial intelligence in oncology. Future Sci OA. 2022;8(4):FSO787. https://doi.org/10.2144/fsoa-2021-0074.
2. Luchini C, Pea A, Scarpa A. Artificial intelligence in oncology: current applications and future perspectives. Br J Cancer. 2022;126:4–9. https://doi.org/10.1038/s41416-021-01633-1.
3. Kaul V, Enslin S, Gross SA. History of artificial intelligence in medicine. Gastrointest Endosc. 2020;92:807–12.
4. Sung H, Ferlay J, Siegel RL, Laversanne M, Soerjomataram I, Jemal A, et al. Global cancer statistics 2020: GLOBOCAN estimates of incidence and mortality worldwide for 36 cancers in 185 countries. CA Cancer J Clin. 2021;71(3):209–49.
5. Cancer incidence in United Arab Emirates, Annual Report of the UAE—National Cancer registry—2021. Statistics and Research Center, Ministry of Health and Prevention. Accessed on 06 June 2024.
6. World Economic Forum. https://www.weforum.org/agenda/2021/07/ai-projects-improving-cancer-screening-outcomes/. Accessed 6 Sept 2022.
7. Shaikh K, Krishnan S, Thanki R. Artificial intelligence in breast cancer early detection and diagnosis. Berlin: Springer; 2021. https://doi.org/10.1007/978-3-030-59208-0.
8. Bishop C. Pattern recognition and machine learning. Berlin: Springer; 2006.
9. Sarker IH. Deep learning: a comprehensive overview on techniques, taxonomy, applications and research directions. SN Comput Sci. 2021;2(6):420. https://doi.org/10.1007/s42979-021-00815-1. Epub 2021 Aug 18. PMID: 34426802; PMCID: PMC8372231.
10. Hopfield JJ. Artificial neural networks. IEEE Circuits Devices Mag. 1988;4:3–10.
11. Kann BH, Thompson R, Thomas CR Jr, Dicker A, Aneja S. Artificial intelligence in oncology: current applications and future directions. Oncology (Williston Park, NY). 2019;33(2):46–53.
12. Chua IS, Gaziel-Yablowitz M, Korach ZT, Kehl KL, Levitan NA, Arriaga YE, Jackson GP, Bates DW, Hassett M. Artificial intelligence in oncology: Path to implementation. Cancer Med. 2021;10:4138. https://doi.org/10.1002/cam4.3935.
13. Abu-Gheida IH, Nijhawan N, Al-Awadhi A, Al-Shamsi HO. General oncology care in the UAE. In: Al-Shamsi HO, Abu-Gheida IH, Iqbal F, Al-Awadhi A, editors. Cancer in the Arab world. Singapore: Springer; 2022. https://doi.org/10.1007/978-981-16-7945-2_19.
14. Blair I, Sharif A. Health and health systems performance in the United Arab Emirates. World Hosp Health Serv. 2013;49(4):12–7.
15. https://www.who.int/news/item/07-02-2000-world-health-organization-assesses-the-world's-health-systems.
16. Al-Shamsi HO. The state of cancer care in the United Arab Emirates in 2022. Clin Pract. 2022;12(6):955–85. https://doi.org/10.3390/clinpract12060101.
17. Artificial Intelligence in GI screening, Sheikh Shakhbout Medical City. https://ssmc.ae/news/ssmc-leverages-artificial-intelligence-in-gi-screening/. Accessed 22 Sept 2022.
18. Healthcare IT News, EMEA. https://www.healthcareitnews.com/news/emea/nyu-and-nyu-abu-dhabi-develop-new-ai-tool-breast-cancer-detection. Accessed 22 Sept 2022.

19. Khaleej Times, Health, Abu Dhabi rolls out new programme to detect, treat breast cancer patients. https://www.khaleejtimes.com/health/abu-dhabi-rolls-out-new-programme-to-detect-treat-breast-cancer-patients. Accessed 22 Sept 2022.
20. Krizhevsky A, Sutskever I, Hinton GE. ImageNet classification with deep convolutional neural networks. In: Pereira F, Burges CJC, Bottou L, Weinberger KQ, editors. Advances in neural information processing systems 25. Curran Associates, Inc; 2012. p. 1097–105. http://papers.nips.cc/paper/4824-imagenet-classification-with-deep-convolutional-neural-networks.pdf.
21. Russakovsky O, Deng J, Su H, et al. ImageNet large scale visual recognition challenge. Int J Comput Vis. 2015;115:211–52.
22. Bird RE, Wallace TW, Yankaskas BC. Analysis of cancers missed at screening mammography. Radiology. 1992;184(3):613–7.
23. Baker JA, Rosen EL, Lo JY, Gimenez EI, Walsh R, Soo MS. Computer-aided detection (CAD) in screening mammography: sensitivity of commercial CAD systems for detecting architectural distortion. Am J Roentgenol. 2003;181(4):1083–8.
24. DOI: 10.1200/EDBK_350652 American Society of Clinical Oncology Educational Book 42 (June 10, 2022) 842–851.
25. Esteva A, Kuprel B, Novoa RA, et al. Dermatologist-level classification of skin cancer with deep neural networks. Nature. 2017;542:115–8.
26. Webster DE, Suver C, Doerr M, et al. The mole mapper study, mobile phone skin imaging and melanoma risk data collected using ResearchKit. Sci Data. 2017;4:170005.
27. Wang P, Xiao X, Brown JRG, et al. Development and validation of a deep-learning algorithm for the detection of polyps during colonoscopy. Nat Biomed Eng. 2018;2:741.
28. Data Science Bowl. 2017. https://www.kaggle.com/c/data-science-bowl-2017. Accessed 3 Dec 2018.
29. de Wit J. 2nd place solution for the 2017 National Data Science Bowl. http://juliandewit.github.io/kaggle-ndsb2017/. Accessed 15 March 2018.
30. Hammack D. DSB2017: code for 2nd place solution to the 2017 National Data Science Bowl. 2018. https://github.com/dhammack/DSB2017. Accessed 12 Jan 2018.
31. Kuan K, Ravaut M, Manek G, et al. Deep learning for lung cancer detection: tackling the Kaggle data science Bowl 2017 challenge. 2017. http://arxiv.org/abs/1705.09435. Accessed 14 Feb 2019.
32. Zhao W, Yang J, Sun Y, et al. 3D deep learning from CT scans predicts tumor invasiveness of subcentimeter pulmonary adenocarcinomas. Cancer Res. 2018;78:6881–9.
33. Ciompi F, Chung K, van Riel SJ, et al. Towards automatic pulmonary nodule management in lung cancer screening with deep learning. Sci Rep. 2017;7:46479.
34. Kang G, Liu K, Hou B, Zhang N. 3D multi-view convolutional neural networks for lung nodule classification. PLoS One. 2017;12:e0188290.
35. Wang S, Zhou M, Gevaert O, et al. A multi-view deep convolutional neural networks for lung nodule segmentation. Conf Proc IEEE Eng Med Biol Soc. 2017;2017:1752–5.
36. Sage Bionetworks. Digital mammography DREAM challenge. http://sagebionetworks.org/research-projects/digital-mammography-dream-challenge/. Accessed 3 Dec 2018.
37. Ribli D, Horváth A, Unger Z, et al. Detecting and classifying lesions in mammograms with deep learning. Sci Rep. 2018;8:4165.
38. Trister AD, Buist DSM, Lee CI. Will machine learning tip the balance in breast cancer screening? JAMA Oncol. 2017;3:1463–4.
39. Kann BH, Aneja S, Loganadane GV, et al. Pretreatment identification of head and neck cancer nodal metastasis and extranodal extension using deep learning neural networks. Sci Rep. 2018;8:14036.
40. Chang K, Bai HX, Zhou H, et al. Residual convolutional neural network for determination of IDH status in low- and high-grade gliomas from MR imaging. Clin Cancer Res. 2018;24:1073–81.
41. Chang P, Grinband J, Weinberg BD, et al. Deep-learning convolutional neural networks accurately classify genetic mutations in gliomas. AJNR Am J Neuroradiol. 2018;39:1201–7.

42. Bibault JE, Giraud P, Durdux C, et al. Deep learning and radiomics predict complete response after neo-adjuvant chemoradiation for locally advanced rectal cancer. Sci Rep. 2018;8:12611.
43. Sun R, Limkin EJ, Vakalopoulou M, et al. A radiomics approach to assess tumour-infiltrating CD8 cells and response to anti-PD-1 or anti-PD-L1 immunotherapy: an imaging biomarker, retrospective multicohort study. Lancet Oncol. 2018;19:1180–91.
44. Bejnordi BE, Veta M, van Diest PJ, et al. Diagnostic assessment of deep learning algorithms for detection of lymph node metastases in women with breast cancer. JAMA. 2017;318:2199–210.
45. Arvaniti E, Fricker KS, Moret M, et al. Automated Gleason grading of prostate cancer tissue microarrays via deep learning. Sci Rep. 2018;8:12054.
46. Coudray N, Ocampo PS, Sakellaropoulos T, et al. Classification and mutation prediction from non-small cell lung cancer histopathology images using deep learning. Nat Med. 2018;24:1559–67.
47. Zhang SM, Wang YJ, Zhang ST. Accuracy of artificial intelligence-assisted detection of esophageal cancer and neoplasms on endoscopic images: a systematic review and meta-analysis. J Dig Dis. 2021;22:318–28.
48. Lamb LR, Lehman CD, Gastounioti A, et al. Artificial intelligence (AI) for screening mammography, from the AI special series on AI applications. AJR Am J Roentgenol. 2022;219:369.
49. Leibig C, et al. Combining the strengths of radiologists and AI for breast cancer screening: a retrospective analysis. Lancet Digit Health. 2022;4(7):e507–19. https://doi.org/10.1016/S2589-7500(22)00070-X.
50. Kim HE, Kim HH, Han BK, Kim KH, Han K, Nam H, Lee EH, Kim EK. Changes in cancer detection and false-positive recall in mammography using artificial intelligence: a retrospective, multireader study. Lancet Digit Health. 2020;2(3):e138–48. https://doi.org/10.1016/S2589-7500(20)30003-0. Epub 2020 Feb 6. PMID: 33334578.
51. Bi WL, Hosny A, Schabath MB, et al. Artificial intelligence in cancer imaging: clinical challenges and applications. CA Cancer J Clin. 2019;69:127–57.
52. Tomaszewski MR, Gillies RJ. The biological meaning of radiomic features. Radiology. 2021;298:505–16.
53. Janssen BV, Verhoef S, Wesdorp NJ, et al. Imaging-based machine-learning models to predict clinical outcomes and identify biomarkers in pancreatic cancer: a scoping review. Ann Surg. 2022;275:560–7.
54. Martin-Gonzalez P, Crispin-Ortuzar M, Rundo L, et al. Integrative radiogenomics for virtual biopsy and treatment monitoring in ovarian cancer. Insights Imaging. 2020;11:94.
55. Calabrese E, Villanueva-Meyer JE, Cha S. A fully automated artificial intelligence method for non-invasive, imaging-based identification of genetic alterations in glioblastomas. Sci Rep. 2020;10:11852.
56. Bera K, Schalper KA, Rimm DL, et al. Artificial intelligence in digital pathology—new tools for diagnosis and precision oncology. Nat Rev Clin Oncol. 2019;16:703–15.
57. Nagpal K, Foote D, Liu Y, et al. Development and validation of a deep learning algorithm for improving Gleason scoring of prostate cancer. NPJ Digit Med. 2019;2:1–10.
58. Saltz J, Gupta R, Hou L, et al. Spatial organization and molecular correlation of tumor-infiltrating lymphocytes using deep learning on pathology images. Cell Rep. 2018;23:181–93.
59. Esteva A, Kuprel B, Novoa RA, et al. Dermatologist-level classification of skin cancer with deep neural networks. Nature. 2017;542(7639):115–8.
60. Yu L, Chen H, Dou Q, et al. Automated melanoma recognition in dermoscopy images via very deep residual networks. IEEE Trans Med Imaging. 2017;36:994–1004.
61. Huang T-T, Lei L, Chen C-HA, et al. A new clinical genomic model to predict 10-year recurrence risk in primary operable breast cancer patients. Sci Rep. 2020;10:1–10.
62. Spratt DE, Zhang J, Santiago-Jiménez M, et al. Development and validation of a novel integrated clinical-genomic risk group classification for localized prostate cancer. J Clin Oncol. 2018;36:581–90.
63. Jiang J, Ding Y, Wu M, et al. Integrated genomic analysis identifies a genetic mutation model predicting response to immune checkpoint inhibitors in melanoma. Cancer Med. 2020;9:8498–518.

64. Dlamini Z, Skepu A, Kim N, Mkhabele M, Khanyile R, Molefi T, Mbatha S, Setlai B, Mulaudzi T, Mabongo M, Bida M, Kgoebane-Maseko M, Mathabe K, Lockhat Z, Kgokolo M, Chauke-Malinga N, Ramagaga S, Hull R. AI and precision oncology in clinical cancer genomics: from prevention to targeted cancer therapies-an outcomes based patient care. Inform Med Unlocked. 2022;31:100965. https://doi.org/10.1016/j.imu.2022.100965.

65. Batch KE, et al. Developing a cancer digital twin: supervised metastases detection from consecutive structured radiology reports. Front Artif Intell. 2022;5:826402.

66. Hamet P, Tremblay J. Artificial intelligence in medicine. Metabolism. 2017;69:S36–40.

67. Aarvik P. Artificial intelligence—a promising anti-corruption tool in development settings. https://beta.u4.no/publications/artificial-intelligence-a-promising-anti-corruption-tool-in-development-settings.pdf. Accessed 10 Sept.

68. Bersanelli M, et al. Methods for the integration of multi-omics data: mathematical aspects. BMC Bioinform. 2016;17(Suppl 2):15.

69. Shao D, Dai Y, Li N, Cao X, Zhao W, Cheng L, Rong Z, Huang L, Wang Y, Zhao J. Artificial intelligence in clinical research of cancers. Brief Bioinform. 2022;17;23(1):bbab523 https://doi.org/10.1093/bib/bbab523. PMID: 34929741; PMCID: PMC8769909.

70. Miotto R, Li L, Kidd BA, Dudley JT. Deep patient: an unsupervised representation to predict the future of patients from the electronic health records. Sci Rep. 2016;6:26094.

71. Chang Y, Park H, Yang HJ, et al. Cancer drug response profile scan (CDRscan): a deep learning model that predicts drug effectiveness from cancer genomic signature. Sci Rep. 2018;8:8857.

72. Mottini C, Napolitano F, Li ZX, et al. Computer-aided drug repurposing for cancer therapy: approaches and opportunities to challenge anticancer targets. Semin Cancer Biol. 2021;68:59–74.

73. Vamathevan J, Clark D, Czodrowski P, et al. Applications of machine learning in drug discovery and development. Nat Rev Drug Discov. 2019;18:463–77.

74. Varghese B, Chen F, Hwang D, et al. Objective risk stratification of prostate cancer using machine learning and radiomics applied to multiparametric magnetic resonance images. Sci Rep. 2019;9:1570.

75. Wang S, Chen A, Yang L, et al. Comprehensive analysis of lung cancer pathology images to discover tumor shape and boundary features that predict survival outcome. Sci Rep. 2018;8:10393.

76. Aubreville M, Knipfer C, Oetter N, et al. Automatic classification of cancerous tissue in laser endomicroscopy images of the oral cavity using deep learning. Sci Rep. 2017;7:11979.

77. Granter SR, Beck AH, Papke DJ. Alphago, deep learning, and the future of the human microscopist. Arch Pathol Lab Med. 2017;141:619–21.

78. Vang YS, Chen Z, Xie X. Deep learning framework for multi-class breast cancer histology image classification. 2018; arXiv: 1802.00931.

79. Cha KH, Hadjiiski L, Chan HP, et al. Bladder cancer treatment response assessment in CT using radiomics with deep-learning. Sci Rep. 2017;7:8738.

80. Wang X, Yang W, Weinreb J, et al. Searching for prostate cancer by fully automated magnetic resonance imaging classification: deep learning versus non-deep learning. Sci Rep. 2017;7:15415.

81. Levine AB, Schlosser C, Grewal J, et al. Rise of the machines: advances in deep learning for cancer diagnosis. Trends Cancer. 2019;5:157–69.

82. Han Z, Wei B, Zheng Y, et al. Breast cancer multi-classification from histopathological images with structured deep learning model. Sci Rep. 2017;7:4172.

83. Morais-Rodrigues F, Silverio-Machado R, Kato RB, et al. Analysis of the microarray gene expression for breast cancer progression after the application modified logistic regression. Gene. 2020;726:144168.

84. Maros ME, Capper D, Jones DTW, et al. Machine learning workflows to estimate class probabilities for precision cancer diagnostics on DNA methylation microarray data. Nat Protoc. 2020;15:479–512.

85. Albaradei S, Napolitano F, Thafar MA, et al. MetaCancer: a deep learning-based pan-cancer metastasis prediction model developed using multi-omics data. Comput Struct Biotechnol J. 2021;19:4404–11.

86. Yamamoto Y, Saito A, Tateishi A, et al. Quantitative diagnosis of breast tumors by morphometric classification of microenvironmental myoepithelial cells using a machine learning approach. Sci Rep. 2017;7:46732.
87. Zeng X, Zhong Y, Lin W, et al. Predicting disease-associated circular RNAs using deep forests combined with positive-unlabeled learning methods. Brief Bioinform. 2020;21:1425–36.
88. Radhakrishnan A, Damodaran K, Soylemezoglu AC, et al. Machine learning for nuclear mechano-morphometric biomarkers in cancer diagnosis. Sci Rep. 2017;7:17946.
89. Kann BH, Hosny A, Aerts HJWL. Artificial intelligence for clinical oncology. Cancer Cell. 2021;39:916–27.
90. Xu H, Anderson K, Grann VR, et al. Facilitating cancer research using natural language processing of pathology reports. Stud Health Technol Inform. 2004;107:565–72.
91. Karimi YH, Blayney DW, Kurian AW, et al. Development and use of natural language processing for identification of distant cancer recurrence and sites of distant recurrence using unstructured electronic health record data. JCO Clin Cancer. 2021;5:469–78.
92. Zeng J, Banerjee I, Henry AS, et al. Natural language processing to identify cancer treatments with electronic medical records. JCO Clin Cancer. 2021;5:379–93.
93. Choi J, Park S, Ahn J. RefDNN: a reference drug based neural network for more accurate prediction of anticancer drug resistance. Sci Rep. 2020;10:1861.
94. Huang C, Clayton EA, Matyunina LV, et al. Machine learning predicts individual cancer patient responses to therapeutic drugs with high accuracy. Sci Rep. 2018;8:16444.
95. Borisov N, Tkachev V, Suntsova M, et al. A method of gene expression data transfer from cell lines to cancer patients for machine-learning prediction of drug efficiency. Cell Cycle. 2018;17:486–91.
96. Yanagisawa K, Toratani M, Asai A, et al. Convolutional neural network can recognize drug resistance of single cancer cells. Int J Mol Sci. 2020;21:3166.
97. Ramsundar B, Eastman P, Walters P, et al. Deep learning for the life sciences: applying deep learning to genomics, microscopy, drug discovery and more. Sebastopol: O'Reilly Media; 2019.
98. Mayr A, Klambauer G, Unterthiner T, et al. DeepTox: toxicity prediction using deep learning. Front Environ Sci. 2016;3:80.
99. Napolitano F, Carrella D, Mandriani B, et al. gene2drug: a computational tool for pathway-based rational drug repositioning. Bioinformatics. 2017;34:1498–505.
100. Kuhn M, Szklarczyk D, Pletscher-Frankild S, et al. STITCH 4: integration of protein–chemical interactions with user data. Nucleic Acids Res. 2014;42:D401–7.
101. Senior AW, Evans R, Jumper J, et al. Improved protein structure prediction using potentials from deep learning. Nature. 2020;577:706–10.
102. Aberle DR, Berg CD, Black WC, et al. The national lung screening trial: overview and study design. Radiology. 2011;258:243–53.
103. Jensen MA, Ferretti V, Grossman RL, et al. The NCI genomic data commons as an engine for precision medicine. Blood. 2017;130:453–9.
104. Eraslan G, Avsec Ž, Gagneur J, et al. Deep learning: new computational modelling techniques for genomics. Nat Rev Genet. 2019;20:389–403.
105. Kurnit KC, Bailey AM, Zeng J, et al. "Personalized cancer therapy": a publicly available precision oncology resource. Cancer Res. 2017;77:e123–6.
106. Yu Y, Wang Y, Xia Z, et al. PreMedKB: an integrated precision medicine knowledgebase for interpreting relationships between diseases, genes, variants and drugs. Nucleic Acids Res. 2019;47:D1090–101.
107. Wang J, Cao H, Zhang JZH, Qi Y. Computational protein design with deep learning neural networks. Sci Rep. 2018;8:6349.
108. Eulenberg P, Köhler N, Blasi T, et al. Reconstructing cell cycle and disease progression using deep learning. Nat Commun. 2017;8:463.
109. Buggenthin F, Buettner F, Hoppe PS, et al. Prospective identification of hematopoietic lineage choice by deep learning. Nat Methods. 2017;14:403–6.
110. Artemov AV, Putin E, Vanhaelen Q, et al. Integrated deep learned transcriptomic and structure-based predictor of clinical trials outcomes. 2016. https://doi.org/10.1101/095653v2.

111. Menden MP, Iorio F, Garnett M, et al. Machine learning prediction of cancer cell sensitivity to drugs based on genomic and chemical properties. PLoS One. 2013;8:e61318.
112. Han Y, Kim D. Deep convolutional neural networks for pan-specific peptide-MHC class I binding prediction. BMC Bioinform. 2017;18:585.
113. Zech JR, Badgeley MA, Liu M, et al. Variable generalization performance of a deep learning model to detect pneumonia in chest radiographs: a cross-sectional study. PLoS Med. 2018;15:e1002683.
114. Lambin P, Roelofs E, Reymen B, et al. 'Rapid learning health care in oncology': an approach towards decision support systems enabling customised radiotherapy. Radiother Oncol. 2013;109:159–64.
115. Ross JS, Waldstreicher J, Bamford S, et al. Overview and experience of the YODA project with clinical trial data sharing after 5 years. Sci Data. 2018;5:180268.
116. London JW. Cancer research data-sharing networks. JCO Clin Cancer Inform. 2018;2:1–3.
117. ORIEN: Oncology Research Information Exchange Network. http://oriencancer.org/. Accessed 3 Dec 2018.
118. Academics and Hospitals. Flatiron Health. https://flatiron.com/academics/. Accessed 3 Dec 2018.
119. Wilkinson MD, Dumontier M, Aalbersberg IJ, et al. The FAIR guiding principles for scientific data management and stewardship. Sci Data. 2016;3:160018.
120. Chavan V, Penev L. The data paper: a mechanism to incentivize data publishing in biodiversity science. BMC Bioinform. 2011;12(suppl 15):S2.
121. Zhu G, Pan C, Bei JX, et al. Mutant p53 in cancer progression and targeted therapies. Front Oncol. 2020;10:595187. https://doi.org/10.3389/fonc.2020.595187.
122. Zhang M, Yang H, Wan L, et al. Single-cell transcriptomic architecture and intercellular crosstalk of human intrahepatic cholangiocarcinoma. J Hepatol. 2020;73(5):1118–30. https://doi.org/10.1016/j.jhep.2020.05.039.
123. Fridman WH, Pagès F, Sautès-Fridman C, Galon J. The immune contexture in human tumours: impact on clinical outcome. Nat Rev Cancer. 2012;12(4):298–306. https://doi.org/10.1038/nrc3245.
124. Klein O, Kee D, Nagrial A, et al. Evaluation of combination nivolumab and ipilimumab immunotherapy in patients with advanced biliary tract cancers: subgroup analysis of a phase 2 nonrandomized clinical trial. JAMA Oncol. 2020;6(9):1405–9. https://doi.org/10.1001/jamaoncol.2020.2814.
125. Study of nivolumab in combination with gemcitabine/cisplatin or ipilimumab for patients with advanced unresectable biliary tract cancer. ClinicalTrials.gov. Updated February 1, 2022. https://bit.ly/3yULD6O. Accessed 18 Apr 2022.
126. Artificial Intelligence—Opportunities in Cancer Research by National Cancer Institute, NIH. https://www.cancer.gov/research/areas/diagnosis/artificial-intelligence. Accessed 12 Sept.

Khalid Shaikh is the founder and CEO of Prognica Labs, a healthcare technology company that specializes in developing AI-powered solutions for breast cancer detection and treatment. He is a serial entrepreneur, author, innovator, and business strategist. He has received numerous awards for his innovations in healthcare.

In addition to his professional commitments, he also gives back to the aspiring entrepreneur community by serving as an advisor and mentor. He has published and lectured extensively on healthcare performance improvement, digitalization, and innovation. He is the author of the books artificial intelligence in breast cancer early detection and diagnosis and several research articles published in peer-reviewed journals.

Dr. Sreelekshmi Bekal is the medical director of Prognica labs and has led the company to develop a breakthrough AI solution for detecting early-stage breast cancer to make it more accurate, affordable, and accessible. Dr. Sreelekshmi holds a BDS degree, and she comes with over 8 years of experience in healthcare innovation. She is an entrepreneur and wants to make cutting-edge medical care available to communities around the world, regardless of their resources. She is very actively promoting and advocating breast cancer awareness, screening, and research to help the community, society, and medical education.

Traditional, Complementary, and Integrative Medicine and Cancer Care in the UAE

Heidi Kussmann

19.1 Traditional, Complementary, and Integrative Medicine and Conventional Cancer Care: Opportunities for Whole Person Cancer Care

Patient use of both conventional and integrative oncology before, during, and after conventional cancer care is increasing [1, 2]. Integrative oncology use can exist in the United Arab Emirates (UAE) within a robust framework built on the foundations of regulation, research, and collaboration. In other parts of the world, integrative oncology is employed by patients who want to combine conventional treatment with other therapies to decrease side effects and hopefully have a better treatment outcome [3, 4]. Globally, cancer incidence affects low- and middle-income countries the most. The provision of evidence-based cancer screening, treatment, and conventional care during survivorship and palliative stages of cancer is limited to patients who have insurance or can afford it privately. In the scope of non-conventional care, there is usually no insurance coverage, and patients must pay out of pocket. In low- and middle-income countries, there is limited health literacy and access to affordable oncology care, and patients are using culturally familiar medicine that can be considered traditional, complementary, or integrative medicine [5]. The UAE has experienced definitive growth in the field of oncology and is a destination for top-tier oncology care. The next logical step is to incorporate integrative oncology to combine the best of all care options for people with cancer. It is important to define the fields of traditional, complementary, and integrative medicine and naturopathic oncology in the context of this chapter.

Since 2002, the World Health Organization has adjusted its terminology to integrative instead of alternative medicine to encourage the adoption of traditional, complementary, and integrative medicines (TCIM) in health care systems around

H. Kussmann (✉)
Emirates Oncology Society, Emirates Medical Association, Dubai, United Arab Emirates

© The Author(s) 2024
H. O. Al-Shamsi (ed.), *Cancer Care in the United Arab Emirates*,
https://doi.org/10.1007/978-981-99-6794-0_19

the world [5]. TCIM can be defined as the collaboration of qualified health professions within the subspecialty of oncology. Within the umbrella term of integrative oncology, there exist over 40 professions specializing in supportive cancer care, such as naturopathic doctors, doctors of acupuncture and traditional Chinese medicine, homeopathic doctors, dieticians and nutritionists, anthroposophical medicine, massage therapy, physical/movement therapy, yoga, meditation, psychology, and pastoral care [6]. Each of these professions has strong competencies and supportive care options for people with cancer. Each has research supporting use in cancer and influencing overall survival, and each profession provides education for the public and interested medical professionals via books, research, and accredited education institutions.

There are many challenges to the inclusion of the integrative oncology professions in conventional cancer care [1, 6–19], for which an incomplete list of priorities is summarized below [5, 6, 20]:

1. Develop a TCIM department within each of the UAE oncology hospitals to coordinate and advance the collaboration and research needed between conventional and integrative oncology professionals.
2. Establish professional and patient education about TCIM and establish guidelines for collaborative care.
3. Provide qualified integrative oncology professionals for both inpatient and outpatient settings across the cancer continuum, from screening and early diagnosis to survivorship and palliative care.
4. Contribute clinically relevant integrative oncology research findings to protocols and guidelines for practice, in addition to the evidence in TCIM's care of treatment-related side effects.
5. Address financial, perceptional, and cultural barriers for oncologists and patients to access TCIM.
6. Set an example for the rest of the world on how to bring integrative medicine to people with cancer that improves the patient experience, quality of life, efficacy, outcomes, and overall survival.

19.2 Naturopathic Oncology Within TCIM

Due to the extensive information published about TCIM and the editorial limitations of this chapter, this chapter will briefly review one of the qualified health care professions in TCIM, that of naturopathic oncology, and the evidence supporting some of the profession-based recommendations that can safely be used with conventional oncology treatments. Naturopathic oncologists are licensed naturopathic doctors (NDs) with additional oncology education and training who receive the board-certified status of Fellow by the American Board of Naturopathic Oncology (FABNO). To obtain a license, one must first attend a post-graduate degree-granting program that is accredited by the Council of Naturopathic Medical Education (www.cnme.org). The Council grants the naturopathic medicine doctorate degree to schools that require in-person (not online), 4-year full-time programs and provide standards-based education that combines natural healing systems (homeopathy, Chinese medicine and acupuncture, nutrition, and botanical medicine) *in addition to* the same curriculum as the medical training of medical schools. There are only seven accredited schools in Canada and the USA that provide this extensive education. After graduation, NDs must take the national exams, i.e., the Naturopathic Physicians Licensing Examinations (NPLEX), in order to qualify for province- or state-based licensing, which includes both written and oral examination processes. Only after becoming licensed as an ND and further meeting the requirements for specialization in oncology as set out by the American Board of Naturopathic Oncology can an ND attempt the board certification examination. Renewal of the FABNO board certification is required every 10 years from the original date of certification.

The naturopathic oncology professional care guidelines [2] detail how naturopathic oncology can work in the field of integrative oncology with patients and their oncologists. It includes understanding the established standards and collaborating with patients and oncology professionals to evaluate and prepare the patient for treatment, to provide patients with information on what to avoid and what to use in cancer treatment specific to the patient and his or her treatment details, as well as helping with the side effects. A naturopathic oncology doctor would evaluate the cancer treatment plan and the patient as a whole for deficiencies and use existing research evidence to support the use of specific recommendations to maximize therapeutic potential while reducing early and late side effect incidence and severity [21, 22]. Naturopathic oncology also screens for and advises patients regarding drug–nutrient and drug–herb interactions. This ensures, from the start of a cancer diagnosis through treatment and beyond, that it provides optimal care individualized to each patient according to their conventional oncology treatment touchpoints. Most often, the collaboration starts between medical doctors and naturopathic doctors at tumor board meetings prior to the initiation of therapy and ensures the best integrative and conventional treatments are agreed upon and delivered as seamlessly as possible.

In the context of integrative oncology, naturopathic oncology does more than address concerns about the average quality of life and the common side effects [23] of conventional cancer care through Natural Health Products (NHPs), diet and lifestyle modifications [24, 25], homeopathy [9, 26–29], botanical medicine [30–33], vitamin and nutrient therapy [34, 35], intravenous (IV) nutrient therapy, immune supportive therapies [36], and other targeted treatments. It has been established for years that these modalities work to address the most common detractors of quality of life, such as fatigue [11], pain [37], nausea [38], weight loss or muscle loss [39], sleep disturbances [40], and peripheral neuropathy [41–43].

Now that they have been defined, it makes logical sense to explore the evidence supporting the use of naturopathic and integrative oncology from the start of diagnosis through survivorship and/or palliative care. The focus of this remaining chapter is on the integration of naturopathic oncology throughout the listed conventional treatments below:

1. Screening and diagnosis.
2. Radiation therapy [44–46].
3. Surgery [12, 47].
4. Cytotoxic chemotherapy [48, 49].
5. Targeted therapy [31, 50].
6. Immunotherapy [33, 51].

Disclaimer: The following are not complete or individualized recommendations from naturopathic oncology. For details on doses and regimens, patients must consult with a naturopathic oncologist, as there may be potential interactions and no benefit. It is strongly advised that patients match their health history and cancer treatment details with the correct timing of integrative treatment and that they obtain risk and benefit information when adding TCIM therapy specific to their cancer type, stage, and location. Assessment must include evaluation and NHP recommendations made by a licensed professional (such as a naturopathic doctor) working in either the broad field of integrative oncology or naturopathic oncology.

19.3   Radiation Therapy

This is intended to provide a foundation for combining the best of integrative oncology and naturopathic oncology to work with radiation therapy. The need for guidance is because, as in all aspects of cancer care, many radiation therapy patients are including additional treatments, procedures, items, supplements, etc. without consulting a naturopathic oncology professional and being fully informed of their risks and benefits [52]. Herein lies a missed opportunity to develop integrative guidelines for this area of treatment and to have qualified naturopathic oncology professionals screen problematic or contraindicated items and advise on evidence-based supportive care [53] before radiation-based scans and during radiation therapy.

Radiation has proven efficacious in the treatment of cancer, alongside surgery and chemotherapy. Radiation induces changes in cancer tumor locations and surrounding tissue that can be irreversible. The short-term (within 3 months) effects of radiation therapy include the intended reduction or eradication of the tumor burden on the patient and the adverse effect of inflammation of the nearby tissues and body parts exposed to radiation. These usually subside a few weeks after the completion of radiotherapy.

Prior to radiation treatment, the whole person and the systems-based assessment outlined earlier for each patient include evaluating the following:

- Coping and mental health.
- Exercise or movement habits.
- Nutritional status and lifestyle factors (i.e., smoking).

- Biomarkers of inflammation and digestion.
- Digestive, microbiome, and elimination functions to reduce the incidence of radiation resistance [54, 55].

There has been research into when and what to eat or take during radiation exposure and treatment [45], but there is little evidence to support the use of radioprotectants, and more robust research is needed to ensure the safety and efficacy of everyone involved [56]. There are naturopathic oncology recommendations for occupational exposure and imaging studies that differ from recommendations for a daily radiation treatment schedule of 3–6 weeks duration, but the recommendations for what to avoid remain the same for both scans and treatment. Naturopathic oncology employs a protective protocol for imaging studies that use radiation to reduce the oxidative stress on healthy cells [21, 22, 57–63]. Protection protocols are also available during scans in general [64] and from various types of radiation-based occupational exposure, with the exception of extremely low magnetic fields or electrical shocks [65] that have no causal relationship to cancer development [58, 59, 62, 66–70]. Naturopathic oncology guidelines in preparation for scans and radiation treatment include the following avoidances:

- Mistletoe homeopathic injections, metformin prescriptions, or berberine supplements and IV vitamin C with PET imaging are not recommended, as they theoretically have the potential to interfere with contrast glucose distribution and uptake.
- Multivitamins, iron, beta-carotene, methionine, cysteine, copper, CoQ10, and topical use of mint and cinnamon all have the potential to increase the radiation effect.
- Ozone therapy has the potential to increase the radiation effect.

The typical preparation protocol for imaging studies includes increasing dietary intake of garlic, turmeric, green tea, pomegranate juice, cordyceps mushrooms, selenium from Brazil nuts, zinc from pumpkin seeds, shellfish, and legumes. Following scans, one can supplement with concentrated turmeric, green tea extract, cordyceps, selenium, and zinc, as well as sauna or exercise/sweating, to aid in detoxification and cellular repair mechanisms. Most contrast media used in imaging are metabolized by the liver and eliminated in urine, so hydration with at least 2 L of clean water immediately after imaging is advised in addition to the above.

Detailed radiation therapy protocols are provided for patients according to their type of cancer, the treatment thus far, and their signs and symptoms. These individualized assessment and treatment protocols are intended to reduce early and late radiation adverse events like mucositis, dermatitis (resulting in desquamation, wounds, non-healing ulcers, and radionecrosis), fatigue, neuropathies, weight loss, and hematological suppression [11, 71–83]. These protocols can include, but are not limited to, IV vitamin C, green tea, fish oils, zinc, vitamin D, astragalus, and probiotics [21, 22]. Details should be determined in collaboration with a qualified naturopathic oncology doctor and the radiation oncologist. Not all effects are preventable or treatable with naturopathic or integrative oncology. It is important to include wound care, oncology physical rehabilitation, surgical intervention, and other similar resources as deemed important by the integrative oncology care team.

The support measures of hyperthermia and hyperbaric oxygen are two forms of evidence-based, non-invasive, integrative oncology support and recovery options for radiation therapy. Hyperthermia has been shown to have promising potential when combined with radiation therapy [12, 44, 84–86]. If hyperthermia is not an option but movement is possible, then patients can exercise for 10–15 minutes before each radiation treatment to elevate their core body temperature, circulation, and oxygen levels in the body and increase the radiation's effects on tumor tissue [87, 88]. Hyperbaric oxygen can also be employed to offset early- and late-onset side effects. For example, it can support healing in oral microcirculation [89], in cranial radiation to reduce hippocampal injury [90], in radiation cystitis [91], and in some cancer types like glioblastoma [92]. Both hyperthermia and hyperbaric oxygen have promise and potential in combination with IV vitamin C and radiation treatment [93]. It is highly recommended that hyperthermia and hyperbaric oxygen treatments be offered at the same location as the oncology treatment to facilitate patient compliance and better outcomes.

Some people can experience late adverse effects from radiation after completion that include the permanent reduction of supportive cells in tissue, free radical production, DNA damage, tissue hypoxia, cell death, endothelial and vascular damage, tissue dysfunction and damage, fibrosis, and reduced quality of life. These can show up months or decades after radiation is completed. It is important to talk to patients and identify if, after radiation completion, there are adverse late effects of radiation and to always refer for appropriate management and corrective treatment.

19.4 Surgery

The majority of cancer surgery is resection-focused: the complete removal of the tumor and any surrounding or affected lymph nodes, with as many clear tissue margins as possible. Biopsy and palliative surgery are also tools to help understand the type and stage of cancer and to alleviate emergent issues (i.e., tumors obstructing circulation or digestion). The effect of cancer surgery and the anesthetic agents used in surgery can be compared to an endurance event for the body. The psychological and physical stress from cancer surgery and anesthesia has both beneficial and adverse effects on cancer growth. For patients, it is important to inhibit stress before and after surgery and to target the biochemical mechanisms and signaling pathways involved to inhibit an excessive stress response [94, 95]. The body produces an optimal immune response in the first few days after surgery, yet within 2–3 weeks, this becomes a suppressed immune response. This translates into post-surgery trauma-induced growth factors and immune suppression-mediated tumor progression. This can predispose to T-cell impairment, cytokine reduction, and postoperative sepsis [96]. If this patient already has diabetes, the combined effects of delays or complications in wound healing and immunosuppression must be addressed for the best results. To support the immune system in all phases of cancer treatment, and especially before surgery, this can be accomplished using any combination of the following integrative therapies and options:

- Mindfulness techniques such as meditation, breathwork, and prayer.
- Music therapy, whether in the form of singing, listening, or playing a musical instrument.
- Exercise, tai chi, yoga, and exercise if movement is possible.
- Anti-anxiolytic botanical medicine options such as L-Theanine, Passionflower (*Passiflora incarnata*), Hops (*Humulus lupulus*), or Valerian root (*Valeriana officinalis, Caprifoliaceae*).
- Acupuncture.
- Art therapy.

In preparation for surgery, the patient would spend some time in the integrative oncology department, and the involved professionals would assess and address the areas of mental wellness, physical fitness, dietary intake, hydration, and digestion. To accomplish changes or improvements before surgery depends on co-existing health conditions, prior treatments, the tumor burden, and the time available before surgery. For example, one patient can have anxiety, depression, hypertension, diabetes, chronic liver disease, and cancer. Each condition needs monitoring and integrative treatment to reach a better or more stable state before surgery to reduce immune suppression and the incidence of wound healing challenges and complications [97–99].

Strong evidence exists for the use of homeopathy, meditation, tai chi, yoga, and music therapy to reduce stress before, in the acute recovery phase, and upon discharge home to improve the outcome and reduce complications in healing [12, 47, 100–102]. Fasting mimicking diet (FMD), short-term fasting (STF), and intermittent fasting (IF) in pre-habilitation/preparing for surgery and again in recovery from surgery [103–105] have big benefits for patient recovery. The benefits are cumulative when one combines exercise with intermittent fasting to maintain muscle mass [97] before surgery and when cleared for physical activity afterwards. In addition to specific diet and exercise recommendations, there are evidence-based options available such as oral nutrients [106–109], IV vitamin C [109–111], hyperthermia [112], L-arginine [108], homeopathy [113], hyperbaric oxygen for wound healing [114, 115], acupuncture [7, 98], and botanical medicine [32, 116] to further support recovery. Earlier conclusions about fish oils being a clotting risk have now been refuted, and the current evidence supports the use of omega-3 fish oils without interfering with blood clotting [117–119].

Both conventional and integrative cancer care providers, as in all cancer care, must have a thorough understanding of what patients are doing and what they are taking to achieve treatment goals. It is imperative that trustworthy, open communication about, screening for, and pausing the intake of all drugs, nutrients, and botanicals that are contraindicated for surgery be completed as part of the pre-operative assessment. There is potential for botanical and nutrient interference with blood clotting and anesthesia. Specific evidence-based examples of items to avoid in the diet and supplement format before surgery include the following: ginkgo (*Ginkgo biloba*), cayenne (*Capsicum annuum*), garlic (*Allium sativum*), ginger (*Zingiber officinalis*), Dan Shen (*Angelica sinensis*), fenugreek (*Trigonella foenum-graecum L.*), vitamin E as alpha-tocopherol, curcumin (*Curcuma longa 90%*), chondroitin, and red clover (*Trifolium pratense*) [21, 22, 53].

19.5   Chemotherapy

Chemotherapy includes both targeted and cytotoxic therapies and now has the additional treatment combination of immunotherapies. In each, there exist integrative oncology options for side effect management and augmentation of the therapeutic effect in balance with patient quality of life goals. The most common side effects of cytotoxic-type chemotherapy are listed below:

- Nervous system: neuropathy, tinnitus, vertigo, depression, anxiety.
- Digestive system: constipation, diarrhea, nausea, reflux, dysgeusia, hepatitis, xerostomia, pancreatitis, anorexia/cachexia.
- Bone marrow: low numbers of red, white (including neutrophils), and platelet cells.
- Cardiac system: high or low blood pressure, fast or slow heart rate, heart muscle damage.
- Reproductive system: effects of hormone blockade: impotence, low libido, early or instant menopause, forgetfulness, hot flashes, and joint inflammation.
- Musculoskeletal system: joint pain and muscle aches; spasms, cramps, and pain; reduced strength from muscle loss; cachexia-induced weight loss; loss of fitness.
- Other side effects include fatigue, scarring, insomnia, recurring or severe infections, hair loss, dry eyes, kidney inflammation, dyslipidemia, and an altered state.

Years of research on human safety and efficacy with specific oncology treatments have been conducted in relation to the numerous IVs, oral nutritional supplements, botanical medicine, or other natural health products (NHPs) consumed by patients or prescribed by integrative oncology professionals. Some examples include:

- The use of curcumin and nicotinamide reduces neuropathy [41, 42].
- Correcting liver damage from 5-FU with vitamin C [120].
- Vitamin B5 improves immunotherapy's anti-cancer effects [121].
- Vitamin C induces positive effects with PARP inhibitors [122].

There is still widespread concern, as well as a perception bias or disregard for this research and its value as an adjunct in cancer care. Understandably, the majority of concerns revolve around cancer efficacy and patient safety. Considering recent efforts to study cancer and NHPs in research, there is an appreciably larger body of evidence delineating what works and what does not work. However, to address this as competently as possible, it will always require more research. In the meantime, where evidence does not yet exist, priority is given to supporting the patient's conventional care plan for the best possible outcome. It is a generally accepted practice that when NHPs conflict with conventional oncology treatment, patients are strongly advised to discontinue the item(s) after the conflicting treatment is completed.

To reassure colleagues, it is key that the TCIM professional doing the recommending and prescribing have a thorough pharmacokinetic and pharmacodynamic understanding of these items as well as be able to screen for interactions and contraindications to maintain safe prescribing. When a patient meets with a qualified professional to discuss their IV, botanical and nutritional supplements, and other items such as juices and protein powders purchased from the internet or on the recommendation of a friend, patient satisfaction is met, harm is avoided, and there may even be a cumulative benefit in improving quality of life and fighting cancer, which is the highest priority goal during treatment. The summary is that the evidence supports dietary and botanical medicine supplements being generally recognized as safe (GRAS) when the professional recommending them can evaluate the patient properly and screen for contraindications. Individual patient pharmacodynamics play a large role in drug interactions; there must be access to screening and items in lab ordering to monitor hepatic and renal function when recommending items that can pose a risk alongside chemotherapy treatments, just as when labs are done before chemotherapy. Oncology is but one health care profession where there is ongoing research and review, bringing new information, findings, and protocols for patient care. Health care professionals can find specific items that are referenced in all major peer-reviewed research publishing websites and applications to individually review in more detail as needed. This is a very time-consuming process and should be part of the responsibility of the prescriber of the IV, dietary, and botanical supplements to best serve the patient and the integrative oncology department. Furthermore, it is recommended that, when recommending NHPs, objective data points for the interactions be monitored.

The amount of research on lifestyle intervention, IV therapeutics, dietary, metabolic, phytochemical, and botanical NHP use with cancer chemotherapies cannot be individually reviewed in this work alone and requires ongoing updating as new studies expand upon current knowledge. There is a lot of preclinical and empirical evidence supporting the use of these strategies and options in cancer care in each of the TCIM professions. For example, the use of fasting during chemotherapy has evidence supporting it due to its synergy with cancer treatment. This is reflected positively in the research on short-term fasting [104, 123, 124], fasting-mimicking diets [24, 103, 105, 125–127], intermittent fasting [104, 128], and ketogenic diets [129]. It is important to evaluate for GI, kidney, heart, and liver function; insulin resistance; metabolic syndrome; or other co-morbidities, as well as to consider the risk of disordered eating; the risk of developing cachexia; and cultural, seasonal, and loco-regional influences on dietary patterns. Choosing and recommending the correct dietary approach to complement conventional treatment needs time for patient adaptation and implementation and does require the use of qualified nutritional oncology professionals. Weight and muscle mass maintenance are linked to longer overall survival, and the proper implementation of dietary plans and exercise therapies can support these twin goals.

During the use of immunotherapies and targeted therapies in cancer, the role of integrative oncology providers can continue with the use of synergists and efficacy inducers. Evaluation of the microbiome is also important due to the proven fact that an intact microbiome endures chemotherapy and provides for a better overall outcome with immunotherapies.

As of this writing, the UAE has resources such as the Zayed Complex for Herbal Research and Traditional Medicine through the Department of Health that provide guidelines to consumers and are in ongoing development to remain current as a resource for patients and professionals alike.

Table 19.1 provides a summary of commonly used TCIM NHPs with evidence supporting chemotherapy, targeted therapies, and immune therapies. One statement that is very commonly found in each study is the call for TCIM NHPs to be incorporated into clinical trials when clinical benefits are found.

Table 19.1 Summary of commonly used traditional, complementary, and integrative medicine natural health products and lifestyle strategies employed with chemotherapy, immunotherapy, and targeted therapies. Note that these should be screened for interaction with the specific chemotherapy using an evidence-informed approach by a naturopathic doctor trained in drug-herb-nutrient interactions

Study or supplement name, reference	Details/findings/recommendations
Meta-analysis, supplement safety, systematic review [130] vitamins, botanical, omega-3 fatty acids	• Among 19 trials including patients with cancer undergoing chemotherapy, most ($n = 18$) of the DS studied (e.g., vitamins, botanicals, omega-3 fatty acids) were found to be safe
Chemotherapy and curcumin [131, 132]	• Overall, treatment with curcumin in combination with paclitaxel was superior to the paclitaxel-placebo combination with respect to ORR and physical performance after 12 weeks of treatment • Curcumin given intravenously caused no major safety issues or reduction in quality of life, and it may be beneficial in reducing fatigue • Advances in knowledge: This is the first clinical study to explore the efficacy and safety of administering curcumin intravenously in combination with chemotherapy in the treatment of cancer patients • Curcumin exerted its anticancer effect by increasing reactive oxygen species (ROS) production, which downregulated the DNA repair protein RAD51, leading to upregulation of γH2AX
Curcumin [133] and radiotherapy and chemotherapy	• Systematic review: curcumin increases the effectiveness of chemotherapy and radiotherapy, which results in improved patient survival and increases the expression of anti-metastatic proteins while reducing their side effects
Prostate Cancer Progression and NHPs [134]	• The included trials involved 3418 prostate cancer patients—a median of 64 men per trial—From 13 countries. Various trials evaluated the use of pomegranate seed, green tea, broccoli, and turmeric; flaxseed, low-fat diet, lycopene, selenium, and coenzyme Q10 • All demonstrated beneficial effects

Table 19.1 (continued)

Study or supplement name, reference	Details/findings/recommendations
Vitamins B5 [121] Vitamin C [135–140] D3 [141, 142]	• In a small cohort of melanoma patients, the plasma levels of vitamin B5 positively correlated with responses to PD-1-targeted immunotherapy, favoring differentiation of CD8+ T cells into IL-22 through fueling mitochondrial metabolism • High doses of vitamin C inhibit the growth of prostate, colon, and pancreatic cancers, as well as mesothelioma cell lines • High-dose vitamin C improved fatigue and reduced nausea, vomiting, and loss of appetite • It is generally regarded as safe, but blood vitamin D levels should be monitored on a regular basis when taking it • Vitamin D has the potential to become a valid adjuvant in the treatment of cancer
Fish oil/omega 3 fatty acids [143] or polyunsaturated fatty acids (PUFAs)	• Preclinical evidence reveals that omega-3 PUFAs, and their metabolites might modulate pivotal pathways underlying complications secondary to cancer • Anti-inflammatory and antinociceptive effects • Agonists of G protein-coupled receptors, namely, GPR40/FFA1 and GPR120/FFA4
Gastrointestinal mucositis and probiotics [144, 145]	• Of the agents studied for the prevention and treatment of gastrointestinal mucositis, the evidence continues to support the use of probiotics containing lactobacillus spp. for the prevention of chemoradiotherapy and radiotherapy-induced diarrhea in patients with pelvic malignancy and hyperbaric oxygen therapy to treat radiation-induced proctitis • Twenty clinical trials published between 1988 and 2020 were included in this review. Seventeen studies (85%) revealed predominantly positive results when using probiotics to reduce the incidence of treatment-related side effects in oncology patients, while three studies (15%) reported no impact in their findings • This study sheds some light on the significance of chemotherapy and radiotherapy in altering the composition of the gut microbiota, where probiotic strains may play an important role in preventing or mitigating treatment-related side effects
Melatonin and NSCLC [146] Melatonin and glioblastoma [147] Melatonin and tumor effects [146]	• Enhance the overall survival rate in non-small-cell lung cancer patients (RR = 2.13; 95% CI, 1.41–3.24; $P = 0.0004$; $I^2 = 0\%$) and various solid tumor patients (RR = 2.31; 95% CI, 1.78–2.99; $P < 0.00001$; $I^2 = 0\%$) • Reduce the incidence of neurotoxicity (RR = 0.30, 95% CI, 0.19–0.45; $P < 0.00001$), thrombocytopenia (RR = 0.23; 95% CI, 0.16–0.33; $P < 0.00001$), and asthenia (RR = 0.43, 95% CI, 0.38–0.49; $P < 0.00001$) during chemotherapy • In a case report of a complete response in glioblastoma, melatonin was combined with octreotide and retinoids, vitamin E, and vitamin C • Melatonin improved the tumor remission rate and overall survival rate while reducing the incidence of chemotherapy side effects

(continued)

Table 19.1 (continued)

Study or supplement name, reference	Details/findings/recommendations
Quality of life and mistletoe [148]	• The mistletoe group showed a trend toward less neutropenia ($p = 0.178$) and improved pain and appetite loss scores ($p < 0.0001$ and $p = 0.047$, respectively), while having a positive, but not significant, impact on other EORTC QLQ-C30 scores • Mistletoe extracts were safe in this clinical study • Neither did subcutaneous injections induce fever nor did they influence the frequency of relapse and metastasis within 5 years. This result suggests that mistletoe extracts had no adverse interactions with the anticancer agents used in this study • Certain chemotherapy side effects were reduced in breast cancer patients who received this complementary treatment
Flavonoids [149, 150]	• Silymarin, crocin, anthocyanidins, apigenin, quercetin, luteolin, genistein, and resveratrol all have individual pathway effects in modulating cancer growth • mTOR signaling showed the most targetable promise with flavonoids in chemo-resistant breast cancer, and further clinical trials are necessary to validate mTOR as a target • Apigenin, baicalein, curcumin, EGCG, genistein, luteolin, oridonin, quercetin, and wogonin repress VegF, NF-kappa B (NF-κB, a proinflammatory transcription factor) and inhibit proinflammatory cytokines such as TNF-α and IL-6 in vitro
Exercise and breast, prostate, or colorectal cancer survivors [151], and cardiotoxicity prevention [152], muscle wasting prevention [39, 153]	• A circuit-based, interval-based aerobic and resistance exercise intervention improved patient-reported sleep quality in breast, prostate, and colorectal cancer survivors • Additionally, this exercise-induced improvement in sleep quality may result in reduced insulin resistance • Exercise is feasible with lung cancer and promotes quality of life and survival
Cancer-related pain [7] Climacteric symptoms [154] Fatigue [40] and acupuncture	• Adding non-pharmacological treatment such as acupuncture to conventional pain management is helpful in addressing a patient's pain level • Acupuncture has promising results in treating the severe side effects of hormone blockade in people with breast cancer • Functional assessment of cancer therapy-fatigue (FACT-F) scores changed significantly with acupuncture
Functional food [155–157], diets [102, 158] and metabolic effects [159], and immuno-oncology [159, 160]	• Cinnamaldehyde downregulates *FAK* signaling in osteosarcoma • Chlorella prevents bone marrow suppression from cisplatin • Beet and carrot juice in large amounts benefit when combined with chlorambucil in the treatment of chronic lymphocytic leukemia • This randomized trial showed improvement in the overall functional life index and improvement in hemoglobin in patients with stage 2 and stage 3 colon cancer treated with yoga, naturopathy, and dietary interventions • It is possible to promote patients' response rate to anti-PD-1 by manipulating the gut bacteria composition of non-responders, thereby achieving long-term progression-free survival • Protein arginine methyltransferases (PRMTs) have an effect in combination with immune checkpoint inhibitors • Dietary fiber was associated with significantly improved progression-free survival in 128 patients with melanoma receiving anti-PD-1 immunotherapy

Table 19.1 (continued)

Study or supplement name, reference	Details/findings/recommendations
Astragalus [33, 161] and curcumin	• Astragalus polysaccharide can promote the activities of macrophages, natural killer cells, dendritic cells, T lymphocytes, B lymphocytes, and microglia and induce the expression of a variety of cytokines and chemokines • Astragalus with curcumin reduced the expression of *FGF2*, *MMP2*, *VEGF*, *HGF*, and *TF* in mouse models of human hepatocellular carcinoma, giving potential for combination therapy
Medicinal mushrooms Agaricus, Coriolus, and Ganoderma [36]	• Beneficial effects on quality of life and a reduction of the adverse effects of conventional treatment. Positive effects on antitumor activity and immune modulation were indicated; more research is needed

19.6 Conclusion

In concluding this chapter, there are over 20 years of research and abundant supporting evidence presented for the use of traditional, complementary, and integrative medicine professionals in the field of integrative oncology. Ongoing comparative clinical effectiveness research and collaborative professional inclusion are needed to further define benefits and validate effects within established oncology treatments. Given the challenges and opportunities for the status of integrative oncology, there is high potential for the UAE to become the world leader in providing integrative oncology. This comes in the form of the development of collaborative research, professional regulation, and the inclusion of safe, effective patient-centered care unified with conventional cancer treatment.

Conflict of Interest The author has no conflict of interest to declare.

References

1. Grant SJ, Hunter J, Seely D, Balneaves LG, Rossi E, Bao T. Integrative oncology: international perspectives. Integr Cancer Ther. 2019;18:1534735418823266.
2. Marsden E, Nigh G, Birdsall S, Wright H, Traub M. Oncology Association of Naturopathic Physicians: principles of care guidelines. Curr Oncol. 2019;26(1):12–8.
3. Mathew E, Muttappallymyalil J, Sreedharan J, John L, John J, Mehboob M, et al. Self-reported use of complementary and alternative medicine among the health care consumers at a tertiary Care Center in Ajman, United Arab Emirates. Ann Med Health Sci Res. 2013;3(2):215–9.
4. Segev Y, Lavie O, Stein N, Saliba W, Samuels N, Shalabna E, et al. Correlation between an integrative oncology treatment program and survival in patients with advanced gynecological cancer. Support Care Cancer. 2021;29(7):4055–64.
5. Mao JJ, Pillai GG, Andrade CJ, Ligibel JA, Basu P, Cohen L, et al. Integrative oncology: addressing the global challenges of cancer prevention and treatment. CA Cancer J Clin. 2022;72(2):144–64.

6. Witt CM, Balneaves LG, Carlson LE, Cohen M, Deng G, Fouladbakhsh JM, et al. Education competencies for integrative oncology-results of a systematic review and an international and interprofessional consensus procedure. J Cancer Educ. 2022;37(3):499–507.

7. Ashby J, Toveg M, Ye H, Rubin LH, Reddy S, Chao MT. The assessment and treatment of inpatient cancer-related pain with acupuncture: development of a manual. Med Acupunct. 2022;34(1):15–23.

8. Bagot JL, Theunissen I, Serral A. Perceptions of homeopathy in supportive cancer care among oncologists and general practitioners in France. Support Care Cancer. 2021;29(10):5873–81.

9. Bagot JL, Legrand A, Theunissen I. Use of homeopathy in integrative oncology in Strasbourg, France: multi-center cross-sectional descriptive study of patients undergoing cancer treatment. Homeopathy. 2021;110(3):168–73.

10. Ben-Arye E, Schiff E, Levy M, Raz OG, Barak Y, Bar-Sela G. Barriers and challenges in integration of anthroposophic medicine in supportive breast cancer care. Springerplus. 2013;2:364.

11. David A, Hausner D, Frenkel M. Cancer-related fatigue-is there a role for complementary and integrative medicine? Curr Oncol Rep. 2021;23(12):145.

12. Diller ML, Master VA. Integrative surgical oncology: a model of acute integrative oncology. Cancer. 2021;127(21):3929–38.

13. Höxtermann MD, Haller H, Aboudamaah S, Bachemir A, Dobos G, Cramer H, et al. Safety of acupuncture in oncology: a systematic review and meta-analysis of randomized controlled trials. Cancer. 2022;128(11):2159–73.

14. Klafke N, Mahler C, von Hagens C, Uhlmann L, Bentner M, Schneeweiss A, et al. The effects of an integrated supportive care intervention on quality of life outcomes in outpatients with breast and gynecologic cancer undergoing chemotherapy: results from a randomized controlled trial. Cancer Med. 2019;8(8):3666–76.

15. Psihogios A, Ennis JK, Seely D. Naturopathic Oncology Care for Pediatric Cancers: a practice survey. Integr Cancer Ther. 2019;18:1534735419878504.

16. Rossi EG, Noberasco C, Picchi M, Di Stefano M, Bosinelli F. Integrative oncology and patients refusing conventional anticancer treatments. Complement Ther Clin Pract. 2022;48:101608.

17. Savaş BB, Märtens B, Cramer H, Voiss P, Longolius J, Weiser A, et al. Effects of an interdisciplinary integrative oncology group-based program to strengthen resilience and improve quality of life in cancer patients: results of a prospective longitudinal single-center study. Integr Cancer Ther. 2022;21:15347354221081770.

18. Schütze T, Längler A, Zuzak TJ, Schmidt P, Zernikow B. Use of complementary and alternative medicine by pediatric oncology patients during palliative care. Support Care Cancer. 2016;24(7):2869–75.

19. Seifert G, Blakeslee SB, Calaminus G, Kandil FI, Barth A, Bernig T, et al. Integrative medicine during the intensive phase of chemotherapy in pediatric oncology in Germany: a randomized controlled trial with 5-year follow up. BMC Cancer. 2022;22(1):652.

20. Ben-Arye E, Schiff E, Zollman C, Heusser P, Mountford P, Frenkel M, et al. Integrating complementary medicine in supportive cancer care models across four continents. Med Oncol. 2013;30(2):511.

21. Parmar G, Kaczor T, Boudreau E. Textbook of naturopathic oncology: a desktop guide of integrative cancer care. 1st ed. Canada: Medicatrix Holdings Ltd.; 2020. p. 540.

22. McKinney N. Naturopathic oncology: an encyclopedic guide for patients and physicians. 2nd ed. Vancouver: Liaison Press; 2012. p. 540.

23. Anderson JG, Taylor AG. Use of complementary therapies for cancer symptom management: results of the 2007 National Health Interview Survey. J Altern Complement Med. 2012;18(3):235–41.

24. Vernieri C, Fucà G, Ligorio F, Huber V, Vingiani A, Iannelli F, et al. Fasting-mimicking diet is safe and reshapes metabolism and antitumor immunity in patients with cancer. Cancer Discov. 2022;12(1):90–107.

25. Brandhorst S. Fasting and fasting-mimicking diets for chemotherapy augmentation. Geroscience. 2021;43(3):1201–16.

26. Arora S, Aggarwal A, Singla P, Jyoti S, Tandon S. Anti-proliferative effects of homeopathic medicines on human kidney, colon and breast cancer cells. Homeopathy. 2013;102(4):274–82.
27. Arora S, Tandon S. DNA fragmentation and cell cycle arrest: a hallmark of apoptosis induced by *Ruta graveolens* in human colon cancer cells. Homeopathy. 2015;104(1):36–47.
28. Bagot JL. Using hetero-isotherapics in cancer supportive care: the fruit of 15 years of experience. Homeopathy. 2016;105(1):119–25.
29. Bell IR, Sarter B, Koithan M, Banerji P, Jain S, Ives J. Integrative nanomedicine: treating cancer with nanoscale natural products. Glob Adv Health Med. 2014;3(1):36–53.
30. Advanced Centre for Treatment Research and Education in Cancer (New Bombay India), Central Council for Research in Ayurveda and Siddha (India). Report on screening [sic] of single herbal drug extracts for potential anti-cancer activity: a joint project with Advanced Centre for Treatment, Research, and Education in Cancer (ACTREC) Kharghar, Navi Mumbai and central Council for Research in Ayurveda & Siddha, Dept. of Ayush, Ministry of Health & family welfare, Govt. of India. New Delhi: Central Council for Research in Ayurveda and Siddha; 2009. p. 70.
31. Amritpal S. Herbal drugs as therapeutic agents. Boca Raton: CRC Press; 2014. p. 214.
32. Hong F, Zhao M, Xue LL, Ma X, Liu L, Cai XY, et al. The ethanolic extract of *Artemisia anomala* exerts anti-inflammatory effects via inhibition of NLRP3 inflammasome. Phytomedicine. 2022;102:154163.
33. Li CX, Liu Y, Zhang YZ, Li JC, Lai J. Astragalus polysaccharide: a review of its immunomodulatory effect. Arch Pharm Res. 2022;45(6):367–89.
34. Porciello G, Montagnese C, Crispo A, Grimaldi M, Libra M, Vitale S, et al. Mediterranean diet and quality of life in women treated for breast cancer: a baseline analysis of DEDiCa multicentre trial. PLoS One. 2020;15(10):e0239803.
35. Kasarla SS, Garikapati V, Kumar Y, Dodoala S. Interplay of vitamin D and CYP3A4 polymorphisms in endocrine disorders and cancer. Endocrinol Metab (Seoul). 2022;37(3):392–407.
36. Jeitler M, Michalsen A, Frings D, Hübner M, Fischer M, Koppold-Liebscher DA, et al. Significance of medicinal mushrooms in integrative oncology: a narrative review. Front Pharmacol. 2020;11:580656.
37. Bao Y, Kong X, Yang L, Liu R, Shi Z, Li W, et al. Complementary and alternative medicine for cancer pain: an overview of systematic reviews. Evid Based Complement Alternat Med. 2014;2014:170396.
38. Pérol D, Provençal J, Hardy-Bessard AC, Coeffic D, Jacquin JP, Agostini C, et al. Can treatment with Cocculine improve the control of chemotherapy-induced emesis in early breast cancer patients? A randomized, multi-centered, double-blind, placebo-controlled phase III trial. BMC Cancer. 2012;12:603.
39. Cortiula F, Hendriks LEL, van de Worp WRPH, Schols AMWJ, Vaes RDW, Langen RCJ, et al. Physical exercise at the crossroad between muscle wasting and the immune system: implications for lung cancer cachexia. J Cachexia Sarcopenia Muscle. 2022;13(1):55–67.
40. Lee B, Kim BK, Kim M, Kim AR, Park HJ, Kwon OJ, et al. Electroacupuncture for treating cancer-related insomnia: a multicenter, assessor-blinded, randomized controlled, pilot clinical trial. BMC Complement Med Ther. 2022;22(1):77.
41. Acklin S, Sadhukhan R, Du W, Patra M, Cholia R, Xia F. Nicotinamide riboside alleviates cisplatin-induced peripheral neuropathy via SIRT2 activation. Neurooncol Adv. 2022;4(1):vdac101.
42. Caillaud M, Thompson D, Toma W, White A, Mann J, Roberts JL, et al. Formulated curcumin prevents paclitaxel-induced peripheral neuropathy through reduction in neuroinflammation by modulation of α7 nicotinic acetylcholine receptors. Pharmaceutics. 2022;14(6):1296.
43. Chen Y, Lu R, Wang Y, Gan P. Shaoyao Gancao decoction ameliorates paclitaxel-induced peripheral neuropathy via suppressing TRPV1 and TLR4 signaling expression in rats. Drug Des Dev Ther. 2022;16:2067–81.
44. Le Guevelou J, Chirila ME, Achard V, Guillemin PC, Lorton O, Uiterwijk JWE, et al. Combined hyperthermia and radiotherapy for prostate cancer: a systematic review. Int J Hyperth. 2022;39(1):547–56.

45. Swarup AB, Barrett W, Jazieh AR. The use of complementary and alternative medicine by cancer patients undergoing radiation therapy. Am J Clin Oncol. 2006;29(5):468–73.
46. Arora R. Herbal radiomodulators: applications in medicine, homeland defence, and space, vol. 17. Oxfordshire: CABI; 2008. p. 332.
47. Abushukur Y, Cascardo C, Ibrahim Y, Teklehaimanot F, Knackstedt R. Improving breast surgery outcomes through alternative therapy: a systematic review. Cureus. 2022;14(3):e23443.
48. Maurici CE, Colenbier R, Wylleman B, Brancato L, van Zwol E, Van den Bossche J, et al. Hyperthermia enhances efficacy of chemotherapeutic agents in pancreatic cancer cell lines. Biomol Ther. 2022;12(5):651.
49. Moukarzel LA, Ferrando L, Dopeso H, Stylianou A, Basili T, Pareja F, et al. Hyperthermic intraperitoneal chemotherapy (HIPEC) with carboplatin induces distinct transcriptomic changes in ovarian tumor and normal tissues. Gynecol Oncol. 2022;165(2):239–47.
50. Rahman M. Biomarkers as targeted herbal drug discovery: a pharmacological approach to nanomedicines. 1st ed. Palm Bay: Apple Academic Press; 2021.
51. Liu W, Chen H, Zhu Z, Liu Z, Ma C, Lee YJ, et al. Ferroptosis inducer improves the efficacy of oncolytic virus-mediated cancer immunotherapy. Biomedicines. 2022;10(6):1425.
52. Dupin C, Arsène-Henry A, Charleux T, Haaser T, Trouette R, Vendrely V. Prevalence and expectations of "alternative and complementary medicine" use during radiotherapy in 2016: a prospective study. Cancer Radiother. 2018;22(6–7):682–7.
53. Brinker F. Herbal contraindications and drug interactions plus herbal adjuncts with medicines. 4th ed. Sandy: Eclectic Medical Publications; 2010. p. 603.
54. Huang R, Xiang J, Zhou P, Vitamin D. Gut microbiota, and radiation-related resistance: a love-hate triangle. J Exp Clin Cancer Res. 2019;38(1):493.
55. Shastri AA, Lombardo J, Okere SC, Higgins S, Smith BC, DeAngelis T, et al. Personalized nutrition as a key contributor to improving radiation response in breast cancer. Int J Mol Sci. 2021;23(1):175.
56. Kim LN, Rubenstein RN, Chu JJ, Allen RJ, Mehrara BJ, Nelson JA. Noninvasive systemic modalities for prevention of head and neck radiation-associated soft tissue injury: a narrative review. J Reconstr Microsurg. 2022;38(08):621–9.
57. Zablotska LB, Lane RSD, Randhawa K. Association between exposures to radon and γ-ray radiation and histologic type of lung cancer in Eldorado uranium mining and milling workers from Canada. Cancer. 2022;128(17):3204–16.
58. Werneth CM, Slaba TC, Huff JL, Patel ZS, Simonsen LC. Medical countermeasure requirements to meet NASA's space radiation permissible exposure limits for a Mars Mission scenario. Health Phys. 2022;123(2):116–27.
59. Weinmann S, Tanaka LF, Schauberger G, Osmani V, Klug SJ. Breast cancer among female flight attendants and the role of the occupational exposures: systematic review and meta-analysis. J Occup Environ Med. 2022;64(10):822–30.
60. Sills WS, Tooze JA, Olson JD, Caudell DL, Dugan GO, Johnson BJ, et al. Total-body irradiation is associated with increased incidence of mesenchymal neoplasia in a radiation late effects cohort of rhesus macaques (Macaca mulatta). Int J Radiat Oncol Biol Phys. 2022;113(3):661–74.
61. Richardson DB, Rage E, Demers PA, Do MT, Fenske N, Deffner V, et al. Lung cancer and radon: pooled analysis of uranium miners hired in 1960 or later. Environ Health Perspect. 2022;130(5):57010.
62. Park DJ, Park S, Ma SW, Seo H, Lee SG, Lee KE. Assessment of risks for breast cancer in a flight attendant exposed to night shift work and cosmic ionizing radiation: a case report. Ann Occup Environ Med. 2022;34:e5.
63. Palmer JD, Prasad RN, Cioffi G, Kruchtko C, Zaorsky NG, Trifiletti DM, et al. Exposure to radon and heavy particulate pollution and incidence of brain tumors. Neuro-Oncology. 2022;25(2):407–17.
64. Foucault A, Ancelet S, Dreuil S, Caër-Lorho S, Ducou Le Pointe H, Brisse H, et al. Childhood cancer risks estimates following CT scans: an update of the French CT cohort study. Eur Radiol. 2022;32(8):5491–8.

65. Jalilian H, Guxens M, Heikkinen S, Pukkala E, Huss A, Eshagh Hossaini SK, et al. Malignant lymphoma and occupational exposure to extremely low frequency magnetic fields and electrical shocks: a nested case-control study in a cohort of four Nordic countries. Occup Environ Med. 2022;79(9):631–6.

66. Beyea J. Implications of recent epidemiological studies for compensation of veterans exposed to plutonium. Health Phys. 2022;123(2):133–53.

67. Boice JD, Cohen SS, Mumma MT, Golden AP, Howard SC, Girardi DJ, et al. Mortality among Tennessee Eastman corporation (TEC) uranium processing workers, 1943–2019. Int J Radiat Biol. 2022;99(2):208–28.

68. De Roo B, Bacher K, Verstraete K. Cervical and lumbar spine imaging after traffic and occupational accidents: evaluation of the use of imaging techniques, cumulative radiation dose and associated lifetime cancer risk. Eur J Radiol. 2022;151:110293.

69. Hunter N, Haylock RGE, Gillies M, Zhang W. Extended analysis of solid cancer incidence among the nuclear industry workers in the UK: 1955–2011. Radiat Res. 2022;198(1):1–17.

70. Kirresh A, White L, Mitchell A, Ahmad S, Obika B, Davis S, et al. Radiation-induced coronary artery disease: a difficult clinical conundrum. Clin Med (Lond). 2022;22(3):251–6.

71. Gkantaifi A, Alongi F, Vardas E, Cuccia F, Hajiioannou J, Kyrodimos E, et al. Honey against radiation-induced oral mucositis in head and neck cancer patients. An umbrella review of systematic reviews and meta-analyses of the literature. Rev Recent Clin Trials. 2020;15(4):360–9.

72. Münstedt K, Momm F, Hübner J. Honey in the management of side effects of radiotherapy- or radio/chemotherapy-induced oral mucositis. A systematic review. Complement Ther Clin Pract. 2019;34:145–52.

73. Talakesh T, Tabatabaee N, Atoof F, Aliasgharzadeh A, Sarvizade M, Farhood B, et al. Effect of nano-curcumin on radiotherapy-induced skin reaction in breast cancer patients: a randomized, triple-blind, placebo-controlled trial. Curr Radiopharm. 2022;15(4):332–40.

74. Zhao H, Zhu W, Zhao X, Li X, Zhou Z, Zheng M, et al. Efficacy of epigallocatechin-3-gallate in preventing dermatitis in patients with breast cancer receiving postoperative radiotherapy: a double-blind, placebo-controlled, phase 2 randomized clinical trial. JAMA Dermatol. 2022;158(7):779–86.

75. Rao S, Kalekhan F, Hegde SK, Rao P, Suresh S, Baliga MS. Serum zinc status and the development of mucositis and dermatitis in head-and-neck cancer patients undergoing curative radiotherapy: a pilot study. J Cancer Res Ther. 2022;18(1):42–8.

76. Garbuio DC, Ribeiro VDS, Hamamura AC, Faustino A, Freitas LAP, Viani G, et al. A chitosan-coated chamomile microparticles formulation to prevent radiodermatitis in breast: a double-blinded, controlled, randomized, phase II clinical trial. Am J Clin Oncol. 2022;45(5):183–9.

77. Tai RZ, Loh EW, Tsai JT, Tam KW. Effect of hyaluronic acid on radiotherapy-induced mucocutaneous side effects: a meta-analysis of randomized controlled trials. Support Care Cancer. 2022;30(6):4845–55.

78. McMillan H, Barbon CEA, Cardoso R, Sedory A, Buoy S, Porsche C, et al. Manual therapy for patients with radiation-associated trismus after head and neck cancer. JAMA Otolaryngol Head Neck Surg. 2022;148(5):418–25.

79. Abu Zaid Z, Kay Neoh M, Mat Daud ZA, Md Yusop NB, Ibrahim Z, Abdul Rahman Z, et al. Weight loss in post-chemoradiotherapy head and neck cancer patients. Nutrients. 2022;14(3):548.

80. Ligibel JA, Pierce LJ, Bender CM, Crane TE, Dieli-Conwright C, Hopkins JO, et al. Attention to diet, exercise, and weight in oncology care: results of an American Society of Clinical Oncology national patient survey. Cancer. 2022;128(14):2817–25.

81. Shahid A, Huang M, Liu M, Shamim MA, Parsa C, Orlando R, et al. The medicinal mushroom *Ganoderma lucidum* attenuates UV-induced skin carcinogenesis and immunosuppression. PLoS One. 2022;17(3):e0265615.

82. Donlon NE, Davern M, O'Connell F, Sheppard A, Heeran A, Bhardwaj A, et al. Impact of radiotherapy on the immune landscape in oesophageal adenocarcinoma. World J Gastroenterol. 2022;28(21):2302–19.

83. O'Leary BR, Houwen FK, Johnson CL, Allen BG, Mezhir JJ, Berg DJ, et al. Pharmacological ascorbate as an adjuvant for enhancing radiation-chemotherapy responses in gastric adenocarcinoma. Radiat Res. 2018;189(5):456–65.

84. Mirzaei S, Iranshahy M, Gholamhosseinian H, Matin MM, Rassouli FB. Urolithins increased anticancer effects of chemical drugs, ionizing radiation and hyperthermia on human esophageal carcinoma cells in vitro. Tissue Cell. 2022;77:101846.

85. Lee Y, Kim S, Cha H, Han JH, Choi HJ, Go E, et al. Long-term feasibility of 13.56 MHz modulated electro-hyperthermia-based thermoradiochemotherapy in locally advanced rectal cancer. Cancers (Basel). 2022;14(5):1271.

86. Yan B, Liu C, Wang S, Li H, Jiao J, Lee WSV, et al. Magnetic hyperthermia induces effective and genuine immunogenic tumor cell death with respect to exogenous heating. J Mater Chem B. 2022;10(28):5364–74.

87. Schumacher O, Luo H, Taaffe DR, Galvão DA, Tang C, Chee R, et al. Effects of exercise during radiation therapy on physical function and treatment-related side effects in men with prostate cancer: a systematic review and meta-analysis. Int J Radiat Oncol Biol Phys. 2021;111(3):716–31.

88. Zaorsky NG, Garrett S, Spratt DE, Nguyen PL, Sciamanna C, Schmitz K. Exercise: a treatment that should be prescribed with radiation therapy. Int J Radiat Oncol Biol Phys. 2022;112(1):96–8.

89. Helmers R, Milstein DMJ, Straat NF, Navran A, Teguh DN, van Hulst RA, et al. The impact of hyperbaric oxygen therapy on late irradiation injury in oral microcirculation. Head Neck. 2022;44(7):1646–54.

90. Hokama Y, Nishimura M, Usugi R, Fujiwara K, Katagiri C, Takagi H, et al. Recovery from the damage of cranial radiation modulated by memantine, an NMDA receptor antagonist combined with hyperbaric oxygen therapy. Neuro-Oncology. 2022;25(1):108–22.

91. Sarrió-Sanz P, Sanchez-Caballero L, Martinez-Cayuelas L, Gori CF, Pacheco-Bru JJ, Nakdali-Kassab B, et al. Efficacy, tolerance and predictors of response to the treatment with hyperbaric oxygen therapy for patients with hemorrhagic radiation cystitis. Arch Esp Urol. 2022;75(4):354–60.

92. Alpuim Costa D, Sampaio-Alves M, Netto E, Fernandez G, Oliveira E, Teixeira A, et al. Hyperbaric oxygen therapy as a complementary treatment in glioblastoma-a scoping review. Front Neurol. 2022;13:886603.

93. Ou J, Zhu X, Chen P, Du Y, Lu Y, Peng X, et al. A randomized phase II trial of best supportive care with or without hyperthermia and vitamin C for heavily pretreated, advanced, refractory non-small-cell lung cancer. J Adv Res. 2020;24:175–82.

94. Wang Y, Qu M, Qiu Z, Zhu S, Chen W, Guo K, et al. Surgical stress and cancer progression: new findings and future perspectives. Curr Oncol Rep. 2022;24(11):1501–11.

95. Santander Ballestín S, Lanuza Bardaji A, Marco Continente C, Luesma Bartolomé MJ. Antitumor anesthetic strategy in the perioperative period of the oncological patient: a review. Front Med (Lausanne). 2022;9:799355.

96. Hogan BV, Peter MB, Shenoy HG, Horgan K, Hughes TA. Surgery induced immunosuppression. Surgeon. 2011;9(1):38–43.

97. Kaibori M, Ishizaki M, Matsui K, Nakatake R, Yoshiuchi S, Kimura Y, et al. Perioperative exercise for chronic liver injury patients with hepatocellular carcinoma undergoing hepatectomy. Am J Surg. 2013;206(2):202–9.

98. Zou X, Yang YC, Wang Y, Pei W, Han JG, Lu Y, et al. Electroacupuncture versus sham electroacupuncture in the treatment of postoperative ileus after laparoscopic surgery for colorectal cancer: study protocol for a multicentre, randomised, sham-controlled trial. BMJ Open. 2022;12(4):e050000.

99. Michael CM, Lehrer EJ, Schmitz KH, Zaorsky NG. Prehabilitation exercise therapy for cancer: a systematic review and meta-analysis. Cancer Med. 2021;10(13):4195–205.

100. Ginsberg JP, Raghunathan K, Bassi G, Ulloa L. Review of perioperative music medicine: mechanisms of pain and stress reduction around surgery. Front Med (Lausanne). 2022;9:821022.
101. Brunet J, Wurz A, Hussien J, Pitman A, Conte E, Ennis JK, et al. Exploring the effects of yoga therapy on heart rate variability and patient-reported outcomes after cancer treatment: a study protocol. Integr Cancer Ther. 2022;21:15347354221075576.
102. Raghunath K, Sumathi C, Rajappa SJ, Mohan MVTK, Kumar U, Shaik U, et al. Impact of naturopathy, yoga, and dietary interventions as adjuvant chemotherapy in the management of stage II and III adenocarcinoma of the colon. Int J Color Dis. 2020;35(12):2309–22.
103. Valdemarin F, Caffa I, Persia A, Cremonini AL, Ferrando L, Tagliafico L, et al. Safety and feasibility of fasting-mimicking diet and effects on nutritional status and circulating metabolic and inflammatory factors in cancer patients undergoing active treatment. Cancers (Basel). 2021;13(16):4013.
104. Hu D, Xie Z, Ye Y, Bahijri S, Chen M. The beneficial effects of intermittent fasting: an update on mechanism, and the role of circadian rhythm and gut microbiota. Hepatobiliary Surg Nutr. 2020;9(5):597–602.
105. Caffa I, Spagnolo V, Vernieri C, Valdemarin F, Becherini P, Wei M, et al. Fasting-mimicking diet and hormone therapy induce breast cancer regression. Nature. 2020;583(7817):620–4.
106. Daher GS, Choi KY, Wells JW, Goyal N. A systematic review of oral nutritional supplement and wound healing. Ann Otol Rhinol Laryngol. 2022;131(12):1358–68.
107. Serrano PE, Parpia S, Simunovic M, Duceppe E, Pinto-Sanchez MI, Bhandari M, et al. Perioperative optimization with nutritional supplements in patients undergoing gastrointestinal surgery for cancer: a randomized, placebo-controlled feasibility clinical trial. Surgery. 2022;172(2):670–6.
108. Angka L, Tanese de Souza C, Baxter KE, Khan ST, Market M, Martel AB, et al. Perioperative arginine prevents metastases by accelerating natural killer cell recovery after surgery. Mol Ther. 2022;30(10):3270–83.
109. Norouzi M, Nadjarzadeh A, Maleki M, Khayyatzadeh SS, Hosseini S, Yaseri M, et al. The effects of preoperative supplementation with a combination of beta-hydroxy-beta-methylbutyrate, arginine, and glutamine on inflammatory and hematological markers of patients with heart surgery: a randomized controlled trial. BMC Surg. 2022;22(1):51.
110. Klimant E, Wright H, Rubin D, Seely D, Markman M. Intravenous vitamin C in the supportive care of cancer patients: a review and rational approach. Curr Oncol. 2018;25(2):139–48.
111. Carr AC, Vissers MC, Cook JS. The effect of intravenous vitamin C on cancer- and chemotherapy-related fatigue and quality of life. Front Oncol. 2014;4:283.
112. Flegar L, Zacharis A, Aksoy C, Heers H, Derigs M, Eisenmenger N, et al. Alternative- and focal therapy trends for prostate cancer: a total population analysis of in-patient treatments in Germany from 2006 to 2019. World J Urol. 2022;40(7):1645–52.
113. Gaertner K, Baumgartner S, Walach H. Is homeopathic. Front Surg. 2021;8:680930.
114. Oley MH, Oley MC, Iskandar AAA, Toreh C, Tulong MT, Faruk M. Hyperbaric oxygen therapy for reconstructive urology wounds: a case series. Res Rep Urol. 2021;13:841–52.
115. Yousef A, Solomon I, Hom DB. Can hyperbaric oxygen salvage a compromised local/regional skin flap? Laryngoscope. 2022;132(10):1892–4.
116. Hussain A, Behl S. Treating endocrine and metabolic disorders with herbal medicines. Hershey: Medical Information Science Reference; 2021.
117. Zhang T, Li G, Duan M, Lv T, Feng D, Lu N, et al. Perioperative parenteral fish oil supplementation improves postoperative coagulation function and outcomes in patients undergoing colectomy for ulcerative colitis. JPEN J Parenter Enteral Nutr. 2022;46(4):878–86.
118. Hanindita MH, Irawan R, Ugrasena IDG, Hariastawa IGBA. Comparison of two lipid emulsions on interleukin-1β, interleukin-8 and fatty acid composition in infants post gastrointestinal surgery: a randomized trial. F1000Res. 2020;9:1168.

119. Liu Z, Ge X, Chen L, Sun F, Ai S, Kang X, et al. The addition of ω-3 fish oil fat emulsion to parenteral nutrition reduces short-term complications after laparoscopic surgery for gastric cancer. Nutr Cancer. 2021;73(11–12):2469–76.
120. Al-Asmari AK, Khan AQ, Al-Masri N. Mitigation of 5-fluorouracil-induced liver damage in rats by vitamin C via targeting redox-sensitive transcription factors. Hum Exp Toxicol. 2016;35(11):1203–13.
121. Bourgin M, Kepp O, Kroemer G. Immunostimulatory effects of vitamin B5 improve anticancer immunotherapy. Onco Targets Ther. 2022;11(1):2031500.
122. Ma Y, Chen P, Drisko J, Khabele D, Godwin A, Chen Q. Pharmacological ascorbate induces 'BRCAness' and enhances the effects of poly(ADP-ribose) polymerase inhibitors against BRCA1/2 wild-type ovarian cancer. Oncol Lett. 2020;19(4):2629–38.
123. Mindikoglu AL, Abdulsada MM, Jain A, Jalal PK, Devaraj S, Wilhelm ZR, et al. Intermittent fasting from dawn to sunset for four consecutive weeks induces anticancer serum proteome response and improves metabolic syndrome. Sci Rep. 2020;10(1):18341.
124. de Groot S, Pijl H, van der Hoeven JJM, Kroep JR. Effects of short-term fasting on cancer treatment. J Exp Clin Cancer Res. 2019;38(1):209.
125. Di Tano M, Raucci F, Vernieri C, Caffa I, Buono R, Fanti M, et al. Synergistic effect of fasting-mimicking diet and vitamin C against KRAS mutated cancers. Nat Commun. 2020;11(1):2332.
126. de Groot S, Lugtenberg RT, Cohen D, Welters MJP, Ehsan I, Vreeswijk MPG, et al. Fasting mimicking diet as an adjunct to neoadjuvant chemotherapy for breast cancer in the multicentre randomized phase 2 DIRECT trial. Nat Commun. 2020;11(1):3083.
127. Salvadori G, Zanardi F, Iannelli F, Lobefaro R, Vernieri C, Longo VD. Fasting-mimicking diet blocks triple-negative breast cancer and cancer stem cell escape. Cell Metab. 2021;33(11):2247–59.e6.
128. Ye Y, Xu H, Xie Z, Wang L, Sun Y, Yang H, et al. Time-restricted feeding reduces the detrimental effects of a high-fat diet, possibly by modulating the circadian rhythm of hepatic lipid metabolism and gut microbiota. Front Nutr. 2020;7:596285.
129. Sebastian S, Paul A, Joby J, Saijan S, Vilapurathu JK. Effect of high-dose intravenous ascorbic acid on cancer patients following ketogenic diet. J Cancer Res Ther. 2021;17(6):1583–6.
130. Zeng Z, Mishuk AU, Qian J. Safety of dietary supplements use among patients with cancer: a systematic review. Crit Rev Oncol Hematol. 2020;152:103013.
131. Saghatelyan T, Tananyan A, Janoyan N, Tadevosyan A, Petrosyan H, Hovhannisyan A, et al. Efficacy and safety of curcumin in combination with paclitaxel in patients with advanced, metastatic breast cancer: a comparative, randomized, double-blind, placebo-controlled clinical trial. Phytomedicine. 2020;70:153218.
132. Wang G, Duan P, Wei Z, Liu F. Curcumin sensitizes carboplatin treatment in triple negative breast cancer through reactive oxygen species induced DNA repair pathway. Mol Biol Rep. 2022;49(4):3259–70.
133. Mansouri K, Rasoulpoor S, Daneshkhah A, Abolfathi S, Salari N, Mohammadi M, et al. Clinical effects of curcumin in enhancing cancer therapy: a systematic review. BMC Cancer. 2020;20(1):791.
134. Hackshaw-McGeagh LE, Perry RE, Leach VA, Qandil S, Jeffreys M, Martin RM, et al. A systematic review of dietary, nutritional, and physical activity interventions for the prevention of prostate cancer progression and mortality. Cancer Causes Control. 2015;26(11):1521–50.
135. Chen P, Yu J, Chalmers B, Drisko J, Yang J, Li B, et al. Pharmacological ascorbate induces cytotoxicity in prostate cancer cells through ATP depletion and induction of autophagy. Anticancer Drugs. 2012;23(4):437–44.
136. Pathi SS, Lei P, Sreevalsan S, Chadalapaka G, Jutooru I, Safe S. Pharmacologic doses of ascorbic acid repress specificity protein (Sp) transcription factors and Sp-regulated genes in colon cancer cells. Nutr Cancer. 2011;63(7):1133–42.

137. Du J, Martin SM, Levine M, Wagner BA, Buettner GR, Wang SH, et al. Mechanisms of ascorbate-induced cytotoxicity in pancreatic cancer. Clin Cancer Res. 2010;16(2):509–20.
138. Takemura Y, Satoh M, Satoh K, Hamada H, Sekido Y, Kubota S. High dose of ascorbic acid induces cell death in mesothelioma cells. Biochem Biophys Res Commun. 2010;394(2):249–53.
139. Welsh JL, Wagner BA, van't Erve TJ, Zehr PS, Berg DJ, Halfdanarson TR, et al. Pharmacological ascorbate with gemcitabine for the control of metastatic and node-positive pancreatic cancer (PACMAN): results from a phase I clinical trial. Cancer Chemother Pharmacol. 2013;71(3):765–75.
140. Drisko JA, Serrano OK, Spruce LR, Chen Q, Levine M. Treatment of pancreatic cancer with intravenous vitamin C: a case report. Anticancer Drugs. 2018;29(4):373–9.
141. Frenkel M, Abrams DI, Ladas EJ, Deng G, Hardy M, Capodice JL, et al. Integrating dietary supplements into cancer care. Integr Cancer Ther. 2013;12(5):369–84.
142. Pandolfi F, Franza L, Mandolini C, Conti P. Immune modulation by vitamin D: special emphasis on its role in prevention and treatment of cancer. Clin Ther. 2017;39(5):884–93.
143. Freitas RDS, Campos MM. Protective effects of omega-3 fatty acids in cancer-related complications. Nutrients. 2019;11(5):945.
144. Bowen JM, Gibson RJ, Coller JK, Blijlevens N, Bossi P, Al-Dasooqi N, et al. Systematic review of agents for the management of cancer treatment-related gastrointestinal mucositis and clinical practice guidelines. Support Care Cancer. 2019;27(10):4011–22.
145. Rodriguez-Arrastia M, Martinez-Ortigosa A, Rueda-Ruzafa L, Folch Ayora A, Ropero-Padilla C. Probiotic supplements on oncology patients' treatment-related side effects: a systematic review of randomized controlled trials. Int J Environ Res Public Health. 2021;18(8):4265.
146. Wang Y, Wang P, Zheng X, Du X. Therapeutic strategies of melatonin in cancer patients: a systematic review and meta-analysis. Onco Targets Ther. 2018;11:7895–908.
147. Di Bella G, Leci J, Ricchi A, Toscano R. Recurrent glioblastoma Multiforme (grade IV—WHO 2007): a case of complete objective response—concomitant administration of somatostatin/octreotide, retinoids, Vit E, Vit D3, Vit C, melatonin, D2 R agonists (Di Bella method). Neuro Endocrinol Lett. 2015;36(2):127–32.
148. Pelzer F, Tröger W, Nat DR. Complementary treatment with mistletoe extracts during chemotherapy: safety, neutropenia, fever, and quality of life assessed in a randomized study. J Altern Complement Med. 2018;24(9–10):954–61.
149. Hussain Y, Khan H, Alam W, Aschner M, Abdullah, Alsharif KF, et al. Flavonoids targeting the mTOR signaling cascades in cancer: a potential crosstalk in anti-breast cancer therapy. Oxidat Med Cell Longev. 2022;2022:4831833.
150. Chen SS, Michael A, Butler-Manuel SA. Advances in the treatment of ovarian cancer: a potential role of anti-inflammatory phytochemicals. Discov Med. 2012;13(68):7–17.
151. Normann AJ, Kang DW, Christopher CN, Norris MK, Dieli-Conwright CM. Improved sleep quality is associated with reduced insulin resistance in cancer survivors undertaking circuit, interval-based exercise cancer. Epidemiol Biomark Prev. 2022;31(7):1509–10.
152. CohenSolal A. Exercise to prevent cardiotoxicity in cancer: ready for implementation? Eur J Prev Cardiol. 2022;29(3):462.
153. Keats MR, Grandy SA, Blanchard C, Fowles JR, Neyedli HF, Weeks AC, et al. The impact of resistance exercise on muscle mass in glioblastoma in survivors (RESIST): protocol for a randomized controlled trial. JMIR Res Protoc. 2022;11(5):e37709.
154. D'Alessandro EG, da Silva AV, Cecatto RB, de Brito CMM, Azevedo RS, Lin CA. Acupuncture for climacteric-like symptoms in breast cancer improves sleep, mental and emotional health: a randomized trial. Med Acupunct. 2022;34(1):58–65.
155. Chu SC, Hsieh YS, Hsu LS, Lin CY, Lai YA, Chen PN. Cinnamaldehyde decreases the invasion and u-PA expression of osteosarcoma by down-regulating the FAK signalling pathway. Food Funct. 2022;13(12):6574–82.

156. Lin SH, Li MH, Chuang KA, Lin NH, Chang CH, Wu HC, et al. Extract prevents cisplatin-induced myelotoxicity. Oxidat Med Cell Longev. 2020;2020:7353618.

157. Shakib MC, Gabrial SG, Gabrial GN. Beetroot–carrot juice intake either alone or in combination with antileukemic drug 'chlorambucil' as a potential treatment for chronic lymphocytic leukemia. Open Access Maced J Med Sci. 2015;3(2):331–6.

158. Chen Y, Lai X. Modeling the effect of gut microbiome on therapeutic efficacy of immune checkpoint inhibitors against cancer. Math Biosci. 2022;350:108868.

159. Dai W, Zhang J, Li S, He F, Liu Q, Gong J, et al. Protein arginine methylation: an emerging modification in cancer immunity and immunotherapy. Front Immunol. 2022;13:865964.

160. Zitvogel L, Derosa L, Kroemer G. Modulation of cancer immunotherapy by dietary fibers and over-the-counter probiotics. Cell Metab. 2022;34(3):350–2.

161. Zhang S, Tang D, Zang W, Yin G, Dai J, Sun YU, et al. Synergistic inhibitory effect of traditional Chinese medicine Astragaloside IV and curcumin on tumor growth and angiogenesis in an orthotopic nude-mouse model of human hepatocellular carcinoma. Anticancer Res. 2017;37(2):465–73.

Dr. Heidi Kussmann ND FABNO is a Canadian and American-licensed naturopathic doctor (ND) and board certified in naturopathic oncology (FABNO) since 2010, with recertification in 2020. She completed her BSc in 1997 at the University of Victoria in British Columbia, Canada, followed by the ND postgraduate degree in 2002 from the Canadian College of Naturopathic Medicine in Toronto, Ontario, Canada. She has been in practice for over 20 years and has always worked in integrative cancer settings with people in all stages of cancer. Dr. Kussmann recently completed the Harvard Medical School Global Scholars Clinical Research Trials Certificate in March 2024. Professional memberships include:

- Massachusetts Society of Naturopathic Doctors, Massacheussetts, USA
- Vermont Association of Naturopathic Physicians, Vermont, USA
- The Association of Clinical Research Professionals, Alexandria, Virginia
- European Society of Medical Oncology, Lugano, Switzerland
- Society of Clinical Research Associates, Chalfont, Pennsylvania
- Society for the Immunotherapy of cancer, Milwaukee, Wisconsin
- Metabolic Terrain Institute of Health, Tucson, Arizona
- Oncology Association of Naturopathic Physicians, Juneau, Alaska
- European Society of Integrative Medicine, Berlin, Germany
- Society for Integrative Oncology, Pepper Pike, Ohio
- Emirates Oncology Society, Dubai, UAE
- Canadian Association for Naturopathic Doctors, Toronto, Ontario, Canada

Ibrahim H. Abu-Gheida, Rana Irfan Mahmood, Fady Geara, and Falah Al Khatib

20.1 Introduction

20.1.1 What Is Radiation Oncology

Radiation oncology is a medical specialty that combines the fields of biology, physics, and medicine to utilize the use of ionizing radiation to treat malignancies, benign tumors, and sometimes functional diseases refractory to conventional treatment(s) [1]. Ionizing radiation can be delivered in the form of high-dose X-rays, photons, neutrons, protons, electrons, and heavy ion particles, such as in carbon therapy [2]. While radiation is delivered most commonly as external beam radiation, sealed radiation sources placed in close proximity to the target are also often used, and this technique is called Brachytherapy [2].

I. H. Abu-Gheida (✉)
Emirates Oncology Society, Emirates Medical Association, Dubai, United Arab Emirates

Burjeel Medical City, Abu Dhabi, United Arab Emirates

R. I. Mahmood
Mediclinic City Hospital, Dubai, United Arab Emirates

F. Geara
Cleveland Clinic, Abu Dhabi, United Arab Emirates
e-mail: gearaF@clevelandclinicabudhabi.ae

F. A. Khatib
Emirates Oncology Society, Emirates Medical Association, Abu Dhabi, United Arab Emirates

Al Zahra Hospital, Dubai, United Arab Emirates
e-mail: fak@eim.ae

© The Author(s) 2024
H. O. Al-Shamsi (ed.), *Cancer Care in the United Arab Emirates*,
https://doi.org/10.1007/978-981-99-6794-0_20

337

20.1.2 Origin of Radiation

While people think that the field of radiation therapy is new, its origin actually dates back more than 130 years, when Wilhelm Conrad Rontgen described X-rays as being used to treat cancer in 1896. Natural radioactivity was discovered by Antoine Henri Becquerel in 1896. Pierre and Marie Curie utilized this concept to treat a pharyngeal carcinoma in 1904 by placing a radioactive substance in close proximity to and into the tumors [3]. The concept of fractionated radiotherapy was established by Regaud and Ferroux in the 1920s, and since then, the radiotherapy field has been in a constant state of progress and development [4]. The goal of modern radiotherapy is to deliver the highest dose required for tumor eradication while exposing normal tissues to the least amount of radiation. The major progress happened with the introduction of the CT scan and the ability to plan and deliver 3D-based treatment. Subsequently, further progress was made in beam calculation and delivery with the introduction of intensity-modulated radiotherapy (IMRT), volumetric modulated arc therapy (VMAT), image-guided radiotherapy (IGRT), stereotactic radiosurgery (SRS), stereotactic body radiotherapy (SBRT), and more recently, stereotactic ablative body radiation (SABR) due to its ablative character- istics [3, 5]. With an improved understanding of radiation biology and the effec- tiveness of heavy particles, protons and carbon ions were introduced with advancements in treatment modalities such as intensity-modulated proton therapy (IMPT) [6]. Our ability to plan and deliver radiotherapy with high precision and accuracy continues to improve due to enhanced data processing, the use of robot- ics, and artificial intelligence.

20.1.3 How Often Is It Used?

Approximately 48–62% of all cancer patients will benefit from radiation ther- apy [7, 8]. However, at least one in four people needing radiotherapy does not receive it [9]. Radiotherapy remains a mainstay in the treatment of cancer. A comparison of the contributions toward cure by the major cancer treatment modalities shows that of those cured, 49% are cured by surgery, 40% by radio- therapy, and 11% by chemotherapy [10]. Radiotherapy can be utilized in the primary setting for definitive treatment, such as prostate cancer, or in combina- tion with other local therapies, such as surgeries. It can also be used to palliate symptoms in advanced cancer cases where the cancer has spread to other organs, causing symptoms such as pain [11]. More importantly, with the advancement in systemic therapy and the introduction of targeted and immune therapies, can- cer is becoming more of a chronic disease, so the likelihood of needing radiation therapy throughout the disease process is increasing. Moreover, with the improvement of radiation treatment planning, patient positioning, and treatment delivery, the radiotherapy dose is being delivered at the target with more sparing of the surrounding normal structures; therefore, the utilization of re-irradiation is also increasing [12].

20.2   The Radiation Oncology Team

A radiation oncology department is composed of multiple different individuals with separate and direct roles and responsibilities working together in harmony to provide this highly complex service, which includes:

1. A radiation oncologist is a certified physician who is responsible for assessing each case, providing consultation for patients, and overseeing the radiotherapy plan and treatment delivery.
2. Radiation oncology medical physicists usually hold a graduate degree in medical physics along with certification and clinical experience. They are responsible for commissioning and acceptance testing of the radiotherapy machines. They work along with radiation oncologists, dosimetrists, and therapists to design and ensure adequate delivery of treatment plans, as well as carry out all the quality assurance measurements needed for plans and machines.
3. Radiation therapy technologists (RTTs) are highly qualified and trained technicians who are responsible for ensuring proper patient identification and operating the machines to provide patients with their treatment [13].
4. Radiation therapy machine engineers are usually biomedical engineers with specialized training in radiotherapy machine installation and maintenance.
5. Radiation oncology nurses have special experience in oncology nursing and understand the workflow and process of radiation. They also assist in procedures such as brachytherapy, IV insertion for simulations, and patient monitoring.

20.3   Equipment Suppliers and Support

Within the United Arab Emirates (UAE), rather than purchasing from the primary source, end users (hospitals and facilities) purchase required radiotherapy machines and equipment through third parties. There are several companies based in the UAE that facilitate it. These companies are well equipped and staffed to facilitate and ease the logistics of acquiring and maintaining an adequate supply of all radiotherapy-related equipment and upgrades. For example, companies like Emitac Solutions, Atlas Medical, Al Zahrawi Group, Al Naghi Group, and several others are known to represent major radiotherapy-related companies and equipment suppliers in the UAE, along with all accessories associated with setting up the department, such as quality assurance equipment and patient positioning devices, etc.

Having third-party companies' mediation has both pros and cons. An advantage would be that end-users do not need to contact many different primary providers, as there are many additional equipment and accessories that might be needed but are not provided by the main primary source. This way, the third party helps facilitate the end-user's requests and ensure delivery, customs clearance, and all other steps taken to ensure delivery in a timely fashion. On the other hand, this might hinder and delay direct communications that end users sometimes need to have with their primary source.

20.4 Rules and Regulations

Setting up any department of radiation oncology in the UAE undergoes vigorous evaluation and inspection led by the Federal Authority of Nuclear Regulation (FANR), the Department of Health for Abu Dhabi, the Dubai Health Authority in Dubai, and the Ministry of Health and Prevention for the Northern Emirates. This includes reviewing all the engineering and shielding preparations, manpower preparations, applied protocols and guidelines for clinical practice, and peer review. No department can import machines and begin operations until all necessary clearances have been obtained. Furthermore, inspecting those facilities that operate is also something done in order to ensure patient and public safety are met. Prior to going live, the authorities require a second independent audit of the departments to ensure the safety of the department and neighboring departments or spaces in the facility. Any potential unsafe practice or condition shall be highlighted and adjusted before any facility can start accepting patients for treatment. All department members must have basic training in radiation safety, and their credentials must be reviewed and approved by the governing entity before they can work in this field. Furthermore, TLDs sensitive to photons and neutrons must be worn by the entire department, and the readings must be sent on a regular basis to the authorities.

20.5 Radiotherapy Population in the UAE

According to the UAE Federal Competitiveness and Statistics Center, the population of the UAE has seen a steady increase since 1971 up until 2020 [14]. That increase in population is related to massive growth and economic stability, along with the opportunity provided. This is reflected in the Ministry of Health and Prevention (MOHAP) cancer registry data, with the latest published records from 2019 indicating a steady increase in cancer incidence [15]. Breast cancer remains the most commonly diagnosed cancer in the UAE, and it is also the most common cancer site treated in radiotherapy departments across the UAE [15].

The rising population of the UAE demands an increase in the number of radiotherapy machines to continue to provide optimal use. ESTRO-QUARTS guidelines suggest, on average, one linear accelerator per 80,000–250,000 people in high-resource countries [16]. The European Coordination Committee of the Radiological Electromedical and Health Care IT Industry (COCIR) endorsed this recommendation of 7 machines per million population [17]. A recently proposed national cancer control plan for the UAE aims to not only focus on cancer screening and prevention but also improve equitable access to high-quality treatment [18]. In 2017, the UAE reported 4299 new cases of cancer, which is expected to rise each year [19]. Given that radiotherapy is expected to be required in 60% of these new cases, 2500 new patients are expected to seek this treatment each year. The real workload is likely to be higher due to the increasing use of re-irradiation. The IAEA recommends one LINAC for 200 patients per year [20], and this workload will necessitate 15 LINACS, not including specialized equipment such as the Cyberknife or Gamma knife. While this calculation is based on registry data from 2017, with a 10% annual increase expected, a total of 25 LINACS would be required to meet the demands by 2025. Reassuringly, Table 20.1 shows that 20 LINACS are either in operation or are being planned from 2023 onwards.

Table 20.1 Overview of operational departments as of October 2022

Emirate/ city	Hospital	Machine 1	Machine 2	Machine 3	Brachy	Operational (as of October 2022)
Abu-Dhabi/ Al Ain	Tawam Hospital	Clinac	Clinac	Tomotherapy	Yes	Yes
Abu Dhabi/ Al Bahia	Gulf International Cancer Center	Trilogy	True Beam	NA	Yes	Yes
Abu Dhabi/ MBZ city	Burjeel Medical City	Versa HD	Versa HD	NA	TBA	Yes
Abu Dhabi City	Mediclinic Airport Road	True Beam	NA	NA	NA	Yes
Dubai	American Hospital Dubai	Trilogy	NA	NA	Yes	Yes
	Mediclinic City hospital	True Beam	NA	NA	Yes	Yes
	Advanced Care Oncology Center	Versa HD	NA	NA	NA	Yes
	Saudi German Hospital	Versa HD	NA	NA	NA	Yes
	NeuroSpinal Hospital	CyberKnife	Tomortherapy	NA	NA	Yes
Ras Al Khaimah	Sheikh Khalifah Specialty hospital	Versa HD	Clinac	ViewRay MR Linac	NA	Yes

Varian/Elekta/Accuray/Others

20.6 Departments of Radiation Oncology in the UAE

Radiation oncology practices have been available since 1979, when Tawam Hospital in Al-Ain became operational. This department has been up and running until our day today, and Tawam is considered to be the largest comprehensive cancer hospital in the UAE. There has been a significant increase in the development and investment in radiotherapy in the UAE since the foundation of oncology practice until now, with an exponential increase in the number of radiotherapy departments opening within the past 2 years. Mafraq Hospital in Abu Dhabi, founded in 1983, was the only department that ended its services in 2007, while another facility was established in the same Emirate. Currently, there are at least 10 centers providing radiation therapy services in the UAE: 4 in Abu Dhabi and Al Ain, 5 in Dubai, and 1 in Ras Al-Khaimah. These centers are currently operating 7 linear accelerators (LINACs), 1 tomotherapy unit, and 2 brachytherapy units in the Abu Dhabi/Al Ain region; 2 linear accelerators and 1 ViewRay MR Linac are in Ras Al-Khaimah; while in Dubai, there are currently 4 linear accelerators, 1 tomotherapy machine, 1 cyberknife machine, and 2 brachytherapy units in use (Fig. 20.1 summarizes the timetable of different radiotherapy departments in the UAE). Most LINACS in the UAE are of recent generation and of new status, with Elekta Versa HD™ and Varian—TrueBeam® being the most widely used in the UAE. Complex intensity-modulated radiation therapy (IMRT), RAPID-ARC, and volumetric modulated arc therapy (VMAT) accompanied by advanced image guidance radiation (IGRT) are used in all of those sites. Cyberknife radiosurgery has been introduced for the first time by a Neuro Spinal Hospital in Dubai. Novalis' full system with BrainLab and Elements software, which is a tool for the delivery of precision radiotherapy and stereotactic radiosurgery (SRS), has been introduced for the first time in the UAE by Burjeel Medical City. Finally, in Ras Al-Khaimah, an MRI-guided radiation therapy

*Source: Federal Competitiveness and Statistics Centre

Fig. 20.1 A sketch of the timetable of different radiation departments in the UAE to date

cancer treatment system that combines magnetic resonance imaging with adaptive radiotherapy has been established (Table 20.1).

Finally, as mentioned earlier, there are a number of relatively high-capacity departments that are soon to become operational in Abu Dhabi, as well as expansion into the upper and northern emirates by the Gulf International Cancer Center. Furthermore, the introduction of new treatment modalities such as Gamma Knife and Particle therapy has been announced as part of future expansion plans in at least two centers.

20.7 Education and Training

Currently, there is no formal or accredited radiation oncology training in the UAE. All practicing radiation oncologists have obtained training and certification from abroad and, thereafter, obtained their license to practice in the UAE. Whether it is through Abu Dhabi's Department of Health, Dubai Health Authority, or the Ministry of Health and Prevention, all physicians' documentation gets thorough verification and assessment before a license is granted. Practicing radiation oncologists in the UAE come from different backgrounds of training. Some went through pure radiotherapy training, while others underwent clinical oncology training, which includes both medical and radiation oncology. According to their current active medical licenses in the UAE, the latter group is permitted to practice either medical oncology or radiation oncology, or both, in the UAE.

Specialized training for radiation oncology medical physicists and radiation therapists (RTTs) is currently not available in the UAE. Most of the currently licensed individuals have obtained their training certification abroad. The UAE licensing health authorities have a rigorous process for verifying degrees and training.

20.8 Professional Societies for Radiation Oncology in the UAE

Currently, there is no official radiation oncology society in the UAE. Most of the radiation oncologists are part of or members of oncology societies, with the largest being the Emirates oncology society, under the umbrella of the Emirates Medical Association. There is currently an ongoing attempt through the MOHAP cancer control group to help establish a UAE-level radiotherapy collaborative group.

20.9 Research in Radiation Oncology

Research is an area of active development in the UAE health care sector and in radiation oncology within the UAE. There are several research-active radiation oncologists practicing in the UAE who participate in the ongoing projects locally and internationally. Some centers are accredited as part of international research collaborative groups. For example, the Cleveland Clinic Abu Dhabi is a National Research Group (NRG) site. Burjeel Medical City is involved in several ongoing research projects as a Novalis center. To our knowledge, Burjeel Medical City Radiotherapy has a prospective cancer registry for which at least one abstract has been submitted and a full paper is pending.

20.10 Conclusion

In conclusion, the future of radiation oncology in the UAE is promising. The UAE's diverse population and nationalities provide a very unique and comprehensive patient population and cohort for research. There seems to be a trend toward an oversupply of radiotherapy machines in the UAE. However, with future plans for increased medical tourism and international patients visiting the UAE for treatment, this option is becoming more viable. Creating collaborative radiotherapy groups across the UAE would be another significant and beneficial step forward for society and patient care. It is important to share experience and knowledge through an open and supportive collaborative network, especially in the era of radiotherapy sub-specialization and site-specific expertise. Furthermore, establishing a unified research governance for radiotherapy centers will allow for the enrollment of more patients in future clinical studies. Creating an accredited training program for medical physicists, dosimetrists, and radiation therapists, which is critical to maintaining a sufficient supply of those difficult to recruit, highly specialized, and qualified individuals. Finally, having more novel approaches and utilizing major developments in artificial intelligence and machine learning could help reduce physician time while standardizing and improving the quality of treatment delivery.

Acknowledgments We would like to thank Ms. Faryal Iqbal from Burjeel Medical City for her assistance with Fig. 20.1 as well as our industrial partners for data verification.

Conflict of Interest The authors have no conflict of interest to declare.

References

1. Berman AT, Plastaras JP, Vapiwala N. Radiation oncology: a primer for medical students. J Cancer Educ. 2013;28(3):547–53.
2. Vollmering K. What are the types of radiation therapy used for cancer treatment? 2021. https://www.mdanderson.org/cancerwise/what-are-the-types-of-radiation-therapy-used-for-cancer-treatment.h00-159461634.html.
3. Connell PP, Hellman S. Advances in radiotherapy and implications for the next century: a historical perspective. Cancer Res. 2009;69(2):383–92.
4. Baskar R, Lee KA, Yeo R, Yeoh KW. Cancer and radiation therapy: current advances and future directions. Int J Med Sci. 2012;9(3):193–9.

5. Fiorino C, Guckemberger M, Schwarz M, van der Heide UA, Heijmen B. Technology-driven research for radiotherapy innovation. Mol Oncol. 2020;14(7):1500–13.
6. Moreno AC, Frank SJ, Garden AS, Rosenthal DI, Fuller CD, Gunn GB, et al. Intensity modulated proton therapy (IMPT)—the future of IMRT for head and neck cancer. Oral Oncol. 2019;88:66–74.
7. Delaney G, Jacob S, Featherstone C, Barton M. The role of radiotherapy in cancer treatment: estimating optimal utilization from a review of evidence-based clinical guidelines. Cancer. 2005;104(6):1129–37.
8. Barton MB, Jacob S, Shafiq J, Wong K, Thompson SR, Hanna TP, et al. Estimating the demand for radiotherapy from the evidence: a review of changes from 2003 to 2012. Radiother Oncol. 2014;112(1):140–4.
9. Borras JM, Lievens Y, Dunscombe P, Coffey M, Malicki J, Corral J, et al. The optimal utilization proportion of external beam radiotherapy in European countries: an ESTRO-HERO analysis. Radiother Oncol. 2015;116(1):38–44.
10. Price PM, Illidge TM, Sikora K. Treatment of cancer. 5th ed. London: Hodder Arnold; 2007.
11. Lustig R. Radiation oncology: introduction. Semin Oncol. 2014;41(6):701.
12. Andratschke N, Willmann J, Appelt AL, Alyamani N, Balermpas P, Baumert BG, et al. European Society for Radiotherapy and Oncology and European Organization for Research and Treatment of Cancer consensus on re-irradiation: definition, reporting, and clinical decision making. Lancet Oncol. 2022;23(10):e469–e78.
13. Abdel-Wahab MN-BA, Olson A, Polo A, Shah MM, Zubizarreta E, Patel S. Radiation oncology in global health. In: Mollura DJ, Lungren MP, editors. Radiology in global health: strategies, implementation, and applications. Springer International Publishing; 2019. p. 349–60.
14. UAE STAT: UAE stat.gov. 2021. https://uaestat.fcsc.gov.ae/vis?lc=en&fs[0]=FCSC%20-%20Statistical%20Hierarchy%2C0%7CPopulation%20Estimates%23POP_ES%23&pg=0&fc=FCSC%20-%20Statistical%20Hierarchy&df[ds]=FCSC-RDS&df[id]=DF_POP&df[ag]=FCSA&df[vs]=2.6.0&pd=1970%2C&dq=...A...&ly[rw]=TIME_PERIOD&ly[cl]=MEASURE%2CGENDER%2CUNIT_MEASURE&vw=tb.
15. Shelpai W. Cancer incidence in United Arab Emirates annual report of the UAE—National Cancer Registry—2019. 2021. www.mohap.gov.ae.
16. Slotman BJ, Cottier B, Bentzen SM, Heeren G, Lievens Y, van den Bogaert W. Overview of national guidelines for infrastructure and staffing of radiotherapy. ESTRO-QUARTS: work package 1. Radiother Oncol. 2005;75(3):349–54.
17. Radiotherapy Age Profile and Density December 2019 Edition. COCIR, the European Coordination Committee of the Radiological, Electromedical and Healthcare IT Industry. 2019. https://www.cocir.org/fileadmin/Publications_2019/19107_COC_Radiotherapy_Age_Profile_web4.pdf.
18. Al-Shamsi HO, Abyad AM, Rafii S. A proposal for a National Cancer Control Plan for the UAE: 2022–2026. Clin Pract. 2022;12(1):118–32.
19. Radwan H, Hasan H, Ballout RA, Rizk R. The epidemiology of cancer in the United Arab Emirates: a systematic review. Medicine (Baltimore). 2018;97(50):e13618.
20. Agency IAE. Planning national radiotherapy services: a practical tool. IAEA human health series no. 14.

Dr. Ibrahim H. Abu-Gheida is the Clinical Director of the department of radiation oncology at Burjeel Medical City. He also serves as a regional Radiological Society of North America (RSNA) committee representative for the Middle East and Africa. Dr. Abu-Gheida completed his undergraduate training, where he earned a Bachelor of Science with honors degree from the American University of Beirut. Following this, Dr. Abu-Gheida completed his Medical School training at the American University of Beirut Medical Center. He continued and joined the Department of Internal Medicine at the American University of Beirut. Then he did his training in the Department of Radiation Oncology at the American University of Beirut Medical Center, where he also served as the chief resident. During his training, Dr. Abu-Gheida completed a Harvard-affiliated NIH-funded research program as well. After his residency, Dr. Ibrahim went to the Cleveland Clinic in Ohio, where he was appointed as an Advanced Clinical Radiation Oncology Fellow. Dr. Abu-Gheida went and joined the University of Texas MD Anderson Cancer Center, where he sub-specialized in treating breast, gastrointestinal, and genitourinary cancers. Dr. Abu Gheida played an instrumental role in establishing and heading the radiation oncology facility and department at Burjeel Medical City. He has chaired and co-chaired multiple international oncology conferences. Dr. Ibrahim has more than 40 peer-reviewed papers in prestigious medical journals, including the American Society of Radiation Oncology official journal - the International Journal of Radiation Oncology Biology and Physics, Nature, the Journal of Clinical Oncology, and several others. He is also the primary author and editor of several book chapters published in prestigious books.

Dr. Rana Mahmood completed specialist clinical oncology training in the UK and has been a consultant for the last 16 years. He had several managerial roles and, most recently, was clinical director for cancer services across East Suffolk and North Essex (ESNEFT NHS Trust). He is a passionate teacher and is currently an associate professor at the MBR University of Medicine and Health Sciences in Dubai. Dr. Mahmood has a special interest in advanced image-guided radiotherapy (IM-IGRT), stereotactic radiotherapy (SABR), and image-adapted high-dose-rate brachytherapy (HDR). He set up an interstitial brachytherapy practice in King Faisal Specialist Hospital, Riyadh, and later at Colchester Hospital, university NHS trust, to benefit a wide population in Essex and Suffolk. He is the first to introduce and the only practitioner to provide gynecological interstitial brachytherapy in the UAE. He has also set up the only service in the UAE to provide prostate HDR brachytherapy and a trans-perineal spacer device. He has been the principal investigator for several landmark radiotherapy trials, including STAMPEDE, RADICALS, CHHiP, PACE, RAIDER, and EMBRACE, and he has extensively published his work.

Dr. Fady Geara is a professor and chairman of the department of radiation oncology at the oncology Institute of the Cleveland Clinic Abu Dhabi. He graduated in medicine from the University of Paris and completed his training in radiation oncology both in France and at the University of Texas MD Anderson Cancer Center in Houston, Texas.

He started his professional career at MD Anderson, where he led innovative research in radiation biology and was a major contributor to the development of clinical radiation oncology programs in head and neck and thoracic radiation oncology. He later moved to the American University of Beirut, where he built a modern, state-of-the-art radiation oncology program. He has recently joined the Cleveland Clinic Abu Dhabi to develop the radiation oncology program at the new oncology institute.

Dr. Geara is a diplomate of the American Board of Radiology in radiation oncology, and a holder of a Ph.D. degree in radiation biology. He is a well-known authority in clinical radiation oncology and program development. He is also a prolific writer, with over 126 peer-reviewed articles published, and a committed teacher, with more than 200 teaching lectures given in many countries on current radiation and general oncology topics.

Dr. Falah Al Khatib graduated from the faculty of medicine at Ain Shamis University in 1969. He completed his training and residency in general and thoracic surgery in Iraq from 1970 to 1975. Later, he changed his specialty to radiotherapy, took DMRT in 1980, and obtained fellowship FFRRCSI in 1984.

He worked as a consultant and chief of radiotherapy at Tawam Hospital between 1986 and 2007 and was involved in the management of more than 12,000 patients.

Dr. Falah helped establish Tawam as the main cancer center in the UAE and in the region, and he started a hospital-based tumor registry in 1986, which later became the foundation for the UAE cancer registry in 1998. During his time at Tawam, he created many multidisciplinary clinics for different cancers, like breast, gynecological, H + N, and thyroid. He has been a member of many national, regional, and international committees and a member of the UAE National Cancer Committee. He has been very active in the public education and cancer awareness programs in the UAE via TV, radio, newspaper articles, interviews, and conferences.

From August 2007 to August 2014, Dr. Falah worked as a consultant clinical oncologist at the Gulf International Cancer Center (GICC) in Abu Dhabi, which is the first private oncology facility in the UAE and the region. This center provides radiotherapy and medical oncology as an outpatient facility and has the first PET/CT scan in the region.

From January 2011 to 31/12/2021, he has worked as a consultant clinical oncologist (radiation oncology and medical oncology) at the Mediclinic City Hospital, Dubai, UAE, first as a part-time and since January 2016 as a full-time consultant.

From February 2022, he is working as part-time consultant clinical oncologist at the Advanced Care Oncology Center and at Al Zahra Hospital, Dubai.

Surgical Oncology in the UAE

21

Faek R. El Jamali, Chafik Sidani, and Stephen R. Grobmyer

21.1 Overview of Cancer Care in the UAE

The healthcare sector in the United Arab Emirates (UAE) has seen a tremendous evolution over the past two decades, commensurate with the overall rapid ascent of the economic standard of the country as a whole. From the first and only hospital that opened in the city of Al Ain in 1960, the country has rapidly moved to its current state of over 150 hospitals and over 150 primary healthcare centers in the short span of 40 years. At least 30 of these centers are tertiary care centers. This rapid evolution in the health sector has been coupled with a dramatic improvement in healthcare outcomes [1, 2].

In parallel with the overall evolution of the healthcare sector in the country, cancer care has similarly gradually evolved over time. Cancer care was provided in individual medical centres based on the interests and expertise of the local medical team. As the individual medical centers evolved, so did the oncology practice within each, and this in turn followed the worldwide evolution of cancer care toward a centralized multidisciplinary approach with the emergence of at least four comprehensive cancer centers in the UAE in 2022.

The private sector in both Dubai and Abu Dhabi plays an important role in the delivery of healthcare services. In the remaining five Emirates, the Ministry of Health is both the regulator and the main provider of most healthcare services. Nonetheless, when it comes to cancer care, it appears that the activity is concentrated in specialized centers.

In the UAE, in 2021, the number of deaths from cancer totaled 975 (506 in males and 469 in females) and accounted for 8.2% of all deaths, regardless of nationality,

F. R. El Jamali (✉) · C. Sidani · S. R. Grobmyer
Cleveland Clinic Abu Dhabi, Abu Dhabi, United Arab Emirates
e-mail: eljamaf@clevelandclinicabudhabi.ae; SidaniS@clevelandclinicabudhabi.ae; GrobmyS@clevelandclinicabudhabi.ae

H. O. Al-Shamsi (ed.), *Cancer Care in the United Arab Emirates*,
https://doi.org/10.1007/978-981-99-6794-0_21

type of cancer or gender. Colon (11.49%), trachea, bronchus, and lung (9.85%), and breast (9.64%) cancers cause most cancer deaths [UAE - National Cancer Registry] [3].

In the Emirate of Abu Dhabi, cancer caused 15.2% of all deaths in the Emirate in 2017. Breast (11.5%), bronchus and lung (8.7%), and colon (8.5%) cancers cause the most cancer deaths [4]. In 2017, cancer-related clinical activity occurred most frequently at the nationally designated cancer center Tawam Hospital (36.8% of the total volume of cancer care), followed by Sheikh Khalifa Medical City (SKMC) at 24.8%, Cleveland Clinic Abu Dhabi at 10.9%, Mafraq Hospital at 10.5%, and NMC Specialty Hospital at 4.8% [4]. These numbers have drastically changed over the last few years with the establishment of Sheikh Shakhbout Medical City (as the fusion between SKMC and Mafraq) and the significant growth of cancer services at the Cleveland Clinic Abu Dhabi and other private hospital systems like VPS Healthcare/Burjeel Medical City.

Assuming cancer surgery is generally considered major surgery, and looking at all the major surgery procedures that were carried out in 2020 in the Emirate of Dubai, we note that there were 71,339 major procedures done [5]. Of these, 44,990 major procedures (63%) were done in the private sector, highlighting the importance of the private sector in the overall delivery of healthcare in that Emirate. One major source of concern is the number of major oncology procedures performed as emergencies. Out of 462 oncology-Diagnosis Related Group (DRG) related cases, 252 were done as an emergency. This can be multifactorial, but it indicates a delayed presentation with advanced disease or disease complications. In the Emirate of Dubai, there were close to 100,000 outpatient visits that were cancer-related in 2020 among 11,525 patients.

Interestingly, there were 135 international patients who received oncology treatment in the emirate of Dubai in the year 2020, representing 41% of all international medical tourism cases for that year {DHA}. 58 such patients came from the UK, 31 from Germany, and 19 from the USA. Only five international patients underwent surgery in Dubai in 2020. On the other hand, the WHO and other sources estimate that the UAE government spent almost a quarter of its total healthcare expenditure in 2010 to send its citizens abroad for medical care [6]. Looking at the DHA data for health expenditure abroad for the emirate of Dubai, 326 patients traveled abroad for medical care with an average cost of 620,000 AED per patient, of which 70% were direct medical expenditures. The United Kingdom was the most popular destination for international medical care, followed by Germany and the United States [5]. This represents a 28% drop from the prior year, which may be related to the global lockdown related to the COVID-19 pandemic or the population's increased confidence in the UAE healthcare sector.

21.2 The Evolution of Surgical Oncology as a Specialty

General surgical oncology has seen a tremendous evolution in the last two decades into a distinct surgical specialty with multidisciplinary care at its core. Training is primarily focused on GI, endocrine, soft tissue, and breast oncology. This evolution

Table 21.1 Significant dates in the evolution of surgical oncology in the United States and the United Arab Emirates

1975	James Ewing society renamed Society of Surgical Oncology
1983	First surgical oncology fellowship training program approved
2003	First breast oncology fellowship program approved
2011	American Board of Surgery approves surgical oncology subspecialty certification
2021	Arab Board of Surgery approves surgical oncology subspecialty certification

has been spearheaded internationally by many bodies, including surgical societies and regulatory boards. Many of these societies have had a profound impact on shaping surgical oncology into the distinct specialty it is today, but perhaps none more than the Society of Surgical Oncology (SSO) in the United States of America due to the prominent position American medicine holds on the world stage.

The Society of Surgical Oncology was established in the USA in 1975 as an evolution of the James Ewing Society, which was made up of alumni who had trained at the Memorial Sloan-Kettering Cancer Center and gathered in New York for both scientific and social purposes [7]. Early surgical oncologists placed great emphasis on the total care of the patient. They often oversee not only the surgical aspect of the treatment but also pathology, radiation therapy, and chemotherapy. With time and as each of these specialties evolved into a separate discipline, the role of the surgical oncologist transitioned into focusing on providing surgical care for the cancer patient within this multidisciplinary team.

The impact of the Society of Surgical Oncology on the definition of the specialty was tremendous. The society sets the training curriculum as well as the formal minimum requirements for training. Surgical oncology training guidelines were formalized in 1978, and criteria for approving surgical oncology training programs were defined and utilized until 2014, when the Accreditation Council for Graduate Medical Education (ACGME) took over the process of reviewing the programs [8].

This focus on surgical oncology as a specialty has allowed the value that a well-trained surgical oncologist can bring to cancer care to become quite tangible and has served as the basis for the growth and expansion of surgical oncology training programs, as well as increased recognition of the specialty worldwide. Surgical oncologists bring with them advanced surgical training, having worked with some of the world's foremost experts in cancer surgery at high-volume centers. They also bring with them the multidisciplinary approach that is at the heart of cancer care in this day and age (Table 21.1).

21.3   Current Status of Surgical Oncology in the UAE

It is a challenging task to try and accurately assess the current status of surgical oncology in the UAE due to the lack of objective data on current surgical oncology practices and outcomes. There are at least four fellowship-trained general surgical oncologists who completed training in the USA, including Faek Jamali, who completed training at the University of Pittsburgh; Yasir Akmal at City of Hope National Medical Center; Stephen Grobmyer at Memorial Sloan Kettering Cancer Center; and Sadir Al Rawi at Rosewell Park Cancer Center. This is a reflection of the fact that the United States only graduates 39 fellows in surgical oncology per year, compared to close to 1500 general surgery graduates per year. This is coupled with the fact that, worldwide, surgical oncology continues to lag behind in terms of becoming a well-recognized and defined specialty. Using Europe as an example, 67% of European countries still do not recognize surgical oncology as a separate discipline [9]. As a result, oncologic surgery has largely been confined to the domain of general surgeons in the UAE. There are also physicians who have completed subspecialty training in surgical oncology, such as Dr. Waleed Hassan (urologic oncology at Memorial Sloan Kettering Cancer Center (MSKCC)), Stephanie Ricci in GYN oncology at Johns Hopkins, Dr. Muhieddine Seoud at Kansas University Medical Center, and Dr. Usman Ahmed in thoracic surgical oncology at MSKCC, to name a few.

John Birkmeyer has focused attention on the benefits of specialization in optimizing outcomes in complex cancer surgery [10]. In his seminal paper, he highlighted the clear association between low hospital volume and high operative mortality for major cancer operations, especially esophagectomy and pancreatectomy. These findings have been further corroborated in a number of additional studies and expanded to include rectal, gastric, hepatobiliary, and many other cancer sites. However, volume is not the only driver of improved outcomes. As demonstrated by Bilimoria et al., the outcomes of cancer surgery are improved across a large number of procedures when performed by a trained subspecialist as compared to surgeons with no specialized training [11], highlighting the value of additional focused training in cancer surgery.

Using rectal cancer as a reference model, it has been amply demonstrated that specialization improves the outcome of rectal cancer surgery. The treatment of rectal cancer has been challenging due to its complexity at multiple levels. The anatomical location of the rectum in the deep pelvis, the risk of injury to nearby organs, the complexities related to re-establishing continuity after proctectomy, and the prevention of leaks are formidable technical challenges, to name a few. This complexity is further compounded when we add ever-growing options for neoadjuvant treatment, including TNT, organ preservation, and early rectal cancer management. This complexity has resulted in important variations in the outcomes of rectal cancer surgery among hospitals and surgeons in Europe. Statistically significant differences in R0 resection rates, postoperative morbidity and mortality rates, and long-term oncologic and functional outcomes are noted as a result of the subspecialization [12–14]. Most notable, the operating surgeon was noted to be an independent risk factor in rectal cancer outcomes [15].

Clearly, then, surgical oncology in the UAE is facing challenges. The major challenge relates to the lack of specialization, with most of the cancer surgery being in the domain of the general surgeon. This is compounded by the overall low volume of cases since the population in the UAE is relatively young and cancer incidence rates are generally low in this subgroup. Additional challenges include a lack of awareness, resulting in failure to follow common cancer screening recommendations [16]. Furthermore, healthcare in the UAE is decentralized, with no clear mechanisms of referral and no regionalization of cases into specific centers based on expertise and/or outcomes. Finally, there are significant variations in care across the cancer centers, with a lack of well-defined quality control mechanisms.

21.4 Future Prospects of Surgical Oncology in the UAE

Despite all the above challenges, the future of healthcare in general and surgical oncology in the UAE is bright. The UAE government's policies have led to an era of stability and prosperity, even when the whole world is suffering from multiple crises. This has led to the Emirates becoming a highly attractive place to work. In addition, the UAE leadership has highlighted healthcare as one of the top areas of investment and growth and has brought the top 2 US health systems (Mayo Clinic and Cleveland Clinic) to the UAE. This has led to a parallel rise in healthcare investment in the private sector. Furthermore, over the last few years, the UAE has seen an influx of highly trained, specialized physicians, including fellowship-trained surgical oncologists.

There are currently five comprehensive cancer centers that are in operation in the UAE (Table 21.2). These are defined as centers that offer hematology and oncology, dedicated surgical oncology, expert pathology, radiation oncology, nuclear medicine, and oncology patient support services [17]. Additionally, there is at least one more state-of-the-art cancer center that is currently under active construction or development at Sheikh Shakhbout Medical City in partnership with the Mayo Clinic.

Given the challenges associated with the training and recruitment of surgical oncologists in the Arab world, the Arab Board of Surgery has approved in 2021 the standards and curriculum for the establishment of fellowship training programs in surgical oncology across the Arab world. The standards and criteria are parallel to those that are set forth by the Society of Surgical Oncology and are stringent in setting the requirements needed to approve a fellowship program and a designated center as a training center. Nonetheless, this move by the Arab Board will lead to the establishment of several surgical oncology fellowship programs across the Arab world, leading to an improved workforce that is locally trained and groomed and an increased awareness of the specialty and the importance and value of specialists rendering complex cancer care. Anyone interested in obtaining further information may contact the Arab Board of Surgical Oncology representative in the UAE, Dr. Faek Jamali (first author), or consult the Arab Board website at www.arab-board.org.

Surgical oncology could benefit from the recognition of surgical oncologists as members of the Emirates Oncology Society, similar to the American Society of Clinical Oncology, which welcomes surgeons into membership. In fact, the recently appointed director of the National Cancer Institute in the USA was a former

Table 21.2 Currently established comprehensive cancer centers in the United Arab Emirates

Hospital	Location	Established	International accreditation	Unique services	Services not offered
Tawam Hospital	Al Ain	1979	JCI accredited breast cancer unit	Brachytherapy Pediatric oncology Palliative care Genetic counselling	Hepatobiliary surgery Bone marrow transplantation
American Hospital Dubai	Dubai	2010	JCI clinical care certification in 2017 National accreditation program for breast centers (NAPBC) in 2015	Acute hematology and bone marrow transplantation Pediatric oncology Palliative care	Genetic counselling
Mediclinic City Hospital Dubai	Dubai	2016	JCI accredited breast cancer unit	Brachytherapy Palliative care Pediatric oncology	Genetic counselling Acute hematology and bone marrow transplantation
Burjeel Medical City	Abu Dhabi	2020	European Society of Medical Oncology (ESMO) designated center of oncology and palliative care	Adult and pediatric bone marrow transplantation Acute hematology Palliative care Cancer research unit	Genetic counselling
Cleveland Clinic Abu Dhabi	Abu Dhabi	2022	Membership NSABP, NRG, JCI accreditation	Gynecologic oncology Urologic oncology Advanced reconstructive services Adaptive radiotherapy General surgical oncology Breast oncology HIPEC program Robotic surgery Genetic counseling: Solid organ transplant IRB and clinical trials program Tumor registry	Pediatrics

president of ASCO. Alternatively, the formation of an Emirati Society of Surgical Oncology or Gulf Region Society of Surgical Oncology could help advance the field in the region and increase recognition of the value of surgical oncologists in cancer care.

Further, the establishment of national quality metrics for the surgical care of patients could also help further the role of surgical oncologists in the care of patients in the UAE. Finally, efforts to establish surgical oncology fellowships within the UAE would help to ensure a future pipeline of caregivers to meet the population's needs.

With healthcare at the center of the future vision of the country, the UAE is poised to become a destination for health tourism, catering to the requirements of nearly a billion people in the Middle East and North Africa (MENA) region. Surgical oncology stands to play an important role in this vision, and the authors hope to see the establishment of the Emirati Society of Surgical Oncology, which will act as a source of collaboration and leadership in steering the future of surgical oncology.

21.5 Conclusion

The healthcare sector in the UAE has seen a tremendous evolution over the past two decades, commensurate with the overall rapid ascent of the economic standard of the country as a whole.

In parallel, the practice of surgical oncology in the UAE is rapidly evolving from the domain of the general surgeon to the realm of specialized cancer centers that provide the highest level of multidisciplinary, state-of-the-art cancer care. This has been the result of the establishment of several state-of-the-art cancer centers across the country as well as the recruitment of high-quality professionals in all fields related to oncology.

With healthcare at the center of the future vision of the country, the UAE is poised to become a destination for health tourism, catering to the requirements of nearly a billion people in the MENA region.

Conflict of Interest The authors have no conflict of interest to declare.

References

1. Bell J. Modern UAE health care: from a mud hut to skyscraper hospitals. The National UAE; 2013. http://www.thenational.ae/news/uae-news/health/modern-uae-health-care-from-a-mud-hut-to-skyscraper-hospitals.
2. Loney T, Aw T, Handysides DG, et al. An analysis of the health status of The United Arab Emirates: the "big 4" public health issues. Glob Heal Action. 2013;1:1–8. https://doi.org/10.3402/gha.v6i0.20100.
3. Cancer incidence in United Arab Emirates, Annual Report of the UAE – National Cancer Registry. Statistics and Research Center, Ministry of Health and Prevention. 2021. Accessed 01 Jun 20224.

4. Department of Health Abu Dhabi. Health Statistics 2017. 2022. https://www.doh.gov.ae/resources/opendata.
5. Dubai Health Authority. Dubai annual health statistical report 2020. Dubai: Dubai Health Authority; 2020. https://www.dha.gov.ae/uploads/032022/Annual%20%20Health%20Statistics%20%20Book%2020202022326320.pdf.
6. Vetter P, Boecker K. Benefits of a single payment system: case study of Abu Dhabi health system reforms. Health Policy. 2012;108(2–3):105–14. https://doi.org/10.1016/j.healthpol.2012.08.009.
7. Copeland E, presidential address. Surgical oncology: a specialty in evolution. Ann Surg Oncol. 1999;6(5):424–32.
8. Balch C, et al. 2015 James Ewing lecture: the 75 year history of the Society of Surgical Oncology—part II: the transitional years (1966-1990). Ann Surg Oncol. 2016;23:358–64.
9. Tyler D. 2022 SSO presidential address: what is a surgical oncologist? Evolution of surgical oncology and the society of surgical oncology in an era of hyperspecialization. Ann Surg Oncol. 2022;29:4005–13. https://doi.org/10.1245/s10434-022-11770-3.
10. Birkmeyer JD, et al. Hospital volume and surgical mortality in the United States. NEJM. 2002;346:1128–37.
11. Bilimoria KY, et al. Effect of surgeon training, specialization and experience on outcomes for cancer surgery: a systemic review of the literature. Ann Surg Oncol. 2009;16:1799–808.
12. McArdle C, Hole D. Impact of variability among surgeons on postoperative morbidity and mortality and ultimate survival. Br Med J. 1991;302(6791):1501–5.
13. Harmon JW, Tang DG, Gordon TA, et al. Hospital volume can serve as a surrogate for surgeon volume for achieving excellent outcomes in colorectal resection. Ann Surg. 1999;230(3):404.
14. Dorrance HR, Docherty GM, O'Dwyer PJ. Effect of surgeon specialty interest on patient outcome after potentially curative colorectal cancer surgery. Dis Colon Rectum. 2000;43(4):492–8.
15. Renzulli P, Laffer UT. Learning curve: the surgeon as a prognostic factor in colorectal cancer surgery. Rectal Cancer Treat. 2005;165:86–104.
16. Al Mansouri A, et al. Colorectal screening in ambulatory healthcare service clinics in Abu Dhabi, United Arab Emirates in 2015-2016. BMC Cancer. 2021;21(1):897.
17. Al Shamsi H. The state of cancer care in The United Arab Emirates in 2022. Clin Pract. 2022;12(6):955–85. https://doi.org/10.3390/clinpract12060101.

Dr. Faek R. El Jamali is Consultant Surgical Oncologist and colorectal surgeon at the Cleveland Clinic Abu Dhabi (CCAD). Dr. El Jamali completed training in general surgery at the University of Connecticut and an SSO-approved surgical oncology fellowship at the University of Pittsburgh, leading to American Board certification. He also completed a fellowship in advanced minimally invasive colorectal surgery at IRCAD/EITS in Strasbourg, France. Dr. EL Jamali joined the American University of Beirut Medical Center in 2002, where he practiced until August 2020, progressing to the rank of tenured Professor of Surgery and Vice Chair for Clinical Affairs before relocating to the UAE to join SSMC as a senior consultant and chief of the division of colorectal surgery at Sheikh Shakhbout Medical City in partnership with Mayo Clinic. Dr. Jamali has joined the ranks of the faculty at CCAD in 2023. His practice focuses on surgical oncology, the management of peritoneal surface malignancies, and minimally invasive/robotic colorectal surgery.

Dr. Chafik Sidani is the section head of colon and rectal surgery at Cleveland Clinic Abu Dhabi (CCAD). He is a clinical assistant professor at the Cleveland Clinic Lerner College of Medicine. He established the colorectal surgery Enhanced Recovery After Surgery (ERAS) program at CCAD. He is an active member of the American Society of Colon and Rectal Surgeons and has authored numerous publications and book chapters. Dr. Sidani completed medical school at the American University of Beirut (AUB). He then completed a 2-year postdoctoral research fellowship in gastrointestinal physiology at Yale University. He completed his training in general surgery at Yale New Haven Hospital and Georgetown University Hospital in Washington, DC, USA. Subsequently, he completed a Colon and Rectal Surgery Fellowship at the University of Minnesota in Minneapolis, Minnesota, USA. He then served as Chief of Colon and Rectal Surgery at the Virginia Hospital Center, Arlington, VA, USA, prior to moving with his wife and three children to Abu Dhabi to join Cleveland Clinic Abu Dhabi.

Dr. Stephen R. Grobmyer completed his fellowship in surgical oncology at Memorial Sloan Kettering Cancer Center. He was previously the Section Head of Surgical Oncology and Director of the Breast Center at the Cleveland Clinic in Cleveland, Ohio. He was and is currently Professor of Surgery, at the Lerner College of Medicine at Case Western Reserve University. In Ohio, he held the Lula Zapis Endowed Chair in Breast Cancer Research. He currently serves as Oncology Institute Chair at the Cleveland Clinic Abu Dhabi in the United Arab Emirates. In 2018, Dr. Grobmyer was elected to the American Surgical Association. He is a member of ASCO, and in 2011, he was selected for participation in the ACSO leadership development program. He currently serves on the editorial board of the Annals of Surgical Oncology, Surgery, the European Journal of Surgical Oncology, Gland Surgery, and the Annals of Breast Surgery. He has published over 200 peer-reviewed manuscripts and 25 book chapters (h-index = 53). He has edited a book on cancer nanotechnology. His research programs focused on breast cancer prevention and treatment have been funded by over $16 million in extramural funding. His research has been featured in *The New York Times*, National Public Radio, NBC Nightly News, and *Vogue Magazine*.

Palliative Care in the UAE

22

Neil A. Nijhawan and Humaid O. Al-Shamsi ⓘ

22.1 Introduction to Palliative Care

Palliative care (PC) refers to comprehensive and proactive care provided to people across all age groups who experience significant suffering related to severe illness, particularly those in the final stages of life. Its primary objective is to enhance the well-being of patients, their families, and their caregivers. Figure 22.1 presents the fundamental principles and goals of PC, as outlined on the World Health Organization's (WHO) website [1] (Fig. 22.1).

PC is a relatively new medical specialty, having only achieved specialty status in the United Kingdom (UK) in 1987 and the United States (USA) in 2006 [2], though the roots of what we now recognise as the modern palliative care movement can be traced back to post-World War 2 (WW2) Britain, when Dr. Cicely Saunders

N. A. Nijhawan (✉)
Consultant in Palliative Medicine, Burjeel Medical City, Abu Dhabi, United Arab Emirates

Khalifa University College of Medicine and Health Sciences,
Abu Dhabi, United Arab Emirates

Emirates Oncology Society, Emirates Medical Association, Dubai, United Arab Emirates

Gulf Medical University, Ajman, United Arab Emirates

H. O. Al-Shamsi
Burjeel Cancer Institute, Burjeel Medical City, Burjeel Holdings, Abu Dhabi,
United Arab Emirates

Ras Al Khaimah Medical and Health Sciences University, Ras Al Khaimah, United Arab Emirates

Gulf Medical University, Ajman, United Arab Emirates

Emirates Oncology Society, Emirates Medical Association, Dubai, United Arab Emirates

College of Medicine, University of Sharjah, Sharjah, United Arab Emirates

Gulf Cancer Society, Alsafa, Kuwait
e-mail: alshamsi@burjeel.com; humaid.al-shamsi@medportal.ca

- Includes, prevention, early identification, comprehensive assessment, and management of physical issues, including pain and other distressing symptoms, psychological distress, spiritual distress, and social needs. Whenever possible, these interventions must be evidence-based.
- Provides support to help patients live as fully as possible until death by facilitating effective communication and helping them and their families determine goals of care.
- Is applicable throughout the course of an illness, according to the patient's needs.
- Is provided in conjunction with disease modifying-therapies whenever needed.
- May positively influence the course of the illness.
- Intends neither to hasten nor postpone death, affirms life, and recognizes dying as a natural process.
- Provides support to the family and the caregivers during the patient's illness and in their own bereavement.
- Is delivered while recognizing and respecting the cultural values and beliefs of the patient and the family.
- Is applicable throughout all health care settings (place of residence and institutions) and at all levels (primary to tertiary).
- Can be provided by professionals with basic palliative care training.
- Requires specialist palliative care with a multiprofessional team for referral of complex cases.

Fig. 22.1 Principles and aims of palliative care [1]

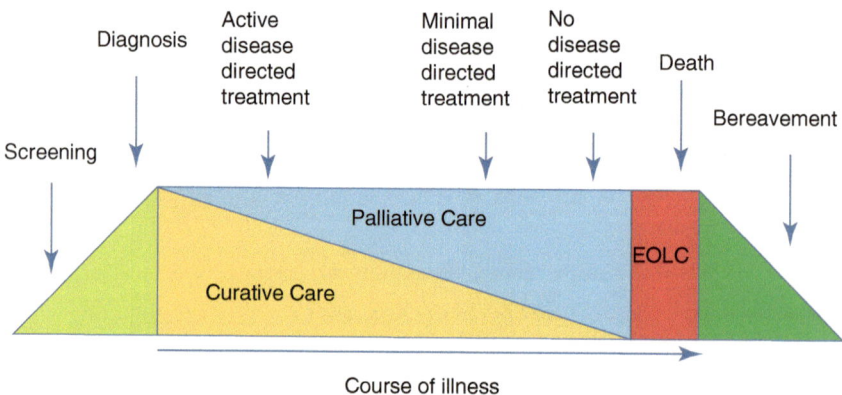

Fig. 22.2 The continuum of palliative and end-of-life care [3]

established the first hospice (St. Christopher's Hospice) in Southeast London in 1967. After visiting St. Christopher's Hospice in 1973, Dr. Balfour Mount helped to create a PC ward at the Royal Victoria Hospital in Montreal, Canada, and is credited with coining the term "palliative care."

Palliative care (PC) is fundamentally a collaborative approach to providing the kind of care one would desire when facing a serious, life-threatening illness, whether it is for oneself or a loved one. PC focuses on addressing the individual's needs, not solely the diagnosis, and it extends across the entire care journey, starting from the moment of diagnosis through the stage when death is approaching and even into the post-bereavement period (Fig. 22.2) [3]. In the past, PC has primarily been associated with the care provided to adult patients in advanced stages of cancer. This narrow focus has led to the misconception that PC is exclusively connected to end-of-life care. Indeed, a growing body of evidence supports the earlier integration of PC with disease-directed treatments for both non-malignant diseases [4] and cancer [5].

Jennifer Temel's ground-breaking research conducted in 2010 [5] showcased the positive impact of early PC intervention for individuals diagnosed with metastatic non-small cell lung cancer. The study revealed notable benefits, including significant improvements in quality of life and mood reported by patients. Interestingly, despite the fact that the group receiving earlier palliative care had less aggressive end-of-life treatments, their median survival time was longer compared to the control group: 11.6 months versus 8.9 months ($P = 0.02$).

The transformation of PC from a philosophy focused solely on end-of-life care to a comprehensive discipline that encompasses the well-being of patients and their families throughout the entire course of illness is clearly evident. This shift is evident in the recent broadening of the concept of supportive care, especially within the field of oncology [6].

The significance of implementing PC in the context of non-malignant illnesses with limited life expectancy cannot be emphasized enough. Given the global rise in aging populations and the corresponding rise in individuals with multiple chronic conditions, the demand for palliative care will inevitably grow [7]. This includes the United Arab Emirates (UAE) as well.

22.2 Palliative Care Provision in the UAE

PC is a relatively recent addition to the medical landscape in the UAE, as depicted in Fig. 22.3. The first PC service in the public health sector was established in 2007 at Tawam Hospital in Al Ain, Abu Dhabi. Initially operating as a consult service alongside their oncology department, it has since evolved into a distinct division within the oncology department. Currently, it offers outpatient clinics, inpatient consultation services, dedicated inpatient beds, and a PC nurse outreach service. Notably, the Tawam PC team remains the only palliative care service within the public health sector. In contrast, the private healthcare sector has witnessed an increase in PC services since 2015. The American Hospital in Dubai, for instance, provides outpatient clinics, inpatient consultation services, and inpatient PC beds, with the majority of their referrals being patients diagnosed with cancer, similar to Tawam.

In September 2019, Mediclinic City Hospital in Dubai introduced its PC service, which presently offers outpatient clinics, inpatient reviews, and a consultation

Fig. 22.3 The UAE palliative care landscape timeline

service. A similar consult service was initiated at Parkview Hospital, a sister facility, in 2020. The most recent expansion in the PC service landscape occurred in March 2020 with the launch of the palliative and supportive care service at Burjeel Medical City (BMC) in Abu Dhabi. Similar to Tawam Hospital, BMC offers a comprehensive PC service, including integrated outpatient clinics with pain medicine and physical medicine and rehabilitation clinics, an inpatient consultation service, dedicated palliative care inpatient beds, and a PC nurse outreach service. Additionally, within the Burjeel healthcare ecosystem, there is a homecare service available, enabling the PC team to provide support for patients in need of palliative and end-of-life care within the comfort of their homes.

22.3 Factors Influencing Palliative Care Provision in the UAE

22.3.1 Philosophical Culture Clash

Cultural attitudes play a significant role in the realm of palliative care (PC), particularly when it comes to addressing cultural variations in end-of-life care. Western medicine emphasizes the autonomy of the individual patient, whereas Middle Eastern cultures place greater importance on the extended family as the primary social institution and decision-maker. Middle Eastern patients often rely heavily on the support of others during times of crisis or illness, as opposed to relying solely on personal coping mechanisms, which is more aligned with Western ideals of individual autonomy and maintaining personal independence. In the UAE, the availability of this extended family support network is not always guaranteed, especially for many expatriate workers.

In the Middle East, Islam is not merely a religious choice but rather a comprehensive way of life that encompasses specific beliefs shaping perspectives on health and illness. Common themes include the beliefs that [8]:

- The destiny of each individual is determined when their soul is brought into existence. While fate is predetermined, one cannot be aware of their own destiny, thus it is advisable to seek God's favor through obedience.
- Illness is perceived as a divine punishment bestowed by God.
- Every aspect of existence is aligned with God's grand design.

These convictions are manifested in an almost fatalistic acceptance of death and illness. However, this perspective is balanced by the belief within Islam that it is a duty to preserve life until God determines its end. This duality creates a common scenario for healthcare providers, where a patient with an advanced progressive illness is rapidly deteriorating with a bleak prognosis, yet the family often insists on exhausting all available options. Discussions regarding maintaining the patient's quality of life by avoiding medically futile interventions are often met with resistance and negativity. The perception is that doctors are abandoning hope, whereas in reality, only God knows the true extent of the prognosis. It is firmly believed that hope must never diminish, as relinquishing hope would mean forfeiting God's assistance. Even if hope appears futile according to Western medical standards, it is believed that hope aids patients in coping with their illness. Consequently, openly discussing the removal of hope from the patient is deemed both tactless and unpardonable [9]. These underlying beliefs typically give rise to contrasting viewpoints among doctors regarding the importance of being transparent and truthful with patients about their illness. Additionally, these beliefs contribute to the elevated expectations held by the general population regarding the effectiveness of medical treatments. It is worth noting that hope is not a fixed concept, as there is evidence suggesting that patients' hopes evolve over time [10]. There is ample evidence to support the fact that patients worldwide frequently choose to wait until their family members have left the room before requesting the doctor to be candid with them and validate their suspicions regarding their illness [11, 12].

The challenge at hand is how to balance the physician's requirement for complete disclosure regarding a patient's condition with the family's desire to protect their loved one and maintain hope. The solution lies in effective communication. While it is crucial to have open and honest discussions with both the patient and their family, it is also important to recognize that not all patients wish to receive all the intricate details of their condition. In primary care consultations, it is common to begin by inquiring about the patient's preferred level of information. This helps determine whether they prefer to have all the specifics or if they are content with knowing just enough. It is beneficial to engage in this conversation in the presence of family members, reassuring them that there are no hidden motives behind the consultation and that it will be conducted according to the patient's or family's preferences and at their desired pace.

22.3.2 Palliative Care Education in the UAE

Regardless of the location, nearly all publications addressing the obstacles to palliative care provision emphasize the attitudes of healthcare providers. The stigma surrounding palliative care teams is pervasive worldwide, including in the UAE, where there remains a prevailing belief that palliative care solely revolves around pain management and end-of-life care [13].

The majority of the existing literature concerning the attitudes of healthcare workers towards palliative care predominantly concentrates on physicians. However, it is the nursing staff who frequently shoulder the responsibility of engaging in challenging conversations with patients and their families [14]. A significant portion of nurses in the UAE are expatriates from India and the Philippines. Many of these nurses are relatively inexperienced, with less than 5 years of clinical experience since completing their qualifications. Despite their limited experience, they frequently find themselves in the challenging position of caring for terminally ill patients and having to navigate emotionally difficult conversations with family members.

Offering specialized training to our nursing colleagues in the UAE presents a challenge. Although there are nursing education practitioner positions available, there is a scarcity of individuals with sufficient palliative care training and experience. In various other regions, nurses with specialized training in palliative care, adept at comprehensive patient assessment and managing physical and psychological symptoms, are integral members of interdisciplinary palliative care teams. These roles, known as clinical nurse specialists (CNS), have been established since the 1990s in the UK and the USA, and their positive impact on patients and families is well-documented. These benefits include enhanced symptom control, psychological support, and increased advocacy for patients and their families [15]. The shortage of doctors with sufficient palliative care training was one of the primary factors that prompted the creation of the clinical nurse specialist (CNS) role. Consequently, two parallel efforts were undertaken: (1) training a group of palliative care CNS professionals and (2) establishing clinical governance structures to support this clinical role. These initiatives resulted in the rapid growth of the CNS role, both in hospitals and community settings worldwide.

However, there are several obstacles hindering the development of this role in the UAE:

- There is a global scarcity of clinical nurse specialists in palliative care [16].
- The hierarchical structure in healthcare within the UAE remains predominantly traditional and physician-centered, where doctors make decisions and nurses follow their orders.
- In the Emirati society, similar to many Middle Eastern cultures, the guidance of the most senior physician is highly respected and followed. Consequently, it may be culturally challenging to accept clinical advice from a nurse, as the nursing profession is traditionally considered subordinate to physicians.
- The migration pattern of nurses in the UAE generally follows a familiar pattern. Inexperienced nurses often come to the UAE for a temporary period of 2–3 years before moving on to countries like the US, Canada, or the UK. Additionally, the continuous influx of nurses from Southeast Asia further diminishes the motivation to provide specialized training for these nurses.

22.3.3 Palliative Care and Opioid Medications

It is understandable that there is a common misconception that PC primarily focuses on pain management, considering that pain is the most prevalent symptom among patients receiving such care. Ensuring proper pain management at the end of life is both the patient's right and the responsibility of healthcare providers. The World Health Organization has a clear stance on this matter: patients have the right to receive appropriate treatment and effective control of their pain, following well-defined guidelines and recommendations [17]. Through our observations, we have learned that unmanaged pain significantly impacts every aspect of an individual's life. Conversely, when pain is adequately controlled, it results in enhanced well-being and improved quality of life for patients and their families. Despite the existence of effective pain relief options, the issue of inadequate cancer pain management persists. According to the World Health Organization (WHO), approximately five billion people reside in countries with limited or no access to pain medications. Among them, an estimated 5.5 million terminally ill cancer patients [18] face inequitable access to medicines, resulting in approximately 80% of them experiencing moderate to severe pain.

In PC, our primary emphasis is on prioritizing pain relief, as this brings about a notable enhancement in the quality of life. By addressing pain, we can improve comfort, foster a better sense of well-being, and promote improved sleep for patients. This, in turn, creates a conducive environment for patients to openly discuss their hopes, fears, and aspirations. While we employ a wide range of interventions, such as non-pharmacological methods and radiation therapy for specific cancers, opioids play a crucial role in managing moderate to severe pain.

Non-specialists often struggle with effectively managing severe pain, which is most commonly experienced by individuals diagnosed with cancer. The WHO guidelines for cancer pain in adults [19] recommend the three-step analgesic ladder (Fig. 22.4), although there is no pharmacological need for Step 2, and compared to weak Step 2 opioids, the benefit from low dose morphine (20–30 mg/24-h PO) is greater and more rapid [20]. Step 2 is necessary due to the challenging and, at times, impossible accessibility of potent (Step 3) opioids in numerous countries. This is despite the fact that morphine, which is relatively inexpensive, is the most readily accessible and extensively studied opioid analgesic.

Based on the Atlas of Palliative Care in the Eastern Mediterranean region, the UAE's median consumption of opioids (excluding methadone) stands at 3.03%, slightly lower than the regional median of 3.27% [21]. Based on data from the International Narcotic Control Board (INCB), opioid consumption in the UAE is lower than anticipated. The defined daily dose for statistical purposes (SDDD) per million inhabitants per day is 162, which falls under the category of inadequate usage, as values below 200 are considered inadequate and values below 100 are considered very inadequate [22].

Two recurring topics found in all publications discussing palliative care provision in the UAE are (1) the prevalent belief that using potent opioids will inevitably result in addiction and (2) challenges related to limited access and prescription restrictions for strong opioids and other frequently used medications.

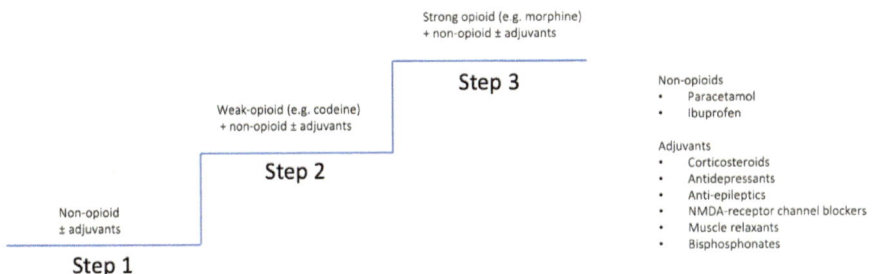

Fig. 22.4 The World Health Organization three-step analgesic ladder [23]

Patients diagnosed with cancer frequently experience intense pain that may require the use of opioids for effective management. However, this situation can become more complicated when physicians feel uneasy about utilizing strong opioids, particularly for severe and complex pain. The apprehension regarding potential opioid abuse or misuse by the patient can actually hinder the successful management of pain.

In reality, the UAE does have access to all the necessary medications for palliative care as recommended by the International Association of Hospice and Palliative Care (IAHPC)—as shown in Table 22.1 [24]. Although methadone is typically associated with the treatment of opioid misuse, the 5 mg tablets of methadone are accessible for utilization in palliative care settings. Moreover, hydromorphone is now accessible in various forms, including immediate and sustained-release oral preparations, as well as an injectable form.

Table 22.1 The IAHPC list of essential medications for palliative care [24]

Medication	Formulation	Indication	UAE availability
Amitriptyline	50 mg tablets	Depression Neuropathic pain	Yes
Bisacodyl	10 mg tablets 10 mg rectal suppositories	Constipation	Yes Yes
Carbamazepine	100–200 mg tablets	Neuropathic pain	Yes
Citalopram	10–20 mg tablets	Depression	Yes
Codeine	30 mg tablets	Pain: mild to moderate Diarrhoea	Yes
Dexamethasone	0.5–4 mg tablets 4 mg/mL injection	Anorexia Nausea and vomiting Neuropathic pain	Yes
Diazepam	2.5–10 mg tablets 5 mg/mL injection 10 mg rectal suppository	Anxiety Muscle relaxant	Yes Yes Yes
Diclofenac	25–50 mg tablets 50–75 mg/3 mL injection	Inflammatory pain	Yes
Diphenhydramine	25 mg tablets 50 mg/mL injection	Antihistamine Motion sickness	
Fentanyl transdermal patch	12.5–100 µg/h	Pain: moderate to severe	Yes
Gabapentin	300–400 mg tablets	Neuropathic pain	Yes
Haloperidol	0.5–5 mg tablets 0.5–5 mg/mL injection	Delirium Nausea and vomiting Terminal restlessness	Yes
Hyoscine butylbromide	10 mg tablets 10 mg/mL injection	Visceral pain Nausea and vomiting Terminal respiratory congestion	Yes
Ibuprofen	200–400 mg tablets	Inflammatory pain	Yes
Levomepromazine	5–50 mg tablets 25 mg/mL injection	Delirium Terminal restlessness	Yes Yes

Table 22.1 (continued)

Medication	Formulation	Indication	UAE availability
Loperamide	2 mg tablets	Diarrhoea	Yes
Lorazepam	0.5–2 mg tablets 2–4 mg/mL injection	Anxiety	Yes Yes
Megestrol acetate	160 mg tablets 40 mg/mL solution	Anorexia	Yes Yes
Methadone	5 mg tablets	Pain: moderate to severe Neuropathic pain	Yes
Metoclopramide	10 mg tablets 5 mg/mL injection	Nausea and vomiting	Yes
Midazolam	1–5 mg/mL injection	Anxiety Terminal restlessness	Yes
Fleet® mineral oil enema			Yes
Mirtazapine	15–30 mg tablets	Depression Anorexia	Yes
Morphine	Immediate release 10–60 mg tablets Immediate release 10 mg/5 mL solution Immediate release 10 mg/mL injection Sustained release 10 mg tablets Sustained release 30 mg tablets	Pain: moderate to severe Dyspnoea	Yes Yes Yes Yes Yes
Octreotide	100micrograms/mL injection	Diarrhoea Vomiting	Yes
Oral rehydration salts		Diarrhoea	Yes
Oxycodone	5 mg tablets	Pain: moderate to severe	Yes
Paracetamol	100–500 mg tablets 500 mg rectal suppositories	Pain: mild to moderate	Yes
Prednisolone (dexamethasone alt)	5 mg tablets	Anorexia	Yes
Senna	8.6 mg tablets	Constipation	Yes
Tramadol	50 mg immediate release tablets/capsules 50 mg/mL injection	Pain: mild to moderate	Yes
Trazodone	25–75 mg tablets	Insomnia	Yes
Zolpidem	5–10 mg tablets	Insomnia	Yes

The introduction of the Unified Electronic Platform [25] (Openjet) in 2019 has simplified the process of monitoring and tracking the prescription and distribution of controlled and narcotic medications, including opioids. This national online prescribing platform requires the use of the patient's national identity card (Emirates ID) with a specialized card reader. This development has enhanced safety for prescribers and decreased the potential for opioid medication misuse. Furthermore, the platform reduces inefficiencies associated with paper prescriptions and ensures accurate monitoring of prescription and medication distribution.

The primary concern lies in the restrictive regulations surrounding the prescription of controlled medications. Specifically:

- Only consultant physicians are authorized to provide a 30-day supply of controlled medications, while other physicians are limited to prescribing between 7- and 14-day supplies.
- Current regulations prohibit the prescription of injectable forms of controlled substances, including opioids and other centrally acting medications, for non-hospitalized patients. This poses a challenge to providing end-of-life care at home for patients who prefer this option, as we are unable to initiate intravenous or subcutaneous infusions of commonly used palliative care medications.

While progress has been made in the UAE to enhance access to frequently used palliative care (PC) medications, there is still a significant amount of work remaining. This includes the crucial step of mandating the inclusion of all commonly used PC medications and their various formulations in the national formulary.

22.3.4 Palliative Care Multidisciplinary Team Approach

Due to its comprehensive approach, PC considers all aspects of a patient's well-being, encompassing physical, psychological, emotional, sociocultural, and spiritual needs. Therefore, a PC assessment typically covers these various domains. To ensure comprehensive care, most palliative care teams strive to be multi-disciplinary in their composition, consisting of professionals such as nurses, doctors, social workers, psychologists, chaplains, physiotherapists, occupational therapists, and complementary therapists. As previously mentioned, the availability of PC-trained nurses and doctors in the UAE has been addressed. While social worker and psychologist roles do exist in the UAE healthcare system, many may not have received specific training in PC.

When patients are faced with a recent diagnosis of a life-limiting illness or are nearing the end of their lives, we acknowledge that addressing their spiritual well-being is just as significant as addressing their clinical concerns. Spiritual health plays a vital role in overall human well-being as it enables individuals to navigate personal existential challenges in different aspects of life, including stressful circumstances, illness, or the presence of death [26]. Attending to the spiritual care needs of patients and their families is essential for enhancing their overall quality of life. While not all patients may adhere to a specific religious affiliation, many grapple with existential questions related to life, illness, and finding meaning in suffering. While palliative care nurses and doctors are typically trained to assess patients' spiritual needs, they cannot replace the role of a chaplain. It is widely acknowledged that patients often require someone who can listen to their concerns, fears, and regrets without judgment [27]. Chaplains recognize the significance of providing a non-judgmental presence that does not necessarily offer advice or solutions but rather creates a space where all emotions and challenges can be acknowledged. They are trained to extend their support to everyone, irrespective of their religious or spiritual beliefs, and their unique perspective adds depth to the medical team by emphasizing the importance of finding meaning in both life and death. Currently, the role of a hospital chaplain is not established in the UAE, although patients are free to practice their chosen faith. However, considering the UAE's population, which primarily consists of expatriates from diverse backgrounds, it becomes crucial to provide culturally appropriate care, especially for patients nearing the end of life.

The role of physiotherapists and occupational therapists within a palliative care (PC) team is of utmost importance. Physiotherapists primarily focus on maximizing movement and comfort, while occupational therapists prioritize optimizing functional abilities. These therapists work collaboratively to help patients engage in daily life activities to the best of their abilities despite limitations or restrictions caused by the progression of their illness. When patients experience sudden limitations in their functional abilities due to illness, it becomes crucial to facilitate a shift in their mindset as part of the rehabilitation process. Therapists initiate the process by conducting assessments with the patient and their family, attentively listening to the patient's narrative, observing their capabilities, and discussing their physical, social, emotional, and spiritual needs. Based on these evaluations, goals are established that align with the patient's desires to continue specific activities and achieve personal milestones. These goals may involve tasks like self-care during showering, getting in and out of bed independently, spending quality time with loved ones, or completing a particular activity. As a result, patients can re-prioritize their daily activities, conserving energy for the activities that hold the most significance to them. This process is dynamic and not fixed, with goals regularly reassessed and adjusted to accommodate the evolving nature of disease progression and the patient's physical capabilities [28].

22.3.5 Advanced Care Planning (ACP)

Advance care planning (ACP) involves the crucial task of making significant decisions regarding the care an individual wishes to receive if they become incapable of expressing their preferences. Although some patients may believe they are too young or in good health to consider creating an advanced care plan, it holds particular importance for those with progressive, life-limiting illnesses. As part of palliative care, we frequently assist patients in this process. ACP entails both legal and personal decision-making to develop a comprehensive plan that can be shared with key individuals, outlining the individual's desires and preferences as they near the end of life. It serves as a means to ensure that the patient's wishes are respected and upheld when they are unable to communicate their choices.

In various regions around the world, the process of advance care planning (ACP) encompasses several considerations, such as an advanced directive and discussions related to cardiopulmonary resuscitation (CPR), preferred place of care (PPC), and preferred place of death (PPD). An advanced directive is a written document that outlines an individual's healthcare preferences to be followed if they are unable to make decisions or express their wishes. Typically, it focuses on situations where the person is terminally ill and specifies which medical treatments they do or do not want healthcare providers to pursue in the absence of their informed consent. For instance, it may indicate preferences regarding the use of ventilators or the insertion of feeding tubes. Additionally, the advanced directive can address any religious or spiritual preferences that the individual wishes to be observed.

While advanced directives do not hold formal or legal recognition within the UAE, it is customary for the medical team to engage in discussions with the patient's relatives regarding the patient's preferences. Within the prevailing cultural attitude toward serious illness in the UAE, the belief is that everything is ultimately in the hands of God. Therefore, there is a perception that attempting to predict, prognosticate, or plan for the future is unnecessary since whatever unfolds will be in accordance with God's will. Consequently, the predominant approach is to continue with all available medical treatments, even those that may be considered medically futile or inappropriate. Unfortunately, when healthcare workers are compelled to prolong a patient's suffering and death by pursuing futile curative treatments, they may experience moral distress. This occurs when there is a conflict between their professional obligations and the knowledge that the interventions being employed are unlikely to yield any meaningful benefit [29].

22.3.6 Allow Natural Death

Cardiopulmonary resuscitation (CPR) is a medical intervention that involves invasive measures, and its original purpose was not to be administered to patients who are in the process of dying from an irreversible condition [30]. Decisions regarding do not attempt cardio-pulmonary resuscitation (DNACPR) serve as a means of communication, indicating when CPR should not be performed on patients either due to their personal wishes or because it is unlikely to succeed. DNACPR decisions play a crucial role in safeguarding patients from potential harm, although they have acquired practical, legal, and emotional implications that extend beyond their original purpose. Doctors often hesitate to initiate discussions about DNACPR due to concerns of causing distress to patients and their families, as well as fears of potential complaints or legal action.

Since August 2016, the implementation of UAE Federal Law No. 4 on Medical Liability has brought about various revisions to the previous Medical Liability law. One significant change is the acceptance of natural death as a permissible outcome for patients diagnosed with terminal illnesses [31]. Under this new law, healthcare professionals are now authorized to withhold cardiopulmonary resuscitation (CPR) from terminally ill or dying patients with incurable illnesses, allowing natural death to occur. However, certain conditions must be met in order to proceed with this approach:

- The patient experiences an irreversible medical condition.
- All available treatment options have been attempted.
- The treatment has been established as ineffective for the specific medical condition.
- The attending doctor recommends against administering CPR to the patient.
- A minimum of three consulting doctors concur that it is in the patient's best interests to allow natural death and withhold CPR (in this situation, the patient's consent, guardian's consent, or custodian's consent is not necessary).

Nevertheless, if a patient explicitly requests CPR, it cannot be denied even if healthcare providers deem it medically futile. Unfortunately, there are no available national-level statistics that illustrate the frequency of implementing the "allow natural death" policy or how often terminally ill patients request full CPR.

Even if a patient and their family agree that allowing natural death aligns with their preferences and desires, significant challenges persist regarding the preferred location for receiving care and passing away. Although official statistics on the place of death are unavailable, it is probable that most patients receive end-of-life care in hospitals. Even patients who express a desire for home-based end-of-life care are unlikely to have their preference fulfilled due to the absence of established community palliative care teams or community-based palliative care units (hospices) within the UAE currently.

22.3.7 Financial Cost of Palliative Care

Whether healthcare services are provided through the government health system or the private sector, the payment for such services is typically facilitated either within a health insurance framework or by patients covering the costs themselves. In the UAE, each emirate has its own regulations regarding medical insurance, with Abu Dhabi and Dubai requiring employers to provide mandatory medical coverage for employees and their dependents. The existing reimbursement system adds complexity to the ability to deliver palliative care (PC) services to patients. Similar to other non-procedural healthcare interventions, PC consultations, including advanced care planning, often face underappreciation, with insufficient value placed on these compassionate, communication-centered procedures [33]. The potential complications can have equally harmful and potentially irreversible effects. Historically, medical insurance providers did not include coverage for palliative care services under their policies. However, this has undergone a transformation with the introduction of specific diagnosis-related group (DRG) codes for palliative care, encompassing both inpatient and outpatient consultations.

The financial benefits of PC have been well-established. Patients who receive PC services, in comparison to those receiving standard care, generally experience a decrease in hospitalizations, shorter hospital stays, fewer admissions to intensive care units, and fewer visits to the emergency department. Notably, PC has been associated with cost savings of US$ 4251 per hospital stay for cancer patients and US$ 2105 per hospital stay for patients with non-cancer illnesses [32].

22.4 Conclusion

Palliative care services in the UAE are continuously developing. As the global population ages and the incidence of cancer and other life-limiting illnesses rises, the demand for comprehensive palliative care will also increase. However, addressing this need in the UAE goes beyond solely recruiting more nurses and doctors, as there is a worldwide shortage of palliative care-trained healthcare professionals. Enhancing palliative care in the UAE will require a multifaceted approach beyond increasing staffing numbers. Our recommendations for improving palliative care in the UAE were highlighted in a previous publication [33], and they have since been refined and diagrammatically represented [34] (Fig. 22.5).

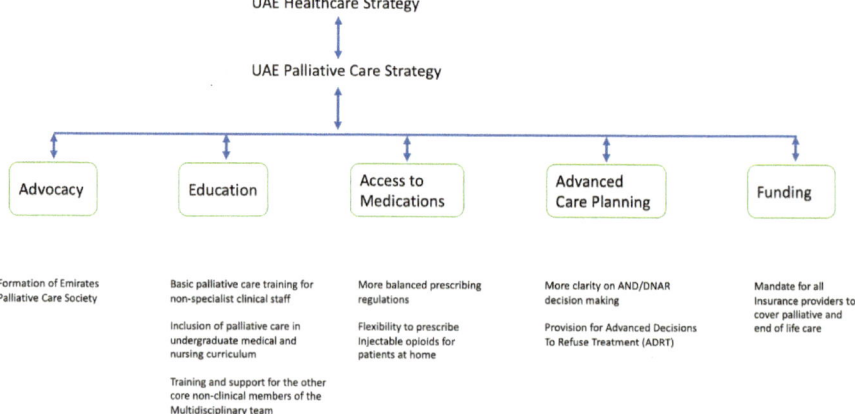

Fig. 22.5 Recommendations for palliative care in the UAE

Conflict of Interest The authors have no conflict of interest to declare.

References

1. https://www.who.int/cancer/palliative/definition/en/.
2. https://aahpm.org/uploads/AAHPM16_Medical_Student_BroWEB.pdf.
3. Myatra SN, Salins N, Iyer S, Macaden SC, Divatia JV, Muckaden M, Kulkarni P, Simha S, Mani RK. End-of-life care policy: an integrated care plan for the dying: a joint position statement of the Indian Society of Critical Care Medicine (ISCCM) and the Indian Association of Palliative Care (IAPC). Indian J Crit Care Med. 2014;18(9):615–35. https://doi.org/10.4103/0972-5229.140155.
4. Kingston AEH, Kirkland J, Hadjimichalis A. Palliative care in non-malignant disease. Medicine. 2020;48(1):37–42. https://doi.org/10.1016/j.mpmed.2019.10.010.
5. Temel JS, et al. Early palliative care for patients with metastatic non-small-cell lung cancer. N Engl J Med. 2010;363:733–42. https://doi.org/10.1056/NEJMoa1000678.
6. https://www.esmo.org/guidelines/supportive-and-palliative-care?page=2.
7. Etkind SN, et al. How many people will need palliative care in 2040? Past trends, future projections and implications for services. BMC Med. 2017;15(1):102. https://doi.org/10.1186/s12916-017-0860-2.
8. Lipson JG, Meleis AI. Issues in health care of middle eastern patients. West J Med. 1983;139(6):854–61.
9. Meleis AI, Jonsen AR. Ethical crises and cultural differences. West J Med. 1983;138:889–93.
10. Daneault S. Ultimate journey of the terminally ill, ways and pathways of Hope. Can Fam Physician. 2016;62(8):648–56.
11. Srivastava guardian article. https://www.theguardian.com/commentisfree/2018/apr/04/should-a-doctor-always-disclose-a-terminal-diagnosis.
12. Al-Alfi N. Palliative care in the United Arab Emirates: a Nurse's perspectives. J Palliat Care Med. 2015;S5:S5–005. https://doi.org/10.4172/2165-7386.1000S5006.
13. Silbermann M, et al. Evaluating palliative care needs in middle eastern countries. J Palliat Med. 2015;18(1):18–25. https://doi.org/10.1089/jpm.2014.0194.
14. Al-Alfi N. Cultural thoughts on palliative care in the UAE. Palliat Med Hosp Care. 2017;SE(1):S51–5. https://doi.org/10.17140/PMHCOJ-SE-1-111.
15. Jack, et al. Impact of the palliative care clinical nurse specialist on patients and relatives: a stakeholder evaluation. Eur J Oncol Nurs. 2002;6(4):236–42. https://doi.org/10.1054/ejon.2002.0214.
16. https://www.healthleadersmedia.com/clinical-care/palliative-care-faces-staffing-shortages-workforce-valley-looms.
17. Carlson CL. Effectiveness of the World Health Organisation cancer pain relief guidelines: an integrative review. J Pain Res. 2016;2016(9):515–34. https://doi.org/10.2147/JPR.S97759.
18. Global Alliance to Pain Relief Initiative (GAPRI). Access to Essential Pain Medicines Brief 2010. http://www.gapri.org/understand-problem.
19. https://www.who.int/cancer/palliative/painladder/en/.
20. Twycross R, et al. Palliative care formulary. 6th ed; 2019. p. 295–6. ISBN 9780857113481.
21. Osman H, Rihan A, Garralda E, Rhee JY, Pons JJ, de Lima L, Tfayli A, Centeno C. Atlas of Palliative Care in the Eastern Mediterranean Region. Houston: IAHPC Press; 2017.
22. INCB. Narcotic drugs estimated world requirements for 2020. 2019. ISBN: 978-92-1-148309-3.
23. https://professionals.wrha.mb.ca/old/professionals/files/PDTip_AnalgesicLadder.pdf.
24. IAHPC list of essential medications. https://hospicecare.com/what-we-do/projects/palliative-care-essentials/iahpc-essential-medicines-for-palliative-care/.
25. https://doh.gov.ae/en/news/Department-of-Health-adopts-Unified-Electronic-Platform-for-narcotic-drugs-and-controlled-medicines.
26. Juskiene V. Spiritual health as an integral component of human wellbeing. Appl Res Health Social Sci Interf Interact. 2016;13(1):11. https://doi.org/10.1515/arhss-2016-0002.
27. https://thedenverhospice.org/what-to-expect-from-a-visit-with-the-hospice-chaplain/.
28. Tavemark S, Hermansson LN, Blomberg K. Enabling activity in palliative care: focus groups among occupational therapists. BMC Palliat Care. 2019;18:17. https://doi.org/10.1186/s12904-019-0394-9.

29. Wolf AT. Palliative care and moral distress in the ICU. An integrative literature review. J Hosp Palliat Nurs. 2016;18(5):405–12. https://doi.org/10.4037/ccn2019645.
30. Fritz, Slowther. Resuscitation policy should focus on the patient, not the decision. Br Med J. 2017;356:J813. https://doi.org/10.1136/bmj.j813.
31. https://www.dha.gov.ae/Asset%20Library/MarketingAssets/20180611/(E)%20Federal%20 Decree%20no.%204%20of%202016.pdf.
32. May P, et al. Economics of palliative care for hospitalized adults with serious illness: a meta-analysis. JAMA Intern Med. 2018;178(6):820–9. https://doi.org/10.1001/ jamainternmed.2018.0750.
33. Nijhawan NA, Al-Shamsi HO. Palliative care in The United Arab Emirates (UAE). In: Laher AGI, editor. Handbook of healthcare in the Arab world. Springer Nature Switzerland; 2021. p. 1–18. https://doi.org/10.1007/978-3-319-74365-3_102-1.
34. Nijhawan NA, Al-Shamsi HO. Experiences and challenges of a new palliative care service in the United Arab Emirates. Palliat Med Hosp Care Open J. 2022;8(2):30–4. https://doi. org/10.17140/PMHCOJ-8-150.

Dr. Neil A. Nijhawan is a UK-trained consultant in Palliative Medicine at Burjeel Medical City. After medical school at Kings College London, he pursued speciality training in Palliative Medicine in London, with rotations in acute general hospitals, domiciliary visits, community hospices, and tertiary oncology centres. Prior to completing his palliative medicine training, Neil returned to his childhood home, Trinidad in the West Indies, to help set up and commission the new Caura Hospital Palliative Care Unit, where he was the Medical Director. This unit was opened in 2014 and provides a comprehensive palliative care service, including a 12-bed inpatient unit, weekly outpatient clinics, and a palliative care consult service at the local university hospital. After completing his specialty training, Neil worked as a consultant in palliative medicine at the Imperial College Healthcare NHS Trust in London, where he was the clinical lead for palliative medicine. His clinical area of interest is symptom control (including pain, nausea, breathlessness, and fatigue) and assistance with complex treatment decision-making at the end of life, and he is often called on to provide an independent second opinion. He is active in palliative care education and palliative care advocacy and is currently the UAE representative to the WHO Eastern Mediterranean Region Palliative Care Expert Network. Neil holds adjunct faculty positions with both Khalifa University and Gulf Medical University where he is Clinical Associate Professor in Hospice& Palliative Medicine.

Prof. Humaid Obaid Al-Shamsi is the Chief Executive Officer of Burjeel Cancer Institute in Abu Dhabi, UAE, President of the Emirates Oncology Society, Lead of the Gulf Cancer Society, Full Professor of Oncology at the Ras Al Khaimah Medical and Health Sciences University, Ras Al Khaimah, UAE, and an Adjunct Professor of Oncology at the College of Medicine, University of Sharjah. He is the first Emirati to be promoted as a professor in oncology in the UAE. He is also the Chairman for Colorectal Cancer in the MENA region, appointed by the prestigious National Comprehensive Cancer Network®. He is also the only member of Lung Cancer Policy Network in the MENA region that aims to advance lung cancer research and screening globally. He is the

Chairman of the Oncology and Hematology Fellowship Training Program for the National Institute for Health Specialties in the United Arab Emirates. He is the only member in GCC in the WIN Consortium which is comprised of organizations representing all stakeholders in personalized cancer medicine globally.

He is board-certified in both internal medicine and oncology from the UK, USA (ABIM), and Canada (FRCPC). He has also been awarded the FRCP (London) in 2023 and FRCP (Glasgow) in 2024. He is the only physician in the UAE with a subspecialty fellowship certification and training in gastrointestinal oncology and the first Emirati to train and complete a clinical post-doctoral fellowship in palliative care. He was an assistant professor at the University of Texas MD Anderson Cancer Center between 2014 and 2017. He has published more than 140 peer-reviewed articles in JAMA Oncology, Lancet Oncology, The Oncologist, BMC Cancer, and many others. His area of expertise includes precision oncology and cancer care in the UAE. In 2016, he published with his group from MD Anderson the JCO paper describing a new distinct subgroup of CRC, NON V600 BRAF-mutated CRC. In 2022, he published the first book about cancer research in the UAE and also the first book about cancer in the Arab world, both of which were launched at Dubai Expo 2020. *Cancer in the Arab World* has been downloaded more than 450,000 times in its first 18 months of publication and is the ultimate source of cancer data in the Arab region. He also published the first comprehensive book about cancer care in the UAE which is the first book in UAE history to document the cancer care in the UAE with many topics addressed for the first time, e.g., neuroendocrine tumors in the UAE. He is passionate about advancing cancer care in the UAE and the GCC and has made significant contributions to cancer awareness and early detection for the public using social media platforms. He is considered as the most followed oncologist in the world with over 300,000 subscribers across his social media platforms (Instagram, Twitter, LinkedIn, and TikTok). In 2022, he was awarded the prestigious Feigenbaum Leadership Excellence Award from Sheikh Hamdan Smart University for his exceptional leadership and research and the Sharjah Award for Volunteering. He was also named the Researcher of the Year in the UAE in 2020 and 2021 by the Emirates Oncology Society.

In May 2024, HH Sheikh Mansour bin Zayed Al Nahyan, Vice President of the United Arab Emirates, awarded him the first place in UAE Nafis program for outstanding leadership in private sector across all business and medical disciplines. Beside his clinical and administrative duties, he is engaged in education and various levels of research training for medical trainees to enhance their clinical and research skills. His mission is to advance cancer care in the UAE and the MENA region and make cancer care accessible to everyone in need around the globe.

Nuclear Medicine in the UAE

Abdulrahim Al Suhaili

23.1 Introduction

On December 2, 1971, the newly formed nation's medical services were truly deplorable. Still, leaders with vision understood that and put in place a system to improve healthcare both horizontally to cover the whole country and vertically by creating high-standard medical facilities. As a result, high-medical facilities such as Mafraq, Tawam, and Al Jazeera in Abu Dhabi; Al Baraha, Al Maktoum, Rashid Hospital, and Latifa Hospital in Dubai; Kuwait Hospital; and Al Qassimi hospitals in Sharjah provide healthcare services in all cities and rural areas.

Nuclear medicine services were added as a new service in Abu Dhabi at Mafraq Hospital in 1979, followed by Tawam Hospital in 1982. Dubai Hospital started in 1983, to the author's knowledge.

All radiopharmaceuticals were imported on a weekly basis; therefore, positron emission tomography (PET) was not available until the first cyclotron was built in Abu Dhabi by the Gulf International Cancer Centre (GICC) in 2009, and then the first PET scan was done. Al Mulla Group established the second cyclotron in Al Nahda, followed by the first PET study in Dubai at an American hospital in 2010. Nowadays, nuclear medicine services for diagnosis and therapy are available in many emirates, whether owned by the government or the private sector.

Radionuclide therapy started in Mafraq and Tawam for high-dose treatment, particularly for hyperthyroidism, thyroid cancer, and neuroblastoma cases. It was a long journey from the early 80s until today, when we have more sophisticated therapies at our fingertips.

A. Al Suhaili (✉)
Department of Nuclear Medicine, Burjeel Medical City, Abu Dhabi, United Arab Emirates
e-mail: abdulrahim.suhaili@burjeelmedicalcity.com

The strategic collaboration between universities and hospitals in the United Arab Emirates (UAE) was blooming.

23.1.1 What Is Nuclear Medicine?

The best description of what nuclear medicine is "the medical specialty that is different than radiology, immunology, cell biology, and physiology, but has a little bit of all of them in one specialty."

There is no single definition for this specialty since it started to be available after the Second World War in 1945. However, the path to using radioactivity in medicine began after Henri Becquerel's discovery of radioactivity in 1897 and Roentgen's discovery of X-rays a year earlier. All attempts by doctors to treat failed due to a lack of understanding of cell biology and molecular behavior [1, 2].

Although radiation exposure can cause cancer, it can also cure it. Oncology was the driving force behind nuclear medicine's development and innovation, and as oncology progressed deeper into the cellular and genetic levels, nuclear medicine followed the same path faster to create molecular imaging and theranostic. In recent years, hybrid systems have solved anatomic issues by having radiology merge with nuclear imaging to create SPECT-CT, PET-CT, and then SPECT-MRI and PET-MRI. Nuclear medicine is used in oncology for early detection, establishing diagnosis, and determining the staging of cancers. It is also used for determining the early response to therapy.

23.2   Nuclear Medicine Services in the UAE

The nuclear medicine services in the UAE are divided into three categories:

- Gamma ray detection, and imaging using gamma camera and SPECT detectors
- Positron emission and dual photon detection tomography (PET and PET-CT)
- Radionuclides therapy, by using alpha and beta particles to induce regional ionization, which causes cell destruction (theranostic) [3].

23.2.1 Gamma-Ray Imaging

Single photon emission computerized tomography (SPECT), and then in the twenty-first century, SPECT merged with CT to create SPECT-CT. It is used in 80% of the procedures performed in nuclear medicine (see scope of services) attached.

Technetium-99m, iodine-131, 123, Ga-67, Tl-201, and In-111 were in use until today. They were used for imaging and functional studies.

Examples:

- **Radioiodine** is taken up by the thyroid gland for manufacturing tri-iodo-thyronine (T3), or tetra-iodo-thyronine (T4), or thyroxine. The speed and amount of uptake can determine the status of thyroid function (normal, hyperthyroidism, or hypothyroidism). The time of radioiodine residency in the thyroid can determine the status of function, like cold, neutral, or hot nodules. Iodine-131 can be used for therapy, which will be discussed later.
- **Gallium-67** was first discovered as a bone agent, but it was found to be accumulated in cancer and inflammatory cells and has since been used in oncology for the localization and spread of cancers. It was used to detect occult cancer and pyrexia of unknown origin (PUO) before the PET era.
- **Thallium-201** chloride is used for myocardial imaging to detect ischemia and infarction, but it was also found to be helpful in differentiating between recurrence and fibrosis in treated masses. It plays an important role in brain tumors.
- **Indium-111** chloride is used in neuroendocrine tumors (NEN and NET) and is labeled as somatostatin receptor agent like octreotide. It was the main imaging modality in neuroblastoma, paraganglioma, medullary thyroid cancer, and GI-NEN. It also opened the door to using it to treat these patients by replacing indium-111 with iodine-131. It is still being used successfully in children with neuroblastoma.

23.2.2 What Are Radiopharmaceuticals?

Pharmaceuticals are chemical compounds that have certain physiological characteristics and can be easily directed to certain targets or images. When a radionuclide is labeled, it will transport it to the same target organ, hence the term "radiopharmaceutical compound".

If the radioactive compound emits gamma rays, then it can be seen by the gamma camera until it reaches the target organ, which allows us to see the uptake percentage, distribution, and viability, for example, of prostate cancer [4].

Prostrate-specific membrane antigen, known as PSMA, is a protein that is found mainly in prostate cancer cells, and when labeled with a gamma or PET compound and then injected into patients with elevated PSA, it will accumulate in these tissues, whether it is a primary disease, a recurrence, or elsewhere, such as metastatic disease. The isotope is the radionuclide, PSMA is the pharmaceutical, and the compound is called radioactive-PSMA. If labeled with a PET tracer like fluorine-18 to form ^{18}F-PSMA or gallium-68 to form ^{68}GA-PSMA.

If the scan shows spread or recurrence and the main treatment for prostate cancer has failed, the fluorine or gallium can be replaced by a beta or alpha emitter like lutetium-177 (beta), actinium-225 or astatine-211 (alpha) for treatment.

Gamma camera imaging is used in detecting skeletal metastasis and localization of the sentinel lymph node (SLN) in most cancers prior to surgery, which revolutionized surgical management and minimized morbidity.

In renal cancer, a gamma camera is used to assess the remaining function of the other kidney prior to nephrectomy or to predict pulmonary function prior to lung resection.

23.2.3 Positron Emission Tomography (PET) in Oncology

The electron (e-) is a negatively charged particle found in all atoms' orbits on earth. It is the source for X-ray imaging. But in nuclear medicine, the source of radiation comes from the *nucleus of the atom*. As a result, it earned the name "nuclear medicine."

But there is another type of electron that lives for a very short period before being annihilated (disappearing) completely. The matter mass is converted to energy, which is used in PET. The electron here does not have a negative charge (negatron-), but a positive charge, and is hence called a "positive electron" (positron+), which can be detected by positron emission tomography (PET) scanners.

PET tracers are designed to target certain organs through the metabolic route. Because of their high demands, cancer cells are always hungry for glucose. If a molecule of glucose is labeled as a PET tracer, it will be consumed by cancer cells at a higher rate than normal tissue. And by modifying the glucose molecule by taking one oxygen atom, it will end up with deoxyglucose (DG). When we label it fluorine-18 (a positron emitter), a new compound will be formed called ^{18}F-fluorodeoxyglucose (^{18}F-FDG). Cancer cells cannot recognize the difference between glucose and deoxyglucose and consume FDG. Once it enters the cancer cells, it will not be able to leave like ordinary glucose and remain there, allowing us to obtain images after the uptake period (about 1 h).

Fluorine-18 can be labeled with other compounds to increase the specificity of ^{18}F-PSMA uptake; however, since fluorine-18 has a 110-min half-life, the production centers (cyclotron) should be located within 2 h from the imaging facility.

The UAE has cyclotrons in a few places in the country (Table 23.1).

All the cyclotrons in the UAE produce mainly ^{18}F-fluorine, and there is a need for a bigger cyclotron that can produce other PET tracers and iodine-123, which is very much needed for imaging the thyroid and other organs using SPECT/CT.

PET imaging can use another source, such as gallium-68, which is produced by a generator that can be milked daily and makes ^{68}Ga available every day within the department, avoiding the logistical problems associated with cyclotrons.

Table 23.1 Functioning cyclotrons' availability in the UAE

S. No	Location
1	Gulf International Cancer Center (GICC) is located in Al Bahia, near Abu Dhabi
2	Monrol cyclotron, located in Al Nahda, Dubai
3	Tawam Mubadala Cyclotron in Al Ain, which was later owned by the Cleveland Clinic

A new field of imaging was opened several years ago by using an alternative to ^{18}F and cyclotron called "fibroblast activation protein," FAP inhibitors are overexpressed in cancer-associated fibroblasts of several tumor entities. FAPI can be detected in various malignant neoplasms and is associated with tumor cell migration, invasion, and angiogenesis.

By targeting FAP, the ^{68}gallium-labeled FAP-inhibitor (^{68}GA-$FAPI$) is developed and used in imaging tumor stroma. ^{68}Ga-FAPI uptake is more specific than ^{18}F-FDG uptake, since the latter has a higher false positive rate than ^{68}GA-FAPI, such as in cases of inflammatory disease, physiologic G.I. uptake, and infected tissue. ^{18}F-FDG has less uptake in certain cancers like well-differentiated hepatocellular carcinoma (HCC), renal cell carcinoma, gastric, and signet ring cell carcinoma, resulting in a high false-negative rate. ^{68}Ga-FAPI is favorable for diagnosing G.I. cancers. ^{68}Ga-FAPI uptake was found to be very high, $SUV\ max > 12$ in sarcoma, esophageal, breast, cholangiocarcinoma, and lung cancers [5].

^{68}Ga-FAPI uptake with $SUV\ max < 6$ was observed in pheochromocytomas, differentiated renal cells, thyroid, adenoid, cystic, and gastric cancers. The average SUV max of hepatocellular, colorectal, head and neck, ovarian, pancreatic, and prostate cancer was intermediate.

Because the ^{68}Ga-FAPI races contain the universal DOTA tracer, the chelator also adds theranostic approach after labeling the ligand with beta emitter, opening up another avenue for a more specific approach to cancer diagnosis and therapy [6, 7].

23.2.4 Theranostic

Because of the creators of newly used therapeutic tracers, nuclear medicine is moving more towards therapy than diagnostics.

The benefits of radiotracer therapy include the ability to focus on the most active part of the cancer while sparing neighboring organs from unnecessary irradiation. The radionuclides used in therapy are either elements, molecules, or compounds.

The most classical element is radioactive iodine-131. In 1939, it was the first to be used in thyroid disease to image the thyroid gland, and it was later used to treat hyperthyroidism. After World War II, it became widely available and was also used to treat thyroid cancer.

Most well-differentiated thyroid cancers can be diagnosed, staged, and treated with radioiodine.

Thyroid surgery is the first line of treatment. Nuclear medicine is used to evaluate the outcome of surgery using postoperative radioiodine imaging. Therefore, it is important to keep the patient away from iodine-rich medications like thyroxine, iodine-based antiseptics, and contrast media used with CT imaging. These can cause delays in management. The American Thyroid Association (ATA) put out guidelines on how to manage thyroid cancer, which are the best guidelines worldwide.

Very rarely, thyroid cancer cells become resistant to the trapping of iodine (radioiodine refractory cancer cells), which is usually associated with a loss of thyroid differentiation features. Such changes correlate with mitogen-activated protein kinase (MAPK), which is found to be higher in tumors with BRAF (B-Raf proto-oncogene) mutations. A tyrosine kinase inhibitor (TKI) was found to be helpful in improving thyroid uptake and therapy.

A patient with high thyroglobulin (TG) and a negative radioiodine whole-body scan can have a positive scan with ^{18}FDG PET. Because these cancers are rare, a PET scan can only be performed after a negative iodine scan. This was given the name "TENIS syndromes" (TG Elevated Negative Iodine Scan). Treatment of these types of cancers with radioiodine is still possible using very high doses of 150–300 mCi, with a good response judged by a continuous fall in TG [9].

23.2.4.1   Trance Arterial Radio Embolization (TARE)

Many cancers were previously treated with interventional procedures, such as embolization of feeding vessels, intra-tumoral chemotherapy, and electromechanical ablation. An alternative approach to covert inoperable hepatic cell carcinoma (HCC) was used with good success and involves the injection of microparticles that are loaded with beta emitters such as yttrium-90 (^{90}Y), aiming to:

1. Convert inoperable liver cancer to an operable disease
2. Increase survival in patients with primary or metastatic liver disease, primarily from GI cancers

This type of treatment needs a good setup and the selection of the proper candidates. It should consist of an oncologist, G.I. surgeon, interventional radiologist, and nuclear medicine physician. The nuclear medicine team should also have a qualified medical physicist, a hot lab, and imaging technology.

TARE is expanding to other cancers but still needs more studies, guidelines, endorsement by regulatory bodies, and insurance reimbursement.

23.3   Equipment and Manpower Availability in the UAE

23.3.1   Equipment Availability in the UAE

Gamma cameras (Table 23.2) and PET scanners (Table 23.3) are available in a few places in the UAE.

Table 23.2 Availability of gamma cameras in the UAE

Emirate	Facility name
Abu Dhabi	Cleveland Clinic Abu Dhabi
	Mediclinic
	Burjeel Medical City (BMC)
	Sheikh Shakhboot Medical City (SSMC)
	Sheikh Khalifa Medical Centre (SKMC)
Al Ain	Tawam Hospital
Dubai	Dubai Hospital
	American Hospital Dubai
	Mediclinic City Hospital
	Clemenceau Medical Center Hospital
	Saudi German Hospital
Sharjah	NMC Royal Hospital
Fujairah	Fujairah Hospital
Ras Al Khaimah	Sheikh Khalifa Specialist Hospital

Table 23.3 Availability of PET scanners in the UAE

Emirate	Facility name
Abu Dhabi	Burjeel Medical City (BMC)
	Gulf International Cancer Centre (GICC) in Al-Bahia
Al Ain	Cleveland Clinic Abu Dhabi (CCAD) (There will be more in the future)
Dubai	American Hospital Dubai
	Mediclinic City Hospital
	Advanced Care Oncology Centre (ACOC)
	Clemenceau Medical Center Hospital
Ras Al Khaimah	Sheikh Khalifa Specialist Hospital

23.3.2 Manpower

Manpower consists of:

23.3.2.1   Nuclear Medicine Physicians
Approximately, the total number of nuclear medicine physicians is around 27 which are currently working in the UAE (Table 23.4).

23.3.2.2   Technical Staff
There is a standard formula that states that the technical staff for each piece of equipment should be at least 1.5 times the number of pieces of equipment plus at

Table 23.4 An estimation of nuclear medicine physicians currently working in the UAE

Total no. of nuclear medicine physicians	Emirates
11	Abu Dhabi and Al Ain
10	Dubai
2	Sharjah
2	Ras Al Khaimah
2	Fujairah

least one medical physicist for each center. The number of technical staff is not fixed, and there is a shortage everywhere.

23.3.2.3 Nursing Staff in Nuclear Medicine

Nurses' availability in each center depends on whether there are outpatient services or admissions for radionuclide therapy. The anticipated number is around 25 nurses.

PET-CT is used mainly for oncology patients for:

- Initial diagnosis, staging, and restaging
- Assessment of response to treatment
- Early detection of a recurrence or metastasis
- Aiding in radiotherapy planning and dosimetry
- Selecting the site of biopsy or finding an occult primary

PET-CT is extremely useful in both pediatric and adult cancers. It changed the way Hodgkin's and non-Hodgkin's lymphomas were managed. It is the main economic player in cancer management by minimizing the use of expensive chemotherapy when the PET-CT scan does not show a good response.

23.4   Scope of Service

Nuclear medicine facilities should be able to offer the following services (to UAE and international patients) a variety of nuclear and molecular scans, and radionuclide therapy.

23.4.1 SPECT-CT Studies

- **Endocrines:** Thyroid scan, parathyroid scan, radioiodine whole-body scan for thyroid cancer, neuroendocrine tumors imaging, and adrenal gland scintigraphy.
- **Gastrointestinal:** Esophageal reflux, transit time, and gastric emptying scintigraphy; Meckel's diverticulum scintigraphy; salivary gland scintigraphy; gastrointestinal tract (GIT) bleeding with red blood cell labeling scintigraphy; hepatobiliary study (HIDA); liver scintigraphy; and Barret's esophagus scan.

- **Cardiovascular System:** Myocardial perfusion study (rest and stress), heart ejection fraction with gated studies for left ventricle function, and myocardial viability study.
- **Urogenital System:** Dynamic renal study, diethylenetriamine pentaacetate (DTPA), mercaptoacetyltriglycine (MAG3), ethylene cysteine (EC) with diuretics, static renal study (DMSA), cysto-scintigraphy for direct and indirect vesicoureteral reflux (VUR), varicocele study, testicular perfusion scintigraphy, assessment of renal function for donors, and assessment of transplant perfusion, function, rejection, and leak.
- **Musculoskeletal Studies:** Bone scintigraphy (3 phase, 2 phase, and whole-body bone scan), joint scintigraphy, Charcot foot scintigraphy, stress fracture, bone mineral densitometry (BMD), body composition study, Fracture Risk Assessment Tool (FRAX), and trabecular bone score (TBS).
- **Inflammation/Infection:** Gallium 67 scintigraphy, white blood cell (WBC) leuko-scan, thallium scintigraphy, and pyrexia of unknown origin (PUO).
- **Miscellaneous Studies**
  - Dacryo-scintigraphy (for lachrymal duct)
  - Lung ventilation and perfusion studies, V/Q ratio, and pulmonary hypertension
  - Lympho-scintigraphy (for upper and lower limbs)
  - Sentinel lymph node scintigraphy
  - Brain scintigraphy (for epilepsy, parkinsonism, dementia, and brain death).

23.4.2 PET-CT Studies

- **Oncology:** Useful in the *staging and restaging* of solid organ malignancies and to search for the unknown primary, response to treatment, and detection of early recurrence.
- ^{18}F-FDG whole body/regional PET-CT, ^{18}F-NaF PET-CT bone scan, ^{18}F-choline, gallium ^{68}Ga-PSMA and DOTA.
- **Non-oncology:** Useful in pyrexia of unknown origin (PUO), epilepsy, dementia, myocardial viability, cardiac sarcoidosis, inflammatory pathologies like sarcoidosis, prosthesis related infections, osteomyelitis, etc.

The average minimum staffing plan for any nuclear medicine department to establish, according to the author's knowledge, is shown in Table 23.5, and Table 23.6 lists the additional equipment needed to establish a nuclear medicine department.

Table 23.5 Minimal staffing plan for a nuclear medicine department to initiate

Staff type	Total number	Staff breakdown
Physician	3	Consultant (1)
		Specialist (1)
		General physician/resident (1)
Medical physicist	1	Consultant physicist (1)
Nurses	3	Nuclear medicine experienced nurses (3)
Clinical Support Staff	5	Senior nuclear medicine technologist (3)
		Nuclear medicine technologist (2)
Administrative Staff	3	Coordinator, receptionist/insurance (3)

Table 23.6 List of the additional equipment needed to establish a nuclear medicine department

S. No	Equipment
1	Dual energy X-ray absorptiometer (for bone mineral density)
2	Technigas—lung ventilation delivery system
3	Radiation (area monitoring system)
4	Dose calibrator
5	Contamination monitor
6	Personal dosimeter
7	Fume hood
8	Processing and reviewing station
9	Archiving computer system
10	Glucose monitoring machine
11	Crash trolley
12	Treadmill and cardiac stress facility
13	Defibrillator, cardiac monitors, oximeters
14	Infusions pumps

23.5   Future Outlook

The future of nuclear medicine is moving towards the molecular level and targeting genes. The standard approach to prostate cancer is surgical, medical, and radiotherapy, depending on the protocol used. Radionuclide therapy with prostate-specific membrane antigen (PSMA) was used when other treatments failed, but the new approach is to start with beta- or alpha-labeled PSMA before surgery, and the results are encouraging.

In breast cancer, an intratumoral single dose of astatine-211 as gold nanoparticles can suppress the growth of tumor tissue strongly without radiation exposure to other organs. Other attempts are still going on, like labeling raloxifene or herceptin with radionuclides for therapy. *HER-2* imaging with 64Cu-DOTA-transluzumab can pick up a very early and small metastasis and subsequently be dealt with very early [10, 11].

The new LU-177 LUTATERA is more effective in treating neuroblastoma than I-131 mIBG [8].

Radionuclides targeted gene therapy as the ultimate direction for the treatment of many cancers and achieving a complete cure.

TARE is expanding for many single or multiple metastases in the liver that originated from intestinal cancers.

23.6 Conclusion

It has been a great achievement in a relatively short period of time since the beginning of nuclear medicine in the UAE by a few pioneers. Now that we have proudly achieved the current level in this promising field, we hope to advance in this direction, as we have in many other fields.

Forty-three years was a hard and difficult but enjoyable mission, particularly when you see patients coming from many neighboring countries for treatment in the UAE.

The ultimate goal is to conduct original research and create a scientific base to make new discoveries that we can share with the rest of the world on a reciprocal basis.

Acknowledgement I would like to thank Dr. Anshu Misra for her support in writing the chapter.

Conflict of Interest The author has no conflict of interest to declare.

References

1. Carlson S. A glance at the history of nuclear medicine. Acta Oncol. 1995;34(8):1095–102. https://doi.org/10.3109/02841869509127236.
2. Wagner HN. A personal history of nuclear medicine. Springer; 2006. https://doi.org/10.1007/b138066#about-this-book.
3. Paez D, Becic T, Bhonsle U, et al. Current status of nuclear medicine practice in the middle east. Semin Nucl Med. 2016;46(4):265–72. https://doi.org/10.1053/j.semnuclmed.2016.01.005.
4. Gutfilen B, Valentini G. Radiopharmaceuticals in nuclear medicine: recent developments for SPECT and PET studies. Biomed Res Int. 2014;2014:426892.
5. Kratochwil C, Flechsig P, Lindner T, et al. ^{68}Ga-FAPI PET/CT: tracer uptake in 28 different kinds of cancer. Radiol Imaging Cancer. 2019;1(1):e194003.
6. Koerber SA, Staudinger F, Kratochwil C, et al. The role of 68Ga-FAPI PET/CT for patients with malignancies of the lower gastrointestinal tract: first clinical experience. J Nuclear. 2020;61(9):1331–6.
7. Siveke JT. Fibroblast activation protein: targeting root of tumor microenvironment. J Nucl Med. 2018;59:1412–4.
8. Parghane RV, Naik C, Talole S, Desmukh A, Chaukar D, Banerjee S, Basu S. Clinical utility of ^{177}Lu-DOTATATE PRRT in somatostatin receptor-positive metastatic medullary carcinoma of thyroid patients with assessment of efficacy, survival analysis, prognostic variables, and toxicity. Head Neck. 2020;42(3):401–16. https://doi.org/10.1002/hed.26024.
9. Vasileiou M, et al. Thyroid disease assessment and management: summary of NICE guidance. BMJ. 2020;368:m41. https://doi.org/10.1136/bmj.m41.
10. Kato H, Huang X, Kadonaga Y, Katayama D. Intratumoral administration of astatine-211-labeled gold nanoparticle for alpha therapy. J Nanobiotechnol. 2021;19(1):223.
11. Al-Ibrahim A, et al. Theranostics in the Arab world. Achiev Chall Jordan Med J. 2022;56(2):188–205. https://doi.org/10.35516/jmj.v56i2.243.

Dr. Abdulrahim Al Suhaili graduated from Ain-Shams Medical School with honors in Egypt. He was SHO at the Institute of Radiology and Nuclear Medicine in Baghdad, Iraq, and then Registrar at the same place. He got a scholarship to Johns Hopkins University (Nuclear Cardiology and Oncology), USA, in 1976. He was a specialist at the Ibn Al-Nafis Cardiovascular Hospital in Baghdad, Iraq. He did his postgraduate from Royal Postgraduate Medical School, University of London, MSc. 1983. Dr. Abdulrahim worked in the nuclear medicine department of Hammersmith Hospital in London, UK, from 1982 to 1984. He worked at Mubarak Al Kabeer Hospital and Amiri Hospital (1984–1987), Kuwait. He was Head of Nuclear Medicine at Tawam Hospital from 1987 to 2004 in Abu Dhabi, UAE. He also worked as an Associate Professor at the UAE University, Al Ain, UAE (1990–2007). He was an Associate Clinical Professor at Dubai Medical School (2004–2019). He was head of nuclear medicine and densitometry at Dubai Hospital from 2004 to 2014. He was the first to start BMD in the UAE in 1995 and the first to teach bone densitometry in the UAE. Dr. Abdulrahim was the founding member of the Pan Arab Osteoporosis Society in 1998 and the Emirates Osteoporosis Society. He was the president of the Pan Arab Osteoporosis Society from 2005 to 2008 and is also the current president of the Emirates Osteoporosis Society. His research, along with that of his colleagues, on osteoporosis among women in the UAE was awarded the Sh. Hamdan Medical Awards in 2006.

Dr. Abdulrahim Ibrahim Al Suhaili has become the Director of Nuclear Medicine and Bone Densitometry in Burjeel Medical City, UAE. He has published 51 articles in US, UK, EU, and UAE journals. He wrote chapters in two books on nuclear medicine. He is a member of the editorial boards of many journals in nuclear medicine and densitometry.

Pediatric Cancer in the UAE

24

Zainul Aaabideen Kanakande Kandy, Ammar Morad, and Eman Taryam Alshamsi

24.1 Introduction

Globally, progress in the field of pediatric oncology is one of the biggest success stories in the oncology field in the last few decades. The 5-year survival rate for most pediatric cancers is now 80–90% (1–3); however, there are very few publications from the United Arab Emirates (UAE) on the survival rate of childhood cancers in the UAE (4–6). The UAE has made remarkable progress in the field of pediatric oncology in the last few decades. The UAE as a country has highly advanced infrastructure and provides a safe environment and very comfortable facilities for both residents and visitors. These attract many visitors to come to the UAE for the holiday and for healthcare. It is the vision of the government to promote medical tourism, and it has tremendous potential to explore and

Z. A. Kanakande Kandy (✉)
Department of Pediatric Oncology Hematology and BMT, Burjeel Medical City, Abu Dhabi, United Arab Emirates
e-mail: zainul.aabideen@burjeelmedicalcity.com

A. Morad
Mediclinic City Hospital, Dubai, United Arab Emirates
e-mail: ammar.morad@mediclinic.ae

E. Taryam Alshamsi
Hematology and Oncology Division, Pediatric Department,
Al Jalila Children's Specialty Hospital, Dubai, United Arab Emirates

© The Author(s) 2024   389
H. O. Al-Shamsi (ed.), *Cancer Care in the United Arab Emirates*,
https://doi.org/10.1007/978-981-99-6794-0_24

establish. Therefore, it is important to reflect on and analyze the past and current challenges and use this insight to plan for the future. The most important domains in pediatric oncology care, as in any other medical field, are service, research, and education, and we need all-round development in all these domains, which will make pediatric oncology in the UAE among the best in the world.

24.2   History of the UAE's Pediatric Oncology Services

Cancer care services for children were initially developed in the public sector but are now available in both the public and private sectors (4, 7, 8). Tawam Hospital, the first hospital in the UAE, was opened by the UAE government in September 1979 in Al Ain, Abu Dhabi, to deliver care to children with cancer (4, 7, 8). Dubai Hospital was established in 1983 in Dubai and also provides pediatric oncology services. The Dubai pediatric hempathology oncology unit moved to Al Jalila Children's specialty hospital in April 2023. It is under the government of Dubai.

Sheikh Khalifa Medical City (SKMC) in Abu Dhabi was the second government hospital in Abu Dhabi to also deliver pediatric cancer care, and it was opened in 2005.

In the private sector, hospitals that provide pediatric cancer care in the UAE include Burjeel Medical City, Abu Dhabi; Royal NMC, Abu Dhabi; American Hospital; and Mediclinic City, Dubai (Tables 24.1 and 24.2).

The first dedicated cancer hospital in the UAE is the Gulf International Cancer Centre (GICC), which was opened in 2007 (7). This hospital, however, does not offer pediatric oncology services.

Table 24.1 Names of a few hospitals providing pediatric oncology in the UAE

Hospital	City
Public	
Tawam Hospital	Al Ain
Sheikha Khalifa Medial City (SKMC)	Abu Dhabi
Al Jalila Children's Hospital	Dubai
Private	
Burjeel Medical City	Abu Dhabi
NMC Hospital Abu Dhabi	Abu Dhabi
American Hospital	Dubai
Mediclinic City Hospital	Dubai
Clemenceau Medical Center (CMC) Hospital	Dubai
NMC Hospital Sharjah	Sharjah

Table 24.2 Incidence of pediatric cancer in the UAE as per National Cancer Registry (NCR) (9)

	2014	2015	2017	2019	2021
New pediatric cancer case age group of 0–14 years	154	165	146	125	154
Male/female	55.2% / 44.8%	57% / 43%	55% / 45%	53.6% / 46.4%	55% / 45%
0–4-year age group	77 (50.0%)	75 (45.5%)	62 (42.5%)	63 (50.4%)	72 (46.8%)
5–9-year age group	41 (26.6%)	48 (29.1%)	50 (34.2%)	24 (19.2%)	39 (25.3%)
10–14-year age group	36 (23.4%)	42 (25.5%)	34 (23.3%)	38 (30.4%)	43 (27.9%)
Leukemia	67 (43.5%)	68 (41.2%)	61 (41.8%)	44 (35.2%)	66 (42.9%)
Brain and CNS	22 (14.3%)	21 (12.7%)	7 (4.8%)	14 (11.2%)	23 (14.9%)
Connective and soft tissue	4 (2.6%)	4 (2.4%)	–	9 (7.2%)	–
Non-Hodgkin lymphoma	11 (7.1%)	15 (9.1%)	10 (6.8%)	9 (7.2%)	13 (8.4%)
Hodgkin's lymphoma	11 (7.1%)	10 (6.1%)	–	–	–
Bone and articular cartilage	5 (3.2%)	3 (1.8%)	–	7 (5.6%)	7 (4.5%)
Kidney & Renal pelvis	6 (3.9%)	10 (6.1%)	11(7.5%)	–	–
Liver and intrahepatic bile ducts	4 (2.6%)	5 (3.0%)	8 (5.5%)	–	–

Source: Ministry of Health and Prevention, Statistics and Research Center, National Disease Registry—UAE National Cancer Registry Report, 2014–2021

24.3   Radiation Oncology Service for Children in the UAE

The first radiation oncology program was started at Tawam Hospital, followed by GICC (7). The first radiation oncology service in the Northern Emirates was started at Sheikh Khalifa Speciality Hospital in Ras Al Khaimah in 2015. The first radiation oncology service in the private sector was started at an American hospital in Dubai. Currently, there are many private hospitals in the UAE that provide radiation services, including Burjeel Medical City (7).

24.4   Pediatric Hematopoietic Stem Cell Transplantation (HSCT)

Stem cell transplantation (SCT) is one of the more advanced treatments. It is curative and lifesaving for many pediatric conditions, including childhood cancer. This service was not available in the UAE until March 2022. This was one of the main reasons for children going abroad. The first allogenic bone marrow transplant (BMT) was successfully done in the UAE in March 2022 at Burjeel Medical City, Abu Dhabi (10). In August 2022, the first BMT in a child with Acute Lymphoblastic Leukemia

(ALL) was successfully completed. The first BMT for Acute Myeloid Leukemia (AML) was done in January 2023. The first haploidentical BMT was done in the UAE in January 2023 (unpublished data on file).

A total of 164 pediatric patients underwent HSCT outside the UAE between 2016 and 2018 (11) including children residing in the UAE. An estimated 200 patients, including non-citizens, need HSCT annually in the UAE. Currently, CAR-T cell therapy and gene therapy are not offered in the UAE.

It is a need of the hour to develop a center of excellence for stem cell transplantation in children, which has previously been a major reason to travel abroad.

24.5 Long-Term Fertility Issue and Sperm Banking

As the outcome for children with cancer improves, the number of adults who are childhood cancer survivors is also increasing. Infertility is one of the long-term side effects of many cancer drugs and radiation treatments. Fertility preservation is becoming increasingly important for these younger patients. This service is available in the UAE; however, lack of awareness about this program is the main obstacle. In addition to this, insurance coverage for this service is also an issue, especially for expatriates.

24.6 Funding for the Pediatric Cancer Care in the UAE

Cancer care in the UAE is expensive, but it is funded by the government for UAE nationals. But for expatriates, it is covered by insurance. There are many expatriates' children, especially in the Northern Emirates, without insurance or whose insurance is inadequate to cover the expense. They do get free treatment through the mandate program available at Tawam Hospital. In effect, all resident children with cancer, irrespective of their nationalities and insurance coverage, get most of the cancer treatment available in the UAE. In addition to this, there are many charities, such as the Red Crescent Society, Sharjah TV, the Child Fund under the umbrella of the Al Jalila Foundation, Rahma, and Friends of Cancer Patients (FOCP), that support children with cancer financially. Recently, BMT services were established in the UAE. Since it is a new service in this country, many insurance companies have not yet recognized it and are not covering the expenses.

24.7 Availability of Cancer Medicines and Investigation Facilities for Treating Children with Cancer in the UAE

Most of the cancer drugs used for treating children with cancer as per the internationally recognized protocols, including the latest Food and Drug Administration (FDA)-approved medications, are available in the UAE. But they are expensive. However, a few medications in syrup form are still unavailable in the UAE.

24.8 Pathology, Molecular Cytogenetics, and MRD (Minimal Residual Disease) Testing in the UAE

Laboratories in the UAE are accredited by the College of American Pathologists (CAP), which standardizes pathology reporting. However, cytogenetic and molecular diagnostic testing for soft tissue sarcoma, Wilms tumors, and brain tumors is not easily available in the UAE. These tests are sent abroad to centers in the USA, Europe, Canada, and India, resulting in significant delays in diagnosis and the start of treatment in many cases. Minimal residual disease (MRD) is another important test for leukemia management. Recently, it was established in the UAE in two hospitals. There is a need for such facilities to be well established in the UAE, which will provide the UAE with all of the services required to provide treatment at an international level.

24.9 Medical and Nursing Team in Pediatric Oncology in the UAE

Consultant pediatric oncologists, junior doctors, pediatric oncology nurses, nurse coordinators, and pediatric oncology advanced nurse practitioners are important human resources in pediatric cancer care.

Delivery of pediatric oncology services in the UAE is led by consultants. Most of the consultants in the UAE were trained either in the USA, Canada, or the UK.

There are an estimated 16 pediatric oncologists in the UAE. All the hospitals that deliver pediatric oncology services have pediatric surgical departments. Pediatric surgeons perform most of the pediatric oncology surgeries. There is no dedicated pediatric radiation oncologist in the UAE.

24.10 Pediatric Oncology Nursing

The pediatric oncology nurse shortage is a challenge in the UAE. Most oncology nurses in the UAE are from India, the Philippines, Jordan, and Lebanon. There are very few pediatric oncology nurses who are UAE nationals.

Furthermore, the role of a pediatric oncology advanced nurse practitioner is not very well established in the UAE. Pediatric oncology advanced nurse practitioners provide a significant contribution in western countries like the UK and the USA.

Developing an advanced nurse practitioner role in oncology nursing in the UAE will foster and improve nursing care for cancer patients and their families. Communication between the nurse and parents is very important in pediatric oncology practice. Therefore, foreign nurses should be encouraged to learn the local Arabic language as a priority.

There are no structured training programs for pediatric oncology nursing in the UAE. As it is mandatory to have attendance at the continuing medical education (CME) for license renewal, pediatric oncology nurses attend the conference with

CME hours. Having a pediatric oncology nursing track in the pediatric oncology annual conference for continuing their education and improving evidence-based nursing practice will help advance their skills.

24.11 Protocol-Based Cancer Treatment in the UAE

Most of the international pediatric cancer centers that treat children with cancer are based on evidence-based protocols. This is one of the main reasons for the improved outcome of pediatric cancer treatment globally (1, 3, 12). These are the Children's Oncology Group (COG), the UKCCLG, the BFM Protocol, etc. (3, 12, 13). Although the UAE does not have national protocols and guidelines, all pediatric oncology centers in the UAE follow either one of the above-mentioned protocols.

24.12 Pediatric Oncology Research in the UAE

The foundation of research in the medical field is accurate knowledge of the epidemiology of diseases. Unfortunately, there is a real paucity of epidemiologic data on pediatric cancers in the UAE (6, 12, 14). After an extensive literature search, the number of publications in PubMed related to pediatric oncology in the UAE is surprisingly low. So far, we could find only 27 publications in the last 50 years (4–8, 11, 14–35). At present, there is no recent publication related to outcomes in children with cancer treated in the UAE except for three published before 2003 (4, 5, 8).

Internationally, pediatric cancer outcomes have improved as a result of the use of uniform guidelines and the very effective enrolment of patients in prospective multicentric clinical trials conducted by professional organizations such as COG and UKCCLG (12, 13). However, the UAE lacks a national pediatric oncology research group.

There are no organized clinical trials related to pediatric oncology in the UAE, and there is a lack of good prospectively published studies on the epidemiology, biology, or outcome of childhood cancers in the UAE.

24.12.1 Cancer Registry

Since 2014, data on the incidence of pediatric cancer in the UAE has been made available through the National Cancer Registry (NCR).

24.13 The Pediatric Oncology Education in the UAE

There is no formal postgraduate program in pediatric hemo-oncology in the UAE. UAE nationals who want to further pursue pediatric oncology after medical graduation go abroad for specialization.

24.13.1 Pediatric Oncology Continuing Medical Education (CME) in the UAE

Continuous medical education of primary care practitioners and pediatricians in the early diagnosis and prompt referral of childhood cancers is a very important step to improving the outcome of pediatric cancer care.

In the last 5 years, there has been an annual Emirates Pediatric Hematology and Oncology conference in the UAE, which has played a big role in the continuous medical education of practicing pediatricians in the UAE by improving their knowledge in pediatric oncology to diagnose cancers in children early. The first Emirates Pediatric Bone Marrow Transplantation Congress was held in the UAE in 2022. Both conferences were attended by renowned international and national speakers.

24.14 Emirates Pediatric Hematology and Oncology Society

Organizations like the Emirates Pediatric Hematology and Oncology Society do not exist as they do among adult oncologists like the Emirates Oncology Society (EOS) and Emirates Hematology Society (EHS). It is the official organization representing adult oncology healthcare providers in the UAE under the Emirates Medical Association's (EMA) umbrella. We recommend establishing such an organization to represent pediatric oncologists in the UAE.

24.15 Support Program for Families with Children on Cancer Treatment

In many countries, there are many support groups to help families with children diagnosed with cancer. In addition to financial support, they have a big role in providing psychological support to the parents during their most difficult time. In the UAE, there are many organizations like the Red Crescent Society, Rahma, and Friends of Cancer Patients (FOCP) that provide lots of support, including financial support, for such families. Many hospitals in the UAE do activities on International Cancer Day in February and Childhood Cancer Awareness Month in September to motivate children with cancer and their families and to raise awareness regarding childhood cancer.

24.16 Medical Tourism for Pediatric Cancer Care: Bringing Revenue to the UAE Rather Than Spending Huge Sums Abroad for Cancer Treatment of Children

The UAE is a popular holiday destination for people all around the world. Considering the infrastructure and very comforting facilities, the UAE can become a very popular destination for healthcare services.

The UAE, as a nation, has a lot of potential to become a hub for medical tourism. Many factors distinguish it from other countries, including its geographical location, the availability of healthcare expertise with US and UK training and experience in the UAE, and the recent establishment of many world-class hospitals, including a pediatric BMT facility and the availability of cancer medicine in the UAE.

But there are many obstacles, including the high price of the cancer treatment and the trust of international patients in the existing healthcare system in the UAE.

On the contrary, the UAE spent vast sums of money outside the UAE for cancer treatment and bone marrow transplantation for children seeking treatment elsewhere.

The USA, the Federal Republic of Germany, the Republic of Singapore, the Republic of Korea (South Korea), the Kingdom of Thailand, and the UK are the most chosen destinations for healthcare tourism for UAE nationals.

There are several sponsoring agencies in the UAE that cater to pediatric cancer care abroad, including all the health authorities (Department of Health, Dubai Health Authority, and Ministry of Health and Prevention), Presidential Affairs offices, the armed forces, police, and charity organizations. The lack of treatment options in the UAE is one of the criteria for sending patients abroad. However, despite the availability of pediatric cancer treatments and a recently opened BMT facility in the UAE, many patients go abroad for the treatment.

24.17 Recommendations for the Progress of Pediatric Oncology Services in the UAE

Cancer Registry: Optimal utilization of the existing national cancer registry through optimal reporting and use of the data for analysis for proper understanding of the current incidence, outcome, and challenges for further development.

Encourage Research Publications: Peer-reviewed journals should be encouraged for all pediatric oncologists. They should be given appropriate support and recognition for their commitment to publishing evidence.

Establish Services: There are services that need to be established as a priority to deliver pediatric oncology services locally at an international standard.

It is a need of the hour to develop a center of excellence for stem cell transplantation in children, which has previously been a major reason to travel abroad.

National Multidisciplinary Tumor (MDT) Board: There is a need for a national MDT board to discuss difficult cases to improve patient care.

Invest in nurse training: Regular nurse training and CME in pediatric oncology will educate and empower the nursing workforce in pediatric oncology.

We recommend that oncology nurses in the UAE receive incentives for their dedication and commitment with a periodic hike in their salary.

Medical Tourism: Reduce residents' going abroad for pediatric oncology treatments and bone marrow transplants that are already available in the UAE by improving their trust in the healthcare system. Promote medical tourism, which attracts more visitors to the UAE for medical treatment.

We recommend that treatment abroad be limited to complex pediatric cancer cases. Promoting public trust in cancer care within the UAE is an important aspect that needs special attention.

24.18 Conclusion

The pediatric oncology services have developed significantly in the UAE since the country was formed. All the children who reside in the country are entitled to get treatment, irrespective of their nationalities and insurance coverage. However, there are areas that need close attention and improvement as a priority.

Conflicts of Interest The authors have no conflict of interest to declare.

References

1. Robison LL, Armstrong GT, Boice JD, Chow EJ, Davies SM, Donaldson SS, et al. The childhood cancer survivor study: a National Cancer Institute-supported resource for outcome and intervention research. J Clin Oncol. 2009;27(14):2308–18.
2. Ward E, DeSantis C, Robbins A, Kohler B, Jemal A. Childhood and adolescent cancer statistics, 2014. CA Cancer J Clin. 2014;64(2):83–103.
3. Allemani C, Matsuda T, Di Carlo V, Harewood R, Matz M, Niksic M, et al. Global surveillance of trends in cancer survival 2000-14 (CONCORD-3): analysis of individual records for 37 513 025 patients diagnosed with one of 18 cancers from 322 population-based registries in 71 countries. Lancet. 2018;391(10125):1023–75.
4. Mpofu C, Revesz T. History and current state of pediatric oncology and hematology in The United Arab Emirates. Pediatr Hematol Oncol. 1996;13(1):1–7.
5. Nawaz A, Matta H, Jacobsz A, Shawis R, Mpofu C, Al-Salem A. Wilms' tumor: the Tawam hospital experience. Ann Saudi Med. 1999;19(3):257–60.
6. Radwan H, Hasan H, Ballout RA, Rizk R. The epidemiology of cancer in The United Arab Emirates: a systematic review. Medicine (Baltimore). 2018;97(50):e13618.
7. Al-Shamsi HO. The state of cancer Care in The United Arab Emirates in 2022. Clin Pract. 2022;12(6):955–85.
8. El-Hayek M, Trad O, Donner M, Hardy D. Pediatric oncology in The United Arab Emirates: the Tawam hospital experience. Med Pediatr Oncol. 2003;41(5):486–7.
9. Cancer incidence in United Arab Emirates, Annual report of the UAE-National Cancer Registry. Statistics and Research Center, Ministry of Health and Prevention.
10. WAM. first pediatric BMT in the UAE 2022. www.wam.ae.
11. Al-Shamsi HO, Abyad A, Kaloyannidis P, El-Saddik A, Alrustamani A, Abu Gheida I, et al. Establishment of the first comprehensive adult and pediatric hematopoietic stem cell transplant unit in The United Arab Emirates: rising to the challenge. Clin Pract. 2022;12(1):84–90.
12. O'Leary M, Krailo M, Anderson JR, Reaman GH, Children's Oncology G. Progress in childhood cancer: 50 years of research collaboration, a report from the Children's Oncology group. Semin Oncol. 2008;35(5):484–93.
13. Bleyer WA. The U.S. pediatric cancer clinical trials programmes: international implications and the way forward. Eur J Cancer. 1997;33(9):1439–47.
14. Khan S, Kambris MEK, AlShamsi ET. Epidemiology of brain tumors in The United Arab Emirates: a National Registry Cross-sectional Study. BMC Neurol. 2020;20(1):301.
15. Alnuaimi E, Al Halabi M, Khamis A, Kowash M. Oral health problems in leukaemic paediatric patients in The United Arab Emirates: a retrospective study. Eur J Paediatr Dent. 2018;19(3):226–32.

16. Bener A, Denic S, Al-Mazrouei M. Consanguinity and family history of cancer in children with leukemia and lymphomas. Cancer. 2001;92(1):1–6.
17. Mahmood S, Revesz T, Mpofu C. Febrile episodes in children with cancer in The United Arab Emirates. Pediatr Hematol Oncol. 1996;13(2):135–42.
18. Revesz T, Mpofu C, Oyejide C. Ethnic differences in the lymphoid malignancies of children in The United Arab Emirates. A clue to aetiology? Leukemia. 1995;9(1):189–93.
19. Revesz T, Pramathan T, Mpofu C. Leukaemia phenotype and ethnicity in children living in The United Arab Emirates. Haematologia (Budap). 1996;28(1):9–12.
20. Razzak HA, Harbi A, Shelpai W, Qawas X. Risk factors of cancer in The United Arab Emirates. Gulf J Oncolog. 2018;1(26):49–57.
21. Shattaf A, Jamil A, Khanani MF, El-Hayek M, Baroudi M, Trad O, et al. Undifferentiated sarcoma of the liver: a rare pediatric tumor. Ann Saudi Med. 2012;32(2):203–5.
22. Ishaqi MK, El-Hayek M, Gassas A, Khanani M, Trad O, Baroudi M, et al. Allogeneic stem cell transplantation for Glanzmann thrombasthenia. Pediatr Blood Cancer. 2009;52(5):682–3.
23. Jumaa PA, Sonnevend A, Pal T, El Hag M, Amith R, Trad O. The molecular epidemiology of Stenotrophomonas maltophilia bacteraemia in a tertiary referral hospital in The United Arab Emirates 2000-2004. Ann Clin Microbiol Antimicrob. 2006;5:32.
24. El Hayek M, Trad O, Jamil A. Vincristine-induced urinary bladder paralysis. J Pediatr Hematol Oncol. 2005;27(5):286–7.
25. El-Hayek M, Trad O, Hardy D, Islam S. The triad of seizures, hypertension, and neuroblastoma: the first described case. J Pediatr Hematol Oncol. 2004;26(8):523–5.
26. El-Hayek M, Lestringant GG, Frossard PM. Xeroderma pigmentosum in four siblings with three different types of malignancies simultaneously in one. J Pediatr Hematol Oncol. 2004;26(8):473–5.
27. Eapen V, Revesz T. Psychosocial correlates of paediatric cancer in The United Arab Emirates. Support Care Cancer. 2003;11(3):185–9.
28. Lestringant GG, Masouye I, El-Hayek M, Girardet C, Revesz T, Frossard PM. Diffuse calcinosis cutis in a patient with congenital leukemia and leukemia cutis. Dermatology. 2000;200(2):147–50.
29. Nawaz A, Mpofu C, Shawis R, Matta H, Jacobsz A, Kassir S, et al. Synchronous bilateral Wilms' tumor. Pediatr Surg Int. 1999;15(1):42–5.
30. Nawaz A, Matta H, Jacobsz A, Mpofu C, Al-Salem A. Unresectable hepatoblastoma: the role of preoperative chemotherapy. Ann Saudi Med. 1999;19(6):553–6.
31. Eapen V, Revesz T, Mpofu C, Daradkeh T. Self-perception profile in children with cancer: self vs parent report. Psychol Rep. 1999;84(2):427–32.
32. Revesz T, Mpofu C, Oyejide C, Bener A. Socioeconomic factors in the families of children with lymphoid malignancy in the UAE. Leukemia. 1997;11(4):588–93.
33. Mpofu C, Sztriha L, Revesz T. Neuroblastoma-associated paraneoplastic syndrome with anti-Hu antineuronal antibodies presenting at the time of recurrence. Pediatr Hematol Oncol. 1996;13(4):369–73.
34. Revesz T, Obeid K, Mpofu C. Severe lactic acidosis and renal involvement in a patient with relapsed Burkitt's lymphoma. Pediatr Hematol Oncol. 1995;12(3):283–8.
35. Revesz T, Mpofu C, Fletcher S, Lytle B, Shawis R, Amirlak I, et al. Progression to anaplasia in bilateral Wilms' tumor. Med Pediatr Oncol. 1994;23(1):40–9.

Dr. Zainul Aaabideen Kanakande Kandy is highly skilled and experienced in paediatric haematology, oncology, and bone marrow transplants. He received his degree in medicine at Calicut University in Kerala, India; subsequently, he took his post-graduate degree in paediatrics from the University of Mumbai .

In 2001, Dr. Zainul moved to the United Kingdom, where he specialized and undertook further training and experience in paediatric haematology, paediatric oncology, and paediatric bone marrow transplantation in various hospitals in the UK.

    Royal Marsden Hospital, London, IK
    University College Hospitals of London UCLH London
    Imperial College London, UK
    Manchester Children Hospital, Manchester, UK
    Alder Hey Children Hospital, Liverpool, UK
    Great North Children Hospital, Newcastle, UK

Dr. Zainul gained his master's degree in paediatric oncology at Birmingham University as well as a certificate in medical education at Manchester University. Subsequently, he completed his International Fellowship in paediatric bone marrow transplantation at Great North Children's Hospital in Newcastle.

He worked previously as a consultant pediatric at the University Hospital of Coventry, Warwickshare, and Royal Oldham Hospital before moving to the UAE.

In the UAE, he had worked at the Department of pediatric hematology and oncology at Tawam Hospital as a consultant, Al Ain, before joining at Burjeel Medical City as head of pediatric hematology, oncology, and BMT.

Aside from his commitment to further advance his experience and knowledge in the fields of paediatrics, paediatric haematology, oncology, and bone marrow transplantation in the United Kingdom, he also published articles and actively supported the community by sharing his knowledge and participating in international conferences across countries.

Dr. Ammar Morad completed his training in pediatric hematology and oncology at Texas Children's Hospital and subsequently practiced as an assistant professor. He has practiced in the USA for the past 35 years, and his most recent position was at Cincinnati Children's Hospital, ranked among the top three pediatric cancer centers in the USA.

He has published extensively in world-renowned journals and taught several generations of future pediatric oncologists. His passion for teaching earned him the title of most outstanding faculty member in the pediatric department at Texas Tech University.

During his tenure, he created and directed the pediatric cancer centre at the Women's and Children's Hospital in Louisiana, where he achieved a 90% cure rate for the 150 cases he has managed according to the latest treatment protocols.

Dr. Morad brings with him a wealth of experience in all aspects of pediatric blood conditions and childhood cancers, backed by access to and collaboration with premier paediatric cancer centers to provide unrivalled, personalized care in Dubai and the UAE.

Dr. Eman Taryam Alshamsi is currently serving as a consultant pediatric hematologist and oncologist at Al Jalila Children's Hospital. She graduated in 2002 from the faculty of medicine and health sciences at the United Arab Emirates University. Following graduation, she trained and practiced at Tawam Hospital UAE in 2005 and as staff from January 2017 until June 2021, when she joined Al Qassimi Women and Children Hospital AQWCH as a consultant in hematology oncology and head of department until July 2023. She acquired the Arab Board diploma of medical specialization in pediatrics and RCPCH. She has been trained and obtained a diploma in clinical research affiliated with Vienna School of Clinical Research.

Later on, she completed a clinical fellowship program in pediatric hematology oncology and bone marrow transplant at Sickkids Hospital in Toronto, Canada, which included full training in pediatric clinical hematology oncology /bone marrow transplant and thrombosis for 3 years and 6 months as a hemostasis fellow. She has also attended an administrative fellowship program in MD – Anderson, Texas, USA.

Dr. Al Shamsi has been part of several on-going collaborative research projects at Tawam and AQWCH hospitals, MBRU, UAEU, and Zayed University with PubMed publications. Author and reviewer for bleeding disorder and pediatric cancer chapters.

She hosted the SIOP Asia conference in Abu Dhabi in April 2019. Dr. Al Shamsi is involved in many national and international committees and activities, including the Scientific Sub-Committee of the International Medical Awards 2023-2024 (Sheikh Hamdan Bin Rashid Al Maktoum Award for medical science), the Transfusion Committee in AQWCH/AQH, the NIHS pediatric hematology oncology accreditation (Head of the Sub-committee), national bone marrow transplants, national cancer screening and prevention, SIOP, and the Women Leader in pediatric oncology SIOP (WLPO). Patient support activity and awareness with CCI and Friends of cancer patients (FOCP) UAE.

Hassan Shahryar Sheikh ⓘ and Kiran Munawar ⓘ

25.1 Introduction

The United Arab Emirates (UAE) is situated in the southeast of the Arabian Peninsula, bordering Oman and Saudi Arabia, and is a member of the Gulf Cooperation Council (GCC) of Arab countries. It has an estimated population of about 9,282,410 in 2020 [1–3]. The World Bank classifies it as a high-income country [4].

In 2017, the age group >60 years does not represent a large share of the population, and the demographics of the UAE are fast changing [5, 6]. Life expectancy at birth in the UAE continues to improve slowly, and the most recent estimates for 2020 are 78 and 81.4 years for males and females, respectively [7]. The projection is that the younger workforce will work their way up the population pyramid. Furthermore, the recent incentives from the government to attract and retain highly skilled expatriate workers and foreign investors with long-term residency and retirement options may increase the number of expatriates living past their retirement age in the UAE. As a result, the geriatric population in the UAE is likely to surge in the next 20–40 years. The World Health Organization (WHO) estimated that countries like the UAE should anticipate a fivefold or greater increase in the proportion of their geriatric population from 2000 to 2050 [8].

H. S. Sheikh (✉)
Sheikh Shakhbout Medical City, Abu Dhabi, United Arab Emirates

Khalifa University, Abu Dhabi, United Arab Emirates
e-mail: hssheikh@ssmc.ae

K. Munawar
St Bartholomew's Hospital, Barts Health NHS Trust, London, UK
e-mail: kiran.munawar@nhs.net

25.1.1 Definition of the Older Individuals

Ageing is commonly measured by chronological age, and, as a convention, a person aged 65 or more is often referred to as "older individual" [9]. The elderly are divided into three groups by the American National Institute for Ageing: young adults (65–74 years old); older adults (75–84 years old); and the oldest adults (over 85 years old) [10]. As per the National Policy on Senior Emiratis, individuals who are 60 years of age or older are considered seniors [11].

The current retirement age for most employees is 60 years. Therefore, the definition of "elderly" in the UAE is here considered to be above the age of 60 years.

25.2 Cancer Incidence in the Elderly (Age ≥60 years) in the UAE

According to the 2021 Annual Report of the UAE National Cancer Registry, a total of 5830 new malignant cases were reported in the country. 25.6% of these cases occurred among Emirati nationals. Females made up 55.1% of these cases, regardless of origin.

When stratified by age groups, 29.18% of all cancer cases occurred in the age ≥60 years, irrespective of gender and origin. As a result, nearly one-third of all cancer cases in the UAE occurred among the elderly. The data indicate the highest incidence of malignant cases in the age groups 40–44 years (12.6%), 50–54 years (11.2%), 45–49 years (11.1%), and 50–59 years (10.8%) 35–39 years (10.3%) as shown in Fig. 25.1(a) [6].

According to gender in the total population, among females, 23% of all new malignant cases were diagnosed in the age range ≥60 years, and among males, 36.2% or one third of all new malignant cases occurred in the age range ≥60 years, as shown in Fig. 25.1(b, c) [6].

People over the age of 60 accounted for 40.4% of new malignant cases among Emirati nationals, compared to 25.1% among non-citizens. The data indicate that the highest incidence of malignant cases in Emirati citizens was observed in the age group 55–59 years (9.6%), as shown in Fig. 25.1(d). Among the Emirati male citizens, almost half (50.5%) of all the new malignant cases were diagnosed in people aged ≥60 years, compared to one-third (33.1%) in the Emirati female citizens. The data show that the highest frequency of cancer was observed among Emirati females in the age group of 40–44 years (11.4%), and among Emirati males in the age group of 70–74 years (10.7%), followed by a second highest frequency in the age group of 60–64 years (10.3%) (Fig. 25.1(e, f), Table 25.1) [6].

The most commonly diagnosed cancers in the UAE population also vary considerably by age group, with particular differences in the cancer types diagnosed in adults' aged ≥60 years compared to the younger population. The breast, colorectal, prostate, and lung were the most frequent solid malignant tumors, respectively,

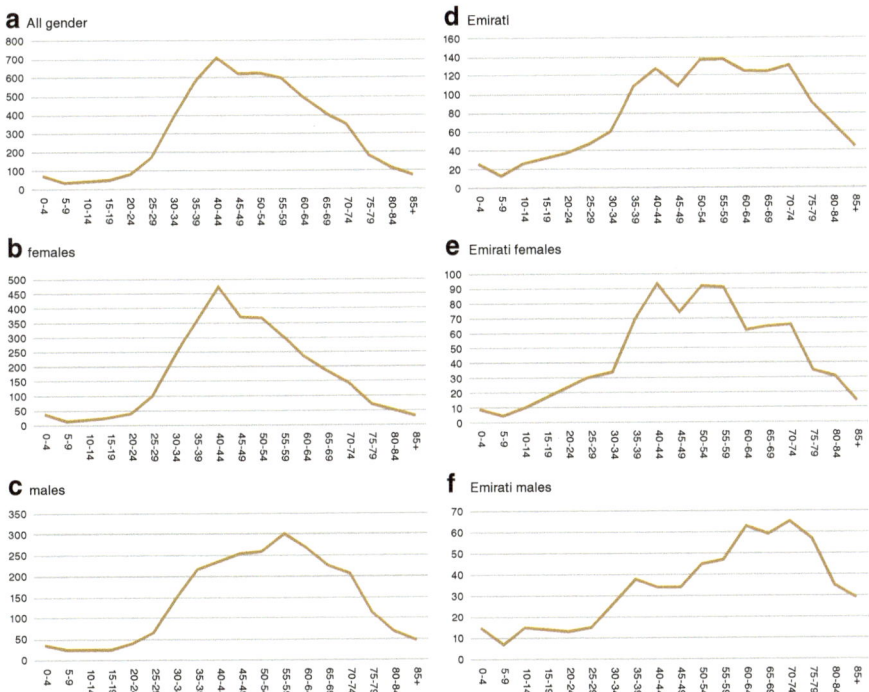

Fig. 25.1 Annual Report of the UAE—National Cancer Registry—2021. Statistics and Research Center, Ministry of Health and Prevention. Adapted from the Cancer Incidence in United Arab Emirates [6]. Age group distribution of malignant cases in UAE among all gender (**a**), females (**b**), males (**c**), Emirati (**d**), Emirati females (**e**), and Emirati males (**f**)

among all populations aged ≥60 years. In the Emirati citizens, the most frequent malignant tumors were breast, colorectal, prostate, lung, and uterus, respectively, in the age ≥60 years. Further specifics of the data by gender, age, and nationality are limited and not available for further analysis. The incidence of cancer burden is expected to rise significantly by 2040 (Table 25.2) [6, 12, 13].

Similar to the cancer incidence, the mortality rate from cancer also increases with age (Fig. 25.2) [2]. According to the Department of Health (DOH) report on Abu Dhabi health statistics 2017, the death rate per 1000 increases significantly starting at around age ≥60 years [2, p. 21].

These statistics clearly demonstrate that older patients with cancer in the UAE constitute a sizable and significant population. Among Emirati citizens, individuals aged ≥60 years carry a high burden of cancer.

Cancer incidence rates in neighbouring GCC countries follow a similar trend of increasing cancer incidence with age, with the age group ≥60 years representing the population with the highest cancer burden (Fig. 25.3) [14].

Table 25.1 Distribution of primary sites (malignant cases) by age group, among all, in 2021, in the UAE

Primary site ICD-10	(0–9)	(10–19)	(20–29)	(30–39)	(40–49)	(50–59)	(60–69)	(70–79)	(80+)
C00–C14 Lip, oral cavity & pharynx	0	1	3	24	42	43	28	7	6
C15 Esophagus	0	0	2	4	7	4	1	6	3
C16 Stomach	0	0	5	15	29	25	33	24	3
C17 Small intestine	0	0	3	4	6	6	4	2	1
C18–C21 Colorectal	1	2	10	69	107	129	112	77	25
C22 Liver and intrahepatic bile ducts	0	0	3	10	8	33	23	31	6
C23, C24 Gallbladder, other and unspecified part of biliary tract	0	0	0	4	9	8	14	7	4
C25 Pancreas	0	0	1	11	15	32	29	14	8
C26 Other and ill-defined digestive organs	0	0	0	0	4	1	1	1	1
C30, C31 Nasal cavity, middle ear, accessory sinuses	0	0	0	1	3	5	3	0	0
C32 Larynx	0	0	0	0	5	10	7	7	0
C34 Bronchus and lung	0	0	4	16	32	62	47	50	20
C37 Thymus	0	0	0	5	4	1	0	0	0
C38 Heart, mediastinum, and pleura	0	0	2	2	1	0	0	1	0
C40–C41 Bone and articular cartilage	1	11	1	6	7	4	2	1	1
C43 Skin melanoma	0	0	6	13	12	15	3	2	0
C44 Skin (Carcinoma)	0	1	7	44	70	77	39	24	11
C45 Mesothelioma	0	0	1	0	2	1	1	1	0
C46 Kaposi sarcoma	0	0	0	1	1	0	1	0	0
C48 Retroperitoneum and peritoneum	2	1	0	2	5	3	2	2	1
C49 Connective and soft tissue	2	7	1	12	10	8	4	2	1
C50 Breast	0	0	13	223	407	275	149	55	17
C51 Vulva	0	0	0	1	1	0	0	0	1
C52 Vagina	0	0	1	0	0	0	2	0	0
C53 Cervix uteri	0	0	5	38	45	34	14	4	1

C54–C55 Uterus	0	0	2	21	37	42	46	24	1
C56 Ovary	0	1	4	19	27	33	12	9	3
C57 Other and unspecified female genital organs	0	1	0	0	0	3	2	0	0
C58 Placenta	0	0	2	0	1	1	0	0	0
C61 Prostate	0	0	0	1	9	56	107	60	18
C62 Testis	2	1	17	30	9	1	0	0	0
C64–C65 Kidney & renal pelvis	10	0	3	19	43	32	27	12	5
C66, C68 Ureter and other urinary organs	0	0	1	0	0	2	1	0	1
C67 Urinary bladder	0	0	1	4	16	24	37	26	18
C69 Eye	1	0	0	1	3	0	0	0	0
C70–C72 Brain & CNS	13	12	9	30	25	29	21	8	2
C73 Thyroid	2	16	69	206	169	94	27	8	4
C74–C75 Other endocrine glands	2	0	3	4	1	1	0	0	0
C76–C80 Unknown or unspecified sites	6	2	1	5	8	13	11	8	7
C81 Hodgkin's lymphoma	2	16	29	21	14	3	2	3	1
C82–C85, C96 Non-Hodgkin lymphoma	7	8	19	35	41	52	32	24	10
C88, C90 Multiple myeloma	0	0	2	3	20	22	27	10	5
C91–C95 Leukemia	59	15	23	56	64	36	28	19	4
Other hematopoietic malignancies	0	0	0	3	11	10	6	2	4
Other malignancy	1	0	0	0	1	0	0	0	0
Grand total	111	95	253	963	1331	1230	905	531	193

Source: Ministry of Health and Prevention, Statistics and Research Center, National Disease Registry—UAE National Cancer Registry Report, 2021

Table 25.2 Distribution of primary sites (malignant cases) by age group, among Emirati, in 2021, in the UAE

Primary site ICD-10	(0–9)	(10–19)	(20–29)	(30–39)	(40–49)	(50–59)	(60–69)	(70–79)	(80+)
C00–C14 Lip, oral cavity & pharynx	0	0	0	5	5	7	5	4	4
C15 Esophagus	0	0	0	1	1	1	0	3	3
C16 Stomach	0	0	2	3	4	4	9	6	2
C17 Small intestine	0	0	1	0	1	1	1	2	0
C18–C21 Colorectal	1	1	2	17	13	39	37	33	17
C22 Liver and intrahepatic bile ducts	0	0	0	1	1	3	4	17	4
C23, C24 Gallbladder, other and unspecified part of biliary tract	0	0	0	0	0	3	3	5	1
C25 Pancreas	0	0	0	1	7	10	4	7	2
C26 Other and ill-defined digestive organs	0	0	0	0	1	0	0	1	0
C30, C31 Nasal cavity, middle ear, accessory sinuses	0	0	0	0	0	1	0	0	0
C32 Larynx	0	0	0	0	2	3	4	3	0
C34 Bronchus and lung	0	0	1	1	7	11	15	14	11
C37 Thymus	0	0	0	1	2	0	0	0	0
C40–C41 Bone and articular cartilage	0	6	1	0	2	1	0	1	1
C43 Skin melanoma	0	0	0	2	1	0	0	0	0
C44 Skin (Carcinoma)	0	0	2	3	3	3	3	4	6
C45 Mesothelioma	0	0	0	0	0	1	0	0	0
C46 Kaposi sarcoma	0	0	0	1	0	0	0	0	0
C48 Retroperitoneum and peritoneum	1	1	0	0	1	1	1	0	0
C49 Connective and soft tissue	1	5	0	2	3	2	2	1	1
C50 Breast	0	0	3	35	55	60	38	22	5
C51 Vulva	0	0	0	0	1	0	0	0	0
C52 Vagina	0	0	0	0	0	0	1	0	0

(continued)

C53 Cervix uteri	0	0	2	0	9	7	3	1	1
C54–C55 Uterus	0	0	0	5	12	12	18	12	1
C56 Ovary	0	1	0	3	4	9	1	3	2
C57 Other and unspecified female genital organs	0	0	0	0	0	1	0	0	0
C61 Prostate	0	0	0	0	0	11	30	21	9
C62 Testis	1	1	3	9	0	1	0	0	0
C64–C65 Kidney & renal pelvis	1	0	1	5	11	5	10	7	2
C66, C68 Ureter and other urinary organs	0	0	0	0	0	1	0	0	1
C67 Urinary bladder	0	0	0	0	4	6	14	14	11
C69 Eye	0	0	0	0	1	0	0	0	0
C70–C72 Brain & CNS	7	8	2	6	3	6	7	4	1
C73 Thyroid	1	10	37	49	50	29	9	4	4
C74–C75 Other endocrine glands	2	0	0	0	0	0	0	0	0
C76–C80 Unknown or unspecified sites	4	1	1	1	2	2	4	3	5
C81 Hodgkin's lymphoma	0	9	12	1	5	1	2	1	1
C82–C85, C96 Non-Hodgkin lymphoma	0	4	7	11	8	14	12	14	7
C88, C90 Multiple myeloma	0	0	0	1	1	9	5	4	2
C91–C95 Leukemia	16	9	6	4	13	9	4	9	2
Other hematopoietic malignancies	0	0	0	0	1	1	3	1	2
Other malignancy	1	0	0	0	0	0	0	0	0
Grand total	36	56	83	168	234	275	249	222	108

Source: Ministry of Health and Prevention, Statistics and Research Center, National Disease Registry—UAE National Cancer Registry Report, 2021

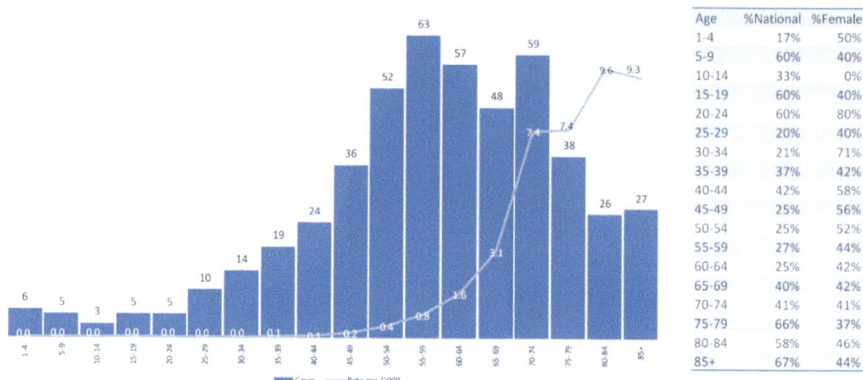

Fig. 25.2 Cancer death cases by age group. Source: Abu Dhabi Health Statistics 2017 [2]

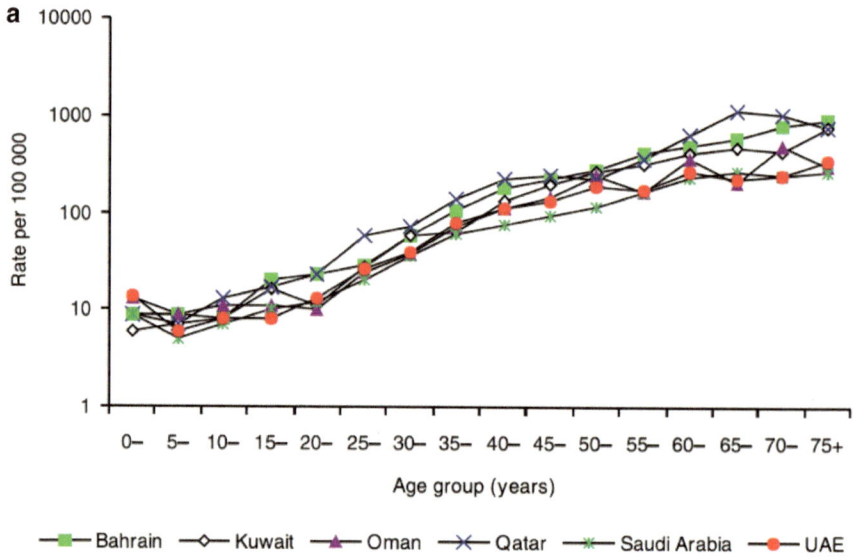

Fig. 25.3 Age-specific incidence rates of all cancers in females (**a**) and males (**b**) in the Gulf Cooperation Council. Source: Incidence of cancer in Gulf Cooperation Council countries, 1998–2001 [14]

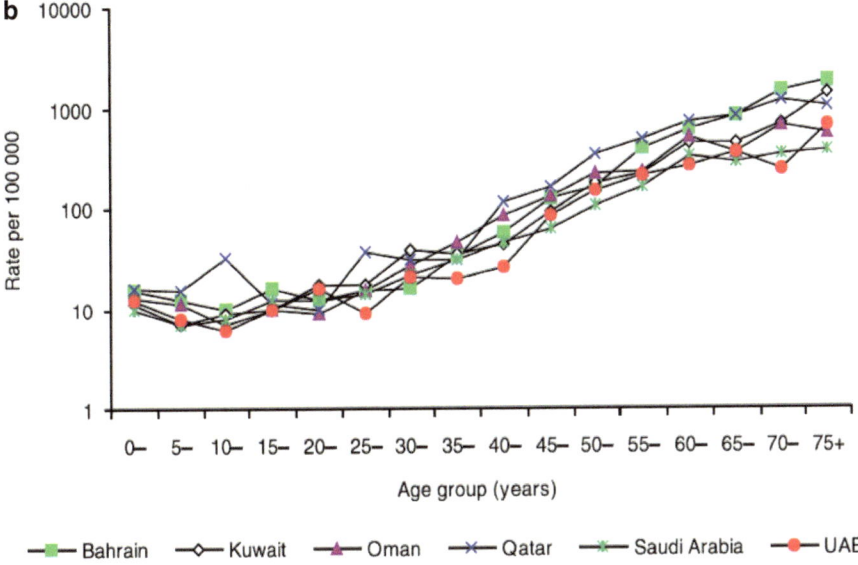

Fig. 25.3 (continued)

25.3 Geriatric Oncology

Geriatrics is a discipline of medicine that deals with the healthcare of older individuals, and geriatric oncology is a sub-discipline of geriatrics that recognizes the uniqueness of older individuals with cancer that requires specialized care and treatment. Cancer is a complex disease that requires a multidisciplinary approach, and special aspects need to be emphasized for the older individuals [15, 16]. The presence of competing comorbidities makes this age group complex. Studies show that only 8% of older patients with cancer have no comorbidities, while up to 55% have three or more co-morbidities. Finding the right balance between overtreatment and undertreatment is challenging yet critical in the clinical decision-making process for older patients with cancer. It is an area of ongoing research to ascertain the priority of care among competing cancers and comorbidities in an older patient with cancer [17, 18].

The field of geriatric oncology has now fully come of age since its beginnings in the 1980s, when American Society of Clinical Oncology (ASCO) President Dr. Kennedy recognized the study of ageing and cancer as a distinct area of interest and unmet need. Since then, it has seen major advancements and recognition as a subspecialty within oncology by several organizations and major cooperative groups. In order to promote awareness, ASCO organized a clinical practice forum in 2000, a symposium during its annual meeting in 2002, and published a document titled "Cancer Care in the Older Patient" as part of their Curriculum Series [15, 16]. Similarly, the International Society of Geriatric Oncology, headquartered in

Switzerland, established various task forces to assess current literature and provide treatment recommendations. In the United States, the National Comprehensive Cancer Network issued practice guidelines for older adult oncology, while the Geriatric Oncology Consortium was founded to initiate clinical trials and raise awareness about challenges faced by elderly patients. The Journal of Clinical Oncology (JCO) released a dedicated series on Geriatric Oncology (GO) in 2007 and subsequently in 2014. These publications aimed to showcase the advancements made by researchers in this field, provide updated evidence-based treatment recommendations for older cancer patients to clinical oncologists, and identify areas of limited knowledge to inspire future research endeavours. Despite being a very fertile area of research and practice, the field of GO is not without its own unique challenges. These challenges can be broadly classified into three categories: (1) establishing a GO clinical service; (2) educating and training personnel; and (3) conducting research in GO. These challenges are being met to varying degrees, depending on the resources of individual countries and organizations. To fulfil this resource disparity from a global oncology perspective, several GO initiatives have been taken across the globe that are revolutionizing the way older adults with cancer are treated [19]. Major oncology organizations have now integrated geriatric oncology (GO) into their global oncology curriculum and have issued guidelines on enhancing clinical practice, training, and research in this field. Notably, the Food and Drug Administration (FDA) is spearheading a global regulatory initiative aimed at expanding the body of evidence for older adults with cancer [19].

25.3.1 Geriatric Oncology in the UAE

A quick look at the global landscape in geriatric oncology will reveal that cancer centers around the world with a dedicated GO service or program are mostly located in high-income countries (HICs), where older adults represent a large share of their populations. GO, like other specialities, is highly resource-dependent, and the presence of skilled personnel and multidisciplinary teams is one of the key resources required to establish a GO program. Unfortunately, skilled personnel and multidisciplinary teams in GO are globally lacking. Therefore, it comes as no surprise that currently no formal GO programs or clinical services exist in the UAE, and provisions for elder care in the UAE remain very limited [5]. There are limited options when it comes to home care programs, typically provided by hospitals or private service providers. In the United Arab Emirates (UAE), for example, there were only 21 licensed geriatricians available in 2020 [5, p. 5], and it is unclear whether they are involved in the treatment of cancer patients. Additionally, there are few healthcare providers with specialized training in this area, and there is a notable absence of local research on ageing and eldercare. Furthermore, there is a scarcity of published studies addressing geriatric oncology issues, and medical students and postgraduate trainees lack a formal educational curriculum on geriatric oncology.

At present, there are only two residential nursing care facilities available for the elderly in the entire country [5]. These facilities are typically considered as a last resort, as specific eligibility criteria must be met by senior citizens seeking care. A recent survey conducted among 2735 UAE residents examined the attitudes of the population towards older individuals, their knowledge and perceptions of elder care, as well as the experiences, expectations, and preferences of Emiratis regarding older age. The findings revealed that the current care system for the elderly in the UAE is not well-developed. Information regarding elder care was severely limited, and respondents were unaware of the emerging challenges associated with the care of older individuals [5].

Moreover, apart from the lack of physical infrastructure, there are also notable changes in social dynamics, particularly the decline of the extended family model and the growing trend towards smaller nuclear families. These transformations will also affect elderly care since the majority of older individuals currently receive care at home from their families or with the assistance of domestic helpers. It is worth noting that UAE nationals have insufficient awareness about the country's increasingly ageing population. Younger individuals are also unaware of the consequences of ageing, while senior Emiratis lack adequate knowledge about maintaining healthy lifestyles, engaging in active pursuits, and understanding the demand for care and support required by the elderly [5].

In the coming years, the demand for elder care is anticipated to increase due to various factors, including shifting demographics within the country, changes in disease patterns and dependency rates, evolving expectations of older individuals, and the changing structure of families. The resources available, at present, are not sufficient to cater to the greater number of older people with cancer that will exist in the future. As a result, it is incumbent on private and public stakeholders in the UAE's healthcare sector to investigate and plan for the development of a sustainable and effective elderly care system capable of meeting the demands of complex medical care such as cancer treatment in the near future.

25.4   Future Directions

Many efforts are underway to advance the care of older individuals with cancer globally. In 2018, an international multidisciplinary working group at the International Society of Geriatric Oncology (SIOG) proposed a comprehensive framework for the global advancement of care for older adults with cancer worldwide [20]. This broad expert consensus, known as the Top Priorities Initiative, addressed four priority domains: education, clinical practice, research, and strengthening collaborations and partnerships.

These 12 priorities, listed in Table 25.3, can serve as the framework for setting up a robust and comprehensive geriatric oncology clinical service and infrastructure, along with the skilled personnel, in the UAE.

Table 25.3 The 12 priorities of the Top Priorities Initiative, proposed by the International Society of Geriatric Oncology (SIOG) in 2018

Education	
Priority 1	Integrate geriatric oncology into training programs for health-care professionals
Priority 2	Provide educational material and activities on geriatric oncology for health-care professionals
Priority 3	Educate the general public about the relevance of providing age-appropriate care for older adults with cancer
Clinical practice	
Priority 4	Implementing models to provide optimal care for older adults with cancer
Priority 5	Develop guidelines for the optimal treatment of older adults with cancer
Priority 6	Establish centers of excellence for delivering clinical care, doing clinical and translational research, and providing educational opportunities
Research	
Priority 7	Improve the relevance of clinical trials to older adults with cancer
Priority 8	Evaluate the benefits of allocated treatments and co-management in improving treatment outcomes for older adults with cancer
Priority 9	Use personalized medicine technologies to improve cancer understanding and management for older adults
Collaborations and partnerships	
Priority 10	Strengthen links between SIOG and the geriatric oncology workforce, international specialized agencies, global and regional professional organizations, policy makers, and patient advocacy groups
Priority 11	Promote the inclusion of specific provisions for delivering evidence-based care for older adults in national cancer control plans
Priority 12	Create global funding mechanisms for professional development and promote research on the interface of cancer and ageing

Source: Priorities for the global advancement of care for older adults with cancer: an update of the International Society of Geriatric Oncology Priorities Initiative [20]

25.5 Conclusion

In conclusion, while the proportion of the older population with cancer in the UAE is modest compared to the younger population, the demographics and social dynamics of the population are fast changing. Cancer remains a significant cause of morbidity and mortality and affects the older population disproportionately. The incidence of cancer is high in older individuals, and they carry a large burden of cancer in the UAE. It is expected that the proportion of older individuals will rise significantly in the near future. As a result, the demand for ageing care for cancer patients is expected to skyrocket in the coming years. Due to a lack of ageing care resources, the country currently lacks formal geriatric oncology clinical services. Education, training, and research in the field of geriatric oncology are also lacking. To provide optimal care to cancer patients over the age of 65, comprehensive planning and resource allocation to establish centers of excellence, training programs, research, and aged care facilities, international collaborations, and partnerships in geriatric oncology are urgently required.

Conflict of Interest The authors have no conflict of interest to declare.

References

1. Blair I, Sharif AA. Population structure and the burden of disease in The United Arab Emirates. J Epidemiol Glob Health. 2012;2(2):61–71.
2. Health Do. Abu Dhabi health statistics 2017. Department of Health; 2018.
3. Al-Shamsi HO, Abu-Gheida IH, Iqbal F, Al-Awadhi A, editors. Cancer in the Arab world. Springer; 2022.
4. UAE Population UAEP, FCSC; 2023. https://fcsc.gov.ae/ar-ae/Pages/Statistics/Statistics-by-Subject.aspx#/%3Ffolder=%D8%A7%D9%84%D8%B3%D9%83%D8%A7%D9%86%D9%8A%D8%A9%20%D9%88%D8%A7%D9%84%D8%A7%D8%AC%D8%AA%D9%85%D8%A7%D8%B9%D9%8A%D8%A9/%D8%A7%D9%84%D8%B3%D9%83%D8%A7%D9%86/%D8%A7%D9%84%D8%B3%D9%83%D8%A7%D9%86%20%D9%88%D8%A7%D9%84%D8%AA%D8%B9%D8%AF%D8%A7%D8%AF%D8%A7%D8%AA%20%D8%A7%D9%84%D8%B3%D9%83%D8%A7%D9%86%D9%8A%D8%A9&subject=%D8%A7%D9%84%D8%B3%D9%83%D8%A7%D9%86%D9%8A%D8%A9%20%D9%88%D8%A7%D9%84%D8%A7%D8%AC%D8%AA%D9%85%D8%A7%D8%B9%D9%8A%D8%A9.
5. Almarabta S, Ridge N. What the UAE population thinks of aging and aged care (Strategic Report No. 5). Sheikh Saud bin Saqr Al Qasimi Foundation for Policy Research; 2021.
6. Cancer incidence in United Arab Emirates annual report of the UAE—National Cancer Registry—2021. Statistics and Research Center, Ministry of Health and Prevention; 2021.
7. https://fcsc.gov.ae/ar-ae/Pages/Statistics/Statistics-by-Subject.aspx#/%3Ffolder=%D8%A7%D9%84%D8%A7%D9%82%D8%AA%D8%B5%D8%A7%D8%AF%D9%8A%D8%A9/%D8%A7%D9%84%D8%AD%D8%B3%D8%A7%D8%A8%D8%A7%D8%AA%20%D8%A7%D9%84%D9%88%D8%B7%D9%86%D9%8A%D8%A9/%D8%A7%D9%84%D8%AD.
8. Hajjar RR, Atli T, Al-Mandhari Z, Oudrhiri M, Balducci L, Silbermann M. Prevalence of aging population in the Middle East and its implications on cancer incidence and care. Ann Oncol 2013;24 Suppl 7:vii11–24.
9. Orimo HIH, Suzuki T, Araki A, Hosoi T, Sawabe M, Orimo H, Ito H, Suzuki T, Araki A, Hosoi T, et al. Reviewing the definition of "elderly". Geriatr Gerontol Int. 2006;6(3):149–58.
10. National Research Council (U.S.). Panel on statistics for an aging population., Gilford DM. The aging population in the twenty-first century: statistics for health policy. Washington, DC: National Academy Press; 1988. xv, 323 p. p.
11. Portal TUAEG. Senior Emiratis—the official portal of the UAE government. The United Arab Emirates' Government Portal. 2022;
12. Cheema S, Maisonneuve P, Lowenfels AB, Abraham A, Doraiswamy S, Mamtani R. Influence of age on 2040 cancer burden in the older population of the Gulf Cooperation Council (GCC) countries: public health implications. Cancer Control. 2021;28:10732748211027158.
13. Ferlay J, Ervik M, Lam F, Colombet M, Mery L, Piñeros M, et al. Global cancer observatory: cancer today. Lyon, France: International Agency for Research on Cancer; 2018.
14. Al-Hamdan N, Ravichandran K, Al-Sayyad J, Al-Lawati J, Khazal Z, Al-Khateeb F, et al. Incidence of cancer in Gulf Cooperation Council countries, 1998-2001. East Mediterr Health J. 2009;15(3):600–11.
15. Lichtman SM, Balducci L, Aapro M. Geriatric oncology: a field coming of age. J Clin Oncol. 2007;25(14):1821–3.
16. Lichtman SM, Hurria A, Jacobsen PB. Geriatric oncology: an overview. J Clin Oncol. 2014;32(24):2521–2.
17. Extermann M, Overcash J, Lyman GH, Parr J, Balducci L. Comorbidity and functional status are independent in older cancer patients. J Clin Oncol. 1998;16(4):1582–7.
18. Repetto L, Venturino A, Vercelli M, Gianni W, Biancardi V, Casella C, et al. Performance status and comorbidity in elderly cancer patients compared with young patients with neoplasia and elderly patients without neoplastic conditions. Cancer. 1998;82(4):760–5.
19. Kanesvaran R, Mohile S, Soto-Perez-de-Celis E, Singh H. The globalization of geriatric oncology: from data to practice. Am Soc Clin Oncol Educ Book. 2020;40:1–9.

20. Extermann M, Brain E, Canin B, Cherian MN, Cheung KL, de Glas N, et al. Priorities for the global advancement of care for older adults with cancer: an update of the International Society of Geriatric Oncology Priorities Initiative. Lancet Oncol. 2021;22(1):e29–36.

Dr. Hassan Shahryar Sheikh is a consultant medical oncologist and hematologist at Sheikh Shakbout Medical City in Abu Dhabi, UAE, and an adjunct associate professor of medicine at Khalifa University. He is a Diplomate of the American Board of Internal Medicine, Geriatrics, Medical Oncology, and Hematology. He graduated from Aga Khan University Medical College in Pakistan and completed residency and fellowship trainings in internal medicine, medical oncology, and hematology at Penn State University Milton S. Hershey Medical Center, Hershey, USA. He also took fellowship training in geriatrics and ageing at the University of Rochester, New York, USA. He served as an Assistant Professor of Medicine at Penn State University for a few years prior to returning to Pakistan, where he served as the Head of Medical Oncology at the Shaukat Khanum Memorial Cancer Hospital in Lahore, Pakistan. Dr. Shahryar is well published in peer-reviewed international journals and is a speaker at many national and regional cancer conferences.

Dr. Kiran Munawar is a Clinical Fellow in Solid tumour Oncology at St Bartholomew's Hospital, Barts Health NHS Trust, London, UK. She received her medical degree from Allama Iqbal Medical College, University of Health Sciences, Lahore, Pakistan. After working as a Medical Officer in Medical Oncology at Shaukat Khanum Memorial Cancer Hospital and Research Centre, Pakistan, she joined Addenbrookes Hospital, Cambridge University Hospitals NHS Foundation Trust, Cambridge, UK, as a Clinical Fellow in Oncology. She aims to pursue clinical work and research in oncology, with an emphasis on melanoma and cutaneous oncology.

Breast Cancer in the UAE

Aydah Al-Awadhi ⓘ, Faryal Iqbal ⓘ, Hampig R. Kourie,
and Humaid O. Al-Shamsi ⓘ

26.1 Introduction

Breast cancer (BC) emerged as the most prevalent cancer globally in 2020, with approximately 2.26 million new BC cases and nearly 685,000 BC-related fatalities reported worldwide during that year [1]. Regarding cancer-related mortality, BC is ranked fifth overall and first among women [1]. Based on the UAE National Cancer Registry 2021 report [2], breast cancer claimed approximately 9.64% of annual cancer-related deaths. Throughout the year, the UAE National Cancer Registry

A. Al-Awadhi
Tawam Hospital, Al Ain, United Arab Emirates

Emirates Oncology Society, Emirates Medical Association, Dubai, United Arab Emirates
e-mail: ayawadhi@seha.ae

F. Iqbal
Burjeel Medical City, Abu Dhabi, United Arab Emirates
e-mail: faryal.iqbal@burjeelmedicalcity.com

H. R. Kourie
Hematology-Oncology Department, Faculty of Medicine, Saint Joseph University,
Beirut, Lebanon
e-mail: hampig.kourie@usj.edu.lb

H. O. Al-Shamsi (✉)
Burjeel Cancer Institute, Burjeel Medical City, Burjeel Holdings, Abu Dhabi,
United Arab Emirates

Ras Al Khaimah Medical and Health Sciences University, Ras Al Khaimah, United Arab Emirates

Gulf Medical University, Ajman, United Arab Emirates

Emirates Oncology Society, Emirates Medical Association, Dubai, United Arab Emirates

College of Medicine, University of Sharjah, Sharjah, United Arab Emirates

Gulf Cancer Society, Alsafa, Kuwait
e-mail: alshamsi@burjeel.com; humaid.al-shamsi@medportal.ca

© The Author(s) 2024
H. O. Al-Shamsi (ed.), *Cancer Care in the United Arab Emirates*,
https://doi.org/10.1007/978-981-99-6794-0_26

recorded a total of 1139 breast cancer cases among the country's population, which constituted 20.3% of all reported malignant cases in 2021.

Based on 2020 data from the World Health Organization (WHO), breast cancer stands as the predominant form of cancer in the UAE, representing an incidence rate of 21.4% (1054 cases). It also holds the unfortunate distinction of being the primary cause of cancer-related deaths in the region [1]. The majority of diagnosed cases are individuals below the age of 50 [3]. The situation represents a significant public health concern in the UAE.

In this chapter, we will examine published articles and research findings from the United Arab Emirates (UAE) to gain a deeper understanding of the characteristics and outcomes associated with breast cancer. Additionally, we will supplement this knowledge with our own experiences in managing breast cancer at renowned cancer facilities. We believe that individuals interested in the status of breast cancer in the UAE will find this review to be thorough and inclusive. Furthermore, we will highlight existing gaps and areas for improvement, emphasizing the importance of resource allocation and future clinical and research endeavors.

26.2   Epidemiology of Breast Cancer in the UAE

According to the latest information available from the UAE National Cancer Registry in 2021, there were 1139 new instances of breast cancer. Out of these cases, 1128 were reported among women, while the remaining eleven were identified in men. These figures account for approximately 20.2% of all cancer cases reported during the same year. The crude incidence rate was calculated at 40.1 per 100,000 for female population, while the age-standardized incidence rate (ASR) was determined to be 52 per 100,000 for female population based on 2021 data. Notably, breast cancer ranked as the most prevalent malignancy among women, comprising 36.9% of all female cancer cases [2]. In 2021, breast cancer ranked third among cancer-related fatalities, accounting for an approximate average of 9.64% of annual cancer deaths [2]. The incidence of male breast cancer is approximately 1%, similar to rates observed in the United States (US) and the United Kingdom (UK). In comparison, the prevalence of male breast cancer in central Africa is reported to be as high as 6% of all breast cancer cases [4, 5].

Historical data on breast cancer in the UAE were limited prior to 2011, when the UAE National Cancer Registry was initially established. Hence, we sought previously unreported historical data from prominent oncology facilities and clinics across the country, spanning various time periods prior to 2011. The aggregate data from reports from the UAE National Cancer Registry that have been published, as well as unpublished sources from clinics during the previous 40 years, are summarized in Fig. 26.1.

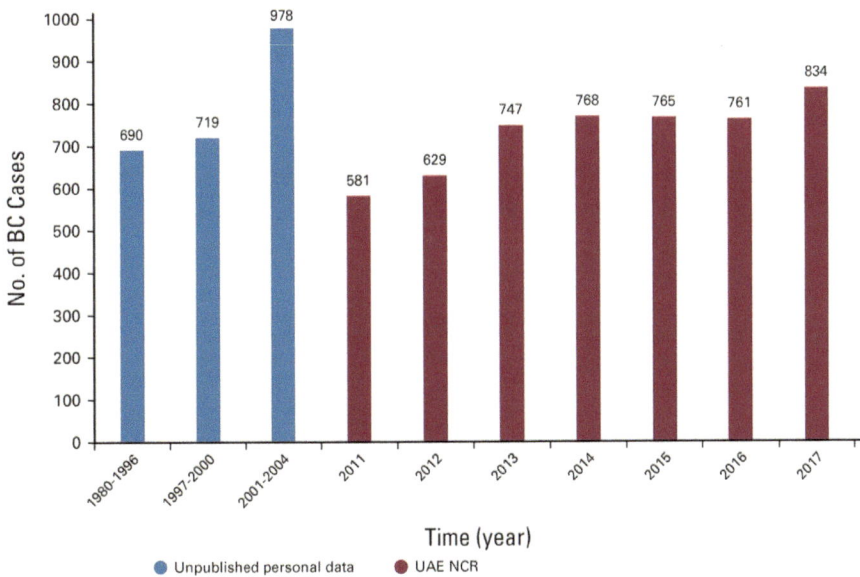

Fig. 26.1 Registered breast cancer cases in the UAE over the past four decades. Source: NCR, National Cancer Registry; UAE, United Arab Emirates

Previous reports have indicated that the average age in Arab nations, as a whole, tends to be around 10 years lower compared to Western nations [6]. The potential causes of this phenomenon include genetic and environmental factors, a relatively younger population in comparison to Western countries, and cultural influences leading to reduced rates of breast cancer screening among elderly Arab women [7].

Based on reports, the typical age of breast cancer diagnosis in the UAE falls within the range of 48–49 years old [3, 8, 9]. In 2021, the majority of breast cancer cases were observed among patients below the age of 60, as depicted in Fig. 26.2a. More specifically, for UAE nationals, individuals aged 50–59 (60 out of 218 cases, accounting for 27.5%) and, for non-UAE citizens, those aged 40–49 (352 out of 921 cases, making up 38.2%) represented the highest proportions [2]. Based on 2021 data, the average age of the population in the UAE is reported to be 32.8 years [10]. The age-specific incidence rates rise steadily from age 25 in females. After age 74, age-specific breast cancer incidence decreases drastically as depicted in Fig. 26.2b [2].

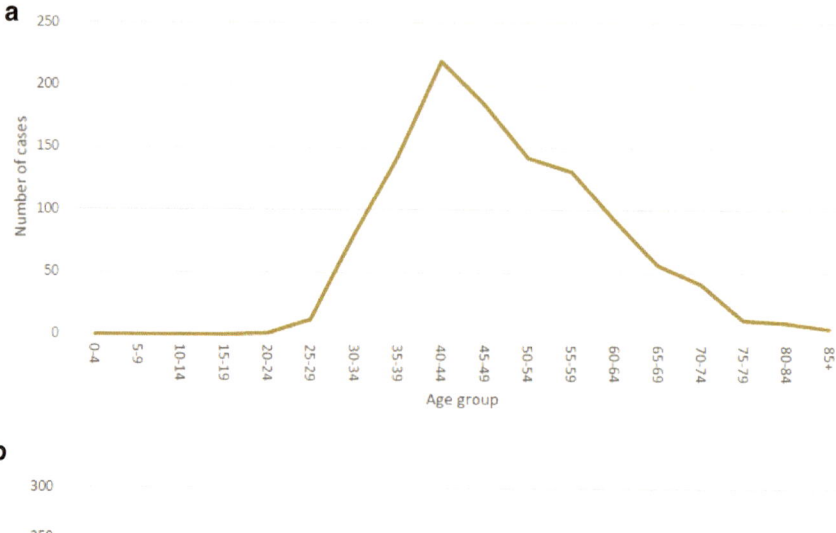

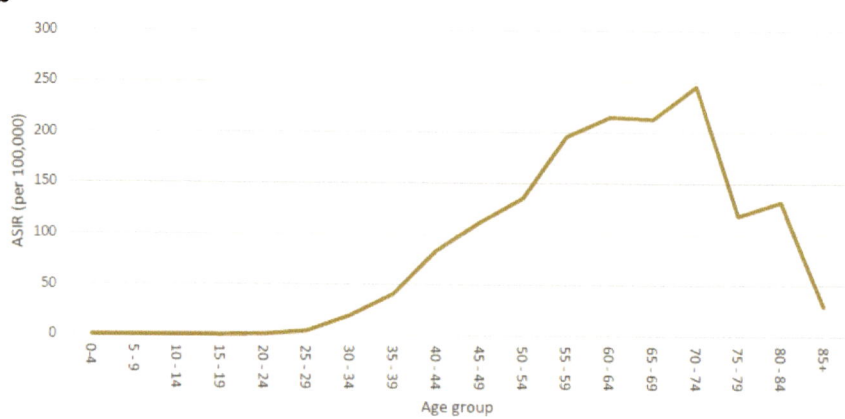

Fig. 26.2 (**a**) Age group distribution of female breast cancer cases in the UAE in 2021. (**b**) Age-specific incidence rate (ASIR) for female breast cancer cases in the UAE in 2021. Source: Ministry of Health and Prevention, Statistics and Research Center, National Disease Registry—UAE National Cancer Registry Report, 2021

26.3 Clinicopathological Features of Breast Cancer

In general, breast cancer in the Arab population exhibits distinct clinicopathological features when compared to the Western world. These include an earlier age of onset, a higher tumor grade, an increased prevalence of HER2 amplification, and a lower occurrence of the luminal subtype [11]. Regrettably, there is a scarcity of available data regarding the clinicopathological characteristics of breast cancer in the UAE. However, a research study carried out by the University of Sharjah examined 94 breast cancer patients from the Northern Emirates of the United Arab Emirates. The findings revealed that the majority of patients (78 out of 94, or 83%) exhibited invasive ductal carcinoma and tested positive for hormone receptors. Additionally, it was observed that women under the age of 40 were more likely than older women to exhibit HER2 overexpression, as determined by immunohistochemistry (IHC) or fluorescence in situ hybridization (FISH), with a statistically significant correlation ($p = 0.007$) [3].

In a separate study involving a group of 78 Arab patients, including 25% from the United Arab Emirates, the median age at diagnosis was reported as 52.3 years, ranging from 37 to 82 years, with 38.5% of individuals being 50 years old. Among this cohort, 82.1% had invasive ductal carcinoma, 19.2% exhibited HER2 overexpression, and 26.9% had triple-negative breast cancer. Notably, at the time of diagnosis, 46.2% of cases were identified as stage IV disease [12].

A different study provided insights into the clinicopathological attributes of 130 breast cancer patients with BRCA1/2 mutations. The average age of these patients was 42.9 years, and approximately 50.7% of them had a positive family history of breast cancer. The majority of patients (66.2%) exhibited stage I/II disease, with invasive ductal carcinoma being the prevalent subtype (81.5%). Additionally, 45.3% of patients had hormone receptor-positive breast cancer [13].

Although comprehensive stage data are unavailable in the UAE Cancer Registry, there is evidence indicating a decline in the incidence of stage IV breast cancer cases compared to other stages. Notably, the proportion of localized disease has shown an increase from 10% in 2011 to 25% in 2017 [2]. This could potentially be attributed to heightened societal awareness regarding the significance of early detection, consequently prompting an increase in breast cancer screening practices.

Conversely, a qualitative study carried out in the UAE focused on 19 breast cancer survivors ranging in age from 35 to 70 years old. These individuals had exhibited delayed medical attention-seeking and diagnosis after experiencing symptoms. The study identified that the primary reasons for such delays were spousal abandonment and the fear of facing social stigmatization [14]. Culture exerts a substantial influence on the decisions made by women in UAE society. The limited understanding of breast cancer signs and symptoms, as well as the absence of regular screening, significantly affects how symptoms are assessed and subsequent choices regarding follow-up diagnostic procedures. Consequently, this can lead to the presentation of advanced-stage disease and a delay in receiving timely treatment [14].

A separate study conducted at Tawam Hospital, a prominent cancer center in the UAE, focused on male breast cancer. The study spanned from 2000 to 2020 and revealed that male breast cancer accounted for 0.75% (28 out of 3733 cases) of all breast cancer cases. The median age at diagnosis was 51, and the majority of patients (26 out of 28) were diagnosed at an early stage, with only 2 cases identified as stage IV disease. Among the male breast cancer patients, 21 out of 28 were diagnosed with hormonal receptor (HR)-positive, or HER-2-negative disease [15].

26.3.1 Somatic and Germline Mutations

There is a scarcity of comprehensive data regarding somatic and germline mutations in breast cancer, specifically within the UAE and other Gulf Cooperation Council

(GCC) nations. However, based on available information, it seems that BRCA1/2 mutations do not play a prominent role in hereditary breast cancer in GCC countries. Currently, there is a lack of available data on the involvement of hereditary genetic mutations in breast cancer, specifically within the UAE [16]. Another study examined the prevalence of germline BRCA mutations in the UAE BC population and found a positive incidence of 130/309 (11.9%). Pathogenic and potentially pathogenic mutations were detected in 34.6% of patients. BRCA2 was the most prevalent gene discovered [13]. During the period between 2015 and 2017, a research study was conducted in the UAE involving 276 individuals diagnosed with hereditary cancer syndromes, with a majority of patients having breast cancer. The majority of participants were from the Middle East and Asia-Pacific regions. Out of the 276 patients, 24 individuals (8.7%) were found to have a pathogenic mutation in either the BRCA1 gene (13 cases), the BRCA2 gene (eight cases), or c.1100delC in the CHEK2 gene (three cases). Additionally, 30 patients (10.8%) had a variant of unknown significance (VUS) detected in one of the three genes [17].

The Arab genome project, initiated by Saudi Arabia and continued by Qatar, Kuwait, and the United Arab Emirates, aims to facilitate the identification of novel breast cancer biomarkers. These biomarkers have the potential to enhance the prognosis and enable targeted therapies for the disease. To improve the treatment and genetic counseling for Arab patients with breast cancer-associated genetic mutations, it is crucial to conduct comprehensive and well-controlled genetic epidemiological studies. These studies will provide accurate estimates of the frequency of genetic variants, including BRCA1/2 and non-BRCA1/2 mutations, among breast cancer patients in the Gulf States.

The National Arab Genome Project in the UAE aims to address this gap in knowledge by leveraging Next-Generation Sequencing (NGS) technology. This project aims to create a comprehensive catalog of mutations that are unique to the Arab population in the UAE. By collecting and analyzing this data, it seeks to enhance our understanding of the Arab genome. The primary objective is to compare the Arab genome with other ethnic groups' genomes, highlighting both similarities and differences. Such comparisons may provide insights into the genetic predisposition to breast cancer within the UAE population [18].

Conversely, there is a lack of available data concerning somatic genetic mutations specific to breast cancer in Arab women.

The only study to date addressing this is a cross-sectional analysis of Arab breast cancer patients who were diagnosed at a single facility between 2000 and 2018 and underwent Ampliseq 46-Gene or 50-Gene next-generation sequencing. It included 78 Arab women, and the somatic mutation rates were discovered by next-generation sequencing to be: NPM1, 2.5%; MPL, 1.3%; JAK2, 2.5%; KIT, 7.7%; KRAS, 3.8%; and NRAS, 3.8%. TP53, 23.1%; ATM, 2.6%; IDH1, 2.6%; IDH2, 3.8%; PTEN, 7.7%; PIK3CA, 15.4%; and APC, 7.7%. This study implies potential differences from estimates for the Western population. These findings demand more extensive epidemiology research that takes into account the changing significance that these variants play in prognostication and individualized treatment [12].

26.4   Breast Cancer Screening Knowledge and Practice

The UAE's "National Guidelines for Breast Cancer Screening and Diagnosis" by the Ministry of Health and Prevention, 2022, advises beginning mammography at age 40 and continuing every 2 years.

Based on a cross-sectional survey conducted on 492 females in the UAE, aged 25 to 45, the level of awareness among respondents regarding basic breast cancer information, including risk factors, warning signs and symptoms, and screening practices, was lower than expected. The study attributed these findings to the inadequate involvement of physicians and health authorities in raising awareness. It recommended addressing this issue by implementing awareness campaigns to bridge knowledge gaps and actively engaging medical professionals in educating both patients and the general public [19]. Additional studies examining awareness of breast cancer and breast self-examination (BSE) have indicated that knowledge regarding risk factors, warning signs and symptoms, and the practice of BSE is relatively limited among individuals in the UAE [20].

Although women expressed a desire for increased year-round breast cancer awareness initiatives and improved screening accessibility, they generally conveyed positive attitudes towards breast cancer screening [21]. Therefore, it remains evident that there is a need to enhance women's awareness of breast cancer in the UAE, with the aim of promoting breast cancer screening. There is a necessity to improve the national screening program by enhancing accessibility and optimizing resource allocation to ensure its effectiveness and focus [22].

In general, there has been a rise in collaborative endeavors by the government, private sectors, nonprofit organizations, charities, and individuals to elevate awareness about breast cancer. These efforts have been particularly prominent during the month of October, recognized globally as Breast Cancer Awareness Month. Some examples of the effort to promote early detection and screening include the awareness campaign "Pink Caravan," organized walks for fundraising and awareness, free screenings, breast health checkups, etc.

26.5   Treatment Modalities

26.5.1 Surgery

Overall, in the UAE, there are a good number of well-trained breast surgeons, some of whom have been dedicated to oncoplastic training.

In the UAE, there is a rapid expansion of oncoplastic and reconstructive breast surgery, despite the limited availability of published techniques and outcomes. Over the past decade, oncoplastic lumpectomy procedures have become the established standard of care as needed. Initially, level I procedures primarily involved the removal of less than 20% of breast volume, allowing for glandular displacement to repair the defect. With time, level 2 treatments became more prevalent, permitting the resection of up to 50% of breast volume through operations such as mammoplasty reduction. These advancements have led to a reduction in positive margin rates and an increase in breast conservation rates. In collaboration with the American Society of Breast Surgeons, the UAE has hosted annual oncoplastic workshops with practical training over the past 5 years. In 2021, the UAE adopted the use of magnetic seed localization technology, a groundbreaking wire-free approach that assists radiologists and surgeons in locating breast abnormalities for tissue excision, to the author's knowledge.

However, based on our observations, we have noticed that total mastectomies, and at times even double mastectomies, are still being performed more frequently than necessary. This trend is more evident among general surgeons who lack specialized oncology training or experience, and particularly when these procedures are carried out against the recommendations of multidisciplinary teams (MDT). To address this issue, potential solutions include education and training initiatives, streamlining reimbursement processes, and implementing mandatory MDT consultations prior to surgical interventions.

In the past, women who anticipated postoperative radiation treatment before 2009 did not undergo immediate reconstruction following mastectomy. The primary approach for delayed reconstruction during that period was the transverse rectus abdominis myocutaneous (TRAM) flap. However, after 2009, instant two-stage breast reconstruction involving an expander-implant technique became available for women scheduled to receive postoperative radiation, to the author's knowledge. As the expansion of tissue expanders caused discomfort for irradiated women, this procedure was swiftly replaced by rapid direct-to-implant reconstruction performed in a single step. The introduction of synthetic or biological advancements in surgical meshes facilitated the shift from two-stage to one-stage reconstruction. Previously, subpectoral implant placement was the standard, but there has been an increasing preference for pre-pectoral insertion. These advancements in breast surgery have significantly reduced the need for delayed reconstruction using autologous flaps. Nonetheless, certain patients may still require delayed reconstruction, and specialized clinics in the UAE are proficient in performing autologous flaps, such as the deep inferior epigastric perforator (DIEP) flap.

26.5.2 Radiotherapy

Radiotherapy plays a crucial role in the comprehensive management of breast cancer, whether it is administered in the early-stage setting after breast-conserving surgery, for postmastectomy radiotherapy (PMRT), or in metastatic scenarios. In the UAE, radiation oncologists are frequently involved in the initial treatment planning for breast cancer patients, often as part of a multidisciplinary team (MDT). However, there are instances in certain locations where their consultation occurs later in the treatment process [23]. Numerous radiation facilities have been established throughout the United Arab Emirates, with additional centers announced for the future. These facilities offer advanced and state-of-the-art care for various types of cancer, including breast cancer [22].

Radiotherapy in the United Arab Emirates has experienced substantial advancements over the last two decades, leading to improved treatment delivery and a reduction in both early and potential long-term side effects. Notably, the current recognized standard of care involves the utilization of 3D-conformal computed tomography-based radiation planning and dosage computation, replacing the traditional 2D X-ray film approach. Forward planning, field-in-field intensity modulated radiation (IMRT), image-guided radiotherapy (IGRT), and volumetric arc radiotherapy (VMAT) are now commonly employed in radiotherapy centers throughout the country. Additionally, sophisticated systems for patient mobility and tracking, such as deep inspiration breath hold (DIBH) and active breath control (ABC) treatment, have proven effective in significantly reducing radiation exposure to the heart, coronary veins, and lungs [24]. This is crucial due to the relatively young age of the breast cancer population in the UAE and the high prevalence of comorbid conditions like diabetes and cardiovascular disease.

The COVID-19 epidemic has hastened the adoption of mild hypofractionated therapy for early-stage breast cancer after breast-conserving surgery [25, 26], as a result of recent technological advancements and level-I evidence data.

Conversely, there is limited utilization of extreme hypofractionated treatment, partial breast radiation, and intraoperative radiotherapy. This may be attributed to patient demographic characteristics and the stage of cancer at the time of diagnosis.

In the context of PMRT, the most frequent fractionation employed in the UAE remains the conventional 5–6 weeks of 2 Gy per fraction; mild hypofractionation is rarely used despite the outcomes of the Chinese hypofractionation trial [27], but concerns surrounding the lymphedema rate have recently been addressed [27].

In metastatic BC, radiation is mostly administered for symptom treatment in a palliative context. Recent years have seen the availability and standardization of stereotactic ablative body radiation (SABR), particularly for individuals with oligometastatic illness [28]. Stereotactic radiosurgery or stereotactic radiation is frequently administered for brain metastases and is preferred, if possible, prior to whole-brain radiotherapy or hippocampal-sparing whole-brain radiotherapy, if required.

26.5.3 Palliative and Supportive Care

Palliative care, which prioritizes the needs of the patient rather than the specific diagnosis, varies in its level of input and involvement depending on the stage of breast cancer. In the UAE, multidisciplinary treatment teams (MDTs) often recommend and consider palliative mastectomy and radiation for the management of fungating chest wall cancers [29]. The recently published guidelines for palliative care by the Department of Breast Medical Oncology at the MD Anderson Cancer Center were developed to specifically address various symptoms, aiming to support oncologists in delivering more customized treatment options [30]. Special attention is given to addressing the management of various symptoms, including pain, breathlessness, fatigue, distress, anxiety, exercise, nutrition, and advance care planning.

These recommendations recognize the global shortage of palliative care profession-als, which is further confirmed by our experience in the UAE. When the initial evaluation highlighting the urgent requirement for palliative care in the UAE was published in 2018, only two hospitals (Tawam Hospital in Abu Dhabi and The American Hospital in Dubai) offered palliative care services [31]. Presently, Burjeel Medical City in Abu Dhabi, Mediclinic City Hospital in Dubai, and soon the Cleveland Clinic Abu Dhabi will also offer palliative care services, either in an inpatient or clinic setting. As these relatively recent additions to the healthcare workforce develop closer collaborations with their oncology counterparts, the early integration of palliative care becomes more feasible. This integration holds the potential to enhance outcomes and improve the quality of life for breast cancer patients and their families. Additionally, regulatory authorities are actively address-ing obstacles to ensure the availability of opioid analgesics, thus promoting better pain management in palliative care settings.

26.5.4 Survivorship Programs

The only study conducted on breast cancer survival in the UAE involved a retro-spective analysis of 988 patients from a single institution, with a follow-up period of 35 months. The study projected 2-year and 5-year survival rates of 97% and 89%, respectively, which were similar to those observed in Western countries like Australia (89.5%) and Canada (88.2%) during the same timeframe [14, 32]. The 5-year survival rate in the UAE is notably impressive when compared to other coun-tries in the same region, such as Qatar (71.95%) and Kuwait (75.2%) [32].

The care provided to cancer survivors is a crucial but often overlooked com-ponent of cancer treatment. It addresses both short-term and long-term complica-tions of treatment, the risk of cancer recurrence, the potential for developing second primary malignancies, adherence to prescribed adjuvant hormonal thera-pies, and recommended lifestyle adjustments such as weight management, physi-cal activity, and exercise [33]. Breast cancer survivorship services and established programs are still in their early stages of development in the UAE, highlighting a significant gap in meeting the needs of survivors. Furthermore, existing pro-grams should share their experiences and challenges to contribute to the develop-ment of best practices in this field.

26.5.5 Social Support Programs

Support groups for cancer patients and their families play a vital role in helping individuals cope with their condition. These groups offer valuable resources such as knowledge, comfort, coping strategies, anxiety reduction, and a platform for shar-ing similar experiences and receiving emotional support. In the UAE, there are sev-eral breast cancer support groups available, including The Cancer Patient Care Society—Rahma, Friends of Cancer, Breast Friends, Moazzara—Emirates

Association for Cancer Support, The Cancer Majlis, Bosom Buddies, and Pure Heart 4 Cancer. Some of these organizations also provide financial assistance for cancer treatment. Additionally, there has been a significant increase in the utilization of social media platforms and WhatsApp support groups, facilitating virtual connections between patients and healthcare providers. The primary challenge lies in the lack of coordination among different support groups, but we believe that collaboration will greatly enhance the impact and effectiveness of these supportive initiatives.

26.5.6 Accreditation of Oncology Centers

The accreditation of oncology practices involves an impartial assessment conducted by external peers to evaluate and enhance the quality of treatment. It serves as a means to promote practice improvement, advance research, teaching, and clinical practice, and establish criteria for the development of services. Moreover, accreditation increases public confidence in the organization. Burjeel Medical City in Abu Dhabi is the first facility in the UAE to receive approval from the European Society of Medical Oncology as an integrated oncology and palliative care center, a distinction it has held since 2021. Additionally, Tawam Hospital has been accredited as a breast center by the Joint Commission International since 2017, followed by Al Zahra Hospital in 2020 and Mediclinic City Hospitals in 2022.

26.5.7 Use of Predictive and Prognostic Genomic Score Testing

The Amsterdam 70-gene profile (MammaPrint) and the Oncotype DX 21-gene (more prevalent in the UAE) recurrence scores are essential in generating a predictive and/ or prognostic signature for some cases to decide the benefit of adjuvant chemotherapy in BC [34, 35]. In a particular study, the Oncotype DX recurrence score assessment had a clear impact on the recommendations for adjuvant treatment [36]. Few insurance companies cover the cost of some of these tests; otherwise, the Emirates Oncology Society urged regulators that these tests be covered in the appropriate clinical environment, given their high price. Recommendations are based on additional clinicopathologic criteria and recurrence estimating techniques (e.g., www.breastrecurrenceestimator.onc.jhmi.edu) when recurrence score testing is not possible.

26.5.8 Cost and Monetary Burden of Cancer Treatments

There is limited available data regarding the cost of breast cancer anticancer therapy in the UAE. The only existing report evaluated the top 20 cancer medications utilized in Abu Dhabi in 2011, which revealed that breast cancer accounted for a significant portion of drug expenditures (8 out of the top 20). The introduction of newer drugs and innovative therapies is expected to contribute to increased expenses for cancer treatment in the United States. To mitigate some of the rising costs, there is a need for greater utilization of biosimilars. While there has been partial adoption of our recommendation to incorporate biosimilars in the formularies of cancer centers in the UAE, further improvements are necessary [37].

26.6   Breast Cancer Research in the UAE

We systematically conducted an extensive literature search to identify articles related to breast cancer in the UAE. On August 8, 2021, we performed a PubMed search using specific terms such as "breast" AND "Cancer* OR oncology* OR malignant* OR tumor OR tumour" AND "emirates OR UAE." This search yielded a total of 203 journal articles authored by individuals from the UAE, with the earliest publication dating back to 2001. The majority of studies fell into the categories of fundamental science or translational research (45.8%) or observational studies (26.1%). Non-data-driven publications, such as reviews, consensus statements, and editorials, accounted for 40 articles (19%). Interestingly, we found only six clinical studies (as shown in Fig. 26.3). It is worth noting that out of the 163 publications with data-driven content, only 62 (38%) were conducted within the UAE. The remaining research was carried out abroad, with many authors having affiliations both in the UAE and international institutions, indicating a significant level of international collaboration [38]. Over time, there has been an overall increase in breast cancer research output, despite the persistence of significant gaps in evidence. The growth of academic institutions and research programs dedicated to molecular and cellular research has contributed to the expansion of fundamental and translational research in the UAE. Observational studies have primarily focused on screening and fundamental epidemiological aspects. Moving forward, it is crucial to prioritize the expansion of national registries and the longitudinal collection of clinically relevant factors. These efforts can provide valuable insights into the clinical and molecular profiles of breast cancer in the country, as well as inform survival measures that can impact management decisions. It is worth noting that there is a noticeable lack of therapeutic clinical studies, which is a shared concern in the broader region.

Fig. 26.3 Breast cancer journal publications from the UAE (2001–2021) by study type

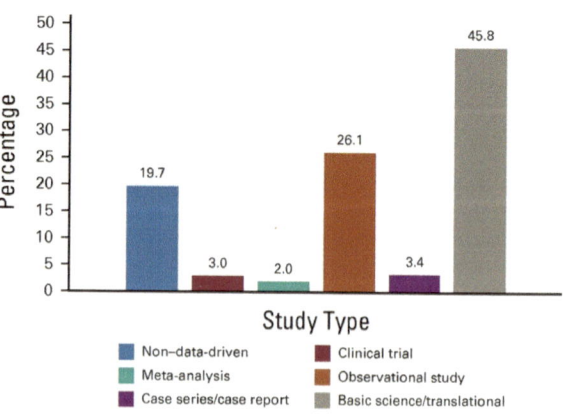

Previous efforts to initiate large-scale randomized studies have been limited, primarily due to challenges in recruiting participants. This highlights the importance of increasing public and healthcare provider awareness regarding the significance of clinical trials in facilitating access to innovative cancer therapies.

As the regulatory and resource infrastructures in the country are well-developed and capable of meeting international standards and requirements, there are several ongoing initiatives to form partnerships with sponsors of clinical trials. These endeavors aim to facilitate the implementation of interventional studies within the country.

26.7   Hereditary Predisposition of Breast Cancer in the UAE

Based on the existing statistics, consanguinity may be associated with a decreased occurrence of breast cancer [39–41]. In Arab women, the occurrence of parental consanguinity may contribute to a lower incidence of breast cancer. This could be attributed to a reduced gene conflict resulting from a higher prevalence of first cousin marriages, particularly those involving the father's brother's son. Additionally, the protective effect of homozygosity for an unknown gene or genes related to breast cancer may further contribute to this phenomenon [39]. In a study conducted among the population of the United Arab Emirates, which compared cases of breast cancer to a control group, it was found that parental consanguinity was associated with a decreased risk of breast cancer in younger women below the age of 50 [41]. In a separate study conducted in the UAE, computer simulations were employed to investigate the impact of mating patterns on the carrier rate of BRCA1/2 mutations over 40 generations. The findings indicated that in a predominantly consanguineous population, the carrier rate of BRCA1/2 mutations decreased at an average rate of 0.022% per 25 years. This decrease occurred six times faster compared to a non-consanguineous population [42].

26.8   Conclusion

Breast cancer (BC) is highly prevalent in the UAE and ranks among the leading causes of cancer-related mortality, posing significant challenges to the healthcare system. The increasing incidence of breast cancer diagnoses can be attributed to both population growth and improved data collection through registries and screening programs. Notably, breast cancer cases in the UAE are commonly observed in individuals in their 40s, which aligns with the relatively young population of the country compared to other developed regions like the United States and the European Union. However, there is a scarcity of studies examining the clinical and genetic risk factors specific to the UAE population, warranting further research in this area. Despite positive attitudes among women, breast cancer screening and general awareness remain insufficient, highlighting the need for increased efforts to improve education and screening programs. The UAE offers management options for breast cancer, including palliative care and survivorship initiatives, which are similar to those available in Western nations. However, there is a noticeable lack of therapeutic clinical trials, although progress is being made in the academic and regulatory sectors to address this gap.

Conflict of Interest The authors have no conflict of interest to declare.

References

1. Sung H, et al. Global cancer statistics 2020: GLOBOCAN estimates of incidence and mortality worldwide for 36 cancers in 185 countries. Cancer J Clin. 2021;71(3):209–49.
2. Cancer incidence in United Arab Emirates, Annual Report of the UAE – National Cancer Registry. Statistics and Research Center, Ministry of Health and Prevention. 2021. Accessed 28 Mar 2024.
3. Bendardaf R, et al. Incidence and Clinicopathological features of breast cancer in the northern emirates: experience from Sharjah breast care center. Int J Women's Health. 2020;12:893–9.
4. Siegel RL, Miller KD, Jemal A. Cancer statistics, 2017. CA Cancer J Clin. 2017;67(1):7–30.
5. Sasco AJ, Lowenfels AB, Pasker-de Jong P. Review article: epidemiology of male breast cancer. A meta-analysis of published case-control studies and discussion of selected aetiological factors. Int J Cancer. 1993;53(4):538–49.
6. Hashim MJ, et al. Burden of breast cancer in the Arab world: findings from global burden of disease, 2016. J Epidemiol Glob Health. 2018;8(1–2):54–8.
7. Donnelly TT, et al. Beliefs and attitudes about breast cancer and screening practices among Arab women living in Qatar: a cross-sectional study. BMC Womens Health. 2013;13:49.
8. Najjar H, Easson A. Age at diagnosis of breast cancer in Arab nations. Int J Surg. 2010;8(6):448–52.
9. Al-Shamsi HO, Alrawi SJ. Breast cancer screening in The United Arab Emirates: is it time to call for a screening at an earlier age? J Cancer Prev Curr Res. 2018;9:123–6.
10. Source: Federal Competitiveness and Statistics Centre 2022. https://fcsc.gov.ae/en-us/Pages/Statistics/Statistics-by-Subject.aspx. Accessed 2 May 2022.
11. Chouchane L, Boussen H, Sastry KS. Breast cancer in Arab populations: molecular characteristics and disease management implications. Lancet Oncol. 2013;14(10):e417–24.
12. Al-Shamsi HO, et al. Molecular spectra and frequency patterns of somatic mutations in Arab women with breast cancer. Oncologist. 2021;26(11):e2086–9.
13. Altinoz A, et al. Clinicopathological characteristics of gene-positive breast cancer in The United Arab Emirates. Breast. 2020;53:119–24.
14. Elobaid Y, et al. Breast cancer presentation delays among Arab and national women in the UAE: a qualitative study. SSM Popul Health. 2016;2:155–63.
15. Al-Awadhi A, et al. Abstract P1-23-02: incidence, clinicopathological features and treatment patterns of male breast cancer (MBC) in a high-volume cancer center in The United Arab Emirates (UAE). Cancer Res. 2022;82(4_Supplement):P1-23-02-P1-23-02.
16. Rahman S, Zayed H. Breast cancer in the GCC countries: a focus on BRCA1/2 and non-BRCA1/2 genes. Gene. 2018;668:73–6.
17. Dawood SS, et al. Analysis of hereditary cancer syndromes in patients from The United Arab Emirates. J Clin Oncol. 2018;36(15_suppl):e13509–9.
18. Al-Ali M, et al. A 1000 Arab genome project to study the Emirati population. J Hum Genet. 2018;63(4):533–6.
19. Younis M, et al. Knowledge and awareness of breast cancer among young women in The United Arab Emirates. Adv Breast Cancer Res. 2016;05:163–76.
20. Hegde P, et al. Breast cancer risk factor awareness and utilization of screening program: a cross-sectional study among women in the northern emirates. Gulf J Oncolog. 2018;1(27):24–30.

21. Abu Awwad D, et al. Women's breast cancer knowledge and health communication in The United Arab Emirates. Healthcare. 2020;8(4):495.
22. Al-Shamsi H, et al. The state of cancer Care in The United Arab Emirates in 2020: challenges and recommendations, a report by The United Arab Emirates oncology task force. Gulf J Oncolog. 2020;1(32):71–87.
23. Shah C, Al-Hilli Z, Vicini F. Advances in breast cancer radiotherapy: implications for current and future practice. JCO Oncol Pract. 2021;17(12):697–706.
24. Nissen HD, Appelt AL. Improved heart, lung and target dose with deep inspiration breath hold in a large clinical series of breast cancer patients. Radiother Oncol. 2013;106(1):28–32.
25. Whelan TJ, et al. Long-term results of hypofractionated radiation therapy for breast cancer. N Engl J Med. 2010;362(6):513–20.
26. Haviland JS, et al. The UK standardisation of breast radiotherapy (START) trials of radiotherapy hypofractionation for treatment of early breast cancer: 10-year follow-up results of two randomised controlled trials. Lancet Oncol. 2013;14(11):1086–94.
27. Wang SL, et al. Hypofractionated versus conventional fractionated postmastectomy radiotherapy for patients with high-risk breast cancer: a randomised, non-inferiority, open-label, phase 3 trial. Lancet Oncol. 2019;20(3):352–60.
28. Palma DA, et al. Stereotactic ablative radiotherapy for the comprehensive treatment of Oligometastatic cancers: long-term results of the SABR-COMET phase II randomized trial. J Clin Oncol. 2020;38(25):2830–8.
29. Cherny NI, Paluch-Shimon S, Berner-Wygoda Y. Palliative care: needs of advanced breast cancer patients. Breast Cancer (Dove Med Press). 2018;10:231–43.
30. Kida K, et al. Optimal supportive Care for Patients with Metastatic Breast Cancer According to their disease progression phase. JCO Oncol Pract. 2021;17(4):177–83.
31. Al-Shamsi HO, Tareen M. Palliative Care in The United Arab Emirates, a desperate need. Palliat Med Care. 2018;5(3):1–4.
32. Allemani C, et al. Global surveillance of cancer survival 1995-2009: analysis of individual data for 25,676,887 patients from 279 population-based registries in 67 countries (CONCORD-2). Lancet. 2015;385(9972):977–1010.
33. Nardin S, et al. Breast cancer survivorship, quality of life, and late toxicities. Front Oncol. 2020;10:864.
34. Cognetti F, Naso G. The clinician's perspective on the 21-gene assay in early breast cancer. Oncotarget. 2021;12(26):2514–30.
35. Lee YJ, et al. A nomogram for predicting probability of low risk of MammaPrint results in women with clinically high-risk breast cancer. Sci Rep. 2021;11(1):23509.
36. Jaafar H, et al. Impact of Oncotype DX testing on adjuvant treatment decisions in patients with early breast cancer: a single-center study in The United Arab Emirates. Asia Pac J Clin Oncol. 2014;10(4):354–60.
37. The National Guidelines for Breast Cancer Screening and diagnosis. United Arab Emirates, Ministry of Health & Prevention; 2014.
38. Lewison G, et al. Cancer research in the 57 organisation of Islamic cooperation (OIC) countries, 2008-17. Ecancermedicalscience. 2020;14:1094.
39. Denic S, Agarwal MM. Breast cancer protection by genomic imprinting in close kin families. BMC Med Genet. 2017;18(1):136.
40. Medimegh I, et al. Consanguinity protecting effect against breast cancer among Tunisian women: analysis of BRCA1 haplotypes. Asian Pac J Cancer Prev. 2015;16(9):4051–5.
41. Denic S, Bener A. Consanguinity decreases risk of breast cancer—cervical cancer unaffected. Br J Cancer. 2001;85(11):1675–9.
42. Denic S, Al-Gazali L. Breast cancer, consanguinity, and lethal tumor genes: simulation of BRCA1/2 prevalence over 40 generations. Int J Mol Med. 2002;10(6):713–9.

Dr. Aydah Al-Awadhi is a consultant medical oncologist who graduated from medical school at the United Arab Emirates University in 2012, then completed her internal medicine residency in the United States and completed her medical oncology and hematology fellowship at the University of Texas, MD Anderson Cancer Center, in 2019. She is American Board Certified in Internal Medicine, Hematology, and Medical Oncology. She was awarded the Emirates Oncology Society (EOS) Member of the Year award for 2021 and the Emirates Oncology Society Women in Oncology of the Year award. She is also currently the scientific chairperson of the Emirates Oncology Society and the chair of the breast cancer working group. Dr. Al-Awadhi has published many articles and book chapters on breast cancer and oncology, with main expertise in breast cancer and sarcoma.

In 2022, he was awarded the prestigious Feigenbaum Leadership Excellence Awards from Sheikh Hamdan Smart University for his exceptional leadership and research, as well as the Sharjah Award for Volunteering. He was also named the Researcher of the Year in the UAE in 2020 and 2021 by the Emirates Oncology Society.

Ms. Faryal Iqbal is the research associate at Burjeel Medical City, Abu Dhabi, United Arab Emirates. She completed her undergraduate studies in molecular biology and biotechnology. Following that, she received a postgraduate qualification in molecular genetics. She co-edited "Cancer in the Arab World," the first extensive book covering cancer care across all Arab countries. The book succeeded significantly, with over 450,000 downloads within just two years. She has several peer-reviewed papers under her name. In September 2023, Ms. Iqbal received the "EOS Research Award" from the prestigious Emirates Oncology Society for her research efforts. Moreover, she assists in different aspects of clinical trial implementation at a research site. Her research interests and publications encompass oncology, hematology, and genetics.

Dr. Hampig Raphael Kourie is an assistant professor in Hematology-oncology department, Faculty of Medicine, Saint Joseph University, Beirut, Lebanon.

Hampig Raphael Kourie gained his medical doctor degree from the Faculty of Medicine of Saint Joseph University in Beirut Lebanon in 2010. He started his fellowship in Hôtel-Dieu de France, Saint Joseph University hospital in Beirut, Lebanon and he continued his medical oncology fellowship in Jules Bordet Institute in Brussels, Belgium from 2014 to 2016. Since September 2016, he worked in the digestive oncology department in Hôpital Européen Georges Pompidou (HEGP) in Paris as a researcher in the Association des Gastro-entérologues Oncologues en France (AGEO). He is also certified from Paris Diderot and Paris Descartes universities in hereditary cancers and digestive oncology. He gained his PhD in genetics from Saint Joseph University of Beirut in 2023 and his MEMS from ESA business school in

2023. He is practicing as hematologist-oncologist in Hôtel Dieu de France, Hôpital Saint Joseph des Soeurs de la Croix and Bellevue Medical Center.

He is actually an ESMO Faculty in the colorectal cancer group. He founded and directed the Middle East Biomarkers course and the Middle East and North Africa GI Oncology Summit. He has more than 200 peer-reviewed articles, mainly in immunotherapy, oncogenomics and digestive oncology fields.

Prof. Humaid Obaid Al-Shamsi is the Chief Executive Officer of Burjeel Cancer Institute in Abu Dhabi, UAE, President of the Emirates Oncology Society, Lead of the Gulf Cancer Society, Full Professor of Oncology at the Ras Al Khaimah Medical and Health Sciences University, Ras Al Khaimah, UAE, and an Adjunct Professor of Oncology at the College of Medicine, University of Sharjah. He is the first Emirati to be promoted as a professor in oncology in the UAE. He is also the Chairman for Colorectal Cancer in the MENA region, appointed by the prestigious National Comprehensive Cancer Network®. He is also the only member of Lung Cancer Policy Network in the MENA region that aims to advance lung cancer research and screening globally. He is the Chairman of the Oncology and Hematology Fellowship Training Program for the National Institute for Health Specialties in the United Arab Emirates. He is the only member in GCC in the WIN Consortium which is comprised of organizations representing all stakeholders in personalized cancer medicine globally.

He is board-certified in both internal medicine and oncology from the UK, USA (ABIM), and Canada (FRCPC). He has also been awarded the FRCP (London) in 2023 and FRCP (Glasgow) in 2024. He is the only physician in the UAE with a subspecialty fellowship certification and training in gastrointestinal oncology and the first Emirati to train and complete a clinical post-doctoral fellowship in palliative care. He was an assistant professor at the University of Texas MD Anderson Cancer Center between 2014 and 2017. He has published more than 140 peer-reviewed articles in JAMA Oncology, Lancet Oncology, The Oncologist, BMC Cancer, and many others. His area of expertise includes precision oncology and cancer care in the UAE. In 2016, he published with his group from MD Anderson the JCO paper describing a new distinct subgroup of CRC, NON V600 BRAF-mutated CRC. In 2022, he published the first book about cancer research in the UAE and also the first book about cancer in the Arab world, both of which were launched at Dubai Expo 2020. *Cancer in the Arab World* has been downloaded more than 450,000 times in its first 18 months of publication and is the ultimate source of cancer data in the Arab region. He also published the first comprehensive book about cancer care in the UAE which is the first book in UAE history to document the cancer care in the UAE with many topics addressed for the first time, e.g., neuroendocrine tumors in the UAE. He is passionate about advancing cancer care in the

UAE and the GCC and has made significant contributions to cancer awareness and early detection for the public using social media platforms. He is considered as the most followed oncologist in the world with over 300,000 subscribers across his social media platforms (Instagram, Twitter, LinkedIn, and TikTok). In 2022, he was awarded the prestigious Feigenbaum Leadership Excellence Award from Sheikh Hamdan Smart University for his exceptional leadership and research and the Sharjah Award for Volunteering. He was also named the Researcher of the Year in the UAE in 2020 and 2021 by the Emirates Oncology Society.

In May 2024, HH Sheikh Mansour bin Zayed Al Nahyan, Vice President of the United Arab Emirates, awarded him the first place in UAE Nafis program for outstanding leadership in private sector across all business and medical disciplines. Beside his clinical and administrative duties, he is engaged in education and various levels of research training for medical trainees to enhance their clinical and research skills. His mission is to advance cancer care in the UAE and the MENA region and make cancer care accessible to everyone in need around the globe.

Humaid O. Al-Shamsi ⓘ, Faryal Iqbal ⓘ, Hampig R. Kourie,
Adhari Al Zaabi ⓘ, Amin M. Abyad,
and Nadia Abdelwahed ⓘ

H. O. Al-Shamsi (✉)
Burjeel Cancer Institute, Burjeel Medical City, Burjeel Holdings, Abu Dhabi,
United Arab Emirates

Ras Al Khaimah Medical and Health Sciences University, Ras Al Khaimah, United Arab Emirates

Gulf Medical University, Ajman, United Arab Emirates

Emirates Oncology Society, Emirates Medical Association, Dubai, United Arab Emirates

College of Medicine, University of Sharjah, Sharjah, United Arab Emirates

Gulf Cancer Society, Alsafa, Kuwait
e-mail: alshamsi@burjeel.com; humaid.al-shamsi@medportal.ca

F. Iqbal
Burjeel Medical City, Abu Dhabi, United Arab Emirates
e-mail: faryal.iqbal@burjeelmedicalcity.com

H. R. Kourie
Hematology-Oncology Department, Faculty of Medicine, Saint Joseph University,
Beirut, Lebanon
e-mail: hampig.kourie@usj.edu.lb

A. A. Zaabi
College of Medicine and health Sciences, Al Seeb, Oman
e-mail: adhari@squ.edu.om

A. M. Abyad · N. Abdelwahed
Burjeel Medical City, Abu Dhabi, United Arab Emirates

Emirates Oncology Society, Emirates Medical Association, Dubai, United Arab Emirates
e-mail: nadia.abdelwahed@burjeelmedicalcity.com

© The Author(s) 2024
H. O. Al-Shamsi (ed.), *Cancer Care in the United Arab Emirates*,
https://doi.org/10.1007/978-981-99-6794-0_27

27.1   Introduction

Colorectal cancer (CRC) is currently the third leading cause of cancer-related death worldwide and the fourth most commonly diagnosed cancer, according to GLOBOCAN 2018 data. The incidence of CRC is steadily increasing, particularly in developing countries. This form of cancer, also known as colorectal adenocarcinoma, typically originates from the glandular epithelial cells in the large intestine. Cancers emerge as a result of a series of genetic or epigenetic mutations that give them a selective advantage [1, 2]. The hyper-proliferative cells that have an abnormal replication and survival boost induce benign adenomas that may evolve into carcinomas and metastasize over the years [1, 3].

Researchers and physicians should empower themselves with a robust understanding of CRC development patterns, genetic and environmental risk factors, and, lastly, the molecular transformation of CRC in such a way that leads to the prevention and treatment of this deadly neoplasm [1].

27.2   Epidemiology of Colorectal Cancer (CRC) in the UAE

27.2.1 Overall Cancer Incidence Rate in the UAE

In 2021, the UAE National Cancer Registry (UAE-NCR) recorded a total of 5830 newly diagnosed cancer cases, both malignant and in situ, affecting individuals of both genders across the UAE population. Of these, malignant cases comprised the majority with 5612 instances (96%), while in situ cases accounted for 218 instances (4%). The data revealed a higher incidence among females [3210 (55.1%)] compared to males [2620 (44.9%)], across all nationalities. Among the total cancer cases, UAE citizens accounted for 1493 (25.6%), while non-UAE citizens represented 4337 (74.3%). The overall crude incidence rate of cancer in 2021 cancer data was 60.5/100,000 for both genders [4].

27.2.2 Incidence of CRC in the UAE

In 2021, CRC was the third most commonly diagnosed malignancy among the UAE population for both genders, accounting for 532 out of 5612 anger cases. Non-UAE citizens in the UAE population represented a higher proportion of these cases, with 372 diagnoses (71.1%), while UAE citizens accounted for 160 cases (30.5%). CRC affected males more significantly, constituting 12.5% of male cancer cases compared to 7% of female cancer cases [4]. Figure 27.1 shows a clear predominance of colorectal cancer in males.

The data from UAE-NCR indicate a steady increase in the incidence of colorectal cancer over the last decade, with the number of cases rising from 377 in 2013 [females: 160; males: 217] to 532 cases in 2021 [females: 213; males: 319] (Fig. 27.2; Table 27.1) [4]. There are many factors to consider while contemplating this data, such as the exponential growth of the UAE population, the improved

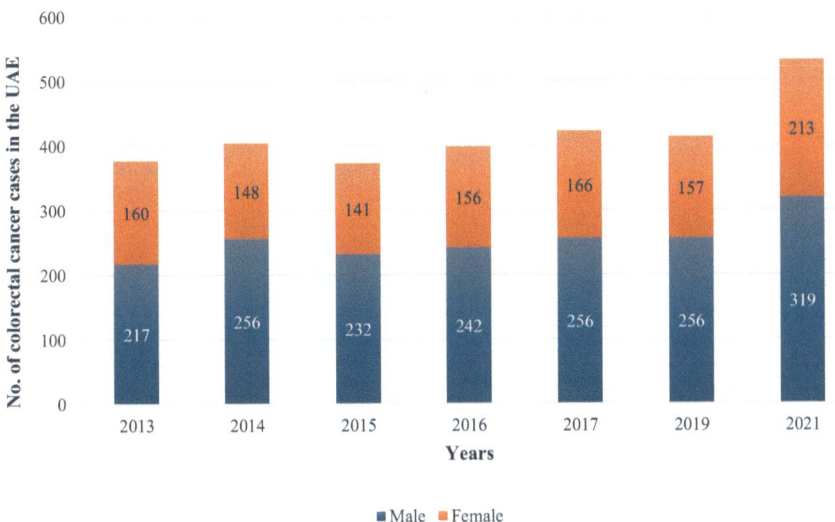

Fig. 27.1 Distribution of colorectal cancer cases by gender, 2013–2021. Source: Ministry of Health and Prevention, Statistics and Research Center, National Disease Registry—UAE National Cancer Registry Report, 2013–2021

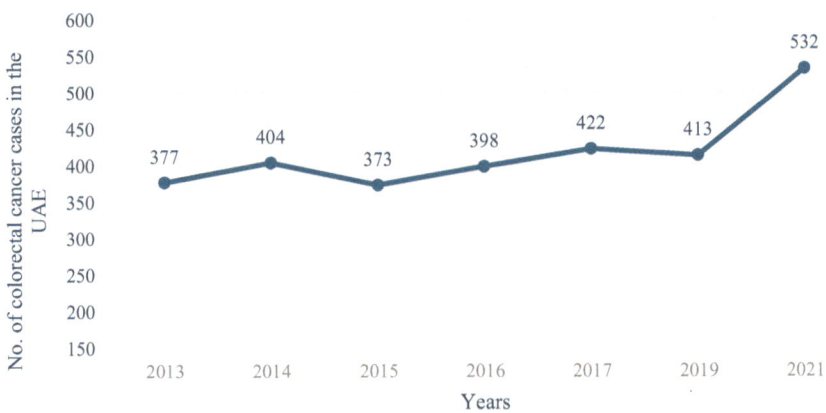

Fig. 27.2 Number of colorectal cancer cases in the UAE. Source: Ministry of Health and Prevention, Statistics and Research Center, National Disease Registry—UAE National Cancer Registry Report, 2013–2021

documentation system through UAE-NCR, and the improvement in population-wide screening and awareness initiatives across the country [5].

Currently, the UAE hosts approximately 200 nationalities who work and reside within the country. The largest non-UAE community hails from India, followed by Pakistan, Bangladesh, other Asian countries, Europe, and Africa [6]. This diversity is reflected in the recent CRC statistics, which show a higher incidence of CRC cases among non-UAE nationals compared to UAE nationals (Fig. 27.3) [4].

Table 27.1 Colorectal cancer demographics among the UAE population, 2013–2021

Year	UAE population (in millions)	Total malignant cases (in numbers)	CRC cancer cases (in numbers)	Percentage (%)	Crude rate CRC cancer cases per 100,000
2013	8.66	3574	377	10.55	–
2014	8.79	3610	404	11.19	4.45
2015	8.93	3744	373	9.96	4.1
2016	9.12	3982	398	9.99	–
2017	9.3	4123	422	10.24	4.5
2019	9.5	4381	413	9.43	–
2021	–	5612	532	9.47	5.7

Source: UAE population: https://fcsc.gov.ae/en-us/Pages/Statistics/Statistics-by-Subject.aspx#/%3Fsubject=Demography%20and%20Social&folder=Demography%20and%20Social/Population/Population

Ministry of Health and Prevention, Statistics and Research Center, National Disease Registry—UAE National Cancer Registry Report. 2013–2021

Fig. 27.3 Number of colorectal cancer cases among the UAE population by nationality, 2013–2021. Source: Ministry of Health and Prevention, Statistics and Research Center, National Disease Registry—UAE National Cancer Registry Report, 2013–2021

27.2.2.1   CRC Cases by Age

The age-standardized incidence rate (ASR) of CRC in the UAE in 2021 was 13/100,000 for males and 12.8/100,000 for females, and the crude incidence rate was 7.6/100,000 for females and 4.9/100,000 for males [4]. Figure 27.4 (a) represents the age group distribution of CRC cases in the UAE, while (b) shows the age-specific incidence rate for CRC cases in the UAE in 2021 [4].

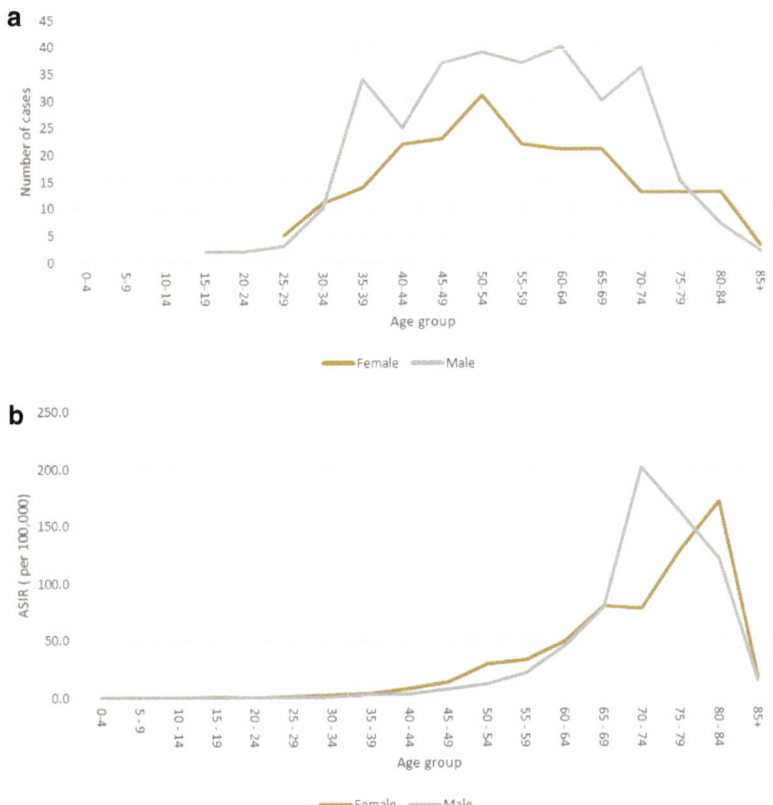

Fig. 27.4 (**a**) Age group distribution of colorectal cancer cases in the UAE in 2021. (**b**) Age-specific incidence rate (ASIR) for colorectal cancer cases in the UAE in 2021. Source: Ministry of Health and Prevention, Statistics and Research Center, National Disease Registry—UAE National Cancer Registry Report, 2021

27.2.3 Reported Stages of CRC Cases in the UAE

The development of colorectal cancer (CRC) typically begins with the noncancerous growth of mucosal epithelial cells, forming structures known as polyps. These polyps can grow very slowly over a period of 10–20 years before potentially becoming malignant [1, 7]. The most common type of polyp is an adenoma, which originates from glandular cells that produce mucus to line the large intestine. While the risk of cancer increases with the growth of these polyps, approximately 10% of adenomas will progress to invasive cancer. Polyps that become invasive are referred to as "adenocarcinomas," which account for 96% of all CRCs [1, 8].

Colorectal cancer arises from the wall of the colon or rectum and has the potential to invade lymphatic vessels or blood vessels, leading to metastasis to distant organs through nearby lymph nodes or the bloodstream. The staging of CRC is

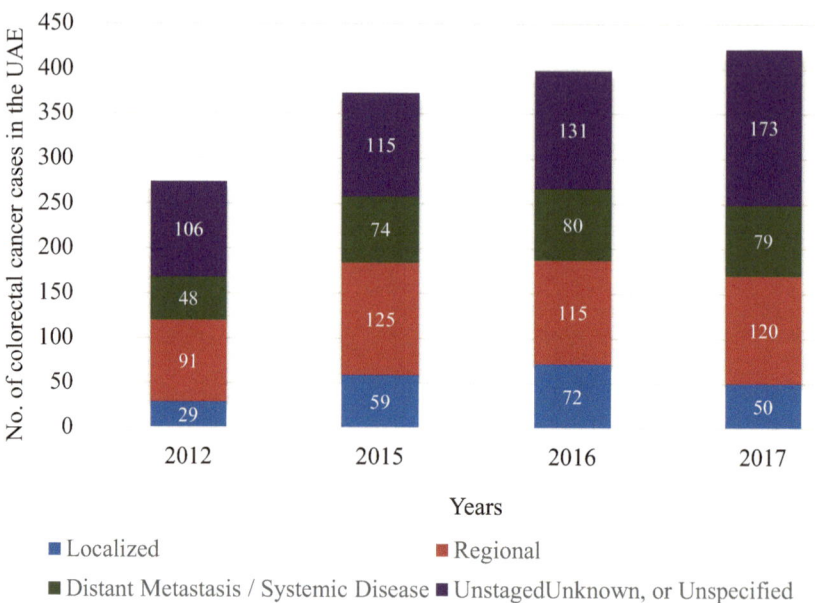

Fig. 27.5 Distribution of colorectal cancer cases by SEER stages in the UAE across the years. Source: Ministry of Health and Prevention, Statistics and Research Center, National Disease Registry—UAE National Cancer Registry Report, 2012–2017

determined by the extent of the invasion, which is crucial for diagnosis. Polyps that have not invaded the wall of the colon or rectum are classified as in situ cancers and are thus not reported as CRCs. Cancers that have penetrated the walls but have not spread beyond them are considered local cancers. In regional cancers, the surrounding lymph nodes or tissues are invaded. Distant cancers have metastasized via the blood stream to distant organs with capillary beds, such as the liver or the lung [1]. The distribution of CRC cases among the UAE population according to the Surveillance, Epidemiology, and End Results (SEER) stage over the years is illustrated in Fig. 27.5 [4].

27.2.4 Mortality Rates of CRC in the UAE

According to the UAE-NCR latest report, malignant neoplasm of the colon was the most lethal cancer, accounting for 11.49% of all cancer deaths, while malignant neoplasm of the rectum ranked the tenth, with 1.33% of cancer deaths [4]. The distribution of mortality from colorectal cancer over the previous years is detailed in Table 27.2.

27.2.5 Trends in the Incidence and Mortality of CRC in the UAE

The trend of CRC incidence and mortality due to malignant neoplasm of the colon and rectum for the years 2017, 2019, and 2021 through UAE-NCR was analyzed (Fig. 27.6) [4]. The chart shows a decrease in both CRC mortality and incidence

Table 27.2 Distribution of malignant colorectal cancer mortality cases in the UAE

Year	Underlying cause of death	Percentage (%)
2014	Malignant neoplasm of colorectal	8.8
2015	Malignant neoplasm of colorectal	10.6
2017	Malignant neoplasm of colon	10.3
	Malignant neoplasm of rectum	1.7
2019	Malignant neoplasm of colon	8.9
	Malignant neoplasm of rectum	1.1
2021	Malignant neoplasm of colon	11.49
	Malignant neoplasm of rectum	1.33

Source: Ministry of Health and Prevention, Statistics and Research Center, National Disease Registry—UAE National Cancer Registry Report, 2014–2021

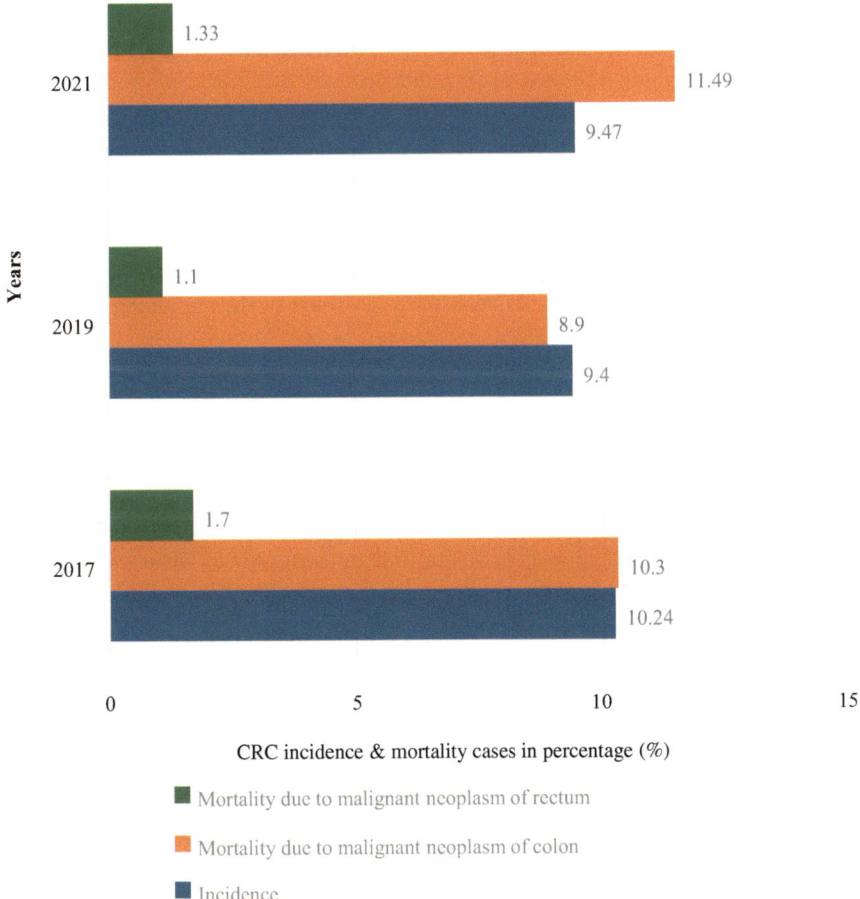

Fig. 27.6 The trend of incidence and mortality cases of CRC in 2017, 2019, and 2021. Source: Ministry of Health and Prevention, Statistics and Research Center, National Disease Registry—UAE National Cancer Registry Report, 2017, 2019, and 2021

among the UAE population. Although the drop is considered insignificant, it shows the country's considerable efforts for optimized screening and awareness in the UAE in order to bring the numbers down.

27.3   Early-Onset Colorectal Cancer in the UAE

Colorectal cancer that is diagnosed before the screening age, which is less than 50 years, is characterized as "Early Onset Colorectal Cancer" (EOCRC). According to the American Cancer Society (ACS), the adult onset of CRC incidence (>50 years) has decreased by 2% per year. Meanwhile, an increase in EOCRC cases of 2% per year has been observed [9, 10]. The predominance of young people at diagnosis, younger than the western population, was observed among all CRC reports across the Arab region. Around 17–38% of CRC patients were younger than 40 years old when diagnosed [10–12].

To date, there is an incomplete understanding of why there has been such an increase in the incidence rate of EOCRC. A crucial challenge in determining the components influencing the rise in the incidence rate of EOCRC is whether EOCRC and LOCRC are equivalent diseases or if there are any distinct underlying biological trails that interact with various risk factors for EOCRC. EOCRC was strongly believed to be closely associated with hereditary familial syndromes or genetic factors. Astonishingly, new evidence contradicts this belief. The discovery of genetic group assessment in a young CRC patient cohort revealed that germline genetic mutations were carried by only one in five of these patients, with approximately 25% having first-degree relatives with CRC. The bulk of the remaining patients are sporadic [10, 13, 14].

27.4   Risk Factors for CRC Development in the UAE

The development of CRC is influenced by both non-modifiable and modifiable risk factors. Non-modifiable factors include personal medical history (such as sex, age, race, history of adenomatous polyps, and inflammatory bowel disease) and family history, which individuals cannot control. On the other hand, modifiable factors are associated with individual habits and lifestyles. By changing these modifiable factors, individuals can reduce their risk of developing CRC [15].

There are several theories regarding the development of colorectal cancer and its association with red meat. One theory suggests that gut microbiomes might play a role in this relationship, impacting the connection between CRC and diet and influencing how red meat affects the progression of CRC [16, 17]. A study found that the disability-adjusted life years (DALYs) per 100,000 and age-standardized mortality rates (ASMRs) associated with diets high in red meat, low in calcium, milk, fiber, and whole grains were higher in men from Palestine, the United Arab Emirates, Jordan, Lebanon, Turkey, and Bahrain compared to the other countries. Similarly, for women, the DALYs per 100,000 and ASMRs related to such diets and their impact on the CRC were higher in Palestine, the United Arab Emirates, Jordan, Lebanon, Qatar, Libya, Afghanistan, Turkey, and Bahrain compared to the other countries [18].

Environmental factors and dietary habits in this region might also increase the risk of CRC. A meta-analysis identified smoking as a significant risk factor for CRC in the Eastern Mediterranean Region (EMRO) region [19, 20].

A large retrospective analysis revealed that a history of Helicobacter pylori infection was modestly but statistically significantly associated with an increased risk of CRC, including fatal cases [21].

The increasing incidence of colorectal cancer, particularly among younger age groups, is largely due to dietary and lifestyle changes. These changes include the adoption of a Westernized diet, higher consumption of animal-source foods, excess body weight, sedentary lifestyles, increased alcohol consumption, smoking, and increased intake of red and processed meats. These shifts are linked to the ongoing socioeconomic development in several Middle Eastern countries [22, 23].

Understanding these risk factors can help in developing comprehensive public health strategies and personalized interventions to reduce CRC incidence in the UAE.

27.5 Screening for CRC in the UAE

Screening is widely recognized as the most effective measure to reduce cancer incidence and mortality rates [24]. The development of CRC can be prevented by detecting and removing precancerous lesions through regular screening [25]. The UAE governmental healthcare system has announced the revised recommendations for performing early screening for CRC in the UAE. The UAE national cancer registry in 2021 has observed the most common CRC cases in males and the third among females [4]. MOHAP has released colorectal cancer screening guidelines in 2023, and they apply to all healthcare providers (facilities and professionals) in the United Arab Emirates providing CRC screening services, including mobile units.

Screening tests for individuals at average risk of colorectal cancer, as specified in Fig. 27.7, are colonoscopy, every 10 years or fecal immunochemical test (FIT), every year. The eligible population must be offered colonoscopy screening as per Fig. 27.7; in case of refusal, the patient should be offered a FIT. Detailed guidelines are given in Appendix O: The National Guideline for Colorectal Cancer Screening and Diagnosis in this book [26].

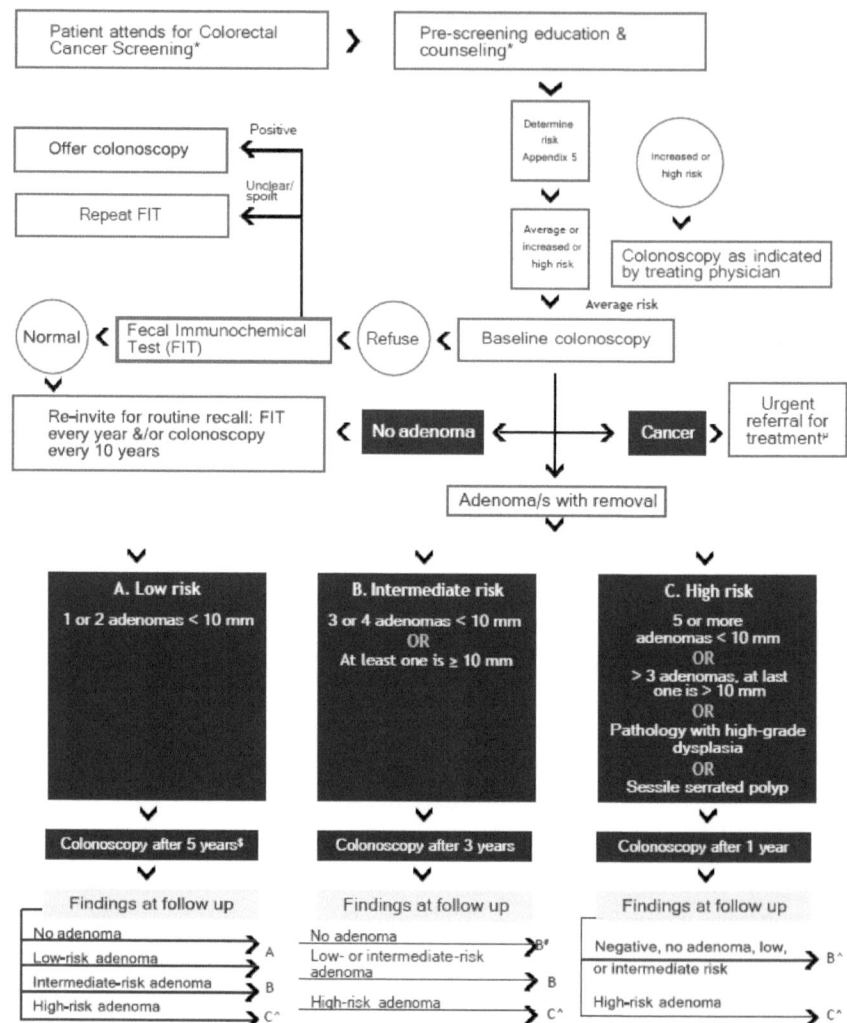

Fig. 27.7 Colorectal cancer screening and diagnosis pathway [26]. *Physician consultation: New patient or existing patient identified during visit for other purpose. ᵘUrgent referral to oncology center within 2 weeks. $Consider age, comorbidity, family history accuracy, and completeness of examination high-risk adenoma C^. #Stop surveillance if there is a further negative result (no adenoma). ^All histopathologically diagnosed cancers should be treated as per colon cancer guidelines. Source: The national guideline for colorectal cancer screening and diagnosis–2023–Ministry of Health and Prevention, UAE

27.6 Conclusion

One of the most common malignancies among Arabs is colorectal cancer (CRC), whose annual incidence rate is sharply rising. Hence, the UAE is also bearing a greater burden from the CRC. The updated guidelines for doing early CRC screening in the United Arab Emirates have been released by the Ministry of Health and Prevention, UAE, in 2023. The UAE should undertake CRC prevention programs, and all infrastructure and resources should be directed toward offering complete cancer care at every stage of the illness. To alleviate the burden of CRC in the nation, research on the cost-effectiveness of high-risk populations or nationwide screening alternatives must be conducted in the UAE.

Conflict of Interest The authors have no conflict of interest to declare.

References

1. Rawla P, Sunkara T, Barsouk A. Epidemiology of colorectal cancer: incidence, mortality, survival, and risk factors. Gastroenterol Rev/Przegląd Gastroenterologiczny. 2019;14(2):89–103.
2. Ewing I, Hurley JJ, Josephides E, Millar A. The molecular genetics of colorectal cancer. Front Gastroenterol. 2014;5:26–30.
3. Vogelstein B, Fearon ER, Hamilton SR, et al. Genetic alterations during colorectal-tumor development. N Engl J Med. 1988;319:525–32.
4. Cancer incidence in United Arab Emirates, Annual Report of the UAE - National Cancer Registry. Statistics and Research Center, Ministry of Health and Prevention.
5. Al-Shamsi HO, Abdelwahed N, Al-Awadhi A, Albashir M, Abyad AM, Rafii S, et al. Breast cancer in The United Arab Emirates. JCO Global Oncology. 2023;9:e2200247.
6. "UAE Fact Sheet. About the UAE." https://u.ae/en/about-the-uae/fact-sheet. Accessed 27 Aug 2022.
7. Stryker SJ, Wolff BG, Culp CE, et al. Natural history of untreated colonic polyps. Gastroenterology. 1987;93:1009–13.
8. Stewart SL, Wike JM, Kato I, et al. A population-based study of colorectal cancer histology in the United States, 1998-2001. Cancer. 2006;107:1128–41.
9. Al-Shamsi HO, et al. Early onset colorectal cancer in the United Arab Emirates, where do we stand? Acta Sci Cancer Biol. 2020;4(11):24–7.
10. Al Zaabi A. Colorectal cancer in the Arab world. In: Al-Shamsi HO, Abu-Gheida IH, Iqbal F, Al-Awadhi A, editors. Cancer in the Arab world. Singapore: Springer; 2022. https://doi.org/10.1007/978-981-16-7945-2_23.
11. Al Zaabi A, Al Shehhi A, Sayed S, Al Adawi H, Al Faris F, Al Alyani O, Al Asmi M, Al-Mirza A, Panchatcharam S, Al-Shaibi M. Early onset colorectal cancer in Arabs, are we dealing with a distinct disease? Cancers (Basel). 2023;15(3):889. https://doi.org/10.3390/cancers15030889. PMID: 36765846; PMCID: PMC9913248
12. Al Zaabi A, Al Harrasi A, Al Musalami A, Al Mahyijari N, Al Hinai K, Al Adawi H, Al-Shamsi HO. Early onset colorectal cancer: challenges across the cancer care continuum. Ann Med Surg (Lond). 2022;22(82):104453. https://doi.org/10.1016/j.amsu.2022.104453. PMID: 36268309; PMCID: PMC9577444
13. Stoffel EM, Koeppe E, Everett J, Ulintz P, Kiel M, Osborne J, et al. Germline genetic features of young individuals with colorectal cancer. Gastroenterology 2018;154(4):897.e1–905.e1. 64.
14. Cavestro GM, Mannucci A, Zuppardo RA, Di Leo M, Stoffel E, Tonon G. Early onset sporadic colorectal cancer: worrisome trends and oncogenic features. Dig Liver Dis. 2018;50:521–32.
15. Hossain MS, Karunuwati H, Jairoun AA, Urbi Z, Ooi J, John A, Lim YC, Kibria KMK, Mohiuddin AKM, Ming LC, Goh KW, Hadi MA. Colorectal cancer: a review of carcinogen-

esis, global epidemiology, current challenges, risk factors, preventive and treatment strategies. Cancers (Basel). 2022;14(7):1732. https://doi.org/10.3390/cancers14071732.

16. Sivasubramanian BP, Dave M, Panchal V, Saifa-Bonsu J, Konka S, Noei F, Nagaraj S, Terpari U, Savani P, Vekaria PH, Samala Venkata V, Manjani L. Comprehensive review of red meat consumption and the risk of cancer. Cureus. 2023;15(9):e45324. https://doi.org/10.7759/cureus.45324.

17. Abu-Ghazaleh N, Chua WJ, Gopalan V. Intestinal microbiota and its association with colon cancer and red/processed meat consumption. J Gastroenterol Hepatol. 2021;36:75–88.

18. Pasdar Y, Shadmani FK, Fateh HL, et al. The burden of colorectal cancer attributable to dietary risk in Middle East and north African from 1990 to 2019. Sci Rep. 2023;13:20244. https://doi.org/10.1038/s41598-023-47647-y.

19. Keivanlou MH, Amini-Salehi E, Joukar F, Letafatkar N, Habibi A, Norouzi N, Vakilpour A, Aleali MS, Rafat Z, Ashoobi MT, Mansour-Ghanaei F, Hassanipour S. Family history of cancer as a potential risk factor for colorectal cancer in EMRO countries: a systematic review and meta-analysis. Sci Rep. 2023;13(1):17457. https://doi.org/10.1038/s41598-023-44487-8. PMID: 37838786; PMCID: PMC10576738.

20. Keivanlou MH, et al. Association between smoking and colorectal cancer in eastern Mediterranean regional office (EMRO): a systematic review and meta-analysis. Saudi J Gastroenterol. 2023;29:204–11. https://doi.org/10.4103/sjg.sjg_163_23.

21. https://www.medpagetoday.com/hematologyoncology/coloncancer/109023

22. https://ascopubs.org/doi/10.1200/EDBK_390520?url_ver=Z39.88-2003&rfr_id=ori:rid:crossref.org&rfr_dat=cr_pub%20%200pubmed

23. Arafa MA, Farhat K. Colorectal cancer in the Arab world—screening practices and future prospects. Asian Pac J Cancer Prev. 2015;16:7425–30.

24. Fayadh MH, Sabih SA, Quadri HA. 8 years observational study on colorectal cancer in UAE. J Coloproctol (Rio de Janeiro). 2019;39:394–5.

25. Cho E, Lee JE, Rimm EB, Fuchs CS, Giovannucci EL. Alcohol consumption and the risk of colon cancer by family history of colorectal cancer. Am J Clin Nutr. 2012;95(2):413–9. https://doi.org/10.3945/ajcn.111.022145. Epub 2012 Jan 4

26. https://mohap.gov.ae/assets/download/2b685966/Colorectal%20Cancer%20guide-line_638511990655953382.pdf.aspx

Prof. Humaid Obaid Al-Shamsi is the Chief Executive Officer of Burjeel Cancer Institute in Abu Dhabi, UAE, President of the Emirates Oncology Society, Lead of the Gulf Cancer Society, Full Professor of Oncology at the Ras Al Khaimah Medical and Health Sciences University, Ras Al Khaimah, UAE, and an Adjunct Professor of Oncology at the College of Medicine, University of Sharjah. He is the first Emirati to be promoted as a professor in oncology in the UAE. He is also the Chairman for Colorectal Cancer in the MENA region, appointed by the prestigious National Comprehensive Cancer Network®. He is also the only member of Lung Cancer Policy Network in the MENA region that aims to advance lung cancer research and screening globally. He is the Chairman of the Oncology and Hematology Fellowship Training Program for the National Institute for Health Specialties in the United Arab Emirates. He is the only member in GCC in the WIN Consortium which is comprised of organizations representing all stakeholders in personalized cancer medicine globally.

He is board-certified in both internal medicine and oncology from the UK, USA (ABIM), and Canada (FRCPC). He has also been awarded the FRCP (London) in 2023 and FRCP (Glasgow) in 2024. He is the only physician in the UAE with a subspecialty fellowship certification and training in gastrointestinal oncology and the first Emirati to train and complete a clinical post-doctoral fellowship in palliative care. He was an assistant professor at the University of Texas MD Anderson Cancer Center between 2014 and 2017. He has published more than 140 peer-reviewed articles in JAMA Oncology, Lancet Oncology, The Oncologist, BMC Cancer, and many others. His area of expertise includes precision oncology and cancer care in the UAE. In 2016, he published with his group from MD Anderson the JCO paper describing a new distinct subgroup of CRC, NON V600 BRAF-mutated CRC. In 2022, he published the first book about cancer research in the UAE and also the first book about cancer in the Arab world, both of which were launched at Dubai Expo 2020. *Cancer in the Arab World* has been downloaded more than 450,000 times in its first 18 months of publication and is the ultimate source of cancer data in the Arab region. He also published the first comprehensive book about cancer care in the UAE which is the first book in UAE history to document the cancer care in the UAE with many topics addressed for the first time, e.g., neuroendocrine tumors in the UAE. He is passionate about advancing cancer care in the UAE and the GCC and has made significant contributions to cancer awareness and early detection for the public using social media platforms. He is considered as the most followed oncologist in the world with over 300,000 subscribers across his social media platforms (Instagram, Twitter, LinkedIn, and TikTok). In 2022, he was awarded the prestigious Feigenbaum Leadership Excellence Award from Sheikh Hamdan Smart University for his exceptional leadership and research and the Sharjah Award for Volunteering. He was also named the Researcher of the Year in the UAE in 2020 and 2021 by the Emirates Oncology Society.

In May 2024, HH Sheikh Mansour bin Zayed Al Nahyan, Vice President of the United Arab Emirates, awarded him the first place in UAE Nafis program for outstanding leadership in private sector across all business and medical disciplines. Beside his clinical and administrative duties, he is engaged in education and various levels of research training for medical trainees to enhance their clinical and research skills. His mission is to advance cancer care in the UAE and the MENA region and make cancer care accessible to everyone in need around the globe.

Ms. Faryal Iqbal is the research associate at Burjeel Medical City, Abu Dhabi, United Arab Emirates. She completed her undergraduate studies in molecular biology and biotechnology. Following that, she received a postgraduate qualification in molecular genetics. She co-edited "Cancer in the Arab World," the first extensive book covering cancer care across all Arab countries. The book succeeded significantly, with over 450,000 downloads within just two years. She has several peer-reviewed papers under her name. In September 2023, Ms. Iqbal received the "EOS Research Award" from the prestigious Emirates Oncology Society for her research efforts. Moreover, she assists in different aspects of clinical trial implementation at a research site. Her research interests and publications encompass oncology, hematology, and genetics.

Dr. Hampig R. Kourie gained his medical doctor degree from the Faculty of Medicine of Saint Joseph University in Beirut Lebanon in 2010. He started his fellowship in Hôtel-Dieu de France, Saint Joseph University hospital in Beirut, Lebanon, and he continued his medical oncology fellowship in Jules Bordet Institute in Brussels, Belgium from 2014 to 2016. Since September 2016, he worked in the digestive oncology department in Hôpital Européen Georges Pompidou (HEGP) in Paris as a researcher in the Association des Gastro-entérologues Oncologues en France (AGEO). He is also certified from Paris Diderot and Paris Descartes universities in hereditary cancers and digestive oncology. He gained his PhD in genetics from Saint Joseph University of Beirut in 2023 and his MEMS from ESA business school in 2023. He is practicing as hematologist-oncologist in Hôtel Dieu de France, Hôpital Saint Joseph des Soeurs de la Croix and Bellevue Medical Center.

He is actually an ESMO Faculty in the colorectal cancer group. He founded and directed the Middle East Biomarkers course and the Middle East and North Africa GI Oncology Summit. He has more than 200 peer-reviewed articles, mainly in immunotherapy, oncogenomics, and digestive oncology fields.

Dr. Adhari Al Zaabi is an MD-PhD and Assistant Professor at Sultan Qaboos University's College of Medicine and Health Sciences. She earned her PhD in cancer prevention from Brigham Young University, USA, and was awarded a prestigious Fellowship in Molecular and Behavioral Cancer Prevention by the National Cancer Institute in USA, 2018.

Her expertise lies in the early detection and prevention of colorectal cancer, particularly early-onset colorectal cancer, and the application of artificial intelligence in healthcare. Adhari has secured numerous research grants, including two consecutive awards from His Majesty's Fund for projects focused on cancer registry automation using machine learning and analyzing the need for a national colorectal cancer screening program.

Recently, she received the Women in Artificial Intelligence Award in Dubai in 2023 and was the first Arab woman to win the International Brain Research Award from France in 2020. In 2021, she earned the Best Poster Award at the prestigious European Society of Medical Oncology conference in Singapore, 2022. Adhari has contributed several chapters to books in areas such as colorectal cancer and artificial intelligence in health.

Dr. Amin M. Abyad earned his medical degree, Bachelor in Medicine and Surgery (MBChB), from Beirut Arab University. After completing his internship, he joined the Internal Medicine Residency at Makassed Hospital in Beirut, Lebanon, which is affiliated with the American University of Beirut Medical Center (AUBMC). Dr. Amin was appointed as Chief Resident of Internal Medicine (2017–2018). Then Dr. Amin started his fellowship in hematology and medical oncology at Makassed Hospital, where he received intensive training in hematology and medical oncology. Dr. Amin joined Burjeel Medical City in July 2021. Dr. Amin is highly interested in malignant hematology and solid malignancies. He has been highly involved in clinical research, being involved in multiple research projects and publishing in multiple peer-reviewed journals. Dr. Abyad believes in patient-centered care, trying to enhance patient outcomes through the application of the latest evidence-based practice and personalized medicine.

Dr. Nadia Abdelwahed with a solid foundation of medical studies, specialized in the branch of medical oncology and added valuable practical experience in the administration of oncology intervention and therapy by addressing cases from all stages and conditions of the disease. Her specialty practice began in 2015 and continued for 3 years, making her an expert in the diagnosis and screening of all cancer types, especially breast cancer, and an expert in cancer treatment methods, qualifying her through the School of Medicine residency program. During this period, she addressed both the inpatient and outpatient categories of patients. After this academic training and practical performance, Dr. Nadia secured a placement as a medical oncologist (specialist) in a leading university oncology hospital in Damascus, Syria. After completing her tenure of office there, she joined another multispecialty center as a medical oncologist from 2018 until 2021. Alongside her postgraduate studies and practical training, Dr. Nadia conducted her independent research studies leading to the master's degree (Neoadjuvant Capecitabine in Rectal Cancer) Research in 2018. In 2017, she passed the European Society of Medical Oncology (ESMO) exam and was certified for a period of 5 years. In 2021, she also excelled at the Immune Oncology Course at Harvard Medical School in the United States, earning a Certificate of Achievement. She is also a recipient of the FRON Prize (Forum of Research in Oncology) for her research work on "Impact of HER-2 Ratio on Efficacy of Trastuzumab in Early Breast Cancer" in 2017. In 2021, Dr. Nadia relocated to the UAE and qualified for the HAAD Licensing Exam as a Specialist in Medical Oncology; soon after, she joined VPS Healthcare at the Burjeel Cancer Institute of the Burjeel Medical City.

Gastric Cancer in the UAE

28

Nadia Abdelwahed, Salem Al Asousi, Faryal Iqbal ⓘ,
Amin M. Abyad, Neil A. Nijhawan, Hampig R. Kourie,
Ibrahim H. Abu-Gheida, Basil Ammori,
and Humaid O. Al-Shamsi ⓘ

N. Abdelwahed · A. M. Abyad · I. H. Abu-Gheida
Emirates Oncology Society, Emirates Medical Association, Dubai, United Arab Emirates

Burjeel Medical City, Abu Dhabi, United Arab Emirates

S. Al Asousi
Al Sabah Hospital, Kuwait City, Kuwait

F. Iqbal · N. A. Nijhawan · B. Ammori
Burjeel Medical City, Abu Dhabi, United Arab Emirates
e-mail: faryal.iqbal@burjeelmedicalcity.com; neil.nijhawan@burjeelmedicalcity.com;
basil.ammori@burjeel.com

H. R. Kourie
Hematology-Oncology Department, Faculty of Medicine,
Saint Joseph University, Beirut, Lebanon
e-mail: hampig.kourie@usj.edu.lb

H. O. Al-Shamsi (✉)
Burjeel Cancer Institute, Burjeel Medical City, Burjeel Holdings, Abu Dhabi, United Arab Emirates

Ras Al Khaimah Medical and Health Sciences University, Ras Al Khaimah, United Arab Emirates

Gulf Medical University, Ajman, United Arab Emirates

Emirates Oncology Society, Emirates Medical Association, Dubai, United Arab Emirates

College of Medicine, University of Sharjah, Sharjah, United Arab Emirates

Gulf Cancer Society, Alsafa, Kuwait
e-mail: alshamsi@burjeel.com; humaid.al-shamsi@medportal.ca

© The Author(s) 2024
H. O. Al-Shamsi (ed.), *Cancer Care in the United Arab Emirates*,
https://doi.org/10.1007/978-981-99-6794-0_28

28.1   Introduction

Gastric cancer remains a major health problem, as it is one of the most aggressive cancer types with high death rates globally [1]. Although its incidence rate has decreased over time in the United States and Western Europe, gastric cancer is still the fifth most common and the third leading cause of cancer death worldwide. This is largely due to the later presentation with a diagnosis of locally unresectable or metastatic gastric cancer, which generally carries a poor prognosis [2]. The incidence rate of gastric cancer varies widely between countries, and its 5-year survival rate varies extensively in Japan, where it reaches 90% vs. only 30% in Europe [3]. This different rate is mostly due to the early routine screening methods with endoscopic evaluation in Japan [4]. Gastric cancer, in its early stages, is located mostly in the antrum of the stomach, with a rate of 57.5% and a lesser curvature at 37.8% [5]. Multiple factors can trigger gastric cancer when combined, like genetic disorders (which make up only 3–5% of the cases), such as hereditary diffuse gastric cancer from mutations in the tumor suppressor gene CDH1, Lynch syndrome, and other genetic defects [6].

Other risk factors related to gastric cancer include an unhealthy lifestyle, including smoking, with a relative risk rate of 1.62 in males, making it one of the most risky factors [7], alcohol abuse, although four prospective large cohort trials provided little support for the association between alcohol intake and gastric cancer [8], processed food, and high salt diets, which cause damage to the gastric mucosa and trigger the carcinogenic pathway of cancer [9]. Obesity with a body mass index (BMI) of more than 30 increases the risk of gastric cancer by 1.5-fold [10]. Infection with Helicobacter pylori is reported to be the strongest risk factor correlated with gastric cancer, which occurs in 3% of H. pylori-infected patients [11]. Gastric cancer is adenocarcinoma in 95% of cases, arising from epithelial cells, and is subdivided, according to Lauren's classification, into the intestinal well-differentiated type, which is mostly sporadic and linked to environmental risk factors, and the diffuse undifferentiated type, which frequently presents metastatic disease and has a poor prognosis [12]. In the Arab World and reviewing the Globocan 2020 database, reviews show that the incidence of gastric cancer is low in this region, and most of the cases are diagnosed at advanced stages, making the 5-year survival rates low when compared to European countries. The figures seem similar: 21.1% in Oman compared to 30% in Europe. Oman had the highest incidence, with a rate of 8.0 per 100,000, while Saudi Arabia and the United Arab Emirates (UAE) had rates of 2.7 and 4.4 per 100,000, respectively [13]. Some retrospective studies for Saudi Arabia in 2017 showed that the mean age was 50 years and the male-to-female ratio was 2.3:1, with most of the tumors located in the body of the stomach [14]. A cohort study of gastric cancer from Gulf Council countries showed that the cancer was located anatomically in 53% of the stomach antrum, with histologic features classified as intestinal type in 61.50% of patients and signet ring morphology in 30.20% of patients. This was the first study to look at the *HER-2* status in relation to gastric cancer in Gulf Cooperation Council (GCC) countries; the findings revealed *HER-2* gene amplification in 20% of GCC gastric patients [14].

In this article, we aim to highlight the updated data on gastric cancer in the UAE and to review and try to understand the risk factors, clinical features, and histologic characteristics of gastric cancer in the UAE. This can help determine if preventive steps can aid in the workup and treatment of such an aggressive disease.

28.2 Epidemiology of Gastric Cancer

28.2.1 Overall Cancer Incidence in the UAE

In 2021, the UAE—National Cancer Registry (NCR) recorded 5830 newly diagnosed cancer cases (both malignant and in situ). Newly diagnosed malignant cases totaled 5612 (96%), and 218 (4%) were in situ cases. Taking all cancer types together, the female gender was diagnosed with cancer more than the male gender; cancer was diagnosed in 3210 (55.1%) females versus 2620 (44.9%) males [15]. Among Emiratis, 1431 cancer cases in 2021 were newly diagnosed as new malignant cases, versus only 62 (4.2%) in situ cases. Cancer case incidents involving non-Emiratis were higher and counted at 4337, with a total of 4181 (96.4%) malignant cases and 156 (3.6%) in situ cases [15].

28.2.2 Gastric Cancer Incidence

Gastric cancer is not among the ten most common cancer types in the UAE. It had a low incidence from 2013 to 2021, with 134 cases out of 5612 (2.38%) newly diagnosed in 2021 (Table 28.1). Among the population in the UAE, individuals who were not UAE citizens represented a larger proportion of gastric cancer cases, totaling 104 cases (77.6%). In contrast, a smaller number of cases, specifically 30 (22.3%), were reported among UAE citizens. These proportions are depicted in Fig. 28.1. Gastric cancer affects men more than women, accounting for 70.8% of cases versus 29% of women. Figure 28.2 shows a clear predominance of gastric cancer in males. The crude incidence rate of gastric cancer per 100,000 people is generally stable and low, from 1.1 in 2014 to 1.4 in 2021 for both genders [15] (Table 28.1).

Table 28.1 Stomach cancer demographics among the UAE population during 2013–2021

Year	UAE population (in millions)	Total malignant cases (in numbers)	Stomach cancer cases (in numbers)	Percentage (%)	Crude incidence rate of stomach cancer cases per 100,000 population
2013	8.66	3574	105	2.94	–
2014	8.79	3610	101	2.79	1.1
2015	8.93	3744	108	2.88	1.2
2016	9.12	3982	111	2.78	–
2017	9.3	4123	95	2.30	1.0
2019	9.5	4381	89	2.03	–
2021	–	5612	134	2.38	1.4

Source: UAE population: https://fcsc.gov.ae/en-us/Pages/Statistics/Statistics-by-Subject.aspx#/%3Fsubject=Demography%20and%20Social&folder=Demography%20and%20Social/Population/Population

Ministry of Health and Prevention, Statistics and Research Center, National Disease Registry—UAE National Cancer Registry Report, 2013–2021

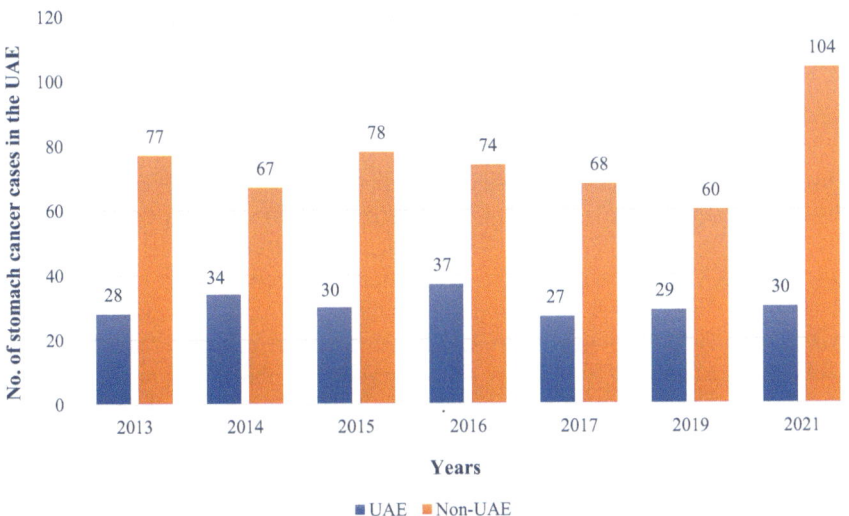

Fig. 28.1 The number of stomach cancer cases (malignant) among the UAE population according to nationality, 2013–2021. Source: Ministry of Health and Prevention, Statistics and Research Center, National Disease Registry—UAE National Cancer Registry Report, 2013–2021

Fig. 28.2 Distribution of stomach cancer cases (malignant) according to gender, 2013–2021. Source: Ministry of Health and Prevention, Statistics and Research Center, National Disease Registry—UAE National Cancer Registry Report, 2013–2021

28.3 Chronicled Data of Gastric Cancer in the UAE, 2013–2021

28.3.1 Incidence

This chapter summarizes the overall data obtained from UAE-NCR reports. The data show steady stability in the occurrence of gastric cancer over the last decade, with 105 cases in 2013 [females: 32; males: 73] to 134 cases in 2021 [females: 39; males: 95] [15] (Fig. 28.3, Table 28.1). There are many factors to consider while preparing this observatory data: the documentation system through UAE-NCR is becoming more evolved, and the screening programs and the awareness campaign are becoming more popular and advanced [16]. This could be attributed to the UAE's rapid population growth, particularly among young people, with over 200 nationalities settled and working in the UAE. The largest population residing in the UAE is from India, Pakistan, Bangladesh, other Asian nations, Europe, and Africa, respectively [17]. Figure 28.4 shows the distribution of stomach cancer cases by the Surveillance, Epidemiology, and End Results (SEER) stages in the UAE across the years.

28.3.2 Mortality

Malignant neoplasm of the stomach is the fifth most common cause of cancer death in both sexes in the UAE, with an estimated average of 4.3% of cancer deaths occurring during the year 2021 [15]. The distribution of mortality cases due to malignant neoplasm of the stomach over the previous years is shown in Table 28.2.

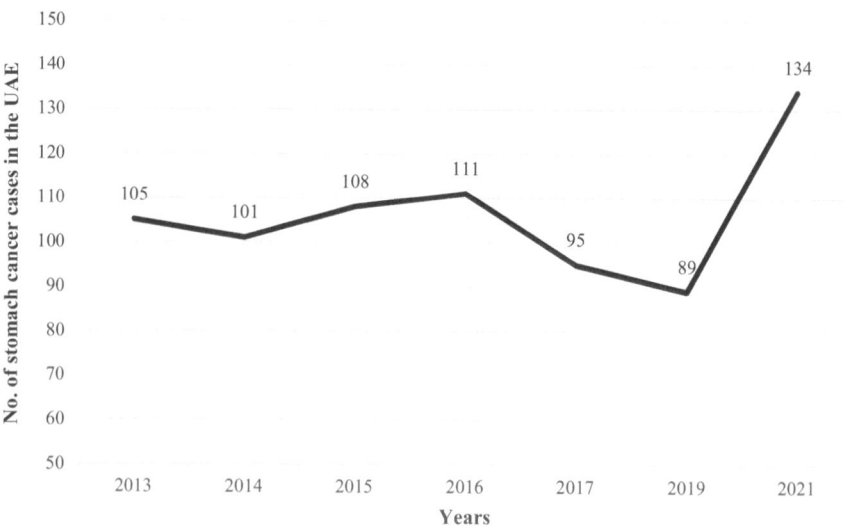

Fig. 28.3 Number of stomach cancer cases (malignant) in the UAE across the years 2013–2021. Source: Ministry of Health and Prevention, Statistics and Research Center, National Disease Registry—UAE National Cancer Registry Report, 2013–2021

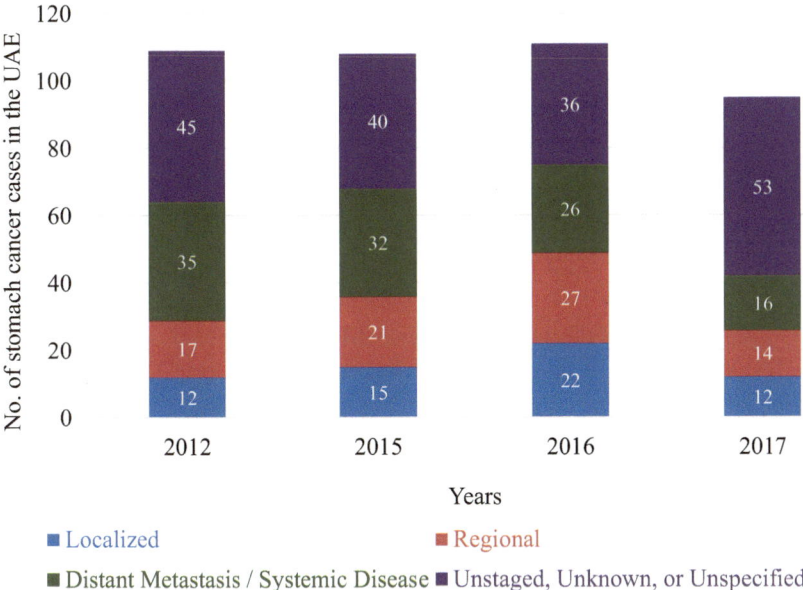

Fig. 28.4 Distribution of stomach cancer cases by SEER stages in the UAE across the years. Source: Ministry of Health and Prevention, Statistics and Research Center, National Disease Registry—UAE National Cancer Registry Report, 2012–2017

Table 28.2 Distribution of malignant neoplasm of stomach mortality cases in the UAE

Year	Percentage (%)
2014	5.4
2015	5.1
2017	5.2
2019	4.7
2021	4.3

Source: Ministry of Health and Prevention, Statistics and Research Center, National Disease Registry—UAE National Cancer Registry Report, 2014–2021

28.4   Risk Factors for Gastric Cancer in the UAE

Helicobacter pylori (H. pylori) is a gram-negative bacteria that affects the stomach, causing chronic inflammation in the gastric mucosa and is the leading cause of stomach cancer [18].

It's the cause of noncardiac gastric cancer in 90%, as this infection stimulates the pathway of dysplasia going through atrophic gastritis and intestinal metaplasia [19].

The prevalence of H. pylori in the UAE has been investigated in a prospective study of healthy people living in the UAE of different nationalities with no symptoms who were found to have these bacteria through stool testing. Infection was found in 41% of the study sample, with more females infected than males, and the median age was in their 30 s [20]. Another major challenge is adhering to follow-up after H. pylori eradication. To provide an illustration, more than two-thirds (71%) of patients infected with H. pylori at a tertiary care center in the UAE failed to attend their scheduled follow-up appointments [21].

Smoking is considered one of the riskiest environmental factors causing cancer in the stomach [22]. The risk increases with duration, reaching 33% when smoking for 40 years [23]. In a cross-sectional study, smoking affects 24.3% of males in the UAE population, mostly Arab residents, and 0.8% of females. Cigarettes have the highest prevalence, at 77.4%. An alarming survey found that Midwakh smoking affects young UAE nationals, with 16% still smoking in the UAE, despite the fact that the prevalence is lower than that in other Arab countries [24].

Metabolic disorders are accused of being risk factors for cancer of the stomach; a cohort study revealed a higher incidence of gastric cancer in the metabolically abnormal obese population. Obesity is associated with a higher prevalence of H. pylori infection, according to multiple studies. Other explanations of obesity's correlation to gastric cancer can be related to gastroesophageal reflux and insulin resistance [25]. Obesity affects non-UAE citizens far more than Emiratis, according to the UAE national diabetes and lifestyle style [26].

A systematic review published in 2015 showed a significant relationship between gastric cancer and a high-salt diet, with a higher risk of injury of 12% for each 5 g/day [27].

A meta-analysis of multiple cohort studies published in 1987–2016 found that high-dose alcohol intake has a higher risk of incidence of gastric cancer, with a 7% increase for each 10 g/day [28].

The national rehabilitation center in the UAE studied alcohol and multiple substance misuse in adults aged 13–18 between 2013 and 2015. Results showed an increase in alcohol intake [29].

28.5 Screening

Screening for gastric cancer in healthy populations is still under investigation. It is preferred for high-risk areas, as in Japan and Korea, but still, there is no unified modality and guidelines regarding when to start and how to screen [30, 31]. In Japan, screening is advised for people aged 50 and above using endoscopy as a modality every 2–3 years or barium imaging annually [32]. Interscreen gaps are still under evaluation in randomized trials. A Korean cohort study found that screening at 2–3-year intervals reduced the risk of advanced care and, as a result, mortality [33].

Prevention of H. pylori infection by routine tests and treatment for this infection in the asymptomatic population has been applied in multiple regions and has shown a decrease in the incidence rate of gastric cancer, especially in first-degree relatives of patients diagnosed with gastric cancer [34, 35].

28.6 Clinicopathological Features of Gastric Cancer in the UAE

A single cross-sectional study involving 96 patients from the GCC, with 26% of them being Emiratis, was the only research that examined the clinicopathological characteristics of gastric cancer in the UAE. This study specifically investigated the anatomical location, histology, and stages of the disease. No other studies have been conducted on this topic in the UAE.

In this particular study, the authors identified 96 patients who were diagnosed with gastric adenocarcinoma, confirmed through histological examination. These patients originated from various countries, including Saudi Arabia (41.6%), the United Arab Emirates (27%), Qatar (14.6%), Kuwait (9.5%), Oman (4.2%), and Bahrain (3.1%). The initial symptoms reported by the patients were epigastric pain in 52% of cases, dyspepsia in 67.7%, weight loss in 72.9%, and melena (blood in the stool) in 7.3%. The median time from the onset of symptoms to diagnosis was 9.3 months, ranging from 2 to 18 months. The median age of diagnosis was 54.5 years, with 40 patients (42%) being under the age of 50. The male-to-female ratio was 1.7, with 61% of the patients being male and 39% female. In 90% of cases, the diagnosis was made before the patients visited MD Anderson, while in 10.4% of cases, the diagnosis was made in the United States. At the time of diagnosis, 76% of cases were already characterized as metastatic. Histological analysis revealed that the intestinal type of gastric cancer was predominant in 61% of cases, while 30% exhibited "signet cell" histology, and 9% had an indeterminate type. Notably, among the patients under the age of 50, the signet ring type was the most common, accounting for 71.4% of cases (30 out of 40). Among the tested samples, 6 out of 28 (20.1%) showed HER-2 amplification.

This study provided the initial report on the clinicopathological characteristics of gastric cancer (GC) in patients from the GCC. Prior to this study, previous reports had separately assessed GC patients in each individual GCC country [36–38]. During a span of 34 years, from 1981 to 2015, our retrospective study examined 96 patients originating from GCC countries who received treatment for gastric carcinoma at the MD Anderson Cancer Center. The majority of these patients came from Saudi Arabia, the UAE, and Qatar. It is noteworthy that 57% of the patients (55 individuals) underwent treatment before the year 2000, which can be attributed to the limited availability of well-established cancer centers in the GCC region during that time.

The average age of the group of patients from the GCC in this study was 54.5 years, ranging from 20 to 80 years. Notably, 40 patients (42%) were younger than 50 years old. This stands in contrast to the average age of males and females in the United States, which was 67.4 years according to the SEER data spanning from 1973 to 2014 [39]. The difference in age between this cohort and the population in the United States is 12.9 years, with the GCC patients being notably younger. This finding aligns with data observed in other Arab countries. For instance, a study conducted in Tunisia analyzed 860 cases of gastric cancer, revealing an average age of 59 years and 27% of the cases being younger than 50 years old [40]. The sex distribution in this study exhibited a ratio of 1.7, indicating a higher prevalence among males. This observation is consistent with a previous study conducted in Oman, which also reported a comparable sex ratio of 1.7 [37] and was lower than the reported study from the Kingdom of Saudi Arabia (KSA) with a male to female ratio of 2.3 [38].

Lauren's classification, developed in 1965, is widely recognized as the primary classification system for gastric carcinoma. This classification categorizes gastric carcinoma into two main types: intestinal and diffuse. These types possess distinct characteristics, including differences in morphology, genetics, clinical presentation, progression patterns, and epidemiology [41]. Within our group of patients, the histology classification revealed that the intestinal type of gastric carcinoma was the most prevalent, accounting for 61.5% of cases. Additionally, 30.2% of cases exhibited a "signet cell" histology, while 8.3% fell under the category of indeterminate GC type. These findings differ from a study conducted in KSA, where 91.5% of gastric carcinoma cases were identified as intestinal type [38]. It is noteworthy that among younger patients, there was a remarkable prevalence of the signet ring type, accounting for 71.4% (30 out of the 40 cases) in individuals below the age of 50. This observation aligns with previous studies indicating that the diffuse type, characterized by signet ring cells, tends to be more prevalent in younger populations. Moreover, this subtype is associated with a poorer prognosis and the highest recurrence frequency (63%) among the four molecular subtypes of gastric carcinoma [42].

Limited information is available regarding the prevalence of HER2 in gastric cancer patients from GCC countries. The only existing study on HER2, as mentioned earlier, was conducted in the KSA and involved only nine patients, all of whom tested negative for HER2 amplification. In our cohort, a total of 28 patients were tested, and 6 of them (20.1%) exhibited HER2 amplification. This represents the first report of the HER2 amplification rate in GCC patients, which aligns with the reported incidence of HER2 amplification in advanced gastric cancer. It has been observed that between 7% and 38% of gastroesophageal adenocarcinomas display amplification and/or overexpression of HER2. Notably, the frequency of overexpression tends to be slightly higher in cancers of the esophagogastric junction (EGJ) compared to those in the stomach (32% versus 21%, respectively). Furthermore, overexpression in the stomach varies based on histological type, with intestinal-type cancers exhibiting a higher prevalence (ranging from 3 to 23%) compared to diffuse-type gastric cancers (ranging from 0 to 6%). Additionally, the degree of differentiation also influences HER2 overexpression, with well and moderately differentiated cancers displaying higher rates compared to poorly differentiated ones [43].

The distribution of gastric cancer (GC) across anatomical locations in our study was as follows: 51 cases (53%) in the antrum, 24 cases (25%) in the body, 9 cases (10%) in the fundus, and 12 cases (12%) in the cardia [14].

28.7 Treatment Modalities for Gastric Cancer in the UAE

Molecular testing is a critical step in the management of almost all cancers, particularly advanced gastric cancer. HER2 testing is widely available in the UAE for advanced gastric cancer. On the other hand, MSI testing, PDL-1 (CPS combined positive score), tumor mutational burden (TMB), neurotrophic tyrosine receptor kinase (NTRK), and next-generation sequencing (NGS) availability vary according to hospital and insurance coverage; many patients in our experience may not have access to these tests due to these limitations.

28.8 Multidisciplinary Team (MDT)

An integrated multidisciplinary approach has become the standard of practice for managing cancer patients at major governmental and private healthcare centers in the UAE [44]. Weekly MDT tumor board meetings take place at major healthcare institutions with the active participation of consultants from various disciplines, including gastroenterology, radiology, nuclear medicine physicians, pathology, gastrointestinal surgeons, medical and radiation oncologists, as well as palliative care physicians. MDT is encouraged by almost all cancer specialists in cancer care in the UAE, but some patients with a gastric cancer diagnosis are treated without MDT input at smaller hospitals and clinics.

28.9 Centralization of Cancer Services

There is no doubt that the centralization of cancer care to create high-volume hospitals, particularly for less common cancers requiring high-risk surgery, such as gastric cancer, can be associated with reduced postoperative mortality and improved overall outcomes [45]. In the Netherlands, as an example, the implementation of centralized gastric cancer surgery in 2012 was associated with significant changes in outcomes, including reductions in cardiac morbidity and 30- and 90-day postoperative mortality and improvements in lymph node retrieval and 2-year overall survival [46]. Furthermore, the centralization of gastric cancer treatment in the Netherlands resulted in the successful introduction of laparoscopic surgery at high-volume centers (6 vs. 40%, $p < 0.01$), which was associated with a significant reduction in hospital stays [47]. In the UAE, the Emirates Oncology Society established an Oncology Task Force in 2019 and recommended the establishment of tertiary cancer centers for better healthcare delivery and improved outcomes [48].

28.10 Neoadjuvant Chemotherapy (NAC)

Although either perioperative or adjuvant chemotherapy improves the survival of patients with non-early operable gastric cancer (stage 1B or higher) [49], neoadjuvant chemotherapy (NAC) followed by D2 gastrectomy has become the more widely practiced standard of care in patients with atypical glandular cells (AGC), including in the UAE. UAE oncologists widely use the FLOT protocol for perioperative chemotherapy for eligible patients [50].

28.11 Surgery

Function-preserving limited gastric resections for EGC, with its low rate of regional lymph node metastases, are increasingly being applied and could include endoscopic, laparoscopic, or combined endoscopy-laparoscopy approaches that offer excellent survival and improved quality of life [51]. In patients with resectable, nonmetastatic AGC, gastrectomy with D2 lymphadenectomy—which involves retrieval of perigastric lymph node stations and those along the branches of the celiac axis—is the standard of care [52]. While D2 gastrectomy is conventionally and more commonly performed by open laparotomy, selected high-volume centers globally adopt minimally invasive approaches (laparoscopic, laparoscopic-assisted, or robotic-assisted). A meta-analysis of eight randomized controlled trials and 22 high-quality non-randomized comparative studies of laparoscopic versus open distal gastrectomy performed by experienced surgeons at high-volume centers for AGC that included 16,029 patients showed benefits in terms of reduction in operative blood loss, serious complications, and hospital stay [53]. The 5-year overall survival of the laparoscopic approach to D2 distal gastrectomy was comparable to that of open surgery, as shown in the multicenter randomized CLASS 0–1 Chinese trial (n = 1056 patients). In the Middle East and the UAE, the minimally invasive approach to D2 gastrectomy for AGC is not widely adopted, yet it is recommended, and the authors advocate for it in keeping with the best surgical practices. The author reported a comparative case-matched controlled study of laparoscopic versus open D2 gastrectomy for AGC where patients were matched for age and extent of resection (total vs. subtotal gastrectomy) and demonstrated significant reductions in intraoperative blood loss and hospital stay (median, 3.0 vs. 7.5 days, $p < 0.001$) while maintaining comparable early oncologic outcomes (median lymph node retrieved, 40.5 vs. 31.5, $p = 0.181$; R0 resection rates 100% vs. 89%, $p = 0.486$) [54].

28.12 Radiation Therapy

As in most of the world, the practice of involving radiation therapy in the treatment of gastric tumors varies across institutions in the UAE and is case-specific. As all cases are typically discussed on tumor boards, the role of the radiation oncologist is definitely vital. For patients presenting with metastatic disease, palliative radiotherapy is usually offered to patients with controlled symptoms. While for non-metastatic diseases, surgery remains the best curative approach. However, for patients with unresectable disease, radiotherapy (with or without chemotherapy) may be used to provide local control benefits [55]. Patients who present with respectable disease require a thorough discussion of their pathological findings (whether these patients received neoadjuvant systemic therapy or not) to determine the indication or not for adjuvant radiation therapy. While surgery is crucial in the treatment of gastric cancer, surgical expertise and the extent of lymph node dissection vary across institutions within the UAE. This reflects on the indications for adjuvant radiation, especially if patients haven't received adequate D2 dissection. For patients who undergo upfront surgery without preoperative treatment, the role of adjuvant radiation is based on pathology findings and extends surgical node dissection. Adjuvant fluoropyrimidine-based concurrent chemoradiotherapy is recommended for patients with more than pT1 any N+ disease, unless primary D2 dissection is made with an R0 resection. Almost all radiotherapy centers within the UAE are equipped with state-of-the-art, advanced machines and have good expertise in treating that cancer. Moreover, most, if not all, departments do have advanced image guidance and gating tools, which allow treatment delivery in the most accurate way possible using 3-D conformal radiation or intensity-modulated radiotherapy (IMRT). More recently, with the shift towards neoadjuvant systemic therapy in patients presenting with locally advanced resectable gastric tumors and especially after the results of the CRITICS trial [56], the utilization of radiation in the adjuvant setting for patients who received neoadjuvant systemic therapy has gone down. However, for patients with high-risk features such as R1 resection, which cannot be re-resected, current practice guidelines, including the National Comprehensive Cancer Network (NCCN) guidelines, which are mostly followed across the UAE, still recommend adjuvant fluoropyrimidine-based concurrent chemoradiotherapy [57]. Finally, the utilization of neoadjuvant radiation therapy for gastric cancer is still not fully endorsed in the UAE, most likely because current guidelines do not state it and the role of this approach is still being evaluated [58].

Therefore, there remains a great deal of controversy in the radiation practice for gastric cancer within the UAE as well as in the rest of the world. Multidisciplinary approaches and case-by-case individualized treatment are key. While surgical expertise for treating gastric cancer varies, this remains the key point in reflecting on the outcome of this patient population in the nation. However, having advanced technologies and good radiotherapy expertise in almost all centers in the UAE makes enrolling patients in ongoing international trials addressing the role of neoadjuvant radiation in gastric cancer quite appealing.

28.13   Systemic Therapy

Anticancer therapies are widely available in the UAE, including chemotherapy, targeted therapies, and immunotherapies. FOLFOX/CAPOX/FOLFIRI are the most widely used chemotherapy combinations for advanced gastric cancer. The addition of anti-HER2 or immunotherapy is also utilized in keeping with the NCCN guidelines. Ramucirumab and paclitaxel are commonly used as second-line therapies in patients with metastatic disease.

28.14 Palliative Care

Most patients with gastric cancer have advanced-stage disease at the point of diagnosis, resulting in a significant symptom burden that adversely affects their quality of life. The early integration of palliative care into the care of patients with cancer is now accepted as standard of care since the landmark Temel trial of 2010 [59], which demonstrated that in patients with advanced non-small-cell lung cancer, early palliative care involvement resulted in not only improved quality of life but also longer survival as compared to patients who received standard care. Gastric cancer is particularly morbid, with poor food tolerance (including early satiety), nausea and/or vomiting, abdominal pain and bloating, bleeding, fatigue, and a low mood commonly reported [60].

Improved interdisciplinary symptom control may increase the chances of a patient's functional status being good enough to tolerate systemic anticancer therapy. As the condition progresses, or in cases of advanced gastric cancer, malignant gastric outlet obstruction (GOO) can dominate. Whether treated via an endoscopic or surgical approach, the aim of GOO treatment is to reduce nausea and vomiting and enable the patient to tolerate oral intake. If it is not possible to restore physiological gastrointestinal tract integrity by surgically bypassing the obstruction, a combination of a gastrostomy tube (to vent the stomach) and a separate distal jejunostomy tube (for enteral nutrition) may be utilized. As patients survive longer with advanced gastric cancer, these palliative approaches are likely to be utilized more commonly.

While we are fortunate within the UAE to have access to both the full complement of modern medical equipment and essential palliative care medications [61], access to palliative care teams remains limited to a few specialist hospitals in the major conurbations of Dubai and Abu Dhabi [62]. Early palliative and supportive care input for patients with gastric cancer may help patients deal with the physical and psychological morbidity of gastric cancer.

28.15 Conclusion

Gastric cancer has a lower incidence rate in the UAE when compared to global rates, but it is the fifth leading cause of cancer-related death. This process helps us understand cancer behavior in the stomach and the mechanisms triggering it. Infectious control of H. pylori bacteria by screening healthy asymptomatic populations, launching screening programs for highly at-risk populations, and launching awareness campaigns to encourage balanced diets and healthy lifestyle rules could prevent and decrease the mortality rate of gastric cancer.

Conflict of Interest The authors have no conflict of interest to declare.

References

1. Torre LA, Siegel RL, Ward EM, Jemal A. Global cancer incidence and mortality rates and trends—an update. Cancer Epidemiol Biomarkers Prev. 2016;25:16–27.
2. Sitarz R, Skierucha M, Mielko J, et al. Gastric cancer: epidemiology, prevention, classification, and treatment. Cancer Manag Res. 2018;10:239–48.
3. Anderson LA, Tavilla A, Brenner H, et al. Survival for oesophageal, stomach and small intestine cancers in Europe 1999-2007: results from EUROCARE-5. Eur J Cancer. 2015;51:2144–57.
4. Asaka M, Mabe K. Strategies for eliminating death from gastric cancer in Japan. Proc Jpn Acad Ser B Phys Biol Sci. 2014;90:251–8.
5. Kim K, Cho Y, Sohn JH, et al. Clinicopathologic characteristics of early gastric cancer according to specific intragastric location. BMC Gastroenterol. 2019;19:24.
6. Colvin H, Yamamoto K, Wada N, Mori M. Hereditary gastric cancer syndromes. Surg Oncol Clin N Am. 2015;24:765–77.
7. Nomura AM, Wilkens LR, Henderson BE, et al. The association of cigarette smoking with gastric cancer: the multiethnic cohort study. Cancer Causes Control. 2012;23:51–8.
8. Ma K, Baloch Z, He TT, Xia X. Alcohol consumption and gastric cancer risk: a meta-analysis. Med Sci Monit. 2017;23:238–46.
9. D'Elia L, Galletti F, Strazzullo P. Dietary salt intake and risk of gastric cancer. Cancer Treat Res. 2014;159:83–95.
10. Hashimoto Y, Hamaguchi M, Obora A, et al. Impact of metabolically healthy obesity on the risk of incident gastric cancer: a population-based cohort study. BMC Endocr Disord. 2020;20:11.
11. Wroblewski LE, Peek RM Jr, Wilson KT. Helicobacter pylori and gastric cancer: factors that modulate disease risk. Clin Microbiol Rev. 2010;23:713–39.
12. Ma J, Shen H, Kapesa L, Zeng S. Lauren classification and individualized chemotherapy in gastric cancer. Oncol Lett. 2016;11:2959–64.
13. Aoude M, Mousallem M, Abdo M, et al. Gastric cancer in the Arab world: a systematic review. East Mediterr Health J. 2022;28:521–31.
14. Humaid O, Al-Shamsi MA, Abu-Gheida I, Iqbal F, Alrawi S, Ajani JA. Clinicopathological features of gastric cancer in a cohort of gulf council countries' patients: a cross sectional study of 96 cases. Journal of Oncology Research Reviews & Reports. 2021;2(2):1–3.
15. Cancer incidence in United Arab Emirates, Annual report of the UAE - National Cancer Registry. Statistics and Research Center, Ministry of Health and Prevention.
16. Koornneef E, Robben P, Blair I. Progress and outcomes of health systems reform in The United Arab Emirates: a systematic review. BMC Health Serv Res. 2017;17:672.
17. Al-Shamsi SR, Bener A, Al-Sharhan M, et al. Clinicopathological pattern of colorectal cancer in The United Arab Emirates. Saudi Med J. 2003;24:518–22.
18. Polk DB, Peek RM Jr. Helicobacter pylori: gastric cancer and beyond. Nat Rev Cancer. 2010;10:403–14.
19. Moss SF. The clinical evidence linking helicobacter pylori to gastric cancer. Cell Mol Gastroenterol Hepatol. 2017;3:183–91.
20. Khoder G, Muhammad JS, Mahmoud I, et al. Prevalence of helicobacter pylori and its associated factors among healthy asymptomatic residents in The United Arab Emirates. Pathogens. 2019;8:44.
21. Waness A, Bismar M, Alasadi M, Elmustafa N, Sharqi KA, Elghul A, et al. Continuity of care challenges in GCC countries: H. pylori eradication as example in a UAE tertiary care center. Int J Med Sci Public Health. 2015;4:1125–31.
22. Trédaniel J, Boffetta P, Buiatti E, et al. Tobacco smoking and gastric cancer: review and meta-analysis. Int J Cancer. 1997;72:565–73.

23. Praud D, Rota M, Pelucchi C, et al. Cigarette smoking and gastric cancer in the stomach cancer pooling (StoP) project. Eur J Cancer Prev. 2018;27:124–33.
24. Razzak HA, Harbi A, Ahli S. Tobacco smoking prevalence, health risk, and cessation in the UAE. Oman Med J. 2020;35:e165.
25. Li Q, Zhang J, Zhou Y, Qiao L. Obesity and gastric cancer. Front Biosci (Landmark Ed). 2012;17:2383–90.
26. Sulaiman N, Elbadawi S, Hussein A, et al. Prevalence of overweight and obesity in United Arab Emirates expatriates: the UAE National Diabetes and lifestyle study. Diabetol Metab Syndr. 2017;9:88.
27. Fang X, Wei J, He X, et al. Landscape of dietary factors associated with risk of gastric cancer: a systematic review and dose-response meta-analysis of prospective cohort studies. Eur J Cancer. 2015;51:2820–32.
28. Han X, Xiao L, Yu Y, et al. Alcohol consumption and gastric cancer risk: a meta-analysis of prospective cohort studies. Oncotarget. 2017;8:83237–45.
29. Al Ghaferi HA, Ali AY, Gawad TA, Wanigaratne S. Developing substance misuse services in United Arab Emirates: the National Rehabilitation Centre experience. BJPsych Int. 2017;14:92–6.
30. Mizoue T, Yoshimura T, Tokui N, et al. Prospective study of screening for stomach cancer in Japan. Int J Cancer. 2003;106:103–7.
31. Pisani P, Oliver WE, Parkin DM, et al. Case-control study of gastric cancer screening in Venezuela. Br J Cancer. 1994;69:1102–5.
32. Hamashima C, Shibuya D, Yamazaki H, et al. The Japanese guidelines for gastric cancer screening. Jpn J Clin Oncol. 2008;38:259–67.
33. Bae JM, Shin SY, Kim EH. Mean sojourn time of preclinical gastric cancer in Korean men: a retrospective observational study. J Prev Med Public Health. 2014;47:201–5.
34. Ford AC, Forman D, Hunt RH, et al. Helicobacter pylori eradication therapy to prevent gastric cancer in healthy asymptomatic infected individuals: systematic review and meta-analysis of randomised controlled trials. BMJ. 2014;348:g3174.
35. Wong BC, Lam SK, Wong WM, et al. Helicobacter pylori eradication to prevent gastric cancer in a high-risk region of China: a randomized controlled trial. JAMA. 2004;291:187–94.
36. Al-Mahrouqi H, Parkin L, Sharples K. Incidence of stomach cancer in Oman and the other gulf cooperation council countries. Oman Med J. 2011;26:258–62.
37. Green R, Northway MG, Buena NB, et al. Gastric carcinoma in the Sultanate of Oman: incidence and distribution. Ann Saudi Med. 1996;16:291–5.
38. Patel RM, Knezevic A, Shenvi N, et al. Association of red blood cell transfusion, anemia, and necrotizing enterocolitis in very low-birth-weight infants. JAMA. 2016;315:889–97.
39. Milano AF. 20-year comparative survival and mortality of cancer of the stomach by age, sex, race, stage, grade, cohort entry time-period, Disease Duration & Selected ICD-O-3 oncologic phenotypes: a systematic review of 157,258 cases for diagnosis years 1973-2014: (SEER*Stat 8.3.4). J Insur Med. 2019;48:5–23.
40. Elghali MA, Gouader A, Bouriga R, et al. Gastric adenocarcinomas in Central Tunisia: evolution specificities through two decades and relation with helicobacter pylori. Oncology. 2018;95:121–8.
41. Machlowska J, Puculek M, Sitarz M, et al. State of the art for gastric signet ring cell carcinoma: from classification, prognosis, and genomic characteristics to specified treatments. Cancer Manag Res. 2019;11:2151–61.
42. Cristescu R, Lee J, Nebozhyn M, et al. Molecular analysis of gastric cancer identifies subtypes associated with distinct clinical outcomes. Nat Med. 2015;21:449–56.
43. Sodhi CP, Shi XH, Richardson WM, et al. Toll-like receptor-4 inhibits enterocyte proliferation via impaired beta-catenin signaling in necrotizing enterocolitis. Gastroenterology. 2010;138:185–96.
44. Al-Shamsi HO, Abyad AM, Rafii S. A proposal for a National Cancer Control Plan for the UAE: 2022-2026. Clin Pract. 2022;12:118–32.
45. Grilli R, Violi F, Bassi MC, Marino M. The effects of centralizing cancer surgery on postoperative mortality: a systematic review and meta-analysis. J Health Serv Res Policy. 2021;26:289–301.

46. van Putten M, Johnston BT, Murray LJ, et al. 'Missed' oesophageal adenocarcinoma and high-grade dysplasia in Barrett's oesophagus patients: a large population-based study. United European Gastroenterol J. 2018;6:519–28.
47. van Putten M, Nelen SD, Lemmens V, et al. Overall survival before and after centralization of gastric cancer surgery in The Netherlands. Br J Surg. 2018;105:1807–15.
48. Al-Shamsi H, Darr H, Abu-Gheida I, et al. The state of cancer Care in The United Arab Emirates in 2020: challenges and recommendations, a report by The United Arab Emirates oncology task force. Gulf J Oncolog. 2020;1:71–87.
49. Coccolini F, Nardi M, Montori G, et al. Neoadjuvant chemotherapy in advanced gastric and esophago-gastric cancer. Meta-analysis of randomized trials. Int J Surg. 2018;51:120–7.
50. Al-Batran SE, Homann N, Pauligk C, et al. Perioperative chemotherapy with fluorouracil plus leucovorin, oxaliplatin, and docetaxel versus fluorouracil or capecitabine plus cisplatin and epirubicin for locally advanced, resectable gastric or gastro-oesophageal junction adenocarcinoma (FLOT4): a randomised, phase 2/3 trial. Lancet. 2019;393:1948–57.
51. Nunobe S, Hiki N. Function-preserving surgery for gastric cancer: current status and future perspectives. Transl Gastroenterol Hepatol. 2017;2:77.
52. Smyth EC, Nilsson M, Grabsch HI, et al. Gastric cancer. Lancet. 2020;396:635–48.
53. Chen X, Feng X, Wang M, Yao X. Laparoscopic versus open distal gastrectomy for advanced gastric cancer: a meta-analysis of randomized controlled trials and high-quality nonrandomized comparative studies. Eur J Surg Oncol. 2020;46:1998–2010.
54. Ammori BJ, Asmer H, Al-Najjar H, et al. Laparoscopic versus open D2 gastrectomy for gastric cancer: a case-matched comparative study. J Laparoendosc Adv Surg Tech A. 2020;30:777–82.
55. Willett CG. Results of radiation therapy in gastric cancer. Semin Radiat Oncol. 2002;12:170–5.
56. Cats A, Jansen EPM, van Grieken NCT, et al. Chemotherapy versus chemoradiotherapy after surgery and preoperative chemotherapy for resectable gastric cancer (CRITICS): an international, open-label, randomised phase 3 trial. Lancet Oncol. 2018;19:616–28.
57. https://www.nccn.org/. Accessed 19 Nov 2022.
58. Ng SP, Leong T. Role of radiation therapy in gastric cancer. Ann Surg Oncol. 2021;28:4151–7.
59. Temel JS, Greer JA, Muzikansky A, et al. Early palliative care for patients with metastatic non-small-cell lung cancer. N Engl J Med. 2010;363:733–42.
60. Harada K, Zhao M, Shanbhag N, et al. Palliative care for advanced gastric cancer. Expert Rev Anticancer Ther. 2020;20:575–80.
61. Nijhawan NA, Al-Shamsi HO. Palliative Care in The United Arab Emirates (UAE). In: Laher I, editor. Handbook of healthcare in the Arab world. Cham: Springer International Publishing; 2020. p. 1–18.
62. Nijhawan NA. An update on the state of palliative care development in The United Arab Emirates. Palliat Med Hosp Care Open J. 2022;8(2):27–9. https://doi.org/10.17140/PMHCOJ-8-149.

Dr. Nadia Abdelwahed With a solid foundation of medical studies, Dr. Nadia Abdelwahed specialized in the branch of medical oncology and added valuable practical experience in the administration of oncology intervention and therapy by addressing cases from all stages and conditions of the disease. Her specialty practice began in 2015 and continued for 3 years, making her an expert in the diagnosis and screening of all cancer types, especially breast cancer, and an expert in cancer treatment methods, qualifying her through the school of medicine residency program. During this period, she addressed both the inpatient and outpatient categories of patients. After this academic training and practical performance, Dr. Nadia secured a placement as a medical oncologist (specialist) in a leading university oncology hospital in Damascus, Syria. After completing her tenure of office there, she joined another multispecialty center as a medical oncologist from 2018 until 2021.

Alongside her postgraduate studies and practical training, Dr. Nadia conducted her independent research studies leading to the master's degree (Neoadjuvant Capecitabine in Rectal Cancer) Research in 2018. In 2017, she passed the European Society of Medical Oncology (ESMO) exam and was certified for a period of 5 years. In 2021, she also excelled at the Immune Oncology Course at Harvard Medical School in the United States, earning a Certificate of Achievement. She is also a recipient of the FRON Prize (Forum of Research in Oncology) for her research work on "Impact of HER-2 Ratio on Efficacy of Trastuzumab in Early Breast Cancer" in 2017. In 2021, Dr. Nadia relocated to the UAE and qualified for the HAAD Licensing Exam as a Specialist in Medical Oncology; soon after, she joined VPS Healthcare at the Burjeel Cancer Institute of the Burjeel Medical City.

Dr. Salem Al-Asousi is a graduate of the Royal College of Surgeons in Ireland. He trained in internal medicine at the University of British Columbia in Vancouver, Canada. He has been practicing since 2012 in both internal medicine and gastroenterology in both government and private hospitals in Kuwait. Salem's main interest is advanced gastrointestinal therapies and the effect of weight reduction/management on overall health.

Ms. Faryal Iqbal is the research associate at Burjeel Medical City, Abu Dhabi, United Arab Emirates. She completed her undergraduate studies in molecular biology and biotechnology. Following that, she received a postgraduate qualification in molecular genetics. She co-edited "Cancer in the Arab World," the first extensive book covering cancer care across all Arab countries. The book succeeded significantly, with over 450,000 downloads within just two years. She has several peer-reviewed papers under her name. In September 2023, Ms. Iqbal received the "EOS Research Award" from the prestigious Emirates Oncology Society for her research efforts. Moreover, she assists in different aspects of clinical trial implementation at a research site. Her research interests and publications encompass oncology, hematology, and genetics.

Dr. Amin M. Abyad earned his medical degree, Bachelor in Medicine and Surgery (MBChB), from Beirut Arab University. After completing his internship, he joined the Internal Medicine Residency at Makassed Hospital in Beirut, Lebanon, which is affiliated with the American University of Beirut Medical Center (AUBMC). Dr. Amin was appointed as Chief Resident of Internal Medicine (2017–2018). Then Dr. Amin started his fellowship in hematology and medical oncology at Makassed Hospital, where he received intensive training in hematology and medical oncology. Dr. Amin joined Burjeel Medical City in July 2021. Dr. Amin is highly interested in malignant hematology and solid malignancies. He has been highly involved in clinical research, being involved in multiple research projects and publishing in multiple peer-reviewed journals.

Dr. Abyad believes in patient-centered care, trying to enhance patient outcomes through the application of the latest evidence-based practice and personalized medicine.

Dr. Neil A. Nijhawan is a UK-trained consultant in Palliative Medicine at Burjeel Medical City. After medical school at Kings College London, he pursued specialty training in Palliative Medicine in London, with rotations in acute general hospitals, domiciliary visits, community hospices, and tertiary oncology centers. Prior to completing his palliative medicine training, Neil returned to his childhood home, Trinidad in the West Indies, to help set up and commission the new Caura Hospital Palliative Care Unit, where he was the Medical Director. This unit was opened in 2014 and provides a comprehensive palliative care service, including a 12-bed inpatient unit, weekly outpatient clinics, and a palliative care consult service at the local university hospital. After completing his specialty training, Neil worked as a consultant in palliative medicine at the Imperial College Healthcare NHS Trust in London, where he was the clinical lead for palliative medicine. His clinical area of interest is symptom control (including pain, nausea, breathlessness, and fatigue) and assistance with complex treatment decision-making at the end of life, and he is often called on to provide an independent second opinion. He is active in palliative care education and palliative care advocacy and is currently the UAE representative to the WHO Eastern Mediterranean Region Palliative Care Expert Network. Neil holds adjunct faculty positions with both Khalifa University and Gulf Medical University where he is Clinical Associate Professor in Hospice & Palliative Medicine.

Dr. Hampig R. Kourie gained his medical doctor degree from the Faculty of Medicine of Saint Joseph University in Beirut Lebanon in 2010. He started his fellowship in Hôtel-Dieu de France, Saint Joseph University hospital in Beirut, Lebanon and he continued his medical oncology fellowship in Jules Bordet Institute in Brussels, Belgium from 2014 to 2016. Since September 2016, he worked in the digestive oncology department in Hôpital Européen Georges Pompidou (HEGP) in Paris as a researcher in the Association des Gastro-entérologues Oncologues en France (AGEO). He is also certified from Paris Diderot and Paris Descartes universities in hereditary cancers and digestive oncology. He gained his PhD in genetics from Saint Joseph University of Beirut in 2023 and his MEMS from ESA business school in 2023. He is practicing as hematologist-oncologist in Hôtel Dieu de France, Hôpital Saint Joseph des Soeurs de la Croix and Bellevue Medical Center.

He is actually an ESMO Faculty in the colorectal cancer group. He founded and directed the Middle East Biomarkers course and the Middle East and North Africa GI Oncology Summit. He has more than 200 peer-reviewed articles, mainly in immunotherapy, oncogenomics and digestive oncology fields.

Dr. Ibrahim H. Abu-Gheida is the Clinical Director of the department of radiation oncology at Burjeel Medical City. He also serves as a regional Radiological Society of North America (RSNA) committee representative for the Middle East and Africa. Dr. Abu-Gheida completed his undergraduate training, where he earned a Bachelor of Science with honors degree from the American University of Beirut. Following this, Dr. Abu-Gheida completed his Medical School training at the American University of Beirut Medical Center. He continued and joined the Department of Internal Medicine at the American University of Beirut. Then he did his training in the Department of Radiation Oncology at the American University of Beirut Medical Center, where he also served as the chief resident. During his training, Dr. Abu-Gheida completed a Harvard-affiliated NIH-funded research program as well. After his residency, Dr. Ibrahim went to the Cleveland Clinic in Ohio, where he was appointed as an Advanced Clinical Radiation Oncology Fellow. Dr. Abu-Gheida went and joined the University of Texas MD Anderson Cancer Center, where he sub-specialized in treating breast, gastrointestinal, and genitourinary cancers. Dr. Abu Gheida played an instrumental role in establishing and heading the radiation oncology facility and department at Burjeel Medical City. He has chaired and co-chaired multiple international oncology conferences. Dr. Ibrahim has more than 40 peer-reviewed papers in prestigious medical journals, including the American Society of Radiation Oncology official journal - the International Journal of Radiation Oncology Biology and Physics, Nature, the Journal of Clinical Oncology, and several others. He is also the primary author and editor of several book chapters published in prestigious books.

Prof. Basil Ammori graduated from Baghdad University medical school in 1986, completed his surgical training in the UK, and obtained the CCST in 2000, and was granted a Hunterian Professorship by the Royal College of Surgeons of England in 2000. Appointed a consultant surgeon, he led the laparoscopic bariatric and HPB (hepato-pancreato-biliary) surgery service in Manchester between 2002 and 2017 and was the first to introduce laparoscopic Whipple's procedure in the U.K. He was appointed as an Honorary Professor at the University of Manchester in 2011 and later joined King Hussein Cancer Center, Amman, Jordan, as a full member in November 2017 as a Laparoscopic Gastrointestinal Oncology Surgeon before moving to Burjeel Hospital, Abu Dhabi, in January 2020. He sat on the editorial boards of a number of medical journals and has over 240 peer-reviewed publications and book chapters.

Prof. Humaid Obaid Al-Shamsi is the Chief Executive Officer of Burjeel Cancer Institute in Abu Dhabi, UAE, President of the Emirates Oncology Society, Lead of the Gulf Cancer Society, Full Professor of Oncology at the Ras Al Khaimah Medical and Health Sciences University, Ras Al Khaimah, UAE, and an Adjunct Professor of Oncology at the College of Medicine, University of Sharjah. He is the first Emirati to be promoted as a professor in oncology in the UAE. He is also the Chairman for Colorectal Cancer in the MENA region, appointed by the prestigious National Comprehensive Cancer Network®. He is also the only member of Lung Cancer Policy Network in the MENA region that aims to advance lung cancer research and screening globally. He is the Chairman of the Oncology and Hematology Fellowship Training Program for the National Institute for Health Specialties in the United Arab Emirates. He is the only member in GCC in the WIN Consortium which is comprised of organizations representing all stakeholders in personalized cancer medicine globally.

He is board-certified in both internal medicine and oncology from the UK, USA (ABIM), and Canada (FRCPC). He has also been awarded the FRCP (London) in 2023 and FRCP (Glasgow) in 2024. He is the only physician in the UAE with a subspecialty fellowship certification and training in gastrointestinal oncology and the first Emirati to train and complete a clinical post-doctoral fellowship in palliative care. He was an assistant professor at the University of Texas MD Anderson Cancer Center between 2014 and 2017. He has published more than 140 peer-reviewed articles in JAMA Oncology, Lancet Oncology, The Oncologist, BMC Cancer, and many others. His area of expertise includes precision oncology and cancer care in the UAE. In 2016, he published with his group from MD Anderson the JCO paper describing a new distinct subgroup of CRC, NON V600 BRAF-mutated CRC. In 2022, he published the first book about cancer research in the UAE and also the first book about cancer in the Arab world, both of which were launched at Dubai Expo 2020. *Cancer in the Arab World* has been downloaded more than 450,000 times in its first 18 months of publication and is the ultimate source of cancer data in the Arab region. He also published the first comprehensive book about cancer care in the UAE which is the first book in UAE history to document the cancer care in the UAE with many topics addressed for the first time, e.g., neuroendocrine tumors in the UAE. He is passionate about advancing cancer care in the UAE and the GCC and has made significant contributions to cancer awareness and early detection for the public using social media platforms. He is considered as the most followed oncologist in the world with over 300,000 subscribers across his social media platforms (Instagram, Twitter, LinkedIn, and TikTok). In 2022, he was awarded the prestigious Feigenbaum Leadership Excellence Award from Sheikh Hamdan Smart University for his exceptional leadership and research and the Sharjah Award for Volunteering. He was also named the Researcher of the Year in the UAE in 2020 and 2021 by the Emirates Oncology Society.

In May 2024, HH Sheikh Mansour bin Zayed Al Nahyan, Vice President of the United Arab Emirates, awarded him the first place in UAE Nafis program for outstanding leadership in private sector across all business and medical disciplines. Beside his clinical and administrative duties, he is engaged in education and various levels of research training for medical trainees to enhance their clinical and research skills. His mission is to advance cancer care in the UAE and the MENA region and make cancer care accessible to everyone in need around the globe.

Humaid O. Al-Shamsi ⓘ, Faryal Iqbal ⓘ, Neil A. Nijhawan,
Hampig R. Kourie, Nadia Abdelwahed,
Ibrahim H. Abu-Gheida, and Basil Ammori

H. O. Al-Shamsi (✉)
Burjeel Cancer Institute, Burjeel Medical City, Burjeel Holdings, Abu Dhabi, United Arab Emirates

Ras Al Khaimah Medical and Health Sciences University, Ras Al Khaimah, United Arab Emirates

Gulf Medical University, Ajman, United Arab Emirates

Emirates Oncology Society, Emirates Medical Association, Dubai, United Arab Emirates

College of Medicine, University of Sharjah, Sharjah, United Arab Emirates

Gulf Cancer Society, Alsafa, Kuwait
e-mail: alshamsi@burjeel.com; humaid.al-shamsi@medportal.ca

F. Iqbal · N. A. Nijhawan · B. Ammori
Burjeel Medical City, Abu Dhabi, United Arab Emirates
e-mail: faryal.iqbal@burjeelmedicalcity.com; neil.nijhawan@burjeelmedicalcity.com; basil.ammori@burjeel.com

H. R. Kourie
Hematology-Oncology Department, Faculty of Medicine, Saint Joseph University, Beirut, Lebanon
e-mail: hampig.kourie@usj.edu.lb

N. Abdelwahed · I. H. Abu-Gheida
Emirates Oncology Society, Emirates Medical Association, Dubai, United Arab Emirates

Burjeel Medical City, Abu Dhabi, United Arab Emirates
e-mail: nadia.abdelwahed@burjeelmedicalcity.com

© The Author(s) 2024
H. O. Al-Shamsi (ed.), *Cancer Care in the United Arab Emirates*,
https://doi.org/10.1007/978-981-99-6794-0_29

29.1 Introduction

Pancreatic cancer originates in the pancreas, with pancreatic adenocarcinoma being the prevalent form. Less common are pancreatic neuroendocrine tumors (NETs). Pancreatic adenocarcinoma arises from the unchecked growth of exocrine cells within the pancreas, whereas endocrine cells, constituting a smaller portion, produce hormones such as insulin and glucagon, pivotal in regulating blood sugar levels. Pancreatic neuroendocrine tumors, on the other hand, originate from the endocrine cells [1]. A critical aspect influencing the classification of pancreatic neoplasms is the degree of cellular differentiation they exhibit. Most pancreatic epithelial neoplasms mirror, to some extent, the normal epithelial cell types found in the pancreas, namely ductal, acinar, and endocrine cells. Over 90% of pancreatic neoplasms demonstrate ductal differentiation, encompassing the prevalent infiltrating ductal adenocarcinoma, along with various cystic and intraductal neoplasms. A smaller subset of pancreatic neoplasms, referred to as 'nonductal,' encompasses endocrine and acinar neoplasms, as well as those with mixed or unclear differentiation patterns. Examples include pancreatic endocrine neoplasms, acinar cell carcinoma, pancreatoblastoma, and solid-pseudopapillary neoplasm [2].

Exocrine elements account for more than 95% of pancreatic malignant neoplasms. The endocrine pancreas gives rise to neoplasms (i.e., pancreatic neuroendocrine [islet cell] tumors), which account for less than 5% of pancreatic neoplasms [3]. Neuroendocrine pancreatic cancer will be discussed in the "Neuroendocrine tumors in the UAE" chapter.

The incidence of pancreatic cancer in the United Arab Emirates (UAE) is similar to that in the US [4]. The pancreatic cancer incidence is relatively low in the UAE. While data on the stage of diagnosis of pancreatic cancer within the UAE are still unpublished, extrapolation from international studies indicates that only 15–20% of those patients are diagnosed at an early stage where the tumor is considered "resectable". Unfortunately, most pancreatic cancer cases are diagnosed at a later stage, when surgery is not a suitable option or the cancer has spread elsewhere (Stage IV).

Pancreatic adenocarcinoma, increasingly prevalent and anticipated to rank as the second most fatal cancer in certain regions, typically manifests at an advanced stage. This advanced presentation significantly contributes to dismal five-year survival rates, ranging from 2% to 9%. Consequently, it stands at the bottom of the list among all cancer types in terms of patient prognostic outcomes [5]. This type of cancer can occur at any age. However, the peak incidence is between 60 and 80 years of age. At the beginning of pancreatic cancer onset, the symptoms are quite nonspecific and progressively worsen over time, including weight loss, fatigue, nausea, malaise, and midepigastric pain that usually radiates to the back [6].

Jaundice is an indication of tumors in the head of the pancreas caused by the constriction of the common bile duct and can be the presenting symptom of these tumors. A tumor may invade the duodenum or stomach, leading to gastric outlet obstruction. Pancreatic cancer is usually diagnosed based on clinical symptoms. As a result, diagnosis is commonly delayed until there is little prospect of a cure. However, if the tumor is detected in those who are predisposed to pancreatic cancer or have a family history of the specified disease, an earlier detection may be conceivable. Approximately 90% of all pancreatic cancer cases are sporadic, with the remaining 10% having a hereditary pattern [6].

This chapter will focus on the clinical presentation, diagnostic evaluation, and staging workup for pancreatic cancer, in addition to pathology, adjuvant and neoadjuvant therapy, surgical management, radiotherapy, and palliative treatment.

29.2 Epidemiology of Pancreatic Cancer

29.2.1 The Global Burden of Pancreatic Cancer

The number of pancreatic cancer-related deaths (466,000) is approximately similar to the number of reported cases, i.e., 496,000, because of its dismal prognosis. It is the seventh-leading cause of cancer deaths in both sexes, with 4.7%. The occurrence and mortality rates are approximately three to four times greater in countries with high Human Development Index (HDI) scores compared to those with low to medium scores, with slightly higher rates observed in men than in women. The highest incidence and mortality rates are observed in Europe, North America, and Australia/New Zealand, with only slight variations in their severity attributed to the bleak prognosis associated with the disease [7, 8]. As of now, the risk factors for pancreatic cancer are not fully comprehended. However, there is evidence indicating that smoking, diabetes, obesity, and certain dietary habits such as high intake of red and processed meats, along with excessive alcohol consumption, are all likely linked to an elevated risk of developing the disease [7, 9].

29.2.2 Data Sources

This is the first systematic effort to comprehensively unify the pancreatic cancer incidence reports over the last decade in the UAE. The code "C25–Pancreas" in the tenth revision of the International Classification of Diseases (ICD-10) was identified as pancreatic cancer. We retrieved the open access data from the Ministry of Health and Prevention's (MOHAP) National Cancer Registry (NCR) for the UAE across the years (2013–2017, 2019, and 2021) that included data on all cancers, including pancreatic cancer, along with gender- and nationality-wise distribution in the country [10].

29.2.3 The UAE Burden of Pancreatic Cancer

During the year 2021, the UAE-NCR recorded 110 pancreatic cancer cases out of a total of 5612 cancer cases, representing 1.96% of all malignant cases in 2021. Non-UAE citizens accounted for a higher proportion of pancreatic cancers in the UAE than UAE citizens, accounting for 79 (71.8%), whereas UAE citizens represented a lower proportion of cases, accounting for 31 (28.1%) (Fig. 29.1). Males were affected by pancreatic cancer at a higher rate (69 [62.7%]) than females (41 [37.2%]) (Fig. 29.2) [11]. Figure 29.3 summarizes data on pancreatic cancer occurrences in the UAE from published UAE-NCR reports over the last decade [10].

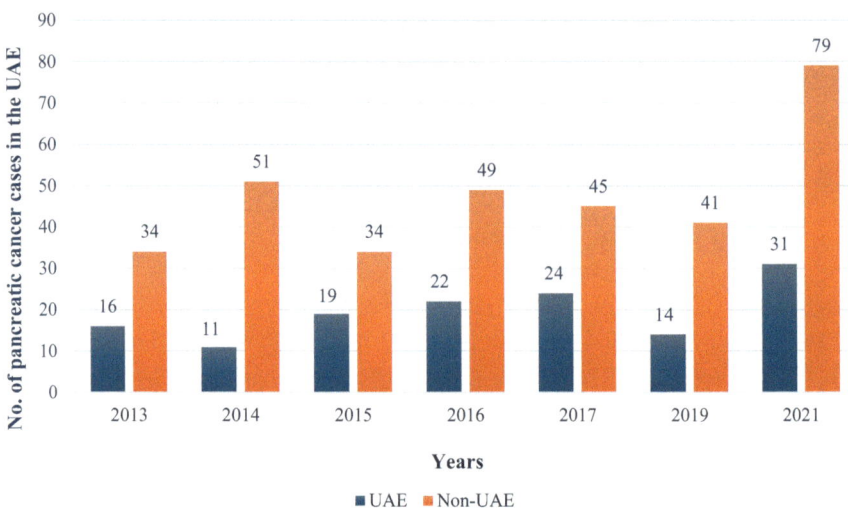

Fig. 29.1 The number of pancreatic cancer cases (malignant) among the UAE population according to nationality, 2013–2021. Source: Ministry of Health and Prevention, Statistics and Research Center, National Disease Registry—UAE National Cancer Registry Report, 2013–2021

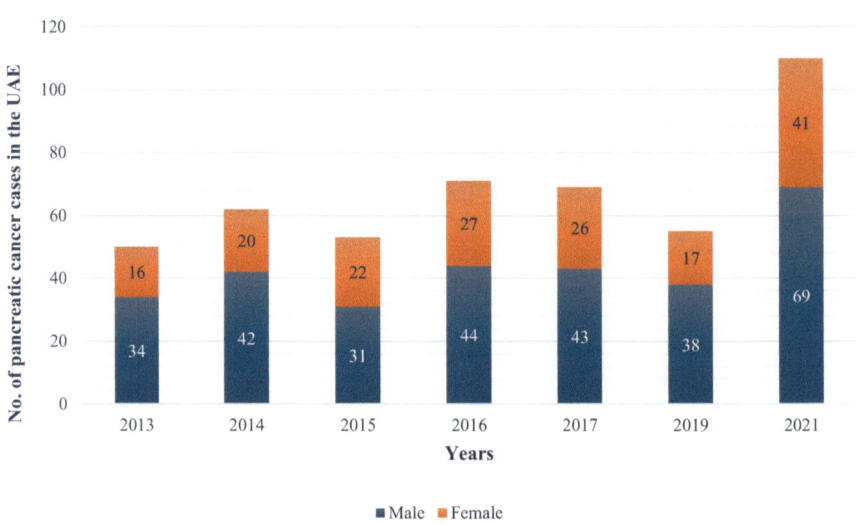

Fig. 29.2 Distribution of pancreatic cancer cases (malignant) according to gender, 2013–2021. Source: Ministry of Health and Prevention, Statistics and Research Center, National Disease Registry—UAE National Cancer Registry Report, 2013–2021

Fig. 29.3 Number of pancreatic cancer (malignant) occurrences in the UAE across the years 2013–2021 Source: Ministry of Health and Prevention, Statistics and Research Center, National Disease Registry—UAE National Cancer Registry Report, 2013–2021

29.3 Surgery for Pancreatic Cancer in the UAE

Presentation with locoregionally advanced and inoperable (30–35%) or metastatic disease (50–55%) remains the most commonly encountered scenario in patients with pancreatic ductal adenocarcinoma, for whom there is no role for palliative resection [12]. However, approximately 15–20% of patients presenting with locoregional disease can be offered surgical resection, which is the only potentially curative treatment, though the majority of these will later relapse [12].

29.3.1 Determining Disease Resectability

A triple-phase (arterial, delayed arterial, and venous) helical computed tomography (CT) of the chest, abdomen, and pelvis with thin slices (the pancreas protocol) is the most commonly applied imaging modality to assess the locoregional disease status and excludes overt metastases in patients with suspected pancreatic ducal adenocarcinoma (PDAC), and for the avoidance of confusion, it is performed before any endoscopic diagnostic or interventional procedures. Magnetic resonance imaging (MRI) with magnetic resonance cholangiopancreatography (MRCP) is an equivalent alternative. Positron emission tomography (PET)-CT is employed selectively in patients with suspected metastases on helical CT or MRI [12]. All the above modalities are available in the UAE, yet their exact utilization in this setting is not standardized, and variation occurs between physicians and hospitals.

While preoperative histological diagnosis is not required and is not always possible in patients with PDAC, the increasing use of neoadjuvant chemotherapy (NAC) with or without radiotherapy protocols in patients with borderline resectable (BR) or resectable disease in the UAE generally requires a prior tissue diagnosis. This is often obtained with an endoscopic ultrasound (EUS)-guided fine needle aspiration biopsy (FNAB), a service that is by and large led by dedicated interventional gastroenterologists. Obtaining a biopsy prior to beginning any cancer treatment in general is widely practiced in order to avoid any medicolegal consequences.

The disease can be considered resectable if the patient's Eastern Cooperative Oncology Group (ECOG) performance status is <3 in the absence of overt metastases and when the major vascular structures, including the superior mesenteric vein (SMV), portal vein (PV), superior mesenteric artery (SMA), celiac axis, and hepatic artery, are clear of the tumor [13]. In this setting, while upfront surgery is the standard of care, the concept of NAC is being tested in clinical trials [14]. On the other hand, neoadjuvant regimens have become the standard of care in patients with BR-PDAC, followed by reevaluation of resectability [14]. There is heterogeneity in the definition of BR-PDAC, as quoted in the National Comprehensive Cancer Network 2021 anatomical criteria [15]. Biological factors suggestive of BR-PDAC were defined at the 20th consensus meeting of the International Association of Pancreatology (IAP) in Sendai, Japan, in 2017, as a tumor potentially resectable anatomically with clinical findings suspicious of, but not proven, distant metastasis, including serum carbohydrate antigen 19-9 (CA19-9) level >500 units/mL or regional lymph node metastasis diagnosed by biopsy or PET-CT [13]. Anatomical features that define unresectable disease include distant metastases, ≥180° SMA encasement, PDAC of the head/uncinate process with celiac artery abutment, inferior vena cava (IVC) involvement, unreconstructible SMV/PV occlusion, aortic invasion, and metastases to distant lymph nodes beyond the locoregional resection field.

29.3.2 Preoperative Biliary Drainage

Preoperative biliary drainage in patients presenting with obstructive jaundice is not routinely recommended and may be associated with an increased incidence of complications and wound infections compared with upfront surgery, as shown in a meta-analysis that included 25 studies (22 retrospective and 3 randomized controlled trials) and 6214 patients [16]. However, endoscopic biliary stenting is indicated in patients with active cholangitis (a rare occurrence in malignant biliary obstruction) and in those with a serum ≥250 μmol/L and is preferred to percutaneous drainage. In this regard, plastic stenting is preferred to covered self-expanding metal stents; the latter were associated with higher intraoperative blood loss and rates of surgery-related morbidity as well as longer postoperative hospital stays in a recent randomized trial of 70 patients with resectable PDAC [17].

29.3.3 Multidisciplinary Team (MDT)

The management of patients with PDAC is discussed at weekly MDT Oncology meetings that include the participation of consultants within the following specialties: gastroenterologists, pathologists, radiologists, nuclear medicine specialists, surgeons, medical and radiation oncologists, and palliative care team representatives. All major cancer centers in the UAE have weekly MDTs. Smaller cancer centers and solo practicing surgeons may not have access to MDT, which may compromise the care of PDAC and other cancers in the UAE.

29.3.4 Centralization of Cancer Services

There is ample evidence that supports the centralization of cancer services, particularly high-risk surgeries such as those for pancreatic cancer, and demonstrates improvements in the quality of the healthcare delivered. In the Netherlands, the centralization of pancreatic surgery resulted in an increase in the resection rate for patients with pancreatic head and periampullary cancer diagnosed in non-pancreatic surgery centers that matched those reported in pancreatic surgery centers and improved overall survival [11]. Centralization enables the establishment of high-volume centers and the delivery of better patient outcomes. In one study from the United States, patients who underwent pancreaticoduodenectomy (PD) for pancreatic cancer at high-volume centers enjoyed improved perioperative outcomes, a reduction in short-term mortality, and better overall survival compared to those treated at low-volume centers [18].

In the UAE, the Oncology Task Force was founded in 2019 under the auspices of the Emirates Oncology Society by practicing cancer care providers with a mandate to fulfill the UAE national agenda, one of whose key performance indicators was to reduce the number of cancer deaths. The Task Force recommended the establishment of tertiary oncology centers that integrate the governmental and private sectors of healthcare providers and that can be easily accessed from all cities within the UAE; these will be supported by multiple affiliated satellite offices equipped with chemotherapy facilities [19].

29.3.5 Neoadjuvant Chemoradiotherapy (NACR)

While upfront surgery is the standard of care for patients with resectable PDAC, the concept of NACR is being tested in clinical trials [20] and is becoming more common in clinical practice in the UAE. On the other hand, neoadjuvant chemotherapy regimens have become the standard of care in patients with borderline resectable PDAC, followed by reevaluation of resectability [20]. Indeed, the PREOPANC Dutch randomized trial showed that neoadjuvant gemcitabine-based chemoradiotherapy followed by surgery and adjuvant gemcitabine improved overall survival

compared with upfront surgery and adjuvant gemcitabine in resectable and borderline resectable pancreatic cancer [21].

29.3.6 Surgery

Surgery is performed within a month of completing the staging investigations
(which may include staging laparoscopy) at most and is conducted by specialty-
trained surgeons. One month from surgery is the most widely practiced approach for
surgery post-neoadjuvant therapy, and up to 6 weeks in the UAE, based on the
patient's recovery post-neoadjuvant therapy.

While traditional open surgery is used to remove exocrine pancreatic cancers, the
global adoption of minimally invasive approaches (laparoscopic, laparoscopic-
assisted, or robotic-assisted) has seen selected surgeons in the UAE offer these techniques at select centres. This is especially true for the less difficult distal
pancreatectomy with *en block* splenectomy, for which there is clear evidence from
the LEOPARD Dutch randomized controlled trial of clinical benefits to patients in
terms of reduced time to functional recovery and rates of delayed gastric emptying,
as well as a better quality of life with the minimally invasive approach compared to
open surgery and without increasing cost [22]. On the other hand, the complexity of
both the resection and the reconstruction in PD and the conflicting evidence from
the three randomized controlled trials regarding the potential advantages of the minimally invasive approaches over open surgery [23] have limited their adoption
among UAE surgeons. Nonetheless, the author has reported two comparative case-
matched controlled trials of laparoscopic versus open PD, one from a U.K. centre
[24] and the other from a Middle Eastern centre [25], and consistently demonstrated
significant reductions in hospital stay while maintaining oncologic outcomes.

29.4   Radiation Oncology

Radiation therapy continues to play an important role for pancreatic cancer patients.
In the palliative setting, where tumors, either the primary tumor or metastatic sites,
are causing significant pain and affecting the patient's quality of life, radiation can
be utilized to alleviate those symptoms [26].

Most cancer centers in the UAE adhere to the National Comprehensive Cancer
Network (NCCN) guidelines, which provide multiple different approaches, including for patients diagnosed with locally advanced or borderline resectable pancreatic
cancer prior to surgery. Radiation options included are concurrent chemoradiotherapy after induction chemotherapy, upfront concurrent chemoradiotherapy, stereotactic ablative body radiotherapy (SABR), or enrollment in a clinical trial [27].
While most radiotherapy departments in the UAE are equipped with the latest radiotherapy machines and upgrades, allowing image guidance and very precise treatment, the utilization of radiation therapy for pancreatic cancer in the UAE remains
lower than estimated. This could be related to multiple factors, including the lack of

a proper referral pattern for those patients, the concern of potentially losing the opportunity to operate on those patients, the lack of radiation oncologists or medical physics expertise in advanced pancreatic cancer radiotherapy techniques, and the lack of information about the presence of experienced gastroenterologists or interventional radiologists who can do the proper fiducial placements needed for image guidance or stereotactic ablative radiotherapy. Finally, and perhaps most importantly, there is a lack of a dedicated center of excellence for treating pancreatic cancer cases in the UAE, which would be ideal given the relatively small number of cases scattered across the country [18]. Having a dedicated center of excellence for nonmetastatic, borderline non-resectable pancreatic tumors would allow patients to access a perhaps more experienced radiation facility, which could have more options and even clinical trials to enroll those patients in and provide the best potential outcome.

All advanced modalities for pancreatic cancer irradiation should be available with the current and near-future expansion of UAE radiotherapy infrastructure and expertise. Work must be done among centers to centralize cases or unresectable nonmetastatic cases that might be a potential candidate for enrollment in a clinical trial testing a novel agent with a novel or advanced radiation technique. Knowledge sharing and collaboration in facing this disease are a must, and in our opinion, dedicated radiation therapy workshops and meetings for such diseases should also be more readily available to keep up with the rapid progress in pancreatic cancer radiotherapy.

29.5 Palliative and Supportive Care for Pancreatic Cancer in the UAE

PDAC carries significant morbidity for patients with a wide spectrum of commonly reported symptoms, including nausea, pain, dyspnea, abdominal distension and bloating, constipation, weight loss, malnutrition, steatorrhea, diarrhea, anxiety, and depression. A thorough supportive care assessment is essential to minimize symptom burden, including dietician input, treatment of pancreatic insufficiency, and a low threshold for consideration of small bowel overgrowth if symptoms do not improve with treatment of exocrine insufficiency. Nausea and vomiting are very common and multifactorial, with multiple potential etiologies, including local and systemic tumor effects, chemotherapy, medications, anxiety, and gastric outlet obstruction.

The abdominal pain of PDAC has an adverse effect on the patient's quality of life, particularly when symptoms do not improve with systemic anticancer therapies. PDAC is recognized as one of the most painful cancers with its characteristic epigastric distribution, but there is more commonly a mixed picture at presentation: a visceral component that is poorly localized, dull, colicky, and associated with nausea and vomiting. Somatic pain, which is often sharp and well localized, and neuropathic pain, which is referred to as pain in the back and is frequently exacerbated by lying flat, are two types of pain.

Somatic pain arises from local invasion and metastasis into the surrounding peritoneum, retroperitoneum, and bones. Visceral pain arises from the infiltration of adjacent organs and the accumulation of ascites in patients with more advanced stages of disease. The neuropathic pain component is attributed to the perineural invasion. Extra-pancreatic nerve plexus invasion is responsible for the neuropathic pain sensation. Similarities in growth factor receptors and adhesion molecules between pancreatic cancer cells and neuronal cells explain the affinity to neural tissue and lead to increased cancer cell proliferation, migration, and invasion along nerve bundles. Nociceptive signals are carried along sympathetic fibers to the celiac plexus nerves and ganglia (T12-L2) and are transmitted via the splanchnic nerves (T5–T12) to the higher centers of the central nervous system [28].

In accordance with the World Health Organization (WHO) analgesic ladder model [29], patients often receive simple analgesics (Step 1) like paracetamol or nonsteroidal anti-inflammatory drugs for mild pain with the addition of adjuvant analgesics like corticosteroids, gabapentinoids (Pregabalin and Gabapentin), tricyclic antidepressants (amitriptyline), and SNRIs (Duloxetine) for clearly neuropathic pain. With increasing severity of pain, weak opioids (Step 2) (Codeine or Tramadol) are trialed before escalating to Step 3 with strong opioids (Morphine, Oxycodone, Hydromorphone, Methadone, and Fentanyl). The NMDA receptor antagonist, ketamine, is also utilized in specialized palliative care units for patients with refractory neuropathic pain despite high-dose opioid therapy [30]. While all these named analgesics are available within the UAE, availability and clinician familiarity with their use vary from one Emirate to another.

When combinations of traditional opioid analgesics (oral or parenteral) fail to provide adequate analgesia, interventional analgesic techniques may prove beneficial. Celiac plexus blockade is the most commonly used intervention and involves the disruption of visceral pain innervation from the pancreas and adjacent structures by an injection of corticosteroids and/or local anesthetic. Celiac blocks can be performed using either a percutaneous CT-guided (PC) or an endoscopic ultrasound (EUS) approach, with some evidence suggesting that the EUS approach provided better quality analgesia than the PC approach [31]. Likewise, celiac plexus neurolysis, which entails permanent destruction of the plexus, has also been demonstrated to be suitable for patients with a short life expectancy of 3–6 months, again with an EUS approach giving better patient-reported outcomes. Celiac plexus blocks are available at a small but growing number of hospitals across the UAE. Implanted intrathecal drug delivery systems may be considered for patients with a longer predicted life expectancy (>6 months) who continue to have poorly controlled pain refractory to high-dose opioids and celiac plexus blocks, but this is only available at a small number of super-specialized facilities and is not covered by the majority of health insurance policies.

The evidence for early palliative care integration into oncology care is now well established [32], but we also have evidence that early palliative care for patients with PDAC is also associated with reduced emergency department admissions and healthcare costs [33].

29.6 Conclusion

In the UAE, treatment of pancreatic ducal adenocarcinoma, whether early or metastatic, is advanced and largely follows NCCN guidelines. One of the major challenges is decentralized surgical care due to the large number of cancer care providers in the UAE, which may affect the surgical outcome in lower-volume centers.

Conflict of Interest The authors have no conflict of interest to declare.

References

1. American Cancer Society (2024) What is pancreatic cancer? https://www.cancer.org/cancer/types/pancreatic-cancer/about/what-is-pancreatic-cancer.html. Accessed on 04 Jun 2024
2. Klimstra DS. Nonductal neoplasms of the pancreas. Mod Pathol. 2007;20(Suppl 1):S94–112.
3. https://www.uptodate.com/contents/clinical-manifestations-diagnosis-and-staging-of-exocrine-pancreatic-cancer. Accessed 31 Oct 2022.
4. https://www.cancer.net/cancer-types/pancreatic-cancer/statistics. Accessed 1 Nov 2022.
5. McGuigan A, et al. Pancreatic cancer: a review of clinical diagnosis, epidemiology, treatment and outcomes. World J Gastroenterol. 2018;24(43):4846–61.
6. Ansari D, et al. Pancreatic cancer: yesterday, today and tomorrow. Future Oncol. 2016;12(16):1929–46.
7. Arnold M, et al. Global burden of 5 major types of gastrointestinal cancer. Gastroenterology. 2020;159(1):335–349.e15.
8. Sung H, Ferlay J, Siegel RL. Global cancer statistics 2020: GLOBOCAN estimates of incidence and mortality worldwide for 36 cancers in 185 countries. CA Cancer J Clin. 2021;71(3):209–49.
9. American Institute for Cancer Research. Food, nutrition, physical activity, and the prevention of pancreatic cancer. Continuous Update Project Report: World Cancer Research Fund/American Institute for Cancer Research, 2012
10. Cancer Incidence in United Arab Emirates. National Disease Registry Section. Annual Report of the UAE-National Cancer Registry (2011–2017). Ministry of Health & Prevention, United Arab Emirates. https://smartapps.moh.gov.ae/ords/f?p=105:511. Accessed 31 Oct 2022.
11. Latenstein AEJ, et al. Effect of centralization and regionalization of pancreatic surgery on resection rates and survival. Br J Surg. 2021;108(7):826–33.
12. Park W, Chawla A, O'Reilly EM. Pancreatic cancer: a review. JAMA. 2021;326(9):851–62.
13. Isaji S, et al. International consensus on definition and criteria of borderline resectable pancreatic ductal adenocarcinoma 2017. Pancreatology. 2018;18(1):2–11.
14. Tonini V, Zanni M. Pancreatic cancer in 2021: what you need to know to win. World J Gastroenterol. 2021;27(35):5851–89.
15. Nappo G, Donisi G, Zerbi A. Borderline resectable pancreatic cancer: certainties and controversies. World J Gastrointest Surg. 2021;13(6):516–28.
16. Scheufele F, et al. Preoperative biliary stenting versus operation first in jaundiced patients due to malignant lesions in the pancreatic head: a meta-analysis of current literature. Surgery. 2017;161(4):939–50.
17. Mandai K, et al. Fully covered metal stents vs plastic stents for preoperative biliary drainage in patients with resectable pancreatic cancer without neoadjuvant chemotherapy: a multicenter, prospective, randomized controlled trial. J Hepatobiliary Pancreat Sci. 2021;29:1185.
18. Lidsky ME, et al. Going the extra mile: improved survival for pancreatic cancer patients traveling to high-volume centers. Ann Surg. 2017;266(2):333–8.
19. Al-Shamsi H, et al. The state of cancer Care in The United Arab Emirates in 2020: challenges and recommendations, a report by The United Arab Emirates oncology task force. Gulf J Oncolog. 2020;1(32):71–87.

20. Muller PC, et al. Neoadjuvant chemotherapy in pancreatic cancer: an appraisal of the current high-level evidence. Pharmacology. 2021;106(3–4):143–53.
21. Versteijne E, et al. Neoadjuvant Chemoradiotherapy versus upfront surgery for resectable and borderline resectable pancreatic cancer: long-term results of the Dutch randomized PREOPANC trial. J Clin Oncol. 2022;40(11):1220–30.
22. de Rooij T, et al. Minimally invasive versus open distal pancreatectomy (LEOPARD): a multicenter patient-blinded randomized controlled trial. Ann Surg. 2019;269(1):2–9.
23. Nickel F, et al. Laparoscopic versus open Pancreaticoduodenectomy: a systematic review and meta-analysis of randomized controlled trials. Ann Surg. 2020;271(1):54–66.
24. Khaled YS, et al. Matched case-control comparative study of laparoscopic versus open Pancreaticoduodenectomy for malignant lesions. Surg Laparosc Endosc Percutan Tech. 2018;28(1):47–51.
25. Ammori BJ, et al. A case-matched comparative study of laparoscopic versus open Pancreaticoduodenectomy. Surg Laparosc Endosc Percutan Tech. 2020;30(3):276–80.
26. Coveler AL, et al. Pancreas cancer-associated pain management. Oncologist. 2021;26(6):e971–82.
27. https://www.nccn.org/professionals/physician_gls/pdf/pancreatic.pdf. Accessed 2 Nov 2022.
28. Moffat GT, Epstein AS, O'Reilly EM. Pancreatic cancer-a disease in need: optimizing and integrating supportive care. Cancer. 2019;125(22):3927–35.
29. Ventafridda V, et al. WHO guidelines for the use of analgesics in cancer pain. Int J Tissue React. 1985;7(1):93–6.
30. Kwon KM, et al. Factors associated with ketamine use in pancreatic cancer patient in a single hospice center. J Hospice Palliat Care. 2016;19(3):249–55.
31. Santosh D, et al. Clinical trial: a randomized trial comparing fluoroscopy guided percutaneous technique vs. endoscopic ultrasound guided technique of coeliac plexus block for treatment of pain in chronic pancreatitis. Aliment Pharmacol Ther. 2009;29(9):979–84.
32. Temel JS, et al. Early palliative care for patients with metastatic non-small-cell lung cancer. N Engl J Med. 2010;363(8):733–42.
33. Bevins J, et al. Early palliative care is associated with reduced emergency department utilization in pancreatic cancer. Am J Clin Oncol. 2021;44(5):181–6.

Prof. Humaid Obaid Al-Shamsi is the Chief Executive Officer of Burjeel Cancer Institute in Abu Dhabi, UAE, President of the Emirates Oncology Society, Lead of the Gulf Cancer Society, Full Professor of Oncology at the Ras Al Khaimah Medical and Health Sciences University, Ras Al Khaimah, UAE, and an Adjunct Professor of Oncology at the College of Medicine, University of Sharjah. He is the first Emirati to be promoted as a professor in oncology in the UAE. He is also the Chairman for Colorectal Cancer in the MENA region, appointed by the prestigious National Comprehensive Cancer Network®. He is also the only member of Lung Cancer Policy Network in the MENA region that aims to advance lung cancer research and screening globally. He is the Chairman of the Oncology and Hematology Fellowship Training Program for the National Institute for Health Specialties in the United Arab Emirates. He is the only member in GCC in the WIN Consortium which is comprised of organizations representing all stakeholders in personalized cancer medicine globally.

He is board-certified in both internal medicine and oncology from the UK, USA (ABIM), and Canada (FRCPC). He has also been awarded the FRCP (London) in 2023 and FRCP (Glasgow) in 2024. He is the only physician in the UAE with a subspecialty fellowship certification and training in gastrointestinal oncology and the first Emirati to train and complete a clinical post-doctoral fellowship in palliative care. He was an assistant professor at the University of Texas MD Anderson Cancer Center between 2014 and 2017. He has published more than 140 peer-reviewed articles in JAMA Oncology, Lancet Oncology, The Oncologist, BMC Cancer, and many others. His area of expertise includes precision oncology and cancer care in the UAE. In 2016, he published with his group from MD Anderson the JCO paper describing a new distinct subgroup of CRC, NON V600 BRAF-mutated CRC. In 2022, he published the first book about cancer research in the UAE and also the first book about cancer in the Arab world, both of which were launched at Dubai Expo 2020. *Cancer in the Arab World* has been downloaded more than 450,000 times in its first 18 months of publication and is the ultimate source of cancer data in the Arab region. He also published the first comprehensive book about cancer care in the UAE which is the first book in UAE history to document the cancer care in the UAE with many topics addressed for the first time, e.g., neuroendocrine tumors in the UAE. He is passionate about advancing cancer care in the UAE and the GCC and has made significant contributions to cancer awareness and early detection for the public using social media platforms. He is considered as the most followed oncologist in the world with over 300,000 subscribers across his social media platforms (Instagram, Twitter, LinkedIn, and TikTok). In 2022, he was awarded the prestigious Feigenbaum Leadership Excellence Award from Sheikh Hamdan Smart University for his exceptional leadership and research and the Sharjah Award for Volunteering. He was also named the Researcher of the Year in the UAE in 2020 and 2021 by the Emirates Oncology Society.

In May 2024, HH Sheikh Mansour bin Zayed Al Nahyan, Vice President of the United Arab Emirates, awarded him the first place in UAE Nafis program for outstanding leadership in private sector across all business and medical disciplines. Beside his clinical and administrative duties, he is engaged in education and various levels of research training for medical trainees to enhance their clinical and research skills. His mission is to advance cancer care in the UAE and the MENA region and make cancer care accessible to everyone in need around the globe.

Ms. Faryal Iqbal is the research associate at Burjeel Medical City, Abu Dhabi, United Arab Emirates. She completed her undergraduate studies in molecular biology and biotechnology. Following that, she received a postgraduate qualification in molecular genetics. She co-edited "Cancer in the Arab World," the first extensive book covering cancer care across all Arab countries. The book succeeded significantly, with over 450,000 downloads within just two years. She has several peer-reviewed papers under her name. In September 2023, Ms. Iqbal received the "EOS Research Award" from the prestigious Emirates Oncology Society for her research efforts. Moreover, she assists in different aspects of clinical trial implementation at a research site. Her research interests and publications encompass oncology, hematology, and genetics.

Dr. Neil A. Nijhawan is a UK-trained consultant in Palliative Medicine at Burjeel Medical City. After medical school at Kings College London, he pursued speciality training in Palliative Medicine in London, with rotations in acute general hospitals, domiciliary visits, community hospices, and tertiary oncology centres. Prior to completing his palliative medicine training, Neil returned to his childhood home, Trinidad in the West Indies, to help set up and commission the new Caura Hospital Palliative Care Unit, where he was the Medical Director. This unit was opened in 2014 and provides a comprehensive palliative care service, including a 12-bed inpatient unit, weekly outpatient clinics, and a palliative care consult service at the local university hospital. After completing his specialty training, Neil worked as a consultant in palliative medicine at the Imperial College Healthcare NHS Trust in London, where he was the clinical lead for palliative medicine. His clinical area of interest is symptom control (including pain, nausea, breathlessness, and fatigue) and assistance with complex treatment decision-making at the end of life, and he is often called on to provide an independent second opinion. He is active in palliative care education and palliative care advocacy and is currently the UAE representative to the WHO Eastern Mediterranean Region Palliative Care Expert Network. Neil holds adjunct faculty positions with both Khalifa University and Gulf Medical University where he is Clinical Associate Professor in Hospice & Palliative Medicine.

Dr. Hampig R. Kourie gained his medical doctor degree from the Faculty of Medicine of Saint Joseph University in Beirut, Lebanon, in 2010. He started his fellowship in Hôtel-Dieu de France, Saint Joseph University Hospital in Beirut, Lebanon, and he continued his medical oncology fellowship in Jules Bordet Institute in Brussels, Belgium, from 2014 to 2016. Since September 2016, he worked in the digestive oncology department in Hôpital Européen Georges Pompidou (HEGP) in Paris as a researcher in the Association des Gastro-entérologues Oncologues en France (AGEO). He is also certified from Paris Diderot and Paris Descartes universities in hereditary cancers and digestive oncology. He gained his PhD in genetics from Saint Joseph University of Beirut in 2023 and his MEMS from ESA Business School in 2023. He is practicing as hematologist-oncologist in Hôtel Dieu de France, Hôpital Saint Joseph des Soeurs de la Croix, and Bellevue Medical Center.

He is actually an ESMO Faculty in the colorectal cancer group. He founded and directed the Middle East Biomarkers course and the Middle East and North Africa GI Oncology Summit. He has more than 200 peer-reviewed articles, mainly in immunotherapy, oncogenomics, and digestive oncology fields.

Dr. Nadia Abdelwahed With a solid foundation of medical studies, Dr. Nadia Abdelwahed specialized in the branch of medical oncology and added valuable practical experience in the administration of oncology intervention and therapy by addressing cases from all stages and conditions of the disease. Her specialty practice began in 2015 and continued for 3 years, making her an expert in the diagnosis and screening of all cancer types, especially breast cancer, and an expert in cancer treatment methods, qualifying her through the school of medicine residency program. During this period, she addressed both the inpatient and outpatient categories of patients. After this academic training and practical performance, Dr. Nadia secured a placement as a medical oncologist (specialist) in a leading university oncology hospital in Damascus, Syria. After completing her tenure of office there, she joined another multispecialty center as a medical oncologist from 2018 until 2021. Alongside her postgraduate studies and practical training, Dr. Nadia conducted her independent research studies leading to the master's degree (Neoadjuvant Capecitabine in Rectal Cancer) Research in 2018. In 2017, she passed the European Society of Medical Oncology (ESMO) exam and was certified for a period of 5 years. In 2021, she also excelled at the Immune Oncology Course at Harvard Medical School in the United States, earning a Certificate of Achievement. She is also a recipient of the FRON Prize (Forum of Research in Oncology) for her research work on "Impact of HER-2 Ratio on Efficacy of Trastuzumab in Early Breast Cancer" in 2017. In 2021, Dr. Nadia relocated to the UAE and qualified for the HAAD Licensing Exam as a Specialist in Medical Oncology; soon after, she joined VPS Healthcare at the Burjeel Cancer Institute of the Burjeel Medical City.

Dr. Ibrahim H. Abu-Gheida is the Clinical Director of the department of radiation oncology at Burjeel Medical City. He also serves as a regional Radiological Society of North America (RSNA) committee representative for the Middle East and Africa. Dr. Abu-Gheida completed his undergraduate training, where he earned a Bachelor of Science with honors degree from the American University of Beirut. Following this, Dr. Abu-Gheida completed his Medical School training at the American University of Beirut Medical Center. He continued and joined the Department of Internal Medicine at the American University of Beirut. Then he did his training in the Department of Radiation Oncology at the American University of Beirut Medical Center, where he also served as the chief resident. During his training, Dr. Abu-Gheida completed a Harvard-affiliated NIH-funded research program as well. After his residency, Dr. Ibrahim went to the Cleveland Clinic in Ohio, where he was appointed as an Advanced Clinical Radiation Oncology Fellow. Dr. Abu-Gheida went and joined the University of Texas MD Anderson Cancer Center, where he sub-specialized in treating breast, gastrointestinal, and genitourinary cancers. Dr. Abu Gheida played an instrumental role in establishing and heading the radiation oncology facility and department at Burjeel Medical City. He has chaired and co-chaired multiple international oncology conferences. Dr. Ibrahim has more than 40 peer-reviewed papers in prestigious medical journals, including the American Society of Radiation Oncology official journal - the International Journal of Radiation Oncology Biology and Physics, Nature, the Journal of Clinical Oncology, and several others. He is also the primary author and editor of several book chapters published in prestigious books.

Professor Dr. Basil Ammori graduated from Baghdad University medical school in 1986, completed his surgical training in the UK, obtained the CCST in 2000, and was granted a Hunterian Professorship by the Royal College of Surgeons of England in 2000. Appointed a consultant surgeon, he led the laparoscopic bariatric and HPB (hepato-pancreato-biliary) surgery service in Manchester between 2002 and 2017 and was the first to introduce laparoscopic Whipple's procedure in the U.K. He was appointed as an Honorary Professor at the University of Manchester in 2011 and later joined King Hussein Cancer Center, Amman, Jordan, as a full member in November 2017 as a laparoscopic gastrointestinal oncology surgeon before moving to Burjeel Hospital, Abu Dhabi, in January 2020. He sat on the editorial boards of a number of medical journals and has over 240 peer-reviewed publications and book chapters.

Hepatocellular Carcinoma (HCC) in the UAE

30

Salman Wahib Srayaldeen
and Mohamed Ahmed Mohamed Elkhalifa

30.1 Introduction

There are two main types of liver cancer based on their etiology and the origin of the tumor: primary and secondary, or metastatic.

Primary liver cancer: Most cases of primary liver cancer are caused by hepatocellular carcinoma (HCC), which accounts for 90% of all cases. In this case, hepatocytes with liver cells are the main cause of cancer. It has been reported that about 10% of primary liver cancer cases are associated with cholangiocarcinoma. The bile ducts of the liver are the first site where cholangiocarcinoma develops.

Metastatic or secondary liver cancer: Secondary liver cancer or metastatic liver cancer occurs when cancer in another organ or body tissue migrates via the circulatory system to the liver. The liver is a common target for metastases from other cancers, including colon, lung, breast, and other organs.

HCC is a more prevalent cancer among individuals who have certain previous medical conditions, for instance, chronic liver disease (CLD) [1], which includes patients who have chronic hepatitis and non-alcohol-related fatty liver disease (NAFLD). However, it can also occur in individuals with no history of chronic liver illness [1, 2].

According to the 2020 statistics of the World Health Organization (WHO), the worldwide incidence of HCC estimation is 9.5 cases per 100,000 people, with over 905,000 new cases reported in 2020 [3], making it less frequent than breast, prostate, lung, cervical, and other GI malignancies. However, the importance of HCC comes from its being considered the third most notable cause of mortality associated with cancer globally, following lung and colorectal malignancies [4], which explains the narrow difference between incidence and prevalence as 5-Year Relative Survival statistics are very low, which represents and constitutes a major global health problem [5, 6].

S. W. Srayaldeen (✉) · M. A. M. Elkhalifa (✉)
Sheikh Khalifa Specialty Hospital, Ras al Khaimah, United Arab Emirates

© The Author(s) 2024
H. O. Al-Shamsi (ed.), *Cancer Care in the United Arab Emirates*,
https://doi.org/10.1007/978-981-99-6794-0_30

30.2 Epidemiology

HCC is allocated as the seventh cancer in the worldwide statistics [7], with 905,677 new cases in 2020, and the estimated number of cancer-related deaths in 2020 was 830,180 cases [8]. In terms of disease prevalence, men are more likely than women; the prevalence sex ratio varies between 2:1 and 4:1, depending on the geographic region [9].

The global incidence varied by population and continent; Asia is thought to have the highest incidence, accounting for approximately 72% of all new cases, followed by Europe (10%) and Africa (8%) [5, 10].

When considering the incidence of HCC according to the geographical distribution, Mongolia has the highest incidence worldwide with about 71.3 cases per 100,000 people, followed by China with 29.2 cases per 100,000 people; however, when comparing the number of populations between these countries, China is considered to have the highest frequency [3, 11].

Italy is the most common country in Europe, with 18.1 per 100,000 people, and Egypt has 28.3 per 100,000 people in Africa [3, 7].

The Middle East and North Africa (MENA) and Gulf countries, excluding Egypt, have the lowest incidence of HCC compared to other countries; for example, Syria and Jordan have almost equal frequencies of HCC at around 2 per 100,000 people [12].

According to GLOBOCAN 2020, in Gulf countries, the highest frequency of HCC is 3.4 per 100,000 people in Saudi Arabia. In terms of the Gulf Cooperation Council's (GCC) countries age-standardized incidence rate (ASR), which was estimated at 4.7 per 100,000 people in 2020 [7], Saudi Arabia ranks first with 5.2 per 100,000 people, followed by Kuwait, and the United Arab Emirates (UAE) ranks last with 2.9 per 100,000 people [3, 11, 13].

The UAE recorded approximately 83 newly diagnosed cases of HCC in 2020 [5, 7], with an incidence of 0.9 per 100,000 people.

The liver cancer demographics are shown in Table 30.1. Figure 30.1 shows the number of liver cancer (malignant) occurrences in the UAE across the years 2013–2021 by UAE- National Cancer Registry (NCR) [14].

The fifth-leading cause of death in the UAE was found to be cancer in the UAE-NCR report, 2021. The number of deaths from all cancers in 2021 totaled 975 (506 in males and 469 in females) and accounted for 8.2% of all deaths, regardless of nationality, type of cancer, or gender [14].

In 2021, the UAE National Cancer Registry reported a total of 5830 newly diagnosed cancer cases. Of these cases, 96% (5612) were invasive cancer cases, while only 4% (218) were in situ cases. These statistics highlight the prevalence of invasive cancer cases in the UAE and the need for continued efforts in cancer prevention and treatment [14].

Cancer incidence rates in the UAE show that colorectal, breast, thyroid, leukemia, and skin cancers are the five most frequently diagnosed types of disease among males as well as females. While colorectal, skin, prostate, leukemia, and Non-Hodgkin Lymphoma (NHL) were the top-ranked cancers among males, breast,

Table 30.1 Liver cancer demographics among the UAE population during 2013–2021

Year	UAE population (in millions)	Total malignant cases (in numbers)	Liver cancer cases (in numbers)	Percentage (%)	Crude incidence rate liver cancer cases per 100,000
2013	8.66	3574	69	1.93	–
2014	8.79	3610	58	1.61	0.64
2015	8.93	3744	68	1.82	0.7
2016	9.12	3982	83	2.08	–
2017	9.3	4123	72	1.75	0.8
2019	9.5	4381	72	1.64	–
2021	–	5612	114	2.0	1.2

Source: UAE population: https://fcsc.gov.ae/ar-ae/Pages/Statistics/Statistics-by-Subject; Ministry of Health and Prevention, Statistics and Research Center, National Disease Registry—UAE National Cancer Registry Report

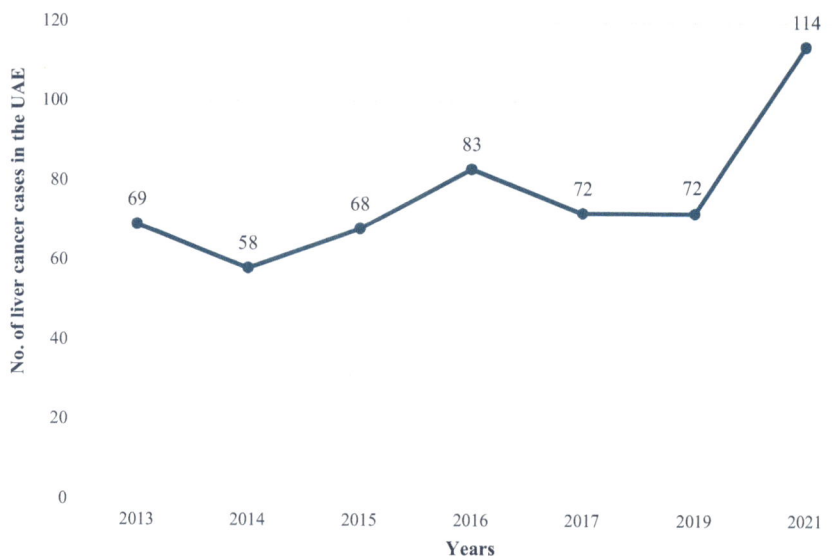

Fig. 30.1 Number of liver cancer (malignant) occurrences in the UAE across the years 2013–2021. Source: Ministry of Health and Prevention, Statistics and Research Center, National Disease Registry—UAE National Cancer Registry Report, 2013–2021

thyroid, colorectal, uterus, and ovary were the top-ranked cancers among females. However, HCC is not among the top ten causes of the common cancers' death [15–17].

According to UAE-NCR data collected in 2021, the number of patients with HCC increased gradually from 69 new cases in 2013 to 114 new cases in all populations in 2021 [14], of which 30 are UAE nationality patients with a gender distribution of 19 males and 11 females, and 84 non-UAE national patients (62 males and 22 females) [14] (Figs. 30.2 and 30.4). Figure 30.3 shows the distribution of liver cancer cases by surveillance, epidemiology, and end results (SEER) stages in the UAE across the years.

In 2020, the UAE reported 83 new cases and 96 (0.49%) deaths of HCC. With a crude rate of 0.84 and an ASR of 2.9 per 100,000 people [7], the incremental incidence of HCC in the UAE could be justified by the increasing incidence of obesity, alcohol consumption, and the aging of the population [18, 19].

In the UAE, the majority of patients at diagnosis are older (over 80 years old) [20], and the majority of cases are advanced or unstaged [15, 21] (Fig. 30.3); the male to female ratio for HCC incidence is 2.7:1. This is higher than the global average of 2.3:1. At an older age (80+ years), the male-to-female ratio is closer to 2:1 [20] (Fig. 30.4). Furthermore, males have a markedly higher risk of dying from HCC than females [9, 22]. In the UAE, chronic hepatitis B or C forms are responsible for and considered the leading risk factor for developing hepatocellular carcinoma, rather than other potential causes [23].

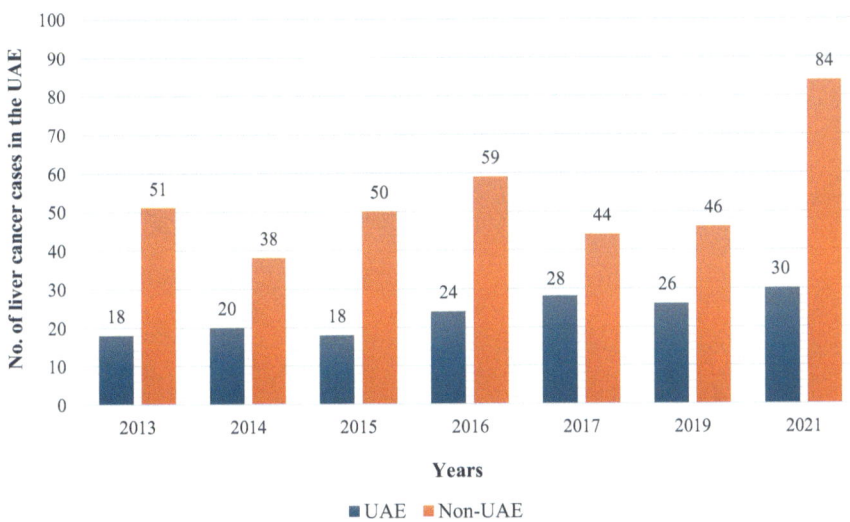

Fig. 30.2 The number of liver cancer cases (malignant) among the UAE population according to nationality, 2013–2021. Source: Ministry of Health and Prevention, Statistics and Research Center, National Disease Registry—UAE National Cancer Registry Report

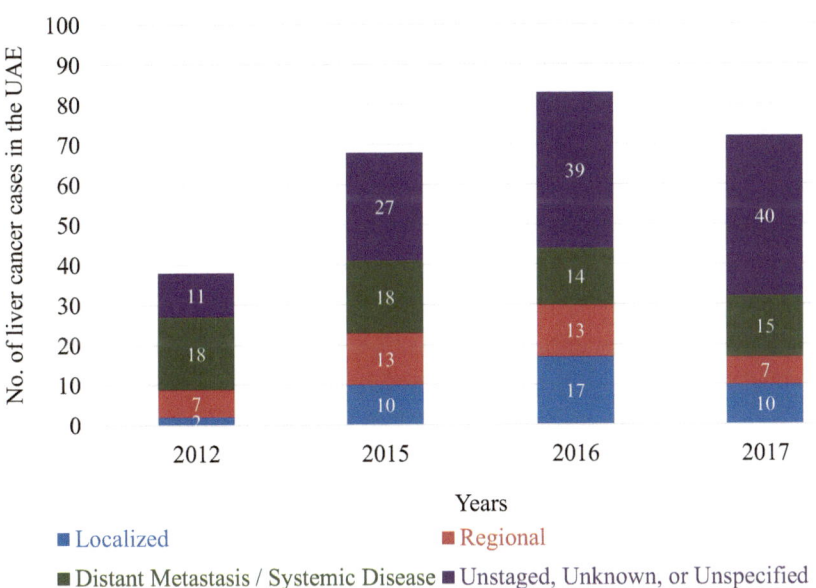

Fig. 30.3 Distribution of liver cancer cases by SEER stages in the UAE across the years. Source: Ministry of Health and Prevention, Statistics and Research Center, National Disease Registry—UAE National Cancer Registry Report, 2012–2017

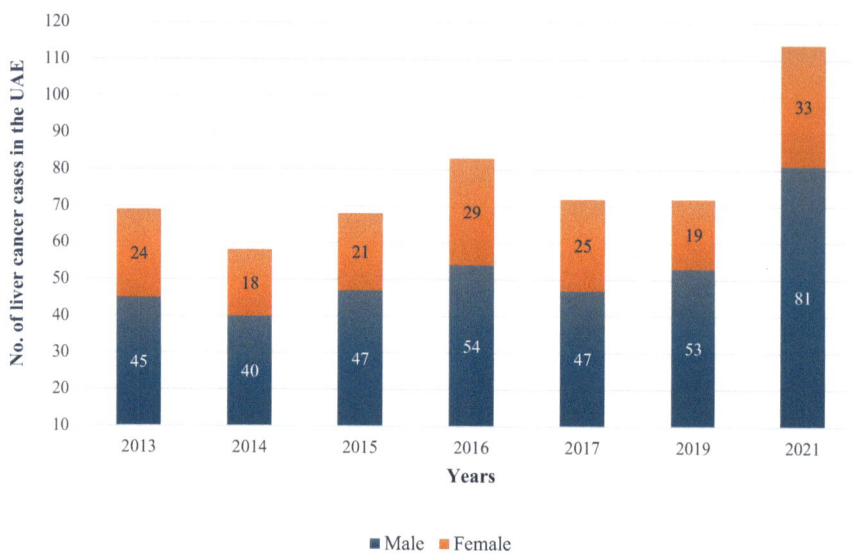

Fig. 30.4 Distribution of liver cancer cases (malignant) according to gender, 2013–2021. Source: Ministry of Health and Prevention, Statistics and Research Center, National Disease Registry—UAE National Cancer Registry Report

30.3   Risk Factors

Factors that increase the risk of primary liver cancer include:

30.3.1 HBV Chronic Infection

Hepatitis B virus (HBV) infection has severe complications, starting with chronic sequelae, which are associated with an increased lifetime risk of developing HCC and cirrhosis of the liver (LC). Limited, extensive research and studies in the UAE examine the burden of HBV-related hepatocellular carcinoma, which poses a significant obstacle to understanding the disease comprehensively [24, 25].

In the UAE, chronic hepatitis B is the most prevalent risk factor for HCC, with 42% in 1990 and 41% in 2017. However, liver cancer rates related to the hepatitis B virus are lower in the UAE than in most western nations [26], such as Spain, the United Kingdom, and the United States [23]. As a result of comprehensive measures to effectively control the spread of disease, beginning with a widespread vaccination program, HBV vaccination programs have been mandatory for all newborns since 1991, and all infants receive four doses of the vaccine at 0 months, 2 months, 4 months, and 6 months. In addition, as part of the visa renewal process, all expatriate residents in the UAE are periodically examined for hepatitis B, along with other infectious diseases such as HIV [16, 17, 25].

The incidence of HBV was 0.23% in the most recent survey of blood donors. As of the 2016 Population Census, an estimated 1.2 million people in the UAE hold Emirati citizenship [25], a percentage of 0.23% would result in an estimated 2760 HBV cases. Assuming a prevalence of 1.5%, as reported among pregnant women in 2000, the country would have 18,000 HBV cases. According to estimations made by medical experts, the prevalence of HBV in the general population of the UAE is likely to range from 1 to 1.5%, leading to an anticipated number of cases ranging from 12,000 to 18,000 cases, as illustrated in and [17, 23].

Chronic hepatitis B virus infection increases the risk of liver-related disorders and end-stage liver diseases, including HCC, liver failure, and cirrhosis [27, 28]. According to the research, most patients with HBV who develop HCC have cirrhosis [29]. HBV infection was also associated with a greater annual incidence of HCC in cirrhotic individuals compared to those without cirrhosis [29].

The quality of life of individuals with chronic hepatitis B declined along with the severity of their disease [27]. Patients with untreated HBV chronic infection have an increased risk of developing hepatocellular carcinoma with increasing chronological age, proportional to both the presence or absence of liver disease and the degree of liver disease progression status [30]. According to recent studies, individuals with cirrhosis due to HBV have a significantly higher risk of developing HCC, with a 31-fold increase compared to those without cirrhosis. The mortality rate is also significantly higher, with a 44-fold increase in mortality for those with cirrhosis [28, 31].

Over the course of 20 years, there has been a significant increase in mortality rates for HCC caused by the HBV in the Arab population worldwide, specifically in the UAE.

The mortality rate of HBV-associated HCC has increased by 137% in the Arab population, leading to the deaths of 6447 Arabs. This increase was twice as high as that observed in other worldwide statistics [32]. Additionally, the population of the UAE experienced an increase in HBV-associated HCC deaths of approximately 10% between 1990 and 2010, with males being affected at a higher rate than females, with a rate of 3.2 per 100,000 males and 1.2 per 100,000 females.

30.3.2 The Dual Infection, Triple-Infection, and HCC

Hepatitis D virus (HDV) infection has been linked to developing severe decompensated chronic viral hepatitis and liver end-stage, which are significant health concerns [33, 34]. However, the replication and generation of complete virion fragments of HDV are dependent on the presence of HBsAg [33]. Because of this, about 5–10% of patients with chronic hepatitis B are also infected with the delta hepatitis, a condition known as "dual infection" due to the simultaneous presence of two different types of hepatitis viruses, the HBV and the HDV [35].

30.3.3 Metabolic Factors, Nonalcoholic Fatty Liver Disease, and HCC

Diabetes and obesity are important risk factors for both hepatitis C virus (HCV) infection and HCC. There are 240,000 cases of diabetes mellitus (D.M.) in the UAE, of which 210,000 are expatriates (77.5%) with a prevalence rate of 15.2% and 47,000,000 are U.A.E. nationals (25.4%) with a prevalence rate of 25.4% [17, 21]. In 2008, cardiovascular risk factors were assessed among adult UAE nationals in Abu Dhabi as a prerequisite for enrollment in national insurance. This examination found that 33% of males and 38% of females were obese [12, 17]. This is equivalent to the United States' obesity rate [12]. In addition, according to the World Health Organization's definition of obesity, 40.2% of the 44,942 UAE students assessed in 2016 were overweight, 24.4% were obese, and 5.7% were very or morbidly obese [12, 16, 17].

Furthermore, diabetes and hepatocellular carcinoma from a researcher's viewpoint. Forty observational studies from the MEDLINE, EMBASE, and Web of Science databases between January 1, 2000, and June 24, 2020, investigate the links between diabetic mellitus (DM), hypertension, dyslipidemia, and obesity and the risk of HCC due to chronic HBV infection [29, 36]. When it came to meta-analysis, only DM had enough studies to make it worthwhile. Diabetes mellitus is a very serious threat [29].

However, to completely comprehend the relationship between antidiabetic drugs, glycemic management, and HCC, additional comprehensive studies are required to draw firm conclusions [36, 37].

30.3.4 Nonalcoholic Fatty Liver Disease

CLD resulting from NAFLD affects a considerable proportion of the young and elderly population [38], and its prevalence is increasing globally [12]. NFLAD occurs in people who do not consume alcohol, distinguishing it from alcoholic liver disease (ALD). Recently, the incidence of NFLAD has been increasing due to metabolic diseases and lifestyle factors, with obesity, type 2 diabetes (T2DM) hyperlipidemia, and metabolic syndrome constituting the most prevalent risk factors for this trend. Nonalcoholic fatty liver (NAFL) is pathologically separate from nonalcoholic steatohepatitis (NASH) [38, 39].

NAFL clearly has a 5% lipid deposit in hepatocyte cells known as "hepatic steatosis" without evidence of hepatocellular injury [38], whereas NASH is distinguished by an aggressive form of fatty liver disease steatosis with inflammation and hepatocyte ballooning with or without liver fibrosis [38]; NASH is the most advanced form of NAFLD and is more likely to lead to decompensated liver disease and HCC [40]. In addition, liver cell cancer, or HCC, is prevalent. Due to nonalcoholic steatohepatitis, an increasing number of HCC diagnoses are being made [38–40].

Furthermore, NASH is the hepatic manifestation of the metabolic syndrome, in which obesity and diabetes are independently related risk factors for HCC [40]. Fifty percent of patients with NASH develop HCC before cirrhosis [41]. Moreover, as a result of chronic hepatitis C (CHC) and chronic hepatitis B (CHB), HCC mortality is decreasing. NAFLD is increasing HCC mortality [40, 42].

The prevalence of NAFLD in the UAE was estimated to be 255,000 cases in 2017 (25%) and 372,000 cases (46%) by 2030 [43], with an overall prevalence rate of 30.2%; the prevalence of NASH was estimated to be 4.1% in 2017 and to increase to 86.0% by 2030 (78,300 cases) [43], with an overall prevalence rate of 30.2%. Moreover, the number of nonalcoholic fatty liver disease cases in the UAE correlated with the high prevalence of type 2 diabetes and obesity, with an average body mass index (BMI) of 28.8; all of these factors, along with the aging of the population, contributed to an increase in NFLAD and, consequently, an increase in the prevalence of HCC [18].

Alswat et al. examined the clinical burden of NAFLD/NASH in Saudi Arabia and the UAE and represented significant increases in advanced liver disease and NASH-related mortality by 2030 [43]. The UAE had the highest prevalence of liver cirrhosis related to NASH (1119.21%), followed by Qatar (776.90%) and Oman (540.67%) as a result of the aforementioned variables [39]. The incidence of decompensated liver cirrhosis and HCC was assessed at 60 cases in 2017 and is projected to increase by 241% to 190 cases in 2030 [43]. The prevalence of HCC is estimated to be 1% [43].

30.3.5 Chronic Infection with HCV

In the UAE, HCV infection is the second most important cause of HCC, with a 27% prevalence rate in 1990 and a 27.3% prevalence rate in 2017 [26].

Infection with the HCV, also known as the hepatotropic RNA virus, is a leading cause of severe hepatic fibrosis and cirrhosis, and it also significantly increases the risk of developing HCC [44]. Morbidity and mortality from HCV-related HCC continue to be high as rates of HCV cirrhosis rise [44].

A comparative analysis of the incidence of hepatocellular carcinoma in America, Europe, Japan, and Latin America shows that hepatocellular cancer is largely attributed to the HCV [45]. In contrast, the HBV is the leading etiology of HCC in most of Asia and Africa [23]. Because of the visa regulation and the screening program provided by the UAE's Ministry of Health and Prevention (MOHAP) and other UAE health facilities to all populations, local and nonlocal, the prevalence of HCV in the UAE is relatively low at 0.1% [46].

Globally, HCV-infected people have a 15- to 20-fold higher significant risk of HCC [44], with an annual incidence of 1–4% in cirrhotic patients older than 30 years [44, 47]. Mortality related to HCV-associated HCC rose by 21.1% during the previous decade, whereas deaths attributable to factors other than HCV and alcohol remained constant [48]. Within 20–30 years, approximately 20% of chronic hepatitis C patients will develop liver cirrhosis, with a significant risk of progressing to HCC. The annual rate of HCC due to HCV exposure is 1–4% [49].

Regarding the association between HCV genotype and the risk of HCC, genotype 3 was associated with an 80% higher risk of HCC than other genotypes [50]. In the UAE, genotype 1 is the most prevalent, followed by genotypes 3 and 4, with genotype 4 being the most widespread in Middle Eastern countries. Among expatriates, genotype 1 was prevalent among Iranians, genotype 4 among Egyptians, and genotype 3 among Pakistanis [51].

As with HBV and HCC, development risk is also connected with lifestyle habits such as smoking, alcohol drinking, and coffee consumption [29, 36]. In addition, HBV, smoking, and alcohol are correlated with the developing incidence of HCC, as demonstrated by numerous studies and articles [37].

It has been proven that alcohol use exacerbates HCV-associated HCC, but coffee consumption may be protective. Multiple studies show that drinking at least one cup of coffee per day reduces the risk of developing HCC [44].

Both a decrease in the pace of hepatic fibrosis development and a decrease in the risk of HCC have been associated with coffee use [52].

Diabetes and obesity are two important risk factors for HCV infection and HCC. Chronic HCV becomes more comorbid with diabetes and obesity, increasing the incidence of HCC by 2–3 folds with diabetes and 1.5–4 times with obesity [44]. Diabetes mellitus-mediated HCC development is likely to involve elevated insulin levels and insulin resistance, which lead to increased inflammation, cellular proliferation, apoptosis inhibition, and the generation of tumor-causing mutations [53], whereas obesity leads to an increase in proinflammatory cytokines, adiponectin, and insulin resistance, all of which are potential mediators of carcinogenesis in HCV-related HCC [44].

Chronic HCV may be treated with either the standard treatment approach based on interferon (IFN) or with direct-acting antiviral agents (DAAs), both of which reduce the risk of HCC [54]. Studies comparing the rates of HCC occurrence and recurrence in patients with HCV-related cirrhosis after DAA vs. interferon (IFN)-based cure found no significant differences [44]. Studies comparing the rates of HCC recurrence in patients who received DAA versus IFN-based regimens found no difference in the rates of recurrence.

Results demonstrated no statistically significant difference in the incidence or recurrence of HCC between patients treated with DAA or IFN [54].

According to a 2017 paper on the epidemiology of hepatitis C in the GCC countries, the prevalence of hepatitis C among nationals was 0.24% (95% CI 0.02–0.63) in the UAE, 0.44% (95% CI 0.29–0.62) in Kuwait, 0.51% (95% CI 0.43–0.59) in Qatar, and 1.65% (95% CI 1.40–1.91) in Saudi Arabia [55].

Bahrain and Oman have no accessible statistics. Among the entire resident populations, HCV prevalence was 0.30% (95% CI 0.23–0.38) in Bahrain, 0.41% (95% CI 0.35–0.46) in Oman, 1.06% (95% CI 0.51–1.81) in Qatar, 1.45% (95% CI 0.75–2.34) in Kuwait, 1.63% (95% CI 1.42–1.84) in Saudi Arabia, and 1.64% (95% CI 0.96–2.49) in the UAE. Expatriate communities, especially those of Egyptian descent, demonstrated a higher incidence [55].

According to WHO statistics, it is estimated that there are 22 million people infected with HCV in the WHO Eastern Mediterranean Region, with around 15.4 million of those individuals suffering from chronic infection. The WHO estimates that the prevalence of HCV in the K.S.A. is between 0.6 and 2%, whereas in the U.A.E. and Kuwait, it is less than 0.5% [46]. The World Health Organization has tasked us with wiping out HBV and HCV by the year 2030.

The UAE has launched a micro-elimination program for the HCV with the aim of providing HCV treatment to all nationalities residing in the seven Emirates. The program has been successful in increasing the accessibility of HCV treatment by making it covered under medical insurance. The main objective of the micro-elimination of HCV program is to reduce the prevalence of HCV infection and the severity of the illness among high-risk populations [46].

To achieve this goal, the UAE government and the Ministry of Health have taken a proactive approach to community engagement and action. Healthcare professionals in blood banks, clinics, hospitals, courts, and rehabilitation facilities have been engaged to help achieve the goal of HCV micro-elimination. The government and the Ministry of Health have also organized a workshop on the micro-elimination of HCV to bring together stakeholders and provide a platform for discussion and collaboration [46].

In conclusion, the UAE's micro-elimination program for HCV is a comprehensive and collaborative effort aimed at reducing the prevalence of HCV infection and the severity of the illness. By increasing access to treatment and engaging healthcare professionals and communities, the program aims to achieve HCV micro-elimination and improve the health of high-risk populations in the UAE. [46].

30.3.6 Cirrhosis

The UAE, followed by Qatar and the Philippines [56], has the most remarkable increase in liver cirrhosis mortality. Furthermore, people with liver cirrhosis are more likely to develop HCC [57]. Fattovich et al. (2004) discovered that approximately 98% of hepatocellular carcinoma patients had liver cirrhosis [58]. In the long term, follow-up studies hypothesized that up to one-third of patients with cirrhosis would develop HCC, with an incidence rate ranging from 1 to 8% per year. Generally, hepatitis B virus-related liver cirrhosis is the best predictor of HCC incidence and mortality, with a significantly increased risk of HCC of around 31-fold and an increased mortality risk of about 44-fold compared to non-cirrhotic individuals [59, 60].

30.3.7 Exposure to Aflatoxins

The fungi Aspergillus flavus and Aspergillus parasiticus create mycotoxins known as aflatoxins, potent hepatotoxins, and potential liver carcinogens [61].

Aflatoxins can infect a wide variety of foods, including cereals, seeds, herbs, and nuts. In warm and humid climates, where mold thrives, the disease is most common [62].

An international study revealed that aflatoxin might play a role in 4.6–28.2% of all HCC cases worldwide. In the same analysis, the rate of HCC attributable to aflatoxin in Eastern Mediterranean nations, including the UAE, is approximately 10%. Unfortunately [41], there are no recent studies about the risk of aflatoxin-related HCC in the UAE. However, a 1999 article on aflatoxin contamination in the UAE demonstrated that the rice stored in some homes in the UAE might contain aflatoxin and increase the risk of HCC [63].

30.3.8 Excessive Alcohol Consumption

Consuming more than a modest quantity of alcohol on a regular basis may cause permanent impairment to the liver and raise the probability of developing liver cancer over time. Excessive alcohol consumption causes alcohol liver disease (ALD) and is responsible for more than half of all HCC cases in the eastern European nation [64]. Moreover, in the UAE, one paper published in 2020 indicates ALD causes HCC by 12.78%, compared to other countries in the GCC and MENA [26]. This percentage is lower than in Sudan and Turkey but higher than in other Gulf countries; another study examining the risk factor of HCC within the WHO Eastern Mediterranean Region (EMR) found ALD to be 22% of the etiology of HCC in the UAE, the highest among neighboring countries [26].

30.4 Clinical Presentation and Diagnosis of HCC in the UAE

Based on the clinical approach, early-stage HCC can be separated from more advanced stages. In the early stages of HCC [2, 65], many patients experience either no symptoms or a variety of nonspecific ones, including mild to moderate pain in the upper abdomen, early satiety (when you feel full when you've eaten less than usual) [22], fatigue, unintentional weight loss, or a discernible lump around the abdomen's upper region. These are the hallmark symptoms of hepatocellular carcinoma in 95% of patients. The appearance of signs and symptoms may vary based on the severity, size, and site of the disease or damage; jaundice may be indicative of severe disease [66]. Patients with HCC may also exhibit hypoglycemia, hypercalcemia, diarrhea, and cutaneous signs such as dermatomycosis and pemphigus foliaceus [66]. Typically, extrahepatic metastases move to the lungs, abdomen lymph nodes, and bones [67] (Fig. 30.5).

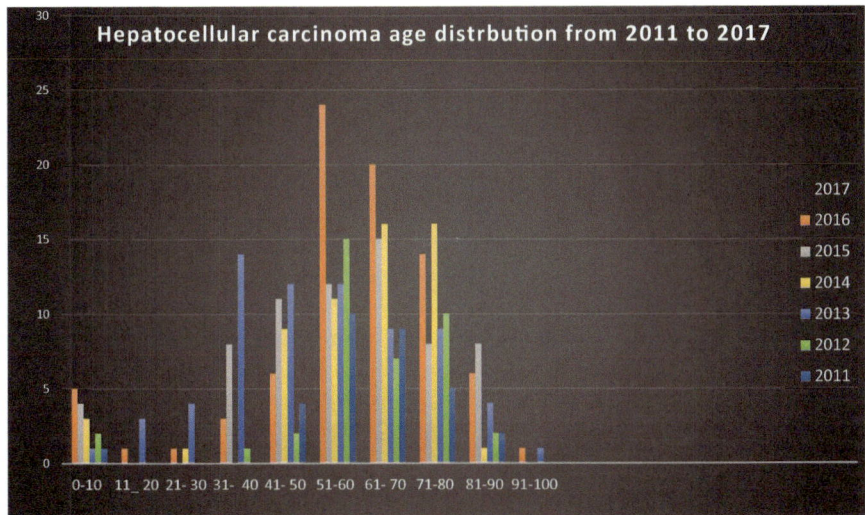

Fig. 30.5 The age distribution of hepatocellular carcinoma in the UAE from 60 to 80 years
Source: Ministry of Health and Prevention, Statistics and Research Center, National Disease
Registry—UAE National Cancer Registry Report

30.5   HCC Management in the UAE

One of the major performance metrics for the UAE National Agenda's pillar of
world-class healthcare is the reduction of cancer-related mortality [12, 17]. To
achieve this objective on the ground, in the UAE, all health authorities have
adopted a comprehensive healthcare system for cancer screening, diagnosis, and
treatment provided to all citizens and residents through its preventive and curative
health services [12], beginning with the federal health authority level in the
Ministry of Health and Prevention healthcare and continuing to the local health
authority level as in Abu Dhabi, the Department of Health (DOH), and Dubai
Health Authority (DHA), through all hospitals in both the public and private sectors, as well as primary health care centers and specialized health centers distributed all across the country [21].

30.6   Hepatocellular Carcinoma Diagnostic Approach

Hepatocellular carcinoma often develops and progresses silently, limiting and challenging its detection in the early stages before the appearance of advanced cancer
[68]. In numerous cases [69], HCC can be diagnosed with noninvasive imaging
techniques, eliminating the need to perform a biopsy. Even in cases where a biopsy

is necessary, imaging continues to be frequently recommended for guidance. HCC staging approaches must consider liver function, tumor burden, and performance status [70, 71] for comprehensive medical plan decision-making.

30.6.1 Hepatocellular Carcinoma's Pathohistological Diagnosis

Pathohistological diagnosis is the most accurate method to identify HCC and its differential diagnoses since the classification of liver cancer is based on morphological features and characteristics [72]. The WHO and the International Consensus Group have collectively come up with cornerstone criteria and recommendations for the diagnosis of HCC, which include pathohistological, histological, and immunobiological methods [73]. Generally, patients with cirrhosis should be diagnosed with HCC using noninvasive criteria, with or without pathological confirmation. Nevertheless, pathology confirmation remains essential for validating the diagnosis of HCC in patients without liver cirrhosis [66]. Patients with a significant possibility of developing HCC undergo evaluation, and the diagnosis is made based on whether the lesion size measurements are less or more than 1 cm [73]. The American Association for the Study of Liver Diseases (ASLD) recommends monitoring lesions less than 1 cm in diameter using ultrasonography (US) every 3–6 months. However, larger than 1 cm in diameter should be monitored with additional diagnostic tools, such as four-phase computed tomography (CT) or magnetic resonance imaging (MRI) [70, 73]. Meanwhile, the European Association for the Study of the Liver (EASL) recommendations classify masses into three categories: less than 1 cm, between 1 and 2 cm, and more than 2 cm in size [68, 72], with each category having its own set of diagnostic criteria. Nonetheless, the Asian Pacific Association for the Study of the Liver (APASL) guideline does not take liver lesion size into account [74].

30.6.2 Biopsy

In the majority of cases, a biopsy is unnecessary since a diagnosis can be made based on radiological features and an increased alpha-fetoprotein (AFP) serum, particularly in individuals with a history of cirrhosis or other risk factors [66, 72, 73].

Liver biopsies have been shown to be 100% specific for diagnosing HCC. However, due to the difficulty of wide-ranging differential diagnosis in highly differentiated hepatocellular tumors, this may not be the case in routine diagnostic biopsies [68]. Reliable and accurate diagnosis of HCC through liver biopsy relies on many factors, including biopsy site, tumor size, and differentiation, in addition to the expertise of the biopsy specialist and the histopathologist [66, 74]. According to reports, it ranges from around 90 to 95% across the board for tumor sizes. In addition, nodules are notoriously challenging to identify pathologically, unlike some other masses and lumps [75].

For patients without chronic liver disease, non-cirrhotic livers are more often impacted by metastases from extrahepatic malignancy than cirrhotic livers; hence, a biopsy is more likely to confirm the diagnosis [76]. The competence of the biopsy expertise and the pathologist is crucial for an accurate diagnosis (Figs. 30.6 and 30.7).

Hepatocellular carcinoma CT and MR imaging reporting and data collection have recently been standardized worldwide (HCC) [67, 72, 73]. However, prior imaging-based criteria showed significant limitations [78]. Therefore, the Liver Imaging Reporting and Data System (LI-RADS) has been created to improve accuracy for borderline patients, ranging from LR-1 (certainly benign) to LR-5 (likely malignant) (inevitably HCC) [70, 72, 78].

30.6.3 Imaging Study

HCC is distinguished from other solid tumors by its unique reliance on imaging modalities as the cornerstone for diagnosis. Notably, even in situations in which a biopsy becomes necessary, imaging techniques are commonly used to guide the procedure [70, 73].

30.6.3.1 Ultrasound Examination

Routine ultrasound (US) has the ability to effectively detect focal liver lesions in their early stages with a high level of sensitivity. Moreover, it can accurately classify these lesions as either cystic or parenchymal and determine whether they

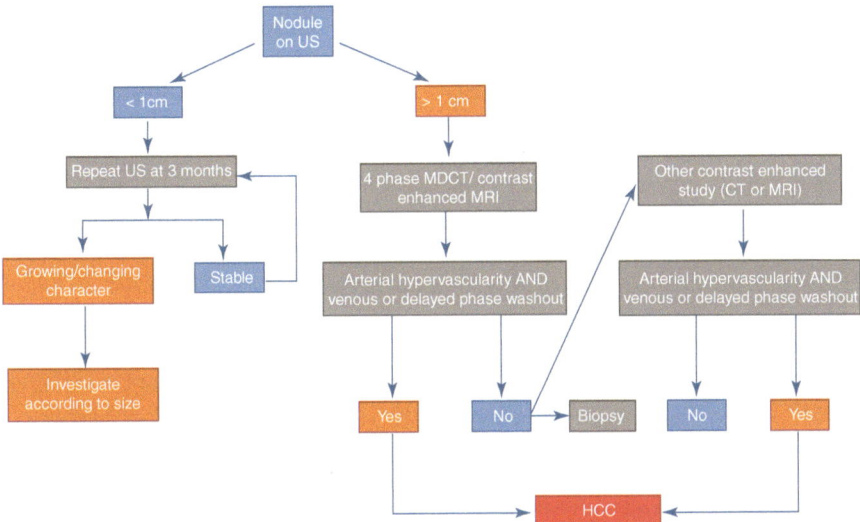

Fig. 30.6 The following text describes the diagnostic algorithm outlined in the American Association for the Study of Liver Diseases (AASLD) guideline for the detection of nodules through ultrasound (US) in patients who are at risk of HCC [77]

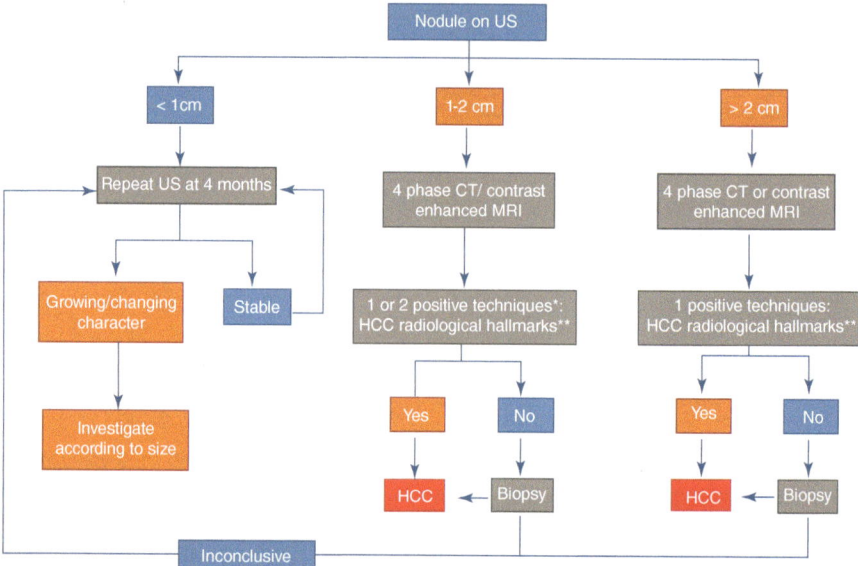

* One imaging technique only recommended in centers of excellence with high-end radiological equipment.
** HCC radiological hallmark arterial hypervascularity and venous/late phase washout.

Fig. 30.7 The diagnostic algorithm outlined in the European Association for the Study of the Liver (EASL) guidelines is designed to aid in the diagnosis of liver nodules detected by ultrasound (US) in patients at risk of HCC [77]

are benign or malignant [66, 75]. In addition, color Doppler flow US imaging provides more information about the nature of the liver mass and its intricate anatomical relationship with essential intrahepatic vascular structures [70, 72]. The sensitivity of ultrasound for HCC detection is 60%, and the specificity is 97%. The sensitivity increased to 79% when combined with the AFP assessment [71, 75]. Contrast-enhanced ultrasonography (CEUS) demonstrates notable efficacy in assessing the microvascular perfusion of hepatic neoplasms while simultaneously providing valuable assistance in the direction of interventional procedures and appraising treatment efficacy.

30.6.3.2 CT and MRI

Patients with abnormal liver US and elevated blood alpha-fetoprotein (AFP) screening values should undergo dynamic contrast-enhanced CT and multimodal MRI. These imaging methods successfully diagnose liver issues by displaying blood flow and tissue characteristics of the liver. By illuminating abnormalities in hepatic blood perfusion and tissue features, these cutting-edge imaging techniques bring attention to liver illnesses. Imaging is helpful for both creating and continuing to monitor the course of treatment [72].

A triphasic contrast CT scan of the liver is better than an ultrasound, with high specificity compared to ultrasound findings. If there is an increased alpha-fetoprotein and an abnormal ultrasound with the focal liver lesion(s), the diagnosis of HCC needs a more advanced workup. A CT four-phase liver (protocol) or contrast-enhanced magnetic resonance imaging of the abdomen is recommended for diagnostic verification [70, 75]. Multimodal MRI detects and diagnoses smaller liver cancers (2 cm in diameter) more effectively than dynamic contrast-enhanced computerized tomography [79]. An MRI of the abdomen has a sensitivity and specificity of 81% and 85%, respectively, compared to 68% and 93% for a CT scan [66, 75].

In recent years, numerous research conducted and meta-analyses on the determination of the diagnostic effectiveness of CTs and MRIs for the identification of small HCC in high-risk patients have been published; MRI has stronger diagnostic efficacy than CT, but there is no consensus recommendation to choose one method over the other [67, 70, 72].

Gadoxetic acid is a gadolinium-based MRI contrast agent used to diagnose liver lesions and abnormalities [70, 72]. The use of gadoxetic acid MRI has been added to the most current revisions of the recommendations issued by the European Association for the Study of the Liver (EASL) [68] and the American Association for the Study of Liver Diseases (AASLD) [73]. HCC has been approved as a noninvasive diagnostic criterion by these prestigious scientific bodies. However, there is not enough evidence to recommend Gadoxetic MRI over extracellular contrast MRI as the gold standard in diagnosis [67].

30.6.3.3 Serum Markers

The concentration of serum alpha-fetoprotein (AFP) is considered the most reliable biomarker for the diagnosis of hepatocellular carcinoma [72]. The fetal liver and yolk sac create a glycoprotein called alpha-fetoprotein (AFP) during pregnancy. AFP levels may rise in noncancerous conditions such as pregnancy, chronic or active liver illnesses, and cancerous conditions such as embryonal tumors of the gonads and HCC [70, 79].

In clinical practice, a 20 ng/mL blood alpha-fetoprotein is often used as a threshold to initiate an evaluation for HCC. However, this detection rate has a sensitivity of less than 60% and a specificity of about 80% when it comes to HCC [80]. In relation to sensitivity and specificity, the diagnostic threshold of 400 nanograms per milliliter (ng/mL) for alpha-fetoprotein (AFP) demonstrates a considerably superior level of accuracy compared to a threshold of 200 ng/mL, regardless of the presence or absence of ultrasound [79]. Based on the findings, it can be concluded that in individuals at a high risk of developing HCC, an AFP serum level exceeding 400 ng/mL serves as a nearly definitive diagnostic criterion for HCC, exhibiting a specificity exceeding 95% [70, 72].

30.7 Intermediate and Advanced Hepatocellular Carcinoma Staging and Treatment Options for Respectable and Unrespectable Patients

The development of hepatocellular carcinoma adds a new level of complexity to the management of chronic liver disease, which already includes cirrhosis, fibrosis, and esophageal varices [81]. A multidisciplinary team (MDT) including interventional radiology, surgery, medical oncology, and radiology is essential for effective treatment of HCC. Next comes staging and prognosis [67, 71]. The poor prognosis for HCC results from the disease itself, its underlying cause, and the degree of impaired liver function [82, 83]. The degree of liver dysfunction is measured by various scores and classifications, including the Child-Pugh classification, the Model for End-Stage Liver Disease (MELD) system, and the albumin-bilirubin (ALBI) classification [82, 83].

The classification ALBI, which includes albumin and bilirubin in its nomogram, was developed to predict the evolution of patients with HCC [67, 82, 83]. However, this ALBI classification has not been widely used in recent years and is separate from the HCC treatment recommendation. Therefore, we used validated scales such as the Eastern Cooperative Oncology Group (ECOG) performance status or the Karnofsky index instead of this system [82, 83].

The characteristics of tumors and the spread of metastasis are described in extensive detail in the Tumor-Node-Metastasis (TNM) Staging System developed by the American Joint Committee on Cancer [69]. One of the most popular methods to simplify HCC treatment is the Barcelona Clinic Liver Cancer (BCLC) staging approach developed at the Barcelona Clinic. According to the BCLC methodology, patients may be classified into one of five HCC stages: 0, A, B, C, or D. The recommendations propose either curative or palliative treatment based on the patient's tumor status (number, size, vascular invasion, extrahepatic location), liver function (Child-Pugh score), and performance status (PS, defined by the Eastern Cooperative Oncology Group scale6). This strategy considers not only the kind of tumor but also the severity of liver disease and the patient's functional status, as shown by tests [68, 69, 82]. Treatment decisions are based on the results of an all-encompassing appraisal of the disease called BCLC staging, which takes into account tumor characteristics, liver function, and overall performance status.

The BCLC system has undergone consistent validation and is highly endorsed for prognosticating and allocating treatment. The Barcelona Clinic Liver Cancer (BCLC) system, initially introduced at the end of the 20th century, stands as the predominant staging framework utilized for liver cancer across numerous international regions, including the UAE and Western nations. BCLC's latest revision, outlined in the January 2022 edition [84] of the Journal of Hepatology, reaffirms its authoritative position in the field.

The first practice change in BCLC 2022 involves the endorsement of treatment stage migration (TSM) [84]. The 2022 BCLC strategy encompasses a sophisticated clinical decision-making module that allows for personalized treatment allocation by considering precise patient and tumor profiles, leveraging local proficiency, and optimizing technical resources. The subsequent modification in this update pertains to the acknowledgement of liver transplantation (LT) as one of the principal objectives. In comparison to the 2018 version [70, 84], where LT was suggested solely for multifocal 3 cm HCCs [69], the current update identifies three arrows pointing towards LT for small multifocal HCCs [84], an intermediate stage, or stage B, of BCLC patients who have successfully undergone downstaging through trans arterial chemoembolization (TACE) or trans arterial radioembolization (TARE), and successful downstaging by TACE. The interventional radiology (IR) team will play a decisive role in achieving the multidisciplinary team's (MDT) objective of increasing the number of transplanted patients [69, 70, 84, 85].

The BCLC framework incorporates variables associated with the tumor's stage, liver function, and symptomatic presence [68]. It establishes the prognosis by categorizing patients into five distinct stages that are indicative of appropriate treatment approaches [70, 84].

Different forms of liver cancer, according to tumor size and invasion of other parts of the liver and vessels, provide different therapeutic opportunities and difficulties. Surgically interventional as resectable liver cancer, transplantable liver cancer, nonsurgically unresectable liver cancer, and metastatic liver cancer [81].

Ablation is the treatment of choice for BCLC 0, while resection and TACE are also viable options. The findings of the LEGACY study [86], which compared TARE to TACE, suggest that TARE is as effective but should only be used for solitary HCCs less than 8 cm [86]. However, resection is favored in BCLC A for tumors larger than 2 cm [69], because of the high incidence of tumor recurrence recorded following ablation.

According to the updated 2022 guidelines from the BCLC, best clinical practice is recommended for patients who are not eligible for LT or have multiple tumors. The BCLC recommends a minimally invasive approach as a possible alternative to liver resection for these patients. Current guidelines also suggest that minimally invasive procedures for liver tumors should be based on tumor size. For HCC that is less than 3 cm, ablation is recommended, whereas tumors larger than 3 cm should undergo TACE [84].

Unfortunately, patients with significant liver dysfunction (Child-Pugh C or B cirrhosis with concomitant decompensations such as chronic or recurrent hepatic encephalopathy (HE), refractory ascites, or spontaneous bacterial peritonitis) or patients whose general health has deteriorated dramatically to the point where they are no longer candidates for LT can only be managed palliatively rather than curative therapy [87, 88].

30.8 Treatment Modality of HCC

30.8.1 Treatment for HCC Includes Surgery

Tumors that arise in a non-cirrhotic liver are effectively treated by surgical resection because substantial resections can be carried out with a reduced risk of complications and a good chance of survival [89]; some underlying causes, such as NFLAD and HBV, can lead to HCC in patients without a cirrhosis background [89], making surgical resection the treatment of choice [68]. That is why assessing liver function and reserve volume before any operation is crucial [68, 89]. However, a meta-analysis by Xu et al. comparing survival outcomes between the combined approach

($n = 197$) and liver resection only ($n = 269$) in patients with primary HCC showed that at 3 years, the overall and disease-free survival were comparable, despite the fact that decompensated cirrhosis is a formal contraindication for a liver transplant [68].

Since no adjuvant treatment has been shown to be effective in preventing recurrences after surgery in HCC, resection is recommended for patients with solitary HCC arising in the non-cirrhotic liver or in a cirrhotic liver with preserved liver function, normal bilirubin, and a hepatic venous pressure gradient of less than or equal to 10 mmHg [90].

High rates of curable recurrence after resection with curative intent warrant close monitoring (evidence high; recommendation strong). Follow-up intervals need to be clearly defined. Three- to four-month intervals are reasonable in the first year [68].

30.8.2 Role of Liver Transplant LT for HCC in the UAE

Theoretically, a liver transplant is the therapy of choice for HCC patients. However, the danger of posttransplant recurrence makes this option limited [67]. Unfortunately, a liver transplant is associated with a significant probability of tumor recurrence; on average, 15–20% of patients transplanted for HCC relapsed within 2 years after transplantation [91], although this percentage varies by center and disease degree [92]. The presence of macroscopic vascular invasion and/or extrahepatic metastases is a contraindication to LT [67].

In the GCC, which includes the UAE, when selecting candidates for LT, professionals use a strict set of criteria known as the Milan criteria [93]. Milan criteria place restrictions on who can undergo a transplant due to their HCC based on the following: there is no evidence of angioinvasion or extrahepatic spread; the tumor's total diameter is less than 5 cm; there are fewer than three tumor foci; and the largest tumor site has a diameter of less than 3 cm [93].

On February 1, 2018, when the first liver transplant from a healthy donor was carried out in the UAE at Cleveland Clinic Hospital (CCAD), Abu Dhabi, LT was officially implemented in the UAE [94, 95].

In May 2017, a decision from the Ministry of Health and Prevention established the legal definition of brain death in the UAE, opening the door for deceased donor organ transplantation. This allowed for the successful introduction of solid organ transplantation [95].

Fourteen liver transplants from living donors and 11 from deceased donors were successfully completed at the Cleveland Clinic Hospital (CCAD), Abu Dhabi, with 16% of patients diagnosed with HCC (four patients). Twenty recipients have had at least a year of follow-up, and both graft and patient survival have been actuarially estimated to be 100% after 1 year. Patients, even those with HCC, have not seen a recurrence. After a median follow-up of 647 days (range, 247–1002), a total of 24 patients and 25 grafts had a 100% and 96% survival rate, respectively [94].

30.8.3 Ablation Treatment of HCC

Ablation techniques are used to treat HCC, and these techniques include radiofrequency ablation (RFA), microwave ablation (MWA), percutaneous ethanol injection (PEI), and others [67].

Percutaneous radiofrequency ablation (RFA) has replaced surgery for small nodules of HCC (less than 2 cm in diameter) [96]. It has become the standard of care for unresectable early HCCs in recent years[97]. When compared to PEI, RF ablation has exhibited superior ablative capacity and survival benefits [98].

A meta-analysis evaluates MWA with RFA based on data from seven studies involving 774 patients: one randomized controlled trial (RCT) and six retrospective studies [96]. Though the overall response rate (ORR) was similar between the two treatment groups (OR 1.01, 95% CI 0.53–1.87, $p = 0.98$), MWA performed better than RFA in the case of larger nodules (OR 0.46, 95% CI 0.24–0.89, $p = 0.02$). Despite the apparent superiority of MWA in larger neoplasms, both percutaneous methods have similar efficacy and a similar 3-year SR after RFA [99, 100].

Thermal ablation with radiofrequency is the gold standard for patients with BCLC-0 and BCLC-A tumors who are not surgical candidates. However, technical considerations (tumor location) and hepatic and extrahepatic patient circumstances warrant consideration of thermal ablation as an alternative to surgical resection for solitary tumors measuring 2–3 cm in size [68].

Radiofrequency ablation in suitable areas may be used as a first-line therapy, even in surgical patients with extremely early-stage HCC (BCLC-0) [68]. When thermal ablation is not viable, especially for tumors, ethanol injection may be used as an alternative [96, 99].

30.8.4 Role of Locoregional Therapy for Patients with HCC in the UAE

The American Association for the Study of Liver Diseases (AASLD) [73], the European Association for the Study of the Liver (EASL) [68], and the Asian Pacific Association for the Study of the Liver (APASL) (98), all recommend TACE for patients with intermediate-stage HCC (BCLC stage B) [68, 69, 73], defined as patients with multinodular disease, performance status 0, Child-Pugh class A or B cirrhosis, and without portal vein invasion or extrahepatic disease [68, 101]. Major contraindications include decompensated cirrhosis and/or multicentric involvement of both liver lobes that prevent selective intervention and severely reduced portal vein blood flow [67]; the most common adverse event of TACE is a post-embolization syndrome, while liver failure, abscesses, ischemic cholecystitis, or even death affect less than 1% of patients. Fever is an indicator of tumor necrosis, and antibiotic prophylaxis does not reduce the risk of infection [102].

A retrospective review of 150 patients with HCC in Qatar revealed that patients who underwent TACE as first-line therapy had an enhanced median survival of 27 months (95% CI 20.3–33.7). In a separate retrospective analysis from a single institution in Kuwait, 12.6% of patients underwent TACE, whereas 8.2% underwent RFA [103, 104].

There are two forms of TACE used in clinical practice: conventional TACE (cTACE) and TACE with drug-eluting beads (DEB-TACE). The conventional method employs cytotoxic drugs such as doxorubicin (the most common cytotoxic drug), mitomycin, or cisplatin, followed by the infusion of lipiodol, an oily radio-opaque agent [102, 105]. In contrast, DEB-TACE employs non-resorbable embolic microspheres loaded with chemotherapy drugs that are capable of sustained release [106].

Based on the results of a meta-analysis conducted by Zou et al., it was found that DEB-TACE has shown a higher rate of complete response and overall survival (OS) compared to conventional TACE [102]. However, there is yet to be a definitive answer as to why DEB-TACE appears to produce better results in both ways. Also, Chen et al. and Han et al. reached the same conclusion that more remarkable OS was associated with DEB-TACE [105]. In contrast, numerous studies have found no difference in overall response (OR) or adverse events between the two techniques [102, 106].

TACE is presently regarded as the treatment of choice for selected individuals with HCC at the intermediate stage (BCLC stage B) [67]. Although TACE is a viable treatment option for HCC, it may not be effective in all patients in order to better determine which patients will benefit from TACE. Medical professionals have developed seven criteria. These criteria were initially used to predict the prognosis of HCC patients undergoing LT and are now used to subclassify patients within the BCLC-B stage, successfully assess, and maximize the potential benefits of TACE for each case.

These criteria include tumor size, number of tumors, tumor marker, and liver function (child bug), considering the heterogeneity of HCC tumors [107]. Current studies suggest that intermediate-stage patients may benefit from TACE if they have Child-Pugh scores of 7, a performance status of 0 (PS 0), a high multinodular tumor burden that is not bulky, and no evidence of extrahepatic illness [108, 109].

Recent research indicates that TACE benefits early-stage HCC patients who are not candidates for surgery or ablation therapy [110]. Additionally, TACE may also be performed before LT to serve as a bridge therapy for candidates on the transplant list, improving and increasing compliance with the Milan criteria leading to better outcomes [111].

Numerous meta-analyses conclude that TACE is associated with increased OS and tumor response in comparison to conservative treatment [110]. However, it is advisable to avoid performing TACE on individuals who have decompensated liver disease, poor renal function, macrovascular invasion of the portal or hepatic vein, or extrahepatic spread. This is according to a study that highlights the potential risks associated with the procedure [68].

Regarding the combined treatment between TACE and other modalities of treatment, there is no benefit of combining TACE with RFA compared to RFA as monotherapy in small HCC tumors less than 3 cm, but patients may achieve benefit with intermediate tumors (3–10 cm) or are ineligible for RFA monotherapy due to tumor location [112, 113]. According to a randomized controlled trial by Peng et al., liver failure, advanced cirrhosis (Child-Pugh C), total bilirubin >3 mg/dL, evidence of extrahepatic disease, full portal vein thrombosis, and uncorrectable coagulopathy are among the few contraindications to TACE therapy [114, 115]. Considered related contraindications include severe atherosclerosis, renal failure, and an allergy to contrast material [116, 117].

Following the initial TACE procedure, each patient must be reevaluated, and because HCCs have predominantly arterial (hyper-) vascularization, modified Response Evaluation Criteria in Solid Tumors criteria (mRECIST) are used. The patient's response to TACE can be characterized as a complete response (CR), a partial response (PR), stable disease (SD), or progressive illness (PD) [118, 119].

30.9   Systemic Therapies Used for Patients with HCC

Before 2007, the United States Food and Drug Administration (FDA) had not authorized any medicine for the treatment of advanced HCC. Sorafenib was the first and only drug licensed for this purpose [120]. With the rapid development and approval of novel molecularly targeted therapies and immune checkpoint inhibitors (ICIs) [121] in the last 3 years, more options have become available for the treatment of advanced HCC. The combination of an immune checkpoint inhibitor (IC) and a vascular endothelial growth factor (VEGF) inhibitor is currently recommended in guideline recommendations as first-line therapy for HCC [122, 123].

Moreover, the positive safety and effectiveness data from the phase III IMbrave150 study represent a significant pivot in the first-line treatment of HCC [121, 124]. The duration of the lines represents the study period, from its actual inception to its eventual FDA registration. Therefore, the therapies in the green boxes are regarded as "second-line," while the treatments in the red boxes are "first-line" [121, 124] (Fig. 30.8).

30.9.1 Molecular Targeted Therapies

30.9.1.1   Sorafenib
Sorafenib is a form of medication that blocks certain signalling pathways responsible for the growth and development of HCC. It primarily targets the Raf-MEK-ERK and VEGFR 1-3, as well as the PDGFR and PDGFR pathways [121]. Heart and Renal Protection (SHARP) research (Sorafenib et al.; NCT00105443) found that sorafenib increased OS and progression-free survival compared to placebo.

Sorafenib has been compared to various TKIs in the literature, including sunitinib and linifanib. However, driving failed to show superiority and was more dangerous than sorafenib [126]. Patients from the GCC area with advanced HCC and Child-Pugh A/B who had failed or were ineligible for local palliative ablation

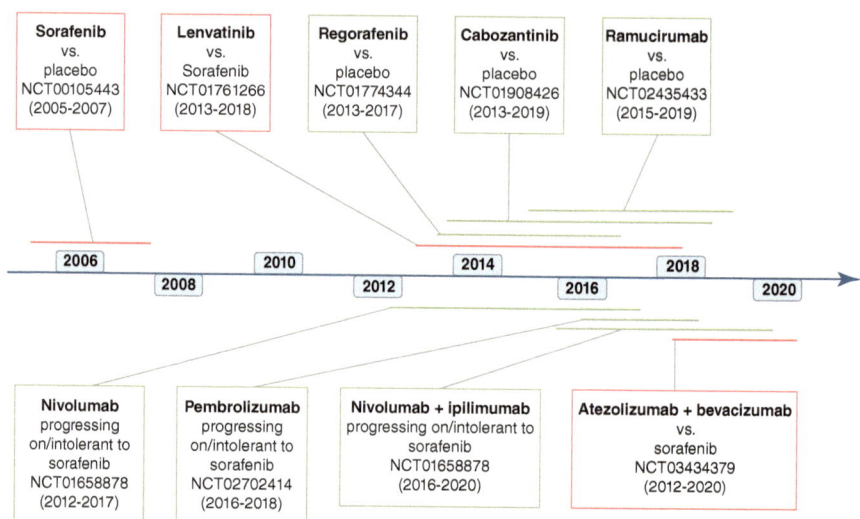

Fig. 30.8 Flowchart of a timeline demonstrates the timing of major clinical trials, including drugs currently approved for advanced HCC. Dual lines on the timeline represent trial initiation and FDA approval. First-line therapies are shown in red boxes, while second-line therapies are shown in green boxes [125]

treatment were included in retrospective research undertaken by Rasul et al. in Qatar, GCC, to evaluate the safety and effectiveness of sorafenib. Sorafenib was safe and improved survival, especially in the high-risk Child-Pugh A [122] group. The two most common sorafenib-related side events were fatigue and skin rash (237).

Furthermore, patients with the least liver damage (Child-Pugh A) had better survival rates when treated with sorafenib than patients with Child-Pugh B and C [123].

The most recent recommendations [121] recommend tyrosine kinase inhibitors (TKIs), sorafenib, or lenvatinib as alternatives for patients who do not meet the criteria for treatment with atezolizumab in combination with bevacizumab.

30.9.1.2 Lenvatinib

Lenvatinib is a potent small-molecule inhibitor that targets and binds to a range of receptors with impressive selectivity. These receptors include retinoic acid, kinase inhibitors, vascular endothelial growth factor (VEGF) 1–3, fibroblast growth factor (FGFR) 1–4, and platelet-derived growth factor (PDGFR). Its ability to bind to multiple receptor types makes it a promising choice for therapeutic interventions targeting these pathways, highlighting its remarkable efficacy.

The efficacy of lenvatinib was evaluated in a primary clinical trial called REFLECT (NCT01761266), which compared lenvatinib with sorafenib as first-line therapy for patients with unresectable HCC and Child-Pugh disease, a liver disease with adequate liver function. The primary objective of the study was to assess OS, while the secondary endpoints included time to progression (TTP) and progression-free survival (PFS). The findings indicated that patients receiving lenvatinib had a median survival time (OS) of 13.6 months, which was significantly higher than the median survival time (OS of 12.0 months, TTP) observed in patients receiving sorafenib.

The REFLECT study showed that the median survival OS with lenvatinib was 13.6 months, which was better than the 12.0 months observed with sorafenib (TTP). Furthermore, the study showed a significantly improved progression-free survival (PFS) of 7.4 months in patients treated with lenvatinib, compared with a PFS of 3.7 months in patients receiving sorafenib. These compelling results were noted and documented in references [126–128]. Subsequently, the FDA granted approval for lenvatinib as a first-line therapy for unresectable hepatocellular carcinoma [128].

30.9.1.3 Regorafenib

Regorafenib, a multi-kinase inhibitor that shares structural similarity with sorafenib, was investigated in a randomized, double-blind, phase III clinical trial called the RESORCE trial (NCT01774344). In this study, regorafenib effectively targeted multiple kinases, including VEGFR2, VEGFR3, PDGFR, FGFR-1, Kit, Ret, and B-Raf [103]. In particular, when administered as second-line therapy to patients previously treated with sorafenib who had failed treatment, regorafenib showed a significant improvement in OS compared with placebo [126]. The adverse effects of regorafenib were comparable to those of placebo. These positive findings from the RESORCE trial [121, 129] now authorize the use of regorafenib as a viable second-line treatment option for patients with sorafenib-resistant advanced HCC when no other treatment alternatives are available.

30.9.1.4 Cabozantinib

Cabozantinib is classified as a small-molecule tyrosine kinase inhibitor that selectively targets and blocks essential tyrosine kinases, which play important roles in tumorigenesis due to their role in cellular processes, including cell growth, proliferation, differentiation, and survival.

In addition, cabozantinib has a dual mechanism of action, blocking tumor angiogenesis by preventing the growth of new blood vessels, thereby reducing the blood supply the tumor needs to grow through VEGFR and effectively interfering with tumor growth and metastasis-limiting activities through MET, resulting in a remarkable antitumor effect [121, 126, 130].

To investigate its efficacy, a phase III clinical trial called CELESTIAL (NCT01908426) was conducted in patients with unresectable HCC who had previously failed treatment with sorafenib and were unresponsive to curative therapies. The results of the study showed significantly longer median OS in patients treated with cabozantinib compared to those receiving a placebo [131].

In January 2019, the FDA issued an authorization for Cabozantinib to be a second-line therapeutic alternative for patients afflicted with advanced HCC who had previously undergone sorafenib treatment. This approval was granted on the grounds of the promising results observed in the CELESTIAL study [132].

30.9.1.5 Ramucirumab

Ramucirumab is a type of medication that belongs to the class of humanized recombinant monoclonal IgG1 antibodies. Its primary mechanism of action is selective binding to VEGFR-2, effectively inhibiting activation of the VEGF pathway [126].

REACH (NCT01140347) is the name of the clinical trial conducted to discover the effects of ramucirumab on the progression of advanced HCC among individuals who were already treated with sorafenib. The participants in the study were all receiving treatment for their disease with sorafenib. Interestingly, the results of this study showed no increase in OS compared to the placebo group [121]. However, patients with advanced HCC and high AFP levels (400 ng/mL) who had previously been treated with sorafenib were enrolled in another study called REACH-2 (NCT02435433) to compare the efficacy of ramucirumab with placebo.

The results of this study indicated that the ramucirumab group had better OS and median progression-free survival [121, 130]. Based on the results of the study REACH-2, the FDA granted approval for the use of ramucirumab as second-line therapy for HCC [133].

30.9.2 Monotherapy with Immune Checkpoint Inhibitors

30.9.2.1 Nivolumab

Nivolumab is an immunotherapeutic agent as a human immunoglobulin G4 antibody that inhibits the signalling pathway of programmed cell death protein 1. Hence, the FDA has given its approval for this method of treatment to be used in various types of tumors, such as metastatic non-small cell lung cancer, esophageal cancer, and advanced renal cell carcinoma [121, 126].

CheckMate 040 is a non-comparative phase I/II (ClinicalTrials.gov Identifier: NCT01658878) conducted to examine the efficacy of nivolumab as a monotherapy. This drug was tested in patients with HCC, specifically patients who had previously been treated with sorafenib.

The study showed an ORR of 20% (42 out of 214 patients, 95% CI: 15–26), with 39 patients showing a partial response (meaning their tumor shrank) and three patients showing a complete response (meaning their tumors disappeared entirely). In addition, a median progression-free survival of 4.0 months was observed in the study, and the overall duration of response was 9.9 months [134].

Therefore, on September 22, 2017, these positive findings led to the FDA granting expedited authorization for administering nivolumab to patients with HCC who had undergone sorafenib treatment for hepatocellular carcinoma [135].

Moreover, results from another randomized, multicentre phase III trial, CheckMate 459 (NCT02576509), were published in Barcelona 2019 European Society for Medical Oncology (ESMO) conference. Patients were randomly selected to be treated with either nivolumab or sorafenib as first-line therapy. The study failed to identify a significant increase in OS. However, the primary outcome suggests that immunotherapy has a potential position as first-line therapy, which could influence the current standard of medical management [136].

30.9.2.2 Pembrolizumab

Pembrolizumab, an immune checkpoint inhibitor that uses a monoclonal antibody against programmed cell death protein 1 (PD-1), underwent investigation in the Keynote 224 trial (ClinicalTrials.gov identifier: NCT02702414) [121, 126]. The efficacy of pembrolizumab in treating these individuals was the focus of the study conducted to understand the outcomes. Eligibility criteria for the HCC trial patients receiving therapeutic sorafenib and with advanced or refractory disease were invited to participate in this experiment to evaluate both the efficacy and risk of pembrolizumab.

The trial findings demonstrated durable responses and favorable progression-free survival (PFS) of 4.8 months, median OS of 12.9 months, and TTP of 4.9 months in the HCC patient population [137].

The keynote trial results led to the FDA's approval of pembrolizumab on November 9, 2018, as a treatment for patients with advanced hepatocellular carcinoma resistant to sorafenib and experiencing disease progression [138].

Building upon the Keynote-224 trial, a phase III double-blind, randomized, controlled trial known as the Keynote-240 was conducted. This study compared pembrolizumab with best supportive care to placebo and best supportive care as second-line therapy for HCC patients previously receiving systemic therapy (250). The study results indicated that pembrolizumab did not show statistically significant improvements in OS and PFS compared to the placebo group [139].

ASCO 2021 added valuable information on both trials. It provided further insights into pembrolizumab's potential as a gold standard therapy for patients with late-stage HCC, supported by recent study data [140].

The Keynote-224 trial's second cohort, which enrolled advanced HCC patients who had not undergone any previous comprehensive treatment, reported an ORR of 16%, a median PFS of 4 months, and a median OS of 17 months. Therefore,

The Keynote 240 trial findings were further elucidated at the 2021 American Society of Clinical Oncology Gastrointestinal (ASCO GI) conference. The updated data demonstrated a statistically significant improvement in the median OS within the pembrolizumab cohort, which reached 13.9 months as opposed to 10.5 months observed in the placebo arm. In addition, the median PFS was 3.3 months and the ORR was 18.3% in the pembrolizumab group compared with 4.4% in the placebo group [141].

30.9.3 Combination Therapy

30.9.3.1 Atezolizumab + Bevacizumab

Atezolizumab, an IgG1 monoclonal antibody, exhibits specificity in binding to PD-L1, disrupting its interaction with PD-1 and reversing the suppression of T-cells.

On the other hand, bevacizumab, a humanized monoclonal antibody that targets vascular endothelial growth factor (VEGF), functions by inhibiting angiogenesis and impeding tumor growth in HCC [121, 126].

This combination therapy has emerged as a promising and preferred first-line treatment approach for advanced HCC, supported by the IMbrave 150 (NCT03434379) clinical trial outcomes. IMbrave 150, a global, open-label, phase III study, involved 501 patients randomly assigned to receive either atezolizumab + bevacizumab or sorafenib. The cohort receiving the combination of atezolizumab and bevacizumab demonstrated a statistically significant extension in both OS and progression-free survival (PFS) when compared to the sorafenib group. However, the median OS for the atezolizumab + bevacizumab group could not be evaluated (NE), while it was 13.2 months in the sorafenib group. Additionally, based on the response evaluation criteria in solid tumors (RECIST) version 1.1 criteria, the combination therapy group demonstrated an ORR of 27.3% compared to 11.9% in the sorafenib group [142].

In May 2020, the utilization of atezolizumab in conjunction with bevacizumab received approval from the FDA for administering first-line therapy to patients diagnosed with advanced HCC.

This decision was based on the compelling safety and efficacy outcomes observed in the IMbrave 150 clinical trial (Finn et al., 2021). Notably, Finn et al. presented updated findings from the IMbrave 150 study during the 2021 ASCO Gastrointestinal Cancer Symposium (p. 169). With an additional 12 months of follow-up, the combination of atezolizumab + bevacizumab demonstrated sustained clinical efficacy benefits compared to sorafenib.

This treatment regimen achieved the longest OS observed in first-line phase III studies and is currently considered the standard of care for advanced HCC patients who have not previously received systemic therapy [143].

30.9.3.2 Nivolumab + Ipilimumab

The combination of an anti-PD-1 antibody called nivolumab and a CTLA-4 inhibitor known as ipilimumab was examined in a study called CheckMate 040. Fifty patients in cohort 4 of CheckMate trial 040 (NCT01658878) received nivolumab plus ipilimumab at a total of four dosages at 3-week intervals. As part of ongoing treatment, patients received 240 mg single dosage of nivolumab every 14 days. This regimen was carefully followed to ensure the continuity and efficacy of the patient's treatment.

According to the study, using RECIST v1.1 criteria yielded an ORR of 32%, with four patients achieving a complete response and 12 patients showing a partial response. Based on the data collected, it was found that the average response time was 17.5 months. As a result, on March 10, 2020, the dose regimen of arms A was approved for advanced HCC patients who had progressed on sorafenib treatment. Recently, the revised CheckMate 040 result was released at ASCO 2021, and it showed that the mOS length had improved to 22.2 months. This improvement resulted in the 24-month OS rate increasing to 46%, while the 36-month OS rate was 42%. Furthermore, recent studies on the combination of nivolumab and ipilimumab as second-line therapy have demonstrated significant improvements in clinical outcomes and survival rates [144].

30.10 Supportive Management

The use of psychoactive medicines, in particular benzodiazepines, for the treatment of psychological distress is leading to an increased risk of falls, accidents, and an impaired mental state in patients with advanced cirrhosis [68].

The use of psychoactive medicines, in particular benzodiazepines, for the treatment of psychological distress is associated with an increased risk of falls, accidents, and an impaired mental state in patients with advanced cirrhosis. Patients with HCC and liver cirrhosis should thus use these drugs with extreme caution [145].

According to the patient's condition, psycho-oncological support and nutritional supplementation are recommended [68]. Patients with cancer are presented with a wide range of physical, nutritional, and psychosocial issues that require a patient-specific, integrated intervention and consistent follow-ups to secure and consolidate the improved quality of life [146].

Because of the poor prognosis of patients with chronic HCC, as defined by the Barcelona Clinic Liver Cancer (BCLC) criteria, with a life expectancy of around 3–4 months [68], the treatment of end-stage illness is limited to symptomatic care, and there is no need for tumor-directed therapy. These patients should get palliative care, which includes pain control, nourishment, and emotional support [145–147].

30.11 Hepatocellular Carcinoma Preventive Measurements in the UAE

30.11.1   Vaccination

All newborns have been given four doses of the HBV vaccine at 0, 2, 4, and 6 months of age as part of mandatory vaccination programs that have been in place since 1991 [25]; furthermore, the high-risk population receives additional HBV vaccine booster shots [25].

30.11.2   Treatment for Viral Hepatitis

Numerous studies demonstrate that treating chronic HBV infection minimizes the incidence risk of HCC [29]. The majority of doctors follow the guidelines of the European Association for the Study of the Liver (EASL) [148]. By limiting disease progression and mortality, the major objective of HBV treatment is to increase survival and quality of life. PEGylated interferon (Peg-IFN-alfa-2), lamivudine (LAM), telbivudine, and adefavir are given much less often than entecavir (ETV) and tenofovir disoproxil fumarate (TDF) in the UAE [23, 25]. In 12 cohort studies and 1 RCT, CHB patients have been examined with entecavir (ETV), lamivudine (LAM), telbivudine (LdT), and/or tenofovir disoproxil fumarate (TDF). The meta-analysis demonstrated that ETV was better than LAM in terms of the incidence of HCC

($p = 0.001$). We discovered no significant difference in HCC incidence between ETV and TDF ($p = 0.08$) [148].

In addition, one study demonstrated the efficacy of tenofovir in lowering HCC incidence relative to entecavir, and there is now a recommendation for the widespread use of tenofovir in chronic hepatitis B patients [149, 150]. By reducing HBV DNA levels, the long-term course of antiviral therapy effectively decreases the incidence of HCC among people with chronic hepatitis B (CHB) [151].

According to a recent study, there is no significant difference between entecavir (ETV) and tenofovir disoproxil fumarate (TDF) treatments when it comes to the risk of HCC development in treatment-naive CHB patients, regardless of preexisting cirrhosis [149]. In conclusion, antiviral treatments may minimize the development of HCC and death in individuals with chronic hepatitis C, especially when a sustained virological response (SVR) is established [152].

30.11.3   Other Medications

30.11.3.1   Statins

The statins class of medications has a major impact on lowering cholesterol levels, thereby decreasing the possibility of cardiovascular disorders, including coronary artery disease and stroke [153].

Meanwhile, elevated cholesterol levels are now recognized as a significant contributor to liver cancer. The potential benefits of statins for preventing liver cancer have been investigated in 24 randomized controlled trials with a total of 59,070 participants.

When comparing statin users with the non-statin group, patients on a statin medication course significantly had a lower chance of developing HCC (risk ratio: 0.55, 95% confidence interval: 0.47–0.61, I2 = 84.39%).

More importantly, the results of the effectiveness of statin therapy were supported by strong data, showing that statins lower the incidence of HCC in people with nonalcoholic fatty liver disease [154] and in those with diabetes and liver cirrhosis. Additionally, statins could potentially lower the incidence of HCC in those with chronic hepatitis B and hepatitis C viruses, specifically individuals with liver cirrhosis [155].

30.11.3.2   Aspirin and Other Nonsteroidal Anti-Inflammatory Drugs

A meta-analysis and comprehensive review were conducted to evaluate the correlation between aspirin as well as NSAIDs, or nonsteroidal anti-inflammatory drugs, and the prevalence of HCC.

The analysis included 19 research projects with a total of 149,000 participants. The results indicated that all categories of aspirin and NSAID treatment, including aspirin only, nonaspirin NSAIDs only, and a combination of both, were associated with a decreased risk of HCC incidence and improved liver-related mortality. The study shows daily aspirin use specifically minimized the probability of liver cancer for those diagnosed with HBV. These findings support the hypothesis that using aspirin and NSAIDs may serve as a protective measure against HCC [156–158]. However, there was no statistically significant variance among aspirin users or non-users in the rate of gastrointestinal bleeding.

30.11.3.3 Metformin

Big data was collected from April 2017 to January 2019 and published in 2020. A combined collection of eight scientific research papers, consisting of four cohort studies and four case-control studies, examines the potential role of metformin therapy in minimizing the incidence of liver cancer. The metformin effect was most pronounced with early-stage HCC or patients at potentially curative tumor stages [159, 160].

30.11.4   Lifestyle Factor

30.11.4.1 Coffee Consumption

Consumption of coffee on a regular basis lowers incidences of both liver disease and HCC.; this statistic, supported by numerous clinical and observational trials, reveals that any coffee consumption declines the probability of developing hepatocellular by 40% [161]. Coffee contains large amounts of antioxidants, suggesting biologic plausibility for the protective effect [161, 162].

30.11.4.2 Diet

Consumption of a healthy diet has been found to have a positive impact on reducing the risk of cancer as well as other chronic diseases such as cardiovascular disease (CVD) [161]. Notably, studies have indicated that the intake of white meat, fish, omega-3 fatty acids, and vegetables is remarkably related to a decreased risk of HCC [162–164], while the consumption of red meat has been linked to an increased risk of HCC [163].

The potential hepatoprotective effects of Vitamin E, functioning as an antioxidant, have been suggested for mitigating oxidative stress-induced damage to the liver and HCC. Patients with liver cirrhosis/fibrosis and some nonalcoholic steatohepatitis [97, 165] who were given vitamin E showed positive clinical outcomes, indicating the beneficial effects of vitamin E supplements [97, 165].

Nevertheless, additional investigation is necessary to substantiate this hypothesis.

30.11.4.3 Physical Activity

Maintaining a healthy weight within the normal range, implementing dietary modifications, and engaging in more physical exercise have all been shown to enhance a person's health and well-being over time.

One study from South Korea examines patients with chronic hepatitis B as a high-risk group for developing hepatocellular carcinoma and the potential prevention role of physical activity. The study analyzed data from 9727 patients' treatment-naive with chronic hepatitis b. found that the engaging low to moderate physical activity).

Specifically, patients with low to moderate physical activity ranging between 500 and 1500 (MET)-min/week demonstrated a significantly lower risk of HCC compared to those who were inactive in both patients with and without cirrhosis.

Moreover, the highest outcomes of preventive measures and the biggest gains from physical activities are shown in obese or high body mass index, male participants, non-diabetes participants, and young groups.

However, those with cirrhosis who continued to engage in vigorous exercise did not achieve similar positive benefits in HCC prevention because high-intensity exercise raises blood testosterone levels, which has been linked to an increased risk of HCC in cirrhotic individuals.

Guidelines for physical activity need more research and investigation. That will be appropriate for each group, taking into account the intensity, nature, duration, and amount of physical exercise.

30.12 Conclusion

The incidence of HCC in the UAE has been stable, with 60–80 cases per year in the UAE between 2013 and 2019, with a potential future increase in incidence with rising rates of obesity, diabetes, and excessive alcohol consumption, as well as hepatitis B virus, hepatitis C virus, and NASH.

The UAE has successfully implemented measures to keep the prevalence of HCV infection low. The widespread availability of screening and treatment facilities, coupled with effective public health campaigns, has helped to detect infected individuals early and provide them with appropriate treatment, thus reducing the overall prevalence of the disease. Additionally, a comprehensive vaccination program targeting high-risk populations, such as healthcare workers, newborns, and mothers, has contributed to the reduction of HBV infection rates in the country.

Despite the success in controlling viral hepatitis, the increasing rates of obesity and diabetes in recent years could potentially lead to an escalation in the incidence of liver cancer in the future.

To effectively address liver cancer prevention and management in the UAE, a comprehensive and interdisciplinary approach is necessary, requiring the collaboration of public health officials and healthcare providers in implementing preventive measures and early detection strategies to reduce the burden of liver cancer in the country. Public health interventions aimed at curtailing the prevalence of hepatitis B and C virus infections have been effective in the UAE.

Furthermore, promoting healthy lifestyles and reducing modifiable risk factors, such as excessive alcohol consumption, obesity, and diabetes, are pivotal in preventing and managing liver cancer in the UAE. It is crucial that the government, healthcare providers, and the public forge smart partnerships to achieve common goals. These partnerships would facilitate the pooling of resources, expertise, and knowledge, enabling the development of tailored interventions that consider the unique cultural, social, and economic factors that influence liver cancer risk and outcomes in the UAE.

The implementation of effective public health initiatives, including educational campaigns, screening programs, and affordable access to treatment, is necessary to lower the incidence and mortality rates of liver cancer further in the UAE.

Conflict of Interest The authors have no conflict of interest to declare.

References

1. Llovet JM, Zucman-Rossi J, Pikarsky E, Sangro B, Schwartz M, Sherman M, Gores G. Hepatocellular carcinoma. Nat Rev Dis Prim. 2016;2(1):16018. https://www.nature.com/articles/nrdp201618. Accessed 6 Aug 2019.
2. Mak LY, Cruz-Ramón V, Chinchilla-López P, Torres HA, NK LC, Rice JP, Foxhall LE, Sturgis EM, Merrill JK, Bailey HH, Méndez-Sánchez N, Yuen MF, Hwang JP. Global epidemiology, prevention, and management of hepatocellular carcinoma. Am Soc Clin Oncol Educ Book. 2018;38:262–79. https://doi.org/10.1200/EDBK_200939.
3. GLOBOCAN database (September 2018). https://gco.iarc.fr/today/home. Accessed 1 July 2020.
4. Zhuo Y, Chen O, Chhatwal J. Changing epidemiology of hepatocellular carcinoma and role of surveillance. In: Hoshida Y, editor. Hepatocellular carcinoma: translational precision medicine approaches. Cham, CH: Humana Press; 2019.
5. Hepatocellular carcinoma I Nature Reviews Disease Primers Primer. 2021. https://www.nature.com/articles/s41572-020-00240-3.
6. Bahardoust M, Sarveazad A, Agah S, Babahajian A, Amini N. Predictors of 5 year survival rate in hepatocellular carcinoma patients. J Res Med Sci. 2019;24(1):86.
7. gco.iarc.fr. Cancer today. https://gco.iarc.fr/today/online-analysis-multibars?v=2020&mode=cancer&mode_population=countries&population.
8. Sung H, et al. Global cancer statistics 2020: GLOBOCAN estimates of incidence and mortality worldwide for 36 cancers in 185 countries. CA Cancer J Clin. 2021;71(3):209–49. https://doi.org/10.3322/caac.21660.
9. Hefaiedh R, et al. Gender difference in patients with hepatocellular carcinoma. Tunis Med. 2013;91(8-9):505–8.
10. GLOBOCAN database (September 2018)—https://gco.iarc.fr/today/home aDOUBLEHYPHEN: GLOBOCAN database (September 2018)—https://gco.iarc.fr/today/home. Accessed 1 July 2020.
11. Estimated age-standardized incidence rates (Asia) in 2020, both sexes, all ages. World Health Organization GLOBOCAN; 2020. https://gco.iarc.fr/today/home. Accessed 29 Aug 2022.
12. Abu-Gheida IH, Nijhawan N, Al-Awadhi A, Al-Shamsi HO. Cancer in the Arab world. 2022. pp 301–19.
13. Albarrak J, Al-Shamsi H. Current status of management of hepatocellular carcinoma in The Gulf Region: challenges and recommendations. Cancers (Basel). 2023;15(7):2001. https://doi.org/10.3390/cancers15072001.
14. Cancer incidence in United Arab Emirates, Annual Report of the UAE—National Cancer Registry. Statistics and Research Center, Ministry of Health and Prevention, United Arab Emirates.
15. MCCR-01-344 Cancer Notification Policy—United Arab Emirates Ministry of Health Central Cancer Registry (MCCR); 2014. https://www.mohap.gov.ae/en/OpenData/Pages/default.aspx. Accessed 1 July 2020.
16. Workbook: non-communicable diseases. Cancer Incidence 2016. Department of Health. https://tableau.doh.gov.ae/views/NonCommunicableDiseases/CancerIncidence?%3AisGuestRedirectFromVizportal=y&%3Aembed=y&%3Atoolbar=no. Accessed 7 Aug 2021.

17. Al-Shamsi H, Darr H, Abu-Gheida I, Ansari J, McManus MC, Jaafar H, Tirmazy SH, Elkhoury M, Azribi F, Jelovac D, et al. The state of cancer care in the United Arab Emirates in 2020: challenges and recommendations, a report by the United Arab Emirates Oncology Task Force. Gulf J Oncolog. 2020;1(32):71–87.

18. Alswat K, Aljumah AA, Sanai FM, et al. Nonalcoholic fatty liver disease burden—Saudi Arabia and United Arab Emirates, 2017–2030. Saudi J Gastroenterol. 2018;24(4):211–9. https://doi.org/10.4103/sjg.SJG_122_18.

19. Jawad Hashim M, Sadaf Rizvi S, Khan G. Hepatocellular carcinoma in the United Arab Emirates. In: Carr BI, editor. Liver cancer in the middle east. Springer; 2021. p. 101–8.

20. World Life Expectancy. HEALTH PROFILE ARAB EMIRATES. https://www.worldlifeexpectancy.com/country-health-profile/arab-emirates. Accessed 23 Mar 2023.

21. Health Statistics 2012, Department of Health Abu Dhabi. https://www.scad.ae/Release%20Documents/Health%202012%20English.pdf. Accessed 1 July 2020.

22. Greten TF. Gender disparity in HCC: is it the fat and not the sex? J Exp Med. 2019;216(5):1014–5. https://doi.org/10.1084/jem.20190441.

23. Hashim MJ, Rizvi SS, Khan G. Hepatocellular carcinoma in The United Arab Emirates. In: Liver cancer in the middle east; 2021. p. 101–8.

24. Sharafi H, Alavian SM. The rising threat of hepatocellular Carcinoma in the Middle East and North Africa region: results from global burden of disease study 2017. Clin Liver Disease. 2019;14(6):219–23.

25. Al Zaabi M, et al. Hepatitis B care pathway in the United Arab Emirates: current situation, gaps, and actions. Eur Med J. 2019.

26. Sharafi H, Alavian SM. The rising threat of hepatocellular carcinoma in the middle east and north Africa region: results from global burden of disease study 2017. Clin Liver Dis (Hoboken). 2020;14(6):219–23.

27. Ioannou GN, Splan MF, Weiss NS, et al. Incidence and predictors of hepatocellular carcinoma in patients with cirrhosis. Clin Gastroenterol Hepatol. 2007;5:938.

28. Quang EV, et al. Epidemiological projections of viral-induced hepatocellular carcinoma in the perspective of WHO global hepatitis elimination. Liver Int. 2021;41(5):915–27.

29. Campbell C, et al. Risk factors for the development of hepatocellular carcinoma (HCC) in chronic hepatitis B virus (HBV) infection: a systematic review and meta-analysis. J Viral Hepat. 2021;28(3):493–507. https://doi.org/10.1111/jvh.13452.

30. Kuang X-J, et al. Systematic review of risk factors of hepatocellular carcinoma after hepatitis B surface antigen seroclearance. J Viral Hepat. 2018;25(9):1026–37.

31. Tan YJ. Hepatitis B virus infection and the risk of hepatocellular carcinoma. World J Gastroenterol. 2011;17(44):4853–7. https://doi.org/10.3748/wjg.v17.i44.4853.

32. Khan G, Hashim MJ. Burden of virus-associated liver cancer in the Arab World, 1990–2010. Asian Pac J Cancer Prev. 2015;16(1):265–70. https://doi.org/10.7314/APJCP.2015.16.1.265.

33. https://www.who.int/news-room/fact-sheets/detail/hepatitis-d#:~:text=Hepatitis%20D%20virus%20(HDV)%20affects,B%20(super%2Dinfection.

34. Negro F. Hepatitis D virus coinfection and superinfection. Cold Spring Harb Perspect Med. 2014;4(11):a021550. https://doi.org/10.1101/cshperspect.a021550.

35. Kamal H, et al. Risk of hepatocellular carcinoma in hepatitis B and D virus co-infected patients: a systematic review and meta-analysis of longitudinal studies. J Viral Hepat. 2021;28(10):1431–42.

36. Shin HS, et al. Impact of diabetes, obesity, and dyslipidemia on the risk of hepatocellular carcinoma in patients with chronic liver diseases. Clin Mol Hepatol. 2022;28(4):773–89. https://doi.org/10.3350/cmh.2021.0383.

37. Kramer JR, et al. Effect of diabetes medications and glycemic control on risk of hepatocellular cancer in patients with nonalcoholic fatty liver disease. Hepatology. 2022;75(6):1420–8. https://doi.org/10.1002/hep.32244. Epub 2021 Dec 19.

38. Ahmad MI, et al. Hepatocellular carcinoma due to nonalcoholic fatty liver disease: current concepts and future challenges. J Hepatocell Carcinoma. 2022;9:477–96. https://doi.org/10.2147/JHC.S344559.

39. Zhai M, Liu Z, Long J, Zhou Q, Yang L, Zhou Q, Liu S, Dai Y. The incidence trends of liver cirrhosis caused by nonalcoholic steatohepatitis via the GBD study. 2017;11(1):5195.

40. Klein S, Dufour J-F. Nonalcoholic fatty liver disease and hepatocellular carcinoma. Hepat Oncol. 2017;4(3):83–98. https://doi.org/10.2217/hep-2017-0013.

41. Liu Y, Wu F. Global burden of aflatoxin-induced hepatocellular carcinoma: a risk assessment. Environ Health Perspect. 2010;118(6):818–24. https://doi.org/10.1289/ehp.0901388.

42. Grgurevic I, et al. Hepatocellular carcinoma in non-alcoholic fatty liver disease: from epidemiology to diagnostic approach. Cancers (Basel). 2021;13(22):5844. https://doi.org/10.3390/cancers13225844.

43. Sanai FM, et al. Clinical and economic burden of nonalcoholic steatohepatitis in Saudi Arabia, United Arab Emirates and Kuwait. Hepatol Int. 2021;15(4):912–21.

44. Axley P, et al. Hepatitis C virus and hepatocellular Carcinoma: a narrative review. J Clin Transl Hepatol. 2018;6(1):79–84. https://doi.org/10.14218/JCTH.2017.00067.

45. Hepatocellular carcinoma: A global view. https://doi.org/10.1038/nrgastro.2010.100.

46. Micro-elimination of Hepatitis C Virus in the Middle East Report of an expert workshop. https://easl-ilf.org/wp-content/uploads/2020/02/Micro-elimination-of-Hepatitis-C-Virus-in-the-Middle-East_Report-of-an-expert-workshop-2.pdf.

47. El-Serag HB. Epidemiology of viral hepatitis and hepatocellular carcinoma. Gastroenterology. 2012;142:1264–1273.e1. https://doi.org/10.1053/j.gastro.2011.12.061.

48. GBD 2015 Mortality and Causes of Death Collaborators. Global, regional, and national life expectancy, all-cause mortality, and cause-specific mortality for 249 causes of death, 1980–2015: a systematic analysis for the Global Burden of Disease Study 2015. Lancet. 2016;388:1459–544. https://doi.org/10.1016/S0140-6736(16)31012-1.

49. Omland LH, Krarup H, Jepsen P, Georgsen J, Harritshøj LH, Riisom K, et al. Mortality in patients with chronic and cleared hepatitis C viral infection: a nationwide cohort study. J Hepatol. 2010;53:36–42. https://doi.org/10.1016/j.jhep.2010.01.033.

50. Kanwal F, Kramer JR, Ilyas J, Duan Z, El-Serag HB. HCV genotype 3 is associated with an increased risk of cirrhosis and hepatocellular cancer in a national sample of U.S. veterans with HCV. Hepatology. 2014;60:98–105. https://doi.org/10.1002/hep.27095.

51. Abro AH, Al-Dabal L, Younis NJ. Distribution of hepatitis C virus genotypes in Dubai, United Arab Emirates. J Pak Med Assoc. 2010;60(12):987–90.

52. Saab S, Mallam D, Cox GA 2nd, Tong MJ. Impact of coffee on liver diseases: a systematic review. Liver Int. 2014;34:495–504. https://doi.org/10.1111/liv.12304.

53. Rao H, Wu E, Fu S, Yang M, Feng B, Lin A, et al. The higher prevalence of truncal obesity and diabetes in American than Chinese patients with chronic hepatitis C might contribute to more rapid progression to advanced liver disease. Aliment Pharmacol Ther. 2017;46:731–40. https://doi.org/10.1111/apt.14273.

54. Waziry R, et al. Hepatocellular carcinoma risk following direct-acting antiviral HCV therapy: a systematic review, meta-analyses, and meta-regression. J Hepatol. 2017;67(6):1204–12. https://doi.org/10.1016/j.jhep.2017.07.025.

55. Mohamoud YA, et al. Epidemiology of hepatitis C virus in the Arabian Gulf countries: systematic review and meta-analysis of prevalence. Int J Infect Dis. 2016;46:116–25. https://doi.org/10.1016/j.ijid.2016.03.012. Epub 2016 Mar 17.

56. Ye F, et al. The burden of liver cirrhosis in mortality: results from the global burden of disease study. Front Public Health. 2022;10:909455. https://doi.org/10.3389/fpubh.2022.909455.

57. Alan Herbst D, Rajender K, Reddy. Risk factors for hepatocellular carcinoma. Clin Liver Dis (Hoboken). 2013;1(6):180–2. https://doi.org/10.1002/cld.111.

58. Singal AG, Lampertico P, Nahon P. Epidemiology and surveillance for hepatocellular carcinoma: new trends. J Hepatol. 2020;72(2):250–61.

59. Thiele M, et al. Large variations in risk of hepatocellular carcinoma and mortality in treatment naïve hepatitis B patients: systematic review with meta-analyses. PLoS One. 2014;9(9):e107177. https://doi.org/10.1371/journal.pone.01071.

60. Liu Y, et al. Viral biomarkers for hepatitis B virus-related hepatocellular carcinoma occurrence and recurrence. Front Microbiol. 2021;12:665201. https://doi.org/10.3389/fmicb.2021.665201.

61. Benkerroum N. Chronic and acute toxicities of aflatoxins: mechanisms of action. Int J Environ Res Public Health. 2020;17(2):423.

62. Hamid AS, et al. Aflatoxin B1-induced hepatocellular carcinoma in developing countries: geographical distribution, mechanism of action and prevention (review). Oncol Lett. 2013;5(4):1087–92. https://doi.org/10.3892/ol.2013.1169.

63. Osman N, et al. Aflatoxin contamination of rice in the United Arab Emirates. Mycotoxin Res. 1999;15(1):39–44. https://doi.org/10.1007/BF02945213.

64. Patel R, Mueller M. Alcoholic liver disease. 2022.

65. Sagnelli E. et al. Epidemiological and etiological variations in hepatocellular carcinoma. 2019;48(1):7–17. https://doi.org/10.1007/s15010-019-01345-y. Epub 2019 Jul 25.

66. Grgurevic I, Bokun T, Salkic NN, Brkljacic B, Vukelić-Markovic M, Stoos-Veic T, et al. Liver elastography malignancy prediction score for noninvasive characterization of focal liver lesions. Liver Int. 2018 Jun;38(6):1055–63.

67. Reig M, Forner A, Ávila MA, Ayuso C, Mínguez B, Varela M, Bilbao I, Bilbao JI, Burrel M, Bustamante J, Ferrer J, Gómez MÁ, Llovet JM, De la Mata M, Matilla A, Pardo F, Pastrana MA, Rodríguez-Perálvarez M, Tabernero J, Urbano J, Vera R, Sangro B, Bruix J. Diagnosis and treatment of hepatocellular carcinoma. Update of the consensus document of the AEEH, AEC, SEOM, SERAM, SERVEI, and SETH. Med Clin (Barc). 2021;156(9):463.e1–463.e30. doi: https://doi.org/10.1016/j.medcli.2020.09.022. English, Spanish. Epub 2021 Jan 16.

68. Galle PR, Forner A, Llovet JM, Mazzaferro V, Piscaglia F, Raoul J-L, Schirmacher P, Vilgrain V. EASL clinical practice guidelines: management of hepatocellular carcinoma. J Hepatol. 2018;69(1):182–236. https://www.sciencedirect.com/science/article/pii/S0168827818302150

69. Zhou J, Sun H, Wang Z, Cong W, Wang J, Zeng M, Zhou W, Bie P, Liu L, Wen T, Han G, Wang M, Liu R, Lu L, Ren Z, Chen M, Zeng Z, Liang P, Liang C, Chen M. Guidelines for the diagnosis and treatment of hepatocellular carcinoma (2019 edition). Liver Cancer. 2020;9(6):682–720. https://www.karger.com/Article/FullText/509424. Accessed 13 Dec 2020.

70. Kefeli A, Basyigit S, Yeniova AO. Diagnosis of hepatocellular Carcinoma. Updates in Liver Cancer; 2017.

71. Attwa MH. Guide for diagnosis and treatment of hepatocellular carcinoma. World J Hepatol. 2015;7(12):1632. https://www.ncbi.nlm.nih.gov/pmc/articles/PMC4483545/. Accessed 23 Jul 2019.

72. Cartier V, Aubé C. Diagnosis of hepatocellular carcinoma. Diagn Interv Imaging. 2014;95(7–8):709–19.   https://www.sciencedirect.com/science/article/pii/S2211568414001958. Accessed 15 Aug 2019.

73. Heimbach JK, Kulik LM, Finn RS, Sirlin CB, Abecassis MM, Roberts LR, Zhu AX, Murad MH, Marrero JA. AASLD guidelines for the treatment of hepatocellular carcinoma. Hepatology. 2017;67(1):358–80. https://aasldpubs.onlinelibrary.wiley.com/doi/full/10.1002/hep.29086. Accessed 23 Jul 2019.

74. Kim T-H, Kim SY, Tang A, Lee JM. Comparison of international guidelines for noninvasive diagnosis of hepatocellular carcinoma: 2018 update. Clin Mol Hepatol. 2019;25(3):245–63.

75. Hepatocellular carcinoma—symptoms, diagnosis and... https://bestpractice.bmj.com.

76. Ronot M, Fouque O, Esvan M, Lebigot J, Aubé C, Vilgrain V. Comparison of the accuracy of AASLD and LI-RADS criteria for the non-invasive diagnosis of HCC smaller than 3 cm. J Hepatol. 2018;68(4):715–23.

77. Kefeli A, Basyigit S, Yeniova AO. Diagnosis of hepatocellular carcinoma. InTech. 2017; https://doi.org/10.5772/64992.

78. Yoon JH, Park J-W, Lee JM. Noninvasive diagnosis of hepatocellular carcinoma: elaboration on Korean Liver Cancer Study Group-National Cancer Center Korea practice guidelines compared with other guidelines and remaining issues. Korean J Radiol. 2016;17(1):7.

79. Zhang J, Chen G, Zhang P, Zhang J, Li X, Gan D, Cao X, Han M, Du H, Ye Y. The threshold of alpha-fetoprotein (AFP) for the diagnosis of hepatocellular carcinoma: a systematic review and meta-analysis. PLoS One. 2020;15(2):e0228857.

80. Lok AS, Sterling RK, Everhart JE, et al. Des-gamma-carboxy prothrombin and alpha-feto-protein as biomarkers for the early detection of hepatocellular carcinoma. Gastroenterology. 2010;138:493.
81. Pinter M, Trauner M, Peck-Radosavljevic M, Sieghart W. Cancer and liver cirrhosis: implications on prognosis and management. ESMO Open. 2016;1(2):e000042. https://esmoopen.bmj.com/content/1/2/e000042?utm_source=TrendMD&utm_medium=cpc&utm_campaign=Journals_TrendMD_KA. Accessed 2 Jan 2020.
82. Johnson PJ, Berhane S, Kagebayashi C, Satomura S, Teng M, Reeves HL, et al. Assessment of liver function in patients with hepatocellular carcinoma: a new evidencebased approach-the ALBI grade. Clin Oncol. 2015;33:550–8.
83. Malinchoc M, Kamath PS, Gordon FD, Peine CJ, Rank J, ter Borg PC. A model to predict poor survival in patients undergoing transjugular intrahepatic portosystemic shunts. Hepatology. 2000;31:864–71.
84. Lucatelli P, Guiu B. 2022 update of BCLC treatment algorithm of HCC: what's new for interventional radiologists? Cardiovasc Intervent Radiol. 2022;45(3):275–6.
85. Reig M, Forner A, Rimola J, Ferrer-Fábrega J, Burrel M, Garcia-Criado A, et al. BCLC strategy for prognosis prediction and treatment recommendation Barcelona Clinic Liver Cancer (BCLC) staging system. The 2022 update. J Hepatol. 2021:681. https://doi.org/10.1016/j.jhep.2021.11.018.
86. Salem R, Johnson GE, Kim E, Riaz A, Bishay V, Boucher E, et al. Yttrium-90 radioembolization for the treatment of solitary unresectable HCC: the LEGACY study. Hepatology. 2021;74(5):2342–52.
87. Lucatelli P, Burrel M, Guiu B, de Rubeis G, van Delden O, Helmberger T. CIRSE standards of practice on hepatic transarterial chemoembolisation. Cardiovasc Intervent Radiol. 2021;44(12):1851–67.
88. Cabibbo G, Enea M, Attanasio M, Bruix J, Craxi A, Camma C. A meta-analysis of survival rates of untreated patients in randomized clinical trials of hepatocellular carcinoma. Hepatology. 2010;51:1274–83.
89. Grazi GL, Cescon M, Ravaioli M, Ercolani G, Gardini A, Del Gaudio M, et al. Liver resection for hepatocellular carcinoma in cirrhotics and noncirrhotics. Evaluation of clinicopathologic features and comparison of risk factors for long-term survival and tumour recurrence in a single centre. Aliment Pharmacol Ther. 2003;17(Suppl 2):119–29.
90. Essa M. Aleassa, Koji Hashimoto. Combining ablation and resection for the treatment of hepatocellular carcinoma: an attempt to expand treatment options. 2020;27(7):2125–6. https://doi.org/10.1245/s10434-020-08232-z.
91. Bodzin AS, Lunsford KE, Markovic D, Harlander-Locke MP, Busuttil RW, Agopian VG. Predicting mortality in patients developing recurrent hepatocellular carcinoma after liver transplantation: impact of treatment modality and recurrence characteristics. Ann Surg. 2017;266:118–25.
92. Morisson-Sarapak K, et al. Late recurrence of hepatocellular carcinoma in a patient 10 years after liver transplantation unrelated to transplanted. Organ Case Rep Oncol. 2021;14:1754–60. https://doi.org/10.1159/000520535.
93. Bruix J, Qin S, Merle P, et al. Regorafenib for patients with hepatocellular carcinoma who progressed on sorafenib treatment (RESORCE): a randomised, double-blind, placebo-controlled, phase 3 trial. Lancet. 2017;389:56–66.
94. Kumar S, Miller CM, Hashimoto K, Quintini C, Kumar A, Balci NC, Pinna AD. Liver transplantation in The United Arab Emirates from deceased and living donors: initial 2-year experience. Transplantation. 2021;105(9):1881–3.
95. Kumar S, Sankari BR, Miller CM, Obaidli AAKA, Suri RM. Establishment of solid organ transplantation in The United Arab Emirates. Transplantation. 2020;104(4):659–63.
96. Facciorusso A, et al. Microwave ablation versus radiofrequency ablation for treatment of hepatocellular carcinoma: a meta-analysis of randomized controlled trials. Cancers (Basel). 2020;12(12):3796. https://doi.org/10.3390/cancers12123796.

97. Vilar-Gomez E, Vuppalanchi R, Gawrieh S, Ghabril M, Saxena R, Cummings OW, Chalasani N. Vitamin E improves transplant-free survival and hepatic decompensation among patients with nonalcoholic steatohepatitis and advanced fibrosis. Hepatology. 2019;71(2):495–509.

98. Germani G, Pleguezuelo M, Gurusamy K, Meyer T, Isgro G, Burroughs AK. Clinical outcomes of radiofrequency ablation, percutaneous alcohol and acetic acid injection for hepatocelullar carcinoma: a meta-analysis. J Hepatol. 2010;52:380–8.

99. Tan W, et al. Comparison of microwave ablation and radiofrequency ablation for hepatocellular carcinoma: a systematic review and meta-analysis. Int J Hyperthermia. 2019;36(1):264–72. https://doi.org/10.1080/02656736.2018.1562571.

100. Vogl TJ, Farshid P, Naguib NN, Zangos S, Bodelle B, Paul J, et al. Ablation therapy of hepatocellular carcinoma: a comparative study between radiofrequency and microwave ablation. Abdom Imaging. 2015;40:1829–37.

101. Heimbach JK, Kulik LM, Finn RS, et al. AASLD guidelines for the treatment of hepatocellular carcinoma. Hepatology. 2018;67:358–80.

102. Raoul L, Forner A, Bolondi L, Cheung TT, Kloeckner R, de Baere T. Updated use of TACE for hepatocellular carcinoma treatment: how and when to use it based on clinical evidence. Cancer Treat Rev. 2019;72:28–36. https://doi.org/10.1016/j.ctrv.2018.11.002.

103. Rasul KI, Al-Azawi SH, Chandra P. Hepatocellular carcinoma in Qatar. Gulf J Oncolog. 2013;1(14):70–5.

104. Shaaban A, Salamah R, Abo Elseu Y, et al. Presentation and outcomes of hepatocellular carcinoma in the Arabian peninsula: a review of a single institution experience in the sorafenib era. J Gastrointest Canc. 2021;52(1):85–9.

105. Meshari A-A. Role of trans-arterial chemoembolization (TACE) in patients with hepatocellular carcinoma. Eur Rev Med Pharmacol Sci. 2022;26(18):6764–71. https://pubmed.ncbi.nlm.nih.gov/36196724/. Accessed 23 Mar 2023.

106. Sieghart W, Hucke F, Peck-Radosavljevic M. Transarterial chemoembolization: modalities, indication, and patient selection. J Hepatol. 2015;62:1187–95. https://doi.org/10.1016/j.jhep.2015.02.010.

107. Wang H, Cao C, Wei X, Shen K, Shu Y, Wan X, Sun J, Ren X, Dong Y, Liu Y, et al. A comparison between drug-eluting bead-transarterial chemoembolization and conventional transarterial chemoembolization in patients with hepatocellular carcinoma: a meta-analysis of six randomized controlled trials. J Cancer Res Ther. 2020;16:243–9. https://doi.org/10.4103/jcrt.jcrt_504_19.

108. Takada R, Fukutake N, Uehara H, et al. Subclassification of patients with intermediate-stage (Barcelona clinic liver cancer stage-B) hepatocellular carcinoma using the up-to-seven criteria and serum tumor markers. Hepatol Int. 2017;11:105–14. https://doi.org/10.1007/s12072-016-9771-0.

109. Kudo M, et al. Subclassification of BCLC B stage hepatocellular carcinoma and treatment strategies: proposal of modified Bolondi's Subclassification (Kinki Criteria). Dig Dis. 2015;33(6):751–8. https://doi.org/10.1159/000439290.

110. Kotsifa E, et al. Transarterial chemoembolization for hepatocellular carcinoma: why, when, how? J Pers Med. 2022;12(3):436. https://doi.org/10.3390/jpm12030436.

111. Kudo M, Ueshima K, Ikeda M, Torimura T, Tanabe N, Aikata H, Izumi N, Yamasaki T, Nojiri S, Hino K, et al. Randomised, multicentre prospective trial of transarterial chemoembolisation (TACE) plus sorafenib as compared with TACE alone in patients with hepatocellular carcinoma: TACTICS trial. Gut. 2020;69:1492–501. https://doi.org/10.1136/gutjnl-2019-318934.

112. Xue T-C, Xie X-Y, Zhang L, Yin X, Zhang B-H, Ren Z-G. Transarterial chemoembolization for hepatocellular carcinoma with portal vein tumor thrombus: a meta-analysis. BMC Gastroenterol. 2013;13:60. https://doi.org/10.1186/1471-230X-13-60.

113. Kim W, Cho SK, Shin SW, Hyun D, Lee MW, Rhim H. Combination therapy of transarterial chemoembolization (TACE) and radiofrequency ablation (RFA) for small hepatocellular carcinoma: comparison with TACE or RFA monotherapy. Abdom Radiol. 2019;44:2283–92. https://doi.org/10.1007/s00261-019-01952-1.

114. Raoul J-L, Sangro B, Forner A, Mazzaferro VM, Piscaglia F, Bolondi L, Lencioni R. Evolving strategies for the management of intermediate-stage hepatocellular carcinoma: available evidence and expert opinion on the use of transarterial chemoembolization. Cancer Treat Rev. 2011;37:212–20. https://doi.org/10.1016/j.ctrv.2010.07.006.

115. Vogl TJ, Gruber-Rouh T. HCC: transarterial therapies—what the interventional radiologist can offer. Dig Dis Sci. 2019;64:959–67. https://doi.org/10.1007/s10620-019-05542-5.

116. Zaitoun MMA, Elsayed SB, Zaitoun NA, Soliman RK, Elmokadem AH, Farag AA, Amer M, Hendi AM, Mahmoud NEM, El Deen DS, et al. Combined therapy with conventional trans-arterial chemoembolization (cTACE) and microwave ablation (MWA) for hepatocellular carcinoma >3–<5 cm. Int J Hyperth. 2021;38:248–56. https://doi.org/10.1080/0265673 6.2021.1887941.

117. Liu C, Li T, He J-T, Shao H. TACE combined with microwave ablation therapy vs. TACE alone for treatment of early- and intermediate-stage hepatocellular carcinomas larger than 5 cm: a meta-analysis. Diagn Interv Radiol. 2020;26:575–83. https://doi.org/10.5152/ dir.2020.19615.

118. Kudo M, Han K-H, Ye S-L, Zhou J, Huang Y-H, Lin S-M, Wang C-K, Ikeda M, Chan SL, Choo SP, et al. A changing paradigm for the treatment of intermediate-stage hepatocellular carcinoma: Asia-Pacific primary liver cancer expert consensus statements. Liver Cancer. 2020;9:245–60. https://doi.org/10.1159/000507370.

119. Müller L, et al. Current strategies to identify patients that will benefit from TACE treatment and future directions a practical step-by-step guide. J Hepatocell Carcinoma. 2021;8:403–19. https://doi.org/10.2147/JHC.S285735.

120. Forner A, Reig M, Bruix J. Hepatocellular carcinoma. Lancet. 2018;391(10127):1301–14.

121. Zhang H, Zhang W, Jiang L, Chen Y. Recent advances in systemic therapy for hepatocellular carcinoma. Biomark Res. 2022;10(1):3.

122. Rasul K, Issameldin A, Elazzazi S, et al. Can we use Sorafenib for advanced hepatocellular Carcinoma (HCC) child Pugh B? Gulf J Oncol. 2015;1(17):82–4.

123. Rasul KI, Al-Azawi SH, Chandra P, Abou-Alfa GK, Knuth A. Status of hepatocellular carcinoma in Gulf region. 2013;2(4):42.

124. Chang YS, Adnane J, Trail PA, Levy J, Henderson A, Xue D, et al. Sorafenib (BAY 43-9006) inhibits tumor growth and vascularization and induces tumor apoptosis and hypoxia in RCC xenograft models. Cancer Chemother Pharmacol. 2007;59(5):561–74.

125. Zhang H, Zhang W, Jiang L, et al. Recent advances in systemic therapy for hepatocellular carcinoma. Biomark Res. 2022;10:3. https://doi.org/10.1186/s40364-021-00350-4.

126. Alqahtani A, et al. Hepatocellular carcinoma: molecular mechanisms and targeted therapies. Medicina (Kaunas). 2019;55(9):526. https://doi.org/10.3390/medicina55090526.

127. Yamamoto Y, Matsui J, Matsushima T, Obaishi H, Miyazaki K, Nakamura K, et al. Lenvatinib, an angiogenesis inhibitor targeting VEGFR/FGFR, shows broad antitumor activity in human tumor xenograft models associated with microvessel density and pericyte coverage. Vasc Cell. 2014;6:18.

128. The primary endpoint was OS and secondary outcomes were TTP and progression free survival (PFS). Median OS for lenvatinib of 13.6 months was non-inferior to sorafenib of 12.3 months. TTP was 7.4 months for lenvatinib and 3.7 months with sorafenib.

129. Cerrito L, Ponziani FR, Garcovich M, Tortora A, Annicchiarico BE, Pompili M, Siciliano M, Gasbarrini A. Regorafenib: a promising treatment for hepatocellular carcinoma. Expert Opin Pharmacother. 2018;19:1941–8. https://doi.org/10.1080/14656566.2018.1534956.

130. Xiang Q, Chen W, Ren M, Wang J, Zhang H, Deng DY, et al. Cabozantinib suppresses tumor growth and metastasis in hepatocellular carcinoma by a dual blockade of VEGFR2 and MET. Clin Cancer Res. 2014;20(11):2959–70.

131. Abou-Alfa GK, Meyer T, Cheng AL, El-Khoueiry AB, Rimassa L, Ryoo BY, Cicin I, Merle P, Chen Y, Park JW, et al. Cabozantinib in patients with advanced and progressing hepatocellular Carcinoma. N Engl J Med. 2018;379:54–63. https://doi.org/10.1056/NEJMoa1717002.

132. FDA Approves Cabozanitab for Hepatocellular Carcinoma. https://www.fda.gov/Drugs/InformationOnDrugs/ApprovedDrugs/ucm629512.htm. Accessed 17 Apr 2019.
133. FDA Approves Ramucirumab for Hepatocellular Carcinoma. https://www.fda.gov/drugs/resources-information-approved-drugs/fda-approves-ramucirumab-hepatocellular-carcinoma. Accessed 5 June 2019.
134. El-Khoueiry AB, Sangro B, Yau T, Crocenzi TS, Kudo M, Hsu C, et al. Nivolumab in patients with advanced hepatocellular carcinoma (CheckMate 040): an open-label, non-comparative, phase 1/2 dose escalation and expansion trial. Lancet. 2017;389(10088):2492–502.
135. FDA grants accelerated approval to nivolumab for HCC previously treated with sorafenib. https://www.fda.gov/drugs/resources-information-approved-drugs/fda-grants-accelerated-approval-nivolumab-hcc-previously-treated-sorafenib. Accessed 22 Sept 2017.
136. Au TPJ, Finn RS, Cheng A-L, Mathurin P, Edeline J, et al. CheckMate 459: a randomized, multi-center phase III study of nivolumab (NIVO) vs sorafenib (SOR) as first-line (1L) treatment in patients (pts) with advanced hepatocellular carcinoma (aHCC) [abstract LBA38_PR]. Ann Oncol. 2019;30:874–5.
137. Zhu AX, Finn RS, Cattan S, Edeline J, Ogasawara S, Palmer DH, Verslype C, Zagonel V, Rosmorduc O, Vogel A, Sarker D, Verset G, Chan SL, Knox JJ, Daniele B, Ebbinghaus S, Ma J, Siegel AB, Cheng A-L, Kudo M. KEYNOTE-224: pembrolizumab in patients with advanced hepatocellular carcinoma previously treated with sorafenib. J Clin Oncol. 2018;36(4_suppl):209–9.
138. Center for Drug Evaluation and Research. FDA grants accelerated approval to pembrolizumab for hepatocellular carcinoma. FDA; 2019. https://www.fda.gov/drugs/fda-grants-accelerated-approval-pembrolizumab-hepatocellular-carcinoma. Accessed 8 Sept 2022.
139. Finn RS, Chan SL, Zhu AX, Knox JJ, Cheng A-L, Siegel AB, Bautista O, Watson P, Kudo M. KEYNOTE-240: randomized phase III study of pembrolizumab versus best supportive care for second-line advanced hepatocellular carcinoma. J Clin Oncol. 2017;35(4_suppl):TPS503.
140. Van Laethem J-L, Borbath I, Karwal M, Verslype C, Van Vlierberghe H, Kardosh A, Zagonel V, Stal P, Sarker D, Palmer DH, Vogel A, Edeline J, Cattan S, Kudo M, Cheng A-L, Ogasawara S, Siegel AB, Chisamore MJ, Wang A, Zhu AX. Pembrolizumab (pembro) monotherapy for previously untreated advanced hepatocellular carcinoma (HCC): phase 2 KEYNOTE-224 study. J Clin Oncol. 2021;39(15_suppl):4074–4.
141. Center for Drug Evaluation and Research. FDA approves atezolizumab plus bevacizumab for unresectable hepatocellular carcinoma. FDA; 2021. https://www.fda.gov/drugs/drug-approvals-and-databases/fda-approves-atezolizumab-plus-bevacizumab-unresectable-hepatocellular-carcinoma. Accessed 23 Mar 2023.
142. Finn RS, Qin S, Ikeda M, Galle PR, Ducreux M, Kim T-Y, Kudo M, Breder V, Merle P, Kaseb AO, Li D, Verret W, Xu D-Z, Hernandez S, Liu J, Huang C, Mulla S, Wang Y, Lim HY, Zhu AX. Atezolizumab plus bevacizumab in unresectable hepatocellular carcinoma. N Engl J Med. 2020;382(20):1894–905.
143. Cheng A-L, Qin S, Ikeda M, Galle PR, Ducreux M, Kim T-Y, Lim HY, Kudo M, Breder V, Merle P, Kaseb AO, Li D, Verret W, Ma N, Nicholas A, Wang Y, Li L, Zhu AX, Finn RS. Updated efficacy and safety data from IMbrave150: atezolizumab plus bevacizumab vs. sorafenib for unresectable hepatocellular carcinoma. J Hepatol. 2021;76(4):862–73.
144. El-Khoueiry TYAB, Kang Y-K, Kim T-Y, Santoro A, Sangro B, Melero I, et al. Nivolumab (NIVO) plus ipilimumab (IPI) combination therapy in patients (pts) with advanced hepatocellular carcinoma (aHCC): long-term results from CheckMate 040. J Clin Oncol. 2021;39(suppl 33):abstr 269.
145. Kumar M, Panda D. Role of supportive Care for Terminal Stage Hepatocellular Carcinoma. J Clin Exp Hepatol. 2014;4:S130–9.
146. Laube R, Sabih A, Strasser SI, Lim L, Cigolini M, Liu K. Palliative care in hepatocellular carcinoma. J Gastroenterol Hepatol. 2020;36(3):618–28.

147. Rakoski MO, Volk ML. Palliative care for patients with end-stage liver disease: an overview. Clin Liver Disease. 2015;6(1):19–21. https://doi.org/10.1002/cld.478.

148. Wang X, Liu X, Dang Z, Yu L, Jiang Y, Wang X, Yan Z. Nucleos(t)ide analogues for reducing hepatocellular carcinoma in chronic Hepatitis B patients: a systematic review and meta-analysis. Gut and Liver. 2020;14(2):232–47.

149. Gu L, Yao Q, Shen Z, He Y, Ng DM, Yang T, Chen B, Chen P, Mao F, Yu Q. Comparison of tenofovir versus entecavir on reducing incidence of hepatocellular carcinoma in chronic hepatitis B patients: a systematic review and meta-analysis. J Gastroenterol Hepatol. 2020;35(9):1467–76.

150. www.who.int. Hepatitis D. https://www.who.int/news-room/fact-sheets/detail/hepatitis-d#:~:text=Hepatitis%20D%20virus%20(HDV)%20affects. Accessed 23 Mar 2023.

151. Zhang Z, Zhou Y, Yang J, Hu K, Huang Y. The effectiveness of TDF versus ETV on incidence of HCC in CHB patients: a meta analysis. BMC Cancer. 2019;19(1):511.

152. Nishibatake Kinoshita M, Minami T, Tateishi R, Wake T, Nakagomi R, Fujiwara N, Sato M, Uchino K, Enooku K, Nakagawa H, Asaoka Y, Shiina S, Koike K. Impact of direct-acting antivirals on early recurrence of HCV-related HCC: comparison with interferon-based therapy. J Hepatol. 2019;70(1):78–86.

153. Islam MM, Poly TN, Walther BA, Yang H-C, Li Y-C. Statin use and the risk of hepatocellular Carcinoma: a meta-analysis of observational studies. Cancers. 2020;12(3):671.

154. Sung F-C, Yeh Y-T, Muo C-H, Hsu C-C, Tsai W-C, Hsu Y-H. Statins reduce hepatocellular carcinoma risk in patients with chronic kidney disease and end-stage renal disease: a 17-year longitudinal study. Cancers. 2022;14(3):825. https://www.ncbi.nlm.nih.gov/pmc/articles/PMC8834435/#:~:text=doi%3A%C2%A010.3390/cancers14030825. Accessed 23 Mar 2023.

155. Yau T, Park J-W, Finn RS, Cheng A-L, Mathurin P, Edeline J, Kudo M, Harding JJ, Merle P, Rosmorduc O, Wyrwicz L, Schott E, Choo SP, Kelley RK, Sieghart W, Assenat E, Zaucha R, Furuse J, Abou-Alfa GK, El-Khoueiry AB. Nivolumab versus sorafenib in advanced hepatocellular carcinoma (CheckMate 459): a randomised, multicentre, open-label, phase 3 trial. The lancet. Oncology. 2022;23(1):77–90. https://pubmed.ncbi.nlm.nih.gov/34914889/. Accessed 20 Feb 2022.

156. Lee T-Y, Hsu Y-C, Tseng H-C, Yu S-H, Lin J-T, Wu M-S, Wu C-Y. Association of Daily Aspirin Therapy with Risk of hepatocellular Carcinoma in patients with chronic Hepatitis B. JAMA Intern Med. 2019;179(5):633.

157. Tan RZH, Lockart I, Abdel Shaheed C, Danta M. Systematic review with meta-analysis: the effects of non-steroidal anti-inflammatory drugs and anti-platelet therapy on the incidence and recurrence of hepatocellular carcinoma. Aliment Pharmacol Ther. 2021;54(4):356–67.

158. Wang S, Yu Y, Ryan PM, Dang M, Clark C, Kontogiannis V, Rahmani J, Varkaneh HK, Salehisahlabadi A, Day AS, Zhang Y. Association of aspirin therapy with risk of hepatocellular carcinoma: a systematic review and dose-response analysis of cohort studies with 2.5 million participants. Pharmacol Res. 2020;151:104585.

159. Cunha V, Cotrim HP, Rocha R, Carvalho K, Lins-Kusterer L. Metformin in the prevention of hepatocellular carcinoma in diabetic patients: a systematic review. Ann Hepatol. 2019;19(3):232–7.

160. Schulte L, Scheiner B, Voigtländer T, Koch S, Schweitzer N, Marhenke S, Ivanyi P, Manns MP, Rodt T, Hinrichs JB, Weinmann A, Pinter M, Vogel A, Kirstein MM. Treatment with metformin is associated with a prolonged survival in patients with hepatocellular carcinoma. Liver Int. 2019;39(4):714–26.

161. Bravi F, Bosetti C, Tavani A, Gallus S, La Vecchia C. Coffee reduces risk for hepatocellular carcinoma: an updated meta-analysis. Clin Gastroenterol Hepatol. 2013;11(11):1413–1421.e1.

162. Kennedy OJ, Roderick P, Buchanan R, Fallowfield JA, Hayes PC, Parkes J. Coffee, including caffeinated and decaffeinated coffee, and the risk of hepatocellular carcinoma: a systematic review and dose–response meta-analysis. BMJ Open. 2017;7(5):e013739. https://bmjopen.bmj.com/content/7/5/e013739. Accessed 12 Jan 2020.

163. Luo J, Yang Y, Liu J, Lu K, Tang Z, Liu P, Liu L, Zhu Y. Systematic review with meta-analysis: meat consumption and the risk of hepatocellular carcinoma. Aliment Pharmacol Ther. 2014;39(9):913–22.
164. Yu J, Liu Z, Liang D, Li J, Ma S, Wang G, Chen W. Meat intake and the risk of hepatocellular carcinoma: a meta-analysis of observational studies. Nutr Cancer. 2022;74(9):3340–50.
165. George ES, Sood S, Broughton A, Cogan G, Hickey M, Chan WS, Sudan S, Nicoll AJ. The association between diet and hepatocellular carcinoma: a systematic review. Nutrients. 2021;13(1):172.

Dr. Salman Wahib Srayaldeen is a highly qualified and experienced medical professional with a strong background in the fields of hematology and medical oncology. He holds a Master's degree in hematology/medical oncology from Damascus University, where he earned an M.D. degree as well with a very good grade level. He has been practicing medicine for over a decade and has held various positions in the medical field, including senior specialist in hematology/medical oncology at the Oncology Center.

Dr. Salman has a strong commitment to his field, as evidenced by his active participation in various professional organizations, including the Emirates Oncology Society (EOS), the Emirates Hematology Society (EHSA), the Syrian Oncology Society (SOA), the European Society of Medical Oncology (ESMO), the European Hematology Society (EHA), and the American Society of Clinical Oncology (ASCO). He has also published more than ten articles in reputable journals, showcasing his research and knowledge in the field.

Dr. Salman has extensive work experience in the fields of medical oncology and hematology. He has been working as a specialist at Sheikh Khalifa Specialty Hospital in the UAE since May 2017, where he provides specialized care to patients diagnosed with cancer and blood disorders. Before that, he held various positions, including general practitioner, specialist in internal medicine/hematology, and resident doctor in a stem cell transplantation unit.

In addition to his work experience, Dr. Salman has also attended numerous workshops and conferences and has several certificates of attendance, showing his commitment to continuously improving his skills and knowledge in the field. He is excellent in reading, writing, and speaking both Arabic and English, and he is good at German.

Overall, Dr. Salman Wahib Sray Aldeen is a highly qualified and experienced medical professional with a strong background in the fields of hematology and medical oncology. He is dedicated to providing high-quality care to his patients and is constantly working to improve his skills and knowledge in the field.

Dr. Mohamed Ahmed Mohamed Elkhalifa is a highly respected healthcare professional with a broad range of expertise and accomplishments in the field. He holds a MBBS from the University of Gezira in Sudan and is licensed by several healthcare organizations. His research interests include liver, skin cancers, medical education development, epidemiology, and health services research, as well as continuous quality improvement in healthcare firms. His work in these areas has earned him recognition through numerous publications, awards, and invitations to deliver keynote presentations at international conferences.

In addition to his medical degree, Dr. Elkhalifa is a Certified Professional in Healthcare Quality (CPHQ), a Certified Professional in Healthcare Risk Management (CPHRM), and a Project Management Professional (PMP) certified professional. He is also a member of the Royal Colleges of Physicians in the UK (MRCP (UK)).

Before his current role, Dr. Elkhalifa served as a Regional Healthcare Accreditation Coordinator (RAC) in Saudi Arabia, where he was responsible for overseeing accreditation processes and ensuring compliance with accreditation standards. His extensive experience in the field and commitment to quality made him a valuable asset to the organization and helped him to make a positive impact on the healthcare industry.

Dr. Mohamed Elkhalifa's broad range of expertise and experience make him a valuable asset to any healthcare organization. His dedication to improving healthcare quality, combined with his excellent team-leading, administrative, and organizational skills, makes him well-suited to take on leadership roles and drive positive change in healthcare firms. He has a strong commitment to improving healthcare outcomes and making a positive impact on the lives of patients and communities.

Ashish V. Chintakuntlawar, Hani Al-Halabi, and Aref Chehal

31.1   Introduction

Head and neck cancer is one of the most common cancers in the world and one of the more prevalent malignancies among males in the United Arab Emirates (UAE) [1]. Cancers of the head and neck comprise a variety of cancers, including mucosal head and neck squamous cell carcinomas (HNSCC), nasopharyngeal carcinomas (NPC), skin cancers (including melanomas, squamous, and basal cell carcinomas), carcinomas of the paranasal sinus, and salivary gland carcinomas. In this chapter, we will mainly focus on mucosal HNSCCs.

Although HNSCC is the term usually used to describe all mucosal head and neck cancers, it is important to note that it comprises a variety of cancers from distinct subsites, including the oral cavity, oropharynx, larynx, and hypopharynx. Although tobacco and alcohol consumption are the main etiologic factors, cancers in these subsites have variable pathophysiology and clinical outcomes, including rates of distant metastasis and overall survival. In the western hemisphere, the rates of human papilloma virus-associated oropharyngeal cancer (HPV-OPSCC) are also changing demographic trends. Management of HNSCC remains a tough clinical challenge all over the world and requires multidisciplinary collaboration amongst

A. V. Chintakuntlawar (✉)
Division of Medical Oncology, Sheikh Shakhbout Medical City,
Abu Dhabi, United Arab Emirates
e-mail: Chintakuntlawar.ashish@mayo.edu

H. Al-Halabi
Gulf International Cancer Center, Abu Dhabi, United Arab Emirates
e-mail: hani@gulficc.ae

A. Chehal
Division of Medical Oncology, Sheikh Shakhbout Medical City,
Abu Dhabi, United Arab Emirates
e-mail: achehal@ssmc.ae

© The Author(s) 2024
H. O. Al-Shamsi (ed.), *Cancer Care in the United Arab Emirates*,
https://doi.org/10.1007/978-981-99-6794-0_31

many specialties to optimally treat the patient. This is essential not only to achieve better survival outcomes but also to reduce long-term treatment-related morbidity in survivors.

In this chapter, we will discuss the epidemiology, pathogenesis, and management of HNSCC. Our goal is to not only describe the current state of affairs as it relates to therapy but also describe the challenges we face in order to define opportunities for improvement.

31.2 Epidemiology

The data regarding the incidence of HNSCC in the UAE is limited and has previously been published in systematic reviews, published articles, or single institution series. However, recently, data from registries has started to become available. Data from the Ministry of Health and Prevention, UAE National Cancer Registry shows that the absolute numbers of HNSCC (lip, oral cavity, pharynx, and larynx combined) are increasing in the UAE (Fig. 31.1) [2]. This is also true for

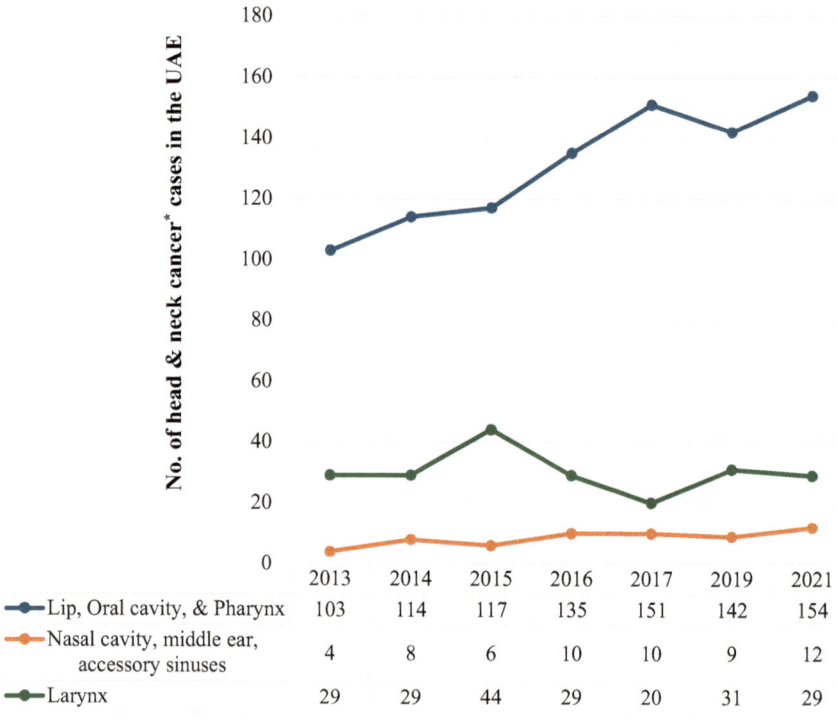

	2013	2014	2015	2016	2017	2019	2021
Lip, Oral cavity, & Pharynx	103	114	117	135	151	142	154
Nasal cavity, middle ear, accessory sinuses	4	8	6	10	10	9	12
Larynx	29	29	44	29	20	31	29

Fig. 31.1 Number of head and neck cancer (malignant) occurrences in the UAE across the years 2013–2021 (Source: Ministry of Health and Prevention, Statistics and Research Center, National Disease Registry—UAE National Cancer Registry Report, 2013–2021).
* Head and neck cancer includes 1. Lip, Oral cavity, & Pharynx, 2. Nasal cavity, middle ear, accessory sinuses, and 3. Larynx

population-adjusted ratios (Table 31.1) and holds true especially for non-UAE residents (Fig. 31.2), males (Fig. 31.3), and oral cavity and oropharyngeal cancers (Fig. 31.1).

The data regarding HPV versus non-HPV oropharyngeal cancers is also lacking in the registry. It will be critical to collect this data and monitor future trends with respect to the native Emirati population as well as the expat population. The UAE is unique with respect to immigration trends, and there are a significant number of expats from both regions, such as the Indian subcontinent and Southeast Asia, where oral carcinomas are common, and from the Western Hemisphere, where HPV-positive OPSCC are on the rise.

Table 31.1 Head and neck cancer demographics among the UAE population during 2013–2021

Year	UAE population (in millions)	Total malignant cases (in numbers)	Head and neck cancer cases (in numbers)	Crude incidence rate of head and neck cancer[a] cases per 100,000 population
2013	8.66	3574	136	–
2014	8.79	3610	151	– Lip, Oral cavity & pharynx: 1.25 – Nasal cavity, middle ear, accessory sinuses: 0.09 – Larynx: 0.32
2015	8.93	3744	167	– Lip, Oral cavity & pharynx: 1.3 – Nasal cavity, middle ear, accessory sinuses: 0.1 – Larynx: 0.5
2016	9.12	3982	174	–
2017	9.3	4123	181	– Lip, Oral cavity & pharynx: 1.6 – Nasal cavity, middle ear, accessory sinuses: 0.1 – Larynx:0.2
2019	9.5	4381	182	–
2021	–	5612	195	– Lip, Oral cavity & pharynx: 1.7 – Nasal cavity, middle ear, accessory sinuses: 0.1 – Larynx:0.3

Source: UAE population: https://fcsc.gov.ae/en-us/Pages/Statistics/Statistics-by-Subject.aspx#/%3Fsubject=Demography%20and%20Social; Ministry of Health and Prevention, Statistics and Research Center, National Disease Registry—UAE National Cancer Registry Report; [a]Head and neck cancer includes 1. Lip, Oral cavity, and Pharynx, 2. Nasal cavity, middle ear, accessory sinuses, and 3. Larynx

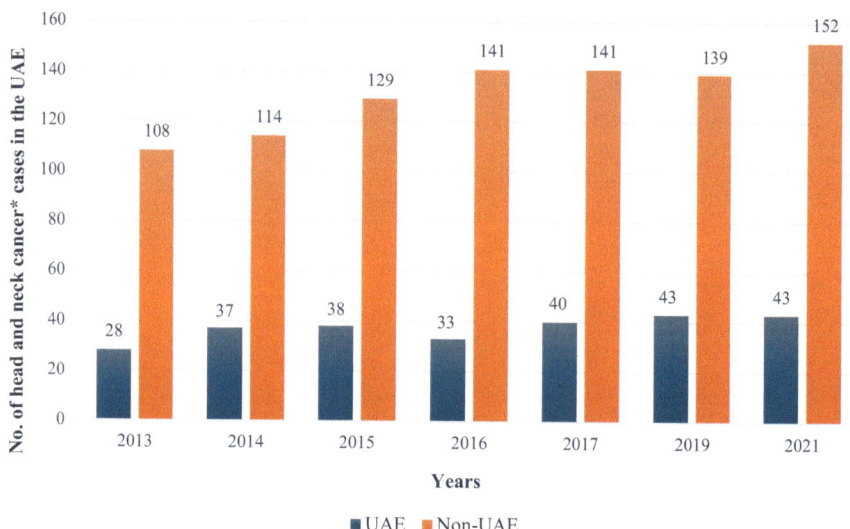

Fig. 31.2 The number of head and neck cancer cases (malignant) among the UAE population according to nationality, 2013–2021 (Source: Ministry of Health and Prevention, Statistics and Research Center, National Disease Registry—UAE National Cancer Registry Report).
* Head and neck cancer includes 1. Lip, Oral cavity, & Pharynx, 2. Nasal cavity, middle ear, accessory sinuses, and 3. Larynx

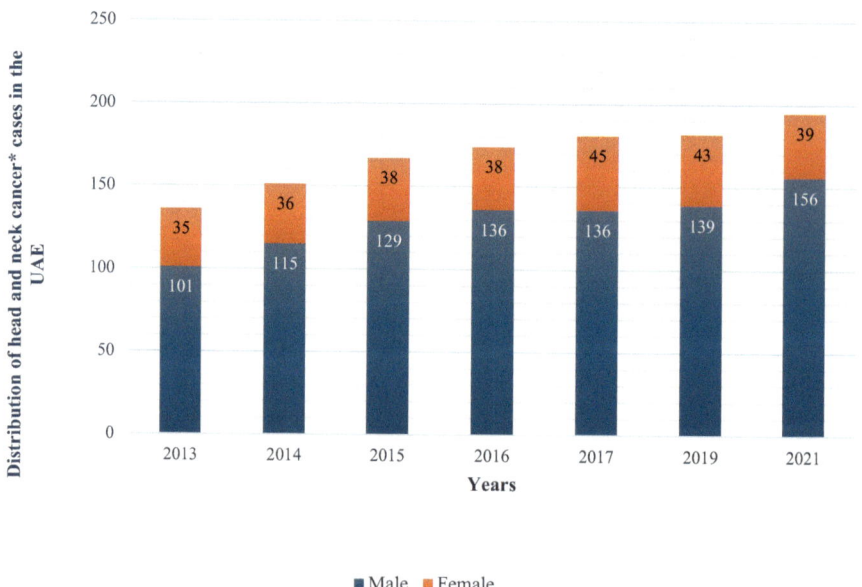

Fig. 31.3 Distribution of head and neck cancer cases (malignant) according to gender, 2013–2021 (Source: Ministry of Health and Prevention, Statistics and Research Center, National Disease Registry—UAE National Cancer Registry Report).
* Head and neck cancer includes 1. Lip, Oral cavity, & Pharynx, 2. Nasal cavity, middle ear, accessory sinuses, and 3. Larynx

31.3 Diagnostic Techniques

The diagnosis of HNSCC requires expert evaluation using endoscopy by a trained and experienced head and neck surgeon. Imaging should be reviewed by a radiologist with expertise in head and neck cancers. A concurrent evaluation by a radiation oncologist and a medical oncologist, or at least a multidisciplinary tumor board evaluation, is necessary for optimal treatment planning. Cross-sectional imaging with CT scanning, MRI, and FDG-PET is also available in the UAE. Access to FDG-PET is limited but is rapidly improving.

31.4 Current Treatment Paradigms

The treatment standards are largely commensurate with National Comprehensive Cancer Network (NCCN) recommendations and Food and Drug Administration (FDA) approvals. Oral cancers are treated with surgical resection, and often microvascular reconstruction is required for locally advanced oral cancers. Although not available in every tertiary care facility, access to microvascular reconstruction surgery is fair in the emirates of Dubai and Abu Dhabi. The availability of experts in microvascular surgery is limited in the northern Emirates. Transoral robotic surgery (TORS) is increasingly being utilized in the surgical resection of OPSCC. Recently, ECOG 3311 and phase 2 and 3 trials from the Mayo Clinic have shown the feasibility and importance of de-escalation therapy via TORS for HPV-OPSCC [3, 4]. On the other hand, the recent randomized trial of TORS versus chemoradiotherapy had to be halted due to unexpected deaths. Therefore, it will be critical to implement the TORS very carefully, and monitoring the outcomes via a credentialing system as in ECOG3311 may be needed [3, 5].

Radiotherapy in the adjuvant or definitive setting (with or without concurrent chemotherapy) is becoming more accessible. Radiotherapy has been available here in the UAE for more than 30 years but was initially limited to one center. In the past 5–10 years, we have witnessed a revolution in the field of radiotherapy with the development of multiple new centers across Abu Dhabi and Dubai. Advanced radiotherapy techniques such as volumetric modulated arc therapy (VMAT), Intensity-modulated radiation therapy (IMRT), and stereotactic radiation are now available to treat HNSCC patients, and older 2D and 3D techniques are being phased out to improve treatment efficacy and patients' quality of life [6]. Proton therapy has shown promise in further improving the therapeutic index of radiotherapy in HNSCC, but as in many parts of the world, it is not yet available in the UAE or the Middle East.

Systemic therapy, administered concurrently with radiotherapy or palliative therapy, is widely available. However, what is lacking is expertise specific to HNSCC and access to clinical trials and novel experimental therapeutics. Cytotoxic chemotherapy agents such as platinum agents, taxanes, and methotrexate are all available. Biologics such as cetuximab [7, 8] and immunotherapeutics such as nivolumab and pembrolizumab are also available, and their use is commensurate with the current guidelines [9, 10]. There are studies from India with innovative strategies keeping in mind the low resource setting [11, 12]. Similar strategies could be tested in the

UAE population with the goal of reducing not only the cost but also the symptom burden on patients, who often work through the therapy.

There are a number of other supportive therapies needed for optimal management of HNSCC patients. Access to dentistry for dental rehabilitation, speech pathology, physical therapy, occupational therapy, dietician services, and audiologic testing is limited and lacks specific expertise. Both access to devices and training related to speech therapy and hearing impairment are limited due to a lack of insurance coverage as well as a shortage of medical professionals.

31.5 Challenges

Even though there has been tremendous progress in terms of diagnostics as well as available modalities of therapy, including chemotherapy, immunotherapy, IMRT, VMAT, and reconstructive surgery, there remain significant challenges that preclude optimal outcomes and the management of long-term morbidities.

Most of the patients with HNSCC in the UAE are from the lower socioeconomic strata and have poor financial and social support. Many patients, especially those with oral tongue, buccal, and oropharyngeal squamous cell carcinoma, belong to the Indian subcontinent. They have extremely limited medical access, which often delays the diagnosis and even therapy. They have almost nonexistent social support and often lose their source of income upon diagnosis, making both therapy and recovery an uphill task for both the patient and the providers. It is very well known that interruptions or delays in therapy are associated with particularly poor outcomes [13, 14]. There are no systematic studies from the UAE to determine how many patients suffer from delays and interruptions during therapy.

There is emerging interventional trial data from India, especially from Tata Memorial Hospital [15–17], but none from the UAE, and there is no comprehensive molecular analysis of these cancers to determine if the genetic drivers are the same.

With a significant number of expats from the western hemisphere, it is likely that we will see more HPV-OPSCC. It is critical that we identify the trends in the local population and the expat population for this particular cancer and pay close attention to the careful descaling of the therapy, preferably in a prospective trial, to reduce morbidity. This makes it essential not to lump them with other HNSCCs and to continue to offer them the same conventional therapies with significant long-term morbidity.

The incidence of depression and anxiety related to therapy and therapy-related morbidity is very high in HNSCC survivors all over the world. HNSCCs is also one of the leading causes of cancer-related suicide, particularly in rural areas [18]. The burden of psychosocial symptoms could be decreased by better access to psychiatry and psychology services as well as patient support groups. Currently, there are no head-and-neck cancer-specific support groups in the UAE.

There is a significant number of patients from the lower socio-economic strata of the UAE population who are disproportionately affected by HNSCCs. This poses greater distress and financial hardship, resulting in delays in diagnosis and therapy and often resulting in interruptions of therapy. All these factors have been shown to negatively affect the survival outcomes of HNSCC. It is critical that we have the means to provide lodging closer to radiotherapy facilities, access to financial support and job security while patients are on therapy and unable to work, and the availability of uninterrupted nutritional support, including tube feedings. All of these are as important as, if not more important than, the pharmacologic therapies we provide to these patients.

31.6 Conclusion

In conclusion, head and neck cancer continues to be a challenge in the UAE, affecting both the native Emirati and expat populations, with tobacco, alcohol, and HPV as emerging etiologies. Access, diagnostic, and therapeutic challenges remain, but steady progress is being made. Clinical trials and ancillary services, including physical, occupational, speech, dental, and psychosocial rehabilitation, need major improvements and engagement.

Acknowledgments The authors are indebted to their patients and families.

Conflict of Interest The authors have no conflict of interest to declare.

References

1. Global Burden of Disease Cancer Colloboration, Kocarnik JM, Compton K, Dean FE, Fu W, Gaw BL, Harvey JD, Henrikson HJ, Lu D, Pennini A, Xu R, Ababneh E, Abbasi-Kangevari M, Abbastabar H, Abd-Elsalam SM, Abdoli A, Abedi A, Abidi H, Abolhassani H, Force LM. Cancer incidence, mortality, years of life lost, years lived with disability, and disability-adjusted life years for 29 cancer groups from 2010 to 2019: a systematic analysis for the global burden of Disease study 2019. JAMA Oncol. 2022;8(3):420–44. https://doi.org/10.1001/jamaoncol.2021.6987.
2. Cancer incidence in United Arab Emirates, Annual Report of the UAE - National Cancer Registry. Statistics and Research Center, Ministry of Health and Prevention (Accessed on 26 Mar 2024).
3. Ferris RL, Flamand Y, Weinstein GS, Li S, Quon H, Mehra R, Garcia JJ, Chung CH, Gillison ML, Duvvuri U, O'Malley BW Jr, Ozer E, Thomas GR, Koch WM, Gross ND, Bell RB, Saba NF, Lango M, Mendez E, Burtness B. Phase II randomized trial of Transoral surgery and low-dose intensity modulated radiation therapy in Resectable p16+ locally advanced oropharynx cancer: an ECOG-ACRIN cancer research group trial (E3311). J Clin Oncol. 2022;40(2):138–49. https://doi.org/10.1200/JCO.21.01752.
4. Ma DJ, Price KA, Moore EJ, Patel SH, Hinni ML, Garcia JJ, Graner DE, Foster NR, Ginos B, Neben-Wittich M, Garces YI, Chintakuntlawar AV, Price DL, Olsen KD, Van Abel KM, Kasperbauer JL, Janus JR, Waddle M, Miller R, et al. Phase II evaluation of aggressive dose De-escalation for adjuvant Chemoradiotherapy in human papillomavirus-associated oropharynx squamous cell carcinoma. J Clin Oncol. 2019;37(22):1909–18. https://doi.org/10.1200/JCO.19.00463.
5. Ferris RL, Flamand Y, Holsinger FC, Weinstein GS, Quon H, Mehra R, Garcia JJ, Hinni ML, Gross ND, Sturgis EM, Duvvuri U, Mendez E, Ridge JA, Magnuson JS, Higgins KA, Patel MR, Smith RB, Karakla DW, Kupferman ME, et al. A novel surgeon credentialing and quality assurance process using transoral surgery for oropharyngeal cancer in ECOG-ACRIN

cancer research group trial E3311. Oral Oncol. 2020;110:104797. https://doi.org/10.1016/j. oraloncology.2020.104797.

6. Rathod S, Gupta T, Ghosh-Laskar S, Murthy V, Budrukkar A, Agarwal J. Quality-of-life (QOL) outcomes in patients with head and neck squamous cell carcinoma (HNSCC) treated with intensity-modulated radiation therapy (IMRT) compared to three-dimensional conformal radiotherapy (3D-CRT): evidence from a prospective randomized study. Oral Oncol. 2013;49(6):634–42. https://doi.org/10.1016/j.oraloncology.2013.02.013.

7. Bonner JA, Harari PM, Giralt J, Azarnia N, Shin DM, Cohen RB, Jones CU, Sur R, Raben D, Jassem J, Ove R, Kies MS, Baselga J, Youssoufian H, Amellal N, Rowinsky EK, Ang KK. Radiotherapy plus cetuximab for squamous-cell carcinoma of the head and neck. N Engl J Med. 2006;354(6):567–78. https://doi.org/10.1056/NEJMoa053422.

8. Vermorken JB, Mesia R, Rivera F, Remenar E, Kawecki A, Rottey S, Erfan J, Zabolotnyy D, Kienzer HR, Cupissol D, Peyrade F, Benasso M, Vynnychenko I, De Raucourt D, Bokemeyer C, Schueler A, Amellal N, Hitt R. Platinum-based chemotherapy plus cetuximab in head and neck cancer. N Engl J Med. 2008;359(11):1116–27. https://doi.org/10.1056/NEJMoa0802656.

9. Burtness B, Harrington KJ, Greil R, Soulieres D, Tahara M, de Castro G Jr, Psyrri A, Baste N, Neupane P, Bratland A, Fuereder T, Hughes BGM, Mesia R, Ngamphaiboon N, Rordorf T, Wan Ishak WZ, Hong RL, Gonzalez Mendoza R, Roy A, Investigators K. Pembrolizumab alone or with chemotherapy versus cetuximab with chemotherapy for recurrent or metastatic squamous cell carcinoma of the head and neck (KEYNOTE-048): a randomised, open-label, phase 3 study. Lancet. 2019;394(10212):1915–28. https://doi.org/10.1016/S0140-6736(19)32591-7.

10. Ferris RL, Blumenschein G Jr, Fayette J, Guigay J, Colevas AD, Licitra L, Harrington K, Kasper S, Vokes EE, Even C, Worden F, Saba NF, Iglesias Docampo LC, Haddad R, Rordorf T, Kiyota N, Tahara M, Monga M, Lynch M, et al. Nivolumab for recurrent squamous-cell carcinoma of the Head and Neck. N Engl J Med. 2016;375(19):1856–67. https://doi.org/10.1056/NEJMoa1602252.

11. Kashyap L, Patil V, Noronha V, Joshi A, Menon N, Jobanputra K, Saha S, Chaturvedi P, Banavali SD, Prabhash K. Efficacy and safety of neoadjuvant chemotherapy (NACT) with paclitaxel plus carboplatin and oral metronomic chemotherapy (OMCT) in patients with technically unresectable oral squamous cell carcinoma (OSCC). Ecancermedicalscience. 2021;15:1325. https://doi.org/10.3332/ecancer.2021.1325.

12. Patil VM, Noronha V, Menon N, Rai R, Bhattacharjee A, Singh A, Nawale K, Jogdhankar S, Tambe R, Dhumal S, Sawant R, Alone M, Karla D, Peelay Z, Pathak S, Balaji A, Kumar S, Purandare N, Agarwal A, et al. Low-dose immunotherapy in head and neck cancer: a randomized study. J Clin Oncol. 2022;41(2):222–32. https://doi.org/10.1200/JCO.22.01015.

13. Karp EE, Yin LX, O'Byrne TJ, Lu LY, Routman DM, Lester SC, Neben Wittich MA, Ma DJ, Price KA, Chintakuntlawar AV, Tasche KK, Price DL, Moore EJ, Van Abel KM. Diagnostic delay in human papillomavirus negative oropharyngeal squamous cell carcinoma. Laryngoscope. 2022;133:1394. https://doi.org/10.1002/lary.30307.

14. Yin LX, Karp EE, Elias A, O'Byrne TJ, Routman DM, Price DL, Kasperbauer JL, Neben-Wittich M, Chintakuntlawar AV, Price KA, Ma DJ, Foote RL, Moore EJ, Van Abel KM. Disease profile and oncologic outcomes after delayed diagnosis of human papillomavirus-associated oropharyngeal cancer. Otolaryngol Head Neck Surg. 2021;165(6):830–7. https://doi.org/10.1177/01945998211000426.

15. D'Cruz AK, Vaish R, Kapre N, Dandekar M, Gupta S, Hawaldar R, Agarwal JP, Pantvaidya G, Chaukar D, Deshmukh A, Kane S, Arya S, Ghosh-Laskar S, Chaturvedi P, Pai P, Nair S, Nair D, Badwe R, Head, & Neck Disease Management, G. Elective versus therapeutic Neck dissection in node-negative Oral cancer. N Engl J Med. 2015;373(6):521–9. https://doi.org/10.1056/NEJMoa1506007.

16. Noronha V, Joshi A, Patil VM, Agarwal J, Ghosh-Laskar S, Budrukkar A, Murthy V, Gupta T, D'Cruz AK, Banavali S, Pai PS, Chaturvedi P, Chaukar D, Pande N, Chandrasekharan A, Talreja V, Vallathol DH, Mathrudev V, Manjrekar A, et al. Once-a-week versus once-Every-3-weeks cisplatin Chemoradiation for locally advanced Head and Neck cancer: a phase III

randomized noninferiority trial. J Clin Oncol. 2018;36(11):1064–72. https://doi.org/10.1200/JCO.2017.74.9457.

17. Noronha V, Patil VM, Joshi A, Mahimkar M, Patel U, Pandey MK, Chandrasekharan A, Dsouza H, Bhattacharjee A, Mahajan A, Sabale N, Agarwal JP, Ghosh-Laskar S, Budrukkar A, D'Cruz AK, Chaturvedi P, Pai PS, Chaukar D, Nair S, et al. Nimotuzumab-cisplatin-radiation versus cisplatin-radiation in HPV negative oropharyngeal cancer. Oncotarget. 2020;11(4):399–408. https://doi.org/10.18632/oncotarget.27443.

18. Osazuwa-Peters N, Barnes JM, Okafor SI, Taylor DB, Hussaini AS, Adjei Boakye E, Simpson MC, Graboyes EM, Lee WT. Incidence and risk of suicide among patients with Head and Neck cancer in rural, urban, and metropolitan areas. JAMA Otolaryngol Head Neck Surg. 2021;147(12):1045–52. https://doi.org/10.1001/jamaoto.2021.1728.

Dr. Ashish V. Chintakuntlawar is a certified medical oncologist by the American Board of Internal Medicine with expertise in treating head and neck cancers as well as endocrine cancers such as thyroid and adrenal carcinomas. His research interests include translational research and early phase clinical trials in head and neck and endocrine cancers. Dr. Chintakuntlawar is a member of the American Society of Clinical Oncology (ASCO), the American Thyroid Association (ATA), the international thyroid oncology group (ITOG), and the American-Australian-Asian adrenal Alliance (A5), an international cooperative group dedicated to research related to adrenal diseases.

Dr. Hani Al-Halabi is an American and Canadian board-certified consultant radiation oncologist, currently practicing at the Gulf International Cancer Center in Abu Dhabi. He has special interests in the management of thoracic and head-and-neck oncology patients and the use of SBRT and SRS for treating primary and metastatic disease. Dr. Al-Halabi is the author of numerous peer-reviewed publications and was the principal investigator of several phase I and II clinical trials. Dr. Al Halabi is an active member of the American Society for Therapeutic Radiation Oncology (ASTRO) and the European Society of Radiation Oncology (ESTRO).

Dr. Aref Chehal is a consultant medical oncologist and hematologist at Sheikh Shakhbout Medical City (SSMC) in Abu Dhabi. Dr. Chehal has more than 20 years of experience in hematology and oncology. He is a recognized and established adjunct professor of medicine at both Khalifa University and Gulf Medical University in the United Arab Emirates.

Thyroid Cancer in the UAE

32

Riyad Bendardaf ⓘ, Iman M. Talaat ⓘ, Noha M. Elemam ⓘ,
and Humaid O. Al-Shamsi ⓘ

R. Bendardaf
Emirates Oncology Society, Emirates Medical Association, Dubai, United Arab Emirates

College of Medicine, University of Sharjah, Sharjah, United Arab Emirates

University Hospital Sharjah, Sharjah, United Arab Emirates
e-mail: riyad.bendardf@uhs.ae

I. M. Talaat
College of Medicine, University of Sharjah, Sharjah, United Arab Emirates

Research Institute for Medical and Health Sciences, University of Sharjah,
Sharjah, United Arab Emirates

Emirates Pathology Society, Dubai, United Arab Emirates
e-mail: italaat@sharjah.ac.ae

N. M. Elemam
College of Medicine, University of Sharjah, Sharjah, United Arab Emirates

Research Institute for Medical and Health Sciences, University of Sharjah,
Sharjah, United Arab Emirates
e-mail: nelemam@sharjah.ac.ae

H. O. Al-Shamsi (✉)
Burjeel Cancer Institute, Burjeel Medical City, Burjeel Holdings, Abu Dhabi, United Arab
Emirates

Ras Al Khaimah Medical and Health Sciences University, Ras Al Khaimah, United Arab Emirates

Gulf Medical University, Ajman, United Arab Emirates

Emirates Oncology Society, Emirates Medical Association, Dubai, United Arab Emirates

College of Medicine, University of Sharjah, Sharjah, United Arab Emirates

Gulf Cancer Society, Alsafa, Kuwait
e-mail: alshamsi@burjeel.com; humaid.al-shamsi@medportal.ca

32.1   Introduction to Thyroid Cancer

The thyroid gland consists of two lobes connected by an isthmus and is located in the neck below the larynx and in front of the trachea. Thyroid cancer (TC) starts as a lump or nodule in the thyroid that is usually asymptomatic. Individuals with previous exposure to high doses of radiation and a genetic history of TC are more prone to thyroid cancer development [1]. Radiotherapy used to treat serious cancers like Hodgkin's disease was found to be associated with an increased risk of developing TC [1]. Another possible cause of TC development is exposure to radioactive iodine released during a nuclear disaster [2].

Although the most common malignancy of the endocrine system is thyroid cancer, it is still very rare, accounting for about 1% of all malignant tumors. However, it has been increasing in frequency in recent years. Thyroid cancer is a malignant tumor of the glandular tissue of the thyroid gland that can be classified based on the differentiation status and affected cell types into differentiated TC (DTC), including papillary, follicular, and Hurthle cell tumors, which make up 95% of TC cases. This is in addition to medullary TC (MTC) and anaplastic TC, as well as some rare subtypes [3].

Papillary thyroid cancer (PTC) is the most prevalent type, especially in well-differentiated TC, accounting for 70–80% of cases. It is the most common form of thyroid cancer to result from radiation exposure. Also, it could occur at any age and is most likely located at the lymph nodes in the neck. However, the prognosis for younger (< 45 years) patients is significantly better than for older (> 45 years) patients [4]. In particular, PTC looks like an irregular solid or cystic nodule within normal thyroid parenchyma. Even though PTC is well-differentiated, it can be quite invasive and spread quickly to other organs as well as the lymphatic system [5].

Follicular thyroid carcinoma (FTC) is the second most frequent thyroid malignancy. On initial presentation, 11% of FTC patients develop metastases outside of the cervical or mediastinal lymph nodes. The age of TC patients affects their life expectancy, as people younger than 45 have a better prognosis. In comparison to papillary cancer, FTC tends to affect older individuals and accounts for 10–15% of all thyroid cancers [6, 7]. Despite having distinct features, the FTC may be invasive. Like papillary cancer, FTC can invade the nearby neck lymph nodes. Additionally, FTC has a higher propensity than PTC to metastasize into blood arteries and spread, particularly to the lungs and bones, where bone metastasis was found to be osteolytic [6, 7].

Neuroendocrine tumor medullary carcinomas of the thyroid (MTC) account for less than 5% of thyroid cancers. It is triggered by the calcitonin-producing C cells of the thyroid, which do not build up radioiodine. Around 75% of cases of MTC are sporadic, while the remaining are associated with multiple endocrine neoplasia type 2 (MEN2), an autosomal-dominant condition caused by mutations in the *RET* proto-oncogene. A test of such a genetic mutation can indicate an initial analysis of the MTC [8]. The clinical outcome of patients is affected by the severity of the disease, tumor biology, and the overall success of the surgical resection [9].

The most aggressive and invasive type of thyroid cancer is anaplastic thyroid carcinoma (ATC), which is also the least likely to respond to therapy. Fortunately, it is a rare type and affects less than 2% of people, but it accounts for 40% of thyroid cancer mortality [10]. According to reports, the 5-year survival rate was less than 10%, and the majority of patients only survived a few months following diagnosis [10]. A fast-growing neck mass is the typical initial sign of ATC in patients, with metastasis in the lungs evident in 50% of the cases at the time of diagnosis [11].

32.2   Thyroid Cancer Classification

Many thyroid tumors are derived from follicular epithelial cells, while only a few arise from calcitonin-secreting C cells. According to the World Health Organization (WHO), follicular cell-derived neoplasms are categorically divided into benign, low-risk, and malignant neoplasms. Another type of classification of thyroid tumors was suggested to be based on the combination of classic histopathology and molecular pathogenesis. For instance, most encapsulated thyroid tumors with follicular cell growth exhibited a RAS-like molecular profile [12, 13]. On the contrary, thyroid tumors with papillary and/or infiltrative growth and nuclear atypia possessed a BRAFV600E-like molecular profile. Thyroid tumors with BRAFV600E and RAS mutations can undergo further genetic changes and result in the progression of high-grade malignancies [14]. The protooncogene RET codes for a transmembrane receptor tyrosine kinase, whose activation could lead to oncogenesis. As mentioned earlier, germline RET mutations result in MEN2 and MTC. RET mutations were reported to be associated with more aggressive diseases in MTC [15]. Less than 10% of differentiated thyroid tumors and anaplastic carcinomas possess RET mutations. When compared to thyroid tumors in older people, RET fusions are more common in thyroid cancers identified in children and young adults. Also, they are more common in patients with previous exposure to environmental radiation [16–21].

32.3   Thyroid Cancer Statistics

On a global scale, an increase in screening methods has resulted in a rise in the diagnosis of thyroid cancer. Also, besides the enhanced diagnosis of early tumors, the greater prevalence of individual risk factors (such as obesity) and the increased exposure to environmental risk factors (iodine levels) have been implicated in the global increase of the TC [22].

According to the Global Cancer Observatory (GLOBOCAN), the age-standardized incidence rate (ASR) worldwide for thyroid cancer was 6.6 per 100,000 in 2020 and 9.1 per 100,000 in 2022. Notably, 586,202 and 821,173 newly diagnosed thyroid cancer cases were reported in 2020 and 2022 across all genders and ages, respectively. Also, the ASR per gender was 13.6 per 100,000 for females and 4.6 per 100,000 for males in 2022. Additionally, the age-standardized mortality rate of thyroid cancer reached 0.57 per 100,000, which accounted for 47,485 cases worldwide [1, 23].

In the United Arab Emirates (UAE), according to the UAE National Cancer Registry (UAE-NCR) 2021, thyroid cancer ranks first among endocrine cancers and

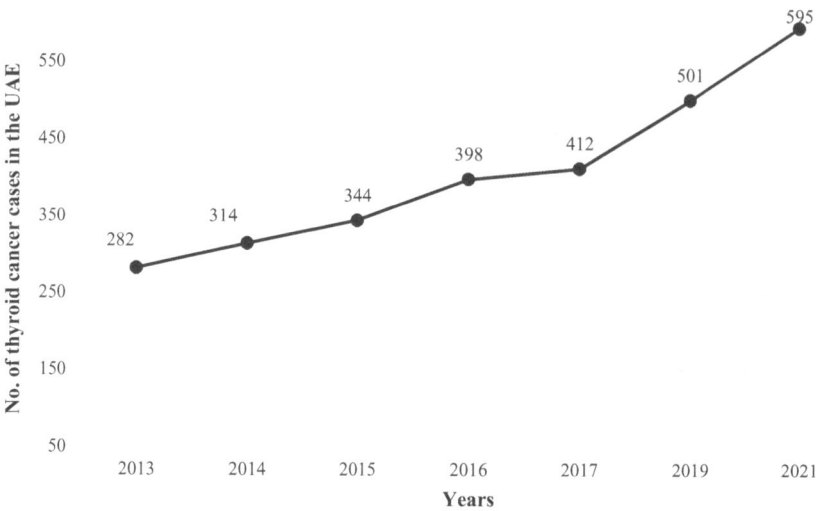

Fig. 32.1 Number of thyroid cancer (malignant) occurrences in the UAE across the years 2013–2021. (Source: Ministry of Health and Prevention, Statistics and Research Center, National Disease Registry—UAE National Cancer Registry Report, 2013–2021)

is the second most prevalent cancer among the UAE population of all genders. Furthermore, thyroid cancer is ranked as the second most prevalent cancer in females and the fourth most common in males. Figure 32.1 shows that the number of thyroid cancer cases rose from 2013 to 2021, where it surged from 7.89% in 2013 to 10.6% in 2021 (Table 32.1) [24].

In the UAE, a total of 3574 cancer cases were diagnosed in 2013, out of which 282 were thyroid cancer (Table 32.1). Interestingly, the percentage of thyroid cancer patients in the UAE reached 10.6% ($n = 595$) from the total number of malignant cancer cases ($n = 5612$) that were newly diagnosed in 2021. Females showed a higher incidence of thyroid cancer, accounting for 77.1% and 74.8% of diagnosed thyroid cancer cases in 2016 and 2019, respectively (Fig. 32.2). In comparison, 5612 cancer cases were diagnosed in 2021. It is worth mentioning that 595 (10.6%) cases were reported to be thyroid cancer, where 402 were UAE nationals (Fig. 32.3). The distribution of thyroid cancer cases by Surveillance, Epidemiology, and End Results (SEER) stages in the UAE across the years is shown in Fig. 32.4 [24].

A study published by Alseddeeqi E. et al. described the incidence of thyroid cancer in Abu Dhabi, the capital of the UAE, from 2012–2015 [25]. 89.9% of TC patients were diagnosed with papillary thyroid cancer, followed by 22.2% of patients with follicular thyroid cancer and 2% of patients with medullary thyroid cancer [25]. Also, the same study reported the sharp increase in the incidence rate noted in 2013 [25], which is similar to the rapid rise seen from 2017 to 2019 (Table 32.1). This could be attributed to an increase in previous exposure to ionizing radiation as well as a rise in iodine deficiency status, especially in females [25, 26]. This emphasizes the significance of having an updated and detailed histological subtype of thyroid cancer in the UAE. According to the UAE 2017 data, 44% of the thyroid cancer cases were localized, while around 20% of the tumors were found to be

Table 32.1 Thyroid cancer demographics among the UAE population during 2013–2021

Year	UAE Population (in millions)	Total malignant cases (in numbers)	Thyroid cancer cases (in numbers)	Percentage (%)	Crude incidence rate of thyroid cancer cases per 100,000 population
2013	8.66	3574	282	7.89	–
2014	8.79	3610	314	8.70	3.46
2015	8.93	3744	344	9.19	3.8
2016	9.12	3982	398	9.99	–
2017	9.30	4123	412	9.99	4.4
2019	9.50	4381	501	11.4	–
2021	–	5612	595	10.6	6.4

Source: UAE population: https://fcsc.gov.ae/en-us/Pages/Statistics/Statistics-by-Subject.aspx#/%3Fsubject=Demography%20and%20Social; Ministry of Health and Prevention, Statistics and Research Center, National Disease Registry—UAE National Cancer Registry Report

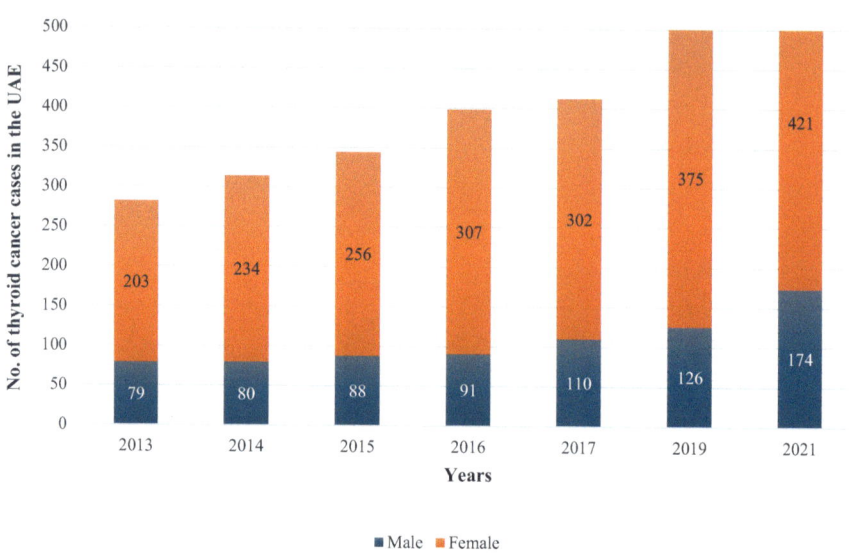

Fig. 32.2 Distribution of thyroid cancer cases (malignant) according to gender, 2013–2021. (Source: Ministry of Health and Prevention, Statistics and Research Center, National Disease Registry—UAE National Cancer Registry Report, 2013–2021)

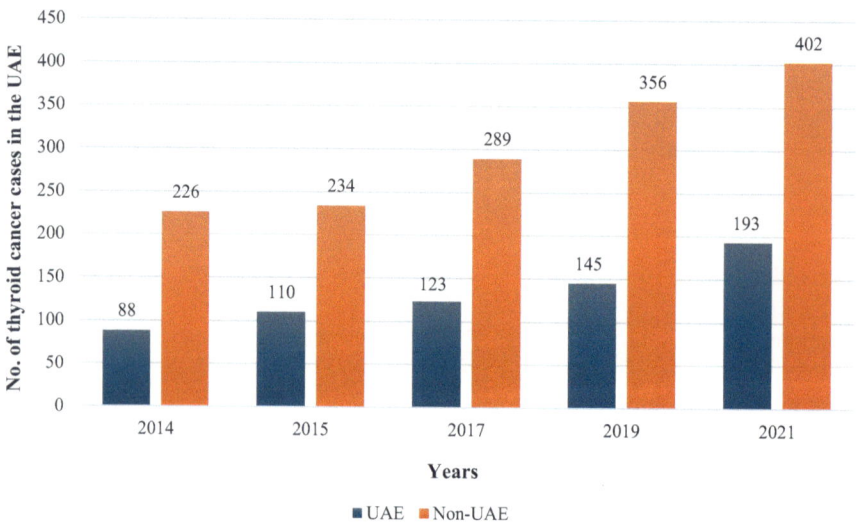

Fig. 32.3 The number of thyroid cancer cases (malignant) among the UAE population according to nationality, 2014–2021. (Source: Ministry of Health and Prevention, Statistics and Research Center, National Disease Registry—UAE National Cancer Registry Report, 2014–2019)

infiltrating the regional lymph nodes. A few cases ($n = 8$) were reported to possess metastatic potential. For instance, 1.9% of thyroid cancer patients were staged as metastatic in comparison to 2.04% in 2017. This could be attributed to early screening through the increased use of diagnostic imaging and surveillance [27–30].

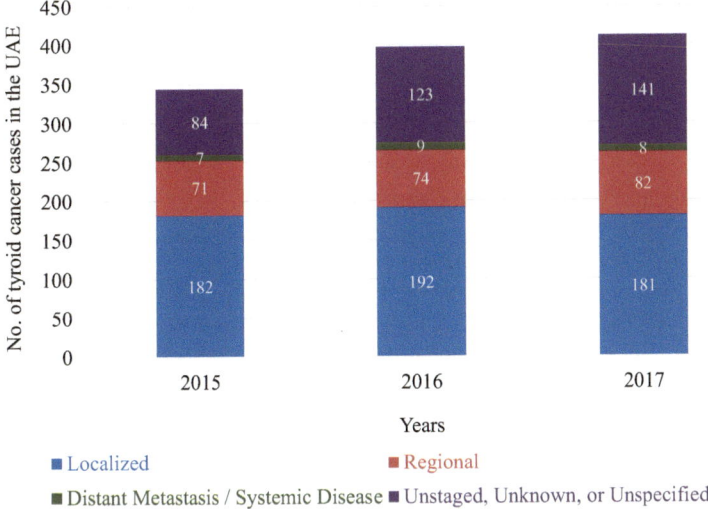

Fig. 32.4 Distribution of thyroid cancer cases by SEER stages in the UAE across the years. (Source: Ministry of Health and Prevention, Statistics and Research Center, National Disease Registry—UAE National Cancer Registry Report, 2015–2017)

32.4 Thyroid Cancer Diagnosis and Treatment

Patients with well-differentiated TC showed a long-term survival rate of around 90%, while patients with poorly differentiated tumors had survival rates below 10%, indicating their resistance to standard treatment options [31, 32]. The known standard therapeutic approaches for radioiodine-refractory and advanced TC are immunotherapy, chemotherapy, and tyrosine kinase inhibitors [33]. Therefore, new research approaches and technologies are necessary in order to discover new targets for therapy for TC.

Kinase inhibitors, such as vandetanib and cabozantinib, are indicated for medullary thyroid cancer, while others, such as sorafenib and lenvatinib, are approved for radioiodine-refractory differentiated thyroid cancer [34–37]. In addition to many other kinases, RET is simultaneously targeted by these multitargeted kinase inhibitors. The safety and longevity of responses to these drugs are partially constrained by toxic side effects that are due to the inhibition of non-RET kinases, such as vascular endothelial growth factor receptor 2 (VEGFR2) [38, 39].

All recommendations include measuring blood thyroid-stimulating hormone (TSH) levels and performing an ultrasonographic assessment of thyroid nodules in order to decide whether a fine needle aspiration biopsy is necessary. Typically, a diagnosis of TC is made by investigating a fine needle aspiration biopsy of a thyroid nodule or a surgically removed nodule [27, 40]. Multiple organizations released guidelines for the diagnosis and/or management of thyroid cancer, such as the American Thyroid Association (ATA) [41, 42], the National Comprehensive Cancer Network (NCCN) [4], and the American Association of Clinical Endocrinologists/

American College of Endocrinology/Associazione Medici Endocrinologi (AACE/ ACE/AME) [43].

Surgery is the main gold-standard treatment for all types of TC, especially differentiated types. The currently recommended treatment involves performing a complete thyroidectomy, which involves extracting the whole thyroid gland and affecting neighboring lymph nodes in the upper chest or neck. Post-surgical resection, patients might be subject to radioactive iodine (^{131}I) and thyrotropin suppression for the destruction of any remnant thyroid cancer cells [4]. However, no curative treatment exists for anaplastic thyroid cancer, as most patients present in an unresectable or metastatic state [31]. All TC patients who have had a thyroidectomy will require thyroid hormone replacement for the rest of their lives. Also, a blood test showing the detection of thyroglobulin might indicate a recurrence of thyroid cancer.

32.5 Conclusion

TC is the most common malignancy of the endocrine system that can be classified based on the differentiation status and affected cell types. Another type of classification of thyroid tumors is based on histopathology and molecular pathogenesis. On a global scale, there is a rise in the incidence of thyroid cancer that could be attributed to the greater prevalence of risk factors such as obesity and exposure to iodine levels. In the UAE, TC ranks first among endocrine cancers and is the second most prevalent cancer among the UAE population. Until now, the standard therapeutic approaches were surgery, radioactive iodine, chemotherapy, tyrosine kinase inhibitors, and immunotherapy. Therefore, further research is needed to discover new targets for therapy for different subtypes of TC.

Conflict of Interest The authors have no conflict of interest to declare.

References

1. Bogović Crnčić T, Ilić Tomaš M, Girotto N, Grbac IS. Risk factors for thyroid cancer: what do we know so far? Acta Clin Croat. 2020;59(Suppl 1):66–72.
2. Wiwanitkit V. Nuclear detonation, thyroid cancer and potassium iodide prophylaxis. Indian J Endocrinol metab. 2011;15(2):96–8.
3. Haugen BR, Alexander EK, Bible KC, Doherty GM, Mandel SJ, Nikiforov YE, et al. 2015 American Thyroid Association management guidelines for adult patients with thyroid nodules and differentiated thyroid cancer: the American Thyroid Association guidelines task force on thyroid nodules and differentiated thyroid cancer. Thyroid. 2016;26(1):1–133.
4. Haddad RI, Bischoff L, Ball D, Bernet V, Blomain E, Busaidy NL, et al. Thyroid carcinoma, version 2.2022, NCCN clinical practice guidelines in oncology. J Natl Compr Canc Netw. 2022;20(8):925–51.
5. Haugen BR, Sawka AM, Alexander EK, Bible KC, Caturegli P, Doherty GM, et al. American Thyroid Association guidelines on the Management of Thyroid Nodules and Differentiated Thyroid Cancer Task Force Review and recommendation on the proposed renaming of encap-

sulated follicular variant papillary thyroid carcinoma without invasion to noninvasive follicular thyroid neoplasm with papillary-like nuclear features. Thyroid. 2017;27(4):481–3.

6. Badulescu CI, Piciu D, Apostu D, Badan M, Piciu A. Follicular thyroid carcinoma—Clinical and diagnostic findings in a 20-year follow up study. Acta endocrinologica. 2020;16(2):170–7.

7. Lloyd RV, Osamura RY. Klöppel Gn, Rosai J. WHO classification of tumours of endocrine organs. 4th edition. Ed. world health O, International Agency for Research on C, editors. International Agency for Research on Cancer IARC: Lyon, France; 2017.

8. Fagin JA, Wells SA Jr. Biologic and clinical perspectives on thyroid cancer. N Engl J Med. 2016;375(11):1054–67.

9. Filetti S, Durante C, Hartl D, Leboulleux S, Locati LD, Newbold K, et al. Thyroid cancer: ESMO clinical practice guidelines for diagnosis, treatment and follow-up†. Ann Oncol. 2019;30(12):1856–83.

10. Cabanillas ME, Zafereo M, Gunn GB, Ferrarotto R. Anaplastic thyroid carcinoma: treatment in the age of molecular targeted therapy. J Oncol Pract. 2016;12(6):511–8.

11. Jonker PK, van Dam GM, Oosting SF, Kruijff S, Fehrmann RS. Identification of novel therapeutic targets in anaplastic thyroid carcinoma using functional genomic mRNA-profiling: paving the way for new avenues? Surgery. 2017;161(1):202–11.

12. Cancer Genome Atlas Research Network. Integrated genomic characterization of papillary thyroid carcinoma. Cell. 2014;159(3):676–90.

13. Yoo SK, Lee S, Kim SJ, Jee HG, Kim BA, Cho H, et al. Comprehensive analysis of the transcriptional and mutational landscape of follicular and papillary thyroid cancers. PLoS Genet. 2016;12(8):e1006239.

14. Baloch ZW, Asa SL, Barletta JA, Ghossein RA, Juhlin CC, Jung CK, et al. Overview of the 2022 WHO classification of thyroid neoplasms. Endocr Pathol. 2022;33(1):27–63.

15. Ciampi R, Romei C, Ramone T, Prete A, Tacito A, Cappagli V, et al. Genetic landscape of somatic mutations in a large cohort of sporadic medullary thyroid carcinomas studied by next-generation targeted Sequencing. iScience. 2019;20:324–36.

16. Santoro M, Papotti M, Chiappetta G, Garcia-Rostan G, Volante M, Johnson C, et al. RET activation and clinicopathologic features in poorly differentiated thyroid tumors. J Clin Endocrinol Metab. 2002;87(1):370–9.

17. Duan H, Li Y, Hu P, Gao J, Ying J, Xu W, et al. Mutational profiling of poorly differentiated and anaplastic thyroid carcinoma by the use of targeted next-generation sequencing. Histopathology. 2019;75(6):890–9.

18. Nikiforov YE, Rowland JM, Bove KE, Monforte-Munoz H, Fagin JA. Distinct pattern of ret oncogene rearrangements in morphological variants of radiation-induced and sporadic thyroid papillary carcinomas in children. Cancer Res. 1997;57(9):1690–4.

19. Ciampi R, Giordano TJ, Wikenheiser-Brokamp K, Koenig RJ, Nikiforov YE. HOOK3-RET: a novel type of RET/PTC rearrangement in papillary thyroid carcinoma. Endocrine-Related Cancer Endocr Relat Cancer. 2007;14(2):445–52.

20. Vanden Borre P, Schrock AB, Anderson PM, Morris JC 3rd, Heilmann AM, Holmes O, et al. Pediatric, adolescent, and young adult thyroid carcinoma harbors frequent and diverse targetable genomic alterations, including kinase fusions. Oncologist. 2017;22(3):255–63.

21. Su X, Li Z, He C, Chen W, Fu X, Yang A. Radiation exposure, young age, and female gender are associated with high prevalence of RET/PTC1 and RET/PTC3 in papillary thyroid cancer: a meta-analysis. Oncotarget. 2016;7(13):16716.

22. Kim J, Gosnell JE, Roman SA. Geographic influences in the global rise of thyroid cancer. Nat Rev Endocrinol. 2020;16(1):17–29.

23. Bray F, Laversanne M, Sung H, et al. Global cancer statistics 2022: GLOBOCAN estimates of incidence and mortality worldwide for 36 cancers in 185 countries. CA Cancer J Clin. 2024; 74(3): 229–63.

24. Cancer incidnece in United Arab Emirates, Annual Report of the UAE - National Cancer registry. Statistics and Research Center, Ministry of Health and Prevention.

25. Alseddeeqi E, Altinoz A, Oulhaj A, Suliman A, Ahmed LA. Incidence of thyroid cancer in Abu Dhabi, UAE: A registry-based study. J Cancer Res Ther. 2023;19(2):321–6.
26. Liu Y, Su L, Xiao H. Review of factors related to the thyroid cancer epidemic. Int J Endocrinol. 2017;2017:5308635.
27. Haugen BR, Alexander EK, Bible KC, Doherty GM, Mandel SJ, Nikiforov YE, et al. 2015 American Thyroid Association management guidelines for adult patients with thyroid nodules and differentiated thyroid cancer: the American Thyroid Association guidelines task force on thyroid nodules and differentiated thyroid cancer. Thyroid. 2015;26(1):1–133.
28. Schmidbauer B, Menhart K, Hellwig D, Grosse J. Differentiated thyroid cancer—treatment: state of the art. Int J Mol Sci. 2017;18(6):1292.
29. Pacini F, Basolo F, Bellantone R, Boni G, Cannizzaro MA, De Palma M, et al. Italian consensus on diagnosis and treatment of differentiated thyroid cancer: joint statements of six Italian societies. J Endocrinol Investig. 2018;41(7):849–76.
30. Amin MB, Greene FL, Edge SB, Compton CC, Gershenwald JE, Brookland RK, et al. The eighth edition AJCC cancer staging manual: continuing to build a bridge from a population-based to a more "personalized" approach to cancer staging. CA Cancer J Clin. 2017;67(2):93–9.
31. Patel KN, Shaha AR. Poorly differentiated thyroid cancer. Curr Opin Otolaryngol Head Neck Surg. 2014;22(2):121–6.
32. Ancker OV, Krüger M, Wehland M, Infanger M, Grimm D. Multikinase inhibitor treatment in thyroid cancer. Int J Mol Sci. 2019;21(1):10.
33. Matrone A, Campopiano MC, Nervo A, Sapuppo G, Tavarelli M, De Leo S. Differentiated thyroid cancer, from active surveillance to advanced therapy: toward a personalized medicine. Front Endocrinol. 2020;10:10.
34. Wells SA Jr, Robinson BG, Gagel RF, Dralle H, Fagin JA, Santoro M, et al. Vandetanib in patients with locally advanced or metastatic medullary thyroid cancer: a randomized, double-blind phase III trial. J Clin Oncol Off J Am Soc Clin Oncol. 2012;30(2):134–41.
35. Elisei R, Schlumberger MJ, Müller SP, Schöffski P, Brose MS, Shah MH, et al. Cabozantinib in progressive medullary thyroid cancer. J Clin Oncol Off J Am Soc Clin Oncol. 2013;31(29):3639–46.
36. Brose MS, Nutting CM, Jarzab B, Elisei R, Siena S, Bastholt L, et al. Sorafenib in radioactive iodine-refractory, locally advanced or metastatic differentiated thyroid cancer: a randomised, double-blind, phase 3 trial. Lancet (London, England). 2014;384(9940):319–28.
37. Schlumberger M, Tahara M, Wirth LJ, Robinson B, Brose MS, Elisei R, et al. Lenvatinib versus placebo in radioiodine-refractory thyroid cancer. N Engl J Med. 2015;372(7):621–30.
38. Brose MS, Bible KC, Chow LQM, Gilbert J, Grande C, Worden F, et al. Management of treatment-related toxicities in advanced medullary thyroid cancer. Cancer Treat Rev. 2018;66:64–73.
39. Cabanillas ME, Habra MA. Lenvatinib: role in thyroid cancer and other solid tumors. Cancer Treat Rev. 2016;42:47–55.
40. Rometo DA, Baranski TJ. Thyroid cancer: what to do after fine needle aspiration. Mo Med. 2011;108(2):93–8.
41. Wells SA Jr, Asa SL, Dralle H, Elisei R, Evans DB, Gagel RF, et al. Revised American Thyroid Association guidelines for the management of medullary thyroid carcinoma. Thyroid. 2015;25(6):567–610.
42. Smallridge RC, Ain KB, Asa SL, Bible KC, Brierley JD, Burman KD, et al. American Thyroid Association guidelines for management of patients with anaplastic thyroid cancer. Thyroid. 2012;22(11):1104–39.
43. Gharib H, Papini E, Garber JR, Duick DS, Harrell RM, Hegedüs L, et al. American Association of Clinical Endocrinologists, American College of Endocrinology, and Associazione Medici Endocrinologi Medical Guidelines For Clinical Practice For The Diagnosis And Management of Thyroid Nodules—2016 update. Endocr Pract. 2016;22(5):622–39.

Prof. Riyad Bendardaf has been a consultant Medical Oncologist at University Hospital Sharjah in the UAE since 2013. Before joining the University Hospital Sharjah, he worked as a Medical Oncology Consultant at the Medical Oncology Unit, Department of Medicine, Benghazi Medical Centre, Benghazi, Libya, and as a Medical Oncology Consultant at the Department of Oncology & Radiotherapy, Turku University Hospital, Turku, Finland, where he initially started as an oncology resident. In 2013, he was appointed as a Clinical Professor in the Clinical Sciences Department, College of Medicine, University of Sharjah, UAE. He worked as a Docent (Associate Professor of Medical Oncology) at the Department of Oncology and Radiotherapy, Turku University Hospital, University of Turku, Finland. He is an active member of different oncology societies, including the European Society of Medical Oncology (ESMO), the American Society of Clinical Oncology (ASCO), the European Society of Digestive Oncology (ESDO), the European School of Oncology (ESO), the Finnish Medical Society (Duodecim), and the European Association of Cancer Research (EACR). He is also on the editorial board of various international peer-reviewed journals. Prof. Bendardaf has more than 60 scientific papers published in international oncology and cancer peer-reviewed journals. He also presented his work at several local and international conferences.

Prof. Iman M. Talaat graduated from the Faculty of Medicine, Alexandria University, Egypt, with the highest honors. She obtained her MSc and Ph.D. degrees in Basic Medical Sciences in Pathology. She is a Professor and Consultant Anatomic Pathologist at the Clinical Sciences Department, College of Medicine, University of Sharjah, UAE, and at the Pathology Department, Faculty of Medicine, Alexandria University, Egypt. She specializes in Anatomic and Cellular Pathology, with her work focused on Cancer Research and Pathogenesis of different neoplastic and non-neoplastic diseases. She is the coordinator of the Immuno-oncology research group at the Research Institute for Medical and Health Sciences, University of Sharjah, UAE. She is a member of the Emirates Medical Association (EMA) and a board member (Cultural Committee Chairperson) of the Emirates Pathology Society (EPS). She is also a member of the American Association of Investigative Pathology (ASIP) and the European Association of Cancer Research (EACR). She is an associate editor and review editor in the pathology section of Frontiers in Medicine and in the breast cancer section of Frontiers in Oncology. She was awarded several grants with more than 100 publications in high-impact factor, peer reviewed journals.

Dr. Noha M. Elemam graduated from the German University of Cairo with a bachelor's degree in Pharmacy and Biotechnology with Highest Honors. She obtained her master's degree with Excellence in Molecular Pharmacology. She was awarded a full-granted scholarship to finally complete her Ph.D. degree with Distinction in Molecular Medicine and Translational Research at the University of Sharjah, along with a dual Ph.D. degree from University of Lübeck, Germany in 2020. Currently, Dr. Noha is a Senior Lecturer of Immunology at the College of Medicine at the University of Sharjah, UAE. She is working on multiple projects in the fields of immunology and viral infections as well as various cancer types such as breast, colon, renal and prostate cancers. Dr. Noha published more than 70 research and review papers in high impact factor journals and participated in around 20 international conferences with oral and poster presentations. Dr. Noha is a young affiliate member of the Mohammed bin Rashed Academy of Scientists (MBRAS). Dr. Noha was selected to be a winner of the L'Oréal-UNESCO For Women in Science Middle East Regional Young Talents 2023 Program-Postdoctoral category. She is an editor and expert reviewer in many prestigious international high impact factor journals. Also, she is a co-inventor in a patent filed in the US patent office, as well as a co-investigator in multiple grants at the University of Sharjah.

Prof. Humaid Obaid Al-Shamsi is the Chief Executive Officer of Burjeel Cancer Institute in Abu Dhabi, UAE, President of the Emirates Oncology Society, Lead of the Gulf Cancer Society, Full Professor of Oncology at the Ras Al Khaimah Medical and Health Sciences University, Ras Al Khaimah, UAE, and an Adjunct Professor of Oncology at the College of Medicine, University of Sharjah. He is the first Emirati to be promoted as a professor in oncology in the UAE. He is also the Chairman for Colorectal Cancer in the MENA region, appointed by the prestigious National Comprehensive Cancer Network®. He is also the only member of Lung Cancer Policy Network in the MENA region that aims to advance lung cancer research and screening globally. He is the Chairman of the Oncology and Hematology Fellowship Training Program for the National Institute for Health Specialties in the United Arab Emirates. He is the only member in GCC in the WIN Consortium which is comprised of organizations representing all stakeholders in personalized cancer medicine globally.

He is board-certified in both internal medicine and oncology from the UK, USA (ABIM), and Canada (FRCPC). He has also been awarded the FRCP (London) in 2023 and FRCP (Glasgow) in 2024. He is the only physician in the UAE with a subspecialty fellowship certification and training in gastrointestinal oncology and the first Emirati to train and complete a clinical post-doctoral fellowship in palliative care. He was an assistant professor at the University of Texas MD Anderson Cancer Center between 2014 and 2017. He has published more than 140 peer-reviewed articles in JAMA Oncology, Lancet Oncology, The Oncologist, BMC Cancer, and many others. His area of expertise includes precision oncology and cancer care in the UAE. In 2016, he published with his group from MD Anderson the JCO paper describing a new distinct subgroup of CRC, NON V600 BRAF-mutated CRC. In 2022, he published the first book about cancer research in the UAE and also the first book about cancer in the Arab world, both of which were launched at Dubai Expo 2020. *Cancer in the Arab World* has been downloaded more than 450,000 times in its first 18 months of publication and is the ultimate source of cancer data in the Arab region. He also published the first comprehensive book about cancer care in the UAE which is the first book in UAE history to document the cancer care in the UAE with many topics addressed for the first time, e.g., neuroendocrine tumors in the UAE. He is passionate about advancing cancer care in the UAE and the GCC and has made significant contributions to cancer awareness and early detection for the public using social media platforms. He is considered as the most followed oncologist in the world with over 300,000 subscribers across his social media platforms (Instagram, Twitter, LinkedIn, and TikTok). In 2022, he was awarded the prestigious Feigenbaum Leadership Excellence Award from Sheikh Hamdan Smart University for his exceptional leadership and research and the Sharjah Award for Volunteering. He was also named the Researcher of the Year in the UAE in 2020 and 2021 by the Emirates Oncology Society.

In May 2024, HH Sheikh Mansour bin Zayed Al Nahyan, Vice President of the United Arab Emirates, awarded him the first place in UAE Nafis program for outstanding leadership in private sector across all business and medical disciplines. Beside his clinical and administrative duties, he is engaged in education and various levels of research training for medical trainees to enhance their clinical and research skills. His mission is to advance cancer care in the UAE and the MENA region and make cancer care accessible to everyone in need around the globe.

Lung Cancer in the UAE

33

Saeed Rafii, Batool Aboud, and Humaid O. Al-Shamsi (ID)

33.1 Introduction

In the United Arab Emirates (UAE), lung cancer is not as common as in other parts of the world; however, the incidence of lung cancer is increasing, and most cases are diagnosed at late stages. Smoking patterns and the unique young population of the UAE could lead to an increasing number of lung cancers in the coming years. Herein, we review the current state of lung cancer in the UAE and make recommendations in order to reduce the incidence of lung cancer and the steps that need to be taken towards early diagnosis.

S. Rafii
Department of Oncology, Mediclinic City Hospital, Dubai, United Arab Emirates

Emirates Oncology Society, Emirates Medical Association, Dubai, United Arab Emirates
e-mail: saeed.rafii@mediclinic.ae

B. Aboud
Department of Oncology, Saudi German Hospital, Ajman, United Arab Emirates

H. O. Al-Shamsi (✉)
Burjeel Cancer Institute, Burjeel Medical City, Burjeel Holdings, Abu Dhabi,
United Arab Emirates

Ras Al Khaimah Medical and Health Sciences University, Ras Al Khaimah, United Arab Emirates

Gulf Medical University, Ajman, United Arab Emirates

Emirates Oncology Society, Emirates Medical Association, Dubai, United Arab Emirates

College of Medicine, University of Sharjah, Sharjah, United Arab Emirates

Gulf Cancer Society, Alsafa, Kuwait
e-mail: alshamsi@burjeel.com; humaid.al-shamsi@medportal.ca

33.2   Lung Cancer Statistics

Lung cancer is a major health challenge worldwide. It is estimated that 2.2 million patients were diagnosed with lung cancer in 2020 [1]. From a mortality perspective, globally, lung cancer accounts for almost 20% of all cancer-related deaths, more than breast and colorectal cancers combined [1].

In the Arab Gulf Cooperation Council (GCC) countries, which include Bahrain, Kuwait, Qatar, Saudi Arabia, Oman, and the UAE, the incidence of lung cancer is significantly lower than in Europe, North America, and some Asian countries. Lung cancer was identified as the seventh most common cancer among men and women in the GCC, with an average number of 505 cases annually, which accounts for 4.6% of total cancer cases each year [2].

According to the UAE National Cancer Registry (UAE-NCR), 231 cases of lung cancer were diagnosed in the UAE in 2021, which accounts for 3.9% of all diagnosed cancers. Lung cancer was the second common cause of cancer death in both sexes, with an estimated average of 96 (9.85%) of cancer deaths per year. Based on this, lung cancer was ranked the seventh most common cancer in men and women in the UAE. In the same year, lung cancer was the sixth most common cancer in men (161 cases; 6.3% of all cancer diagnosed in men) and the tenth most common in women (70 cases; 2.3% of all cancer diagnosed in women) [3].

33.3   Lung Cancer Risk Factors in the UAE

33.3.1 Smoking

Tobacco smoking poses significant health risks ranging from cardiovascular disease to stroke, chronic obstructive airway disease, and an increased risk of a variety of cancers. Lung cancer is possibly the most well-known cancer associated with tobacco smoking, although smoking is also a major risk factor in other malignancies such as bladder and head and neck cancers. It is estimated that tobacco-associated mortality reaches 8.7 million people worldwide, of which 7.4 million deaths are attributed to direct tobacco use [4]. Lung cancer is possibly the most well-known cancer associated with tobacco smoking. Lung cancer is the third most common cause of death related to smoking after ischemic heart disease and chronic obstructive pulmonary disease (COPD) [5].

33.3.1.1   Smoking and Lung Cancer Risk
Smoking is the greatest risk factor and is responsible for more than 85% of all lung cancers. The smoke in tobacco contains over 7000 chemicals and 60 different carcinogenic substances that are toxic and known to be cancer-producing. It is estimated that 1.4 million people died worldwide due to tobacco-related lung cancer in 2019 [6]. With increasing rates of smoking in the Middle East, the incidence of the disease is on the rise. In 2015, the UAE reported 2569 deaths attributable to smoking, with a mortality cost attribution of approximately US$ 1.7 billion [7].

As well as cigarettes, other traditional methods of tobacco smoking, such as midwakh and shisha, are commonly practiced in the UAE. These methods are sometimes used to add other products, such as aromatic leaves, herbs, or flavored oils, to tobacco. Such products are popular among youth and are commonly used during social and family gatherings. In one study, cigarette smoking was the most common form of tobacco use in the UAE (77.4% of smokers), followed by midwakh (15.0%), shisha (waterpipe) (6.8%), and cigars (0.66%) [8]. The emerging new electronic nicotine delivery systems (ENDS), including the e-cigarette and vaping, will increase the burden of lung cancer attributed to the use of these new tools. What we do know is that e-cigarettes contain 15 times the amount of formaldehyde found in traditional cigarettes and that this cancer-causing chemical is associated with an increased risk of lung, oral, and bladder cancer. E-cigarettes produce a number of dangerous chemicals, including acetaldehyde, acrolein, and formaldehyde [9]. These aldehydes can cause lung disease as well as cardiovascular disease.

33.3.2 Air Pollution

Although smoking is still the single most common risk factor for lung cancer, non-smokers (who never smoked) are also at risk of developing lung cancer due to polluted air. Outdoor pollution is thought to be responsible for over 300,000 lung cancer deaths [10]. Recent research, which was recently presented at the European Society of Medical Oncology (ESMO) 2022 annual congress, showed that exposure to fine 2.5 µm particulate matter ($PM_{2.5}$) triggers the inflammatory mediator interleukin-1β, which promotes carcinogenesis [11]. To our knowledge, and at the time of writing this manuscript, there is no study showing an association between air quality and lung cancer in the UAE or the region. The UAE National Air Emissions Inventory Project published its final results in 2019 [12]. The report identifies emissions related to industrial processes and product use (IPPU), energy, and transport as the major sources of $PM_{2.5}$ in the country [12]. Such emissions mainly arise from the metal and mineral industries, including aluminium, iron, steel, and cement production; stationary combustions to produce energy; oil and gas production; road transport, including passenger cars and heavy transportation; and the wear of tires and brake pads.

33.4 Preventative Measures for Lung Cancer in the UAE

The UAE government has made tremendous commitments in order to reduce the health burden of smoking-associated disease through legislation. Tobacco control is one of the most important priorities for health authorities in UAE.

Since the UAE's ratification of the WHO Framework Convention on Tobacco Control in November 2005, the Ministry of Health has developed an integrated strategy to combat this epidemic through the National Tobacco Control Program. It issued legislation supporting tobacco control measures, adopted international best practices, and consequently, reducing tobacco consumption became one of the most important indicators of the national agenda for the UAE Vision 2021. Federal Law No.15 of 2009 forbids and penalizes the sale of tobacco products to those under the age of 18. The law prohibits the sale of tobacco products near schools and places of worship. The sale of sweets that resemble tobacco products, automatic vending equipment and devices for tobacco distribution inside the country, tobacco advertising, and smoking in closed public spaces are also prohibited [13]. Each emirate may have specific rules regarding smoking in public places. For example, the Emirate of Sharjah banned all kinds of smoking in public areas in 2008, and Dubai Municipality does not allow smoking shisha in parks, beaches, and all other public recreational areas. The UAE's National Tobacco Control Committee has indicated its intention to implement a complete smoking ban in public areas in the country [13].

The UAE introduced an excise tax at a rate of 72% on tobacco products in 2017 and regulations for nicotine electronic products in 2019 in order to curb smoking, combat diseases, and prevent the spread of electronic cigarettes and analog products in the markets and outlets in the country [14]. According to the National Health Survey 2018, the prevalence of tobacco use has been reduced from 11% in 2010 to 9% in 2018 [13].

The Ministry of Health and Prevention (MoHAP) held a workshop in April 2021 to develop a national policy to promote healthy lifestyles, with a focus on three national indicators, including the prevalence of smoking any tobacco products.

Promoting a healthy lifestyle and reducing the rate of smoking is one of the national key performance indicators of the UAE National Agenda, as set forth in many documents, including the 2030 Agenda for Sustainable Development [15]. The UAE aims to reduce tobacco consumption from 21.6 percent to 15.7 percent among men and from 1.9 percent to 1.66 percent among women by the year 2021 [13]. Such measures are recognized by the WHO, which has named the UAE as one of the highest-achieving countries that enforce bans on tobacco advertising, promotion, and sponsorship [16]. However, much more needs to be done in order to reduce the burden of smoking-related disease in general and lung cancer in particular.

The UAE government has recognized the importance of improving air quality as a key development priority in the UAE Vision 2021. The UAE government aims to work on the following areas in order to improve the air quality in the country [17]:

- Defining the national standards for air pollution and compliance control.
- Implementation of the transition to a green economy.
- Encouraging the use of clean energy in different fields.
- The sustainability of the transport sector.
- The development of an air quality control network and the reliance on intelligent technologies and solutions for monitoring types of pollutants.

The UAE and the 2030 Agenda for Sustainable Development, which underpins the national sustainable development pillars, recognize the importance of prevention and "seek to reduce cancer and lifestyle-related diseases in order to ensure longer, healthier lives for citizens and residents" [15]. Two of the national key performance indicators are the prevalence of smoking and the rate of deaths from cancer. This underlines the commitment of the UAE to reducing cancer mortality by promoting preventative measures.

33.5 Lung Cancer Screening Program in the UAE

Despite being ranked as the seventh most common cancer, lung cancer is the second leading cause of cancer-related death, accounting for 9.85% of cancer-related mortalities in both men and women in the UAE [3]. This data highlights the fact that the majority of lung cancer cases are diagnosed at late stages when curative therapy options are no longer available. It is estimated that around 80% of lung cancer patients in the UAE are diagnosed at advanced stages [18, 19]. It is well known that the advanced stage at diagnosis carries a poor survival outcome for patients [20]. While 5-year survival for stage IA1 is estimated to be around 92%, no patient with stage IVb lung cancer is expected to live 5 years after diagnosis [21]. Recognizing the importance of early detection of lung cancer and its positive impact on reducing mortality, in 2017, the Department of Health (DoH) in Abu Dhabi launched a lung screening service based on a low-dose CT scan for high-risk individuals aged 55 to 75 with the following risk factors [22]:

- 30 pack-year history of smoking and/or tobacco cessation for less than 15 years.
- 20 pack-year history of tobacco use and/or tobacco cessation for less than 15 years and one additional risk factor.
- 20-year history of water pipe (shisha) and/or dokha, medwakh, and/or all other forms of smoked tobacco use.

This screening service is based on opportunistic recruitment and not on a call-and-recall system. As such, there is currently no official data on the uptake rate of lung cancer screening in the UAE. In August 2022, the Lung Cancer Policy Network (a global multi-stakeholder initiative set up by the Lung Ambition Alliance) published a document titled "Lung cancer screening: learning from implementation" which outlines valuable lessons learned from lung cancer screening programs around the world [23].

Because of the UAE's smoking pattern, eligibility criteria in the UAE may need to be adjusted to target younger, high-risk individuals. Additionally, instead of opportunistic recruitment, targeted outreach may be a better strategy in order to reduce barriers to participation in the screening program. Also, in order to ensure the success of the screening program, an effective referral pathway for patients with abnormal findings is highly important [18, 23, 24]. Lastly, it is of the utmost importance that the screening program be integrated into the health system and covered by the insurance companies.

33.6 Treatment of Lung Cancer in the UAE

Much investment has been made in the UAE in order to achieve world-class healthcare services. As such, the healthcare system in the country is one of the best in the region and among the best worldwide. Various diagnostic and therapeutic resources are available in the UAE to diagnose and treat lung cancer effectively. Many public and private healthcare facilities and hospitals are equipped with diagnostic imaging hardware such as X-rays, CT scans, MRI machines, and PET CT scanners.

Currently, more than 30 cancer centers and clinics and at least four comprehensive cancer centers are operating across the UAE [25, 26].

There is a significant number of specialists, including pulmonologists, radiologists, thoracic surgeons, and medical and radiation oncologists, working in the UAE, both in the public and private sectors, many with specialized training in western countries [26].

Emirates Oncology Society and Emirates Thoracic Society are among the active medical societies in the UAE that promote public awareness, education, and local research in lung cancer [27, 28].

33.6.1 Access to Medicine in the UAE

Access to medicine in the UAE is comparable to many modern western countries. Many modern anticancer therapies, including chemotherapy, the latest targeted therapies, and immunotherapies, are available in the UAE. Approval of cancer medicine in the UAE is among the fastest in the world, which reduces the registration period to around 30 days [29]. In 2020, the UAE became the first country in the Middle East to approve osimertinib as an adjuvant treatment for early non-small lung cancer. The UAE is the leading country in the GCC in the approval of immune checkpoint inhibitors for cancer therapy [29].

33.7 Conclusion and Recommendations

Lung cancer is a deadly disease that incurs a large health and economic burden. A significant proportion of lung cancer patients are diagnosed at advanced stages, when curative treatment is no longer possible. A low-dose CT scan is proven to be effective in identifying lung cancer at earlier stages. We commend the DoH for identifying the need for lung cancer screening and implementing a screening program. In order to have an effective lung cancer screening program that fits the needs of the UAE population, we recommend:

1. Adjusting eligibility criteria for lung cancer screening based on local criteria and population composition. In addition, we recommend a national level of research on the rate of lung cancer among never smokers in order to explore the need for expanding eligibility criteria beyond age and smoking.

2. Instead of an opportunistic recruitment strategy, we recommend lung cancer screening programs that are based on proactive identification of at-risk groups, with particular attention to disadvantaged individuals, including those with less understanding of the importance of screening or those with language barriers.

3. We recommend paying particular attention to cultural barriers that may hinder the participation of women in the lung cancer screening program.

4. We appreciate the UAE government's efforts to reduce smoking rates in the country and encourage a comprehensive, inclusive screening program that supports smoking cessation as well as an integrative referral pathway for patients with abnormal findings and is covered by insurance companies.

Conflict of Interest The authors have no conflict of interest to declare.

References

1. Ferlay J, Ervik M, Lam F, et al. 2020. Global cancer observatory: cancer today. https://gco.iarc.fr/today
2. Al-Lawati J, et al. Epidemiology of lung cancer in Oman: 20-year trends and tumor characteristics. Oman Med J. 2019;34(5):397–403.
3. Cancer incidence in the United Arab Emirates, Annual Report of the UAE- National Cancer Registry-2021. Statistics and Research Center, Ministry of Health and Prevention. (Accessed on 25 Mar 2024).
4. He H, Pan Z, Wu J, Hu C, Bai L, Lyu J. Health effects of tobacco at the global, regional, and National Levels: results from the 2019 global burden of disease study. Nicotine Tob Res. 2022;24(6):864–70. https://doi.org/10.1093/ntr/ntab265.
5. https://tobaccoatlas.org/challenges/health-effects/
6. GBD 2019 Tobacco Collaborators. Spatial, temporal, and demographic patterns in prevalence of smoking tobacco use and attributable disease burden in 204 countries and territories, 1990–2019: a systematic analysis from the Global Burden of Disease Study 2019. Lancet. 2021;397(10292):2337–60. https://doi.org/10.1016/S0140-6736(21)01169-7. Erratum in: Lancet. 2021 Jun 19;397(10292):2336. PMID: 34051883; PMCID: PMC8223261
7. Nagi MA, Riewpaiboon A, Thavorncharoensap M. Cost of premature mortality attributable to smoking in the Middle East and North Africa. East Mediterr Health J. 2021;27(10):974–83. https://doi.org/10.26719/emhj.21.028.
8. Al-Houqani M, Ali R, Hajat C. Tobacco smoking using Midwakh is an emerging health problem--evidence from a large cross-sectional survey in the United Arab Emirates. PLoS One. 2012;7(6):e39189.
9. Jensen RP, et al. Hidden formaldehyde in e-cigarette aerosols. N Engl J Med. 2015;372(4):392–4.
10. Liu X, Mubarik S, Wang F, Yu Y, Wang Y, Shi F, Wen H, Yu C. Lung cancer death attributable to long-term ambient particulate matter ($PM_{2.5}$) exposure in east Asian countries during 1990-2019. Front Med (Lausanne). 2021;8:742076. https://doi.org/10.3389/fmed.2021.742076.
11. Gourd E. New evidence that air pollution contributes substantially to lung cancer. Lancet Oncol. 2022 Oct;23(10):e448. https://doi.org/10.1016/S1470-2045(22)00569-1.
12. https://www.moccae.gov.ae/assets/download/fa2f8dd4/Air%20Emissions%20Inventory%20Report.pdf.aspx.
13. https://u.ae/en/information-and-services/health-and-fitness/tobacco-provisions#:~:text=15%20of%202009%20regarding%20Tobacco,schools)%2C%20health%20and%20sports%20facilities.
14. https://u.ae/en/information-and-services/finance-and-investment/taxation/excise-tax#:~:text=Rate%20of%20excise%20tax&text=100%20per%20cent%20on%20tobacco,cent%20on%20electronic%20smoking%20devices.

15. https://fcsc.gov.ae/en-us/Documents/UAE%20SDGs%20%E2%80%93%20Executive%20 Summary%20%E2%80%93%20VNR%202018%20EN.PDF.
16. https://www.who.int/teams/health-promotion/tobacco-control/global-tobacco-report-2021.
17. https://www.moccae.gov.ae/en/knowledge-and-statistics/air-quality.aspx.
18. Al-Shamsi HO, Jaffar H, Mahboub B, Khan F, Albastaki U, Hammad S, Zaabi AA. Early diagnosis of lung cancer in The United Arab Emirates: challenges and strategic recommendations. Clin. Pract. 2021;11:671–8. https://doi.org/10.3390/clinpract11030082.
19. Jaafar H, Mohieldin A, Mohsen R, Al Farsi A, Maarraou A, Al-Nassar M, Diaeddine T, El Shourbagy D, Dawoud EA. Epidermal growth factor receptor (EGFR) positive non-small-cell lung carcinoma (NSCLC) patients in the Gulf region: current status, challenges, and call for action. J Cancer Prev Curr Res. 2020;11:130–4.
20. Knight SB, Crosbie PA, Balata H, Chudziak J, Hussell T, Dive C. Progress and prospects of early detection in lung cancer. Open Biol. 2017;7:170070.
21. Oldstraw P, Chansky K, Crowley J, et al. The IASLC lung cancer staging project: proposals for revision of the TNM stage groupings in the forthcoming (eighth) edition of the TNM classification for lung cancer. J Thorac Oncol. 2016;11:39.
22. https://www.doh.gov.ae/-/media/Feature/Resources/Standards/Lung-Cancer-Screening-Service-Specifications_Publish.ashx.
23. https://www.lungcancerpolicynetwork.com/app/uploads/Lung-cancer-screening-learning-from-implementation.pdf.
24. Al-Shamsi HO, Abyad AM, Rafii S. A proposal for a National Cancer Control Plan for the UAE: 2022-2026. Clin Pract. 2022;12(1):118–32. https://doi.org/10.3390/clinpract12010016.
25. https://cdn.who.int/media/docs/default-source/country-profiles/cancer/are-2020.pdf?sfvrsn =c932b567_2&download=true.
26. Al-Shamsi HO. The state of cancer care in The United Arab Emirates in 2022. Clin Pract. 2022;12:955–85. https://doi.org/10.3390/clinpract12060101.
27. https://eos-uae.com.
28. https://www.etsociety.ae/.
29. https://www.iqvia.com/-/media/iqvia/pdfs/mea/white-paper/oncology-market-trends-in-gcc-countries.pdf.

Dr. Saeed Rafii is a board-certified consultant medical oncologist. After completion of his primary medical degree, he was trained in internal medicine, followed by subspecialty training in medical oncology in two of the most prestigious cancer hospitals in the UK, Queen Elizabeth Hospital Birmingham and the Royal Marsden Hospital, London. He then completed a clinical fellowship in early-phase clinical trials at the Royal Marsden Hospital, London, and received his CCT (certificate of completion of training) from the UK general medical council. Dr. Rafii was subsequently appointed as an associate professor and consultant in medical oncology at the University of Manchester and the Christie Hospital, where he helped establish and expand the experimental cancer medicine centre. He then moved to University College London Hospital and the Oxford Cancer Network as a consultant medical oncologist.

He has extensive expertise in clinical trials and has been chief, principal or co-investigator on over 100 early and late-phase oncology clinical trials. Dr. Rafii also holds a PhD and a postdoctoral fellowship in molecular cancer genetics. He is a member of the Royal College of Physicians of the UK, the European Society of Medical Oncology (ESMO), and the American Association for Clinical Oncology (ASCO). In 2018, he was elected as Fellow of the Royal College of Physicians of UK (FRCP) for his outstanding medical and research activities.

Dr. Batool Aboud is a specialist in medical oncology. She received her master's degree in medical oncology from Damascus University and is a member of the Syrian Board of Medical Oncology. She practices at the Saudi German Hospital in Ajman, UAE, since 2020, and is a lecturer at Ajman University, where she teaches fifth-year medical school students.

Prof. Humaid Obaid Al-Shamsi is the Chief Executive Officer of Burjeel Cancer Institute in Abu Dhabi, UAE, President of the Emirates Oncology Society, Lead of the Gulf Cancer Society, Full Professor of Oncology at the Ras Al Khaimah Medical and Health Sciences University, Ras Al Khaimah, UAE, and an Adjunct Professor of Oncology at the College of Medicine, University of Sharjah. He is the first Emirati to be promoted as a professor in oncology in the UAE. He is also the Chairman for Colorectal Cancer in the MENA region, appointed by the prestigious National Comprehensive Cancer Network®. He is also the only member of Lung Cancer Policy Network in the MENA region that aims to advance lung cancer research and screening globally. He is the Chairman of the Oncology and Hematology Fellowship Training Program for the National Institute for Health Specialties in the United Arab Emirates. He is the only member in GCC in the WIN Consortium which is comprised of organizations representing all stakeholders in personalized cancer medicine globally.

He is board-certified in both internal medicine and oncology from the UK, USA (ABIM), and Canada (FRCPC). He has also been awarded the FRCP (London) in 2023 and FRCP (Glasgow) in 2024. He is the only physician in the UAE with a subspecialty fellowship certification and training in gastrointestinal oncology and the first Emirati to train and complete a clinical post-doctoral fellowship in palliative care. He was an assistant professor at the University of Texas MD Anderson Cancer Center between 2014 and 2017. He has published more than 140 peer-reviewed articles in JAMA Oncology, Lancet Oncology, The Oncologist, BMC Cancer, and many others. His area of expertise includes precision oncology and cancer care in the UAE. In 2016, he published with his group from MD Anderson the JCO paper describing a new distinct subgroup of CRC, NON V600 BRAF-mutated CRC. In 2022, he published the first book about cancer research in the UAE and also the first book about cancer in the Arab world, both of which were launched at Dubai Expo 2020. *Cancer in the Arab World* has been downloaded more than 450,000 times in its first 18 months of publication and is the ultimate source of cancer data in the Arab region. He also published the first comprehensive book about cancer care in the UAE which is the first book in UAE history to document the cancer care in the UAE with many topics addressed for the first time, e.g., neuroendocrine tumors in the UAE. He is passionate about advancing cancer care in the UAE and the GCC and has made significant contributions to cancer awareness and early detection for the public using social media platforms. He is considered as the most followed oncologist in the world with over 300,000 subscribers across his social media platforms (Instagram, Twitter, LinkedIn, and TikTok). In 2022, he was awarded the prestigious Feigenbaum Leadership Excellence Award from Sheikh Hamdan Smart University for his exceptional leadership and research and the Sharjah Award for Volunteering. He was also named the Researcher of the Year in the UAE in 2020 and 2021 by the Emirates Oncology Society.

In May 2024, HH Sheikh Mansour bin Zayed Al Nahyan, Vice President of the United Arab Emirates, awarded him the first place in UAE Nafis program for outstanding leadership in private sector across all business and medical disciplines. Beside his clinical and administrative duties, he is engaged in education and various levels of research training for medical trainees to enhance their clinical and research skills. His mission is to advance cancer care in the UAE and the MENA region and make cancer care accessible to everyone in need around the globe.

Saladin Sawan [ID], Faryal Iqbal [ID], and Humaid O. Al-Shamsi [ID]

34.1 Introduction

The United Arab Emirates (UAE) was formed as a constitutional federation of seven emirates: Abu Dhabi, Dubai, Sharjah, Ajman, Umm Al Quwain, Ras Al Khaimah, and Fujairah, which came together as one state in December 1971. It is located in the Arabian Peninsula's southeast [1]. According to the United Nations Development Programme (UNDP) in their most recent Human Development Report 2020, the UAE is distinguished as the foremost nation in the Arab world with a "Very High Human Development Index." It holds the 31st position among a total of 189 countries worldwide [2, 3]. The Federal Competitiveness and Statistics Centre published demographic data for the UAE, showing a total population of 9.5 million in 2019, with 3.2 million females (33.7%) [4, 5].

S. Sawan (✉)
University of Manchester, Manchester, UK
e-mail: saladin.sawan@manchester.ac.uk

F. Iqbal
Burjeel Medical City, Abu Dhabi, United Arab Emirates
e-mail: faryal.iqbal@burjeelmedicalcity.com

H. O. Al-Shamsi
Burjeel Cancer Institute, Burjeel Medical City, Burjeel Holdings, Abu Dhabi, United Arab Emirates

Ras Al Khaimah Medical and Health Sciences University, Ras Al Khaimah, United Arab Emirates

Gulf Medical University, Ajman, United Arab Emirates

Emirates Oncology Society, Emirates Medical Association, Dubai, United Arab Emirates

College of Medicine, University of Sharjah, Sharjah, United Arab Emirates

Gulf Cancer Society, Alsafa, Kuwait
e-mail: alshamsi@burjeel.com; humaid.al-shamsi@medportal.ca

© The Author(s) 2024
H. O. Al-Shamsi (ed.), *Cancer Care in the United Arab Emirates*,
https://doi.org/10.1007/978-981-99-6794-0_34

34.2   Cancer Care in the UAE

With the objective of accessing medical data while safeguarding patient confidentiality, the Ministry of Health and Prevention (MOHAP) initiated the creation of the "UAE National Cancer Registry" (UAE-NCR). This registry, designed for both UAE nationals and expatriates, serves as a population-based record of cancer cases in the country [6]. The UAE National Cancer Registry (UAE-NCR) is the exclusive and reliable resource for acquiring accurate information regarding cancer incidence and mortality rates in different regions of the country. It offers valuable insights into the data gathered from all healthcare providers throughout the UAE.

34.2.1 Overall Cancer Incidence Rate in the UAE

In 2021, the UAE-NCR recorded 5830 newly diagnosed cancer cases (malignant and in-situ) in both genders. Out of them, the number of malignant cases was 5612 (96%), whereas 218 (4%) were in situ cases. The cancer affected a greater number of women than men; the number of affected males was 2620 (44.9%), whereas the number of females diagnosed with cancer was 3210 (55.1%) in 2021 [6]. Taken into account the proportion of female and male population, the crude incident rate of cancer in 2021 was 108.7/100,000, 39.5/100,000, and 60.5/100,000 in female, male, and overall crude incidence rates for both genders, respectively. The total number of overall newly diagnosed cancer cases in 2021 was divided according to UAE citizenship: 1493 cases were newly diagnosed with cancer among UAE citizens, whereas 4337 newly diagnosed cancer cases among non-UAE citizens were reported [6]. Hence, 25.6% of newly diagnosed cancers affected Emirati citizens [6].

34.3   Gynecologic Malignancies in the UAE

Any cancer that initiates in a female's reproductive organs is broadly termed "Gynecologic Cancer".

1. Uterine cancer initiates in the uterus.
2. Cervical cancer initiates in the cervix.
3. Ovarian cancer initiates in the ovaries.
4. Vaginal cancer initiates in the vagina.
5. Vulvar cancer initiates in the Vulva [7].

34.4   Gynecologic Cancer Incidence Rate in the UAE

In 2021, there were a total of 490 gynecologic cancer cases (including both malignant and in-situ) among the population of the UAE, out of a total of 3210 newly diagnosed cancer cases in women, representing 15.2% of the total. The data presented in Table 34.1 demonstrates that non-UAE citizens within the UAE population had a higher number of gynecologic cancers, specifically 367 cases, compared to UAE citizens, who accounted for 123 cases [6].

According to the "Cancer Incidence in the United Arab Emirates: Annual Report of the UAE-National Cancer Registry 2021," the UAE population experiences the highest number of malignant cases in the cervix uteri, uterus, and ovaries within the age groups of 40–49, 60–69, and 50–59, respectively. Table 34.2 presents the breakdown of gynecologic cancer cases by age group for the entire UAE population. Furthermore, Table 34.3 displays the distribution of malignant gynecologic cancers by age group, specifically among UAE citizens, while Table 34.4 illustrates the distribution among non-UAE citizens [6].

Table 34.1   The sum of gynecologic cancers out of the total number of newly diagnosed cancer cases among the UAE population according to primary site (malignant and in situ) and nationality, 2021

Primary site	UAE	Non-UAE	Total
All invasive cancers (malignant cases)			
C53 cervix uteri	23	118	141
C54-C55 uterus	60	113	173
C56 ovary	23	85	108
Non-invasive cancers (in-situ cases)			
D06 carcinoma in situ of cervix uteri	17	51	68
Total	**123**	**367**	**490**

Source: Ministry of Health and Prevention, Statistics and Research Center, National Disease Registry—UAE National Cancer Registry Report, 2021

Table 34.2   Primary site (malignant) distribution of gynecologic cancers by age group among all, (UAE and non-UAE Citizens), 2021

Primary Site	0–9	10–19	20–29	30–39	40–49	50–59	60–69	70–79	80+	Total
C53 cervix uteri	0	0	5	38	45	34	14	4	1	141
C54-C55 uterus	0	0	2	21	37	42	46	24	1	173
C56 ovary	0	1	4	19	27	33	12	9	3	108

Source: Ministry of Health and Prevention, Statistics and Research Center, National Disease Registry—UAE National Cancer Registry Report, 2021

Table 34.3 Primary site (malignant) distribution of gynecologic cancers by age group among UAE citizens, 2021

Primary Site	0–9	10–19	20–29	30–39	40–49	50–59	60–69	70–79	80+	Total
C53 cervix uteri	0	0	2	0	9	7	3	1	1	23
C54-C55 uterus	0	0	0	5	12	12	18	12	1	60
C56 ovary	0	1	0	3	4	9	1	3	2	23

Source: Ministry of Health and Prevention, Statistics and Research Center, National Disease Registry—UAE National Cancer Registry Report, 2021

Table 34.4 Primary site (malignant) distribution of gynecologic cancers by age group among non-UAE citizens, 2021

Primary Site	0–9	10–19	20–29	30–39	40–49	50–59	60–69	70–79	80+	Total
C53 cervix uteri	0	0	3	38	36	27	11	3	0	118
C54-C55 uterus	0	0	2	16	25	30	28	12	0	113
C56 ovary	0	0	4	16	23	24	11	6	1	85

Source: Ministry of Health and Prevention, Statistics and Research Center, National Disease Registry—UAE National Cancer Registry Report, 2021

34.5 Recent Incidence of the Most Common Gynecologic Cancers in Among the UAE Population

34.5.1 Cervix Uteri

This particular form of cancer primarily affects women over the age of 30 [8]. The total number of malignant cervix uteri cases among the UAE population in 2021 is presented in Table 34.1. It ranked as the fifth most prevalent cancer among women in the UAE. Specifically, there were 141 cases of cervix uteri cancer, accounting for 4.6% of all female cancer diagnoses in 2021. The stage distribution of cervix uteri cancer cases in 2017, as documented by the UAE-National Cancer Registry (UAE-NCR), is depicted in Fig. 34.1. Table 34.2 highlights that the age group of 40–49 had the highest number of reported cervix uteri cases in 2021 according to the UAE-NCR [6].

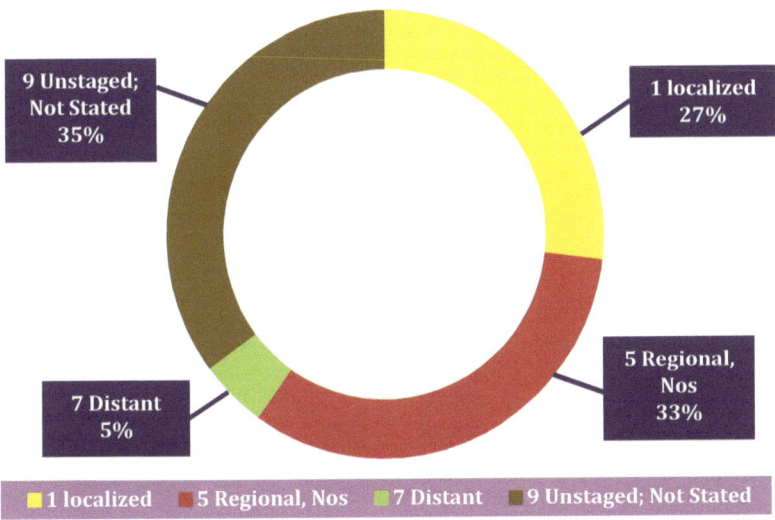

Fig. 34.1 Stage distribution of cervix uteri cancer cases among the UAE population, 2017. (Source: Ministry of Health and Prevention, Statistics and Research Center, National Disease Registry—UAE National Cancer Registry Report, 2017)

34.5.2 Uterus

Among the population of the UAE, uterus cancer is the prevailing gynecologic cancer. The total count of malignant uterus cancer cases among the UAE population in 2021 is provided in Table 34.1. It ranked as the fourth most prevalent cancer among women in the UAE. In 2021, there were 173 reported cases of uterus cancer, constituting 5.7% of all cancer diagnoses in females. As indicated by Table 34.2, the UAE National Cancer Registry (UAE-NCR) recorded the highest number of uterus cancer cases within the age group of 60–69 in 2021 [6].

34.5.3 Ovary

Ovarian cancer is a type of cancer that develops in the ovaries or in adjacent regions such as the fallopian tubes and the peritoneum [9]. Mutations in the BRCA1 and BRCA2 genes, as well as those associated with Lynch syndrome, have the potential to increase the risk of ovarian cancer in women [9]. Table 34.1 provides an overview of the total number of malignant ovary cancer cases within the UAE population in 2021. It ranked as the seventh most prevalent cancer among women in the UAE. Specifically, there were 108 reported cases of ovary cancer, representing 3.5% of all female cancer diagnoses in 2021. According to Table 34.2, the UAE National Cancer Registry (UAE-NCR) documented the highest number of ovarian cancer cases in the age group of 50–59 [6].

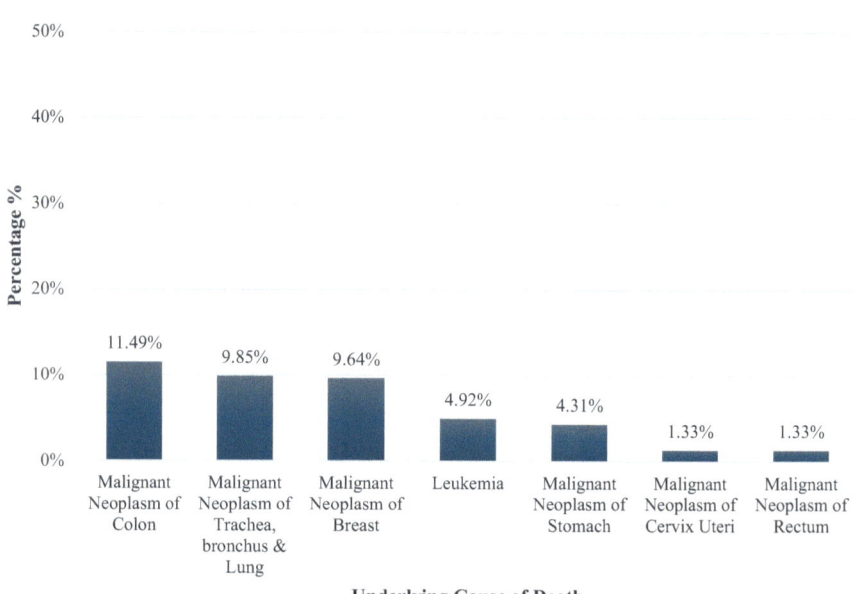

Fig. 34.2 Distribution of cancer mortality rates by type in the UAE in 2021. (Source: Ministry of Health and Prevention, Statistics and Research Center, National Disease Registry—UAE National Cancer Registry Report, 2021)

34.6 Mortality Rate in the UAE Due to Gynecologic Cancer in 2021

Cancer is the fifth-leading cause of death in the UAE [6]. Malignant neoplasm of the cervix uteri contributed to 1.33% of all cancer-related deaths within the UAE population in 2021. It held the sixth position in terms of mortality rates among the UAE population during that year. Figure 34.2 presents the breakdown of malignant cancer deaths by cancer type in the UAE, as reported by the UAE National Cancer Registry (UAE-NCR) in 2021 [6].

34.7 Overall Malignant Gynecologic Cancers in the UAE

Based on the report from the UAE National Cancer Registry (UAE NCR) 2021, uterus cancer has the highest number of overall malignant gynecologic cancer cases, i.e., 173, while the second-highest is cervix uteri, with 141 cases in 2021, and the third is ovarian cancer, with 108 cases. Figure 34.3 provides an overview of the total count of malignant gynecologic cancer cases in the UAE from 2013 to 2021 [6].

Figure 34.4 shows the trendline of total malignant gynecologic cancer cases among UAE population from the year 2013–2019 [6].

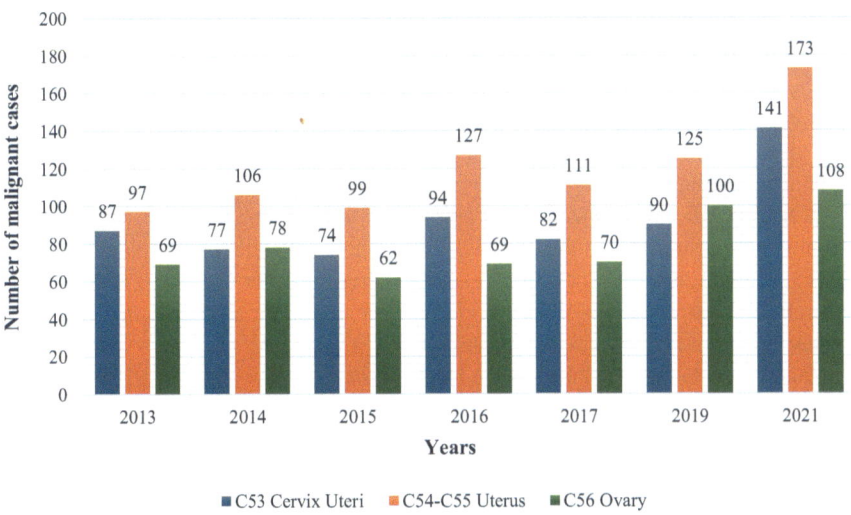

Fig. 34.3 Overall malignant gynecologic cancer cases in the UAE for the years 2013–2021. (Source: Ministry of Health and Prevention, Statistics and Research Center, National Disease Registry—UAE National Cancer Registry Report, 2013–2021)

Fig. 34.4 The trendline of total malignant gynecologic cases in the UAE population between 2013 and 2021. (Source: Ministry of Health and Prevention, Statistics and Research Center, National Disease Registry—UAE National Cancer Registry Report, 2013–2021)

34.8   Cervical Cancer Prevention and Control Program

In the UAE, cervix uteri ranks as the fifth most common cancer among women [6]. Almost 99% of cervical cancer cases are attributed to Human Papillomavirus (HPV) infection in the cervix area. Vaccinations are highly effective in preventing HPV infection among females [10].

The cervical cancer awareness campaign is a component of the Department of Health's larger initiative titled "Live healthily & simply check" campaign. This particular campaign spans six months, starting from October 2017 and concluding in March 2018. It coincides with the global observance months dedicated to raising awareness about cancer prevention initiatives [11].

DOH launched the "Cancer Wave Health Promotion Project" in 2012. The objective of the campaign was to raise community awareness about the importance of regular screening and early detection of the top three cancers, namely breast, cervical, and colorectal [12].

Friends of Cancer Patients (FoCP) collaborated with the United Nations Population Fund (UNFPA) to create guidance programs for the Ministry of Health and partner agencies, aimed at developing and updating their programs for preventing and controlling cervical cancer. FoCP organized its inaugural forum titled "Turning the Tide on HPV and Cervical Cancer" in January 2019, followed by the second forum "Accelerating Action on HPV and Cervical Cancer" in January 2021, in Sharjah, UAE [13].

Due to these significant and commendable measures taken to prevent cervical cancer, the UAE has observed a decline in the number of cervix uteri cases in recent years.

Figure 34.5 shows the trendline of newly diagnosed carcinoma in situ of the cervix uteri for the years 2015, 2016, 2017, 2019, and 2021 [6]. This illustrates the decrease in the number of cervix uteri (one of the most common gynecologic cancers in the UAE).

The UAE needs to set up more screening and prevention campaigns for the sake of other common gynecologic cancers in the UAE, i.e., uterus and ovary cancers among the female population. Physicians have a vital role to play in promoting community awareness campaigns among women in the country. These campaigns aim to mitigate the risk of developing cancer and promote the adoption of healthy lifestyles. It is crucial for physicians to actively engage in these efforts.

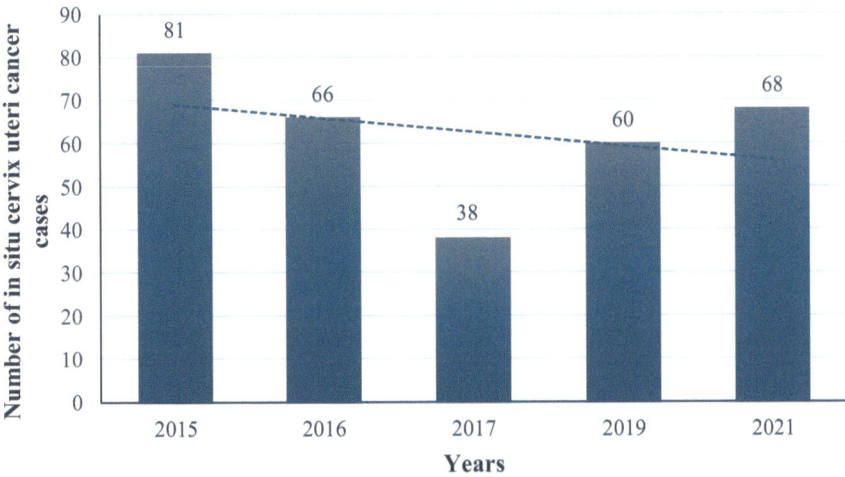

Fig. 34.5 The trendline of D06 carcinoma in situ of the cervix uteri for the years 2015, 2016, 2017, 2019, and 2021. (Source: Ministry of Health and Prevention, Statistics and Research Center, National Disease Registry—UAE National Cancer Registry Report)

34.9 Conclusion

The UAE boasts a sophisticated healthcare system. The significantly high rate of female cancer incidence in the UAE has prompted health authorities to take proactive measures in providing gynecologic cancer care services. These initiatives include the establishment of the National Cancer Registry, the implementation of an HPV vaccination drive, organizing workshops and training for healthcare professionals, conducting awareness campaigns, and launching screening campaigns for females nationwide. To further enhance the multidimensional care for gynecologic cancer patients in the UAE, it is important to foster collaborations with other countries. Such collaborations would help in bringing together various cancer specialties and advanced technologies available within the country.

Conflict of Interest The authors have no conflict of interest to declare.

References

1. Al-Shamsi H, Darr H, Abu-Gheida I, Ansari J, McManus MC, Jaafar H, Al-Khatib F. The state of cancer Care in The United Arab Emirates in 2020: challenges and recommendations, a report by The United Arab Emirates oncology task force. Gulf J Oncolog. 2020;1(32):71–87.
2. The 2020 Human Development Report. United Nations Development Programme (UNDP). 2020. http://hdr.undp.org/sites/default/files/hdr2020.pdf.
3. UAE ranks first in Arab World in Human Development Report 2020. Ministry of Foreign Affairs and International Cooperation, United Arab Emirates. 2020. https://www.mofaic.gov.ae/en/mediahub/news/2020/12/16/16-12-2020-uae-human-development.

4. UAE population, Federal Competitiveness and Statistics Centre. Available at: https://fcsc.gov. ae/en-us/Pages/Statistics/Statistics-by-Subject.aspx#/%3Fyear=&folder=Demography%20 and%20Social/Population/Population&subject=Demography%20and%20Social. Accessed 26 Mar 2023.
5. The United Arab Emirates' Government portal, Population and demographic mix: https://u. ae/en/information-and-services/social-affairs/preserving-the-emirati-national-identity/ population-and-demographic-mix. Accessed 26 Mar 2023.
6. Ministry of Health and Prevention, Statistics and Research Center, National Disease Registry— UAE National Cancer Registry Report.
7. https://www.cdc.gov/gynecologic-cancer/about/index.html. (Accessed on 23 Mar 2024).
8. https://www.cdc.gov/cervical-cancer/about/index.html. (Accessed on 23 Mar 2024).
9. https://www.cdc.gov/ovarian-cancer/about/index.html. (Accessed on 23 Mar 2024).
10. Malboomers et al. "Human papillomavirus is a necessary cause of invasive cervical cancer worldwide" https://doi.org/10.1002/(SICI)1096-9896(199909)189:1%3C12:: AID-PATH431%3E3.0.CO;2-F.
11. https://www.doh.gov.ae/en/news/late-stage-diagnosis-ofcervical-cancer).
12. https://gulfnews.com/uae/government/campaign-against-cancer-launched-in-abu-dhabi-1.1088422. (Accessed on 23 Mar 2024)
13. https://www.focp.ae/our-programs/cervical-cancer/. (Accessed on 23 Mar 2024).

Dr. Saladin Sawan is currently a consultant surgeon in gynaecolgoy oncology in Manchester University NHS Foundation Trust in England. He had worked recently in the United Arab Emirates when he gained an insight into the burden of gynaecological cancers in the UAE.

Saladin is a vice-chair of the Scientific Advisory Committee, the Royal College of Obstetricians and Gynaecologists, UK. He is an honorary lecturer, The University of Manchester, UK.

Ms. Faryal Iqbal is the research associate at Burjeel Medical City, Abu Dhabi, United Arab Emirates. She completed her undergraduate studies in molecular biology and biotechnology. Following that, she received a postgraduate qualification in molecular genetics. She co-edited "Cancer in the Arab World," the first extensive book covering cancer care across all Arab countries. The book succeeded significantly, with over 450,000 downloads within just two years. She has several peer-reviewed papers under her name. In September 2023, Ms, Iqbal received the "EOS Research Award" from the prestigious Emirates Oncology Society for her research efforts. Moreover, she assists in different aspects of clinical trial implementation at a research site. Her research interests and publications encompass oncology, hematology, and genetics.

Prof. Humaid Obaid Al-Shamsi is the Chief Executive Officer of Burjeel Cancer Institute in Abu Dhabi, UAE, President of the Emirates Oncology Society, Lead of the Gulf Cancer Society, Full Professor of Oncology at the Ras Al Khaimah Medical and Health Sciences University, Ras Al Khaimah, UAE, and an Adjunct Professor of Oncology at the College of Medicine, University of Sharjah. He is the first Emirati to be promoted as a professor in oncology in the UAE. He is also the Chairman for Colorectal Cancer in the MENA region, appointed by the prestigious National Comprehensive Cancer Network®. He is also the only member of Lung Cancer Policy Network in the MENA region that aims to advance lung cancer research and screening globally. He is the Chairman of the Oncology and Hematology Fellowship Training Program for the National Institute for Health Specialties in the United Arab Emirates. He is the only member in GCC in the WIN Consortium which is comprised of organizations representing all stakeholders in personalized cancer medicine globally.

He is board-certified in both internal medicine and oncology from the UK, USA (ABIM), and Canada (FRCPC). He has also been awarded the FRCP (London) in 2023 and FRCP (Glasgow) in 2024. He is the only physician in the UAE with a subspecialty fellowship certification and training in gastrointestinal oncology and the first Emirati to train and complete a clinical post-doctoral fellowship in palliative care. He was an assistant professor at the University of Texas MD Anderson Cancer Center between 2014 and 2017. He has published more than 140 peer-reviewed articles in JAMA Oncology, Lancet Oncology, The Oncologist, BMC Cancer, and many others. His area of expertise includes precision oncology and cancer care in the UAE. In 2016, he published with his group from MD Anderson the JCO paper describing a new distinct subgroup of CRC, NON V600 BRAF-mutated CRC. In 2022, he published the first book about cancer research in the UAE and also the first book about cancer in the Arab world, both of which were launched at Dubai Expo 2020. *Cancer in the Arab World* has been downloaded more than 450,000 times in its first 18 months of publication and is the ultimate source of cancer data in the Arab region. He also published the first comprehensive book about cancer care in the UAE which is the first book in UAE history to document the cancer care in the UAE with many topics addressed for the first time, e.g., neuroendocrine tumors in the UAE. He is passionate about advancing cancer care in the UAE and the GCC and has made significant contributions to cancer awareness and early detection for the public using social media platforms. He is considered as the most followed oncologist in the world with over 300,000 subscribers across his social media platforms (Instagram, Twitter, LinkedIn, and TikTok). In 2022, he was awarded the prestigious Feigenbaum Leadership Excellence Award from Sheikh Hamdan Smart University for his exceptional leadership and research and the Sharjah Award for Volunteering. He was also named the Researcher of the Year in the UAE in 2020 and 2021 by the Emirates Oncology Society.

In May 2024, HH Sheikh Mansour bin Zayed Al Nahyan, Vice President of the United Arab Emirates, awarded him the first place in UAE Nafis program for outstanding leadership in private sector across all business and medical disciplines. Beside his clinical and administrative duties, he is engaged in education and various levels of research training for medical trainees to enhance their clinical and research skills. His mission is to advance cancer care in the UAE and the MENA region and make cancer care accessible to everyone in need around the globe.

Mohammed Shahait, Hosam Al-Qudah,
Layth Mula-Hussain ⓘ, Ibrahim H. Abu-Gheida,
Thamir Alkasab, Ali Thwaini, Rabii Madi,
Humaid O. Al-Shamsi ⓘ, Syed Hammad Tirmazy,
and Deborah Mukherji

M. Shahait · R. Madi · D. Mukherji (✉)
Department of Urologist, Clemenceau Medical Center, Dubai, United Arab Emirates
e-mail: mohammed.shahait@cmcdubai.ae; rabii.madi@cmcdubai.ae;
deborah.mukherji@cmcdubai.ae

H. Al-Qudah
Department of Urology, Fakeeh University Hospital, Dubai, United Arab Emirates

L. Mula-Hussain
Radiation Oncology, Dalhousie University, Halifax, NS, Canada
e-mail: layth.mula-hussain@dal.ca; lmulahussain@aol.com

I. H. Abu-Gheida
Emirates Oncology Society, Emirates Medical Association, Dubai, United Arab Emirates

Burjeel Medical City, Abu Dhabi, United Arab Emirates

T. Alkasab
Uro-Oncology Specialist, Alzahra Hospital, Dubai, United Arab Emirates
e-mail: Thamir.Alkasab@azhd.ae

A. Thwaini
Department of Urologist, Mohammed Bin Rashid University of Medicine and Health
Sciences, Dubai, United Arab Emirates
e-mail: Ali.thwaini@mediclinic.ae

H. O. Al-Shamsi
Burjeel Cancer Institute, Burjeel Medical City, Burjeel Holdings, Abu Dhabi,
United Arab Emirates

Ras Al Khaimah Medical and Health Sciences University, Ras Al Khaimah, United Arab Emirates

Gulf Medical University, Ajman, United Arab Emirates

Emirates Oncology Society, Emirates Medical Association, Dubai, United Arab Emirates

College of Medicine, University of Sharjah, Sharjah, United Arab Emirates

Gulf Cancer Society, Alsafa, Kuwait
e-mail: alshamsi@burjeel.com; humaid.al-shamsi@medportal.ca

S. H. Tirmazy
Department of Oncology, Dubai Hospital, Dubai, United Arab Emirates
e-mail: SHTirmazy@dha.gov.ae

© The Author(s) 2024
H. O. Al-Shamsi (ed.), *Cancer Care in the United Arab Emirates*,
https://doi.org/10.1007/978-981-99-6794-0_35

35.1   Introduction

The genitourinary malignancies encompass kidney and bladder cancer, diagnosed
in both men and women; prostate cancer, testicular germ cell tumors, and penile
cancer, diagnosed exclusively in men.

Since the establishment of the United Arab Emirates-national cancer registry
(UAE-NCR), it has captured cases of prostate cancer, kidney and renal pelvis cancer, bladder cancer, and germ cell tumors in men, but not penile cancer cases due to
the rarity of this disease. Prostate cancer ranked as the second most common malignancy in men, Kidney and Renal pelvis ranked ninth, and bladder cancer ranked as
the tenth most common malignancy detected in men in the 2021 UAE-NCR report.
The majority of the GU malignancies registered in this time period were diagnosed
in expats [1] Table 35.1.

Table 35.1 Genitourinary cancer demographics among the UAE population during 2013–2021

Year	UAE population (in millions)	Total malignant cases (in numbers)	GU cancer[a] cases (in numbers)	Percentage (%)	Crude incidence rates of GU cancer[a] cases per 100,000 population
2013	8.66	3574	369	10.32	–
2014	8.79	3610	425	11.77	Prostate: – Testis: – Kidney and renal pelvis: 1.06 Urinary bladder: 1.35
2015	8.93	3744	390	10.41	Prostate: – Testis: – Kidney and renal pelvis: 0.9 Urinary bladder: 1.1
2016	9.12	3982	402	10.09	–
2017	9.3	4123	394	9.5	Prostate: – Testis: – Kidney and renal pelvis: 0.9 Urinary bladder: 1.2
2019	9.5	4381	437	9.9	–
2021	–	5612	588	10.4	Prostate: – Testis: – Kidney and renal pelvis: 1.6 Urinary bladder: 1.4

Source: UAE   population: https://fcsc.gov.ae/en-us/Pages/Statistics/Statistics-by-Subject.
aspx#/%3Fsubject=Demography%20and%20Social; Ministry of Health and Prevention, Statistics
and Research Center, National Disease Registry—UAE National Cancer Registry Report
[a] GU cancer cases include: (1) Prostate (2) Testis (3) Kidney and renal pelvis (4) Urinary bladder

35.2 Prostate Cancer in the UAE

According to Global cancer statistics 2020: GLOBOCAN, the age-standardized rate (ASR) of new prostate cancer cases in the UAE is 13.4 cases per 100,000 population per year, while worldwide it is 30.7 cases per 100,000 population per year. Moreover, the ASR of prostate cancer-related mortality in the UAE is 3.4 cases per 100,000 population per year, compared to the worldwide rate of 7.7 cases per 100,000 population per year. The mortality-to-incidence ratio of prostate cancer in the UAE is similar to that in the world (25%), but it is higher compared to the USA (16.6%) [2].

The differences in the incidence and mortality rates between the UAE and USA might be explained by several factors, such as the age of the population, genetic predisposition, PSA screening penetrance, diet, accessibility to novel medications, and stage at diagnosis [3–5]. For example, 16.8% of the USA population is older than 65 years, compared to 1.9% of the UAE population. In the UAE, there is no established prostate cancer screening program at the national level; even more opportunistic PSA testing is not widespread due to a lack of awareness, and in most instances, it is not covered by private insurance.

To better understand the genomic diversity of prostate cancer among different ancestries, Albawardi et al. compared the mutational profiles of Mediterranean patients with prostate cancer treated at Tawam Hospital to those of prostate cancer patients from different ancestries and found that the tumors of patients of ME ancestry had fewer gene-level copy number aberrations than those in men of other ancestries [6]. In addition, somatic amplification of the glutathione S-transferase family on chromosome 1 (GSTM1, GSTM2, GSTM5) and the IQ motif-containing family on chromosome 3 (IQCF1, IQCF2, IQCF13, IQCF4, IQCF5, IQCF6) was noted more frequently in patients of ME ancestry. The findings of this study are in line with other studies that showed differences in clonal evolution between the Middle Eastern and Western populations [4].

A recent bibliometric analysis of prostate cancer research in the Arab world showed that most Arab-based studies did not involve collaborations and were categorized as low-level evidence [7]. However, prostate cancer health providers in the UAE are actively involved in regional consensus on prostate cancer management, the translation and validation of quality-of-life questionnaires, and multi-institutional studies [8, 9].

Patients diagnosed with prostate cancer in the UAE have good access to diagnostic and treatment modalities and physicians with sub-specialist training. Specialist imaging modalities such as multi-parametric MRI and PET-PSMA are examples of diagnostic modalities. Treatment modalities available for localized disease include robotic surgery, external beam radiation therapy, brachytherapy, and theranostics (including lutetium-PSMA). Access to novel systemic therapies, germline, and somatic genetic testing is excellent.

Recommendations for improving prostate cancer outcomes in the UAE

1. Implementation of population-based PSA screening as well as improving public awareness of prostate cancer symptoms and the importance of screening, including earlier screening, in men with a family history of prostate cancer.

2. Training and awareness for primary healthcare physicians.
3. Implementation of fast-track referral pathways for patients with suspected cancer that include targets.
4. Improved the data documentation process at the hospital level on prostate cancer diagnosis, stage at diagnosis, and treatment outcome data.
5. Encourage multidisciplinary management of prostate disease by sub-specialist teams to ensure optimal patient outcomes.
6. Encourage multidisciplinary management of treatment-related toxicity for men treated with systemic therapy, including cardiovascular and bone health.
7. Early access to specialist palliative care for men with advanced disease.
8. Engagement in context-specific research with improved data collection on population genetics, barriers to screening, patient preferences for treatment, and treatment-related toxicity.

35.3 Bladder Cancer in the UAE

According to Global Cancer Statistics 2020: GLOBOCAN, the age-standardized rate (ASR) of new bladder cancer cases in the UAE is 6.4 cases per 100,000 population per year, while worldwide it is 5.6 cases per 100,000 population per year. Moreover, the ASR of bladder cancer-related mortality in the UAE is 2.7 cases per 100,000 population per year, compared to the worldwide rate of 1.9 cases per 100,000 population per year [2]. The high incidence of bladder cancer in the UAE compared to the rest of the world might be explained by the fact that a substantial number of expats are originally from countries with high bladder cancer incidence, such as Lebanon and Egypt [10].

Although bladder cancer is a disease of the elderly, as the average age at diagnosis of bladder cancer in the USA is 73 years, most bladder cancer cases in the UAE are in patients younger than 65 years [10]. The occurrence of cancer at a younger age in Arab countries compared to western countries is well described in different malignancies that include breast and colorectal cancer [11, 12]. This observation is very crucial for practicing physicians in the UAE, as western guidelines for microscopic hematuria stratify patients based on age according to the age-standardized rate observed in their population [13]. As such, physicians in the UAE should be vigilant about adopting these guidelines into their practice.

There are several identified modifiable risk factors for bladder cancer, such as obesity and smoking [14, 15]. The UAE has a comparable obesity rate compared to the USA, with a trend toward having a higher rate of childhood obesity [16]. Another risk factor is smoking; the overall smoking rate among the UAE population is 18.20%, with a high male-to-female ratio of 44:1 [17]. The highest prevalence was reported in males aged 20–39 years. Several forms of tobacco consumption are observed in the UAE, which include cigarette smoking (77.4%), followed by 15.0% midwakh use (a small pipe used for smoking tobacco), 6.8% waterpipe use, and 0.66% cigar use [18]. The UAE Government's efforts to curb smoking consisted of bundled policies, which include but are not limited to raising the cost of smoking through taxation, mounting sustained social marketing campaigns, and ensuring that health professionals routinely advise smokers to stop smoking, accompanied by behavioral and pharmacological support for cessation and smoking cessation clinics.

Patients in the UAE have excellent access to all diagnostic and treatment modalities for bladder cancer, including sub-specialist surgical expertise, external beam radiation, and recently approved systemic therapies, including immune checkpoint inhibitors.

Recommendations for improving bladder outcomes in the UAE are as follows:

1. Improved the data documentation process at the hospital level, on bladder cancer diagnoses, risk factors, stage at diagnosis, and treatment outcome data.
2. Advocate for improved tobacco control policies and improved public awareness of bladder cancer risks and symptoms.
3. Encourage multidisciplinary management of bladder cancer by sub-specialist teams to ensure optimal patient outcomes.
4. Early access to specialist palliative care for men with advanced disease.
5. Engagement in context-specific research with improved data collection on bladder cancer risk factors, treatment response, barriers to early diagnosis, and patient preferences for treatment.

35.4 Kidney Cancer in the UAE

According to Global cancer statistics 2020: GLOBOCAN, the age-standardized rate (ASR) of new kidney cancer cases in the UAE is 2.3 cases per 100,000 population per year, while worldwide it is 4.6 cases per 100,000 population per year. Moreover, the ASR of kidney cancer-related mortality in the UAE is 0.95 cases per 100,000 population per year, compared to the worldwide rate of 1.8 cases per 100,000 population per year [2]. The number of kidney cancer cases diagnosed in the UAE is low, and it is difficult to infer any conclusions; however, the widespread use of axial imaging led to an increase in the number of early-stage cases between 2013 and 2017.

Patients in the UAE have excellent access to diagnostic and treatment modalities, including sub-specialist expertise in laparoscopic and robotic surgery for localized disease. Patients have access to all recently approved systemic therapies, including targeted therapy and immune checkpoint inhibitors for use in adjuvant and advanced disease settings.

Recommendations for improving kidney cancer outcomes in the UAE are as follows:

1. Improved the data documentation process at the hospital level on kidney cancer diagnoses, risk factors, stage at diagnosis, and treatment outcome data.
2. Encourage multidisciplinary management of kidney cancer by sub-specialist teams to ensure optimal patient outcomes.
3. Early access to specialist palliative care for men with advanced disease.
4. Engagement in context-specific research with improved data collection on kidney cancer risk factors, treatment response, and patient preferences for treatment.

35.5 Testicular Cancer in the UAE

According to Global Cancer Statistics 2020: GLOBOCAN, the age-standardized rate (ASR) of new testicular cancer cases in the UAE is 0.49 cases per 100,000 population per year, while worldwide it is 1.8 cases per 100,000 population per year. Moreover, the ASR of kidney cancer-related mortality in the UAE is 0.02 cases per 100,000 population per year, compared to the worldwide rate of 0.22 cases per 100,000 population per year [2]. The incidence and mortality of testicular cancer in the UAE follow the global pattern of low incidence and mortality of testicular cancer observed in Asia [19]. Patients have access to fertility clinics both in the public and private sectors.

Recommendations for improving testicular cancer outcomes in the UAE are as follows:

1. Improved the data documentation process at the hospital level on testicular cancer diagnoses, risk factors, stage at diagnosis, and treatment outcome data.
2. Encourage multidisciplinary management of testicular cancer by sub-specialist teams to ensure optimal patient outcomes. Due to the rarity of the disease and excellent treatment outcomes with appropriate treatment, it is recommended that there be cross-institutional collaboration and discussion of all advanced cases.
3. Access to fertility centers and insurance coverage for this young patient population.
4. Engagement in context-specific research with improved data collection on treatment outcomes and long-term data to monitor rates of relapse.

35.6 Conclusion

The incidence of GU malignancies in the UAE is expected to rise significantly due to population and demographic change in the coming decade. There is an urgent need to establish population-based screening for prostate cancer and raise public awareness regarding the risk factors and symptoms of GU cancers. Other recommendations include strengthening the UAE cancer registry to include treatment outcome data, mechanisms to audit adherence to guidelines, multidisciplinary sub-specialist care for all patients regardless of location or financial resources, early access to specialist palliative care for patients with advanced disease, and the promotion of context-specific research collaboration.

Conflict of Interest The authors have no conflict of interest to declare.

References

1. Cancer incidence in United Arab Emirates, Annual Report of the UAE - National Cancer Registry. Statistics and Research Center, Ministry of Health and Prevention.

2. Sung H, Ferlay J, Siegel RL, Laversanne M, Soerjomataram I, Jemal A, Bray F. Global cancer statistics 2020: GLOBOCAN estimates of incidence and mortality worldwide for 36 cancers in 185 countries. CA Cancer J Clin. 2021;71(3):209–49.
3. Hilal L, Shahait M, Mukherji D, Charafeddine M, Farhat Z, Temraz S, Khauli R, Shamseddine A. Prostate cancer in the Arab world: a view from the inside. Clin Genitourin Cancer. 2015;13(6):505–11.
4. Abdelsalam RA, Khalifeh I, Box A, Kalantarian M, Ghosh S, Abou-Ouf H, Lotfi T, Shahait M, Palanisamy N, Bismar TA. Molecular characterization of prostate cancer in middle eastern population highlights differences with Western populations with prognostic implication. J Cancer Res Clin Oncol. 2020;146(7):1701–9.
5. Mohammed S. Prostate cancer management in the Middle East. World J Urol. 2020;38(8):2063–4.
6. Albawardi A, Livingstone J, Almarzooqi S, Palanisamy N, Houlahan KE, Awwad AA, Abdelsalam RA, Boutros PC, Bismar TA. Copy number profiles of prostate cancer in men of middle eastern ancestry. Cancers. 2021;13(10):2363.
7. Ali AH, Awada H, Nassereldine H, Zeineddine M, Sater ZA, El-Hajj A, Mukherji D. Prostate cancer in the Arab world: bibliometric review and research priority recommendations. Arab J Urol. 2022;20(2):81–7.
8. Mukherji D, Youssef B, Dagher C, El-Hajj A, Nasr R, Geara F, Rabah D, Al Dousari S, Said R, Ashou R, Wazzan W. Management of patients with high-risk and advanced prostate cancer in the Middle East: resource-stratified consensus recommendations. World J Urol. 2020;38(3):681–93.
9. Awad MA, Hallgarth L, Barayan GA, Shahait M, Abu-Hijlih R, Farkouh AA, Azhar RA, Alghamdi MM, Bugis A, Yaiesh S, Aldousari S. Creation and validation of the harmonized Arabic version of the expanded prostate cancer index composite for clinical practice (EPIC-CP). Arab J Urol. 2022;20(2):88–93.
10. Bray F, Ferlay J, Soerjomataram I, Siegel RL, Torre LA, Jemal A. Global cancer statistics 2018: GLOBOCAN estimates of incidence and mortality worldwide for 36 cancers in 185 countries. CA Cancer J Clin. 2018;68(6):394–424.
11. Najjar H, Easson A. Age at diagnosis of breast cancer in Arab nations. Int J Surg. 2010;8:448–52. https://doi.org/10.1016/j.ijsu.2010.05.012.
12. AlZaabi A. Colorectal cancer in the Arab world. In: Cancer in the Arab World. Singapore: Springer; 2022. p. 363–79.
13. Wollin T, Laroche B, Psooy K. Canadian guidelines for the management of asymptomatic microscopic hematuria in adults. Can Urol Assoc J. 2009;3(1):77–80.
14. Teleka S, et al. Association between blood pressure and BMI with bladder cancer risk and mortality in 340,000 men in three Swedish cohorts. Cancer Med. 2021;10(4):1431–8.
15. Hemminki K, et al. Incidence trends in bladder and lung cancers between Denmark, Finland and Sweden may implicate oral tobacco (snuff/snus) as a possible risk factor. BMC Cancer. 2021;21(1):1–7.
16. Sulaiman N, Elbadawi S, Hussein A, Abusnana S, Madani A, Mairghani M, Alawadi F, Sulaiman A, Zimmet P, Huse O, Shaw J, Peeters A. Prevalence of overweight and obesity in United Arab Emirates expatriates: the UAE National Diabetes and lifestyle study. Diabetol Metab Syndr. 2017;9:88. https://doi.org/10.1186/s13098-017-0287-0.
17. Commar A, et al. WHO global report on trends in prevalence of tobacco smoking 2000-2025. In: WHO global report on trends in prevalence of tobacco smoking 2000-2025. 2nd ed; 2018.
18. Razzak HA, Harbi A, Ahli S. Tobacco smoking prevalence, health risk, and cessation in the UAE. Oman Med J. 2020;35(4):e165. https://doi.org/10.5001/omj.2020.107.
19. Huang J, et al. Worldwide distribution, risk factors, and temporal trends of testicular cancer incidence and mortality: a global analysis. Eur Urol Oncol. 2022;5:566.

Dr. Mohammed Shahait is a doctor of urology and minimally invasive surgery with extensive experience in the management of complex genitourinary tumors and robotic surgery. Dr.Shahait has performed the first robotic surgery in the Kingdom of Jordan. After receiving his medical degree from the Jordanian University of Science and Technology, he completed his residency at the division of urology and kidney transplant at the American University of Beirut Medical Center (AUBMC). Dr. Shahait afterwards completed a 2-year fellowship in Endourology laparoscopic and robotic surgery at the University of Pittsburgh Medical Center, followed by a prestigious 1-year fellowship in advanced robotic surgery at the University of Pennsylvania. He later joined the King Hussein Cancer Center in Amman, Jordan as a consultant in uro-oncology. Dr. Shahait's research focus is on the outcomes of patients undergoing robot-assisted radical prostatectomy for prostate cancer and has more than 45 peer-reviewed articles.

Dr. Hosam Al-Qudah is a consultant urologist with fellowship training in the United States of America. He has been on the Jordanian and Arab Boards of Urology since 2003 and 2004.

He started his residency training in 1998 at the Jordan University of Science and Technology and the American University Hospital of Beirut. After that, he worked in private practice for one year and then moved to the United States of America to receive subspecialty training in reconstructive urology from Wayne State University from July 2004 until June 2005., In July 2005, he moved to do transplant surgery at the University of Maryland, which is one of the pioneering programs in transplant, especially in laparoscopic donor nephrectomy. After that, he did his third fellowship in uro-oncology and advanced laparoscopy at Moffitt Cancer Center in Tampa, Florida, from July 2006 until June 2007.

After finishing this training, he was appointed as an assistant professor of urology at the Jordan University of Science and Technology (King Abdullah University Hospital). At the end of 2008, he moved to Saudi Arabia and worked as a consultant urologist and transplant surgeon at Saad Specialist Hospital for almost 8 years. During these years, he was the surgeon in charge of the transplant service in the hospital.

On July 24th, 2016, he joined Alzahra Hospital in Dubai as head of the division of urology and consultant urologist. During these years, he has performed a good number of complicated oncology and laparoscopic cases.

On June 1st, 2021, he moved to Fakeeh University Hospital to lead the urology service in this hospital. It is a state-of-the-art new hospital that aims to raise health standards in Dubai.

He did the first laparoscopic donor nephrectomy, the first laparoscopic partial nephrectomy, and the first laparoscopic radical prostatectomy in Jordan.

Dr. Layth Mula-Hussain was born, raised and gained his medical degree from the University of Mosul, Ninevah – Iraq. He travelled to Jordan to do a residency in radiation oncology at the King Hussein Cancer Center, then to Germany to do an M.Sc. in advanced oncology at Ulm University. Then he completed four years of clinical fellowships in Canada. Besides many professional memberships, he is a lifetime ESTRO Fellow & IAHPC member, ESCO Graduate, Certified Clinical Investigator, Fellow and the "Regional Adviser for Eastern Canada" of the Royal College of Physicians of Edinburgh, and an ESO Ambassador by the European School of Oncology.

During his 20+ years in oncology, Dr. Mula-Hussain served as a consultant physician in radiation oncology at the King Hussein Cancer Center in Jordan, Zhianawa Cancer Center in Iraq, and Sultan Qaboos Comprehensive Cancer Centre in Oman. He is currently an attending physician at the Cape Breton Cancer Centre, an assistant professor at Dalhousie University in Nova Scotia, Canada, and a visiting professor at Ninevah University in Iraq.

Dr. Mula-Hussain was the founding director of Iraq's first radiation oncology certification board program (2013–2017). He acted as an Expert within imPACT IAEA teams for Pakistan (2013) & Syria (2022), a member of the ASCO International Affairs Steering Committee (2016–2019), and a reviewer in the IAEA curriculum for radiation oncology education and training (2024). He authored / co-authored 80+ manuscripts/ books/ books' chapters, did 100+ scientific presentations, and his efforts were cited over a thousand times with an H-index of 17 "https://www.researchgate.net/profile/Layth-Mula-Hussain".

Dr. Ibrahim H. Abu-Gheida is the Clinical Director of the department of radiation oncology at Burjeel Medical City. He also serves as a regional Radiological Society of North America (RSNA) committee representative for the Middle East and Africa. Dr. Abu-Gheida completed his undergraduate training, where he earned a Bachelor of Science with honors degree from the American University of Beirut. Following this, Dr. Abu-Gheida completed his Medical School training at the American University of Beirut Medical Center. He continued and joined the Department of Internal Medicine at the American University of Beirut. Then he did his training in the Department of Radiation Oncology at the American University of Beirut Medical Center, where he also served as the chief resident. During his training, Dr. Abu-Gheida completed a Harvard-affiliated NIH-funded research program as well. After his residency, Dr. Ibrahim went to the Cleveland Clinic in Ohio, where he was appointed as an Advanced Clinical Radiation Oncology Fellow. Dr. Abu-Gheida went and joined the University of Texas MD Anderson Cancer Center, where he sub-specialized in treating breast, gastrointestinal, and genitourinary cancers. Dr. Abu Gheida played an instrumental role in establishing and heading the radiation oncology facility and department at Burjeel Medical City. He has chaired and co-chaired multiple international oncology conferences. Dr. Ibrahim has more than 40 peer-reviewed papers in prestigious medical journals, including the American Society of Radiation Oncology official journal - the International Journal of Radiation Oncology Biology and Physics, Nature, the Journal of Clinical Oncology, and several others. He is also the primary author and editor of several book chapters published in prestigious books.

Dr. Thamir Alkasab is a distinguished uro-oncologist at Al Zahra Hospital Dubai. He graduated as a uro-oncologist from Princess Margret Cancer Center/University of Toronto in 2016. Renowned for his expertise in urological cancer surgery, he specializes in using robotic/laparoscopic tools for precise interventions. Dr. Thamir has authored many publications in the uro-oncology field. His commitment to advancing treatment options and improving outcomes for cancer patients underscores his dedication to excellence in healthcare.

Dr. Ali Thwaini is a consultant urologist in Dubai with a special interest in urological cancers, namely renal cancers. He was the lead for renal cancers in the Belfast health and social care trust (Belfast City Hospital). His main skills are in advanced laparoscopy, renal cancer, and renal reconstruction procedures. He is also an adjunct clinical professor at Mohammed bin Rashid University in Dubai and an honorary clinical lecturer at Queen's University, Belfast.

Dr. Rabii Madi is a urologic oncologist with 20 years of academic and clinical experience in urologic oncology and minimally invasive surgery. He is serving as the Director of Urology and Robotic Surgery at CMC-Dubai. He was the director of robotic surgery at Case Western Reserve in Cleveland before moving to Georgia and heading the urologic oncology and robotic surgery programs at Augusta University Health for 9 years.

Dr. Madi is among the most experienced robotic surgeons in the USA. He has performed more than 2300 robotic surgeries and has more than 50 peer-reviewed publications and 120 presentations in national and international meetings. He has pioneered new techniques and approaches in urologic oncology and robotic surgery.

Prof. Humaid Obaid Al-Shamsi is the Chief Executive Officer of Burjeel Cancer Institute in Abu Dhabi, UAE, President of the Emirates Oncology Society, Lead of the Gulf Cancer Society, Full Professor of Oncology at the Ras Al Khaimah Medical and Health Sciences University, Ras Al Khaimah, UAE, and an Adjunct Professor of Oncology at the College of Medicine, University of Sharjah. He is the first Emirati to be promoted as a professor in oncology in the UAE. He is also the Chairman for Colorectal Cancer in the MENA region, appointed by the prestigious National Comprehensive Cancer Network®. He is also the only member of Lung Cancer Policy Network in the MENA region that aims to advance lung cancer research and screening globally. He is the Chairman of the Oncology and Hematology Fellowship Training Program for the National Institute for Health Specialties in the United Arab Emirates. He is the only member in GCC in the WIN Consortium which is comprised of organizations representing all stakeholders in personalized cancer medicine globally.

He is board-certified in both internal medicine and oncology from the UK, USA (ABIM), and Canada (FRCPC). He has also been awarded the FRCP (London) in 2023 and FRCP (Glasgow) in 2024. He is the only physician in the UAE with a subspecialty fellowship certification and training in gastrointestinal oncology and the first Emirati to train and complete a clinical post-doctoral fellowship in palliative care. He was an assistant professor at the University of Texas MD Anderson Cancer Center between 2014 and 2017. He has published more than 140 peer-reviewed articles in JAMA Oncology, Lancet Oncology, The Oncologist, BMC Cancer, and many others. His area of expertise includes precision oncology and cancer care in the UAE. In 2016, he published with his group from MD Anderson the JCO paper describing a new distinct subgroup of CRC, NON V600 BRAF-mutated CRC. In 2022, he published the first book about cancer research in the UAE and also the first book about cancer in the Arab world, both of which were launched at Dubai Expo 2020. *Cancer in the Arab World* has been downloaded more than 450,000 times in its first 18 months of publication and is the ultimate source of cancer data in the Arab region. He also published the first comprehensive book about cancer care in the UAE which is the first book in UAE history to document the cancer care in the UAE with many topics addressed for the first time, e.g., neuroendocrine tumors in the UAE. He is passionate about advancing cancer care in the UAE and the GCC and has made significant contributions to cancer awareness and early detection for the public using social media platforms. He is considered as the most followed oncologist in the world with over 300,000 subscribers across his social media platforms (Instagram, Twitter, LinkedIn, and TikTok). In 2022, he was awarded the prestigious Feigenbaum Leadership Excellence Award from Sheikh Hamdan Smart University for his exceptional leadership and research and the Sharjah Award for Volunteering. He was also named the Researcher of the Year in the UAE in 2020 and 2021 by the Emirates Oncology Society.

In May 2024, HH Sheikh Mansour bin Zayed Al Nahyan, Vice President of the United Arab Emirates, awarded him the first place in UAE Nafis program for outstanding leadership in private sector across all business and medical disciplines. Beside his clinical and administrative duties, he is engaged in education and various levels of research training for medical trainees to enhance their clinical and research skills. His mission is to advance cancer care in the UAE and the MENA region and make cancer care accessible to everyone in need around the globe.

Dr. Syed Hammad Tirmazy is a consultant clinical oncologist and head of oncology at Dubai Hospital, DHA. He has substantive experience in managing adult solid cancers and has been involved in multiple international clinical trials. Before moving to Dubai in 2016, he worked as a consultant clinical oncologist in the West Midlands region, UK. After completing his graduation, he moved to the UK in 2003. He initially completed internal medicine training and MRCP and subsequently attained training in clinical oncology and worked in internationally renowned oncology centers in the UK, including Queen Elizabeth Hospital Birmingham. He has a master's in clinical oncology from the University of Birmingham, an FRCR in Clinical Oncology from RCR London, and a CCT in Clinical Oncology. He also attained FRCP Glasgow.

Dr. Deborah Mukherji completed her advanced specialty training in Medical Oncology at Guys and St Thomas's NHS Foundation Trust, London UK and was awarded a Post-Graduate Diploma in Oncology from the Institute of Cancer Research, University of London, in 2011. On completion of her advanced specialty training, Dr. Mukherji joined the Royal Marsden Hospital London specializing in clinical research and drug development with a focus on prostate cancer. Dr. Mukherji joined the faculty of the American University of Beirut, Lebanon, in June 2012 and is currently an associate professor of clinical medicine. She joined Clemenceau Medical Center Dubai, UAE as a consultant medical oncologist in October 2022.

Sarcoma in the UAE

36

Aydah Al-Awadhi [iD] and Philipp Berdel

36.1 Background

According to the World Health Organization (WHO), soft tissue sarcoma is an uncommon and rare type of malignancy in adults, accounting for approximately 2% of all malignancies (primarily malignant bone tumors account for only 0.2% in adults and more than 3% in children and adolescents). By definition, primary bone and soft tissue sarcomas are non-epithelial malignancies. These are characterized by rapid, locally infiltrative growth and show a high risk of metastasis. The most common primary malignant bone tumors are osteosarcomas (approximately 35%), chondrosarcomas (25%), and tumors of the Ewing sarcoma group (16%) [1, 10].

According to the 2020 WHO Classification of Tumors—Soft Tissue and Bone Tumors, bone sarcoma (BS) and soft tissue sarcoma (STS) encompass roughly 100 distinct pathologic entities, many of which are ultrarare (incidence 1 per million) [2, 3]. Sarcomas are uncommon, and as a result, they frequently lack information about their epidemiology, biology, natural history, prognostic and predictive indicators, and sensitivity to standard treatment. This makes diagnosis and clinical decision-making difficult.

A. Al-Awadhi (✉)
Emirates Oncology Society, Emirates Medical Association, Dubai, United Arab Emirates

Tawam Hospital, Al Ain, United Arab Emirates
e-mail: ayawadhi@seha.ae

P. Berdel
Tawam Hospital / STMC, Al Ain, United Arab Emirates

United Arab Emirates University, Al Ain, United Arab Emirates
e-mail: phberdel@seha.ae

© The Author(s) 2024
H. O. Al-Shamsi (ed.), *Cancer Care in the United Arab Emirates*,
https://doi.org/10.1007/978-981-99-6794-0_36

36.2 Burden of Sarcoma in the UAE

As expected, due to the rarity of the disease, there is limited information on STS and bone sarcomas in the United Arab Emirates (UAE). In this chapter, we aim to shed light on all available published data on sarcoma in the UAE and discuss available services for STS and bone tumors, as well as the unmet needs.

According to the UAE National Cancer Registry Report (UAE-NCR) of 2021, connective and soft tissue sarcomas in adult patients accounted for 47 cases out of the total 5612 cancer cases diagnosed in 2021 across the country, representing 0.83% of all malignancies [4]. In the same report, bone and cartilage sarcomas were documented to account for 34 cases out of all cancer cases in 2021, accounting for 0.6% of all cancer cases. In the pediatric population, UAE-NCR reported 4.5% of bone and articular cartilage sarcomas.

Both bone (BS) and soft tissue sarcomas (STS) should ideally be treated at specialized facilities from the start because they require a comprehensive and sophisticated therapeutic strategy involving pathologists, radiologists, specialized surgeons, radiation oncologists, and medical oncologists [5].

Despite recent advances in pathology, there are still diagnostic difficulties that must be acknowledged. Given the rarity of the disease, the significant forms of sarcomas with dynamically changing nomenclature, the morphological heterogeneity within the same class of sarcoma and different types of sarcomas, the pathological overlap between benign and musculoskeletal tumors, and the availability of adequate biopsy material for diagnosis, the task of pathology in sarcoma diagnosis is complex [6]. Research on sarcomas has advanced significantly, and molecular techniques such as next-generation sequencing have been utilized to identify the subtypes of sarcomas [7, 8]. However, these molecular methods are not available in the UAE and are often sent by large cancer centers abroad to be tested. A second opinion from an anatomic pathologist specializing in musculoskeletal tumors should be required prior to making a final treatment decision [7]. For diagnostic, predicative, and prognostic purposes, specialized pathology centers are beginning to incorporate molecular diagnostics into their histopathological reports. Since there is no highly specialized sarcoma pathologist available in the UAE, many centers have collaborated with large centers abroad, mainly in the United States, for a second pathologist's opinion [9].

To guarantee that all patients should receive the best care possible from a multidisciplinary team at a specialized facility. Usually, highly specialized orthopedic oncology surgeons operate on patients with bone and soft tissue sarcomas, and only a few with this expertise are available in the UAE, like in Tawam Hospital, the largest tertiary cancer center in the UAE. For the resection of sarcomas, a wide resection (R0) with a resection margin outside the reactive zone of the tumor (a safety margin within the cancellous bone of 3 cm) and entrainment of the existing biopsy channel are required [10].

Also, only a few oncologists have dedicated, specialized sarcoma training and interest. Therefore, strong cooperation within the community of physicians treating sarcoma patients outside and within a sarcoma center is needed. The reconstruction of the bones and joints after bone sarcoma resection is usually done by the implantation of megaprostheses. It is worth mentioning that in some highly specialized centers, the post-operative reconstruction of the bony defect can be done with recycled bone grafts, e.g., by extracorporeal irradiation (ECI) or liquid nitrogen, when appropriate for the management of primary malignant bone tumors. Usually, National Comprehensive Cancer Network (NCCN) guidelines direct the management of STS and bone sarcomas in the UAE, and when chemotherapy or targeted agents are needed, the standard regimens per National Comprehensive Cancer Network (NCCN) guidelines are utilized. High-dose ifosfamide of more than 12 g/m^2 with mesna per dose is rarely utilized, partly due to a lack of experience with toxicity.

36.3 Conclusion

In conclusion, there is a great unmet need for expertise in the management of different types of sarcomas. Therefore, strong cooperation among the physicians treating sarcoma patients outside and within a sarcoma center is needed.

Conflict of Interest The authors have no conflict of interest to declare.

References

1. Siegel RL, et al. Cancer statistics, 2022. CA Cancer J Clin. 2022;72(1):7–33.
2. Miller KD, et al. Cancer statistics for adolescents and young adults, 2020. CA Cancer J Clin. 2020;70(6):443–59.
3. Italiano A, et al. Clinical effect of molecular methods in sarcoma diagnosis (GENSARC): a prospective, multicentre, observational study. Lancet Oncol. 2016;17(4):532–8.
4. Cancer incidence in United Arab Emirates, Annual Report of the UAE - National Cancer Registry, 2021. Statistics and Research Center, Ministry of Health and Prevention. (Accessed on 28 Mar 2024).
5. Siegel GW, et al. The multidisciplinary management of bone and soft tissue sarcoma: an essential organizational framework. J Multidiscip Healthc. 2015;8:109–15.
6. Liegl-Atzwanger B. The role of pathology in sarcoma. Memo - Mag Eur Med Oncol. 2020;13(2):159–63.
7. Fletcher C, et al. WHO classification of tumours of soft tissue and bone: WHO classification of tumours, vol. 5. World Health Organization; 2013.
8. Szurian K, Kashofer K, Liegl-Atzwanger B. Role of next-generation sequencing as a diagnostic tool for the evaluation of bone and soft-tissue tumors. Pathobiology. 2017;84(6):323–38.
9. Randall RL, et al. Errors in diagnosis and margin determination of soft-tissue sarcomas initially treated at non-tertiary centers. Orthopedics. 2004;27(2):209–12.
10. Berdel P. In: Ruchholtz S, Wirtz DC, editors. Orthopädie und Unfallchirurgie essentials. Intensivkurs zur Weiterbildung. 4th ed. Stuttgart New York: Georg Thieme Verlag; 2021. p. 169–218.

Dr. Aydah Al-Awadhi is a consultant medical oncologist who graduated from medical school at the United Arab Emirates University in 2012, then completed her internal medicine residency in the United States and completed her medical oncology and hematology fellowship at the University of Texas, MD Anderson Cancer Center, in 2019. She is American Board certified in internal medicine, hematology, and medical oncology. She was awarded the Emirates Oncology Society (EOS) Member of the Year Award for 2021 and the Emirates Oncology Society Women in Oncology of the Year Award. She is also currently the scientific chairperson of the Emirates Oncology Society and the chair of the Breast Cancer Working Group. Dr. Al-Awadhi has published many articles and book chapters on breast cancer and oncology, with her main expertise in breast cancer and sarcoma.

Prof. Dr. Philipp Berdel FRCS received his medical degree at the University of Bonn in Germany and then started his residency as an orthopaedic surgeon in 1997 in the Department of Orthopaedic and Trauma Surgery of the University Hospital Aachen, Germany.

After finishing his training as an orthopaedic surgeon in 2005 and fellowships in orthopaedic rheumatology (2006) and pediatric orthopaedic surgery (2008), he received training authorization in both fields in 2008 and 2010, respectively.

In 2011, he reached the highest level of orthopaedic training in Germany, Advanced Orthopaedic Procedures, from the Medical Board Nordrhein, Germany.

Since 2013, he has lived and worked in Al Ain, Abu Dhabi, UAE.

In 2017, he became one of only 45 certified orthopedic oncology surgeons on the German Medical Board (out of 13.400 orthopedic surgeons).

Since 2017, he has been an adjunct professor in the Department of Surgery, College of Medicine and Health Science, United Arab Emirates University (UAEU); since February 2021, he has been a full professor.

He is a member of the German Orthopaedic Society (DGOOC) and of its scientific working group for orthopaedic oncology. On May 1, 2015, he joined the Emirates Medical Association/Emirates Orthopaedic Society as well.

He is a Fellow of the Royal College of Surgeons of England (FRCS) and holds the German, GMC (UK), DOH (Abu Dhabi), and MOH (Oman) licenses.

Neuroendocrine Tumors (NETs) in the UAE

37

Aydah Al-Awadhi ⓘ and Humaid O. Al-Shamsi ⓘ

37.1 Introduction

Neuroendocrine neoplasms (NEN) are uncommon and can develop anywhere in the body, most commonly in the gastrointestinal tract, lung, and pancreas [1, 2]. Although most neuroendocrine tumors (NETs) are neuroendocrine tumors (NETs) with a slow-progressing disease biology, 10–20% of NENs are neuroendocrine carcinomas (NECs), which are highly proliferative tumors. Neuroendocrine tumors occur at a rate of 2.5 to 5 per 100,000 per year [3].

The approach to neuroendocrine neoplasms usually includes extensive clinical assessment, both clinically and chemically, radiological imaging, nuclear scans, and pathological evaluation to reach a definitive diagnosis and treatment plan.

A. Al-Awadhi
Emirates Oncology Society, Emirates Medical Association, Dubai, United Arab Emirates
e-mail: ayawadhi@seha.ae

H. O. Al-Shamsi (✉)
Burjeel Cancer Institute, Burjeel Medical City, Burjeel Holdings, Abu Dhabi, United Arab Emirates

Ras Al Khaimah Medical and Health Sciences University, Ras Al Khaimah, United Arab Emirates

Gulf Medical University, Ajman, United Arab Emirates

Emirates Oncology Society, Emirates Medical Association, Dubai, United Arab Emirates

College of Medicine, University of Sharjah, Sharjah, United Arab Emirates

Gulf Cancer Society, Alsafa, Kuwait
e-mail: alshamsi@burjeel.com; humaid.al-shamsi@medportal.ca

37.2 NETs Prevalence in the UAE

In a survey conducted by Al-Shamsi et al. through the Emirates Oncology Society (EOS) in September 2021, this is the first survey to assess the current burden of NET in the United Arab Emirates (UAE). The survey was distributed to oncologists, surgeons, and endocrinologists through direct emails and WhatsApp groups. The estimated number of invited participants was 110 physicians who potentially treat neuroendocrine tumors. The survey aimed to get a snapshot of the status and burden of NET in the UAE at the time of the survey, which may also reflect the overall burden of NET in the UAE.

Forty-three respondents (39%) completed the survey: 31 were medical oncologists, 3 were general surgeons, 4 were radiation oncologists, 2 were endocrinologists, and only 1 was a gastroenterologist.

Thirty-one respondents (72.1%) had active patients with neuroendocrine tumors at the time of the survey. Sixteen (37.2%) had between 1 and 3 NET patients at the time of the survey, and 11 (25.6%) had 4–6 NET patients. Six (14%) had seven to nine patients, and only four had more than ten patients, which may reflect the referral patterns to some oncologists or centers.

Thirty-one respondents (73.8%) indicated that GI NET was the most common NET in their practice. Six (14.3%) respondents selected lung and two (4.8%) selected gynecological NETs as the most common NETs in their practices.

The pancreas was the top answer for the most common gastro-entero-pancreatic NET, with pancreas 22 (30.6%), stomach 17 (23.6%), small intestine 13 (18.1%), appendix 12 (16%), and lastly, colon 8 (11.1%). According to the responses, the following NETs are the least common in their practice, with the rectum being the least common, followed by the stomach, and then the colon.

Forty-one (95.3%) of the respondents agreed that general practitioners, gastroenterologists, and general surgeons need more education about NETs to better diagnose NETs. 31 (72.1%) reported delays in the diagnosis of their NET patients.

The most used modalities in imaging NETs were Dota-PET with 26 (29.2%), followed by octreotide scan with 22 (24.7%), CT with 17 (19.1%), and PET/CT with 14 (15.7%). Twenty-three (53.5%) respondents reported ordering chromogranin and 5-HIAA as their initial workup for NET patients. Twenty-two (51.2%) agreed that the number of NET patients is increasing over time in their practice.

Thirty-five (81.4%) reported not having access to the NET multidisciplinary tumor board (MDT) in their hospital or practice. 23 respondents (54.8%) reported they needed more experience using peptide receptor radionuclide therapy (PRRT) for the management of NETs.

Unfortunately, there is no data to illustrate the prevalence of NEN in the UAE. An IRB-approved study conducted in a single high-volume center (Tawam hospital) found that from 2010 to 2020, approximately 149 patients were diagnosed with NEN, the majority of whom originated in the GI tract ($N = 67$), followed by the lung ($N = 43$), pancreas ($N = 23$), cervix ($N = 9$), and others with unknown primary origin ($N = 7$) [4]. Interestingly, most of the patients were also metastatic at presentation

(102/149). The median age at diagnosis was 56. Females comprised 51 patients of the total population, whereas men were 98. Interestingly, a good number of patients presented and were diagnosed with carcinoid syndrome ($N = 63$) [4].

Also, in another study conducted by the University of Sharjah and published in 2014 that investigated the clinicopathological characteristics of appendiceal carcinoid in Sharjah, out of the 964 patients who underwent surgery for acute appendicitis from January 2010 through December 2010 in a single center, 9 were found to have an appendiceal carcinoid [5]. The median age is 28.7 years, with male incidence being higher than female incidence. The incidence reported in this study is apparently higher than that reported from other regions in the Gulf countries [5].

In terms of diagnostic workup, most of the modalities are available in large cancer centers, including octreotide SPECT, CT, and SSTR-PET. Also, peptide receptor radionuclide therapy (PRRT) with 177 Lu-DODATE is available in limited centers in the UAE. Other liver-directed modalities, including TACE and TARE, are available as well, but in fewer centers based on the availability of experienced interventional radiologists.

Most of the medications indicated for use in metastatic neuroendocrine tumors or carcinomas are available in the UAE, from chemotherapy to targeted medications and hormonal therapy. There is, however, a high unmet need in terms of public awareness of this tumor, which the Emirates Oncology Society (EOS), in collaboration with different companies and institutions, is working hard to shed light on. For example, EOS and Ipsen, which is a global biopharmaceutical company, have created the largest awareness ribbon displaying the zebra print associated with the condition's awareness activity, which measures 4.8 m^2, breaking the previous record.

Finally, more research is needed to better understand the prevalence and characteristics of NEN in the UAE, as well as to raise public awareness of this disease entity, its implications, and its clinical presentation.

37.3 Conclusion

Despite their increasing incidence, NETs are considered to be rare tumors. There is no published data about the prevalence of NETs in the UAE. In a survey in 2021 for oncologists in the UAE, 43 respondents completed the survey. Thirty-one respondents (72.1%) had active patients with neuroendocrine tumors at the time of the survey. Thirty-one respondents (73.8%) indicated that GI NET was the most common NET in their practice. Six respondents (14.3%) selected lung and two (4.8%) selected gynecological NETs as the most common NETs in their practices. This is the first study to address the potential burden of NETs in the UAE. More education for family physicians, endocrinologists, and gastroenterologists in the UAE is needed to facilitate early diagnosis. More research is needed to assess the burden of NET in the UAE.

Conflict of Interest The authors have no conflict of interest to declare.

References

1. Yao JC, Hassan M, Phan A, Dagohoy C, Leary C, Mares JE, et al. One hundred years after "carcinoid": epidemiology of and prognostic factors for neuroendocrine tumors in 35,825 cases in the United States. J Clin Oncol. 2008;26(18):3063–72.
2. Sorbye H, Strosberg J, Baudin E, Klimstra DS, Yao JC. Gastroenteropancreatic high-grade neuroendocrine carcinoma. Cancer. 2014;120(18):2814–23.
3. Oronsky B, Ma PC, Morgensztern D, Carter CA. Nothing but NET: a review of neuroendocrine tumors and carcinomas. Neoplasia. 2017;19(12):991–1002.
4. Cancer Registry, Tawam Hospital, SEHA, Al Ain, United Arab Emirates.
5. Anwar K, Desai M, Al-Bloushi N, Alam F, Cyprian FS. Prevalence and clinicopathological characteristics of appendiceal carcinoids in Sharjah (United Arab Emirates). World J Gastrointest Oncol. 2014;6(7):253–6.

Dr. Aydah Al-Awadhi is a consultant medical oncologist who graduated from medical school at the United Arab Emirates University in 2012, then completed her internal medicine residency in the United States, and completed her medical oncology and hematology fellowship at the University of Texas, MD Anderson Cancer Center, in 2019. She is American Board-certified in internal medicine, hematology, and medical oncology. She was awarded the Emirates Oncology Society (EOS) Member of the Year Award for 2021 and the Emirates Oncology Society Women in Oncology of the Year Award. She is also currently the scientific chairperson of the Emirates Oncology Society and the chair of the Breast Cancer Working Group. Dr. Al-Awadhi has published many articles and book chapters on breast cancer and oncology, with her main expertise in breast cancer and sarcoma.

Prof. Humaid Obaid Al-Shamsi is the Chief Executive Officer of Burjeel Cancer Institute in Abu Dhabi, UAE, President of the Emirates Oncology Society, Lead of the Gulf Cancer Society, Full Professor of Oncology at the Ras Al Khaimah Medical and Health Sciences University, Ras Al Khaimah, UAE, and an Adjunct Professor of Oncology at the College of Medicine, University of Sharjah. He is the first Emirati to be promoted as a professor in oncology in the UAE. He is also the Chairman for Colorectal Cancer in the MENA region, appointed by the prestigious National Comprehensive Cancer Network®. He is also the only member of Lung Cancer Policy Network in the MENA region that aims to advance lung cancer research and screening globally. He is the Chairman of the Oncology and Hematology Fellowship Training Program for the National Institute for Health Specialties in the United Arab Emirates. He is the only member in GCC in the WIN Consortium which is comprised of organizations representing all stakeholders in personalized cancer medicine globally.

He is board-certified in both internal medicine and oncology from the UK, USA (ABIM), and Canada (FRCPC). He has also been awarded the FRCP (London) in 2023 and FRCP (Glasgow) in 2024. He is the only physician in the UAE with a subspecialty fellowship certification and training in gastrointestinal oncology and the first Emirati to train and complete a clinical post-doctoral fellowship in palliative care. He was an assistant professor at the University of Texas MD Anderson Cancer Center between 2014 and 2017. He has published more than 140 peer-reviewed articles in JAMA Oncology, Lancet Oncology, The Oncologist, BMC Cancer, and many others. His area of expertise includes precision oncology and cancer care in the UAE. In 2016, he published with his group from MD Anderson the JCO paper describing a new distinct subgroup of CRC, NON V600 BRAF-mutated CRC. In 2022, he published the first book about cancer research in the UAE and also the first book about cancer in the Arab world, both of which were launched at Dubai Expo 2020. *Cancer in the Arab World* has been downloaded more than 450,000 times in its first 18 months of publication and is the ultimate source of cancer data in the Arab region. He also published the first comprehensive book about cancer care in the UAE which is the first book in UAE history to document the cancer care in the UAE with many topics addressed for the first time, e.g., neuroendocrine tumors in the UAE. He is passionate about advancing cancer care in the UAE and the GCC and has made significant contributions to cancer awareness and early detection for the public using social media platforms. He is considered as the most followed oncologist in the world with over 300,000 subscribers across his social media platforms (Instagram, Twitter, LinkedIn, and TikTok). In 2022, he was awarded the prestigious Feigenbaum Leadership Excellence Award from Sheikh Hamdan Smart University for his exceptional leadership and research and the Sharjah Award for Volunteering. He was also named the Researcher of the Year in the UAE in 2020 and 2021 by the Emirates Oncology Society.

In May 2024, HH Sheikh Mansour bin Zayed Al Nahyan, Vice President of the United Arab Emirates, awarded him the first place in UAE Nafis program for outstanding leadership in private sector across all business and medical disciplines. Beside his clinical and administrative duties, he is engaged in education and various levels of research training for medical trainees to enhance their clinical and research skills. His mission is to advance cancer care in the UAE and the MENA region and make cancer care accessible to everyone in need around the globe.

Shahrukh Hashmi (ORCID)

38.1 Introduction

Hematologic malignancies, like solid tumors, are an extremely heterogeneous group of diseases. Apart from chemotherapy, immunotherapy, surgery, and radiation therapy, hematologic malignancies have an added arm of cellular therapies (bone marrow transplant, CAR-T cells, and gene therapy), making the patient's journey very complex. The classification of hematologic malignancies is also very problematic, as many different systems exist concurrently and different countries utilize different classifications, which in turn affects the overall national statistics of diseases with respect to incidence rate. Some of the systems existents are the Gall and Mallory classification, Rappaport classification, Kiel classification, Lukes-Collins classification, Working Formulation, Revised European-American classification (REAL), French-American-British system, World Health Organization (WHO) classification 2001, and World Health Organization (WHO) classification 2008, revised in 2016. Though earlier classification systems were based on tissue architecture and the cytologic appearances of neoplastic cells, the current era of genomics is driving the newer classification systems.

Since there are more than 200 countries in the world, it is imperative to know which classification is being used in a particular region. In 2016, the WHO classification was updated and expanded upon the use of objective diagnostic criteria (e.g., particularly for genomics) for clearly defined entities and is now widely accepted in some countries. However, unfortunately, it takes years for the diffusion of

S. Hashmi (✉)
Mayo Clinic, Rochester, MN, USA

Department of Health, Abu Dhabi, United Arab Emirates

Khalifa University, Abu Dhabi, United Arab Emirates
e-mail: hashmi.shahrukh@mayo.edu

© The Author(s) 2024
H. O. Al-Shamsi (ed.), *Cancer Care in the United Arab Emirates*,
https://doi.org/10.1007/978-981-99-6794-0_38

innovation or updated practice in healthcare, and thus one needs to be careful when comparing the epidemiology of hematologic malignancies if different classification systems are being used.

38.2 UAE Statistics

Bearing in mind the critically important differences in the classification of hematologic malignancies, let's focus on the paradigm of hematologic malignancies in the United Arab Emirates (UAE), which is a country of approximately ten million people. Fortunately, the data in the UAE on hematologic malignancies is assimilated based on the latest WHO classification, and the last published data was the sixth annual report of the UAE National Cancer Registry, in which the UAE National Cancer Registry (UAE-NCR) records demographic, cancer, staging, clinical, and treatment information for all cancers diagnosed in the UAE in accordance with internationally accepted registration and coding standards, which are WHO-based unless otherwise specified [1–3]. For UAE and non-UAE citizens in the country, all malignant and in situ cases diagnosed in the UAE during 1 January–31 December 2021 were notified and registered to the UAE National Cancer Registry [1].

There is a striking difference in the incidence rates of invasive cancers between the North American statistics and those of the UAE. The ratio of solid cancers to hematologic malignancies is tilted, with a higher proportion of hematologic malignancies in the UAE. As per the 2021 data of the UAE National Cancer Registry by the Ministry of Health and Prevention (MOHAP), leukemias and non-Hodgkin's lymphomas (NHL) are the fourth and eigth leading cancers in the UAE, respectively. This contrasts with the data from the registry in the United States, where only one hematologic malignancy is in the top 10 cancers [1, 2].

This trend of a higher proportion of hematologic malignancies in the UAE is also found in geographically nearby locations, e.g., in Saudi Arabia, where the last report by the Saudi Cancer Registry (2016) indicates the presence of the NHL and leukemias as the fifth and sixth most common malignancies, respectively.

The current data cannot explain why there is a disparity in a higher incidence of hematologic incidence.

38.3 Healthy Worker Effect in the UAE with Respect to Hematologic Malignancies

Median age (in years) in the UAE is 32.8. The main reason is that 65% of the population is between 25 and 54 years old and represents mainly the workers coming from different countries for work and the majority of them return to their native countries at the time of retirement. Thus, traditional methods of projecting cancer incidence rates have an inherent bias because not only are cancers that are prevalent at a younger age different in the overall sample, but it is also impossible to predict trends given expatriates belong to over 200 countries, thus causing a healthy worker effect [4, 5].

38.4 Consanguinity

In the current genomics era, there is an increased propensity for the diagnosis of genetic conditions predisposing to hematologic malignancies. With targeted next-generation sequencing (NGS) for hematologic cancers and immunodeficiencies, a wide variety of inheritable syndromes are being diagnosed that cause various hematologic malignancies, especially leukemias; e.g., in a large study by the Center for International Blood and Marrow Transplant Research (CIBMTR), the post- and pre-transplant samples of patients with myeloid malignancies were evaluated for inheritable mutations, and though the sample was consistent of adults and elderly patients, 4% of the patients were found to have the SBDS mutation, which is the hallmark of Shwachman-Diamond syndrome (SDS).

Given that Arab populations and some expatriate populations that are prevalent in the UAE (particularly Pakistan, Saudi Arabia, and Nigeria) have a high rate of consanguinity, a higher incidence of inheritable hematologic malignancies is estimated, particularly for leukemias, which can arise from a number of inherited (autosomal recessive) conditions.

38.5 Diagnostic Issues and the Current Paradigm

Since a critically important decision in high-risk hematologic malignancies is hematopoietic stem cell transplantation (HSCT), chimeric antigen receptor T cells (CAR-T cells), and gene therapy, one must have a clear picture of which patients not to take for these very complex and high-risk procedures. For decision-making, one needs a targeted NGS panel, cytogenetics, fluorescence in situ hybridization (FISH) panel, and minimal residual disease (MRD) for assessments of acute leukemias, multiple myeloma, lymphomas, and myelodysplastic syndromes [6].

Some real-world issues that are prevalent in the UAE for diagnostics are the unavailability of these genomic testing panels. If they are available at a particular institution, the denial of approval for these tests is commonly observed in hematologic malignancies, especially if these are send-out tests (out of the country). Some of the testing has been the standard of care for decades, e.g., MRD testing for B-cell acute lymphoblastic leukemia (B-ALL), and unfortunately, it is unavailable in the majority of the hospitals in the UAE as an in-house test.

Similarly, most large healthcare institutions still (as of the writing of this in 2022) do not perform in-house NGS myeloid or lymphoid panels, which is a critical piece of information when deciding on HSCT. For example, some of the known driver mutations in acute leukemias have an almost 0% chance of cure with chemotherapy alone and require an immediate transplant. Unlike solid cancers, the morphology of the cancer and cytopathology have become less important in hematologic malignancies, where the primary driver clone may drive the progression of cancer, but over time, there could be many sub-clones that become dominant and proliferate, and thus genomics is the primary driver of the oncogenesis in hematologic cancers.

38.6 Management Issues

Cellular therapies are not new in the treatment paradigm of hematologic malignancies, as HSCT was performed in the 1950s. About a decade ago, the first reports of successful CAR-T cells were published. In 2017, the first CAR-T cells were approved for leukemias and subsequently for lymphomas in the United States. Since then, at least six CAR-T products have been approved, but all for hematologic malignancies (B-ALL, NHL, and multiple myeloma). As of now, neither the US-FDA nor the EMA have approved a CAR-T product for solid cancers, and the role of autologous HSCT (auto-SCT) is limited to a few solid cancers (germ cell tumors, Wilms tumors, and neuroblastomas). Lastly, gene therapy has been approved by the EMA and the FDA for certain hematologic conditions (e.g., hemophilia and thalassemia), and there is no known role for gene therapy for solid cancers currently. All three treatment modalities—HSCT, CAR-T cells, and gene therapy— require extensive infrastructure in the form of a transplantation unit, dedicated and skilled personnel, and extensive quality management systems. Furthermore, these therapies are expensive, with average costs significantly higher than the annual costs of immunotherapies, chemotherapies, radiotherapies, or oncologic surgeries. For example, CAR-T cell therapy in the United States typically costs between USD 375,000 and 500,00, an allogeneic HSCT is typically priced between USD 400,000 and 750,000, and gene therapy pricing is above a million dollars per treatment [7].

In the UAE, at least four facilities are licensed by the regulatory authorities to perform HSCTs (three in Abu Dhabi and one in Dubai). These are likely going to be the facilities that are also going to be involved in future CAR-T cell therapies and gene therapy. The coverage issues for both the local population and the expatriate population need to be solved in order to provide smooth and optimal care for patients' journeys with hematologic malignancies.

Acute leukemias (particularly acute lymphoblastic leukemia and acute myeloid leukemias) and Burkitt lymphoma are true medical emergencies and need urgent admission and treatment. There are few centers that perform "acute leukemia induction" for acute leukemias, and these centers are present in Dubai, Abu Dhabi, and Al Ain. There is a networked ecosystem for the referral of these patients to these large facilities managing acute leukemias. However, post-leukemia induction, the insurance coverage for both local patients and expatriates needs to improve for pre-transplant requisites, e.g., HLA typing for both the recipient and the donor.

According to the 2021 UAE cancer registry report, there were 304 cases of leukemia and 228 cases of NHL diagnosed in the preceding year [1]. Some cases could be either due to non-reporting bias or emigration/immigration bias. Nonetheless, only a minority of these received HSCT, and data is only available for UAE citizens who traveled abroad for the receipt of the transplant. Now better systems are being placed to track the patient journal, i.e., cancer survivorship, which begins from the time of diagnosis until death, and thus it is imperative that long-term data, both pre- and post-transplant, be captured.

A robust ecosystem of research is being developed in the UAE, which is imperative for hematologic malignancies given the complexity of cellular therapies. CAR-T clinical trials and various stem cell therapy products will likely be tested in human trials, which will help improve both the clinical outcomes and the mortality statistics.

Some issues pertaining specifically to the hematologic malignancies in the UAE are presented in Table 38.1.

Table 38.1 Hematologic malignancies in the UAE

	Current situation	Issues	Mitigation strategies
Statistics tracking	Eighty-five percent are expatriates with majority of them being 20–65 years of age	Immigration and emigration bias Healthy worker effect	Age-adjusted rates with the removal of healthy worker bias
Diagnosis of hematologic malignancies	(a) Most hematologic malignancies require cytogenetics, FISH, and targeted NGS panels for decision-making about treatment (b) Acute leukemias and myeloma need MRD testing post-induction for treatment decisions	(a) Absence of targeted NGS panels (lymphoid and myeloid) in the majority of the hospitals in the UAE (b) Absence of MRD testing post-induction chemotherapy for acute leukemias	(a) Develop in-house capabilities for both NGS panels and also for MRD testing for acute leukemias (b) Influence insurance companies to approve the standard diagnostics for hematologic malignancies, which include genomic testing
Cellular therapies	(a) Bone marrow transplant (hematopoietic stem cell transplant, HSCT) plays a vital role in cure of most of the hematologic malignancies (b) CAR-T cells are being used as a curative treatment in some cancers and as a bridge to transplant for some cancers (c) Gene therapy is being used to cure certain hematologic conditions	(a) HSCT has been started in the UAE, but there is no FACT[a]- or JACIE[b]- accredited center in the country yet (b) No CAR-T cell therapy facility is ready to start commercial CAR-T cell therapy (c) Gene therapy infrastructure is not available	(a) Gain accreditation through the FACT[a]/ JACIE[b] for the HSCT programs in the UAE (b) Start the CAR-T program by establishing robust infrastructure in the UAE, including a state-of-the-art quality management system (QMS)

(continued)

Table 38.1 (continued)

	Current situation	Issues	Mitigation strategies
Hematologic malignancies survivorship	(a) For hematologic malignancies and HSCT, long-term follow-up clinics are required	(a) Absence of dedicated cancer survivorship and BMT survivorship clinics	(a) Establishment of dedicated loss to follow-up (LTFU) or survivorship clinics for leukemia, lymphoma, and myeloma (b) Establish a multidisciplinary clinic for HSCT/ BMT and graft-versus-host disease (GVHD)
Clinical trials	(a) Given the poor prognosis of certain hematologic malignancies, novel drug testing is necessary to improve outcomes (b) the etiology of most of the hematologic malignancies is unknown	(a) No in-house (by UAE investigators) interventional clinical trial has started in the field of hematologic malignancies (b) Gene editing trials for various hematologic conditions have not started yet	(a) Establishing a state-of-the-art clinical trials unit (b) Establish funding mechanisms specifically for hematologic malignancies and stem cell therapies (including CAR-T cells and gene therapy)

[a]*FACT* Foundation for the Accreditation of Cellular Therapy
[b]*JACIE* Joint Accreditation Committee ISCT-Europe & EBMT

38.7   Conclusion

The presence of hematologic malignancies in the UAE is proportionately (compared to solid cancers) higher than that in the United States. The exact causes are unknown. Moreover, the complex management of hematologic malignancies includes stem cell transplantation, CAR-T cells, and gene therapies, besides the traditional chemotherapies and immunotherapies. We need to pay attention to immigration and emigration bias, besides the healthy worker effect in the UAE. Lastly, now there is a great movement toward the establishment of cellular therapy centers in the UAE, which include the only potentially curable treatments for many cancers.

Conflict of Interest   The author has no conflict of interest to declare.

References

1. Cancer Incidence in United Arab Emirates Annual Report, 2021, Statistics and Research Center, Ministry of Health and Prevention (Accessed on 19March2024).
2. https://seer.cancer.gov/report_to_nation/statistics.html#new.
3. Khoury JD, Solary E, Abla O, Akkari Y, Alaggio R, Apperley JF, Bejar R, Berti E, Busque L, Chan JK, Chen W. The 5th edition of the World Health Organization classification of haematolymphoid tumours: myeloid and histiocytic/dendritic neoplasms. Leukemia. 2022;36(7):1703–19.
4. https://fcsc.gov.ae/en-us/Pages/Statistics/Statistics-by-Subject.aspx#/%3Ffolder=Demography%20and%20Social/Population/Population&subject=Demography%20and%20Social.
5. Kirkeleit J, Riise T, Bjørge T, Christiani DC. The healthy worker effect in cancer incidence studies. Am J Epidemiol. 2013;177(11):1218–24.
6. El Achi H, Kanagal-Shamanna R. Biomarkers in acute myeloid leukemia: leveraging next generation sequencing data for optimal therapeutic strategies. Front Oncol. 2021;11:3997.
7. Gene therapies should be for all. Nat Med. 2021. 27, 1311.

Dr. Shahrukh Hashmi is the Director of Research at the Department of Health, Abu Dhabi, UAE. He graduated from Baqai Medical University, Pakistan, as the top graduate, receiving a gold medal from the president of Pakistan. He received a master's degree (MPH) from Yale University, Connecticut, USA. He is a diplomat of the American Board of Preventive Medicine/public health, internal medicine, and hematology. He is a Professor of Medicine at the Mayo Clinic, Rochester, Minnesota, US. He started the Mayo Clinic's first BMT survivorship program. He has served as PI or co-PI on many industry-sponsored and NIH-sponsored trials. His research interests include premature aging, artificial intelligence, and GVHD. He is also involved in stem cell therapeutics, particularly in regenerative hematology. He has authored >300 articles in peer-reviewed journals, including *JAMA*, *Lancet*, and *The NEJM*, and has an H-index of>50, and has >10,000 citations. He chairs many national or international professional committees or groups, including being the chair of the Worldwide Network for Blood and Marrow Transplant's Nuclear Accident Committee (Geneva, Switzerland), the founding chair of the American Society for Blood and Marrow Transplant Society's Survivorship SIG (Chicago, Illinois), and the co-chair of the Center for International Blood and Marrow Transplant Registry's (CIBMTR) Health Services Committee (Milwaukee, Wisconsin).

Hematopoietic Stem Cell Transplantation (HSCT) in the UAE

Humaid O. Al-Shamsi ⓘ, Amin M. Abyad,
Zainul Aaabideen Kanakande Kandy, Biju George,
Mohammed Dar-Yahya, Panayotis Kaloyannidis,
Amro El-Saddik, Shabeeha Rana, and Charbel Khalil ⓘ

H. O. Al-Shamsi (✉)
Burjeel Cancer Institute, Burjeel Medical City, Burjeel Holdings, Abu Dhabi,
United Arab Emirates

Ras Al Khaimah Medical and Health Sciences University, Ras Al Khaimah, United Arab Emirates

Gulf Medical University, Ajman, United Arab Emirates

Emirates Oncology Society, Emirates Medical Association, Dubai, United Arab Emirates

College of Medicine, University of Sharjah, Sharjah, United Arab Emirates

Gulf Cancer Society, Alsafa, Kuwait
e-mail: alshamsi@burjeel.com; humaid.al-shamsi@medportal.ca

A. M. Abyad
Emirates Oncology Society, Emirates Medical Association, Dubai, United Arab Emirates

Department of Oncology, Burjeel Medical City, Abu Dhabi, United Arab Emirates

Z. A. Kanakande Kandy
Department of Oncology, Burjeel Medical City, Abu Dhabi, United Arab Emirates
e-mail: zainul.aabideen@burjeelmedicalcity.com

B. George · P. Kaloyannidis
Department of Haematology, Burjeel Medical City, Abu Dhabi, United Arab Emirates

Department of Bone Marrow Transplant and Cell Therapy, Burjeel Medical City,
Abu Dhabi, United Arab Emirates
e-mail: biju.george@burjeelmedicalcity.com; panagiotis.kalogiannidis@burjeelmedicalcity.com

M. Dar-Yahya
Department of Bone Marrow Transplant and Cell Therapy, Burjeel Medical City,
Abu Dhabi, United Arab Emirates
e-mail: mohammed.hamid@burjeelmedicalcity.com

A. El-Saddik
Department of Oncology, Burjeel Medical City, Abu Dhabi, United Arab Emirates

Department of Bone Marrow Transplant and Cell Therapy, Burjeel Medical City,
Abu Dhabi, United Arab Emirates
e-mail: amro@burjeelholdings.com

S. Rana
Genesis Healthcare centre, Dubai, United Arab Emirates
e-mail: Shabeeha@doctors.org.uk

C. Khalil
Department of Bone Marrow Transplant and Cell Therapy, Burjeel Medical City,
Abu Dhabi, United Arab Emirates

Reviva Regenerative Medicine center, Middle East Institute of Health University Hospital,
Bsalim, Lebanon

School of Medicine, Lebanese American University, Beirut, Lebanon
e-mail: charbel.khalil@burjeelmedicalcity.com

39.1 Introduction

Hematopoietic stem cell transplantation (HSCT) has become an essential and poten-tially life-saving or curative choice for various non-cancerous and cancerous blood-related conditions, solid tumors, and immune system disorders [1]. Over time, there has been a gradual shift in the perception of hematopoietic stem cell transplantation (HSCT) as a treatment option. Initially viewed as a last resort, it has now become an integral part of the international treatment guidelines for both malignant and non-malignant diseases mentioned earlier. The decision to recommend HSCT involves a meticulous evaluation of the risks and benefits by a team of experts. Factors consid-ered include the disease stage, duration, presence of other medical conditions, indi-vidual patient characteristics, transplant protocol, donor type, source of stem cells, conditioning regimen, and more. This comprehensive assessment helps determine whether the benefits of HSCT outweigh those of non-transplant strategies for each patient [1, 2]. Proficiency and knowledge in HSCT play a crucial role in the imple-mentation of cellular therapy and gene therapies, which hold immense potential for revolutionizing the landscape of treatment choices for a wide range of illnesses [3–5]. The effectiveness of any HSCT or cellular therapy program relies heavily on the creation and implementation of a well-structured and specialized unit comprising experienced multidisciplinary teams who adhere to the highest global standards. In this section, we will discuss the advancements made in HSCT services within the United Arab Emirates (UAE) and share our valuable insights and knowledge gained from establishing the first all-encompassing pediatric and adult HSCT service in the country. This service has been specifically designed to address the needs of UAE citi-zens, residents, and individuals from neighboring countries who encounter similar difficulties in accessing HSCT treatments.

39.2 Historical Background of the Unmet Need for HSCT in the UAE

In December 1971, the UAE was formed as a union of seven Emirates. With its rapid economic growth, the UAE has emerged as one of the world's fastest-growing economies, thanks to visionary leadership, strategic planning, significant investments, and exceptional infrastructure development. The healthcare sector has also experienced remarkable progress in recent decades, marked by the establishment of advanced governmental and private healthcare institutions in collaboration with regional and international partners. However, despite these advancements, a significant number of local patients still travel abroad, particularly for cancer treatment [6]. Over the past few years, numerous pediatric and adult patients from the UAE have sought medical treatment overseas, primarily in countries such as the USA, Germany, Korea, the UK, and Singapore. These patients have received care for both cancerous and non-cancerous ailments, with the expenses being covered by the UAE government [7]. A similar situation can be observed with expatriate residents, as they frequently choose to go back to their home countries to receive medical treatment. This decision is influenced by a range of logistical and economic factors, such as inadequate insurance coverage or loss of employment due to illness [8].

Given the prevalence and occurrence of both non-cancerous and cancerous blood-related diseases in the UAE, there is a pressing need for local approaches like HSCT. To address this, the Ministry of Health and Prevention (MOHAP) took the initiative to establish the UAE National Cancer Registry (UAE-NCR) in 2014. This population-based cancer registry encompasses data from both UAE citizens (local Emirati patients) and non-UAE citizens (expatriate residents) [9]. The most recent annual report released by the UAE National Cancer Registry (UAE-NCR) in 2022 provides data on newly diagnosed cases in the year 2019. This includes information on 220 cases of leukemia (comprising 168 expatriate and 52 Emirati patients), 42 cases of multiple myeloma (32 expatriate and 10 Emirati patients), 215 cases of non-Hodgkin's lymphomas (151 expatriate and 64 Emirati patients), and 76 cases of Hodgkin's lymphomas (51 expatriate and 25 Emirati patients) [9]. Regarding non-cancerous reasons for HSCT, the UAE exhibits notably elevated rates of various hereditary blood disorders that have been historically prevalent in the region. These disorders, including thalassemia, impose a substantial burden on public health and healthcare resources in terms of both prevalence and the medical care required [10, 11].

Before the year 2020, the lack of comprehensive HSCT centers in the UAE meant that seeking transplant services abroad became unavoidable. It is estimated that hundreds of patients, both adults and pediatrics, travel abroad annually to seek HSCT treatment. This situation presents several challenges, including limited coverage, logistical difficulties, high costs, and the need for meticulous peri- and post-HSCT procedural management. It not only impacts the psychological well-being of patients and their families but also disrupts the essential continuity of care. Many returning patients often have an inadequate understanding of their medical condition and the treatment they received abroad, and they may lack important medical documentation and reports. These factors can potentially lead to treatment interruptions or discrepancies between local and overseas medical teams due to communication gaps that commonly exist [11, 12]. The overall quality of care for patients is frequently affected by difficulties arising from cultural disparities, ethical and religious considerations, and language barriers, which can compromise their experiences [15]. We have observed that medical tourism is made more complex by factors such as restricted mobility and the added risk of COVID-19 infection during travel. These circumstances necessitate stringent travel arrangements, which are particularly concerning for an already immunocompromised patient population that requires continuous medical care. Therefore, the COVID-19 pandemic and travel restrictions have emphasized the critical need for locally based and self-sufficient HSCT programs within the UAE [12].

Considering all these factors, we have come to realize the significant demand for establishing a comprehensive HSCT unit that can cater to both adult and pediatric patients within the UAE. This would address the unmet needs of a substantial number of patients who can benefit from HSCT without the need to travel abroad. Such a local unit would bring various advantages in terms of clinical care, logistics, and cost-effectiveness. Additionally, it would provide a convenient option for patients from neighboring countries facing similar challenges, offering them the opportunity to seek HSCT in a nearby country that has consistently welcomed foreigners and facilitated their transition and integration.

Currently, there are 3 centers that provide HSCT for the UAE: 2 in Abu Dhabi and 1 in Dubai.

39.3 Abu Dhabi Stem Cell Center (ADSCC)

In July 2020, the Abu Dhabi Stem Cell Center completed the first autologous HSCT case in the UAE [13, 14].

39.4 Conceptualization and Journey Through the Establishment of the Comprehensive HSCT Program at Burjeel Medical City

Our objective was to create a fully fledged HSCT unit for both adults and children, adhering to international standards set by the European Society for Blood and Marrow Transplantation (EBMT). This initiative aligns with the vision of the Abu Dhabi Department of Health and the UAE government, aiming to deliver top-quality healthcare to UAE citizens and residents. Additionally, our goal is to establish Abu Dhabi and the UAE as a burgeoning center for global medical tourism, with a specific focus on cancer care.

The HSCT unit at Burjeel Cancer Institute within Burjeel Medical City (BMC) was established as part of a comprehensive cancer center. BMC's inception took place in 2015, positioning it as the flagship medical facility of Burjeel Holding Healthcare (formerly VPS Healthcare). Spanning over 1.1 million square feet in the capital city of Abu Dhabi, BMC has undergone significant development in the past two years. The oncology department has transformed into a state-of-the-art facility that provides comprehensive cancer care, encompassing medical and radiation oncology, nuclear medicine, surgical oncology, and the first palliative care service in

Abu Dhabi city. A team of highly skilled experts with international training and expertise is dedicated to delivering these specialized services [11]. It is the only center in the UAE that has received a European Society of Medical Oncology (ESMO) designation as an integrated oncology and palliative care services provider [11].

The HSCT initiative at BMC commenced in February 2021, marked by the establishment of a taskforce comprising experienced clinical hematologists and oncologists specializing in adult and pediatric HSCT. The taskforce also included a chief nursing officer, oncology and hematology pharmacists, a human resources representative, pediatric and adult infectious disease specialists, a quality director, a laboratory director, and international advisors with prior involvement in setting up HSCT services in Saudi Arabia and Italy [16]. The initial phase involved evaluating the available workforce and infrastructure, as well as determining any resource requirements. Our primary focus was to augment the manpower by filling any gaps to ensure the full participation and commitment of the experts directly involved in the HSCT project [11].

A specialized unit was constructed, featuring 13 private inpatient rooms equipped with high-efficiency particulate absorbing (HEPA) filters. To oversee the management of the HSCT unit, experienced pediatric and adult HSCT nurse specialists were recruited, with expertise in handling stem cell products, infection control, and administering chemotherapy. Dedicated outpatient clinic rooms were established to facilitate pretransplant evaluations, consultations, patient education, and follow-up, overseen by a knowledgeable HSCT coordinator. Additionally, clinical care pathways were developed in collaboration with various medical departments, including an active emergency room with a capacity of 14 beds, an intensive care unit, pulmonary, gastroenterology, cardiology, and adult and pediatric infectious disease departments. Representatives from these units were designated to ensure alignment with the acute and chronic care requirements of HSCT patients. To address infection control and manage infectious complications related to HSCT, the expertise of a US-trained specialist in solid and HSCT infectious diseases was enlisted [11].

A dedicated collection room on the HSCT floor was equipped with a newly purchased apheresis machine to facilitate stem cell collection. To oversee the apheresis and processing of stem cells, a PhD clinical scientist with expertise in stem cell therapies was appointed. The CD34+ cell count is performed on-site using flow cytometry, and the collected cells are stored in a specialized refrigerator within our facility, maintained at a temperature range of 2–6 °C. Currently, efforts are underway to establish a complete stem cell laboratory and cellular therapy unit for stem cell processing and cryopreservation. In the meantime, the cryopreservation of stem cell products is outsourced to an externally accredited laboratory [11].

In accordance with health regulations in the UAE, hospital-based blood banks are not permitted. Instead, the central Abu Dhabi blood bank ensures round-the-clock availability of irradiated blood products for HSCT patients [11].

The laboratory, imaging facilities (including the PET scanner), interventional radiology services, and radiation department at BMC met all the necessary criteria to complete the remaining requirements for the launch of the HSCT service. With these additions, the HSCT program at BMC successfully fulfilled the recommended criteria for establishing a HSCT program, as outlined by the Worldwide Network for Blood and Marrow Transplantation workshop [11, 17].

The operations of the HSCT unit are regulated by institutional guidelines and protocols, supported by standard operating procedures (SOPs). To ensure adherence to these protocols, regular audits of HSCT procedures and patient treatment outcomes are carried out. Daily rounds are conducted by the HSCT specialist or consultant, while a multidisciplinary team, including specialists such as infectious disease specialists, nutritionists, psychologists, nurses, and other necessary personnel, conducts comprehensive weekly audits [11]. In August 2021, the HSCT unit at BMC obtained the approval of the Abu Dhabi Department of Health after achieving a perfect score of 100% in the audit process [11].

39.5   Adult HSCT Cases Completed at BMC

To initiate our efforts, we chose to prioritize autologous HSCT for multiple myeloma (MM) as our starting point. This decision was based on the fact that MM is the most common indication for transplantation and offers a relatively straightforward protocol, which aids in medical staff training and adaptability. We began by assessing ten cases and adopted the strategy of selecting candidates with the lowest risk for HSCT. Ultimately, two cases were scheduled for October 2021 [11]. The first patient was a 46-year-old Sudanese male who had been diagnosed with MM over 16 years ago. The second patient, a 53-year-old male from Lebanon, was referred to our facility after recently being diagnosed with high-risk MM. Following initial treatment with daratumumab and the VRd protocol, the patient was intended to undergo autologous HSCT after achieving the first remission [11]. However, due to the financial crisis and lack of insurance coverage in Lebanon, the patient could no longer receive autologous transplantation as an intensification treatment [18, 19]. Ultimately, both patients underwent a non-cryopreserved autologous HSCT procedure using melphalan conditioning [11].

To the authors' knowledge, the HSCT program at BMC was the first center in the UAE to complete an HSCT using the cryopreservation technique. The median engraftment time for multiple myeloma was 11 days, and the median engraftment time for lymphoma was 10 days (unpublished data on file).

To the authors' knowledge, the first adult allogeneic (haploidentical) HSCT case in the UAE was also completed at BMC in September 2022. The case was for an AML case involving a 27-year-old male referred from KSA to our facility with poor disease characteristics after receiving intensive remission induction therapy. Unfortunately, due to the financial crisis and lack of insurance coverage, the patient was no longer able to receive his allogeneic transplant. Although the patient did not have a full-match sibling donor, which is considered the optimal donor, the only donor available was his 24-year-old brother with a 50% match, so we have decided to perform the first allo-haploidentical stem cell transplant. The patient received 5.6×10^6/kg CD34 cells, engrafted on day 19, but subsequently expired on Day + due to idiopathic pneumonia syndrome. Since then, four more adult patients have undergone an allogeneic stem cell transplant (unpublished data on file).

Over 100 patients who underwent HSCT from the initiation of the transplant in 2021 at BMC to December 2023 were included in the adult and pediatric categories. Out of 100 patients, most underwent autologous transplantation; others underwent allogeneic transplantation during this period. Patients who underwent allogeneic transplants were haploidentical, full-matched, and related stem cell transplants. The most common malignancies for which patients underwent HSCT at our center were multiple myeloma, followed by Hodgkin's lymphoma, non-Hodgkin lymphoma, amyloidosis, thalassemia, germ cell tumors, chronic myeloid leukemia, acute myeloid leukemia, myelodysplastic syndrome, neuroblastoma, and premature immunodeficiency. The majority of patients who underwent stem cell transplants were in complete remission at the time of the transplant. Acute graft versus host disease (GvHD) was rarely observed in our patients. The majority of patients survived post-transplant (unpublished data on file).

39.6 Pediatric HSCT Cases Completed at BMC

The pediatric HSCT service at BMC is the first and only pediatric HSCT service in the UAE. The service started in March 2022.

The median day of engraftment for thalassemia is 29 days. The median day of engraftment for sickle cell anemia is also 29 days. Severe combined immunodeficiency had 11 days of engraftment. Primary immunodeficiency (PID) had 32 days for engraftment (unpublished data on file).

39.7 HSCT-BMC Outreach Activity and Building Community Trust

In partnership with local print and digital media outlets and designated medical societies, the BMC-HSCT team launched several campaigns to enhance public awareness about hematologic malignancies, their diagnosis, and available treatments. These initiatives aimed to inform the local community about advancements in cancer care within the UAE and to emphasize the availability of HSCT services locally. To reach hematologists and oncologists across the UAE and the GCC region, we directly communicated and shared information about our HSCT program, while also establishing a hotline referral and transfer system. Additionally, we initiated the first comprehensive UAE-wide HSCT weekly virtual multidisciplinary team (MDT) meeting, facilitating discussions on potential HSCT cases and fostering the exchange of experiences and expertise among the oncology and hematology communities in the UAE [11].

In 2023, and to the authors' knowledge, the BMT program at BMC was the first center to obtain membership in the European Society for Bone Marrow Transplantation in the UAE.

39.8 American Hospital Dubai (AHD)

The American Hospital Dubai launched the city's first comprehensive autologous stem cell transplant program in Dubai. The program received a detailed Dubai Health Authority inspection and approval in September 2021. In October 2021, the program performed the first autologous BMT on one of BMC's Nigerian dialysis-dependent high-risk myeloma patients in October 2021. The standard operating procedures, policies, and protocols adapted at American Hospital are based on international standards with reference to specific disease-related entities. So far, the program has performed 10 autologous stem cell transplantations. The program has been moving forward successfully, with excellent patient feedback and satisfaction.

The BMT team includes a consultant hematologist as clinical program director, a specialist hematologist, a BMT quality manager, an apheresis team leader, dedicated transplant-trained nurses, and specialized laboratory staff for stem cell collection, processing, cryopreservation, and thawing prior to transplant. The cases are discussed and agreed upon by the transplant multidisciplinary team, which includes international participation in the form of virtual attendance. The cases performed so far are within standard guidelines for multiple myeloma and relapsed lymphoma.

Being one of the most comprehensive cancer centers in Dubai, American Hospital is ideally suited for medical tourism and is fulfilling this role with great success. Having a specialized service like stem cell transplantation at American Hospital Dubai, which caters to Dubai and the Northern Emirates within the home country, has also opened easy access for UAE nationals and residents to avail themselves of this service without having to travel far and beyond. More recently, Dubai hosted a formal Dubai Health Authority inspection for accreditation, with extremely positive feedback.

39.9 Conclusion

This review outlines the progress that the HSCT programs have made in the UAE over the last two years. From no programs providing HSCT service in the UAE to three programs in ADSCC, BMC, and the AHD. The BMC-HSCT program currently leads HSCT in the UAE, with the most allogeneic and autologous HSCT procedures performed. It is the only program that also provides pediatric HSCT in the UAE, and currently, as of May 2023, it is the only program providing allogenic adult HSCT. This progress would not have been possible without very strong support from the health regulators in the UAE and the UAE government to enhance access to all modalities for cancer care for UAE citizens and residents. The next chapter of the HSCT programs in the UAE is to gain more experience and get international accreditations, e.g., FACT accreditations. The HSCT providers should also continue to gain the trust of patients from the local community and neighboring countries.

Conflict of Interest The authors have no conflict of interest to declare.

References

1. Duarte RF, Labopin M, Bader P, et al. Indications for haematopoietic stem cell transplantation for haematological diseases, solid tumours & immune disorders: current practice in Europe, 2019. Bone Marrow Transplant. 2019;54:1525–52.
2. Watts MJ, Linch DC. Optimisation & quality control of cell processing for autologous stem cell transplantation. Br J Haematol. 2016;175:771–83.
3. Sagoo P, Gaspar HB. The transformative of potential of HSC gene therapy as a genetic medicine. Gene Ther. 2021;30:197.
4. Huang R, Li X, He Y, et al. Recent advances in CAR-T cell engineering. J Hematol Oncol. 2020;13:86.
5. Elverum K, Whitman M. Delivering cellular and gene therapies to patients: solutions for realizing the potential of the next generation of medicine. Gene Ther. 2020;27:537–44.
6. Al-Shamsi HO, Al-Hajeili M, Alrawi S. Chasing the cure around the globe: medical tourism for cancer care from developing countries. J Glob Oncol. 2018;4:1–3.
7. Itihad: 600 Million Dirhams annually for cancer patient treatment abroad. 2014. Accessed 18 Oct 2021. http://www.alittihad.ae/details.php?id=90507&y=2014.
8. Al-Shamsi HO. Expatriates with cancer diagnosis in The United Arab Emirates, hanging on the edge. J Cancer Prev Curr Res. 2018;9(5):257–8. https://doi.org/10.15406/jcpcr.2018.09.00364.
9. Cancer incidence in United Arab Emirates annual report of the UAE—National Cancer Registry. 2017. Ministry of Health & prevention, United Arab Emirates.
10. Abu-Shaheen A, Heena H, Nofal A, et al. Epidemiology of thalassemia in gulf cooperation council countries: a systematic review. Biomed Res Int. 2020;2020:1509501.
11. Al-Shamsi HO, Abyad A, Kaloyannidis P, et al. Establishment of the first comprehensive adult / pediatric hematopoietic stem cell transplant unit in the UAE: rising to the challenge. Clin Pract. 2022;12:84–90.
12. Al-Shamsi HO, Abu-Gheida I, Rana SK, et al. Challenges for cancer patients returning home during SARS-COV-19 pandemic after medical tourism—a consensus report by the emirates oncology task force. BMC Cancer. 2020;20:641.
13. http://www.tradearabia.com/news/HEAL_370921.html.
14. https://adscc.ae/news/first-ever-bone-marrow-transplant-conducted-in-uae.
15. Surbone A. Cultural aspects of communication in cancer care. Recent Results Cancer Res. 2006;168:91–104.
16. Al-Hashmi H, Alsagheir A, Estanislao A, et al. Establishing hematopoietic stem cell transplant programs; overcoming cost through collaboration. Bone Marrow Transplant. 2020;55:695–7.
17. Pasquini MC, Srivastava A, Ahmed SO, et al. Worldwide network for blood and marrow transplantation recommendations for establishing hematopoietic cell transplantation program, part I: minimum requirements and beyond. Biol Blood Marrow Transplant. 2019;25:2322–9.
18. Devi S. Economic crisis hits Lebanese health care. Lancet. 2020;395:548.
19. Shallal A, Lahoud C, Zervos M, Matar M. Lebanon is losing its front line. J Glob Health. 2021;11:03052.

Prof. Humaid Obaid Al-Shamsi is the Chief Executive Officer of Burjeel Cancer Institute in Abu Dhabi, UAE, President of the Emirates Oncology Society, Lead of the Gulf Cancer Society, Full Professor of Oncology at the Ras Al Khaimah Medical and Health Sciences University, Ras Al Khaimah, UAE, and an Adjunct Professor of Oncology at the College of Medicine, University of Sharjah. He is the first Emirati to be promoted as a professor in oncology in the UAE. He is also the Chairman for Colorectal Cancer in the MENA region, appointed by the prestigious National Comprehensive Cancer Network®. He is also the only member of Lung Cancer Policy Network in the MENA region that aims to advance lung cancer research and screening globally. He is the Chairman of the Oncology and Hematology Fellowship Training Program for the National Institute for Health Specialties in the United Arab Emirates. He is the only member in GCC in the WIN Consortium which is comprised of organizations representing all stakeholders in personalized cancer medicine globally.

He is board-certified in both internal medicine and oncology from the UK, USA (ABIM), and Canada (FRCPC). He has also been awarded the FRCP (London) in 2023 and FRCP (Glasgow) in 2024. He is the only physician in the UAE with a subspecialty fellowship certification and training in gastrointestinal oncology and the first Emirati to train and complete a clinical post-doctoral fellowship in palliative care. He was an assistant professor at the University of Texas MD Anderson Cancer Center between 2014 and 2017. He has published more than 140 peer-reviewed articles in JAMA Oncology, Lancet Oncology, The Oncologist, BMC Cancer, and many others. His area of expertise includes precision oncology and cancer care in the UAE. In 2016, he published with his group from MD Anderson the JCO paper describing a new distinct subgroup of CRC, NON V600 BRAF-mutated CRC. In 2022, he published the first book about cancer research in the UAE and also the first book about cancer in the Arab world, both of which were launched at Dubai Expo 2020. *Cancer in the Arab World* has been downloaded more than 450,000 times in its first 18 months of publication and is the ultimate source of cancer data in the Arab region. He also published the first comprehensive book about cancer care in the UAE which is the first book in UAE history to document the cancer care in the UAE with many topics addressed for the first time, e.g., neuroendocrine tumors in the UAE. He is passionate about advancing cancer care in the UAE and the GCC and has made significant contributions to cancer awareness and early detection for the public using social media platforms. He is considered as the most followed oncologist in the world with over 300,000 subscribers across his social media platforms (Instagram, Twitter, LinkedIn, and TikTok). In 2022, he was awarded the prestigious Feigenbaum Leadership Excellence Award from Sheikh Hamdan Smart University for his exceptional leadership and research and the Sharjah Award for Volunteering. He was also named the Researcher of the Year in the UAE in 2020 and 2021 by the Emirates Oncology Society.

In May 2024, HH Sheikh Mansour bin Zayed Al Nahyan, Vice President of the United Arab Emirates, awarded him the first place in UAE Nafis program for outstanding leadership in private sector across all business and medical disciplines. Beside his clinical and administrative duties, he is engaged in education and various levels of research training for medical trainees to enhance their clinical and research skills. His mission is to advance cancer care in the UAE and the MENA region and make cancer care accessible to everyone in need around the globe.

Dr. Amin M. Abyad earned his medical degree, Bachelor in Medicine and Surgery (MBChB), from Beirut Arab University. After completing his internship, he joined the Internal Medicine Residency at Makassed Hospital in Beirut, Lebanon, which is affiliated with the American University of Beirut Medical Center (AUBMC). Dr. Amin was appointed as Chief Resident of Internal Medicine (2017–2018). Then Dr. Amin started his fellowship in hematology and medical oncology at Makassed Hospital, where he received intensive training in hematology and medical oncology. Dr. Amin joined Burjeel Medical City in July 2021. Dr. Amin is highly interested in malignant hematology and solid malignancies. He has been highly involved in clinical research, being involved in multiple research projects and publishing in multiple peer-reviewed journals.

Dr. Abyad believes in patient-centered care, trying to enhance patient outcomes through the application of the latest evidence-based practice and personalized medicine.

Dr. Zainul Aaabideen Kanakande Kandy is highly skilled and experienced in paediatric haematology, oncology, and bone marrow transplants. He received his degree in Medicine at Calicut University in Kerala, India; subsequently, he took his post-graduate degree in paediatrics from the University of Mumbai.In 2001, Dr. Zainul moved to the United Kingdom, where he specialized and undertook further training and experience in paediatric haematology, paediatric oncology, and paediatric bone marrow transplantation in various hospitals in the UK such asRoyal Marsden Hospital, London, UK;University College Hospitals of London, UCLH, London;Imperial College London, UK; Manchester Children Hospital, Manchester, UK; Alder Hey Children Hospital, Liverpool, UK; and Great North Children Hospital, Newcastle, UK.

Dr. Zainul gained his master's degree in paediatric oncology at Birmingham University as well as a certificate in medical education at Manchester University. Subsequently, he completed his International Fellowship in paediatric bone marrow transplantation at Great North Children's Hospital in Newcastle.He worked previously as a consultant pediatric at University Hospital of Coventry and Warwickshare and Royal Oldham Hospital before moving to the UAE.In the UAE, he had worked at the department of pediatric hematology and oncology at Tawam Hospital as a consultant, Al Ain, before joining at Burjeel Medical City as head of pediatric hematology, oncology and BMT.Aside from his commitment to further advance his experience and knowledge in the field of paediatrics, paediatric haematology, oncology, and bone marrow transplantation in the United Kingdom, he also published articles and actively supported the community by sharing his knowledge and participating in international conference across countries.

Dr. Biju George earned his medical degree, Bachelor in Medicine and Surgery (MBBS), from Christian Medical College, Vellore, India. After completing his internship, he completed his internal medicine residency and his higher specialty training in haematology at the same hospital. He had additional training in Westmead Hospital Sydney for 3 years in the field of unrelated and haploidentical stem cell transplantation and helped establish the haploidentical programme in CMC Vellore. He has published more than 250 articles in international peer-reviewed journals and is a reviewer for journals such as Transplant and Cellular Therapy, Bone Marrow Transplantation, and British Journal of Haematology. Dr. George joined Burjeel Medical city in September 2022 and heads the adult Bone Marrow Transplant programme.

Mohammed Dar-Yahya is a highly skilled nursing professional with a rich educational and professional background. He began his journey in healthcare by earning a Bachelor of Science in Nursing from the University of Jordan (2004–2008). Shortly thereafter, Mohammed joined the King Hussein Cancer Center, where he served from 2008 to 2021. During his tenure, he worked as a registered nurse in the bone marrow transplant unit and progressed through various roles, eventually becoming a nurse manager.

Committed to advancing his expertise, Mohammed completed his Master's degree in Palliative Nursing Care between 2016 and 2019. In 2021, he took a significant step by joining Burjeel Medical City. There, he has been instrumental in establishing the first bone marrow transplant unit in the UAE, Abu Dhabi, showcasing his leadership and pioneering contributions to the field of nursing.

Dr. Panayotis Kaloyannidis has joined Burjeel Medical City as specialist hematologist in the adult hematology and bone marrow transplant department. He has 25 years of experience, mainly in the field of malignant hematological diseases and their treatment with conventional chemotherapy and cellular therapies, including hemopoietic stem cell transplantation. Since 1998 he has been actively involved in more than 2000 hematopoietic stem cell transplants (autologous and allogeneic) from any source of hematopoietic cells (bone marrow, peripheral blood, and umbilical cord blood). In 2012–2013, he was trained in cellular therapies with chimeric antigen receptors T cells (CAR T cells) and specific T cells against viral infections and relapse post allogeneic stem cell transplantation. His expertise and areas of interest include diagnosis and treatment of malignant hematological diseases, hematopoietic stem cell transplantation, post-transplantation complications (graft vs. host disease, infections, relapsed disease), cellular therapies, and immunotherapy for malignant hematological diseases. Dr. Kaloyannidis has held a senior Consultant Hematologist position at the Department of Hematology and Bone Marrow Transplantation Unit at the "G. Papanicolaou" General Hospital in

Thessaloniki, Greece, from 1998 to 2014 and as a senior Consultant Hematologist at the Adult Oncology/Hematology and SCT department at King Fahad Specialist Hospital-Dammam, S. Arabia from June 2014 to March 2023. Dr. Kaloyannidis is a member of the Hellenic Society of Hematology, the European Society for Bone Marrow Transplantation (EBMT), and the Saudi Society of Blood and Marrow Transplantation (SSBMT). His areas of expertise include,

- Diagnosis and treatment of malignant hematological diseases
- Hematopoietic stem cell transplantation and post-transplantation complications (graft vs. host disease, infections, relapsed disease).
- Cellular therapies, and immunotherapy for malignant hematological diseases.

Dr. Amro El-Saddik is the Oncology Excellence Lead and the Cell Therapy Program Director at Burjeel Holdings. His role is to lead the team that will make Burjeel Medical City a Center of Excellence in Oncology and Hematology, and a state of the art in patients' healing. He is also responsible for bringing high-quality oncology products to be locally manufactured and packaged in LIFEPharma, aiming to help healthcare authorities and regulators provide more health for the money and keep head room for innovation.He worked as the Oncology Senior Medical Manager for the Gulf States at Pfizer from July 2018 until January 2021. And prior to that, he was a Global Medical Manager for Cell and Gene Therapy at GSK, responsible for the Middle East and North Africa since June 2016. He is also an assistant professor of clinical pathology at Mansoura University, Egypt.

Dr. Shabeeha Rana is a distinguished Consultant Hematologist, trained in the United Kingdom. Her commitment to excellence extends to her dual passions for teaching and clinical research in haematological disease management. With an impressive track record spanning over two decades, Dr. Shabeeha has solidified her reputation as the foremost haematologist in the field, specializing in myeloma, lymphoma, leukemia, and stem cell transplantation, earning the trust of patients and colleagues alike in the United Arab Emirates.

Dr. Shabeeha's career includes contributions to the healthcare landscape of Dubai, such as Dubai's bone marrow and haematopoietic stem cell transplantation program; she has left an indelible mark on the medical community. Notably, she established the first center for Autologous Stem Cell Transplantation at the prestigious American Hospital Dubai, raising the standard of care for patients in the region.In recognition of her outstanding achievements, Dr. Shabeeha has been selected as the first representative from the Gulf region to participate in the European Haematology Association Annual Meeting in 2023, an esteemed distinction within the "Women in Lymphoma" group. Her leadership and contributions continue to shape the future of haematology, making her a trailblazer in her field and a trusted advocate for patients seeking the highest standards of medical care.

Dr. Charbel Khalil received his PhD in cellular therapy and regenerative medicine from Saint Joseph University in Lebanon. I pursued a fellowship in cellular therapy at the University of Paris Diderot and was appointed as a full-time supervisor in the department of cellular therapy at the Institute Gustave Roussy (IGR)—Villejuif/University of Paris XI. I also hold a Master of Science in biobanking from Luxembourg University and a Master in wound management from ICW Germany, a Master in cellular therapy and regenerative medicine from Saint Joseph University, a university diploma in cellular therapy from Paris Diderot, Paris, and he is currently doing a diploma in Immuno-Oncology at Paris XI (Ecole du Cancer—Paris, France). He is the cellular therapy director at the Reviva stem cell platform for the Research and Applications Center-Middle East Institute of Health University Hospital. Reviva has been considered the leading stem cell treatment and research center in the Middle East. It holds the first national cord blood bank, cellular therapy, reconstructive medicine, and regenerative medicine for clinical trials, and it won the Innovation Award in 2014.

Dr. Charbel contributed to the establishment and development of the center's infrastructure and services and supervised the accreditation process in collaboration with Hospital Saint Louis—Paris, France, for JACIE international regulation. This made me an expert in the establishment and direction of any stem cell research center and cord blood bank. He contributed again to the establishment of the first bone marrow transplantation in the United Arab Emirates at Burjeel Medical City in Abu Dhabi and the establishment of the bone marrow transplantation unit at the Lebanese American University Rizk Hospital in Lebanon. He is an assistant professor at the Lebanese American University of Beirut School of Medicine, Beirut, Lebanon.

Dr. Charbel has also been a teacher and researcher at Saint Joseph University of Beirut, Faculty of Dental Medicine and Pharmacy. He's also an active member of the European, Eastern Mediterranean, and Lebanese societies for blood and marrow transplantation and the French society for stem cell and research. He has published and contributed to more than 20 articles, mentored students in the Master's program at Paris XIII, and lectured in universities across Lebanon.

Cancer Survivorship Programs in the UAE

40

Aydah Al-Awadhi ⓘ, Ramanujam A. Singarachari, and Rita A. Sakr ⓘ

40.1 Introduction

Due to demographics, greater early detection, and improved treatment, there are more cancer survivors today. However, current models of care, which are mostly provided by specialists, fall short of meeting the physical, emotional, and supportive care needs of cancer survivors. A survivorship care plan includes tailored check-ups and routine follow-ups to ensure adequate health maintenance to address the physical, psychosocial, and possible long-term effects of the disease and the treatments on the patient [1–3].

40.2 Cancer Survivorship Programs in the UAE

In the United Arab Emirates (UAE), cancer survivorship programs and dedicated clinics have yet to be established as an integral part of caring for cancer patients. To the best of our knowledge, there is no established certified survivorship program or clinic for oncology patients in the UAE. However, some institutions are in the process of facilitating a dedicated track for cancer survivors. Most of the post-treatment surveillance is conducted at their regular follow-up clinics, with possible limitations

A. Al-Awadhi (✉)
Tawam Hospital, Al Ain, United Arab Emirates

Emirates Oncology Society, Emirates Medical Association, Dubai, United Arab Emirates
e-mail: ayawadhi@seha.ae

R. A. Singarachari
Mediclinic Airport Road Hospital, Abu Dhabi, United Arab Emirates

R. A. Sakr
Emirates Oncology Society, Emirates Medical Association, Dubai, United Arab Emirates

King's College Hospital London, Dubai, United Arab Emirates

© The Author(s) 2024
H. O. Al-Shamsi (ed.), *Cancer Care in the United Arab Emirates*,
https://doi.org/10.1007/978-981-99-6794-0_40

due to funding and insurance coverage for screening tests. Regular follow-up for those patients includes a full history and physical examinations, laboratory and imaging investigations as needed, counseling for lifestyle modifications, and psychosocial support.

Ideally, survivorship clinics should be carried out in the primary care setting, but there is also a lack of dedicated training programs for primary care physicians to master oncology survivorship tracking and health maintenance. This is all based on international guidelines [1].

40.2.1 Cancer Survivors' Role in Survivorship Programs

Many cancer survivors in the UAE have participated in their own efforts and joined cancer societies and hospitals to raise awareness about the disease through campaigns, as well as to share their experiences in the local newspaper, magazine, TV, social media, and radio and support other cancer patients either directly or through established support groups.

40.3 Challenges and Recommendations

Finally, cancer survivors' numbers are likely to increase in the next few years, and nationally dedicated guidelines and strategies on long-term survivorship planning and related issues for cancer survivors of various ages have yet to be developed. Efforts by cancer care professionals and non-profit organizations to promote survivor awareness are on the horizon. To optimize holistic cancer care delivery, a national survivorship strategy tailored to cancer, cultural, and population characteristics is required.

40.4 Conclusion

With the anticipated rise in the numbers of cancer survivors in the country due to improved early detection, treatment options, and outcomes, it becomes important to establish a tailored cancer survivorship program to ensure the physical and psychosocial well-being of the patients and mitigate the long-term consequences of disease and treatment.

Conflict of Interest The authors have no conflict of interest to declare.

References

1. Shapiro CL. Cancer survivorship. N Engl J Med. 2018;379(25):2438–50.
2. van Kalsbeek RJ, Mulder RL, Skinner R, Kremer LCM. The concept of cancer survivorship and models for long-term follow-up. Front Horm Res. 2021;54:1–15.
3. Hill RE, Wakefield CE, Cohn RJ, Fardell JE, Brierley ME, Kothe E, et al. Survivorship care plans in cancer: a meta-analysis and systematic review of care plan outcomes. Oncologist. 2020;25(2):e351–e72.

Dr. Aydah Al-Awadhi is a consultant medical oncologist who graduated from medical school at the United Arab Emirates University in 2012, then completed her internal medicine residency in the United States and completed her medical oncology and hematology fellowship at the University of Texas, MD Anderson Cancer Center, in 2019. She is American Board certified in internal medicine, hematology, and medical oncology. She was awarded the Emirates Oncology Society (EOS) Member of the Year Award for 2021 and the Emirates Oncology Society Women in Oncology of the Year Award. She is also currently the scientific chairperson of the Emirates Oncology Society and the chair of the Breast Cancer Working Group. Dr. Al-Awadhi has published many articles and book chapters on breast cancer and oncology, with her main expertise in breast cancer and sarcoma.

Dr. Ramanujam A. Singarachari received his post-doctoral medical degree DM (oncology-medical) from the Regional Cancer Treatment and Research Centre—Cancer Institute (WIA), based in Chennai, India. He is also certified by the European Society of Medical Oncology, the Federation of Royal Colleges of Physicians of the United Kingdom, and the Association of Cancer Physicians—Medical Oncology MRCP UK.

He has a vast experience of 24 years in training and working in a number of eminent and reputable cancer centers in India, New Zealand, and the United Arab Emirates. In the last 20 years in the United Arab Emirates, he worked as a medical oncologist at large hospitals like Sheikh Khalifa Medical City (SKMC) and Sheikh Shakhbout Medical City (SSMC), jointly operated by Cleveland Clinic, Mayo Clinic (USA), and SEHA (a government healthcare provider in the UAE).

His areas of interest are GI cancers, breast cancers, CNS cancers, palliative care, and cancer survivorship.

Dr. Rita A. Sakr is a French/European board-certified, American fellow, consultant breast oncoplastic surgeon, and Ob-gynecologist with over 20 years of surgical, clinical, academic, and research experience in the field. After training in both France and the USA, Dr. Rita Sakr was appointed consultant breast oncoplastic surgeon and Ob-gynecologist in several institutions, including the Institut Curie, Paris; the Memorial Sloan Kettering Cancer Centre, New York; and Assistance Publique—Hôpitaux de Paris. She also held the position of associate professor and co-head of the Breast and Gynecology Care Unit for high-risk patients at Pierre & Marie Curie/Sorbonne University, France. She later relocated to Dubai as a consultant breast surgeon and Ob-gynecologist in the Dr. Sulaiman Al-Habib Hospital and then the American hospital.

Dr. Rita Sakr is a member of various clinical societies such as the American Society of Clinical Oncology, the American Association for Cancer Research, the Lebanese Society of Obstetrics and Gynecology, the Collège National des Gynécologues et Obstétriciens Français, Société Française de Sénologie et Pathologie Mammaire, the New York Academy of Sciences, the European Society of Surgical Oncology, the Emirates Oncology Society, the Europa Donna, etc. Dr. Rita Sakr has more than 100 publications and abstracts in peer-reviewed journals and international conferences and has been awarded many international recognitions, such as the American Association for Cancer Research Award and the San Antonio Breast Cancer Symposium Award.

In addition to her wide surgical, clinical, and academic experience, Dr. Rita Sakr has a passion for patient advocacy and has been actively involved in multiple associations for breast cancer survivorship in Europe, the USA, and the UAE.

Suggested Quality Control Measures for Cancer Care in the UAE

41

Humaid O. Al-Shamsi [ID]

41.1 Introduction

Cancer care in the UAE has evolved dramatically over the last few decades [1]. There are nearly 30 cancer centers that provide everything from basic cancer care to more advanced and complex cancer therapy [1, 2]. In the UAE, there are nearly 90–100 medical oncologists, hematologists, radiation oncologists, and surgical oncologists from various backgrounds and trainings in oncology [2]. This leads to variation in cancer care without consistency or quality measures.

In this chapter, we will discuss the practical recommendations to improve the quality of cancer care in the following domains: pathology, medical oncology, genomic and molecular testing, radiation, surgical oncology, insurance limit and renewal, cancer screening, and other general recommendations (Fig. 41.1).

H. O. Al-Shamsi (✉)
Burjeel Cancer Institute, Burjeel Medical City, Burjeel Holdings, Abu Dhabi,
United Arab Emirates

Ras Al Khaimah Medical and Health Sciences University, Ras Al Khaimah, United Arab Emirates

Gulf Medical University, Ajman, United Arab Emirates

Emirates Oncology Society, Emirates Medical Association, Dubai, United Arab Emirates

College of Medicine, University of Sharjah, Sharjah, United Arab Emirates

Gulf Cancer Society, Alsafa, Kuwait
e-mail: alshamsi@burjeel.com; humaid.al-shamsi@medportal.ca

© The Author(s) 2024
H. O. Al-Shamsi (ed.), *Cancer Care in the United Arab Emirates*,
https://doi.org/10.1007/978-981-99-6794-0_41

Fig. 41.1 The potential domains of cancer care quality in the UAE

41.2 Cancer Care Quality Council and Cancer System Quality Index

Oncologists in the UAE have diverse training backgrounds. Research has demonstrated that receiving specialized training in oncology enhances the results of cancer treatment [3–5]. Ensuring the quality of cancer care in the UAE is a complex task due to the need for standardized delivery. To meet this challenge, it is necessary to establish a cancer system quality index that can guarantee the excellence of care provided [1]. When considering generic quality indicators, it is important to prioritize specific metrics that encompass various aspects of cancer care. These metrics may include waiting times for surgeries, utilization of chemotherapy, 30-day mortality rate after initiating chemotherapy, planning of radiation treatment, adoption of a multidisciplinary approach, advance care planning, end-of-life care, and documentation of cancer care, including pain management.

Our suggestion is to create an autonomous Advisory Cancer Care Quality Council responsible for overseeing the quality of care provided in both government and private hospitals throughout the UAE. We propose adopting a cancer system quality index that bears resemblance to the well-established index implemented in Ontario, Canada, since 2002, recognized as one of the longest-standing indexes of its kind [6–8]. Before adopting and implementing the Ontario cancer system quality index or any other similar quality index, it is crucial to assess its suitability for our healthcare system. It is recommended to foster collaboration between the newly established UAE cancer system quality index and other regional or international organizations focused on cancer care quality. This collaboration aims to facilitate the exchange of knowledge and experiences in order to enhance the overall quality of cancer care [1].

Additionally, we propose the implementation of a centralized multidisciplinary tumor board, which can convene either on-site or virtually on a weekly basis. This tumor board should receive approval and oversight from the Department of Health (DOH), Dubai Health Authority (DHA), and the Ministry of Health and Prevention (MOHAP). It is essential to establish a mandatory rule that requires every confirmed cancer case to be thoroughly discussed and documented at an accredited multidisciplinary tumor board. Subsequent management decisions should be based on the recommendations made during these meetings [1].

41.3 General Recommendations

- Establishment of quality control and auditing for cancer in the UAE in all disciplines: medical oncology, radiation, surgery, pathology, imaging, and palliative care.
- Limit cancer treatment to accredited cancer centers that meet minimum requirements (availability of core MDT members, e.g., medical oncology, radiation therapy, surgical oncology, and palliative care).
- Accreditation of centers providing cancer-related treatments.
- Cancer centers should follow internationally accredited guidelines [9, 10].
- A multidisciplinary team (MDT) recommendation for all newly diagnosed cancer cases. The major challenge is that many cancer cases are misdiagnosed or inaccurately treated, leading to poor outcomes. Regulators must officially recognize MDT, which consists of a consultant (not a specialist) medical oncologist, a consultant radiation oncologist, a consultant general surgeon, a consultant pathologist trained in oncology, and a preferred surgical oncologist.
- The approved MDT must also agree to accommodate other oncologists so they have access to their MDT, as many small hospitals do not have full oncology services and must be paired with an approved MDT to treat cancer patients.
- Insurance only approves surgery for cancer or suspected cancer if MDT recommendations are provided, except in emergency situations. (This cannot be listed as there are many situations in which listing conditions may cause limitations in access to emergency surgeries).
- Adoption of international cancer treatment guidelines in Dubai: We recommend the following:
  - The National Comprehensive Cancer Network® (NCCN) is an alliance consisting of 31 prominent cancer centers that operate as a not-for-profit organization. Their primary focus lies in patient care, research, and education. The NCCN is committed to enhancing and enabling the provision of high-quality, effective, efficient, and easily accessible cancer care, with the ultimate goal of improving the quality of life for patients [10].

– European Society for Medical Oncology (ESMO): The ESMO Clinical Practice Guidelines offer a collection of recommendations crafted by renowned experts and grounded in evidence-based medicine research. These guidelines serve as a valuable resource, equipping patients with a range of suggestions to aid them in making informed decisions about their optimal care choices [9].

41.4 Pathology

- Only an accredited oncology-designated lab can diagnose cancer; rare types of cancer, like sarcoma, must be double-read by two pathologists.
- The regulators need to assign a designated pathology laboratory for cancer diagnosis.

41.5 Medical Oncology

- Chemotherapy options must be given to patients. For example, locally advanced breast cancer with *HER2* positive for treatment with neoadjuvant chemotherapy can be treated with 6 cycles of chemotherapy, but instead many are treating patients with 16 cycles. Both are acceptable, but many patients are not informed about the lighter option with similar efficacy.
- Auditing of all the chemotherapy protocols and deliveries should be done by oncologists and oncology pharmacists.
- Flag system should be implemented for excessive use of chemotherapy in the last 2 weeks before death.

41.6 Genomic and Molecular Testing

- Mandate OncotypeDx© coverage for all breast cancer patients who are clinically eligible as per the NCCN guidelines. This will reduce the use of chemotherapy in early breast cancer [11].
- Mandate insurance coverage for molecular testing for cancers as per the NCCN guidelines.

41.7 Radiation

- Quality control and auditing should be done for indications of radiation.
- Change the payment method from per fraction (per day) to per site; this will reduce the excessive use of higher fractions and side effects.
- All radiotherapy patient plans must be peer-reviewed by two radiation oncologists; an exception will be made for urgent or emergency cases. All IMRT/VMAT

plans must have a QA done by a physicist and verified by another physicist with a minimum pass rate of 95%, a tolerance of 3% for the 3 mm gamma pass rate, and for SBRT, 3% between 1 and 2 mm depending on PTV expansion. All 3D plans must undergo RadCalc with a maximum accepted difference of 3%, unless for small fields or off-axis treatment. All plans with a passing QA rate should be checked, and plans signed and PDFs signed should be checked by the radiation therapists before radiotherapy is delivered. All LINACs must undergo daily, weekly, monthly, and yearly QA (this is standard). Independent external audits for LINACs must be conducted before going live and on a yearly basis. For brachytherapy, daily, monthly, quarterly, and yearly quality assurance should be done. CT/MRI-based treatment planning should be done for all patients when possible. Radiation safety and emergency training for all radiotherapy and nuclear medicine staff are mandatory. Image guidance therapy should be used whenever and as much as possible. CT simulation and daily, monthly, and annual QA should be done.

41.8 Surgical Oncology

- All surgeries are approved by MDT except emergency surgeries, and this must be audited regularly.
- Surgical oncology must be done by a trained surgeon in oncology.
- Rectal surgeries must be done by colorectal surgeons.
- Sarcoma surgeries must be done by orthopedic surgeons with special training in oncology.
- Appropriate reimbursement should be ensured for minimally invasive surgery when it is an evidence-based standard of care (for example, robotic-assisted surgery for prostate cancer).

41.9 Palliative Care

- Make palliative care available in all cancer centers and mandatory.
- Insurance coverage should be ensured for palliative home care and hospices.
- Opioid infusion pump approval should be used in the outpatient setting.

41.10 Insurance Limit and Renewal

- Develop a mechanism to cover cancer patients who have reached their financial limit as cancer treatment becomes more expensive.
- Renewal of insurance for cancer patients, as many patients are unable to renew their insurance due to the high premium requested if diagnosed with cancer. The increase should be limited to, say, 20%.

41.11 Cancer Screening

- Establish a cancer screening task force.
- Review and update the current screening guidelines.
- Ensure insurance or third-party coverage for recommended screening tests.
- Educate the GPs and the public about screening by forming pathways for screening.

41.12 Other

- The use of off-label medications should be audited regularly. Patients should be aware of and fully consent to off-label drugs.
- Promote clinical trials in the UAE and make approval and access to international trials easier.

41.13 Conclusion

Cancer care in the UAE has undergone significant advancements in recent decades. However, despite this progress, there is a noticeable absence of implemented quality indicators to ensure the consistent and standardized delivery of cancer treatment across the country. One of the primary challenges stems from the varying levels of experience and training among oncologists who come from diverse backgrounds to practice in the UAE. While some adhere to the NCCN guidelines, others continue to practice independently, potentially impacting the outcomes for cancer patients. In this chapter, we examine recommendations aimed at improving the quality of cancer care in the UAE. Firstly, we recommend the establishment of an independent Advisory Cancer Care Quality Council, responsible for monitoring the quality of care provided in both government and private hospitals throughout the UAE. Additionally, we suggest adopting a cancer system quality index and conducting regular audits of cancer centers to ensure adherence to quality measures. These measures should encompass all aspects of cancer care, including radiation, surgical procedures, palliative care, and others. It is crucial to emphasize quality control measures that address the current cancer screening programs and enhance accessibility for patients.

Conflict of Interest The author has no conflict of interest to declare.

Acknowledgements We thank Dr. Ibrahim Abu-Gheida for his help in writing the manuscript.

References

1. Al-Shamsi H, Darr H, Abu-Gheida I, Ansari J, McManus MC, Jaafar H, Tirmazy SH, Elkhoury M, Azribi F, Jelovac D, et al. The state of cancer care in the United Arab Emirates in 2020: challenges and recommendations, a report by the United Arab Emirates Oncology Task Force. Gulf J Oncolog. 2020;1(32):71–87.
2. Al-Shamsi HO, Abyad A, Kaloyannidis P, El-Saddik A, Alrustamani A, Abu Gheida I, Ziade A, Dreier NW, Ul-Haq U, Joshua TLA, et al. Establishment of the first comprehensive adult and pediatric hematopoietic stem cell transplant unit in the United Arab Emirates: rising to the challenge. Clin Pract. 2022;12(1):84–90.
3. Smith JA, King PM, Lane RH, Thompson MR. Evidence of the effect of 'specialization' on the management, surgical outcome and survival from colorectal cancer in Wessex. Br J Surg. 2003;90(5):583–92.
4. Vernooij F, Heintz P, Witteveen E, van der Graaf Y. The outcomes of ovarian cancer treatment are better when provided by gynecologic oncologists and in specialized hospitals: a systematic review. Gynecol Oncol. 2007;105(3):801–12.
5. Wu MF, Li J, Lu HW, Wang LJ, Zhang BZ, Lin ZQ. Impact of the care provided by gynecologic oncologists on outcomes of cervical cancer patients treated with radical hysterectomy. Onco Targets Ther. 2016;9:1361–70.
6. Reddeman L, Foxcroft S, Gutierrez E, Hart M, Lockhart E, Mendelsohn M, Ang M, Sharpe M, Warde P, Brundage M, et al. Improving the quality of radiation treatment for patients in Ontario: increasing peer review activities on a jurisdictional level using a change management approach. J Oncol Pract. 2016;12(1):81–2, e61–70.
7. Wild C, Patera N. Measuring quality in cancer care: overview of initiatives in selected countries. Eur J Cancer Care (Engl). 2013;22(6):773–81.
8. Anas R, Stiff J, Speller B, Foster N, Bell R, McLaughlin V, Evans WK. Raising the bar: using program evaluation for quality improvement. Healthc Manage Forum. 2013;26(4):191–5.
9. https://www.esmo.org/guidelines. Accessed 7 Sept 2022.
10. https://www.nccn.org/guidelines/category_1. Accessed 7 Sept 2022.
11. Schneider JG, Khalil DN. Why does oncotype DX recurrence score reduce adjuvant chemotherapy use? Breast Cancer Res Treat. 2012;134(3):1125–32.

Prof. Humaid Obaid Al-Shamsi is the Chief Executive Officer of Burjeel Cancer Institute in Abu Dhabi, UAE, President of the Emirates Oncology Society, Lead of the Gulf Cancer Society, Full Professor of Oncology at the Ras Al Khaimah Medical and Health Sciences University, Ras Al Khaimah, UAE, and an Adjunct Professor of Oncology at the College of Medicine, University of Sharjah. He is the first Emirati to be promoted as a professor in oncology in the UAE. He is also the Chairman for Colorectal Cancer in the MENA region, appointed by the prestigious National Comprehensive Cancer Network®. He is also the only member of Lung Cancer Policy Network in the MENA region that aims to advance lung cancer research and screening globally. He is the Chairman of the Oncology and Hematology Fellowship Training Program for the National Institute for Health Specialties in the United Arab Emirates. He is the only member in GCC in the WIN Consortium which is comprised of organizations representing all stakeholders in personalized cancer medicine globally.

He is board-certified in both internal medicine and oncology from the UK, USA (ABIM), and Canada (FRCPC). He has also been awarded the FRCP (London) in 2023 and FRCP (Glasgow) in 2024. He is the only physician in the UAE with a subspecialty fellowship certification and training in gastrointestinal oncology and the first Emirati to train and complete a clinical post-doctoral fellowship in palliative care. He was an assistant professor at the University of Texas MD Anderson Cancer Center between 2014 and 2017. He has published more than 140 peer-reviewed articles in JAMA Oncology, Lancet Oncology, The Oncologist, BMC Cancer, and many others. His area of expertise includes precision oncology and cancer care in the UAE. In 2016, he published with his group from MD Anderson the JCO paper describing a new distinct subgroup of CRC, NON V600 BRAF-mutated CRC. In 2022, he published the first book about cancer research in the UAE and also the first book about cancer in the Arab world, both of which were launched at Dubai Expo 2020. *Cancer in the Arab World* has been downloaded more than 450,000 times in its first 18 months of publication and is the ultimate source of cancer data in the Arab region. He also published the first comprehensive book about cancer care in the UAE which is the first book in UAE history to document the cancer care in the UAE with many topics addressed for the first time, e.g., neuroendocrine tumors in the UAE. He is passionate about advancing cancer care in the UAE and the GCC and has made significant contributions to cancer awareness and early detection for the public using social media platforms. He is considered as the most followed oncologist in the world with over 300,000 subscribers across his social media platforms (Instagram, Twitter, LinkedIn, and TikTok). In 2022, he was awarded the prestigious Feigenbaum Leadership Excellence Award from Sheikh Hamdan Smart University for his exceptional leadership and research and the Sharjah Award for Volunteering. He was also named the Researcher of the Year in the UAE in 2020 and 2021 by the Emirates Oncology Society.

In May 2024, HH Sheikh Mansour bin Zayed Al Nahyan, Vice President of the United Arab Emirates, awarded him the first place in UAE Nafis program for outstanding leadership in private sector across all business and medical disciplines. Beside his clinical and administrative duties, he is engaged in education and various levels of research training for medical trainees to enhance their clinical and research skills. His mission is to advance cancer care in the UAE and the MENA region and make cancer care accessible to everyone in need around the globe.

Appendix A: The National Guideline for Breast Cancer Screening and Diagnosis

© The Editor(s) (if applicable) and The Author(s), under exclusive license to
Springer Nature Singapore Pte Ltd. 2024
H. O. Al-Shamsi (ed.), *Cancer Care in the United Arab Emirates*,
https://doi.org/10.1007/978-981-99-6794-0

UNITED ARAB EMIRATES
MINISTRY OF HEALTH & PREVENTION

THE NATIONAL GUIDELINE FOR
BREAST CANCER SCREENING AND DIAGNOSIS

3RD EDITION 2023

National Breast Cancer Screening Task Force

Dr. Huda Obaid Al Abdouli—Head of Task Force
Coordinator of National Cancer Screening Program—MOHAP
Dr. Buthaina Bin Belaila—Member
Head of Non-communicable Disease and Mental Health—MOHAP
Prof. Humaid Obaid Al Shamsi—Member
Director of the Burjeel Cancer Institute
President of the Emirates Oncology Society
Dr. Mona Al Ayyan—Member
Consultant General Surgeon/Breast Surgeon
Head of Surgery Department
Saqr Hospital Director
Dr. Lamia Safieldin—Member
Comprehensive Screening Specialist
Department of Health, Abu Dhabi
Dr. Asma Saeed Khammas—Member
Consultant Radiologist and Interventional Breast Imager—EHS
Head of Breast Imaging Unit—Fujairah Hospital
Dr. Nehad Kazim Albastaki—Member
Consultant Breast Radiology
Private Healthcare Sector—Abu Dhabi
Dr. Ola Aldafrawy—Member
Consultant Family Medicine
Dubai Health Authority
Dr. Aydah Al Awadhi
Consultant Physician
TWM-Oncology Clinic-TWM-Medical Affairs
Tawam Hospital
Scientific Chairperson of Emirates
Oncology Society

---

1. Purpose

1.1. To stipulate the service requirements to deliver the National Breast Cancer Screening Program in the United Arab Emirates.

1.2. To set out the minimum Clinical Care Standards and frequency for breast cancer screening as per international evidence-based guidelines.

1.3. To set out the case mix, eligibility criteria and data reporting requirement for breast cancer screening.

1.4. To ensure the population receives quality and safe care and timely referral for diagnosis and/or treatment where appropriate.

2. Scope

2.1. This guideline applies to all healthcare providers (facilities and professionals) in the United Arab Emirates, providing, breast cancer screening and assessment and diagnosis services; including mobile units.

3. Definitions

3.1. **Case mix**: Refer to all females, 40–69 years, determined as eligible for breast cancer screening services, in accordance with the criteria detailed in this guideline.

3.2. **Screening mammograms**: Are carried out for healthy women, who have no symptoms of breast cancer and negative clinical breast examination.

3.3. **Diagnostic mammograms**: Are performed to evaluate a breast complaint or abnormality detected by clinical breast examination or routine screening mammogram.

3.4. **Clinical breast examination (CBE)**: Is an exam conducted by healthcare professional and involves inspection and palpation of all breast tissue including lymph nodes basins.

3.5. **Breast awareness**: Women, 20 years and older, should be encouraged and educate on how to conduct breast self-exam to become aware of the feeling and shape of their breasts, and to report any changes immediately to their healthcare provider.

3.6. **Breast assessment and diagnosis**: Further imaging, clinical breast exam and needle biopsy. The aim of assessment is to obtain a definitive and timely diagnosis of all potential abnormalities detected during screening.

3.7. **First degree relatives**: Parents, siblings, and children
   Second degree relatives: Grandparents, aunts, uncles, nieces, nephews, grandchildren, and half siblings
   Third degree relatives: Great-grandparents, great-aunts, great-uncles, great-grandchildren, and first cousins (refer to Appendix D)

3.8. **Initial screening**: First screening examination of individual women within the screening program, regardless of the organisational screening round in which women are screened.

3.9. **Subsequent screening**: All screening examinations of individual women within the screening program following an initial screening examination, regardless of the organisational screening round in which women are screened. There are two types of subsequent screening examinations:

   3.9.1. Subsequent screening at the regular screening interval, i.e. in accordance with the routine interval defined by the screening policy (SUBS-R).

   3.9.2. Subsequent screening at irregular intervals, i.e. those who miss an invitation to routine screening and return in a subsequent organizational screening round (SUBS-IRR).

4. Duties for Healthcare Providers

All licensed healthcare providers facilities and professionals engaged in providing breast cancer screening and diagnosis services must:

4.1. Provide clinical services and patient care in accordance with this guideline and in accordance with Policies and Standards, Laws and Regulations of the United Arab Emirates; including developing effective recording systems, maintaining confidentiality, privacy and security of patient information.

4.2. Comply with the Federal requirements; laws and policies for patient education and consent. The licensed provider must provide appropriate patient education and information regarding the screening test and must ensure that appropriate patient consent is obtained and documented on the patient's medical record.

4.3. Comply with Federal requirements; laws, policies and standards on managing and maintaining patient medical records, including developing effective recording systems, maintaining confidentiality, privacy and security of patient information.

4.4. Comply with Federal requirements; laws, policies and standards for Information Technology ("IT") and data management, electronic patient records and disease management systems, sharing of screening and diagnostic test, and where applicable pathology results.

4.5. Comply with relevant policies on cultural sensitivity; in particular, providers must ensure:

4.5.1. That only female radiographers, mammographers or technologists are allowed to perform mammographic examination for women.

4.5.2. That the timing of screening appointment for women seeking the service is not delayed beyond 15 working days, due to the limited number of same sex appropriately licensed professionals.

4.5.3. Where delays are likely to occur due to limited availability of same sex licensed professionals at the employing facility, or where there is no female radiographer, that the provider communicates this to the patient and refers/recommends that the patient seeks screening services from another provider.

4.6. Comply with MOHAP requests to inspect and audit records and cooperate with authorized auditors as required.

4.7. Collect and submit data on screening visits and outcomes, as per Appendix B, to the National Cancer Screening Registry; at MOHAP.

4.8. Comply with Federal laws, policies and standards on cancer case reporting and report all confirmed screening-detected cancers to the National Cancer Registry at MOHAP.

5. Enforcement and Sanctions

5.1. Healthcare providers, payers and third-party administrators must comply with the terms and requirements of this guideline. MOHAP may impose sanctions in relation to any breach of requirements under this guideline.

6. Payment for Screening and Follow Up of Breast Cancer

6.1. Eligibility for reimbursement under the Health Insurance Scheme must be in accordance, with local insurance laws for each Emirate.

7. Standard 1: Clinical Services Specifications

7.1. **Breast Cancer Screening Service**
All licensed healthcare facilities providing breast cancer screening services must:

7.1.1. Follow best practice for breast cancer screening and diagnosis care pathways and recommendation of breast cancer screening per Appendices C and D.

7.1.2. Adhere to the clinical performance indicators and timelines for referral in accordance with Appendix E; and ensure availability of evidence of compliance with these indicators.

7.1.3. Comply with requirement of breast screening unit, detailed in Appendix F.

7.1.4. Have an approved referral protocol for referral of women with screen detected abnormalities for further breast assessment unit or treatment.

7.1.5. Establish and maintain record of mammogram outcomes, audit program to follow up positive mammography assessments and to correlate pathology results with the interpreting physician's findings.

7.1.6. Assign a breast cancer facility program coordinator/director who will be accountable to:

7.1.6.1. Report and submit screening visits and outcome data, specified in Sect. 4.

7.1.6.2. Establish internal audit policies and procedures and conduct regular audits, monitoring and evaluation to demonstrate compliance with this guideline and other associated regulatory policies and standards.

7.2. **Breast Assessment and Diagnosis Services**

7.2.1. Breast assessment and diagnosis services must be carried out in Diagnostic Breast Assessment Unit.
These unit must:

7.2.2. Comply with the requirements of Diagnostic Breast Assessment Unit, described in Appendix F.

7.2.3. Comply with breast cancer screening and diagnosis care pathways, clinical quality indicators, and timelines for referral in accordance with Appendices B and E.

7.2.4. Have approved written protocols for the screening assessment and diagnosis; that clearly define the methods of assessment and the diagnostic pathways for all possible assessment outcomes.

7.2.5. Women who require further assessment must be managed in accordance with internationally best practices and recommended guidelines such those of the National Health System Breast Screening Program (NHSBSP) clinical guidelines for breast cancer screening assessment or the National Comprehensive Cancer Network NCCN Breast Cancer Screening and Diagnosis.

7.2.6. Establish internal audit procedures to demonstrate compliance with this guideline and other associated regulatory policies and standards.

7.3. **All licensed healthcare professionals participating in breast cancer screening and diagnosis must**:

7.3.1. Have knowledge of the principles of breast cancer screening, assessment, diagnosis and management.

7.3.2. Participate in continuing medical education and take part in any recognized external quality assessment schemes concerning radiologists and radiographers; to allocate 40% of annual recommended CME for breast imaging.

7.3.3. Conduct breast cancer risk assessment. Detailed history, such as that described in, Appendix B, must be evaluated and completed, each time a woman visits for screening. The purpose of this is to identify risk status, as per risk categories specified in Appendix C and referral women to appropriate screening tests.

7.3.4. Inform all individuals of the procedures and expected time frame to be screened and to receive results.

7.3.5. Ensure that the outcome of screening for breast cancer is reviewed by a multi-disciplinary team involving a full range of specially trained professionals including a radiologist, radiographer, pathologist, surgeon, nurse counselor and medical oncologist/radiotherapist.

7.3.6. Follow up and timely referral of women with abnormal results to further assessment or treatment.

8. Standard 2: Recruitment for Screening

Women eligible for breast cancer screening may be recruited by the healthcare facilities, through the following:

8.1. Targeted invitation

8.1.1. All facilities providing breast cancer screening and diagnosis services must establish an invitation system to ensure identification, successful participation and retaining of eligible population.

8.1.2. Targeted invitation may be established via an electronic or manual invitation system.

8.2. Opportunistic

8.2.1. Physician consultation for related or unrelated reason

8.2.2. Engagement in a health promotion campaign

9. Standard 3: Breast Cancer Screening

9.1. Breast cancer screening must be provided in accordance with the breast screening and diagnosis care pathway as provided in Appendix B, including the following activities:

9.1.1. History and risk assessment

9.1.2. Screening mammogram

9.2. Periodical screening must be carried out as specified in breast cancer screening recommendations in Appendix C.

9.3. Detailed history, such as that described in Appendix B, must be evaluated and completed by the screening facility nurse, each time a woman visits for screening. The purpose of that is to identify patient at increased risk and determine the appropriate screening tests.

9.4. Screening mammography must involve two X-ray images for each breast; craniocaudal (CC) and mediolateral oblique (MLO).

9.5. Digital breast tomosynthesis is recommended as adjunct to screening mammogram. for women with high mammographic breast density and to be considered for women at increased risk in accordance with Appendix C (Reference: European Commission Initiative on Breast Cancer (ECIBC). https://ecibc.jrc. ec.europa.eu/recommendations/. 23 Mar 2023.)

9.6. Women must be provided with (oral and written) education and information, regarding benefits, risk and limitation of breast cancer screening, and about the screening test, associated procedures and expected time frames to receive results.

9.7. Adequate attention must be given to the level of literacy, diversity and linguistic requirements of different populations.

10. Standard 4: Breast Assessment and Diagnosis

10.1. Breast cancer assessment and diagnosis must be provided in accordance with the clinical care pathway and timelines for referral (Appendices B and C).

10.2. Women with abnormal mammogram, who require further assessment and diagnosis must be recalled/referred to Diagnostic Breast Assessment Unit within 15 working days of screening mammogram.

10.3. Assessment and diagnostic work up of screen detected abnormality is best achieved using the triple assessment:

10.3.1. Imaging; usually diagnostic mammography and ultrasound

10.3.2. Clinical examination

10.3.3. Image-guided needle biopsy for histological examination, if indicated

10.3.4. Cytology alone must not be used to obtain a non-operative diagnosis of breast cancer

10.3.5. Clinical examination is mandatory for every woman with a confirmed mammographic or ultrasound abnormality that needs needle biopsy and for all women recalled because of clinical signs or symptoms

10.4. Clinical examination is not mandatory for women whose further imaging is entirely **normal**.

10.5. Core needle biopsy must be performed under image guidance.

10.6. A clip must be placed at site of biopsy during the procedure of needle sampling to identify the lesion/s location; especially in non-palpable lesions.

10.7. Results of assessments must be evaluated and considered by a multidisciplinary team (MDT). Particular attention must be given to address radiology-pathology correlation.

10.8. Early recall for repeat mammography either in screening or diagnostic settings is not recommended and must never be used as a substitute for inexpert or inadequate assessment.

10.9. Early recall rate must be recorded, monitored and audited.

11. Standard 5: Reporting of Screening Mammogram

11.1. Double reading of screening mammogram is mandatory. Mammograms must be interpreted by two independent radiologists.

11.2. In case of discordant opinions between two radiologists, either consensus or preferably arbitration using a third expert screening radiologist can be carried out.

11.3. The final assessment must be reported using the FDA-approved Breast Imaging, Reporting and Data System (BI-RADS®) Final Assessment Categories as described in Appendix G.

11.4. Screening mammograms that require additional assessment tools/imaging should be rated as BI-RADS® 0, 4 or 5 depending on initial evaluation and readers experience.

11.5. Only after full assessment, with additional imaging and/or comparison with prior mammogram; BI-RADS® 1, 2, 4 or 5 can be assigned.

11.6. One final mammogram report to be issued, A synoptic breast imaging report must be used by radiologists containing at least the following information:

    11.6.1. Interpreting physicians' names

    11.6.2. Date of examination

    11.6.3. Patient identification

    11.6.4. Reason for examination

    11.6.5. Breast density

    11.6.6. Description of significant imaging lesions: mammographic characteristics of the lesion; location (in quadrants); distance from the nipple (in mm); and size (maximum diameter in mm)

    11.6.7. Final assessment (BI-RADS®)

    11.6.8. Detailed recommendations should be included in the report

12. Standard 6: Screening Outcomes

12.1. All women must be informed about the results of screening within 3 weeks (15 working days) from date of screening mammogram.

12.2. Women with screening mammogram of normal/benign (BI-RADS® 1/2), are discharged to routine screening. Screening frequency will follow recommendation specified in Appendix G.

12.3. If a woman requires further assessment for abnormal screening mammogram (BI-RADS® 0) or clinical breast exam, referral must be to a Diagnostic Breast Assessment Unit within 5 days of screening mammogram result.

12.4. Women must be notified with assessment results within 5 days of assessment tests.

12.5. At the end of the screening, women must be provided with one final written mammogram report.

12.6. It is the responsibility of the radiologist/referring physician (at the screening or assessment facilities) to inform women regarding her screening and assessment results. Also, send feedback to referring physician at the primary healthcare clinic.

Appendix B: Breast Cancer Screening Pathways

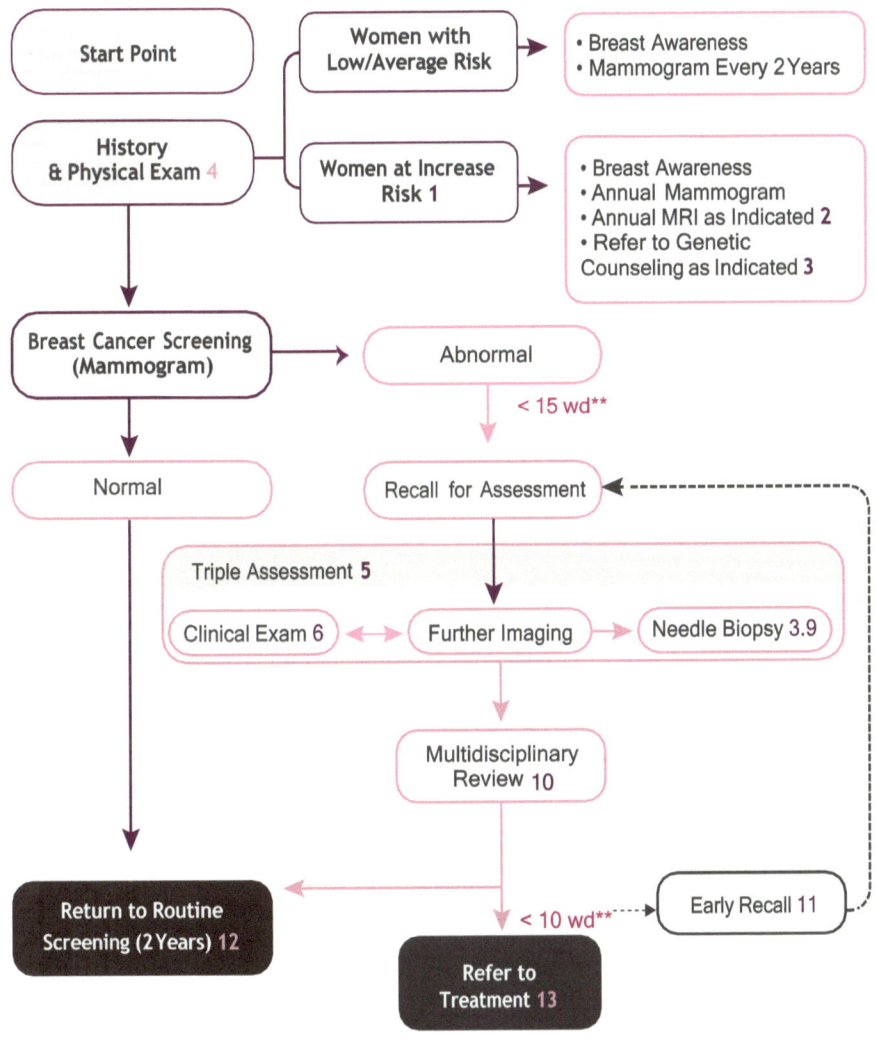

Key

1. Women at increased risk of breast cancer are defined in Appendix C of the standard for the screening and diagnosis of breast cancer.
2. Indication for MRI is stipulated in Appendix C of the standard for the screening diagnosis of breast cancer.
3. Criteria for referral to genetic counselor is detailed in Table C.2.
4. Women with the following criteria should be excluded from screening mammogram: pregnant, breast feeding, had bilateral mastectomy, and had recent mammogram within 24–12 months, under the age of 40, unless she is at increased risk.
5. Triple assessment must be performed in diagnostic breast assessment unit. Requirement of a diagnostic breast assessment unit is detailed in Appendix B.
6. Clinical examination is mandatory for every woman with a confirmed mammographic or ultrasound abnormality that needs needle biopsy.
7. Further imaging usually involves further diagnostic mammography and/or ultrasound.
8. Needle biopsy should be performed under image guidance. Clip placement is done at the time of core needle biopsy to identify lesion locations.
9. Cytology should no longer be used alone to obtain a non-operative diagnosis of breast cancer.
10. Result of assessments are recommended to be discussed by a multidisciplinary team. Women must be informed about results within 5 working days.
11. Early recall is exceptional screening outcome and should be monitored and audited.
12. Screening frequency will follow recommendation specified in Appendix C.
13. Referral of histologically confirmed cancer cases to treatment must be made within 10 working days, following diagnosis.

References

1. NCCN Clinical Practice Guidelines in Oncology, Breast Cancer Screening and Diagnosis. V.1.2022.
2. NHS Clinical Guidelines for Breast Cancer Screening Assessment, NHSBSP publication no. 49.
3. The National Health System (NHS) Cancer Screening Programmes. Technical guidelines for magnetic resonance imaging for the surveillance of women at higher risk of developing breast cancer, NHSBSP publication no. 68.
4. The National Comprehensive Cancer Network (NCCN) Clinical Practice Guidelines in Oncology, genetic/familial high-risk assessment: breast and ovary.

Appendix C: National Breast Cancer Screening Recommendation (Table C.1)

Women at Increased Risk/High Risk

A woman is considered at higher risk of developing breast cancer if she has one or more of the following criteria:

- Previous history of breast cancer.
- Previous treatment with chest radiation at age younger than 30.
- Lobular carcinoma in situ (LCIS) or atypical ductal hyperplasia (ADH) or atypical lobular hyperplasia (ALH), on previous breast biopsy.
- Strong family history or genetic predisposition.
  Criteria of personal or family history of a woman to be categorized as high risk and to follow the high-risk protocol:
  A woman is considered at higher risk of developing breast cancer if she has one or more of the following criteria:

Table C.1 A summary of the national breast cancer screening recommendation

Screening category	Age	Screen assessment tools
Women at average risk	40–69 years 70 years and above	• Mammogram every 2 years • Self-referred
Women at increased/high risk	Age of initiation is individualized according to risk (Table C.2)	• Annual mammogram screening • Begin 10 years prior to the youngest affected family member but not prior to 25 years • Consider tomosynthesis • Annual MRI screening—as indicated not prior to 25 years • Referral to genetic counselor for strong familial/genetic predisposition

Adapted from: NCCN Clinical Practice Guidelines in Oncology. Breast Cancer Screening and Diagnosis. V.1.2022

Table C.2 National screening recommendations for women at increased risk

Previous treatment with chest radiation at a young age (between age of 10 and 30)	Age <25 years	• Screening begin 10 years after radiotherapy
	Age ≥25	• Annual mammography screening (begin 10 years after radiotherapy but not prior to age of 30 years) • Consider tomosynthesis • Annual MRI screening (begin 10 years after radiotherapy but not prior to age of 25 years)
Strong family history or genetic predisposition[a]	Age <25 years	• Referral to genetic risk assessment
	Age ≥25 years	• Annual mammography screening 10 years before the youngest family member but 25 • Youngest family member but not prior to 10–25
Previous history of breast cancer		Surveillance protocol • Annual mammography screening
Lobular carcinoma in situ (LCIS) or atypical ductal hyperplasia (ADH) or atypical lobular hyperplasia (ALH) on previous breast biopsy		Annual mammogram screening: • To begin at diagnosis of LCIS or ADH/ALH but not prior to age of 25 years • Consider annual MRI screening to begin at diagnosis of LCIS or ADH/ALH but not prior to age of 25 years

NCCN Clinical Practice Guidelines in Oncology. Breast Cancer Screening and Diagnosis. V.1.2022
[a] Screening and assessment of women with genetic/familial high risk are individualized and should be in accordance with recognized international guidances such as NCCN guideline

Personal History
- History of breast cancer
- History of ovarian cancer
- Gene mutation: BRCA1, BRCA2, TP 53, or PTEN mutation
- Previous treatment with chest radiation at age younger than 30
- Lobular carcinoma in situ (LCIS) or atypical ductal hyperplasia (ADH) or atypical lobular hyperplasia (ALH), on previous breast biopsy

Family History
- One first degree female relative with
  - Breast cancer diagnosed <50 years
  - Ovarian cancer at any age
  - Bilateral breast cancer where the first diagnosed <50 years
- Two or more first degree relatives, with breast cancer
- One of first-degree or second-degree relative diagnosed with breast cancer or ovarian cancer at any age
- One first-degree male relative with breast cancer at any age
- Having a first-degree relative with gene mutation (BRCA1, BRCA2, TP 53, or PTEN)

References

1. NICE, familial breast cancer. V.1.2022.
2. NCCN Clinical Practice Guidelines in Oncology. Breast Cancer Screening and Diagnosis. V.1.2022.

Criteria of use of MRI as adjunct to mammogram for high-risk women:

- Having BRCA1, 2 mutation
- Having a first-degree relative with BRCA1, 2 mutation
- Received chest radiation between age 10 and 30
- Carry or have a first-degree relative who carries mutation in TP 53 or PTEN genes

Criteria to merit referral for genetic risk evaluation:

- Ovarian cancer
- Male breast cancer
- Personal history of three or more of the following (especially if diagnosed before age of 50 and can include multiple primary cancers in the same individual)
- Breast cancer
- Colon cancer
- Diffuse gastric cancer
- Pancreatic cancer
- Prostate cancer
- Thyroid cancer
- Brain tumors
- Endometrial cancer
- Brain tumors
- Adrenocortical carcinoma
- Melanoma
- Sarcoma
- Leukemia
- Kidney cancer
- Hamartomata's polyps of GI tract

An individual with a breast cancer diagnosis meeting any of the following:

- A known mutation in a cancer susceptibility gene within the family (BRAC1/2, TP 53, or PTEN)
- Early age-onset breast cancer ≤45
- Triple negative (ER-, PR-, HER-) breast cancer and age ≤60
- Two breast cancer primaries in a single individual
- Breast cancer at any age
    - >One close blood relative with breast cancers 50 years
    - >One close blood relative (first or second or third degree) with invasive ovarian cancer at any age
    - >Two close blood relative with breast cancer, prostate cancer and/or pancreatic cancer at any age

- Personal history of pancreatic cancer at any age

An individual with no personal history of cancer but with:

- A close relative with any of the following:
  - >Two breast cancer primaries in a single individual
  - >Two different individuals with breast cancer primaries from the same side of the family maternal or paternal with at least one diagnosed before 50
  - Ovarian cancer
  - Male breast cancer
- First-or second-degree relative with breast cancer ≤45 years
- Family history of three or more of the following (especially if diagnosed before age of 50 and can include multiple primary cancers in the same individual)
  - Breast cancer
  - Colon cancer
  - Diffuse gastric cancer
  - Pancreatic cancer
  - Prostate cancer
  - Thyroid cancer
  - Brain tumors
  - Endometrial cancer
  - Brain tumors
  - Adrenocortical carcinoma
  - Melanoma
  - Sarcoma
  - Leukemia
  - Kidney cancer
  - Hamartomata's polyps of GIT

N. B. Maternal and paternal sides of the family should be considered independently for familial pattern of cancer. First degree: mother, sister, daughter, brother, and father—second degree: grandmother, aunt, niece, and nephew.

References

1. NCCN Clinical Practice Guidelines in Oncology. Breast Cancer Screening and Diagnosis. V.1.2022.
2. The National Comprehensive Cancer Network (NCCN) Clinical Practice Guidelines in Oncology. Genetic/familial high-risk assessment: breast and ovary. V.1.2022.

Appendix D: Pedigree: First-, Second-, and Third-Degree Relatives of Proband

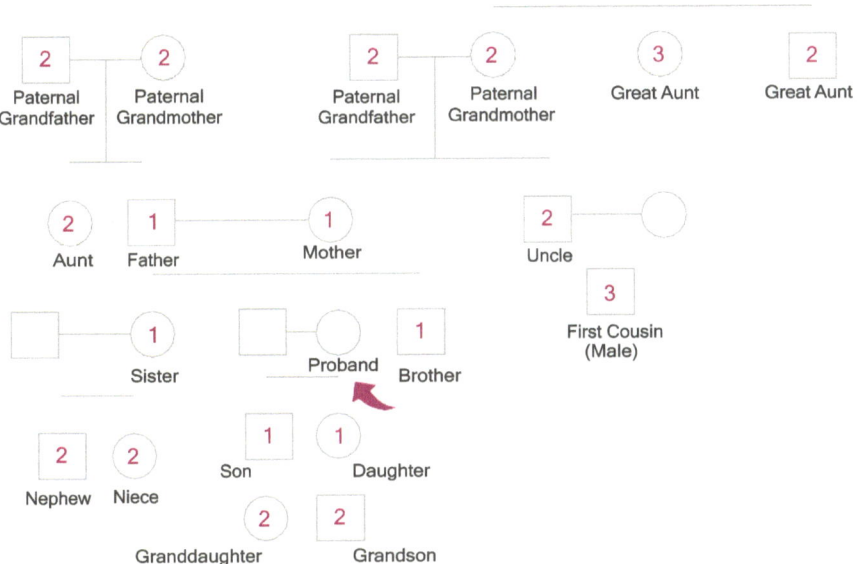

First-degree relatives: parents, siblings, and children

Second-degree relatives: grandparents, aunts, uncles, nieces, nephews, grandchildren, and half siblings

Third-degree relatives: great-grandparents, great-aunts, great-uncles, great-grandchildren, and first cousins

Appendix E: National Breast Cancer Screening Clinical Performance Indicators

Clinical quality indicators	Definition	Calculation	Acceptable level		Desirable level
1. Participation rate	Percentage of women 40–74 years who have a screening mammogram (calculated biennially) as a proportion of the eligible population	Number of women screened at least once (per 2 years period)/target population [(first- and second-year populations averaged from census/forecast)] × 100	>70%		>75%
2. Retention Rate	The estimated percentage of women 40–74 years who are re-screened within 30 months of their previous screen	Kaplan–Meier method[a]	Auditable outcome		>75%
3. Technical repeat rate	Proportion of women undergoing a technical repeat screening examination	[Number of women undergoing a technical repeat/ Number of women screened] × 100	<3%		<1%
4. Abnormal recall rate	Proportion of women recalled for further assessment	[Number of recalls due to abnormal screens/Number of women screened] × 100	At initial screening	Auditable outcome	<7–10%
			At subsequent screening	Auditable outcome	<5–7%
5. Early recall rate	Proportion of women undergoing a technical repeat screening examination	[Number of subjected for early recall/Number of women screened] × 100	<1%		0%
6. Positive predictive value	Proportion of abnormal cases with completed follow-up found to have breast cancer	[Number of screen detected/Number of abnormal screens with complete work-up] × 100	At Initial screening		>5%
			At subsequent screening		>6%

H. O. Al-Shamsi (ed.), *Cancer Care in the United Arab Emirates*, https://doi.org/10.1007/978-981-99-6794-0

Clinical quality indicators	Definition	Calculation	Acceptable level		Desirable level
7. Invasive cancer detection rate	Number of invasive cancers detected per 1000 screens	[Number of invasive cancers detected/ Number of screens] × 1000	At Initial screening		>5 per 1000
			At subsequent screening		>3 per 1000
8. In situ cancer detection rate	Number of ductal carcinomas in situ (DCIS) detected per 1000 screens	[Number of DCIS detected/Number of screens] × 1000	At initial screening		>0.4 per 1000
			subsequent screening		>0.4 per 1000
9. Invasive cancer tumor size	Proportion of invasive screen-detected cancers that are <10 mm in size	[Number of invasive tumor $\leq$10 mm/ Total number of invasive tumors] × 100	Initial screening		20% $\geq$25%
			Subsequent screening		$\geq$25% $\geq$30%
10. Invasive cancer detection rate	Number of women with diagnosis of invasive breast cancer after a normal screening within 12 and 24 months of screen date	[Number of cancers detected in the 0–12-month interval after a normal screening episode/ Total person-years at risk (0–12 months post screen)] × 10,000	Within the first year (0–11 months)		<Per 10,000
			Within the second year (12–23 months)		12 per 10,000
11. Time interval	– Screening mammography and result within 15 working days (wd)			95%	>95%
	– Screening and offered assessment within 5 working days (wd)			90%	>90%
	– Assessment and issuing of results within 5 working days (wd)			90%	>90%
	– Non-operative (needle) biopsy and result 5 working days (wd)			>90%	100%

[a] Refer to Ref. [2] for calculation

References

1. European guidelines for quality assurance in breast cancer screening and diagnosis. Update Mar 2023.
2. Public Health Agency of Canada. Report from the Evaluation Indicators Working Group. Guidelines for monitoring breast screening program performance. Update 2022.

Appendix F: Requirement for Breast Screening and Diagnosis Services

---

Requirement for Breast Screening Unit

1. General
    1.1. Assign a screening program director/coordinator who will be in charge of overall performance, quality assurance of the unit and will be responsible for submitting data on screening visits and outcomes to MOHAP.
    1.2. Perform at least 1000 mammograms a year.
    1.3. Be able to perform risk assessment, physical examinations, and screening mammogram.
    1.4. Monitor data and feedback of results. Keep a formal record of mammogram results, assessment processes, and outcomes.
2. Invitation system
    2.1. Operate a successful personalized invitation system and/or a promotional campaign as well as an organized system for re-inviting all previously screened women.
3. Mammography equipment
    3.1. Specifications must meet recognized standards such as the MQSA final rule published by the FDA.
    3.2. Subject to regular radiographic and physicist quality-controlled tests, in concordance with MQSA rule.
    3.3. Equipment must be maintained and serviced in accordance with the manufacturer's guidelines and service specifications, records must be maintained by providers.
4. Radiographers
    4.1. Radiographers, mammographers, or technologists performing the mammographic examination must have had at least 40 h of training specific to the radiographic aspects of mammography
    4.2. Regularly participate in external quality assessment schemes and radiographic update courses.

5. Radiologists
   5.1. Must have at least 60 h of training specific to mammography.
   5.2. Must read mammograms from a minimum of 1000 screening mammograms annually. Have centralized reading or, in a case of a decentralized programmer, centralized double.
   5.3. This radiologist must take full responsibility for the image quality of the mammograms reported and ensure that where necessary images are repeated until they are of satisfactory standard. The number of all repeated examinations should be recorded.
6. Referral, assessment and feedback
   6.1. Keep a formal record of mammogram results, assessment processes, referrals, and outcomes.
   6.2. Maintain record of mammogram results, referrals, assessment processes, and outcomes.
   6.3. Have an approved protocol for referral of women with screen detected abnormalities to diagnostic breast assessment unit.

Requirement for a Breast Assessment/Diagnostic Unit

1. General
   1.1. Perform at least 2000 mammograms a year.
   1.2. Be able to perform physical examinations and ultrasound examinations as well as the full range of radiographic procedures. Provide cytological examination and/or core biopsy.
   1.3. Sampling under radiological (including stereotactic) or sonographic guidance.
   1.4. Monitor data and feedback of results.
   1.5. Keep a formal record of mammogram results, assessment processes, and outcomes.
2. Physic-technical
   2.1. Have dedicated equipment specifically designed for application in diagnostic mammography, e.g., mammography system with magnification ability and dedicated processing, and be able to provide adequate viewing conditions for mammograms.
   2.2. Have dedicated ultrasound and stereotactic system and needle biopsy device for preoperative tissue diagnosis.
   2.3. Comply with specifications of recognized standards such as the MQSA final rule published by the FDA.
3. Radiographers
   3.1. The radiographers, technologists, or other members of staff performing the mammographic examination must have had at least 40 h of training specific to the radiographic aspects of mammography and regularly participate in external quality assessment schemes and radiographic update courses. These persons must be able to perform good quality mammograms. There should be a nominated lead in the radiographic aspects of quality control.

4. Radiologists
    4.1. Employ a trained radiologist, i.e., a person who has had at least 60 h of training specific to mammography and who in volume reads at least 1000 mammograms per year.
5. Pathology support
    5.1. Have organized and specialist cyto / histopathological support services.
6. Multidisciplinary activities
    6.1. Participate in multidisciplinary communication and review meetings with others responsible for diagnostic and treatment services.

Appendix G: BI-RADS® Final Assessment Categories

CPT— evaluation code	BI-RADS score	Description	Definition
3340F	0	Incomplete. Need additional imaging	The mammogram or ultrasound didn't give enough information to make a clear diagnosis; follow-up imaging is necessary and/or prior mammogram for comparison
3341F	1	Negative	Negative, continue biannual screening mammography (for women 40 and older)
3342F	2	Benign	Benign (non-cancerous) finding, same statistics and plan of follow-up as level 1. This category is for cases that have a finding that is characteristically benign such as cyst of fibro adenoma
3343F	3	Probably benign	Probably benign finding, there is less than 2% chance if cancer, additional examinations done to clear the situation at once
3344F	4	Suspicious 4A AB 4C	Suspicious abnormality. Findings do not have the classic appearance of malignancy. But are sufficiently suspicious to justify recommended biopsy. Carry 2–95% chance of being malignant finding. 4A: finding with a low suspicion of being cancer (>2% and ≤10%) 4B: finding with an intermediate suspicion of being cancer (>10% and ≤50%) 4C: finding of moderate concern of being cancer but not as high category 5 (>50% and <95%)

CPT— evaluation code	BI-RADS score	Description	Definition
3345F	5	Highly suggestive of malignancy	Highly suggestive of malignancy. Classic sign of cancer is seen on the mammogram. All category 5 abnormalities typically receive biopsy and if the biopsy results are benign, the abnormality usually receives re-biopsy since the first biopsy may not have sampled the correct area. Depending on how category 4 and 5, the percentage of category 5 abnormalities that will be cancer may vary between 75% and 99%
3350F	6	Known biopsy proven malignancy	Lesions known to be malignant that are being imaged prior to definitive treatment; assure that treatment is completed

References

1. NCCN breast cancer risk reduction V.1.2023.
2. NICE, familial breast cancer, V.3. 2023.
3. NCCN Clinical Practice Guidelines in Oncology. Breast Cancer Screening and Diagnosis. V.1.2022.
4. The National Comprehensive Cancer Network (NCCN) Clinical Practice Guidelines in Oncology. Genetic/familial high-risk assessment: breast and ovary. V.1.2022.
5. European guidelines for quality assurance in breast cancer screening and diagnosis, Nov 2022.
6. International Agency for Research on Cancer Handbook Working Group, vol. 33(3), 2 Mar 2021
7. Guidelines for Monitoring Breast Screening Program Performance, vol. 20. 2020. p. 795.

Appendix H: The National Guideline for Cervical Cancer Screening and Diagnosis

UNITED ARAB EMIRATES
MINISTRY OF HEALTH & PREVENTION

THE NATIONAL GUIDELINE FOR
CERVICAL CANCER SCREENING AND DIAGNOSIS

3RD EDITION 2023

National Cervical Cancer Screening Task Force

Dr. Huda Obaid Al Abdouli—Head of Task Force
Coordinator of National Cancer Screening Program—MOHAP
Dr. Buthaina Bin Belaila—Member
Head of Non-communicable Disease—MOHAP
Head of National Cancer Screening Program
Prof. Humaid Obaid Al Shamsi—Member
Director of the Burjeel Cancer Institute
President of the Emirates Oncology Society
Dr. Lamia Safieldin—Member
Comprehensive Screening Specialist
Department of Health—Abu Dhabi
Dr. Saad Ghazal Aswad—Member
Chair of Department
General-Obs/Gyno Clinic—Medical Affairs
Tawam Hospital
Dr. Wafa Adel Albayati—Member
Consultant and Head of Department Ob/Gyn
Khorfakan Hospital, EHS
Dr. Suad Hashim Ahmad—Member
Consultant Family Physician
Dubai Health Authority
Dr. Kauser Mansoor Baig—Member
Sr. Consultant in Obstetrics and Gynecology
Prof. and Head of the Department
Sharjah University Hospital
Dr. Saba Alsayari—Member
Consultant in Obstetrics and Gynecology
Head of the Department—Dubai
Dr. Sahar Ibrahim Yassa—Member
Specialist Obstetrics and Gynecology—EHS
The National Cervical Cancer Screening Program

1. Purpose

1.1. This guideline mandates the clinical service specifications and data reporting for National Cervical Cancer Screening Program in the UAE.

1.2. It specifies the clinical care pathway and minimum service standards and specifications to ensure that women screened for cervical cancer receive quality and safe care and timely referral for diagnosis and/or treatment.

2. Scope

2.1. This guideline applies to all healthcare providers (facilities and professionals) licensed in UAE and providing cervical cancer screening services.
2.2. Participating healthcare providers should offer the following services as applicable based on their license category:
    2.2.1. Risk assessment and physical examination.
    2.2.2. Specimen collection and preparation of adequate cervical smear.
    2.2.3. Handling and reporting of cervical smears.
    2.2.4. Follow-up and referral.
2.3. Follow reporting terminologies defined as per Appendix I.

3. Duties of the Healthcare Providers

All licensed healthcare providers; facilities and professionals engaged in providing cervical cancer screening services must:

3.1. Provide clinical services and patient care in accordance with this guideline and in accordance with Policies and Standards, Laws and Regulations of the United Arab Emirates; including developing effective recording systems, maintaining confidentiality, and privacy and security of patient information.
3.2. Comply with the federal requirements; laws and policies for patient education. The National Cervical Cancer Screening Program and consent. The licensed provider must provide appropriate patient education and information regarding the screening test and must ensure that appropriate patient consent is obtained and documented on the patient's medical record.
3.3. Comply with Federal requirements, laws, policies, and standards on managing and maintaining patient medical records, including developing effective recording systems, maintaining confidentiality, privacy and security of patient information.
3.4. Comply with Federal requirements; laws, policies, and standards for Information Technology ("IT") and data management, electronic patient records and disease management systems, sharing of screening and diagnostic test, and where applicable pathology results.
3.5. Comply with MOHAP requests to inspect and audit records and cooperate with authorized auditors as required.
3.6. Collect and submit data on screening visits and outcomes, to the National Cancer Screening Registry, at MOHAP.
3.7. Comply with Federal laws, policies, and standards on cancer case reporting and report all confirmed screening-detected cancers to the National Cancer Registry at MOHAP.

4. Enforcement and Sanctions

4.1. Healthcare providers must comply with the terms and requirements of these guidelines MOHAP may impose sanctions in relation to any breach of requirements under this guideline.

5. Payment for Screening and Follow-Up of Cervical Cancer Screening

5.1. Eligibility for reimbursement under the Health Insurance Scheme must be in accordance, with local insurance laws for each Emirate.

6. Standard 1: Clinical Service Specifications

6.1. Screening facilities responsible for providing screening services must:
  6.1.1. Be licensed according to licensing policies and regulations of the United Arab Emirates.
  6.1.2. Fulfill the eligibility criteria for a cervical cancer screening facility as per clinical best practices in accordance with Appendix J.
  6.1.3. Comply with the cervical cancer screening care pathways, clinical quality indicators, and timelines for referral in accordance with Appendices K–M respectively.
  6.1.4. Assign a screening program coordinator responsible for submitting data on screening visits and outcomes to MOHAP and who will fulfill the responsibilities in accordance with Appendix N.
  6.1.5. Collect and submit data on screening visits and outcomes within 3 weeks of the screening date to MOHAP.
  6.1.6. Report all screen-detected cancer cases to MOHAP, through Cancer Case Notification Form.
  6.1.7. Maintain records for screening tests and outcomes.
  6.1.8. Establish internal audit procedures to demonstrate compliance with these guidelines and other associated regulatory policies and standards.
  6.1.9. Ensure the availability of evidence of compliance with the Cervical Cancer Screening Program Clinical Quality Indicators specified in Appendix L including:
    6.1.9.1. Collection and preparation of adequate cervical smear.
    6.1.9.2. Handling and transporting of specimens to labs assigned by MOHAP to deliver the service.
  6.1.10. Have an approved protocol for referral of women with abnormal results or physical examination to a diagnostic or treatment center.
6.2. Laboratories providing screening services must:
  6.2.1. Be licensed according to licensing policies and regulations of the United Arab Emirates.
  6.2.2. Comply with the applicable elements of the clinical quality indicators in accordance with Appendix M and ensure availability of evidence of compliance with these indicators; such as laboratory records required for accreditation purposes.
  6.2.3. Establish internal audit procedures to demonstrate compliance with this guideline and with other associated regulatory policies and standards.
  6.2.4. Develop, implement, and monitor policies and standard operating procedures for management of smears in accordance with International

Clinical Laboratory standards including: processing, workload, storage, documentation, and reporting.

6.2.5. Attain accreditation by an internationally credible body recognized by MOHAP such as CAP, IOS 15189(2007), JCI/Lab.

6.2.6. Participate in an international external proficiency test by all personnel involved in screening and reporting Pap test.

6.3. Healthcare professionals involved in providing cervical cancer screening services must:

6.3.1. Be licensed according to licensing policies and regulations of the United Arab Emirates.

6.3.2. Comply with the clinical standards detailed in this guideline to provide the most appropriate care, taking responsibility for deciding the best care options for managing cervical cancer cases.

6.3.3. Provide women with culturally and socially relevant education on women's health and with information (oral and written) regarding the screening benefits and limitations of cervical screening, potential outcomes, and next steps that may be required for care management.

6.3.4. Participate in continuing medical education (CME).

7. Standard 2: Screening Test

7.1. Papanicolaou test, (also called Pap test) is the standard test for screening for cervical cancer.

7.2. Liquid-based cytology (LBC) is the accepted standard method for Pap test specimen collection.

7.3. HPV test, as co-testing, for women aged 30 years and above (only internationally approved test is accepted).

8. Standard 3: Frequency of Screening

8.1. The frequency of repeat screening for average-risk, symptom-free women is:

8.1.1. Every 3 years for women aged 25–29 years.

8.1.2. Every 5 years for women aged 30–65 years.

8.2. Women who are immune-compromised due to disease or medication.

8.2.1. Annual screening.

9. Standard 4: Eligibility for Screening

9.1. All sexually active women, symptom-free, aged 25–65 years old (married, divorced, widowed) residing in the UAE, are eligible criteria for screening apply.

9.2. Women are excluded from screening if:

9.2.1. They have received a total hysterectomy for benign indications.

9.2.2. They are over 65 years (provided that the last two previous smears: US cervical were negative).

9.3. Women who have had subtotal hysterectomy (preserving the cervix), or hysterectomy due to cervical cancer or precancerous condition should continue to have cervical screening.

9.4. Screening recommendations remain the same regardless of whether or not they have received the HPV vaccination.

10. Standard 5: Recruitment to Screening

Recruitment of eligible women for screening can be made through:

10.1. Targeted invitation from eligible screening facilities.

10.2. Opportunistic by:

10.2.1. Approaching women who are enrolled in other existing screening programs; e.g. breast cancer.

10.2.2. Physician consultation for related or unrelated reason.

10.2.3. As an outcome of a health promotion campaign.

11. Standard 6: Risk Assessment and Physical Examination

11.1. Women must receive adequate information regarding the screening, Pap test procedure and expected outcomes and timeframe to receive results.

11.2. Detailed history, must be taken to assess risk and frequency of repeating screening, including at least:

11.2.1. Menstrual status (LMP, hysterectomy, pregnant, postpartum, use of contraceptive or hormone therapy).

11.2.2. Previous screening, results of screening, (negative, abnormal, or positive) and any previous treatment, (biopsy, chemotherapy, radiotherapy, or surgery).

11.2.3. Immune-compromised status due to diseases (including HIV) or medication.

11.3. Full clinical examination must be performed including visual inspection of the cervix.

12. Standard 7: Specimen Collection and Preparation of Adequate Pap Test

12.1. The following categories of licensed healthcare physicians are eligible to perform a Pap test:

12.1.1. Licensed gynecologists and obstetricians.

12.1.2. Physicians are already privileged to do so by their institution.

12.2. Eligible physicians must:

    12.2.1. Complete the required form with relevant clinical information in accordance with Standard 6 including any clinical findings e.g. abnormal bleeding or visible lesions etc.

    12.2.2. Collect and manage specimens in accordance with the facility internal policies and procedures for:

        12.2.2.1. Labelling

        12.2.2.2. Storage

        12.2.2.3. Transportation

    12.2.3. Smear-taking must be avoided in the following circumstances and women must be advised when to return for a Pap test:

        12.2.3.1. Menstruation

        12.2.3.2. Vaginal inflammation/infection

        12.2.3.3. Pregnancy (unless a previous smear was abnormal and in the interim the woman becomes pregnant, then the follow-up smear must not be delayed)

13. Standard 8: Cytology Smear Management and Reporting

Clinical laboratories handling and reporting of cytology specimens and cytology smears testing must:

13.1. Manage cervical cytology smears and perform the cytopathology testing as indicated in 6.2.4 and in accordance with laws, regulation, and Clinical Laboratory Standards.

13.2. Make final reports of cervical cytology smear using the Bethesda System (The Bethesda System for Reporting Cervical Cytology).

13.3. The report must be verified by a pathologist for all abnormal and reactive cases, while negative cases can be verified by licensed cytotechnologists using standard synoptic reporting format and containing minimum elements consistent with those of internationally reputable accrediting bodies. The report must include at least the following details:

    13.3.1. Patient's name.

    13.3.2. Age/date of birth.

    13.3.3. Menstrual status (LMP, hysterectomy, pregnant, postpartum, and hormone therapy).

    13.3.4. Relevant clinical information; such as if the patient had previously positive test or had other types of cancer, etc.

    13.3.5. Specimen description (source).

13.4. Reports for specimen adequacy and cytological findings must be returned to the referring physician at the screening center within 8 working days of receiving the specimen.

13.5. The reporting pathologist is the professional responsible for informing the referring physician of the positive cancer results.

13.6. MOHAP may, at its discretion, conduct third-party independent quality assurance testing of laboratories providing cervical smear laboratory test service. Where it does so, providers must comply with MOHAP's direction and cooperate with the MOHAP appointed party.

14. Standard 9: Screening Outcomes and Referrals

14.1. All women must be notified in writing about their results.

14.2. It is the responsibility of the physician at the screening facility to notify and provide a written report to a woman regarding her screening results within 15 working days (3 weeks) of the date of specimen taken.

14.3. If the test outcome is normal the woman is discharged to routine screening as per frequency mentioned in this guideline.

14.4. If the test outcome is unsatisfactory, it must be repeated within 6–12 weeks, treating infection, if present, as indicated.

14.5. If the Pap test outcome is abnormal or positive for intraepithelial lesion or malignancy, the woman's test is managed according to Appendix K.

14.6. If a suspicious visible abnormality is identified during visualization of cervix; the woman must be referred immediately to a Gynecologist Oncologist without receipt of her test results.

14.7. If a woman requires referral for colposcopy or treatment they must be referred to an appropriately licensed healthcare professional, privileged to provide the specialty/oncology services, patients must be seen within the timeframe specified in Appendix M.

14.8. All colposcopy services should be carried out by accredited colposcopist and if the facility has no accredited doctor then arrangements should be made to refer the patient to facility with accredited colposcopist.

Appendix I: Definitions

Term	Definition
The Bethesda system (TBS)	Is a system reporting for cervical or vaginal cytological diagnoses, used for reporting Pap smear results. The name comes from the location (Bethesda, Maryland) of the conference that established the system of reporting
HPV	Human papilloma virus
HPV co-testing	Is a test is done along with the Pap test in women aged 30 years and above, to screen for a high-risk HPV viral type. Only internationally approved test is accepted
ASC-US	Atypical squamous cells of undetermined significance. It is a finding of abnormal cells in the tissue that lines the outer part of the cervix
ASC-H	Suspicious for high-grade dysplasia
LGSIL or LSIL	Low-grade squamous intraepithelial lesion
HGSIL or HSIL	High-grade squamous intraepithelial lesion
AIS	Adenocarcinoma in situ
AGC	Atypical glandular cells

Appendix J: Eligibility Criteria for a Facility to Participate National Cervical Cancer Screening Program

1. **General**
   In addition to the requirements of this standard, the healthcare facility must fulfill the following criteria:
   1.1. Plan capacity to match the demand for screening and the facility capacity
   1.2. Allocate appointment slots for cervical cancer screening linked to the online booking system (when available)
   1.3. Have available adequate equipment to provide safe and quality screening
      1.3.1. Send cervical cytology smears only to licensed Laboratories that meet the requirements of this standard.
      1.3.2. Ensure patient privacy, comfort, and confidentiality at all times.
2. **Human resources**
   2.1. The core team must include at least:
      2.1.1. A program coordinator.
      2.1.2. A licensed physician, gynecologist, or obstetrician, physician privileged to deliver cervical screening care and services.
      2.1.3. A licensed nurse for each clinic with a minimum of 2 years of experience in gynecology or obstetric nursing.
   2.2. Training of licensed health professionals must be delivered using CME/CPD courses accredited by CME department including:
      2.2.1. For physicians; training for Pap smear taking in accordance with international evidence-based training standards and guidelines.
3. **Registration as screening facilities**
   Facilities meeting cervical cancer screening requirements should consider following points:
   3.1. Establish communication with cancer control team
   3.2. Fill service provision form
   3.3. Return filled form back to cancer control team
   3.4. Wait until receive confirmation from cancer control team
   3.5. Receive username and password for data reporting after orientation session with cancer team
   3.6. Commence screening and reporting of screening data to MOHAP

© The Editor(s) (if applicable) and The Author(s), under exclusive license to
Springer Nature Singapore Pte Ltd. 2024
H. O. Al-Shamsi (ed.), *Cancer Care in the United Arab Emirates*,
https://doi.org/10.1007/978-981-99-6794-0

Appendix K

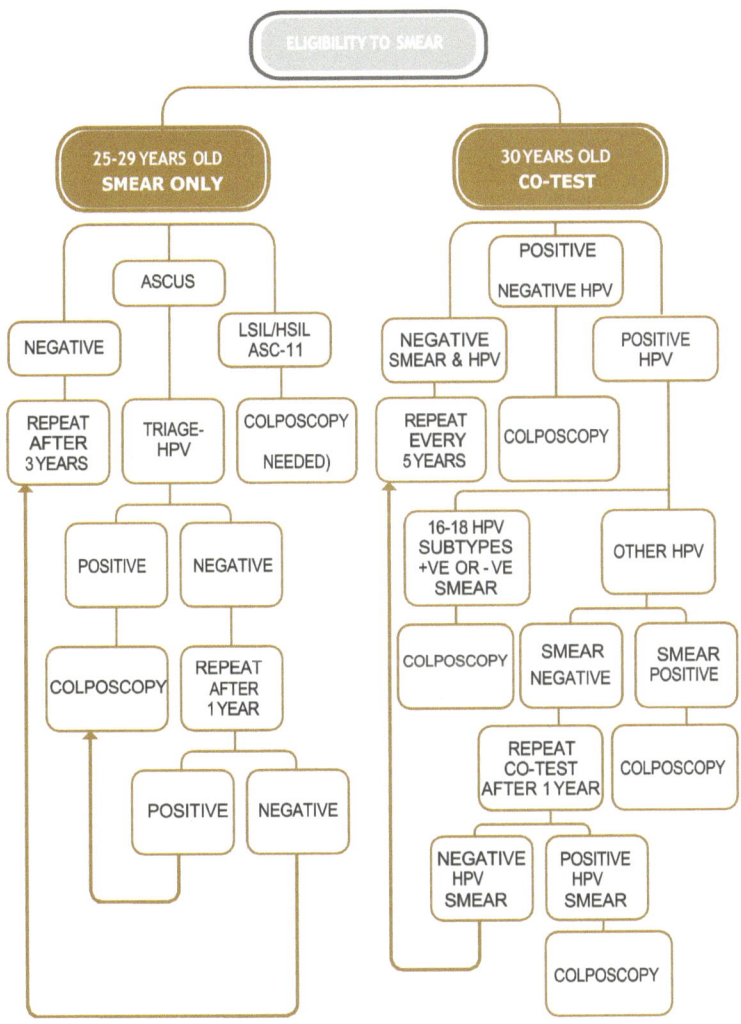

Appendix L

Quality indicators		Acceptable level	Desirable level
Coverage			
Retention rate	Percentage of eligible women re-screened with 3 years after a negative Pap test in a 12-month period	40%	50%
Specimen adequacy unsatisfactory proportion	Percentage of Pap tests that are reported as unsatisfactory in a 12-month period	4.7%	1.3%
Screening test results negative	Percentage of women by their most severe Pap test result in a 12-month period	90%	97%
Cytology turn around time 2 weeks	The average time from the date the specimen is taken to the date the finalized report is issued over a 12-month period	>80%	>90%
Time to colposcopy	Percentage of women with a positive Pap test (HSIL+/ASC-H) who had follow-up colposcopy within 9, 6, and 12 months subsequent to the index Pap test	80%	88%

Quality indicators		Acceptable level	Desirable level
Follow-up			
Biopsy rate	Percentage of women with a positive screening test result (HSIL+/ ASC-H) who received a histological diagnosis in a 12-month period	To be determined	11%
Cytology-histology agreement	Proportion of positive Pap tests with histological work-up found to have a pre-cancerous lesion or invasive cervical cancer in a 12-month period A		
Outcome indicators			
Pre-cancer detection rate	Number of pre-cancerous lesions detected per 1000 women who had a Pap test in a 12-month period		7.1 per 1000

Appendix M: Cervical Cancer Screening Program—Timeframes for Appointments

Cytological pattern	Priority	Appointment
HSIL or ASC-H	Urgent	1–2 weeks
LSIL/ASC-US	Routine	2–6 weeks

H. O. Al-Shamsi (ed.), *Cancer Care in the United Arab Emirates*,
https://doi.org/10.1007/978-981-99-6794-0

Appendix N: Responsibilities of the Facility Cancer Screening Program Coordinator

The healthcare facility cervical cancer screening program coordinator must:

1.1. Be a licensed healthcare professional.
1.2. Have comprehensive and high-quality knowledge in cervical cancer as a disease and its prevention.
1.3. Be responsible for:
  1.3.1. Recruitment of eligible women.
  1.3.2. Follow-up and tracking of screening results to ensure the timeliness and completeness of follow-up.
  1.3.3. Assessing relationships between planned care and approved protocols for care.
  1.3.4. Assessing women's needs for support to remove barriers to screening and follow-up.
  1.3.5. Developing and promoting recall systems that include reminders to patients as appropriate.
  1.3.6. Submitting data on screening visit and outcomes to MOHP via the (cancer screening e-notification system).

References

1. Curry SJ, Krist AH, Owens DK, Barry MJ, Caughey AB, Davidson KW, et al. Screening for cervical cancer: U.S. Preventive Services Task Force recommendation statement. U.S. Preventive Services Task Force. JAMA. 2018;320:674–86. Available at: https://jamanetwork.com/journals/jama/fullarticle/2697704. Retrieved 12 Apr 2021.
2. Saslow D, Solomon D, Lawson HW, Killackey M, Kulasingam SL, Cain J, et al. American Cancer Society, American Society for Colposcopy and Cervical Pathology, and American Society for Clinical Pathology screening guidelines for the prevention and early detection of cervical cancer. Am J Clin Pathol. 2012;137:516–42. Available at: https://academic.oup.com/ajcp/article/137/4/516/1760450. Retrieved 27 Sept 2022.

3. American College of Obstetricians and Gynecologists. Updated guidelines for management of cervical cancer screening abnormalities. Practice Advisory. Washington, DC: American College of Obstetricians and Gynecologists; 2020. Available at: https://www.acog.org/clinical/clinical-guidance/practice-advisory/articles/2020/10/updated-guidelines-for-management-of-cervical-cancer-screening-abnormalities. Retrieved 12 Apr 2021.

4. Perkins RB, Guido RS, Castle PE, Chelmow D, Einstein MH, Garcia F, et al. 2019 ASCCP risk-based management consensus guidelines for abnormal cervical cancer screening tests and cancer precursors. 2019 ASCCP Risk-Based Management Consensus Guidelines Committee [published erratum appears in J Low Genit Tract Dis 2020;24:427]. J Low Genit Tract Dis. 2020;24:102–31. Available at: https://journals.lww.com/jlgtd/Fulltext/2020/04000/2019_ASCCP_Risk_Based_Management_Consensus.3.aspx. Retrieved 12 Apr 2021.

5. Fontham ET, Wolf AM, Church TR, Etzioni R, Flowers CR, Herzig A, et al. Cervical cancer screening for individuals at average risk: 2020 guideline update from the American Cancer Society. CA Cancer J Clin. 2020;70:321–46. Available at: https://acsjournals.onlinelibrary.wiley.com/doi/10.3322/caac.21628. Retrieved 12 Apr 2021.

6. Kim JJ, Burger EA, Regan C, Sy S. Screening for cervical cancer in primary care: a decision analysis for the US Preventive Services Task Force. JAMA. 2018;320:706–14. Available at: https://jamanetwork.com/journals/jama/fullarticle/2697702. Retrieved 12 Apr 2021.

7. Centers for Disease Control and Prevention. HPV-associated cervical cancer rates by race and ethnicity. Available at: https://www.cdc.gov/cancer/hpv/statistics/cervical.htm. Retrieved 2 Aug 2021.

8. Buskwofie A, David-West G, Clare CA. A review of cervical cancer: incidence and disparities. J Natl Med Assoc. 2020;112:229–32. Available at: https://www.sciencedirect.com/science/article/abs/pii/S0027968420300432. Retrieved 3 Mar 2022.

9. Yeh PT, Kennedy CE, de Vuyst H, Narasimhan M. Self-sampling for human papillomavirus (HPV) testing: a systematic review and meta-analysis. BMJ Glob Health. 2019;4:e001351. Available at: https://gh.bmj.com/content/4/3/e001351.long. Retrieved 12 Apr 2021.

10. Elam-Evans LD, Yankey D, Singleton JA, Sterrett N, Markowitz LE, Williams CL, et al. National, regional, state, and selected local area vaccination coverage among adolescents aged 13–17 years—United States, 2019. MMWR Morb Mortal Wkly Rep. 2020;69:1109–16. Available at: https://www.cdc.gov/mmwr/volumes/69/wr/mm6933a1.htm. Retrieved 5 Jun 2022.

11. Agénor M, Pérez AE, Peitzmeier SM, Borrero S. Racial/ethnic disparities in human papillomavirus vaccination initiation and completion among U.S. women in the post-affordable care act era. Ethn Health. 2020;25:393–407. Available at: https://www.tandfonline.com/doi/abs/10.1080/13557858.2018.1427703.

12. Johnson NL, Head KJ, Scott SF, Zimet GD. Persistent disparities in cervical cancer screening uptake: knowledge and sociodemographic determinants of Papanicolaou and human papillomavirus testing among women in the United

States. Public Health Rep. 2020;135:483–91. Available at: https://journals.sagepub.com/doi/10.1177/0033354920925094.

13. Human papillomavirus vaccination. ACOG Committee Opinion No. 809. American College of Obstetricians and Gynecologists. Obstet Gynecol. 2020;136:e15–21. Available at: https://journals.lww.com/greenjournal/Fulltext/2020/08000/Human_Papillomavirus_Vaccination__ACOG_Committee.48.aspx.

Appendix O: The National Guideline for Colorectal Cancer Screening and Diagnosis

UNITED ARAB EMIRATES
MINISTRY OF HEALTH & PREVENTION

THE NATIONAL GUIDELINE FOR
COLORECTAL CANCER SCREENING AND DIAGNOSIS

3RD EDITION 2023

National Colorectal Cancer Screening Task Force

Dr. Huda Obaid Al Abdouli—Head of Task Force
Coordinator of National Cancer Screening Program
Ministry of Health and Prevention
Dr. Buthaina Bin Belaila—Member
Head of Non-communicable Disease
Ministry of Health and Prevention
Prof. Humaid Obaid Al Shamsi—Member
Director of the Burjeel Cancer Institute
President of the Emirates Oncology Society
Dr. Sara Al Bastaki—Member
Consultant Colorectal Surgeon Founder and President of Emirates Society of Colon
and Rectal Surgery
Sheikh Khalifa Medical City
Dr. Lamia Safieldin—Member
Comprehensive Screening Specialist
Department of Health, Abu Dhabi
Dr. Maryam Al Khatry—Member
Consultant Gastroenterology—Member
Obaidullah Hospital—Emirates Health Services—EHS President of the Emirates
Gastroenterology Society
Dr. Sameer Al Awadhi—Member
Consultant Gastroenterology—Member
Dubai Health Authority—DHA
Dr. Thomas Cherukara—Member
Consultant Gastroenterology—Member
Kuwait Hospital—Emirates Health Services—EHS
Dr. Taha Kadir—Member
Consultant Colorectal Surgeon
Private Sector
Dr. Makki Fayadh—Member
Consultant Gastroenterology
Private Sector
Dr. Salim Awadh—Member
Consultant Gastroenterology
Private Sector

1. Purpose

1.1. To stipulate the service requirements to deliver the National Colorectal Cancer
(CRC) Screening Program in the United Arab Emirates.
1.2. To set out the minimum Clinical Care Standards and frequency for CRC
screening as per international evidence-based guidelines.

1.3. To set out the case mix, eligibility criteria and data reporting requirements for colorectal cancer screening.

1.4. To ensure the population receives quality and safe care and timely referral for diagnosis and/or treatment where appropriate.

2. Scope

2.1. This guideline applies to all healthcare providers (facilities and professionals) in the United Arab Emirates, providing CRC screening services, including mobile units.

3. Definitions

3.1. **Colorectal cancer screening**: Means looking for polyps or cancer in the colon and rectum in people who have no symptoms of the disease. CRC screening includes the following services:

    3.1.1. Colorectal cancer screening services

    3.1.2. Colorectal cancer assessment and follow-up

3.2. **Colonoscopy**: Colonoscopy is the endoscopic examination of the large bowel and the distal part of the small bowel with a Charge Coupled Device (CCD), a camera or a fiber optic camera, on a flexible tube passed through the anus. It can provide a visual sight to detect adenomatous polyps and cancer diagnosis (e.g., ulceration and polyps). It also grants the opportunity for biopsy or removal of suspected colorectal cancer lesions.

3.3. **Fecal immunochemical test (FIT)**: **FIT** is a test that investigates by using antibodies to detect blood in the stool sample for signs of cancer.

3.4. **Case mix**: Includes **males and females aged 40–75 years** determined eligible for colorectal cancer screening services, in accordance with the criteria detailed in this guideline. For people ages 76–85, the decision to be screened should be based on a person's preferences and healthcare professional judgment considering life expectancy, overall health, **and** prior screening history.

4. Duties for Healthcare Providers

All licensed healthcare providers, facilities, and professionals engaged in providing CRC screening services must:

4.1. Provide clinical services and patient care in accordance with this guideline and in accordance with laws and regulations, policies, and standards of the United Arab Emirates, including developing effective recording systems, maintaining confidentiality, privacy, and security of patient information.

4.2. Comply with the federal requirements, laws, and policies for patient education and consent. The licensed provider must provide appropriate patient education and information regarding the screening test and must ensure that appropriate patient consent is obtained and documented on the patient's medical record.

4.3. Comply with federal requirements; laws, policies, and standards on managing and maintaining patient medical records, including developing effective recording systems, maintaining confidentiality, privacy, and security of patient information.

4.4. Comply with federal requirements; laws, policies, and standards for Information Technology (IT) and data management, electronic patient records and disease management systems, sharing of screening and diagnostic test, and where applicable pathology results.

4.5. Comply with MOHAP requests to inspect and audit records and cooperate with authorized auditors as required.

4.6. Collect and submit data on screening visits and outcomes, as per Appendix P, to the National Cancer Screening Registry at MOHAP.

4.7. Comply with federal laws, policies, and standards on cancer case reporting and report all confirmed screening-detected cancers to the National Cancer Registry at MOHAP.

5. Enforcement and Sanctions

5.1. Healthcare providers, payers, and third party administrators must comply with the terms and requirements of this guideline. MOHAP may impose sanctions in relation to any breach of requirements under this guideline.

6. Payment for Screening and Follow-Up of Colorectal Cancer

6.1. Eligibility for reimbursement under the health insurance scheme must be in accordance, with local insurance laws for each Emirate.

7. Standard 1: Clinical Service Specifications

7.1. All licensed healthcare screening facilities scheme providing colorectal cancer screening services must:

7.1.1. Follow best practice for colorectal cancer screening as per Appendix P.

7.1.2. Adhere to the clinical performance indicators and timelines in accordance with Appendix Q.

7.1.3. Coordinate referral of individuals with positive screening for further assessment or treatment with diagnostic and oncology centers and develop an agreed protocol and clear process for referrals.

7.1.4. Maintain records for screening tests, outcomes, and clinical performance indicators.

7.1.5. Assign a colorectal cancer facility program coordinator who will be accountable to:

    7.1.5.1. Reports and submits screening visits and outcome data, specified in Sect. 4.

    7.1.5.2. Establish internal audit policies and procedures and conduct regular audits, monitoring and evaluating to demonstrate compliance with this guideline and other associated regulatory policies and standards.

7.1.6. Endoscopy unit providing colorectal cancer screening must meet the criteria for a competent unit infrastructure, equipment, and personnel, as per Appendix R.

7.1.7. Have an approved protocol for referral of individuals with screen detected abnormalities for further assessment or treatment.

7.2. **All licensed laboratories providing** diagnostic histopathology and genetic testing services must:

7.2.1. Have in place the systems, policies, and operating procedures in accordance with the requirements of relevant policies and laboratory standards.

7.2.2. Use specimen identification and labeling in accordance with relevant policies and standards and industry best practices.

7.2.3. Establish internal audit policies and procedures and conduct regular audits, monitoring and evaluating to demonstrate compliance with this guideline and other associated regulatory policies and standards.

7.2.4. Laboratory should be accredited by an internationally credible accrediting body such as CAP, ISO 15189 (2007), JCI/Lab for colorectal cancer.

7.2.5. MOHAP may, at its discretion, conduct third-party independent quality assurance testing of laboratories providing colorectal cancer screening test service. Where it does so, providers must comply with the direction and cooperate with the appointed party.

7.2.6. Labs performing FIT test must:

    7.2.6.1. Follow the manufacturer's instructions for use of the FIT testing kit.

    7.2.6.2. Use an explicit definition for cut-off levels for hemoglobin concentration.

    7.2.6.3. Make provision to record the information concerning the actual amount of hemoglobin, both for tests classified as negative and for those classified as positive.

    7.2.6.4. Labs performing genetic testing must have organized and specialist cyto/histopathological support services who can demonstrate compliance with related policies and laboratory standards.

7.3. **All licensed healthcare professionals** participating in colorectal cancer screening must conduct CRC risk assessment. Detailed history must be evaluated and completed, each time an individual visits for screening. The purpose of this is to identify individuals' risk status, as per risk categories specified in Appendix S, and referral to appropriate screening tests.

7.3.1. Obtain informed patient consent prior to screening. Where consent is granted or refused, the treating physician must document and retain signed consent forms on individuals' medical records.

7.3.2. Inform all individuals of the procedures and expected timeframe to be screened and to receive results.

7.3.3. Ensure that the outcome of screening for colorectal cancer is reviewed by a multi-disciplinary team including gastroenterologist, colorectal surgeon, gastrointestinal oncologist, pathologist, radiologist, physician, and a nurse.

7.3.4. Follow-up and timely referral of individuals with abnormal results to treatment.

8. Standard 2: Screening Tests and Frequency

8.1. Screening tests for individuals at average risk of colorectal cancer, as specified in Appendix S, are as follows:

8.1.1. Colonoscopy, every 10 years.

8.1.2. Fecal Immunochemical Test (FIT) every year.

8.1.3. Eligible population must be offered colonoscopy screening as per Appendix P, in case of refusal, the patient should be offered a FIT.

9. Standard 3: Recruitment for Screening

Population eligible for colorectal cancer screening may be recruited by the health care facilities, through the following:

9.1. Recruitment for screening

9.1.1. All CRC screening facilities must establish an invitation system to ensure identification, successful participation, and retaining of eligible population.

9.1.2. Targeted invitation may be established via an electronic or manual invitation system.

9.2. Opportunistic

9.2.1. Physician consultation for related or unrelated reason.

9.2.2. Engagement in a health promotion campaign.

10. Standard 4: Screening with Colonoscopy

10.1. Pre-colonoscopy assessment

10.1.1. Pre-colonoscopy documentation must include:

10.1.1.1. Patient demographics.

10.1.1.2. Anticoagulant and antiplatelet use.

10.1.1.3. History of diabetes mellitus and use of insulin.

10.1.1.4. Presence of implantable defibrillators or pacemakers.

10.1.1.5. Previous gastrointestinal procedures, including surgeries.

10.1.2. Assessment of patient risk: Physical status of the patient must be documented in accordance with the American Society of Anesthesiology (ASA), Appendix T.

10.1.3. ASA class 3 or higher is at higher risk for cardiopulmonary events, and appropriate measures must be taken in this respect.

10.1.4. Colonic cleansing: Type of bowel preparation must be documented including documentation of careful preparation in accordance with international standards and guidelines. Written instructions to be given to the patient concerning colonic cleansing and need to be mentioned in the report.

10.1.5. Inadequate bowel preparations must not exceed 10% of examinations.

10.2. Colonoscopy procedure

10.2.1. Facility-specific policies and procedures must be in place for the following:

10.2.1.1. Colonoscopy decontamination including infection control.

10.2.1.2. Sedation of patient, considering the patient's status and preferences and recording of all sedation methods and outcomes; consider involving anesthesia service in patients with significant comorbidities such as patients with ASA 3, 4, and 5 (Appendix T).

10.2.1.3. Patient support and comfort, including positioning during the colonoscopy.

10.2.2. To achieve high-quality colonoscopy examination, complete intubation of the colon and careful inspection of the mucosa during withdrawal is necessary.

10.2.2.1. If a complete colonoscopy is not achieved, imaging for documentation of incomplete intubation may be necessary and reasons must be clearly documented.

10.2.2.2. Auditable photo documentation of colonoscopy completion must be available including a panoramic image of the appendiceal orifice, ileocecal valve, and cecum with a video clip with a respective image.

10.2.2.3. Documentation of completion of rectal retroflexion (retroflexion of the endoscope during colonoscopy to increase diagnostic yield) must be recorded.

10.2.2.4. Withdrawal times of the colonoscopy from cecum to anus must be documented and must be not less than 6 min (when no biopsies or polypectomies are performed). The times to be documented include when:

10.2.2.4.1. Endoscope is inserted into the rectum.

10.2.2.4.2. Withdrawal from cecum was started.

10.2.2.4.3. Endoscope is withdrawn completely.

10.2.2.5. A record of the actual model and instrument number used must be maintained by the unit staff to track procedure volume, problems, and infection transmission and instrument repairs.

10.2.2.6. Any adverse clinical events (fall in blood pressure, unplanned reversal of sedation medications, oxygen desaturation, etc.) that occur during colonoscopy as well as all serious events (perforation, bleeding requiring blood transfusion, and/or surgery) must be documented with hard copies attached to the colonoscopy report and reported in accordance with standards for adverse events management and reporting.

10.3. Post-colonoscopy procedures

10.3.1. Patients must be contacted 24 h post-procedure or on the next working day to monitor any complications; this contact must be documented.

10.3.2. Before colonoscopy each patient must receive instructions about management of any potential adverse events following discharge and must be informed that complications may occur within 1–4 weeks post-procedure.

10.3.3. A contact number must be provided to the patient for this purpose and documented in the patient records.

10.3.4. Post-procedure complications must be tracked over a 30-day interval after a colonoscopy.

10.3.5. Discharge instruction form should be given to patient instructing him to call endoscopy unit or the gastroenterology physician on call or to come to ER in case there is any abdominal pain or any complication or concerns after the procedure. Patients should sign this form acknowledging that he understood the post-colonoscopy and the pre-discharge instructions.

10.4. Colonoscopy findings and reporting

10.4.1. Avoid using vague terms to describe polyps in the report.

10.4.2. An estimation of the size and dimension of all polyps must be documented, terms such as "large" or "small" must not be used.

10.4.3. Tattoos preferably be placed for all lesions >10 mm and those with concerning appearance for cancer to mark the location of colon lesions for repeat colonoscopy or surgery.

10.4.4. Lesions that are too large be safely removed must be biopsied and a tattoo injection performed in the vicinity of the lesion and not into the lesion.

10.4.5. Specimen identification and labeling must be in accordance with relevant clinical laboratory standards and industry best practices.

10.4.6. Procedures and protocols for adequate specimen collection, handling, labeling, and reporting must be in accordance with relevant clinical laboratory standards and must be communicated to clinical staff and other clients who are involved in the procedures for processing of colorectal specimens.

10.4.7. Each facility must develop a patient colonoscopy report form, retained on the patient's medical record and made available to auditors. A recommended sample of a standard report is provided.

10.4.8. A standard colonoscopy report must include at least the following information:

10.4.8.1. Patient demographics and history

10.4.8.2. Assessment of patient risk and comorbidity

10.4.8.3. Procedure indications

10.4.8.4. Procedure: technical description

10.4.8.5. Colonoscopy findings

10.4.8.6. Interventions/unplanned events

10.4.8.7. Assessment

10.4.8.8. Follow-up plan

10.4.8.9. Pathology

11. Standard 5: Screening with Fecal Immunochemical Test (FIT)

11.1. FIT test must be offered where the **average risk** patient refuses the screening colonoscopy.

11.2. Patient must be provided with clear and simple instructions regarding collection of sample.

11.3. No drug or dietary restriction is required for FIT, and only one stool sample is needed.

11.4. The quality of the sample must be reproducible and representative of the stool, to be of the required volume and be adequately preserved.

11.5. The samples must be analyzed without delay and kept cool to avoid further sample denaturation and a potential increase in false negative results; and the proportion of unacceptable tests received for measurement must not exceed 3% of all kits received; less than 1% is desirable.

12. Standard 6: Screening Outcomes and Referrals

12.1. At the end of the screening, the screening unit must provide the individuals with a written report with a clear instruction on follow-up plan and next steps, including referral for treatment or next screening dates. Also, send feedback to the referring physician at the primary or ambulatory healthcare clinic.

12.2. It is the sole responsibility of the colonoscopist (in case of screening colonoscopy), or the referring physician (in case of FIT) to inform the individuals with their results and next steps.

12.3. The time between completion of a screening test and receipt of results by the participant must be less than 15 working days (acceptable standard >90% within 15 days).

12.4. Screening with colonoscopy

12.4.1. In case of normal results, negative for polyps, individuals must be re-invited for screening in accordance with the frequencies specified in Sect. 8.

12.4.2. In case of presence of adenoma, colonoscopy must be repeated in accordance with Appendix P.

12.4.3. Adenoma detection rate must be monitored and audited. It is limited to screening colonoscopies; surveillance procedures and repeat endoscopic procedures are excluded.

12.4.4. Individuals with a positive colonoscopy, cancer, must be urgently referred for treatment, within 2 weeks of receiving colonoscopy report.

12.4.5. The time interval between a positive colonoscopy (cancer) and definitive management must be monitored. (Acceptable standard ≥95% of cases must be no more than 31 days).

12.4.6. Death within 30 days after colorectal cancer screening, attributed to complications caused by colonoscopy, must be recorded by e-notification.

12.5. Screening with FIT test:

12.5.1. Individuals with a negative test result are re-invited for screening as per frequencies specified in Sect. 8.

12.5.2. Individuals with a positive test result must be urgently referred for follow-up colonoscopy within 15 working days.

12.5.3. The FIT test must be repeated if results are unclear or spoilt in accordance with Appendix P.

Appendix P: Colorectal Cancer Screening and Diagnosis Pathway

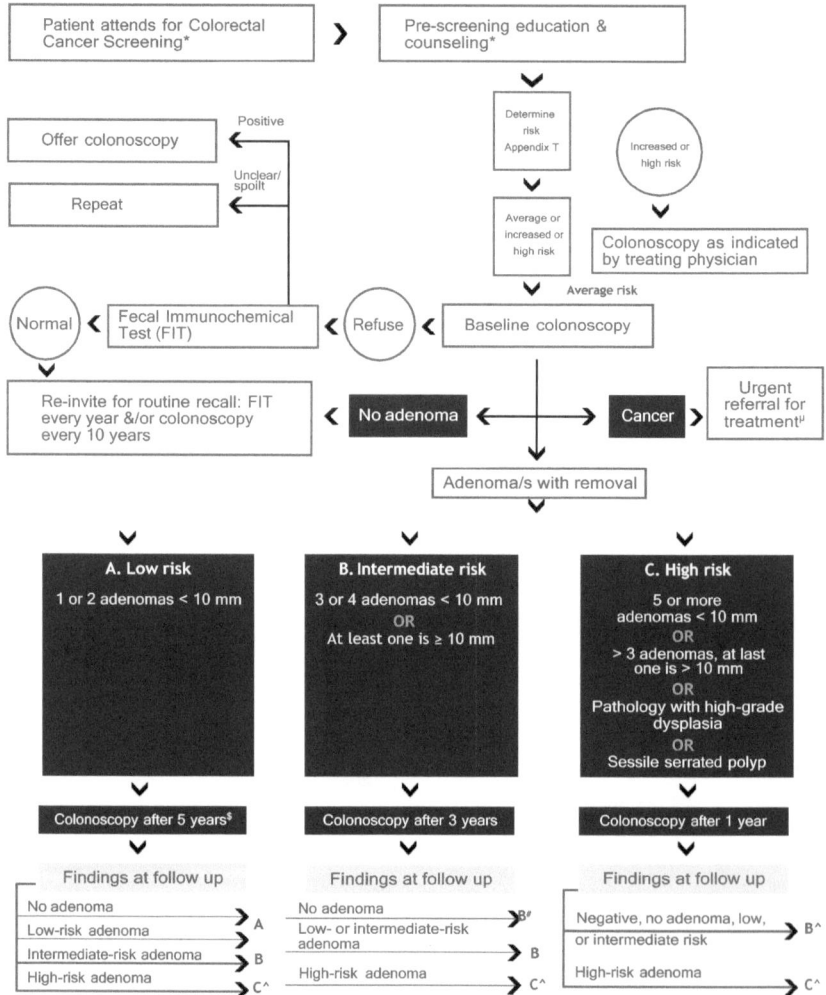

Patient attends for Colorectal Cancer Screening* ❯ Pre-screening education & counseling*

Determine risk Appendix T

Increased or high risk

Offer colonoscopy ← Positive

Repeat ← Unclear/spoilt

Average or increased or high risk

Colonoscopy as indicated by treating physician

Normal ← Fecal Immunochemical Test (FIT) ← Refuse ← Baseline colonoscopy — Average risk

Re-invite for routine recall: FIT every year &/or colonoscopy every 10 years ← No adenoma ⟷ Cancer ❯ Urgent referral for treatment‖

Adenoma/s with removal

A. Low risk
1 or 2 adenomas < 10 mm

B. Intermediate risk
3 or 4 adenomas < 10 mm
OR
At least one is ≥ 10 mm

C. High risk
5 or more adenomas < 10 mm
OR
> 3 adenomas, at last one is > 10 mm
OR
Pathology with high-grade dysplasia
OR
Sessile serrated polyp

Colonoscopy after 5 years$

Colonoscopy after 3 years

Colonoscopy after 1 year

Findings at follow up

No adenoma	→ A
Low-risk adenoma	
Intermediate-risk adenoma	→ B
High-risk adenoma	→ C^

Findings at follow up

No adenoma	→ B#
Low- or intermediate-risk adenoma	→ B
High-risk adenoma	→ C^

Findings at follow up

| Negative, no adenoma, low, or intermediate risk | → B^ |
| High-risk adenoma | → C^ |

Appendix Q: Colorectal Cancer Clinical Performance Indicator

Indicator	Acceptable level	Desirable level
Screening uptake (participation) rate	>45%	>65%
Minimum number of screening colonoscopies undertaken annually by each screening colonoscopist	>150 per annum	>250
Inadequate FIT rate (proportion of people screened with one or more FIT returned none of which were adequate)	<3%	<1%
Maximum time between screening FIT test and receipt of result should be 7 days from sample's dispatch	>90%	>95%
Rate of referral to follow-up colonoscopy after positive FIT test (detects cancer)	90%	>95%
Maximum time between referral after positive screening FIT test and conducting follow-up colonoscopy should be within 31 working days	>90%	>95%
Cecal intubation rate (CIR). Follow-up and screening colonoscopies to be recorded separately (unadjusted CIR with video recorded and photographic evidence)	>90%	>95%
Adenoma detection rate (ADR)	≥35% of colonoscopies	Auditable outcome
Cancer detection rate	≥2 per 1000 screened by FIT ≥11 per 100 colonoscopies	Auditable outcome
Withdrawal time in negative colonoscopies (withdrawal from cecal pole to anus)	≥6 min	
Polyp retrieval rate (retrieval of polypectomy specimens for histological analysis per colonoscopist)[b]	>90% per 100 polyps excised	>95% per 100 polyps excised
Rate of high-grade neoplasia reported by pathologists in a colonoscopy screening program	<5%	
Rate of high-grade neoplasia reported by pathologists in a FIT screening program	<10%	

Indicator	Acceptable level	Desirable level
Endoscopic complications of colonoscopy screening programs	Bleeding <1:150 Perforation <1:1000	
Post polypectomy perforation rate	<1:500	Auditable outcome
Time interval between positive colonoscopy and start of definitive management within 31 days	>90%	>95%

[a] Excellent: no or minimal solid stool and only clear fluid requiring suctionAdequate: collections of semi-solid debris that are cleared with washing/suctionInadequate: solid or semi-solid debris that cannot be cleared effectively

[b] Numerator: number of polyps with histological tissue retrieved for analysisDenominator: number of polyps recorded during lower GI endoscopies

References

1. NHS Cancer Screening Programmes. Quality assurance guidelines for colonoscopy. NHS BCSP Publication No. 6 Feb 2011.
2. European guidelines for quality assurance in colorectal cancer screening and diagnosis. 1st ed. 2010.

Appendix R: Colorectal Cancer Screening Endoscopy Unit Infrastructure, Equipment and Personnel

Endoscopy unit infrastructure and equipment must:

1. Include facilities for adequate pre-colonoscopy assessment, recovery, and be designed to allow efficient patient flow.
2. Match the demand with respect to unit capacity (e.g., equipment and personnel).
3. Provide video-endoscopes with high resolution and image enhancement that facilitate focal application of the dye for the detection and assessment of high-risk colorectal lesions and documentation.
4. Provide adequate supply of accessories suited to the endoscopic interventions undertaken and documentation.
5. Provide properly maintained resuscitation equipment in the endoscopy rooms and recovery areas.
6. Conduct a regular review of all the functioning and cleansing of the colonoscopies. The review should be available at all times in the unit including infection control.
7. Plan capacity that matches demand for screening. Referral to colonoscopy to be within 31 days from a positive FIT test (detects the presence of occult blood in the fecal sample).
8. Referral to colonoscopy to be within 31 days from positive FIT test (detects the presence of occult blood in the fecal sample).

Criteria colorectal cancer screening core team to include:

All members in the colorectal cancer core team should participate in regular multidisciplinary team meetings to discuss each patient with colorectal cancer.

1. At least two gastroenterologists: each conduct a volume of minimum 150 per colonoscopist per year with a cecal completion rate of >90%.
2. Nurse: Two nurses trained to provide support, assistance, information, and advice to every patient. An in-depth understanding of colorectal cancer (diagnosis, treatment, prognosis, staging, and importance of stage at diagnosis), an in-depth understanding of the colorectal screening process (including screening theory and particularly the potential benefits and harms of screening, and the prime importance of quality assurance), and advanced communication skills.
3. Regular training and evaluation for colorectal cancer screening core team according to international guideline.

© The Editor(s) (if applicable) and The Author(s), under exclusive license to Springer Nature Singapore Pte Ltd. 2024
H. O. Al-Shamsi (ed.), *Cancer Care in the United Arab Emirates*,
https://doi.org/10.1007/978-981-99-6794-0

Appendix S: Risk Assessment for Colorectal Cancer

No low risk.

Average age risk:

1. Age ≥40
2. No history of inflammatory bowel disease
3. Negative family history
4. No history of adenoma or colorectal cancer

Increased risk:

1. Personal history of adenoma, sessile serrated polyp (SAP),[1] colorectal cancer, inflammatory bowel disease
2. Positive family history of first or second degree relative with colorectal cancer (screening recommendations vary depending on family history, begin screening at an age approximately 10 years earlier than the age at which the youngest person in family was diagnosed with colorectal polyps or cancer).

High risk:

1. Family history of a hereditary colorectal cancer syndrome such as familial adenomatous polyposis (FAP) or Lynch syndrome (also known as hereditary nonpolyposis colon cancer or HNPCC).
2. Polyposis syndromes (Classical Familial Adenomatous Polyposis (FAP1-), Attenuated Familial Adenomatous Polyposis (AFAP1-), MYH associated

---

[1] Increased risk based on personal history of adenoma(s)/sessile serrated polyp(s) found at colonoscopy:

(a) Low-risk adenoma: ≤2 polyps, <1 cm, tubular.
(b) Advanced or multiple adenomas: high-grade dysplasia, ≥1 cm, villous (>25% villous), between 3 and 10 polyps (fewer than 10 polyps in the setting of a strong family history or younger age (<40 years) may sometimes be associated with an inherited polyposis syndrome).
(c) More than 10 cumulative adenomas (fewer than 10 polyps in the setting of a strong family history or younger age (<40 years) may sometimes be associated with an inherited polyposis syndrome).
(d) Incomplete or piecemeal polypectomy (ink lesion for later identification) or polypectomy of large cancer.

© The Editor(s) (if applicable) and The Author(s), under exclusive license to
Springer Nature Singapore Pte Ltd. 2024
H. O. Al-Shamsi (ed.), *Cancer Care in the United Arab Emirates*,
https://doi.org/10.1007/978-981-99-6794-0

Polyposis (MAP1-), Peutz-Jeghers syndrome (PJS1-), Juvenile polyposis syndrome (JPS1-), hyperplastic polyposis syndrome (HPP1-))

Colorectal cancer screening and surveillance, in high-risk disease family group

High-risk disease groups	Screening procedure	Time of initial screen	Screening procedure and interval
Colorectal cancer			
	Consultation, CT, LFTs, and colonoscopy	Colonoscopy within 6 months of resection only if colon evaluation pre-op is incomplete	CT liver scan within 2 years post-op. Colonoscopy 5 yearly until co-morbidity outweighs
Colonic adenomas			
Low risk	1–2 adenomas, both <1 cm	Colonoscopy	5 years or no surveillance
Intermediate risk	3–4 adenomas, OR at least one adenoma ≥1 cm	Colonoscopy	Every 3 years
High risk	≥5 adenomas or ≥3 with at least one ≥1 cm piecemeal polypectomy	Colonoscopy Colonoscopy or flexi-si (depending on polyp location)	Yearly 3 months consider open surgical resection if incomplete healing of polypectomy scar
Ulcerative colitis and Crohn's colitis			
Low risk	Extensive colitis with no inflammation or left-sided colitis or Crohn's colitis of <50% colon	Pancolonic dye spray with targeted biopsy. If no dye spray then 2–4 random biopsies every 10 cm	Every 10 years from onset of symptoms
Intermediate risk	Extensive colitis with mild active disease or post-inflammatory polyps or family history of colorectal cancer in a FDR <50 years Extensive at least moderate colitis or stricture in past 5 years or dysplasia in past 5 years (declining surgery) or PSC or OLT for PSC) or colorectal cancer in a FDR <50 years Ureterosigmoidostomy		After surgery by 10 years
Acromegaly			
Acromegaly		Colonoscopy	At 40 years

Colorectal cancer screening and surveillance in moderate risk disease family groups: initial screening 10 years earlier than the youngest affected FDM

Family history	Screening procedure	Screening interval
One first-degree relative with CRC or advanced adenoma diagnosed before 60 years of age, or two first-degree relatives diagnosed at any age	Colonoscopy	Start at 40 years of age or 10 years younger than the earliest diagnosis in the patient's family, whichever comes first; colonoscopy should be repeated every 5 years
One first-degree relative with CRC or advanced adenoma diagnosed at 60 years or older, or two second-degree relatives with CRC	Colonoscopy	Start screening colonoscopy at 40 years of age; colonoscopy should be repeated every 10 years
One second- or third-degree relative with CRC	Colonoscopy	Average-risk screening (e.g., start at 40)
Individuals who have Crohn disease with colonic involvement or ulcerative colitis. Screening-repeated every 1–3 years	Colonoscopy	Colonoscopy screening should begin 8–10 years after the onset of symptoms
In individuals with hereditary nonpolyposis colorectal cancer	Colonoscopy	Colonoscopy should begin at 25 years of age and be repeated annually
Individuals with adenomatous polyposis syndromes	Colonoscopy	Colonoscopy between 10 and 20 years of age and be repeated every 1–2 years
Individuals with Peutz-Jeghers syndrome. If results are, negative, testing-repeated every 3 years	Colonoscopy	Esophagogastroduodenoscopy, colonoscopy, and video capsule endoscopy should begin at 8 years of age. If results are negative, testing-repeated every 3 years
Individuals with sessile serrated adenomatous polyposis	Colonoscopy	Colonoscopy should begin as soon as the diagnosis is established and be repeated annually

Affected relatives who are first-degree relatives of each other AND at least one is a first-degree relative of the patient.

- Combinations of three affected relatives in a first-degree kinship include parent and aunt/uncle and/or grandparent; OR 2 siblings/1 parent; OR 2 siblings/1 offspring. Combinations of two affected relatives in a first-degree kinship.
- Include a parent and grandparent, or >2 siblings, or >2 children, or child + sibling. Where both parents are affected, these count as being within the first-degree kinship.
- Clinical genetics referral recommended.
- Centers may vary depending on capacity and referral agreements. Ideally, all such cases should be flagged systematically for future audit on an emirate level.

Colorectal cancer screening and surveillance in high-risk disease family groups

Family history categories[a]	Screening procedure	Age at initial screen	Screening interval and procedure
At-risk HNPCC (fulfils modified Amsterdam criteria, or untested FDR of proven mutation carrier)	MMR gene testing of affected relative colonoscopy ± OGD	Colonoscopy from age 25 years OGD from age 40 years or screening 10 years earlier than the youngest affected FDM	Colonoscopy every 18–24 months (OGD every 2 ears from age 40 years)
MMR gene carrier	Colonoscopy ± OGD		
At-risk FAP (member of FAP family with no mutation identified)	APC gene testing of affected relative colonoscopy	Puberty Flexible approach Important making allowance for variation in maturity	Annual colonoscopy or until aged 30 years Thereafter 3–5 yearly until 60 years proctocolectomy or colectomy if positive
Fulfils clinical FAP criteria, or proven APC mutation carrier opting for deferred surgery prophylactic surgery normally strongly recommended	Colonoscopy Colonoscopy/ OGD	Usually at diagnosis Otherwise puberty Flexible approach important for making allowance for variation in maturity	Recommendation for procto-colectomy and pouch/colectomy before age 30 years Cancer risk increases dramatically age >30 years
			Twice yearly colonoscopy
FAP postcolectomy and IRA	Colonoscopy OGD	After surgery OGD from age 30 years	Colonoscopy every 3 years forward and side-viewing OGD
FAP post procto-colectomy and pouch	DRE and pouch endoscopy Forward and side-viewing OGD	After surgery OGD from age 30 years	Annual exams alternating Flexible/rigid pouch Endoscopy every 3 years Forward and side-viewing OGD

Family history categories[a]	Screening procedure	Age at initial screen	Screening interval and procedure
MUTYH-associated polyposis (MAP)	Genetic testing Colonoscopy ± OGD	Colonoscopy from age 25 years OGD from age 30 years	Mutation carriers should be counselled about the available limited evidence; options include prophylactic colectomy and ileorectal anastomosis; or biennial colonoscopy surveillance Every 3–5 years gastro-duodenoscopy
FDR with MSI-H colorectal cancer and IHC shows loss of MSH2, MSH6 or PMS2 expression	Colonoscopy ± OGD	Colonoscopy from age 25 years OGD from age 40 years	Colonoscopy every 2 years (with OGD aged >40 years)
MLH1 loss and MSI specifically excluded (MLH1 loss in elderly patient with right sided tumor is usually somatic epigenetic event)			
Peutz-Jeghers syndrome	Genetic testing of affected relative Colonoscopy ± OGD	Colonoscopy from age 25 years OGD from age 25 years Small bowel MRI/enteroclysis	2 yearly colonoscopy/consider colectomy and IRA for colonic cancer Small bowel VCE or MRI/enteroclysis 2–4 yearly OGD 2 yearly
Juvenile polyposis	Genetic testing of affected relative Colonoscopy ± OGD	Colonoscopy from age 15 years OGD from age 25 years	Every 2 years colonoscopy and OGD. Extend interval >35 years

1. The Amsterdam criteria for identifying HNPCC are three or more relatives with colorectal cancer:
   - One patient a first degree relative of another.
   - Two generations with cancer.
   - One cancer diagnosed under the age of 45 or other HNPCC-related cancers, e.g., endometrial, ovarian, gastric, upper urothelial, and biliary tree.

2. Clinical genetics referral and family assessment required, if not already in place or if clinical genetics did not initiate referral.
3. FAP, familial adenomatosis polyposis; FDR, first-degree relative (sibling, parent, or child) with colorectal cancer; HNPCC, hereditary non-polyposis colorectal cancer; IHC, immunohistochemistry of tumor material from affected proband; MSI-H, micro-satellite instability high (two or more MSI markers show instability); OGD, esophagogastroduodenoscopy endoscopy; VCE, video capsule endoscopy.

Appendix T: American Society of Anesthesiology Classification System

Class	Description
1	Patient has no organic, physiologic, biochemical, or psychiatric disturbance (healthy, no comorbidity).
2	Mild-moderate systemic disturbance caused either by the condition to be treated surgically or by other pathophysiologic processes (mild-moderate condition, well controlled with medical management; for examples include diabetes, stable coronary artery disease, stable chronic pulmonary disease).
3	Sever, systemic disturbance or disease from whatever cause, even though it may not be possible to define the degree of disability with finality (disease or illness that severely limits normal activity and may require hospitalization or nursing home care; examples include severe stroke, poorly controlled congestive heart failure, or renal failure).
4	Severe systemic disorder that is already life threatening, not always correctable by the operation (examples include coma, acute myocardial infarction, respiratory failure requiring ventilator, support renal failure requiring urgent dialysis, bacterial sepsis with hemodynamic instability).
5	The moribund patient who has little chance of survival.

References

1. Updated guidelines on the management of colon cancer were published on February 1, 2022 by the American Society of Colon and Rectal Surgeons (ASCRS).
2. de Kanter C, Dhaliwal S, Hawks M. Colorectal cancer screening: updated guidelines from the American College of Gastroenterology. Am Fam Physician. 2022;105(3):327–9.
3. Lin JS, Perdue LA, Henrikson NB, Bean SI, Blasi PR. Agency for Healthcare Research and Quality (US); 2021.

Appendix U: DOH Standard for Center of Excellence in Hematopoietic Stem Cell Transplantation (HSCT) Services for Adults and Pediatrics

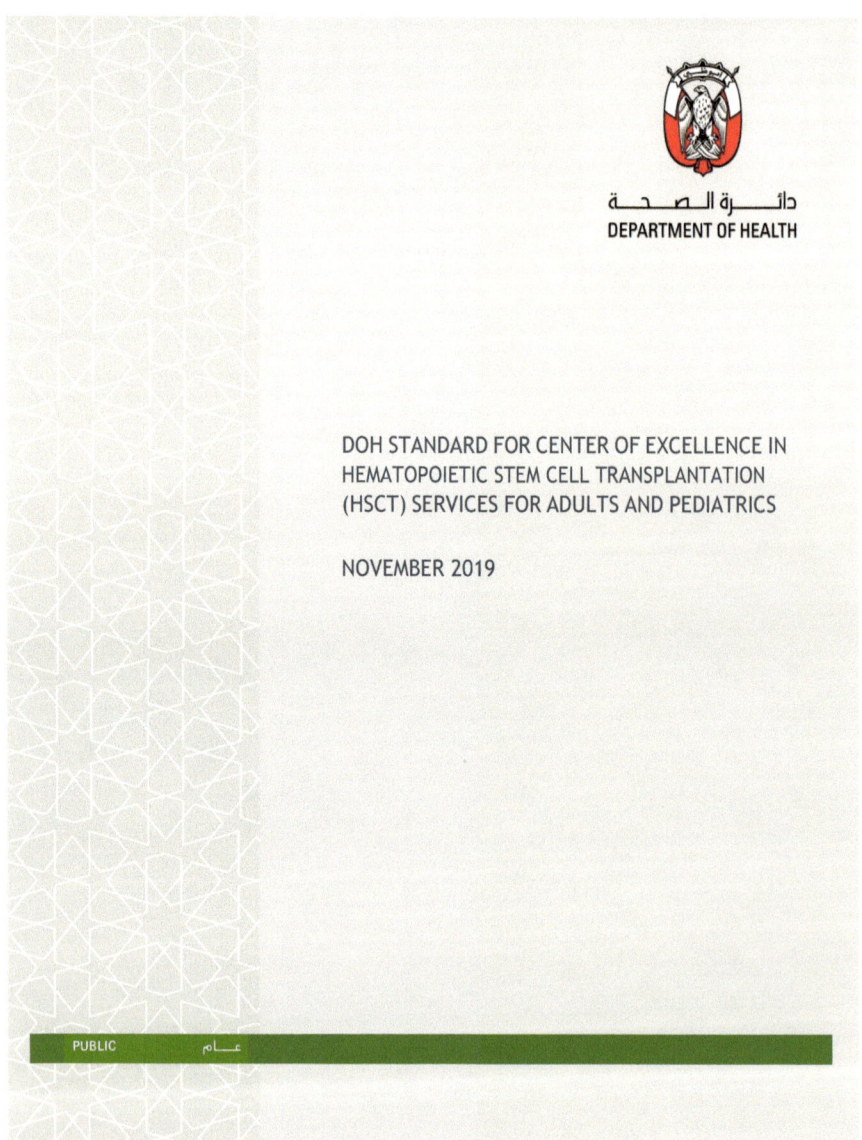

دائـــرة الـصــحــة
DEPARTMENT OF HEALTH

DOH STANDARD FOR CENTER OF EXCELLENCE IN
HEMATOPOIETIC STEM CELL TRANSPLANTATION
(HSCT) SERVICES FOR ADULTS AND PEDIATRICS

NOVEMBER 2019

PUBLIC   عــام

https://doh.gov.ae/-/media/66699F4904F642EF8C3432F69E10084A.ashx

دائـــــرة الـــصـــحـــة
DEPARTMENT OF HEALTH

Document Title:	DOH Standard for Centers of Excellence in Hematopoietic Stem Cell Transplantation (HSCT) Services for Adults and Pediatrics
Document type	Standard
Document Ref. Number:	DOH/HPI/COEHSCT/0.9/2019
Effective Date:	November 2019
Previous versions	None
Document Owner:	Healthcare Planning and Investment
Applies to:	All DOH Licensed Healthcare Providers who seek to be recognized as Centers of Excellence in Hematopoietic Stem Cell Transplantation Services
Classification:	• Public

Note: Read this Standard in conjunction with related UAE Laws, DOH Policies, Standards and Manuals including but not limited to:
- DOH Clinical Privileging Framework Standard
- DOH Data Standards and Procedures
- DOH Standard on Human Subjects Research

1. Purpose

1.1. This Standard defines the service specifications and minimum requirements for healthcare providers to be designated by DOH as Hematopoietic Stem Cell Transplantation Centers of Excellence (COE) in the Emirate of Abu Dhabi.

1.2. The Standard defines the eligibility criteria for all HSCT services in line with DOH Standard for Centers of Excellence—DOH/SD/COE/0.9, evidence based and international guidelines.

2. Definitions

2.1. Centers of Excellence (COE): Specialized and distinguished programs within DOH licensed Healthcare facilities, which can provide an exceptionally high level of expertise and multidisciplinary resources centered on particular service lines and/or services and delivered in a comprehensive, interdisciplinary fashion to achieve the best patient outcomes possible.

2.2. Foundation for the Accreditation of Cellular Therapy—Joint Accreditation Committee (FACT-JACIE): Foundation for the Accreditation of Cellular Therapy (FACT-JACIE) identifies and establishes standards for high quality medical and laboratory practice in cellular therapies.

2.3. Autologous: Derived from and intended for the same individual.

2.4. Allogenic: The biologic relationship between genetically distinct individuals of the same species.

2.5. Pediatrics Patient: Pediatric age as defined by DOH.

3. Abbreviations

3.1. HSCT: Hematopoietic stem cell transplantation

3.2. HEPA: High efficiency particulate air "an efficiency standard of air filter"

3.3. HLA: Human leukocyte antigen

4. Scope

This standard applies to all healthcare providers, public and private, licensed by DOH who seek to qualify as a "Center of Excellence" in Hematopoietic Stem Cell Transplantation (HSCT) services.

5. Implementation Arrangements

DOH shall:

5.1. Ensure that the requirements set out in this Standard are met through its regulatory powers and where necessary, set out further regulatory measures to address the current and future health system needs for developing HSCT Centers of Excellence.

5.2. Ensure that the COE comply with Federal Law and DOH regulations.

5.3. Develop Jawda key performance indicators (Jawda KPI's) to monitor the HSCT COE's performance.

Healthcare providers shall:

5.4. Meet the requirements as set out by DOH in this standard along with the DOH Standard for Centers of Excellence—DOH/SD/COE/0.9 to qualify as a "Center of Excellence" in HSCT services.

5.5. Have in place their own operational guidelines, policies, and procedures.

5.6. Contribute to eliminating International Patient Care (IPC) transfers related to HSCT services.

6. Duties for Healthcare Providers

6.1. The COE in HSCT has to ensure equal access to all patients based on medical needs. The designated COE in HSCT must:

    6.1.1. Ensure and provide evidence that their practices reflect updated international best practices.

6.1.2. Document and monitor quality and safety of clinical care and outcomes of surgical and non-surgical intervention performed on patients, and make these available to DOH for auditing, as and when requested to do so.

6.1.3. Provide records of HSCT related Jawda—Quality Metrics to DOH inspectors.

6.1.4. Maintain Accreditation by a recognized International Accreditation body aligned with DOH and COE and report the findings to DOH (see Appendix V).

6.1.5. Aim to achieve recognized international accreditation in HSCT within 2–5 years.

6.1.6. Follow the clinical and regulatory requirements of this Standard irrespective of provision of COE services to patients who opt not to use health insurance coverage (pre-authorization for coverage and health insurance does not apply in this case).

7. Hematopoietic Stem Cell Transplantation (HSCT) COE Service Requirements and Specifications

7.1. Facilities

7.1.1. Healthcare facilities seeking the COE designation in HSCT should ensure the availability of:

7.1.1.1. A designated inpatient unit that minimizes airborne microbial contamination ideally high efficiency particulate air filtration (HEPA) with positive pressure or laminar airflow for allogeneic transplants.

7.1.1.2. Provisions for prompt evaluation and treatment of patients with complications on a 24-h basis.

7.1.1.3. Access to stem cell lab services that is having international accreditation for stem cell harvest, enumeration; processing and cryopreservation shall be available within the vicinity. The stem cell laboratory shall conform to the National Standards of stem cell procurement, storage, and allocation.

7.1.1.4. Centers performing allogenic HSCT shall have access to HLA-testing laboratory with the capability to carry out DNA-based HLA typing. This HLA-Laboratory shall seek international accreditation.

7.1.1.5. Laboratory support with availability of microbiological tests, monitoring of drug levels, chimerism study, and histopathology services is important. The pathologist shall have experience in the histopathological interpretations of graft versus host disease.

7.1.1.6. A transfusion service to provide irradiated blood products on a 24-h basis.

7.1.1.7. A pharmacy to provide essential medications on a 24-h basis.

7.1.1.8. A radiotherapy service shall be available within the vicinity.

7.1.1.9. Supportive services including specialists in the field of radiology, intensive care, neurology, nephrology, respiratory medicine, gastroenterology, cardiology, and infectious disease shall be available for consultations.

7.2. Healthcare professionals

Healthcare facilities seeking the COE designation in HSCT must fulfil the following requirements related to healthcare professionals:

7.2.1. Valid DOH license in their specialty.

7.2.2. The Head of Clinical Transplant Services shall be a consultant who has at least 1-year specific training in HSCT.

7.2.3. The adult HSCT transplant center shall have at least one physician certified in Internal Medicine and accredited in Hematology or Medical Oncology or Immunology.

7.2.4. Centers performing pediatric transplants shall have at least one physician certified in pediatrics and accredited in hematology/oncology or immunology.

7.2.5. The transplant nurses must have appropriate certification in the management of HSCT patients.

7.2.6. Other supportive staff members shall include a transplant coordinator, pharmacy staff, dietary staff, social worker and physiotherapy staff, and a data manager.

7.3. HSCT services provided in an authorized healthcare facility seeking recognition as a COE in HSCT must include a range of integrated clinical services surgical and non-surgical intervention for its patients in accordance with this Standard including the requirements of FACT-JACIE.

8. Performance Management

Health care provider including those providing pharmacological, surgical, and non-surgical intervention will be required to ensure the following from their services and management systems:

8.1. Are capable of Tracking Performance, including trends in Clinical Quality/ Outcomes for patients by documenting the related JAWDA—Quality Metrics (https://www.doh.gov.ae/resources/jawda-abu-dhabi-healthcare-quality-index).

8.2. Provide seamless care in partnership with other providers, including primary care and hospitals, as required for holistic patient care.

8.3. A Center of Excellence in HSCT shall be required to maintain volumes of greater than or equal to ten (10) new patients/year for autologous transplantation and ten (10) new patients/year for allogeneic transplantation.

9. Clinical Research and Education

9.1. The COE must demonstrate a commitment to education, research, and training focusing on HSC and HSCT-related sciences.

10. Data Management

10.1. Data collection
    10.1.1. Clinical programs shall submit clinical outcomes and specified registry data elements to a national or international database in alignment with related DOH standards.
10.2. The clinical program shall define staff responsible for collecting data maintaining the database.
    10.2.1. Defined data management staff should participate in continuing education annually.

11. Payment Mechanism

11.1. The appropriate compensation model will be adjusted with additional cost associated with clinical leadership, research, education, and technology.

12. Enforcement and Sanctions

12.1. DOH may impose sanctions in relation to any breach of requirements under this standard in accordance with chapter on complaints, investigations, regulatory action, and sanctions, the DOH healthcare regulator manual.

Appendix V: Accredited of HSCT Program

- Foundation for the Accreditation of Cellular Therapy (FACT-JACIE) is the only accrediting organization that addresses all quality aspects of cellular therapy treatments:
  1. Clinical care.
  2. Donor management.
  3. Cell collection.
  4. Cell processing.
  5. Cell storage and banking.
  6. Cell transportation.
  7. Cell administration.
  8. Cell selection.
  9. Cell release.
- Staffing requirements for the clinical unit shall meet the FACT requirements:
  1. Clinical program director
  2. Attending physicians
  3. Mid-level practitioners
  4. Clinical coordinator
  5. Nurses
  6. Consulting specialists
  7. Quality management supervisor/data management
  8. Support services staff (dietitian, psychology, social service)
- The support service staffs are as follows:
  1. Pharmacy staff knowledgeable in the use and monitoring of pharmaceuticals used by the clinical program.
  2. Dietary staff capable of providing dietary consultation regarding the nutritional needs of the transplant recipient, including enteral and parenteral support, and appropriate dietary advice to avoid food-borne illness.
  3. Social services staff.
  4. Psychology services staff.
  5. Physical therapy staff.
  6. Data management staff.

H. O. Al-Shamsi (ed.), *Cancer Care in the United Arab Emirates*, https://doi.org/10.1007/978-981-99-6794-0

References

1. DOH Standard for Centers of Excellence—DOH/SD/COE/0.9.
2. DOH Healthcare Regulator Manual Version 1.0.
3. FACT-JACIE International Standards for Hematopoietic Cellular Therapy Product Collection, Processing, and Administration.
4. National guidelines for hemopoietic stem cell therapy published 2009 and updated 2011. ISBN: 978-983-3433-62-9 (need to get permission to use).
5. NHS standards contract for hematopoietic stem cell transplantation adult NHS commissioning Board. 2013.
6. HSCT charter for KSU (Hematopoietic Stem cell Transplant (HSCT) Program at the Oncology Centre, King Saud University Medical City (KSUMC)).

Appendix W: Cervical Cancer Screening Program Requirements

DEPARTMENT OF HEALTH

CERVICAL CANCER
SCREENING PROGRAM
REQUIRMENTS

MARCH 2022

PUBLIC   عــــام

https://www.doh.gov.ae/-/media/51BDF280150B4AD481064B8E945BDB1D.ashx

DEPARTMENT OF HEALTH

Document Title:	Cervical Cancer Screening Program Requirements		
Document Ref. Number:	DOH/CCSC/SD/1.1	Version:	1.2
Last Reviewed:	March 2021	Next Review:	March 2024
Approval Date:	3rd, March 2022	Effective Date:	16th, March 2022
Document Owner:	Cancer Prevention and Control Section, NCD, Abu Dhabi Public Health Center		
Applies to:	All Licensed Healthcare Providers in the Emirate of Abu Dhabi participating in DOH's Cervical Cancer Screening Program		
Classification:	• Public		
This Standard should be read in conjunction with related UAE laws, DOH standards, policies, and circulars including but not limited to:	• Federal laws on Medical Liability and the Practice of Human Medicine • DOH Providers Manual-Chapter on Standards of Care • DOH Professionals Manual-Chapter on Professional Duties and Obligations • DOH Regulator Manual-Chapter on Data Management • DOH Standard for Clinical Privileging Framework • DOH Standard for Adverse Events reporting and Management • Federal Charter on Patient's Rights and Responsibilities • DOH Standard for Complaints Management in Healthcare Facilities • DOH Data Management Policy and DOH Data Standards • DOH Standard on Telemedicine • DOH Policy on Cultural Sensitivity		

1. Purpose

1.1. This document mandates the clinical service specifications and data reporting for DOH's Cervical Cancer Screening Program in the Emirate of Abu Dhabi.

1.2. It specifies the clinical care pathway and minimum service specifications to ensure that women screened for cervical cancer receive quality and safe care and timely referral for diagnosis and/or treatment.

2. Scope

2.1. These program specifications apply to all healthcare providers (facilities, laboratories, professionals) licensed by DOH in the Emirate of Abu Dhabi who are participating in DOH's Cervical Cancer Screening Program.
2.2. Participating healthcare providers are to provide the following services as applicable based on their license category:
    2.2.1. Risk assessment and physical examination
    2.2.2. Specimen collection and preparation of adequate cervical smear
    2.2.3. Handling and reporting of cervical smears
    2.2.4. Follow-up and referral
2.3. Follow reporting terminologies defined by DOH as per Appendix X.

3. General Duties of the Health Care Providers

All licensed and eligible healthcare providers participating in DOH's Cervical Cancer Screening Program must:

3.1. Provide clinical services and patient care in accordance with DOH Policies and Standards, and the laws and regulations of the Emirate of Abu Dhabi.
3.2. Submit data to DOH via e-Claims in accordance with the DOH Reporting of Health Statistics Policy and as set out in the DOH Data Standards and Procedures (found online at https://www.doh.gov.ae/en/Shafafiya/dictionary).
3.3. Comply with relevant DOH Policies and Standards with special attention to:
    3.3.1. Policies and standards on Patient Education and Consent: The licensed provider must provide appropriate patient education and information regarding the screening test and must ensure that appropriate patient consent is obtained and documented on the patient's medical record.
    3.3.2. Policies and standards on managing patient medical records including developing effective recording systems, maintaining patient records, confidentiality, and privacy and security of patient information; educating patients on services provided; and satisfying the requirements of patient informed consent and patient rights and responsibilities charter.
    3.3.3. DOH data standards and procedures.
3.4. Comply with DOH's requests to inspect and audit records and cooperate with DOH authorized auditors as required by DOH.
3.5. Comply with requirements for information technology (IT) and data management including sharing of screening/diagnosis and where applicable, pathology results, electronic patient records and disease management systems.

4. Enforcement and Sanctions

4.1. Healthcare providers, professionals, and laboratories participating in the DOH's Cancer Screening Program must comply with the terms and requirements of this document.

4.2. Health care providers must comply with the DOH Standard Provider Contract.

4.3. DOH may impose sanctions in relation to any breach of requirements under this program specification in accordance with the [DOH *Policy on Inspections, Complaints, Appeals and Sanctions*].

5. Payment Mechanism

Eligibility for reimbursement under the Health Insurance scheme is as follows:

5.1. For Thiqa holders, reimbursement must be consistent with the DOH Standard for Thiqa Preventive List of Interventions available at www.doh.gov.ae.

5.2. For non-Thiqa holders, payment must be consistent with the individual's health insurance product/plan.

6. Cervical Cancer Screening Program Specifications for Facilities

Facilities participating in DOH's Cervical Cancer Screening Program must:

6.1. Be licensed by the DOH.

6.2. Fulfill the eligibility criteria for a cervical cancer screening program participating facilities in accordance with Appendix Y and approved by DOH as eligible cervical cancer screening program facilities.

6.3. Comply with DOH cervical cancer screening care pathways, clinical quality indicators, and timelines for referral in accordance with Appendices Z–AB, respectively.

6.4. Assign a screening program coordinator responsible for submitting data on screening visits and outcomes to DOH, who will fulfill the responsibilities in accordance with Appendix AC.

6.5. Collect and submit data on screening visits and outcomes within 3 weeks of the screening date, to DOH through the Electronic Cancer Screening Notification (e-*cancer notification*) that can be accessed at: https://www.doh.gov.ae/en/e-services.

6.6. Report all screen-detected cancer cases to DOH, through Cancer Case Notification Form, of the e-*cancer notification*, specified in 4.2.

6.7. Maintain records for screening tests and outcomes.

6.8. Establish internal audit procedures to demonstrate compliance with this document and other associated regulatory policies and standards.

6.9. Ensure availability of evidence of compliance with the Cervical Cancer Screening Program Clinical Quality indicators specified in Appendix AA including:

6.9.1. Collection and preparation of adequate cervical smear.

6.9.2. Handling and transporting of specimens to DOH licensed clinical laboratories.

6.9.3. Have an approved protocol for referral of women with abnormal results or physical examination to a diagnostic or treatment centers.

7. Cervical Cancer Screening Program Specifications for Laboratories

Laboratories participating in DOH Cervical Cancer Screening Program must:

7.1. Be licensed by the DOH.
7.2. Comply with the applicable elements of the Screening Program and other DOH clinical quality indicators in accordance with Appendix AA and ensure availability of evidence of compliance with these indicators such as laboratory records required for accreditation purposes.
7.3. Comply with the DOH Clinical Laboratory Standards.
7.4. Attain accreditation by an international body recognized by the DOH such as, CAP-ISO 15189(2007), or JCI/Lab.
7.5. Participate in an international external proficiency test by all personnel involved in screening and reporting Pap test.
7.6. Establish internal audit procedures to demonstrate compliance with this program specification and with other associated regulatory policies and standards.
7.7. Develop, implement, and monitor policies and standard operating procedures for management of smears in accordance with DOH clinical laboratory standards.

8. Cervical Cancer Screening Program Specifications for Healthcare Professionals

Health professionals participating in the DOH's Cervical Cancer Screening Program must be:

8.1. Licensed by the DOH.
8.2. Comply with the clinical standards detailed in this program specification to provide the most appropriate care, taking responsibility for deciding the best care options for managing cervical cancer cases.
8.3. Provide women with culturally and socially relevant education on women's health and with information (oral and written) regarding the screening benefits and limitations of cervical screening, potential outcomes, and next steps that may be required for care management.
8.4. Participate in continuing medical education (CME) in accordance with DOH requirements.

9. Cervical Cancer Screening Program Screening Tests and Frequency

9.1. Screening tests:
   9.1.1. Papanicolaou test (also called Pap test). Liquid-based cytology (LBC) is the accepted standard method for Pap test specimen collection.
   9.1.2. HPV test, as co-testing, for women age of 30 years and above. Only FDA HPV approved tests are the accepted test for screening.

9.2. Frequency of screening:

The frequency for repeat screening for average risk, symptom free women is as follows:

9.2.1. Every 3 years for women aged 25–29 years.

9.2.2. Every 5 years for women 30–65 years.

9.2.3. Annually for women who are immune-compromised due to disease or medication.

10. Cervical Cancer Screening Program Service Specifications

10.1. Eligibility for screening criteria

10.1.1. All sexually active women (past or present), symptom free, aged 25–65 years residing in the Emirate of Abu Dhabi, except where exclusion criteria for screening apply.

10.1.2. Women are excluded from screening if:

10.1.2.1. They have received a total hysterectomy for benign indications.

10.1.2.2. They are over 65 years, (if the last three previous smears were negative).

10.1.3. Women who have had subtotal hysterectomy (preserving the cervix) or hysterectomy due to cervical cancer or precancerous condition should continue to have cervical screening.

10.1.4. Screening recommendations remain the same regardless of whether or not they have received the HPV vaccination.

10.2. Recruitment to screening

Recruitment of eligible women for screening can be made through: Targeted invitation from the eligible screening facilities.

10.2.1. Opportunistic by:

10.2.1.1. Approaching women who are enrolled in other existing screening programs; e.g., breast cancer.

10.2.1.2. Physician consultation for related or unrelated reason.

10.2.1.3. As an outcome of a health promotion campaign.

10.3. Risk assessment and physical examination

10.3.1. Women must receive adequate information regarding the screening, Pap test procedure, and expected outcomes and timeframe to receive results.

10.3.2. Detailed history must be taken to assess risk and frequency of repeating screening, including at least:

10.3.3. Menstrual status (last menstrual period, hysterectomy, pregnant, postpartum, use of contraceptive, or hormone therapy).

10.3.4. Previous screening, results of screening, (negative, abnormal, or positive), and any previous treatment (biopsy, chemotherapy, radiotherapy, or surgery).

10.3.5. Immune-compromised status due to diseases (including HIV) or medication.

    10.3.6. Full clinical examination must be performed including visual inspection of the cervix.

10.4. Specimen collection and preparation of adequate Pap test

    10.4.1. The following categories of DOH licensed healthcare physicians are eligible to perform a Pap test:

        10.4.1.1. Licensed gynecologists, obstetricians, and family medicine physicians.

        10.4.1.2. Physicians already privileged to do so by their institution.

    10.4.2. Eligible physicians must:

        10.4.2.1. Complete the required form with relevant clinical information in accordance with Article 9.1.2.

        10.4.2.2. Collect and manage specimens in accordance with the facility internal policies and procedures and DOH standards.

    10.4.3. Smear taking must be avoided in the following circumstances and women must be advised when to return for a pap test:

        10.4.3.1. Menstruation, blood loss, breakthrough bleeding.

        10.4.3.2. Vaginal inflammation/infection.

        10.4.3.3. Pregnancy (unless a previous smear was abnormal and in the interim the woman becomes pregnant, then the follow-up smear must not be delayed).

10.5. Cytology smear handling and reporting

Clinical laboratories handling and reporting of cytology specimens and cytology smears testing must:

    10.5.1. Manage cervical cytology smears and perform the cytopathology testing as indicated and in accordance with the specifications of the DOH Clinical Laboratory Standards, including without limitation "Processes for Laboratory Specialties" and "Cytopathology of the HAAD Clinical Laboratory Standards."

    10.5.2. Make final reports of cervical cytology smear using the Bethesda System (The Bethesda System for Reporting Cervical Cytology).

    10.5.3. The report must be verified by a pathologist for all abnormal and reactive cases, while negative cases can be verified by licensed cytotechnologist using standard synoptic reporting format and containing minimum elements consistent with those of internationally reputable accrediting bodies.

    10.5.4. The report must minimally include the following details:

        10.5.4.1. Patient's name.

        10.5.4.2. Age/date of birth.

        10.5.4.3. Menstrual status (LMP, hysterectomy, pregnant, postpartum, hormone therapy).

        10.5.4.4. Relevant clinical information such as if the patient has previously had a positive test or had other types of cancer, etc.

        10.5.4.5. Specimen description (source).

10.5.5. The reporting pathologist is the professional responsible for confirming the positive cancer results.

10.5.6. Reports for specimen adequacy and cytological findings must be returned to the referring physician at the screening center within 8 working days of receiving the specimen.

10.5.7. The DOH may, at its discretion, conduct third-party independent quality assurance testing of laboratories providing cervical smear laboratory test service. Where it does so, providers must comply with DOH's direction and cooperate with the DOH appointed party.

10.6. Screening outcomes and referrals

10.6.1. All women must be notified in writing of the result of their screening tests.

10.6.2. It is the responsibility of the physician at the screening facility to notify and provide a written report to a woman regarding her screening results within 21 days (3 weeks) of the date of specimen taken.

10.6.3. If the test outcome is normal, the woman is discharged to routine screening as per frequency mentioned in this document.

10.6.4. If the test outcome is unsatisfactory, it must be repeated within 6–12 weeks, treating infection, if present, as indicated.

10.6.5. If the Pap test outcome is abnormal (cytology showed intraepithelial lesions or malignancy and/or HPV positive), the woman's test should be managed according to Appendix Z.

10.6.6. If a suspicious visible abnormality is identified during visualization of the cervix, the woman must be referred immediately to a gynecologist oncologist without receipt of her test results.

10.6.7. If a woman requires referral for colposcopy or treatment, the physician must make the referral to an appropriately DOH licensed healthcare professional, privileged to provide the specialty/oncology service. Timelines for referral should be in compliance with Appendix AB.

Appendix X: Definitions

Term	Definition
The Bethesda system (TBS)	Is a system for reporting cervical or vaginal cytological diagnoses, used for reporting Pap smear results. The name comes from the location (Bethesda, Maryland) of the conference that established the system of reporting
HPV	Human papilloma virus
HPV co-testing	Is a test is done along with the Pap test in women aged 30 years and above, to screen for a high-risk HPV viral types. Only FDA approved test is accepted
ASC-US	Atypical squamous cells of undetermined significance. It is a finding of abnormal cells in the tissue that lines the outer part of the cervix
ASC-H	Suspicious for high grade dysplasia
LGSIL or LSIL	Low-grade squamous intraepithelial lesion
HGSIL or HSIL	High-grade squamous intraepithelial lesion
AIS	Adenocarcinoma in situ
AGC	Atypical glandular cells

Appendix Y: Eligibility Criteria for a Facility to Participate in DOH's Cervical Cancer Screening Program

1. General

In addition to the requirements of this program specification, the healthcare facility must fulfill the following criteria:

   1.1. Plan capacity to match the demand for screening and the facility capacity.

   1.2. Allocate appointment slots for cervical cancer screening linked to the DOH online booking system (when available).

   1.3. Have available adequate equipment to provide safe and quality screening:

      1.3.1. Send cervical cytology smears only to DOH licensed laboratories that meet the requirements of this program specification.

      1.3.2. Ensure patient privacy, comfort, and confidentiality at all times.

2. Human resources

   2.1. The core team must include at least:

      2.1.1. A program coordinator.

      2.1.2. A licensed physician, family medicine physician, gynecologist or obstetrician, physician privileged to deliver cervical screening care and services.

      2.1.3. A licensed nurse for each clinic with a minimum of 2 years of experience in gynecology or obstetric nursing.

   2.2. Training of licensed health professionals must be delivered using CME/CPD courses accredited by DOH CME department at https://www.doh.gov.ae/en/programs-initiatives/meed including

      2.2.1. For physicians; training for Pap smear taking in accordance with international evidence-based training standards and guidelines.

3. Registration as DOH screening facilities

Facilities meeting DOH cervical cancer screening requirements should follow DOH facilities' registration process:

   3.1. Register through DOH facility registration website.

Appendix Z: Cervical Cancer Screening Care-Pathway

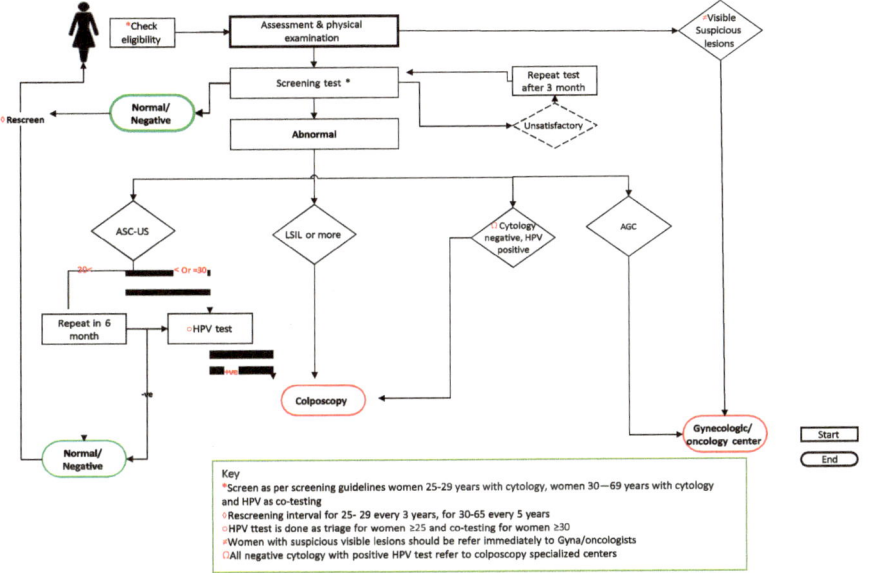

Appendix AA: Cervical Cancer Screening Program—Clinical Quality Indicators

Quality indicator		Acceptable level	Desirable level
Coverage			
Retention rate	Percentage of eligible women re-screened within 3 years after a negative Pap test in a 12-month period	40%	50%
Cytology performance indicators			
Specimen adequacy unsatisfactory proportion	Percentage of Pap tests that are reported as unsatisfactory in a 12-month period	4.7%	1.3
Screening test results Negative	Percentage of women by their most severe Pap test result in a 12-month period	90%	97%
System capacity indicators			
Cytology turnaround time 2 weeks	The average time from the date the specimen is taken to the date the finalized report is issued over a 12-month period	>80%	>90%
Time to colposcopy	Percentage of women with a positive Pap test (HSIL+/ASC-H) who had follow-up colposcopy within 3, 6, 9 and 12 months subsequent to the index Pap test	80%	88%
Follow-up			
Biopsy rate	Percentage of women with a positive screening test result (HSIL+/ASC-H) who received a histological diagnosis in a 12-month period	To be determined	11%
Cytology-histology agreement	Proportion of positive Pap tests with histological work-up found to have a pre-cancerous lesion or invasive cervical cancer in a 12-month period A	To be determined	
Outcome indicators			
Pre-cancer detection rate	Number of pre-cancerous lesions detected per 1000 women who had a Pap test in a 12-month period	To be determined	7.1 per 1000

© The Editor(s) (if applicable) and The Author(s), under exclusive license to Springer Nature Singapore Pte Ltd. 2024
H. O. Al-Shamsi (ed.), *Cancer Care in the United Arab Emirates*,
https://doi.org/10.1007/978-981-99-6794-0

Appendix AB: Cervical Cancer Screening Program—Timeframes for Appointments

Cytological pattern	Priority	Appointment
HSIL or greater	Urgent	1–2 weeks
LSIL	soon	2–4 weeks
ASC-US/ASC-H	Routine	4–8 weeks

© The Editor(s) (if applicable) and The Author(s), under exclusive license to
Springer Nature Singapore Pte Ltd. 2024
H. O. Al-Shamsi (ed.), *Cancer Care in the United Arab Emirates*,
https://doi.org/10.1007/978-981-99-6794-0

Appendix AC: Responsibilities of the Facility Cancer Screening Program Coordinator

The healthcare facility cervical cancer screening program coordinator must:

1.1. Be a licensed healthcare professional.
1.2. Have comprehensive and high-quality knowledge in cervical cancer as a disease and its prevention.
1.3. Be responsible for:
    1.3.1. Recruitment of eligible women.
    1.3.2. Follow-up and tracking of screening results to ensure the timeliness and completeness of follow-up.
    1.3.3. Assessing relationships between planned care and approved protocols for care.
    1.3.4. Assessing women's needs for support to remove barriers to screening and follow-up.
    1.3.5. Developing and promoting recall systems that include reminders to patients as appropriate.
    1.3.6. Submitting data on screening visit and outcomes to DOH via the (cancer screening e-notification system).

References

1. NHS Cervical Screening Program. https://www.bsccp.org.uk/assets/file/uploads/resources/NHSCSP_20_Colposcopy_and_Programme_Management_(3rd_Edition)_(2).pdf.
2. The American Cancer Society guidelines for the prevention and early detection of cervical cancer. https://www.cancer.org/cancer/cervical-cancer/prevention-and-early-detection/cervical-cancer-screening-guidelines.html.
3. Performance monitoring for cervical cancer screening programs in Canada. http://www.phac-aspc.gc.ca/cd-mc/cancer/pmccspc-srpdccuc/pdf/cervical-eng.pdf.
4. Ontario cervical screening guidelines. https://www.cancercare.on.ca/common/pages/UserFile.aspx?fileId=13104.
5. NCCN cervical cancer screening guidelines. https://www.cancercare.on.ca/common/pages/UserFile.aspx?fileId=13104.

6. ACOG-cervical cancer screening guidelines. https://www.acog.org/Patients/FAQs/Cervical-Cancer-Screening.

7. Implementation of cancer screening in the European Union. 2017. https://ec.europa.eu/health/sites/health/files/major_chronic_diseases/docs/2017_cancerscreening_2ndreportimplementation_en.pdf, https://www.ncbi.nlm.nih.gov/pmc/articles/PMC2826099/.

8. European guidelines for quality assurance in cervical cancer screening. http://screening.iarc.fr/doc/ND7007117ENC_002.pdf.

Appendix AD: DOH Colorectal Cancer Screening Program Specifications

DEPARTMENT OF HEALTH

DOH COLORECTAL CANCER SCREENING PROGRAM SPECIFICATIONS

SEPTEMBER 2019

PUBLIC   عــام

https://doh.gov.ae/-/media/980F1D8087214B6D9E630B457B85E4B9.ashx

دائـــــرة الــصــــحــة
DEPARTMENT OF HEALTH

Document Title:	DOH Colorectal Cancer Screening Program Specifications		
Document Ref. Number:	DOH/ADCPH/CRCS/ PS/1.0	Version:	1.0
Approval Date:	04 September 2019	Effective Date:	05 September 2019
Last Reviewed:	August 2019	Next Review:	October 2020
Revision History:	Version 0.9 July, 2012		
Document Owner:	Public Health and Research Division		
Applies to:	All DOH Licensed Healthcare Providers		
Classification:	• Public		

1. Purpose

1.1. To set out the minimum service specifications for DOH Colorectal Cancer Screening Program (CRCSP) through identification of the following:

   1.1.1. Duties of the participating healthcare providers.

   1.1.2. Clinical and administrative service specifications.

   1.1.3. Risk assessment and eligibility criteria.

   1.1.4. Data reporting requirements.

2. Scope

2.1. All DOH licensed healthcare providers and professionals who are participating in DOH's Comprehensive Screening Program for Adults and providing CRC screening services including mobile units.

3. Definitions and Abbreviations

3.1. **Colorectal Cancer Screening Program (CRCSP)** in the Emirate of Abu Dhabi and for the purpose of this document includes the following services:

   3.1.1. Colorectal cancer screening services.

   3.1.2. Colorectal cancer assessment and follow-up.

   3.1.3. Familial and genetic high-risk assessment.

3.2. **Colonoscopy** is the endoscopic examination of the large bowel and the distal part of the small bowel with a Charge Coupled Device (CCD) camera or a fiber optic camera on a flexible tube passed through the anus. It can provide a visual sight to detect adenomatous polyps and cancer diagnosis (e.g., ulceration and polyps). It also gives the opportunity for biopsy or removal of suspected colorectal cancer lesions.

3.3. **Fecal Immunochemical Test (FIT)** is a test that investigates the stool sample for signs of cancer.

3.4. **Case mix** includes males and females aged 40–75 years determined eligible for colorectal cancer screening services, in accordance with the criteria detailed in these program specifications:

    3.4.1. People who are symptomatic and in good health and do not have any problems.

    3.4.2. For adults aged 76–85 years, the decision to screen should be-individualized, considering the patient's overall health prior screening history. The decision will be based on the healthcare professional judgment and the individual's preference.

    3.4.3. People over 85 should no longer get colorectal cancer screening.

3.5. **CT**: Computed tomography.

3.6. **LFTs**: Liver function tests.

3.7. **OLT**: Orthoptic liver transplant.

3.8. **PSC**: Primary sclerosing cholangitis.

3.9. **CRC**: Colorectal cancer.

4. Duties of CRCSP Participating Healthcare Providers

All DOH licensed healthcare providers (facilities and professionals) engaged in CRCSP must:

4.1. Follow the pathway for CRC Screening as per Appendix AE.

4.2. Adhere to the clinical performance indicators and timelines in accordance with Appendix AF.

4.3. Include a CRC risk assessment and refer individuals to appropriate screening test based on risk categories as per Appendix AG.

4.4. Assign a CRCSP coordinator/director who will be accountable to:

    4.4.1. Report and submit screening outcome data to DOH specified in Sect. 4.

    4.4.2. Maintain records for screening tests and outcomes.

    4.4.3. Establish internal audit policies and procedures and conduct regular audits, monitoring and evaluating to demonstrate compliance with these program specifications and other associated regulatory policies and standards.

4.5. Has an endoscopy unit that meets the criteria for a competent unit infrastructure, equipment and personnel, as per Appendix AH.

4.6. Comply with DOH direct or DOH third-party independent quality assurance testing of laboratories providing CRC screening test service.

4.7. Establish an invitation system to ensure successful participation of eligible population. The system may be manual or electronic.

4.8. Provide clinical services and patient care in accordance with these program specifications and in accordance with DOH policies and standards including data submission and cancer E-notification through DOH Data Dictionary Website.

4.9. Provide clinical services and patient care in accordance with laws and regulations of the UAE and the Emirate of Abu Dhabi.

Laboratory services requirements:

DOH licensed healthcare providers engaged in CRCSP must provide laboratory services that:

4.10. Perform the CRC screening laboratory tests in accordance with the requirements and specifications of the CRCSP.

4.11. Use specimen identification and labelling in accordance with DOH Clinical Laboratory Standards and industry best practices.

4.12. Is accredited by an internationally accrediting body for CRC screening tests.

4.13. Ensure that laboratory performing the FIT test:

4.13.1. Follow the manufacturer's instructions for use of the FIT testing kit.

4.13.2. Use an explicit definition for cut-off levels for hemoglobin concentration.

4.13.3. Make provision to record the information concerning the actual amount of hemoglobin, both for tests classified as negative and for those classified as positive.

4.13.4. Employ licensed professionals who are privileged and have evidenced their ability to undertake different types of fecal occult blood test and in-depth understanding of the technology required to perform the fecal occult blood test.

4.14. Ensure that laboratory performing genetic testing must have a specialist cyto/histopathological support services.

Healthcare professional requirements:

DOH licensed healthcare providers engaged in CRCSP must:

4.15. Have a multi-disciplinary team that includes a gastroenterologist, colorectal surgeon, gastrointestinal oncologist, pathologist, radiologist, medical, and a nurse to review the outcomes of screening for colorectal cancer.

4.16. All healthcare professionals participating in colorectal cancer screening must:

4.16.1. Obtain informed patient consent prior to screening.

4.16.2. Where consent is granted or refused, the treating physician must document and retain signed consent forms on patients' medical records.

4.16.3. Inform all patients of the procedures and expected timeframe to be screened and to receive results.

Reporting requirements:

In addition to the routine e-Claims data, DOH licensed healthcare providers (facilities and professionals) engaged in CRCSP must:

4.17. Collect and submit to DOH data on screening visits and outcomes within 2 weeks of the screening date, through the cancer screening form of the cancer E-notification system on DOH Data Dictionary Website.

4.18. Report all screening-detected cancers to DOH, using cancer case notification form (Cancer Notification Appendix AR) available on DOH Data Dictionary Website.

5. CRCSP Risk Assessment and Eligibility Criteria

5.1. **Risk assessment**

    5.1.1. A detailed medical history of the screening candidate, such as that described in the cancer screening form available at DOH e-services website or as per Appendix AU should be taken to determine if the candidate's risk category is average, increased or high as per Appendix AG.

    5.1.2. Colorectal Screening Program-Screening individuals at increased risk for CRC:

        5.1.2.1. Individuals with increased risk for CRC should be referred for assessment and screening to specialized centers or hospitals that have specialized multidisciplinary teams providing gastroenterology, genetic counseling, genetic testing, surgery, oncology, and preferably psychosocial support services.

        5.1.2.2. All individuals at increased risk must be managed in accordance with Appendices AI, AJ, and international best practices and guidelines such as NICE Guidelines for colorectal cancer screening and surveillance in moderate and high-risk groups and NICE guidelines for colonoscopy surveillance for the prevention of colorectal cancer in people with ulcerative colitis, Crohn's disease or adenomas.

    5.1.3. Colorectal Screening Program-Familial/Genetic High-Risk Assessment:

        5.1.3.1. All individuals with suspected or confirmed genetic/familial high risk as per Appendices AK and AL should be referred only to specialized centers or hospitals that have specialized multidisciplinary teams providing gastroenterology, genetic counseling, genetic testing, surgery, oncology, and preferably psychosocial support services for the purpose of assessment, genetic counselling, and mutation analysis of relevant genes where appropriate.

        5.1.3.2. Individuals with gene mutation must be managed in accordance with international best practices and guidelines.

        5.1.3.3. Genetic counseling is highly recommended when genetic testing is offered and after disclosure of results.

        5.1.3.4. Genetic counseling can be given by a genetic counselor, medical geneticist, oncologist, surgeon, oncologist, or other healthcare professional with expertise and experience in genetic counseling who is privileged by the facility to provide counseling.

        5.1.3.5. A list of recommended genetic tests is available in Appendix AL.

5.2. **Screening tests and frequency**

    5.2.1. For individuals at average risk of colorectal cancer:

        5.2.1.1. Eligible population must be offered the two options for colonoscopy screening as per Appendix AE or FIT test in case of colonoscopy refusal.

5.2.1.2. Frequency of screening should be as follows:

5.2.1.2.1. Colonoscopy screening, every 10 years.

5.2.1.2.2. Fecal Immunochemical Test (FIT) every year (annually).

5.2.2. For individuals at increased or high risk of colorectal cancer:

5.2.2.1. Colonoscopy screening should be offered.

5.2.2.2. The frequency and age of initiation should be individualized for each person as described in Appendices AI–AK.

5.2.2.3. Further investigations, genetic testing, and counselling should be pursued for individuals with suspected familial/genetic high risk as per Appendices AK and AL.

5.2.2.4. All healthcare providers must utilize evidence based-practice for CRC screening as per Appendices AG and AM–AU.

5.2.3. For individuals who will go through CRCSP with Fecal Immunochemical Test (FIT), they must be provided with clear and simple instructions regarding collection of the sample.

5.3. **Recruitment for screening**

Individuals eligible for CRC screening may be recruited in healthcare facilities through one of the following:

5.3.1. Targeted invitation which may be established via an electronic or manual invitation system call and recall initiative with Daman.

5.3.2. Opportunistic:

5.3.2.1. New physician consultation for related or unrelated reason.

5.3.2.2. Engagement in a health promotion campaign.

5.4. **Online booking to screening appointment**

All DOH Licensed Healthcare Providers (Facilities and Professionals) engaged in CRCSP:

5.4.1. Are encouraged use optional online booking system for screening appointments in order to facilitate access to screening services available at DOH website.

5.4.1.1. Facilities that utilize the provided appointment system must nominate a person to access the online booking schedule by which secure access will be provided by DOH.

5.4.2. Are encouraged to provide flexible timeslots to enable this functionality.

5.4.3. Ensure that they meet the available appointments to receive patients.

5.5. **Outcomes and referrals**

5.5.1. At the end of the screening procedure, in case of positive outcome, the screening unit must provide the candidate with a copy of the DOH e-cancer screening referral form (Appendix AT).

5.5.2. The time between completion of a screening test and receipt of results by the participant must be less than 15 working days (acceptable standard >90% within 15 days).

5.5.3. Screening with colonoscopy

       5.5.3.1. In case of normal results, (i.e., negative for polyps) individuals must be re-invited for screening in accordance with the frequencies specified in point 5.2.

       5.5.3.2. In case of the presence of adenoma, colonoscopy must be repeated in accordance with Appendix AE.

       5.5.3.3. Adenoma detection rate must be monitored and audited. It is limited to screening colonoscopies and surveillance procedures. Repeat endoscopic procedures are excluded.

       5.5.3.4. The time interval between a positive colonoscopy (cancer) and beginning of definitive management must be less than 31 working days (acceptable standard $\geq 95\%$ of cases must be no more than 31 days).

       5.5.3.5. Death within 30 days after colorectal cancer screening, attributed to complications caused by colonoscopy, must be recorded by e-notification.

   5.5.4. Screening with FIT test (not applicable for increased and high-risk group)

       5.5.4.1. Patients with a negative test result should be re-invited for screening as per frequencies specified in point 5.2.

       5.5.4.2. Patients with a positive test result must be scheduled for follow-up colonoscopy within 31 days of referral.

       5.5.4.3. The FIT test must be repeated if results are unclear or spoilt in accordance with Appendix AL.

6. Enforcement and Sanctions

6.1. Healthcare providers participating in DOH Comprehensive Screening Program—Colorectal Screening Program, payers and third-party administrators must comply with the terms and requirements of these program specifications.

6.2. DOH may impose sanctions in relation to any breach of requirements under these program specifications in accordance with the Chapter on Inspections, Complaints, Appeals and Sanction, the Healthcare Regulatory Manual.

7. Payment for Screening and Follow-Up of Colorectal Cancer

7.1. Eligibility for reimbursement under the Health Insurance Scheme must be in accordance with the Standard Provider Contract and as applicable by the Thiqa Prevention List, DOH Mandatory Tariff and associated Claims and Adjudication Rules and the Coding Manual. All documents are available from the DOH website in Data Dictionary.

Appendix AE: Colorectal Cancer Screening—Care-Pathway

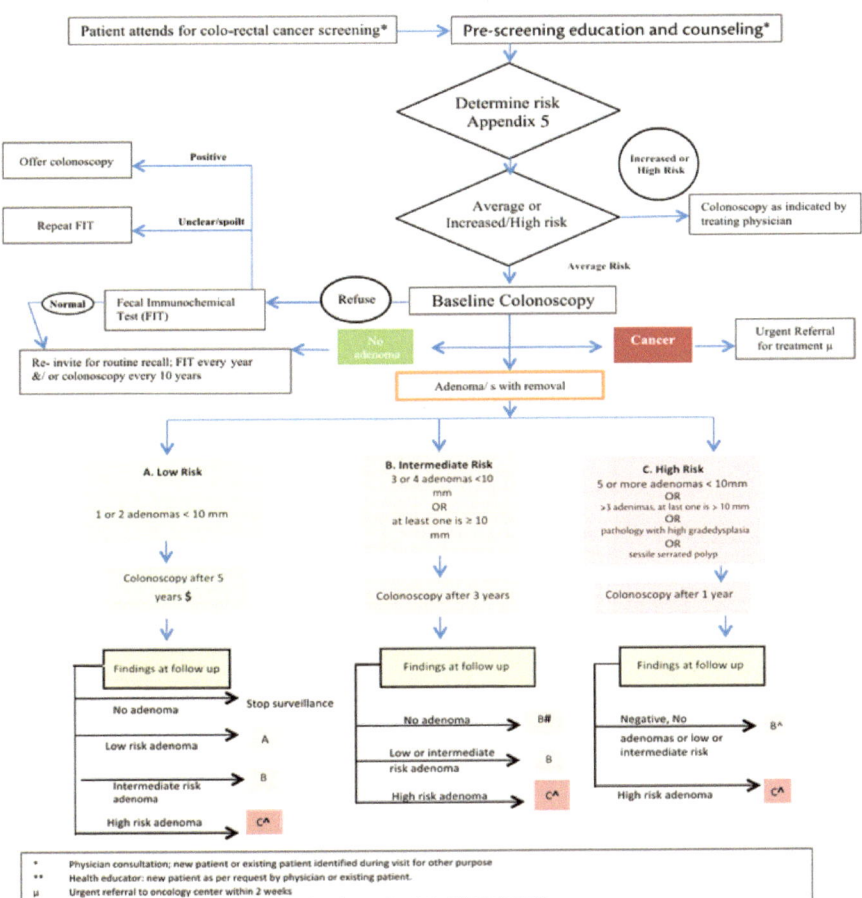

Appendix AF: Colorectal Cancer Clinical Performance Indicators[1]

Indicator	Acceptable level	Desirable level
Screening uptake (participation) rate	>55%	>70%
Minimum number of screening colonoscopies undertaken annually by each screening colonoscopist	>150 per annum	>250
Inadequate FIT rate (proportion of people screened with one or more FIT returned none of which were adequate)	<3%	<1%
Maximum time between screening FIT test and receipt of result should be 7 days from sample's dispatch	>90%	100%
Rate of referral to follow-up colonoscopy after positive FIT test (detects cancer)	95%	>100%
Maximum time between referral after positive screening FIT test and conducting a follow-up colonoscopy should be within 31 days	>90%	>95%
Cecal intubation rate (CIR). Follow-up and screening colonoscopies to be recorded separately (unadjusted CIR with video recorded and photographic evidence)	>90%	>95%
Adenoma detection rate on males	25%	Auditable outcome
Adenoma detection rate on females	15%	Auditable outcome
Cancer detection rate	≥2 per 100 screened by FIT ≥11 per 100 colonoscopies	Auditable outcome
Withdrawal time in negative colonoscopies (withdrawal from cecal pole to anus)	≥6 min	Auditable outcome
Polyp retrieval rate (retrieval of polypectomy specimens for histological analysis per colonoscopist)[b]	>90% per 100 polyps excised	>95% per 100 polyps excised
Rate of high-grade neoplasia reported by pathologists in a FIT screening program	<10%	

[1] NHS Cancer Screening Programs quality assurance guidelines for colonoscopy. European guidelines for quality assurance in colorectal cancer screening and diagnosis.

Indicator	Acceptable level	Desirable level
Endoscopic complications of colonoscopy screening programs	Bleeding <1:150 Perforation <1:1000	
Post polypectomy perforation rate	<1:500	Auditable outcome
Time interval between positive colonoscopy and start of definitive management within 31 days	>90%	>95%

Bowel cleansing should be audited

Proposed standard: At least more than 90% of examinations should be rated as "adequate" bowel cleansing or excellent

[a] Excellent: no or minimal solid stool and only clear fluid requiring suctionAdequate: collections of semi-solid debris that are cleared with washing/suctionInadequate: solid or semi-solid debris that cannot be cleared effectively

[b] Numerator: number of polyps with histological tissue retrieved for analysisDenominator: number of polyps recorded during lower GI endoscopies

Appendix AG: Pre-colonoscopy Risk Assessment for Colorectal Cancer

Average risk:

1. Age ≥40
2. No history of adenoma or colorectal cancer
3. No history of inflammatory bowel disease
4. Negative family history

Increased risk:

1. Personal or family history of adenoma, sessile serrated polyp (SAP),[1] colorectal cancer, inflammatory bowel disease.
2. Positive family history of first- or second-degree relative with colorectal cancer (screening recommendations vary depending on family history).

High risk:

1. Hereditary nonpolyposis colorectal cancer (HNPCC).
2. Family history of polyposis syndromes (Classical Familial Adenomatous Polyposis (FAP-1), Attenuated Familial Adenomatous Polyposis (AFAP-1), MYH associated polyposis (MAP-1), Peutz-Jeghers syndrome (PJS-1), Juvenile Polyposis Syndrome (JPS-1), Hyperplastic Polyposis Syndrome (HPP-1 Cowden syndrome, Li-Fraumeni syndrome).

---

[1] Increased risk is based on personal history of adenoma(s)/sessile serrated polyp(s) found at colonoscopy:

(a) Low-risk adenoma: ≤2 polyps, <1 cm, tubular.
(b) Advanced or multiple adenomas: high-grade dysplasia, ≥1 cm, villous (>25% villous), between 3 and 10 polyps (fewer than 10 polyps in the setting of a strong family history or younger age (<40 years) may sometimes be associated with an inherited polyposis syndrome).
(c) More than 10 cumulative adenomas (fewer than 10 polyps in the setting of a strong family history or younger age (<40 years) may sometimes be associated with an inherited polyposis syndrome).
(d) Incomplete or piecemeal polypectomy (ink lesion for later identification) or polypectomy of large cancer.

Appendix AH: Colorectal Cancer Screening Endoscopy Unit Infrastructure, Equipment and Personnel

Endoscopy unit infrastructure and equipment must:

1. Include facilities for adequate pre-colonoscopy assessment, recovery and be designed to allow efficient patient flow.
2. Match the demand with respect to unit capacity (e.g., equipment and personnel).
3. Provide video-endoscopes that facilitate focal application of the dye for the detection and assessment of high-risk colorectal lesions.
4. Provide adequate supply of accessories suited to the endoscopic interventions undertaken.
5. Provide properly maintained resuscitation equipment in the endoscopy rooms and recovery areas.
6. Conduct regular reviews of all the functioning and cleansing of the colonoscopies. Results of the reviews should be available at all times in the unit.
7. Plan capacity that matches demand for screening. Referral to colonoscopy to be within 31 days from a positive FIT test (detects the presence of occult blood in the fecal sample).

Criteria colorectal cancer screening core team to include:

All members in the colorectal cancer core team should participate in regular multidisciplinary team meetings to discuss each patient with colorectal cancer.

1. At least 2 gastroenterologists: licensed by DOH, colonoscopy volume minimum 150 per colonoscopist per year with a cecal completion rate of >95%.
2. Nurse: Two nurses trained to provide support, assistance, information, and advice to every patient. An in-depth understanding of colorectal cancer (diagnosis, treatment, prognosis, staging, and importance of stage at diagnosis), an in-depth understanding of the colorectal screening process (including screening theory and particularly the potential benefits and harms of screening, and the prime importance of quality assurance), and advanced communication skills.

Appendix AI: DOH Recommendation for Colorectal Cancer Screening and Surveillance, Increased and High-Risk Disease Family Group

High-risk disease groups	Screening procedure	Time of initial screen	Screening procedure and interval
Colorectal cancer			
	Consultation, CT, LFTs and colonoscopy	Colonoscopy within 6 months of resection only if colon evaluation pre-op is incomplete	CT liver scan within 2 years post-op. Colonoscopy 5 years until co-morbidity outweighs
Colonic adenomas			
Low risk	1–2 adenomas, both <1 cm	Colonoscopy	5 years or no surveillance
Intermediate risk	3–4 adenomas, or at least one adenoma ≥1 cm	Colonoscopy	Every 3 years
High risk	≥5 adenomas or ≥3 with at least one ≥1 cm piecemeal polypectomy	Colonoscopy Colonoscopy or flexi-si (depending on polyp location)	Yearly 3 months consider open surgical resection if incomplete healing of polypectomy scar
Ulcerative colitis and Crohn's colitis			
Low risk	Extensive colitis with no inflammation or left-sided colitis or Crohn's colitis of <50% colon	Pancolonic dye spray with targeted biopsy. If no dye spray, then 2–4 random biopsies every 10 cm	Every 10 years from onset of symptoms
Intermediate risk	Extensive colitis with mild active disease or post-inflammatory polyps or family history of colorectal cancer in a FDR <50 years Extensive at least moderate colitis or stricture in past 5 years or dysplasia in past 5 years (declining surgery) or PSC or OLT for PSC) or colorectal cancer in a FDR <50 years Ureterosigmoidostomy		After surgery by 10 years

High-risk disease groups	Screening procedure	Time of initial screen	Screening procedure and interval
Acromegaly			
Acromegaly	–	Colonoscopy	At 40 years

Appendix AJ: DOH Recommendations for Colorectal Cancer Screening and Surveillance in Moderate Risk Disease Family Groups

Initial screening 10 years earlier than the youngest affected FDM

Family history	Screening procedure	Screening interval
One first-degree relative with CRC or advanced adenoma diagnosed before 60 years of age or two first-degree relatives diagnosed at any age	Colonoscopy	Start at 40 years of age or 10 years younger than the earliest diagnosis in the patient's family, whichever comes first; colonoscopy should be repeated every 5 years
One first-degree relative with CRC or advanced adenoma diagnosed at 60 years or older or two second-degree relatives with CRC	Colonoscopy	Start screening colonoscopy at 40 years of age; colonoscopy should be repeated every 10 years
One second- or third-degree relative with CRC	Colonoscopy	Average-risk screening (e.g., start at 40)
Individuals who have Crohn disease with colonic involvement or ulcerative colitis. Screening repeated every 1–3 years	Colonoscopy	Colonoscopy screening should begin 8–10 years after the onset of symptoms
In individuals with hereditary nonpolyposis colorectal cancer	Colonoscopy	Colonoscopy should begin at 25 years of age and be repeated annually
Individuals with adenomatous polyposis syndromes	Colonoscopy	Colonoscopy between 10 and 20 years of age and be repeated every 1–2 years

© The Editor(s) (if applicable) and The Author(s), under exclusive license to
Springer Nature Singapore Pte Ltd. 2024
H. O. Al-Shamsi (ed.), *Cancer Care in the United Arab Emirates*,
https://doi.org/10.1007/978-981-99-6794-0

Family history	Screening procedure	Screening interval
Individuals with Peutz-Jeghers syndrome. If results are, negative, testing repeated every 3 years.	Colonoscopy	Esophagogastroduodenoscopy, colonoscopy, and video capsule endoscopy should begin at 8 years of age. If results are negative, testing-repeated every 3 years
Individuals with sessile serrated adenomatous polyposis	Colonoscopy	Colonoscopy should begin as soon as the diagnosis is established and be repeated annually

- Affected relatives who are first-degree relatives of each other AND at least one is a first-degree relative of the patient.
- Combinations of 3 affected relatives in a first-degree kinship include: parent and aunt/uncle and/or grandparent OR 2 siblings/1 parent; OR 2 siblings/1 offspring. Combinations of 2 affected relatives in a first-degree kinship.
- Include a parent and grandparent, or >2 siblings, or >2 children, or child + sibling. Where both parents are affected, these count as being within the first-degree kinship.
- Clinical genetics referral recommended.
- Centers may vary depending on capacity and referral agreements. Ideally, all such cases should be flagged systematically for future audit on an emirate level.

Appendix AK: DOH Summary of Recommendations for Colorectal Cancer Screening and Surveillance in High-Risk Disease Family Groups

Family history categories[a]	Screening procedure	Age at initial screen	Screening interval and procedure
At-risk HNPCC (fulfils modified Amsterdam criteria, or untested FDR of proven mutation carrier)	MMR gene testing of affected relative colonoscopy ± OGD	Colonoscopy from age 25 years OGD from age 40 years or screening 10 years earlier than the youngest affected FDM	Colonoscopy every 18–24 months (OGD every 2 years from age 40 years)
MMR gene carrier	Colonoscopy ± OGD		
At-risk FAP(member of FAP family with no mutation identified)	APC gene testing of affected relative colonoscopy	Puberty Flexible approach Importance of making allowance for variation in maturity	Annual colonoscopy or until aged 30 years Thereafter 3–5 yearly until 60 years proctocolectomy or colectomy if positive
Fulfils clinical FAP criteria, or proven APC mutation carrier opting for deferred surgery Prophylactic surgery normally strongly recommended	Colonoscopy Colonoscopy/ OGD	Usually at diagnosis otherwise puberty Flexible approach Importance of making allowance for variation in maturity	Recommendation for proctocolectomy and pouch/colectomy before age 30 years Cancer risk increases dramatically age >30 years
			Twice yearly colonoscopy
FAP post-colectomy and IRA	Colonoscopy OGD	After surgery OGD from age 30 years	Colonoscopy Every 3 years forward and side-viewing OGD
FAP post-procto colectomy and pouch	DRE and pouch endoscopy Forward and side-viewing OGD	After surgery OGD from age 30 years	Annual exams alternating
			Flexible/rigid pouch
			Endoscopy every 3 years
			Forward and side-viewing OGD

Family history categories[a]	Screening procedure	Age at initial screen	Screening interval and procedure
MUTYH-associated polyposis (MAP)	Genetic testing Colonoscopy ± OGD	Colonoscopy from age 25 years OGD from age 30 years	Mutation carriers should be counselled about the available limited evidence Options include prophylactic colectomy and ileorectal anastomosis; or biennial colonoscopy surveillance Every 3–5 years gastro-duodenoscopy
FDR with MSI-H colorectal cancer and IHC shows loss of MSH2, MSH6, or PMS2 expression	Colonoscopy ± OGD	Colonoscopy from age 25 years OGD from age 40 years	Colonoscopy every 2 years (with OGD aged >40 years).
MLH1 loss and MSI specifically excluded (MLH1 loss in elderly patient with right-sided tumor is usually somatic epigenetic event)			
Peutz-Jeghers syndrome	Genetic testing of affected relative Colonoscopy ± OGD	Colonoscopy from age 25 years. OGD from age 25 years Small bowel MRI/enteroclysis	Twice yearly colonoscopy/consider colectomy and IRA for colonic cancer Small bowel VCE or MRI/enteroclysis 2–4 yearly OGD twice yearly.
Juvenile polyposis	Genetic testing of affected relative Colonoscopy ± OGD	Colonoscopy from age 15 years OGD from age 25 years	Every 2 years colonoscopy and OGD. Extend interval >35 years

1. The Amsterdam criteria for identifying HNPCC are three or more relatives with colorectal cancer:
   - One patient a first-degree relative of another.
   - Two generations with cancer.
   - One cancer diagnosed under the age of 45 or other HNPCC-related cancers, e.g., endometrial, ovarian, gastric, upper urothelial, and biliary tree.
2. Clinical genetics referral and family assessment required, if not already in place or if clinical genetics did not initiate referral.
3. FAP, familial adenomatosis polyposis; FDR, first-degree relative (sibling, parent, or child) with colorectal cancer; HNPCC, hereditary non-polyposis colorectal cancer; IHC, immunohistochemistry of tumor material from affected proband; MSI-H, micro-satellite instability-high (two or more MSI markers show instability); OGD, esophagogastroduodenoscopy endoscopy; VCE, video capsule endoscopy.

Appendix AL: Genetic Test

Available genetic tests for the patient or his/her affected family member(s) that may be recommended by the cancer genetics professional based on the assessment.

Disease	Reasonable gene
Lynch syndrome/hereditary non-polyposis colorectal cancer (HNPCC)	Genes responsible: MLH1, MSH2, MSH6, PMS2
	<OMIM 114500, 120435, 120436, 276300, 609309, 600678, 600259
Familial adenomatous polyposis (FAP)	APC
	<OMIM 175100
Peutz-Jeghers syndrome	LKB1
	<OMIM 175200
Juvenile polyposis	SMAD4, BMPR1A (Juvenile polyposis)
	<OMIM 174900
Rare subtype hereditary mixed juvenile/adenomatous polyposis	Locus on chr15q (GREM1 or SGNE1 may be responsible)
	<OMIM 601228
MUTYH associated polyposis (MAP)	MUTYH <OMIM 608456

Appendix AM: Pre-colonoscopy Assessment

1. Pre-colonoscopy documentation must include:
   - 2.1. Patient demographics
   - 2.2. Anticoagulant use
   - 2.3. History of diabetes mellitus and use of insulin
   - 2.4. Presence of implantable defibrillators or pacemakers
   - 2.5. Previous gastrointestinal procedures
2. Assessment of patient risk: physical status of the patient must be documented in accordance with the American Society of Anesthesiology (ASA) (Appendix AN).
3. ASA class 3 or higher is at higher risk for cardiopulmonary events and appropriate measures must be taken in this respect.
4. Colonic cleansing: Type of bowel preparation must be documented including documentation of careful preparation in accordance with international standards and guidelines. Written instructions to be given to patient concerning colonic cleansing.
5. Inadequate bowel preparations must not exceed 10% of examinations.

Appendix AN: Pre-colonoscopy Assessment—American Society of Anesthesiology Classification System

Class	Description
1	Patient has no organic, physiologic, biochemical, or psychiatric disturbance (healthy, no comorbidity)
2	Mild-moderate systemic disturbance caused either by the condition to be treated surgically or by other pathophysiologic processes (mild-moderate condition, well controlled with medical management; examples include diabetes, stable coronary artery disease, stable chronic pulmonary disease)
3	Severe, systemic disturbance or disease from whatever cause, even though it may not be possible to define the degree of disability with finality (disease or illness that severely limits normal activity and may require hospitalization or nursing home care; examples include severe stroke, poorly controlled congestive heart failure, or renal failure)
4	Severe systemic disorder that is already life threatening, not always correctable by the operation (examples include, acute myocardial infarction, respiratory failure requiring ventilator support, renal failure requiring urgent dialysis, bacterial sepsis with hemodynamic instability)
5	The moribund patient who has little chance of survival

H. O. Al-Shamsi (ed.), *Cancer Care in the United Arab Emirates*,
https://doi.org/10.1007/978-981-99-6794-0

Appendix AO: Colonoscopy Procedure

1. Facility specific Policies and Procedures must be in place for the following:
   1.1. Colonoscopy decontamination including infection control in accordance with: http://www.healthdesign.com.au/haad.hfg/Full_Index/haad_b_day_surgery_procedure_unit.pdf
   1.2. Sedation of patient, considering patient status and preferences and recording of all sedation methods and outcomes; consider involving anesthesia service in patients with significant comorbidities such as patients with ASA 3, 4, and 5 (Appendix AM).
   1.3. Patient support and comfort, including positioning during the colonoscopy.
2. To achieve high-quality colonoscopic examination, complete intubation of the colon and careful inspection of the mucosa during withdrawal are necessary.
   2.1. If a complete colonoscopy is not achieved, imaging for documentation of incomplete intubation may be necessary and reasons must be clearly documented.
   2.2. Auditable photo documentation of colonoscopy completion must be available; including a panoramic image of the appendiceal orifice, ileo-cecal valve and cecum, or a video clip with a respective image.
   2.3. Documentation of completion of rectal retroflexion (retroflexion of the endoscope during colonoscopy to increase diagnostic yield) must be recorded.
   2.4. Withdrawal times of the colonoscopy from cecum to anus must be documented and must be not less than 6 min (when no biopsies or polypectomies are performed). The times to be documented include when:
   - Endoscope is inserted into the rectum.
   - Withdrawal from cecum was started.
   - Endoscope is withdrawn completely.

   A record of the actual model and instrument number used must be maintained by the unit staff to track procedure volume, problems, and infection transmission and instrument repairs.

H. O. Al-Shamsi (ed.), *Cancer Care in the United Arab Emirates*,
https://doi.org/10.1007/978-981-99-6794-0

2.5. Any adverse clinical events (fall in blood pressure, unplanned reversal of sedation medications, oxygen desaturation, etc.) that occur during colonoscopy as well all serious events (perforation, bleeding requiring blood transfusion, and/or surgery) must be documented and attached to the colonoscopy report (Appendix AS) and reported in accordance with DOH Standard for Adverse Events Management and Reporting.

Appendix AP: Post-colonoscopy Procedures

1. Patients must be contacted 24 h post-procedure or on the next working day to monitor any complications; this contact must be documented.
2. Patients must receive instructions about management of any potential adverse events following discharge and must be informed that complications may occur within 1–4 weeks post procedure.
3. A contact number must be provided to the patient for this purpose and documented in the patient records.
4. Post-procedure complications must be tracked over a 30-day interval after a colonoscopy.
5. Discharge instruction form should be-given to patient instructing him to call endoscopy unit or the gastroenterology physician on call or to come to ER in case, there is any abdominal pain or any complication or concerns after the procedure. Patient should sign this form acknowledging that he understood the post colonoscopy and the pre-discharge instructions.

Appendix AQ: Colonoscopy Findings
Colonoscopy Findings

1. If use of vague terms to describe polyps in the report should be avoided.
2. An estimation of the size and dimension of all polyps must be documented. Terms such as "large" or "small" must not be used
3. Tattoos must be placed for all lesions ≥10 mm and those with concerning appearance for cancer to mark the location of colon lesions for repeat colonoscopy or surgery as per international updated recommendations and best practice.
4. Lesions that are too large to be safely-removed must be biopsied and a tattoo injection performed in the-vicinity of the lesion and not into the lesion.
5. Specimen identification and labelling must be in accordance with DOH Clinical Laboratory Standards and industry best practices.
6. Procedures and protocols for adequate specimen collection, handling, labelling, and reporting must be in accordance with DOH Clinical Laboratory Standard and must be-communicated to clinical staff and other clients who are involved in the procedures for processing of colorectal specimens.

Appendix AR: Colonoscopy Report

1. Each facility must develop a patient colonoscopy report form, retained on the patient's medical record and made available to DOH auditors.
2. The report must include at least the following information:
   - 2.1. Patient demographics and history.
   - 2.2. Assessment of patient risk and comorbidity.
   - 2.3. Procedure indications.
   - 2.4. Procedure: technical description.
   - 2.5. Colonoscopy findings.
   - 2.6. Interventions/unplanned events.
   - 2.7. Assessment.
   - 2.8. Follow-up plan.
   - 2.9. Pathology.

© The Editor(s) (if applicable) and The Author(s), under exclusive license to
Springer Nature Singapore Pte Ltd. 2024
H. O. Al-Shamsi (ed.), *Cancer Care in the United Arab Emirates*,
https://doi.org/10.1007/978-981-99-6794-0

Section 1: Patient demographics

Name	Date of birth	Gender	MRN	Associated diagnosis
		Male		
		Female		

Section 2: Colonoscopy procedure

Informed consent signed	Indication for colonoscopy	ASA risk
Yes		
No		

Date	Time (24 hrs.)	Physical exam conducted	Patient, procedure confirmed
		yes no	yes no

Performed by	Preparation and technique

Bowel preparation

Protocol

		Antibiotic prophylaxis given?	
Quality		yes	no

Monitoring during procedure

Premedication	Patient position
Endoscope model	Endoscope manufacturer

Procedure tolerated?	Complications (state any):		
yes no			

Section 3: Colonoscopy findings

Route of entry	Extent of examination

If incomplete examination, state reason

Method of verifying extent	Duration of colonoscopy withdrawal (minutes)

Rectal retro flexion performed.	Cecum identified by
yes no	Photographic of appendicular orifice? yes no

Positive Findings	Finding
yes no	

Extent	Severity

Characterized by

Consistent with

Removal method

Mass	Number	Size (mm, mention all)

Location(s)

Descriptors

Removal method	Retrieve
Photo documentation attached.	Other findings?
Yes no	
Discussed with patient?	
yes no	

Appendix A5: Example of—Techniques for Colonoscopic—Tattooing Protocol

Indications	Equipment	Procedure
• Prior to surgery to localise pathology • To mark lesions for endoscopic surveillance ➢ DO NOT TATTOO RECTAL LESIONS as they disrupt surgical planes. ➢ There is no need to tattoo lesions in the caecum, however, if in doubt, then place a tattoo	• Primed variceal injection needle with 10ml syringe filled with normal saline • 5ml syringe filled with Spot® (or 0.9ml sterilised Black (India) ink made up to 5ml with normal saline)	• Direct needle at an angle to mucosa • Raise a bleb using 1-2ml of saline • Swap to syringe filled with Spot® or India ink • Inject 1ml into the bleb to create tattoo • Swap to syringe filled with saline and flush ink out with 1ml saline before removing needle • Repeat process for 3 tattoos

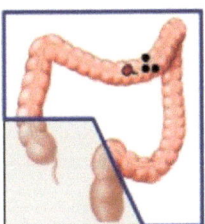

Place 3 tattoos **DISTAL** to lesion

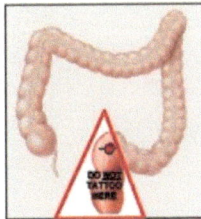

DO NOT place tattoo below 20cm but clearly record distance of LESION from anal verge

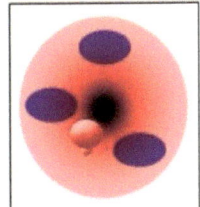

Place tattoos 120° apart **AS CLOSE TO LESION AS POSSIBLE** but separate from it

REMEMBER: TO DOCUMENT HOW MANY TATTOOS WERE PLACED AND THE POSITION RELATIVE TO THE LESION

v14 August 2011

© St Mark's Hospital 2011

Appendix AT: E-Notification Cancer Screening Referral Form

Cancer Screening ID:	
First Name:	
Last Name :	
Date of Birth:	
Emirate I.D.	
Colonoscopy Report:	
Recommended next step:	
FIT Report :	
Recommended next step:	
Referred to :	
Date of referral:	

Appendix AU: Cancer Screening Data Requirement—Screening Visit Outcome

Patient Information			
First Name		Emirates ID Number	
Middle Name		Medical File Number	
Last Name		DOB (age in years)	
Gender	Male/ Female	City of residence	Emirate of residence
Nationality		Residence number	
Marital status		Mobile number	
BMI		Other number	

Personal health history			
Personal history	Inflammatory bowel disease	Y / N	
	Colon adenoma/Polyp	Y / N	
	Colorectal Cancer	Y / N	

Inherited syndrome: HNPCC	Y / N
Inherited syndrome: FAP	Y / N
* Family history of colorectal ,breast, ovarian, prostate, kidney and pancreatic cancer in first or second degree relative?	Y / N

* If positive family history of cancer refer to genetic testing if applicable

Screening History			
Registration status	New	Method of recruitment	Invited for screening
			Walk in
	Registered		With opportunistic
Last screening test performed anywhere			
Colonoscopy (date)			
FIT (date)			

Colorectal Cancer Screening			
A. Screening with Colonoscopy			
Colonoscopy done	Yes/ No	Date of Colonoscopy	
Biopsy done	Yes/ No	Was cecum reached	Yes/ No
Complications of colonoscopy	Yes/ No	If yes, specify:	
Colonoscopy report	• Incomplete exam (the reason) • Invasive cancer • Advanced adenoma • **More than 10 adenomas** • 10 - 3 adenomas • 2-1 tubular adenomas < 1cm • Hyperplastic polyp • Unknown histology • No pathology • Serrated sessile polyp		
Date patient notified with report		Recommended Next Step	• Screening in normal interval • Refer for treatment
Referred to other hospital	Yes/No	Date of referral	
B. Screening with FIT			
FIT done	Yes/No	Date of FIT	
FIT report	• Inadequate • Positive • Negative		
Date patient notified with report	Yes/No	Recommended Next Step	• Screening in normal interval • Repeat test • Refer for colonoscopy
Referred to other hospital	Yes/No	Date of referral	

Route of entry	Extent of examination
If incomplete examination, state reason	
Method of verifying extent	Duration of colonoscopy withdrawal (minutes)
Rectal retro flexion performed? ■ Yes ■ No	Cecum identified by Photographic of appendicular orifice? ■ Yes ■ No
Positive Findings Yes No	Finding:
Extent	Severity
Characterized by	
Consistent with	
Removal method	

Mass	Number	Size (mm, mention all)
	Location(s)	
	Descriptors	

Removal method	Retrieval
Video / Photo documentation attached? ■ Yes ■ No	Other findings?
Discussed with patient? ■ Yes ■ No	

Section 4: Impression and plan

Diagnosis
■ Colo-rectal carcinoma ■ Other (state other)
Course
Follow up
Counseled

Appendix AV: JAWDA KPI Quarterly Guidelines for Hematopoietic Stem Cell Transplant (HSCT) Service Providers

Jawda HSCT Quality Performance Indicators

دائــــرة الــــصـــحــة

DEPARTMENT OF HEALTH

JAWDA KPI Quarterly Guidelines for Haematopoietic Stem Cell Transplant (HSCT) Service Providers

January 2020

Executive Summary

The Department of Health—Abu Dhabi (DOH) is the regulatory body of the health-care sector in the Emirate of Abu Dhabi and ensures excellence in healthcare for the community by monitoring the health status of its population.

The Emirate of Abu Dhabi is experiencing a substantial growth in the number of hospitals, centers, and clinics. This is ranging from school clinics and mobile units to internationally renowned specialist and tertiary academic centers. Although, access and quality of care have improved dramatically over the last couple of decades, mirroring the economic upturn and population boom of Emirate of Abu Dhabi; however, challenges remain in addressing further improvements.

The main challenges that are presented with increasingly dynamic population include an aging population with increased expectation for treatment, utilization of technology and diverse workforce leading to increased complexity of healthcare provision in Abu Dhabi. All of this results in an increased and inherent risk to quality and patient safety.

DOH has developed dynamic and comprehensive quality framework in order to bring about improvements across the health sector. This guidance relates to the quality indicators that DOH is mandating the quarterly reporting against by the operating general and specialist hospitals in Abu Dhabi.

The guidance sets out the full definition and method of calculation for patient safety and clinical effectiveness indicators. For enquiries about this guidance, please contact jawda@DoH.gov.ae

This document is subject for review and therefore it is advisable to utilize online versions available on the DOH at all times.

Published: January 2020 Version 1

About This Guidance

The guidance sets out the definitions and reporting frequency of JAWDA hematopoietic stem cell transplantation (HSCT) facilities performance indicators. The Department of Health (DoH), with consultation from local and international expertise of hematopoietic stem cell transplantation (HSCT), has developed hematopoietic stem cell transplantation performance indicators that are aimed for assessing the degree to which a provider competently and safely delivers the appropriate clinical services to the patient within the optimal period of time.

The JAWDA KPI for hematopoietic stem cell transplantation patients in this guidance includes measures to monitor morbidity and mortality in patients undergoing hematopoietic stem cell transplantation. Healthcare providers are the most qualified professionals to develop and evaluate quality of care for kidney disease patients; therefore, it is crucial that clinicians retain a leadership position in defining performance among hematopoietic stem cell transplantation healthcare providers.

Who is this guidance for?

All DoH licensed healthcare facilities providing hematopoietic stem cell transplantation services in the emirate of Abu Dhabi.

How do I follow this guidance?

Each provider will nominate one member of staff to coordinate, collect, monitor, and report hematopoietic stem cell transplantation services quality indicators data as per communicated dates. The nominated healthcare facility lead must in the first instance e-mail their contact details (if different from previous submission) to JAWDA@doh.gov.ae and submit the required quarterly quality performance indicators through JAWDA online portal.

What are the regulation related to this guidance?

- Legislation establishing the health sector.
- As per DoH Policy for Quality and Patient Safety issued January 15, 2017, this guidance applies to all DoH licensed hospital healthcare facilities in the emirate of Abu Dhabi in accordance with the requirements set out in this standard.
- DOH standard on stem cell therapies and products and regenerative medicine.

Hematopoietic Stem Cell Transplantation Quality Performance Indicators

Type: Hematopoietic stem cell transplantation   **Number: BMT001**

KPI description (title)	Percentage of patients with successful engraftment
Domain	Patient centered
Sub-domain	Clinical outcome
Definition	Percentage of patients with successful engraftment
Calculation	**Numerator**: Number of patients where engraftment was successful (successful defined as neutrophil count of ($>0.5 \times 10^9$ L) for 3 consecutive days by day plus 28) **Denominator**: Total number of patients transplanted in the first 6 months of the previous 7 month reporting period
Reporting frequency	Quarterly
Unit measure	Percentage
International comparison if available	Specialized Services Quality Dashboards—Blood and infection metric definitions for 2019/20
Desired direction	Higher is better
Data source	• Centrally collected claim data (KEH) • Patient medical record

Type: Hematopoietic stem cell transplantation   **Number: BMT002**

KPI description (title)	Percentage of patients dying within 100 days of autologous transplant
Domain	Patient centered
Sub-domain	Clinical outcome
Definition	Percentage of patients dying within 100 days of *autologous* transplant
Calculation	**Numerator**: Number of patients in denominator who died within 100 days of *autologous* transplant **Denominator**: Total number of *autologous* transplants in the first 365 days of the previous 465-day reporting period

KPI description (title)	Percentage of patients dying within 100 days of autologous transplant
Reporting frequency	Quarterly
Unit measure	Percentage of BMT died within 100 days of *autologous* transplant
International comparison if available	Specialized Services Quality Dashboards—Blood and infection metric definitions for 2019/20
Desired direction	Lower is better

Type: Hematopoietic stem cell transplantation   **Number: BMT003**

KPI description (title)	Percentage of patients alive at 1 year post autologous transplant
Domain	Patient centered
Sub-domain	Clinical outcome
Definition	Percentage of patients alive at 1 year post *autologous* transplant
Calculation	**Numerator**: Number of patients in denominator alive 1 year after *autologous* transplant **Denominator**: Total number of *autologous transplants* in the first 12 months of the previous 24-month reporting period
Reporting frequency	Annual
Unit measure	Percentage of survival at 1 year after *autologous transplant*
International comparison if available	Specialized Services Quality Dashboards—Blood and Infection metric definitions for 2019/20
Desired direction	Higher is better
Data source	• Centrally collected claim data (KEH) • Patient medical record

Type: Hematopoietic stem cell transplantation   **Number: BMT004**

KPI description (title)	Percentage of patients dying within 100 days of allogenic transplant
Domain	Patient centered
Sub-domain	Clinical outcome
Definition	Percentage of patients dying within 100 days of *allogenic* transplant
Calculation	**Numerator**: Number of patients in denominator who died within 100 days of *allogenic* transplant **Denominator**: Total number of *allogenic* transplants in the first 365 days of the previous 465-day reporting period
Reporting frequency	Quarterly
Unit measure	Percentage of BMT died within 100 days of *allogenic* transplant
International comparison if available	Specialized Services Quality Dashboards—Blood and infection metric definitions for 2019/20
Desired direction	Lower is better
Data source	• Centrally collected claim data (KEH) • Patient medical record

KPI description (title)	Percentage of patients dying within 100 days of autologous transplant
Data source	• Centrally collected claim data (KEH) • Patient medical record

Type: Hematopoietic stem cell transplantation   Number: BMT005

KPI description (title)	Percentage of patients alive at 1 year post allogenic transplant
Domain	Patient centered
Sub-domain	Clinical outcome
Definition	Percentage of patients alive at 1 year post *allogenic* transplant
Calculation	**Numerator**: Number of patients in denominator alive 1 year after *allogenic* transplant **Denominator**: Total number of *allogenic* transplants in the first 12 months of the previous 24-month reporting period
Reporting frequency	Annual
Unit measure	Overall survival at 1 year
Exclusions	Percentage of survival at 1 year after *allogenic* transplant
International comparison if available	Specialized Services Quality Dashboards—Blood and infection metric definitions for 2019/20
Desired direction	Higher is better
Data source	• Centrally collected claim data (KEH) • Patient medical record chart

Appendix AW: DOH Lung Cancer Screening Service Specifications

دائـــرة الـصــحــة
DEPARTMENT OF HEALTH

DOH LUNG CANCER SCREENING SERVICE SPECIFICATIONS

December 2018

PUBLIC   عـــام

https://www.doh.gov.ae/-/media/Feature/Resources/Standards/Lung-Cancer-Screening-Service-Specifications_Publish.ashx

دائـــــرة الـصــــحــة
DEPARTMENT OF HEALTH

Document Title:	DOH Lung Cancer Screening Service Specifications		
Document Ref. Number:	PH/NCD/LCSC/SR/0.9	Version:	0.9
Approval Date:	20/12/2018	Effective Date:	20/12/2018
Document Owner:	Cancer Prevention and Control Section, NCD, Public Health Division		
Applies to:	Licensed Healthcare Providers in the Emirate of Abu Dhabi participating in the DOH's Lung Cancer Screening Program		
Classification:	• Public		

1. Purpose

1.1. This document sets the service specifications for DOH's lung cancer screening program in the emirate of Abu Dhabi.
1.2. It specifies the minimum service specifications to ensure that high-risk candidates screened for lung cancer receive quality and safe care and timely referral for diagnosis and/or treatment.

2. Scope

These specifications apply to healthcare providers (facilities, professionals, and laboratories) licensed by DOH in the emirate of Abu Dhabi who are participating in DOH's Lung Cancer Screening Program and specify:
2.1. Terms used in DOH Lung Cancer Screening Program.
2.2. Recommended roles, responsibilities, and services offered by healthcare providers (facilities and professionals) participating in DOH's Lung Cancer Screening Program across the emirate of Abu Dhabi.
2.3. Lung cancer screening case mix.

3. Definitions

	Category	Definition
3.1	Lung cancer screening	The process for the early detection of lung cancer. It includes recruitment of individuals at a high risk of developing lung cancer, counseling of these individuals and low-dose computer topography aided screening by a multidisciplinary team

	Category	Definition
3.2	LDCT scan	A procedure that uses low-dose computer topography (LDCT) radiation to make a series of very detailed pictures of areas inside the body in a spiral path. The procedure is also called a low-dose helical CT scan
3.3	Case mix	High-risk candidates for lung cancer, except where exclusion criteria for LDCT apply
3.4	Lung cancer screening risk assessment	An estimate of the likelihood of developing lung cancer in asymptomatic candidate based on age, total cumulative exposure to tobacco smoke, years since quitting tobacco and additional risk factor for lung cancer other than second-hand smoking
3.5	Additional risk factor lung cancer other than second-hand smoking	Include cancer history, lung disease history as chronic obstructive pulmonary disease or pulmonary fibrosis, family history of lung cancer, radon exposure and occupational exposure of silica, cadmium, asbestos, arsenic, beryllium, chromium, diesel fumes, and nickel
3.6	Pack-year	A way to measure the amount a candidate has smoked over a specific period
3.7	Informed and shared decision-making	A documented process of mutual decision-making involving eligible candidate and lung screening healthcare provider and before any decision is made to initiate lung cancer screening including the following elements: • Willingness to undergo follow-up diagnostic testing and treatment • The importance of adherence to lung cancer screening schedule • Lung cancer screening potential benefits (reduce the risk of dying from lung cancer) • Lung cancer screening potential limitations and harms (false-positive and false-negative results, over diagnosis, incidental findings, and radiation exposure) • Adherence to tobacco cessation counseling and treatment
3.8	False-positive result	Positive screening with a completed negative work-up or follow-up of at least 12 months with no diagnosis of lung cancer
3.9	False-negative result	Negative screening associated with diagnosis of lung cancer within 12 months of baseline examination
3.10	Over diagnosis	The detection of indolent lung cancer that would not have become clinically apparent
3.11	Incidental findings	Results that arise outside the original purpose of lung cancer early detection
3.12	Multi-disciplinary team	A team responsible for individualized and evidence-based management of candidates with positive lung cancer screening results. It consists of radiologists, pulmonologists, thoracic surgeons, oncologists, pathologists, family physicians, and nurses

	Category	Definition
3.13	Tobacco cessation intervention	Tobacco cessation counseling and treatment for more than 10 min at PRIMARY CARE CLINCS visit including brief advice, set up quitting date, offer pharmacological agents treatment, offer tobacco cessation specialist appointment, and enforce maintaining tobacco abstinence if former tobacco user

4. General Duties of the Health Care Providers

All licensed and eligible healthcare providers participating in DOH's Lung Cancer Screening Program must:

4.1. Submit data to DOH via e-claims in accordance with the DOH Reporting of Health Statistics Policy and as set out in the DOH Data Standards and Procedures (found online at www.haad.ae/datadictionary).

4.2. Comply with relevant DOH policies and standards.

4.3. Comply with DOH's requests to inspect and audit records and cooperate with DOH authorized auditors as required by DOH.

4.4. Comply with requirements for information technology (IT) and data management including sharing of screening/diagnosis and where applicable, pathology results, electronic patient records, and disease management systems.

5. Lung Cancer Screening Program-Facilities Specifications

In order to be designated as DOH lung cancer screening center, the facility should obtain DOH approval prior to offering the services by completing the service provision form (refer to Appendix AZ)

5.1. All DOH licensed healthcare providers (facilities and professionals) engaged in DOH lung cancer screening program must comply with general regulations governing health care facilities and specific regulations related to these standards (refer to Appendices AX and AY).

5.2. DOH designated lung cancer screening center should adhere to DoH lung cancer screening program performance indicators (refer to Appendix BA).

5.3. A lung cancer screening center should assign a program coordinator/director who will be accountable to:

5.3.1. Report screening and screening outcome data to DOH.

5.3.2. Notify screened candidates of their screening results within the expected timeframe.

5.3.3. Ensure the candidate's enrollment in tobacco cession program.

5.3.4. Assure clear and communicated process for the management of positive cases either within the same facility or in another facility approved by DOH to participate in the Lung Cancer Screening Program.

5.3.5. Maintain records for screening tests and outcomes.

5.3.6. Ensure that the candidate is provided with the right information regarding the screening, assessment, follow-up care, and ensure that the candidate' informed consent form is obtained and documented.

5.3.7. Ensure program key performance indicators are met and records are kept for audits purposes.

5.3.8. Coordinate and organize CME training for the healthcare providers involved in screening program for quality assurance.

6. Case Mix-Eligibility Criteria and Recruitment

6.1. Inclusion/eligibility criteria:

6.1.1. High-risk candidates of lung cancer aged 55–75 years with:

6.1.1.1. 30 Pack-year history of smoking, and/or tobacco cessation <15 years (refer to Appendix BB for calculation of pack-year tobacco use).

6.1.1.2. 20 Pack-year history of tobacco use, and/or tobacco cessation <15 years and one additional risk factor.

6.1.1.3. 20 Year history of water pipe (shisha) and/or dokha, medwakh and/or all other forms of smoked tobacco use.[1]

6.2. Recruitment-eligible candidates for lung cancer screening might be recruited through one of the following:

6.2.1. Opportunistic recruitment.

6.2.2. Online subscription: booking a screening appointment is available at www.haad.ae/simplycheck.ae.

6.3. Exclusion criteria-potential reasons to exclude eligible candidates from screening may include the following:

6.3.1. Metallic implants or devices in the chest or back, such as pacemakers or Harrington fixation rods.

6.3.2. Personal history of lung cancer.

6.3.3. Requirement for home oxygen supplementation.

6.3.4. Unexplained weight loss of more than 7 kilograms in the 12 months prior to eligibility assessment.

6.3.5. Pneumonia or acute respiratory infection treated with antibiotics in the last 12 weeks.

6.3.6. Chest CT examination in the 12 months prior to eligibility assessment.

6.3.7. Patient is not a good candidate for surgical treatment.

---

[1] This category was not addressed in the international guidelines; it was added to be assessed in the pilot phase of the program, due to the popularity of this form of tobacco use among smokers in emirate of Abu Dhabi.

7. DOH Lung Cancer Screening Pathway (Refer to Appendix AY)

Lung cancer screening center should follow the DOH Lung Cancer Screening Pathway:

7.1. Physician in lung cancer screening facility should obtain and document informed consent from eligible candidate to participate in lung cancer screening program.

7.2. All eligible candidates enrolled in a screening program should receive smoking cessation interventions.

7.3. Tobacco cessation program physicians liaising with lung cancer screening centers should:

7.3.1. Collect data and complete the DOH E-cancer screening form jointly (refer to Appendix BC).

7.3.2. Ensure the candidate was provided proper education and information regarding the screening benefits and limitation, assessment, follow-up care, ensure that candidate' informed consent is obtained and documented.

7.4. Healthcare providers at the Lung Cancer Screening Center should:

7.4.1. Inform the candidate about the date and method of receiving screening results.

7.4.2. Report screening outcomes to DOH through the e-notification system.

7.5. In case of negative results, the candidate should:

7.5.1. Have the next lung cancer screening appointment scheduled as per the screening criteria.

7.5.2. Be encouraged to continue following up with the tobacco cessation clinic/center.

7.6. In case of positive results:

7.6.1. The screening results should be assessed and discussed by a multidisciplinary team prior to referral of the candidate to a treatment facility (refer to Appendix BD for the roles and responsibilities of the multidisciplinary team).

7.6.2. The candidate must be referred to a treatment facility.

7.6.3. The program coordinator must report confirmed cancer cases to DOH using the Cancer Case Notification Form of the Cancer Surveillance e-notification.

8. DOH Lung Cancer Screening Protocol

8.1. Screening centers should develop a documented protocol used for lung cancer screening to include image production, image reading, screening frequency, follow-up of scan results, and management of positive cases and communication of results.

8.2. The protocols may be reviewed and updated based on evidence-based best practices recommended by the National Comprehensive Cancer Network (NCCN), the American College of Radiology (ACR), or equivalent.

8.3. Double reading of screening LDCT[2] is required:

    8.3.1. LDCT should be interpreted by two independent radiologists.

    8.3.2. In case of discordant opinions between two radiologists, either consensus or preferably arbitration using a third expert screening radiologist can be carried out.

8.4. Lung cancer imaging report must be-completed by radiologist, containing at least the following information:

    8.4.1. Interpreting physicians' names.

    8.4.2. Date of examination.

    8.4.3. Patient identification.

    8.4.4. Description of significant imaging lesions.

    8.4.5. Final assessment (Lung RADS).

    8.4.6. Recommended next steps.

8.5. Final assessment report should be prepared/completed using Lung-RAD assessment categories or equivalent (refer to Appendix BE).

9. Screening and Tobacco Cessation

Screening should not be viewed as an alternative to tobacco cessation:

9.1. Candidates currently using tobacco should be informed of their continuing risk of developing lung cancer.

9.2. Candidates currently using tobacco that are willing to participate in lung cancer screening program should receive tobacco cessation intervention parallel to lung cancer screening.

9.3. Healthcare provides should increase attention to all forms of tobacco use including shisha, medwakh, and others.

9.4. Healthcare providers are encouraged to use their best clinical judgment in assessing the risk of lung cancer for shisha and medwakh (refer to Appendix BB, for pack-years approximation regarding other forms of tobacco use).

10. Cessation of Screening for Lung Cancer

Yearly lung cancer screening should cease when the candidate being screened:

10.1. Turns 76 years old.

10.2. Has not smoked in 15 years.

10.3. Develops a health problem that makes him or her unwilling or unable to have surgery if lung cancer is found.

---

[2] The only approved and recommended screening tool for lung cancer is low dose computed topography CT scan (LDCT).

11. Data Collection and Submission

Facilities participating in DOH Lung Cancer Screening Program should submit to DoH:

11.1. Via e-claims in accordance with the DOH Reporting of Health Statistics Policy and as set out in the DOH Data Standards and Procedures (www.DOH. ae/datadictionary).

11.2. Data on screening visits, outcomes within 2 weeks of the screening date, through the cancer screening form of the cancer surveillance e-notification system found on the DOH website at: http://www.DOH.ae/DOH/tabid/1084/ Default.aspx.

11.3. Report all screen-detected cancers to DOH, through cancer case notification form, of the cancer surveillance e-notification, available from: https://bpm-web.DOH.ae/usermanagement/login.aspx.

12. Payment Mechanism

Eligibility for reimbursement under the health insurance scheme is as follows:

12.1. For Thiqa holders, reimbursement must be consistent with the DOH (previously HAAD) Standard for Thiqa Preventive List of Interventions available at www.haad.ae.

12.2. For non-Thiqa holders, payment must be consistent with the individual's health insurance product/plan.

13. Enforcement and Sanctions

13.1. Healthcare providers and professionals participating in the DOH's (previously HAAD) Cancer Screening Program must comply with the:

13.1.1. Terms and requirements of this standard.

13.1.2. DOH (Previously HAAD) standard provider contract.

13.2. DOH may impose sanctions in relation to any breach of requirements under this standard in accordance with the [DOH (Previously HAAD) *Policy on Inspections, Complaints, Appeals and Sanctions*].

Appendix AX: Related Regulations

Written orders for lung cancer LDCT screenings must be appropriately documented in the beneficiary's medical record and must contain the following information:

- Date of birth.
- Actual pack-year smoking history (number).
- Current smoking status, and for former smokers, the number of years since quitting smoking.
- A statement that the beneficiary is asymptomatic (no signs or symptoms of lung cancer).
- The provider identifier (license number) of the ordering practitioner.

Appendix AY: Care Pathway

Lung cancer screening care-pathway

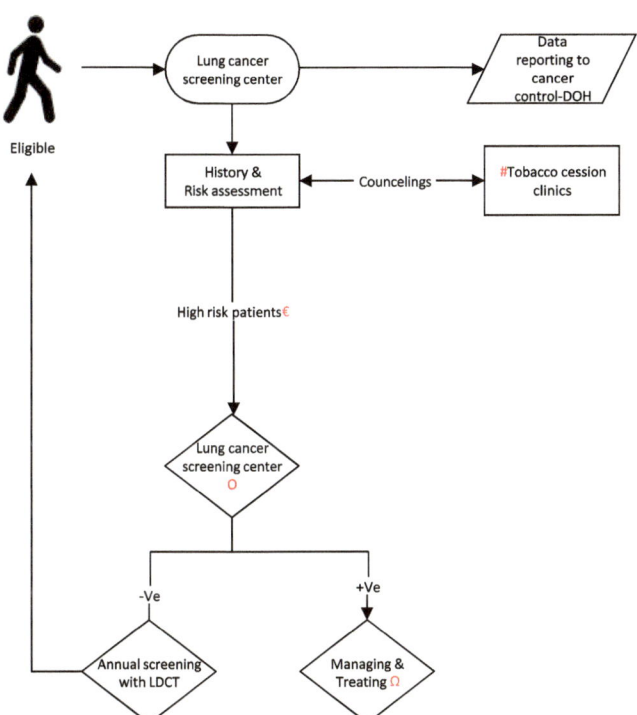

Appendix AZ: Lung Cancer Screening Centre Application Form

Facility informations							
Name of the facility			Address/Street:				
City/Town:			Pod. Box:				
						Tick if appropriate	
Follow screening protocol according to DOH standards or lung cancer screening specifications							
We will submit e-claims							
We will submit data on screening visits and outcomes through cancer e-notification, found at: http://www.DOH.ae/DOH/tabid/871/Default.aspx							
We will comply with DOH standards for Lung Cancer Screening and Diagnosis, found at http://www.DOH.ae/DOH/tabid/820/Default.aspx							
Screening specified appointment slots	Time	Days of the week					
		Saturday	Sunday	Monday	Tuesday	Wednesday	Thursday
Lung cancer screening clinic	A.M.						
	P.M.						
LDCT scanning	A.M.						
	P.M.						
	Name			Mobile	Office landline	E-mail	
Designated program coordinator							
Facility medical director							
Facility administrator							

Filled in by _____   Signature: _____

Date: _____

Please fill in and return to Cancer Prevention and Control Section, Public Health and Research Division DOH, email: ldeen@doh.gov.ae

© The Editor(s) (if applicable) and The Author(s), under exclusive license to Springer Nature Singapore Pte Ltd. 2024
H. O. Al-Shamsi (ed.), *Cancer Care in the United Arab Emirates*,
https://doi.org/10.1007/978-981-99-6794-0

Appendix BA: Lung Cancer Screening Performance Indicators

Performance indicators	Definition	Calculation	Acceptable level	Desirable level
Participation rate	Percentage of subjects who have a screening LDCT as a proportion of the population at risk	[Number of screened subjects/Population at risk] × 100	>10%	>12%
Retention rate	Percentage of LDCT negative screened subjects who follow-up screening	[Number of annual follow-up screening examinations/Baseline screening examinations] × 100	>50%	>70%
Abnormal recall rate	Proportion of screened subjects recalled for further assessment (clinical follow-up screening examinations)	[Number of clinical follow-up screening examinations/Baseline screening examinations] × 100	<25%	<20%
False positive rate	Percentage of positive screening without definitive diagnosis within 1 year	[Number of positive screening without definitive diagnosis within 1 year/True negative results] × 100	<27%	<16%
		True negative results = initial negative results + of positive screening without definitive diagnosis		
Lung cancer detection rate	Number of detected lung cancer cases per 1000 screening examinations	[Number of detected lung cancer cases (per 1-year period)/Total number of screening CT] × 1000	0.5%	−1%
Stage 1 lung cancer rate	Percentage of lung cancer patients diagnosed with stage 1	[Number of stage 1 lung cancer detected/Total number of lung cancer detected] ×100	>65%	>80%
Incidental findings rate	Percentage of findings requiring follow-up for diseases other than lung cancer	[Number of detected non-lung cancer related abnormality/Total screened population] × 100	>5%	>7.5%
Tobacco quit rates among participates in the program	Percentage screened persons who have quit tobacco after participation in lung cancer screening program	[Number of tobacco quit rates among participates/Total tobacco users] × 100	>12%	>20%

Performance indicators	Definition	Calculation	Acceptable level	Desirable level
Diagnostic interval	Maximum time between LDCT screening and receipt of result		≥90% within 7 working days	

Appendix BB: Calculation of Pack-Year Tobacco Use

It is calculated by multiplying the number of packs of cigarettes smoked per day by the number of years the person has smoked.

- Number of pack-years = (packs smoked per day) × (years as a smoker) or
- Number of pack-years = (number of cigarettes smoked per day × number of years smoked)/20

Appendix BC: DOH E-Cancer Screening Form

For information only, web-based "Cancer Screening Form" must be completed, available on http://www.haad.ae/DOH/tabid/1084/Default.aspx.

Appendix BD: Lung Cancer Screening Multi-disciplinary Team Composition

Lung cancer screening multi-disciplinary team members and their roles	
Radiologists	• Have documented training in diagnostic radiology and radiation safety • Involvement in the supervision and interpretation of at least 300 chest computed tomography acquisitions in the past 3 years • Direct supervision LDCT screening process • Reporting the results of LDCT scans for lung cancer screening • Take full responsibility for the quality of the LDCT report • Ensure minimum radiation exposure to screening participants
Oncologists (preferred thoracic oncologist)	Participate in developing follow-up and treatment plan
Thoracic surgeons	• Experience in minimally invasive techniques, VATS procedures, and VATS lobectomy with complete staging through lymph, adenectomy • Reporting on surgical outcomes
Pathologists	• Expertise in cytopathology and pulmonary pathology to report increased number of cytologies, biopsies, and other procedures that result from LDCT screening. Expertise includes lung cancer biomarker testing performed by immunohistochemistry. Double reporting of all lung cancer diagnosis • Use current international standards (Royal College of Pathologists, UK or College of American Pathologists Cancer Protocols) for pathology reporting of cytology/small biopsy, resections, and standardized data collection for future development of the screening program, tumor registries, research, audit and clinical trials. http://www.cap.org/ShowProperty?nodePath=/UCMCon/Contribution%20Folders/WebContent/pdf/cp-lung-17protocol-4000.pdf) • Tissue triage, conservation and referral to molecular pathology for lung cancer predictive biomarker testing or other studies • Timely communication of results at multidisciplinary meetings, correlation of radiology and pathology results, determining appropriate and timely management for patients with lung cancer • Quality assurance—Participation in proficiency testing of lung cancer predictive biomarker tests, relevant subspecialty schemes, and audit of reports

Lung cancer screening multi-disciplinary team members and their roles	
Pulmonologist s	• Evaluate lung nodules • Perform non-surgical bronchoscopy, image-guided biopsy, Endobronchial Ultrasound (EBUS) services
Family physicians	• Recruiting for the program • Explain to participants harms and benefits of lung screening • Collect data • Reporting to DOH
Nurses and support staff	Assist patients with coordination of their care within the continuum

Appendix BE: Lung Cancer Screening RAD Reporting Lung-RADS™ Version 1.0 Assessment Categories April 28, 2014

Category descriptor	Category descriptor	Primary category	Management
Incomplete	–	0	Additional lung cancer screening CT images and/or comparison to prior chest CT examinations are needed
Negative	No nodules and definitely benign nodules	1	Continue annual screening with LDCT in 12 months
Benign appearance or behavior	Nodules with a very low likelihood of becoming a clinically active cancer due to size or lack of growth	2	
Probably benign	Probably benign finding(s)—short-term follow-up suggested; includes nodules with a low likelihood of becoming a clinically active cancer	3	6-month LDCT
4A	3-month LDCT; PET/CT may be used when there is a ≥8 mm solid component	Findings for which additional diagnostic testing and/or tissue sampling is recommended	Suspicious
4B		Chest CT with or without contrast, PET/CT and/or tissue sampling depending on the probability of malignancy and comorbidities. PET/CT may be used when there is a ≥8 mm solid component	
Significant—other		S	
Prior lung cancer		C	

References

1. National Comprehensive Cancer Network. Lung cancer screening guideline, version 1.2015.
2. American College of Radiology. Lung CT screening reporting and data system, Lung-RADS™. 2014.
3. U.S. Preventive Services Task Force recommendation statement: screening for lung cancer, USPSTF. 2013.
4. U.S. Preventive Services Task Force recommendation statement: screening for lung cancer, USPSTF. 2016.
5. National Lung Screening Trial Research Team, Aberle DR, Adams AM, et al. Reduced lung-cancer mortality with low dose computed tomographic screening. N Engl J Med. 2011;365:395–409.
6. The American College of Chest Physicians Lung Cancer Standards. 3rd ed.
7. Medicare coverage for cancer prevention and early detection.
8. Dataset for histopathological reporting of lung cancer. Royal College of Pathologists UK.
9. Protocol for the examination of specimens from patients with primary non-small cell carcinoma, small cell carcinoma, or carcinoid tumor of the lung, College of American Pathologists, Cancer Protocol Templates.

Appendix BF: Standards for Oncology Services: Version (1)

Standards for Oncology Services

<u>Version (1)</u>

Issue date: 2/4/2024

Effective date: 2/6/2024

Health Policies and Standards Department

Health Regulation Sector (2024)

800342 (DHA) | dha.gov.ae | @dha_dubai | Dubai Health Authority

Introduction

The Health Regulation Sector (HRS) plays a key role in regulating the health sector. HRS is mandated by the Dubai Health Authority (DHA) Law No. (6) of the year 2018 with its amendments pertaining to DHA, to undertake several functions including but not limited to:

- Developing regulation, policy, standards, guidelines to improve quality and patient safety and promote the growth and development of the health sector.
- Licensure and inspection of health facilities as well as healthcare professionals in ensuring compliance to best practice.
- Managing patient complaints and assuring patient and physician rights are upheld.
- Governing the use of narcotics, controlled and semi-controlled medications.
- Strengthening health tourism and assuring ongoing growth.
- Assuring management of health informatics, e-health and promoting innovation.

The Standards for Oncology Services aims to fulfil the following overarching priorities of Dubai Health Sector Strategy 2026:

- Pioneering human-centered health system to promote trust, safety, quality, and care for patients and their families.
- Make Dubai a lighthouse for healthcare governance, integration, and regulation.
- Pioneering prevention efforts against non-communicable diseases.
- Foster healthcare education, research, and innovation.

Executive Summary

Dubai Health Authority (DHA) is pleased to present the DHA Standards for Oncology Services, which aims to improve the quality of oncology services in healthcare facilities.

This regulation places an emphasis on services' requirements with a focus on quality of services and safety of patients and healthcare professionals based on the international standards of best practices in this domain, while taking into consideration the local and federal laws. Therefore, this document provides a base for the Health Regulation Sector (HRS) to assess the oncology services provided in the Emirate of Dubai and to ensure a safe and competent delivery of services.

It will also assist oncology service providers in developing their quality management systems and in assessing their own competence to ensure compliance with DHA regulatory requirements and the United Arab Emirates (UAE) federal laws.

Definitions

Antineoplastic: Meaning anti-cancer therapy and cytotoxic (cell-killing) therapy. Includes immunotherapy, hormonal therapy, targeted therapy, and chemotherapy.

Cancer: Defined as a term for diseases in which abnormal cells divide without control and can invade nearby tissues.

CT simulation: Shall be defined as a CT procedure in which the specific pathology is localized within the patient, who is placed in a precise and reproducible position, for use in treatment planning for radiation therapies. CT simulation utilizes a conventional CT scanner outfitted with specific simulation hardware and software.

External radiation therapy: Shall be defined as the use of high-energy penetrating wave or particle beams used to damage or destroy cancerous cells. External radiation therapy may also be used as a form of treatment for some non-cancerous diseases and is frequently delivered on a recurring outpatient basis. High-energy beams do not leave the patient "radioactive," and there are no concerns about exposure of the patient to other persons post-treatment. See Linear Accelerator.

Healthcare professional: Shall be defined as healthcare personal working in healthcare facilities and required to be licensed as per the applicable laws in United Arab Emirates.

Hospice: Shall be defined as a facility or program designed to provide a caring environment for meeting the physical and emotional needs of the terminally ill.

Intensity modulated radiation therapy (IMRT): Shall be defined as an advanced external beam radiation therapy, which utilizes computer images to match radiation to the size and shape of a tumor. Using multiple smaller beams from different angles and of varying intensities, IMRT varies the shape of the radiation delivered to the treatment area, minimizing damage to surrounding healthy tissue. See Stereotactic Radiosurgery.

Internal radiation therapy: Shall be defined as the use of low-level radioactive implants or "seeds" to deliver radiation to local tissue structures. Frequently implanted in tumors, the radioactive decay damages or destroys the immediately surrounding tissue. Implants are specifically chosen to match the prescribed radiation dose necessary to damage the tumor while protecting the surrounding healthy tissues. Radioactive implants are placed surgically. Depending upon the implant's intensity, patients may be "radioactive" for a period of time post-implantation and may need to remain in hospital, segregated from others until the radioactive decay reduces the strength of the implant.

Licensure: Shall be defined as issuing an official permission to operate a health facility to an individual, government, corporation, partnership, Limited Liability Company (LLC), or other form of business operation that is legally responsible for the facility's operation.

Linear accelerator (Linac): Shall be defined as a device, which produces and delivers high-energy beams, which, in the hospital setting, is used to damage or destroy targeted tissues or structures, frequently cancerous tumors, within the patient's body. See Stereotactic Radiosurgery.

Multidisciplinary team: MDT in oncology is defined as the cooperation between different specialized professionals involved in cancer care with the overarching goal of improving treatment efficiency and patient care.

Oncology: Shall be defined as a branch of medicine that specializes in the diagnosis and treatment of cancer. It includes medical oncology (the use of chemotherapy, hormone therapy, and other drugs to treat cancer), radiation oncology (the use of radiation therapy to treat cancer), and surgical oncology (the use of surgery and other procedures to treat cancer).

Palliative: Shall mean an approach that improves the quality of life of patients and their families facing the problem associated with life-threatening illness, through the prevention and relieving of suffering by means of early identification and impeccable assessment and treatment of pain and other problems, physical, psychosocial, and spiritual.

Palliative care: Refers to patient- and family-centered care that optimizes quality of life by anticipating, preventing, and treating suffering.

Patient: Shall be defined as any individual who receives medical attention, care, or treatment by any healthcare professional or admitted in a health facility.

Picture archiving and communication system (PACS): Shall be defined as the digital capture, transfer, and storage of diagnostic images. A PACS system consists of workstations for interpretation, image/data producing modalities, a web server for distribution, printers for file records, image servers for information transfer and holding, and an archive of off-line information. A computer network is needed to support each of these devices.

Precision oncology: Aims to deliver the right cancer treatment to the right patient at the right dose and the right time.

Radiation therapy: Shall be defined as use of high-energy radiation to shrink tumors and kill cancer cells. X-rays, gamma rays, and charged particles are types of radiation used for cancer treatment. The radiation may be delivered by a machine outside the body (external beam radiation therapy), or it may come from radioactive material placed in the body near cancer cells (internal radiation therapy, also called brachytherapy).

Stereotactic radiosurgery: Shall be defined as the process by which radiation beams are projected to the tumor or target area from multiple points of origin. This allows relatively high radiation doses to the target area while exposing the surrounding tissues to significantly lower levels of radiation energy. Stereotactic radiosurgery equipment is available in both frame-based systems for treatment of head and neck, and frameless systems, which can treat any anatomic area.

Supervised area: Shall be defined as any area not already designated as a controlled area but where occupational exposure conditions need to be kept under review even although specific protection measures and safety provisions are not normally needed.

Surgical oncology: Shall be defined as a specialized area of oncology that engages surgeons in the cure and management of cancer.

Treatment planning: Shall be defined as following precise identification of the position, size, and shape of a tumor or target area, typically through MR, PET/CT, SPECT/CT, or CT based simulation, the optimal means of radiation therapy is planned in which the precise radiation doses are delivered to target areas while minimizing the radiation exposure to adjacent and surrounding tissues. This plan is typically mapped out three dimensionally and computer plotted to guide radiation therapy/radiosurgery.

Abbreviations

ACLS Advanced Cardiac Life Support
BLS   Basic Life Support
DHA   Dubai Health Authority
DM   Dubai Municipality
ECG   Electrocardiography
EMT   Emergency Medical Technician
FANR Federal Authority Nuclear Regulation
HRS   Health Regulation Sector
ICU   Intensive Care Unit
IPPV Intermittent positive pressure ventilation
MDT   Multidisciplinary Team
PALS Pediatric Advanced Life Support
PPE   Personal Protection Equipment
QAP   Quality Assurance Program
RN   Registered Nurse
UAE   United Arab Emirates
UPS   Uninterrupted Power Supply

1. Background

1.1. Oncology services provide diagnoses, treatment, and follow-up for cancer in adults using chemotherapy, hormonal therapy, biological therapy, targeted therapy, and immunotherapy. Those services include but are not limited to breast cancer screening, bowel cancer screening, upper or lower GI endoscopy and bronchoscopy, comprehensive tumor board, genomic testing, surgery, radiotherapy, and second opinion.

2. Scope

2.1. All DHA licensed facilities that provide oncology services.

3. Purpose

3.1. To assure provision of the highest levels of safety and quality for oncology services in Dubai Health Authority (DHA) licensed health facilities.

4. Applicability

4.1. DHA licensed healthcare professionals and health facilities providing oncology services.

5. Standard 1: Registration and Licensure Procedures

5.1. All health facilities providing oncology services shall adhere to the United Arab Emirates (UAE) Laws and Dubai regulations.
5.2. Health facilities aiming to provide oncology services shall comply with the DHA licensure and administrative procedures available on the DHA website https://www.dha.gov.ae.
5.3. Licensed health facilities opting to add oncology services shall inform Health Regulation Sector (HRS) and submit an application to HRS to obtain permission to provide the required service.
5.4. Oncology services shall only be provided in one of the following facilities:
    5.4.1. Hospital/unit attached to a hospital
    5.4.2. Day surgical center
    5.4.3. Cancer treatment center
    5.4.4. Breast unit
    5.4.5. Outpatient clinic

6. Standard 2: Health Facility Requirements

6.1. The health facility should meet the health facility requirement as per the DHA Health Facility Guidelines (HFG).
6.2. A comprehensive oncology service shall consist of the following:
(Note: If the applicant provides a single oncology service, then only the relevant requirements should be considered).
    6.2.1. Reception and waiting areas
    6.2.2. Consultation and examination rooms
    6.2.3. Diagnostic imaging services
    6.2.4. Radiotherapy services
    6.2.5. Mould room
    6.2.6. Treatment planning room
    6.2.7. Chemotherapy services
    6.2.8. Surgical care
    6.2.9. Intensive Care Unit (ICU)
    6.2.10. Palliative care
    6.2.11. Acute hematology service
    6.2.12. Bone marrow transplant
    6.2.13. Pediatric oncology hematology service
    6.2.14. Nuclear medicine

6.2.15. Interventional radiology

6.2.16. Oncology pharmacy with aseptic chemotherapy preparation area

6.2.17. Histopathology

6.2.18. Fertility preservation service

6.2.19. Inpatient rooms

6.2.20. Outpatient holding area

6.2.21. Clinical laboratory and blood services

6.2.22. Support areas for oncology care

6.2.23. Staff areas including staff station, staff change areas, etc.

6.2.24. Meeting room where the multidisciplinary team gets together to discuss cases

6.3. The health facility should install and operate equipment required for provision of the proposed services in accordance to the manufacturer's specifications.

6.4. The health facility shall ensure easy access to the health facility and treatment areas for all patient groups.

6.5. The health facility design shall provide assurance of patients and staff safety.

6.6. The health facility shall have appropriate equipment and trained healthcare professionals to manage critical and emergency cases.

6.7. The health facility should develop the following policies and procedure, but not limited to:

6.7.1. Patient acceptance criteria

6.7.2. Patient assessment and admission

6.7.3. Patient education and informed consent

6.7.4. Patient health record

6.7.5. Infection control measures and hazardous waste management

6.7.6. Incident reporting

6.7.7. Patient privacy

6.7.8. Medication management

6.7.9. Emergency action plan

6.7.10. Patient discharge/transfer

6.8. The health facility shall provide documented evidence of the following:

6.8.1. Appropriate storage and preparation of chemotherapy, targeted therapy, and immunotherapy medicine

6.8.2. Transfer of critical/complicated cases when required

6.8.3. Patient discharge

6.8.4. Clinical laboratory services

6.8.5. Equipment maintenance services

6.8.6. Multidisciplinary decision-making and management of patients

6.8.7. Laundry services

6.8.8. Medical waste management as per Dubai Municipality (DM) requirements

6.8.9. Housekeeping services

6.9. The health facility shall maintain charter of patients' rights and responsibilities posted at the entrance of the premise in two languages (Arabic and English).

6.10. The health facility shall have in place a written plan for monitoring equipment for electrical and mechanical safety, with monthly visual inspections for apparent defects.

6.11. The health facility shall ensure it has in place adequate lighting and utilities, including temperature controls, water taps, medical gases, sinks and drains, lighting, electrical outlets, and communications.

7. Standard 3: Healthcare Professionals Requirements

7.1. Medical oncologist

    7.1.1. A medical oncologist is a highly trained specialist who is responsible for the diagnosis and treatment of patients with cancer. They must be assisted by a competent team to provide effective treatment.

7.2. Radiation oncologist

7.3. Radiation therapist

7.4. Surgical oncologist

    7.4.1. Including specialization in colorectal, upper GI, hepatobiliary, breast oncoplastic, urology, GYN oncology, thoracic surgery, head and neck surgery, and neurosurgery

7.5. Oncology nurses

7.6. Chemotherapy nurses

7.7. Oncology pharmacist

7.8. Oncology social worker

7.9. Radiation technician

7.10. Radiation physicist

7.11. Pathologist

7.12. Hematologist

7.13. Lab technician

7.14. Nutritionist

7.15. Physical therapist

7.16. Palliative care specialist

7.17. Healthcare professionals with sub-specialty or specialty oncology training from a DHA approved institution

7.18. Nuclear medicine specialists

7.19. Chemotherapy unit includes (but not limited to):

    7.19.1. Internal medicine consultant/specialist present at the facility at all times

    7.19.2. Medical oncologist

    7.19.3. Clinical pharmacist

    7.19.4. Specialty nurse—oncology

    7.19.5. Palliative care physician

7.20. Multidisciplinary team:

    7.20.1. All Cancer Care Centers must have a multidisciplinary team with a minimum membership including diagnostic radiologists, pathologists, surgical oncologist, radiation oncologists, and medical oncologists to achieve high levels of quality care to manage the disease.

7.20.2. The multidisciplinary team may include physicians ranging from primary care providers to specialists in all oncology disciplines. In addition, care requires input from many other clinical and allied-health professionals including nursing, social work, genetics, nutrition, rehabilitation, and others.

7.20.3. Multidisciplinary team must meet on a regular basis to discuss the management of patients who are diagnosed with cancer.

7.20.4. The multidisciplinary team is responsible for goal setting, planning, initiating, implementing, evaluating, and improving all cancer-related activities in the program.

7.21. Diagnostic imaging unit:

7.21.1. Diagnostic radiologist

7.21.2. Radiologist

7.21.3. Radiographer

7.21.4. Magnetic Resonance Imaging (MRI) technologist

7.21.5. Sonographer

7.21.6. Interventional radiology service

7.22. For radiation therapy unit, the clinical use of ionizing radiation is a complex process involving highly trained personnel in a variety of interrelated activities that include:

7.22.1. Radiation oncologist

(a) There should be one (1) radiation oncologist for each 35–45 patients under treatment at the facility.

7.22.2. Physicist:

(a) There should be one physicist present for each center

(b) A therapist with specialized training in dosimetry, a "Dosimetrist," may render additional support

7.22.3. Radiotherapy technologist

(a) Two technologists are required for the operation of each treatment machine.

(b) An additional technologist will also be present with special training in simulation techniques.

7.22.4. Mould room technician

7.22.5. Nuclear medicine technologist

7.22.6. Specialty nurse—Oncology

(a) A nurse with special competence and skills required for the management of oncology patients.

7.22.7. Support personnel

(a) Personnel will be present to attend to the needs of the patients and the facility in the general categories of administration, compiling of documentation, scheduling, etc.

(b) Additional staff may be required for transcription, mold fabrication, and other tasks as identified by the facility.

7.23. Surgical oncology unit includes (but not limited to):
    7.23.1. Anaesthesiologist
    7.23.2. Surgical oncologist
    7.23.3. Specialty nurse—Oncology
    7.23.4. Anesthesia technologist
    7.23.5. Anesthesia technician
7.24. Pediatric oncology unit
    7.24.1. Pediatric oncologist
    7.24.2. Pediatric hematologist
    7.24.3. Pediatric surgeon/surgical oncologist (as per 7.20.12)
    7.24.4. Pediatric transfusion medicine
    7.24.5. Registered nurse
    7.24.6. Pediatric nurse
    7.24.7. The medical staff at such a facility is composed of a multidisciplinary team of a primary care pediatrician, pediatric medical subspecialists, and pediatric surgical specialist like hematologists/oncologists, surgeons, urologists, neurologists, neurosurgeons, orthopedic surgeons, radiation oncologists, pathologists, child life specialists, and diagnostic radiologists. These physicians and nurse practitioners, pediatric nurses, social workers, pharmacists, nutritionists, and other allied-health professionals shall care for the child or adolescent with cancer.
    7.24.8. Pediatric hematologist/oncologist is the coordinator for the diagnosis and treatment of most children and adolescents with cancer. He/she must be assisted by a competent team to provide effective treatment that can comprise of:
        (a) Pediatric oncology nurses who are certified in chemotherapy, knowledgeable about pediatric protocols, and experienced in the management of complications of therapy.
        (b) Rehabilitation pediatric physical and mental rehabilitation services including pediatric physiatrists.
        (c) Social workers and access to support groups.
        (d) Pediatric nutrition expert.
    7.24.9. Radiologists with specific expertise in the diagnostic imaging of infants, children, and adolescents.
    7.24.10. Radiation oncologist trained and experienced in the treatment of infants, children, and adolescents.
    7.24.11. Pediatric surgeons/urologist, surgical specialists with pediatric expertise (i.e., training and certification, if available) in neurosurgery, orthopedics, ophthalmology, otolaryngology, etc.
    7.24.12. Pediatric subspecialists available to participate actively in all areas of the care of the child with cancer, including anaesthesiology, intensive care, infectious diseases, cardiology, neurology, endocrinology and metabolism, genetics, gastroenterology, child and adolescent psychiatry, nephrology, and pulmonology.

7.24.13. A pathologist experienced in pediatric oncology is an essential member of the multidisciplinary team at the pediatric oncology center.

7.25. Clinical laboratory:

7.25.1. Anatomic and clinical pathologist

7.25.2. Cytopathologist

7.25.3. Hematopathology

7.25.4. Pediatric pathologist

7.26. Support staff that the facility may have are as follows:

7.26.1. Nursing staff

7.26.2. Biomedical engineer

7.26.3. Quality assurance officer

7.26.4. IT support staff

7.26.5. Pharmacist

7.26.6. Therapist (Physiotherapist, Occupational Therapist, Speech Therapist)

7.26.7. Social workers

7.26.8. Clinical psychologist

7.26.9. Dieticians

7.26.10. Wig fitters

7.26.11. Emergency medical technician advances (paramedic)

7.27. Physicians:

7.27.1. A suitably qualified DHA licensed consultant oncologist/physician shall be nominated as medical director of the oncology center who shall be responsible for overall management of the facility.

7.27.2. A DHA licensed consultant pediatric oncologist must be associated with the facility in case pediatric oncology services are provided (children from birth to eighteen (18) years of age, this age could be extended to twenty-one (21) years of age as per the American Cancer Society).

7.27.3. The paediatric oncologist must be present when pediatric oncology services are provided.

7.27.4. The oncologist must ensure adequate monitoring of patients during treatment, and subsequent aftercare.

7.27.5. The oncologist shall be contactable at all times to render emergency care.

7.27.6. In the event that the oncologist on duty is unable to fulfil his/her full responsibility to the patients of the oncology center, he/she must arrange for a similarly qualified physician to be responsible for the total care of the patients in the facility.

7.27.7. The medical director is ultimately responsible in ensuring that the monitoring and safety devices and resuscitation equipment are in proper working condition at all times.

7.27.8. The need for treatment and choice of modality shall be based on MDT recommendation, sound clinical principles, internationally recognized guidelines, and thorough clinical evaluation of medical condition and comorbid by the attending oncologist.

7.27.9. The attending oncologist may recommend to the end stage cancer patient the modality that is best suited to him/her. This shall be based on the patient's, other comorbid conditions, ability to comply with treatment, available family support, and other social factors.

7.27.10. The patient shall be allowed to make a fully-informed choice of modality, after receiving adequate counseling from his/her oncologist on the different modalities available and the modality that is most appropriate for the patient's need.

7.27.11. There shall be a documented Quality Assurance Program (QAP) to ensure quality patient care through objective and systematic monitoring, evaluation, identification of problems and action to improve the level and appropriateness of care. The QAP shall include:

(a) Documented policies and procedures related to the safety while conducting all patient care activities.

(b) Documented regular biannual reviews of the policies and procedures.

(c) Documented reviews of deaths, accidents, complications, and injuries arising from treatment.

7.28. Nursing staff:

7.28.1. Nurses with specialized knowledge and skills shall provide oncology nursing care.

7.28.2. The nurse in charge must be a qualified DHA licensed Registered Nurse (RN), with at least two (2) years of experience in oncology.

7.28.3. The ratio of trained RNs/patients shall be 1:3 at a given time.

7.28.4. All the nurses shall have an Oncology Nursing Society (ONS) certification and maintain Continuous Professional Development (CPD) by attending ONS programs.

7.28.5. There shall be at least one (1) nurse with a minimum of six (6) months of training or experience/training to be physically present at the oncology center at all times to monitor the patients throughout the treatment/procedure, to be available to deal with any emergencies that may arise and to alert the oncologist when necessary.

7.28.6. The attending RN is responsible for the general checkup of the patient including vital statistics and recording the initial assessment in the medical records.

7.28.7. All RNs shall hold current BLS and ACLS certifications.

7.29. Biomedical engineer:

7.29.1. Employ a biomedical engineer or have contracts with the manufacturers of the equipment for regular monitoring and maintaining equipment.

7.30. Radiation safety officer:

7.30.1. Uses ionizing radiations for medical use may be required to have a Radiation Protection Program (RPP).

7.31. Quality assurance officer:

    7.31.1. The quality assurance officer will monitor the quality improvement program activity and report the findings to the cancer committee at least annually and recommend corrective action if activity falls below the annual goal or requirement.

7.32. Pharmacist

    7.32.1. A DHA licensed pharmacist shall be in charge of maintaining the medicines and solutions that will be administered to patients with a minimum of one (1) year experience in chemotherapy preparation.

7.33. Therapist (Physiotherapist, Occupational Therapist, Speech Therapist):

    7.33.1. DHA licenses healthcare professionals to support the cancer treatment offered at the facility.

7.34. Clinical psychologist:

    7.34.1. At least one (1) DHA licensed clinical psychologist to help people who are having difficulty coping with cancer or cancer treatment.

7.35. Dietician:

    7.35.1. At least one (1) dietician shall maintain progress notes of all patients treated in the facility.

7.36. Medical social worker:

    7.36.1. There shall be some medical social workers associated with oncology center.

    7.36.2. The medical social workers shall be involved in psychosocial evaluation, case work counseling of patients and families, group work, evaluate and facilitate rehabilitation, team care planning, and collaboration, facilitate community agency referral, and improve communication with treating team.

    7.36.3. The social workers are required to maintain notes of the patients.

7.37. Infection control nurse:

    7.37.1. To perform regular audits, conducts surveillance of cultures and insures best practice for patient access.

8. Standard 4: Diagnostic Imaging Requirements

8.1. The diagnostic imaging services may include the following:

    8.1.1. Conventional radiography (X-ray unit)

    8.1.2. Ultrasound

    8.1.3. MRI

    8.1.4. Digital mammography

    8.1.5. Sonography

    8.1.6. CT: PET CT imaging and SPECT/CT

    8.1.7. For detailed information, please refer to Diagnostic Imaging Services Regulation on the DHA website www.dha.gov.ae.

    8.1.8. Diagnostic imaging services must comply with the FANR laws and regulations regarding the use of ionizing radiation and radioactive materials. For further information regarding FANR, law and regulations please visit FANR website www.fanr.gov.ae.

9. Standard 5: Radiation Requirements

9.1. The facility layout shall be planned in accordance with the local radiation safety regulations and internationally accepted radiation safety standards and in consultation with the radiation oncologist, physicist, and equipment manufacturer.

9.2. The room design, construction, and shielding shall be as per FANR and the manufacturers.

9.3. The radiation unit may have an inpatient facility for frail patients, patients traveling long distances and the occasional patient who has severe reactions to any of the treatments administered in the facility (a bed for every 10 patients).

9.4. The radiation therapy unit shall:

    9.4.1. Be located on the ground floor or lower floors of the oncology center to accommodate the weight of the equipment and ease of installation and replacement.

    9.4.2. Ensure properly designed rigid support structures located above the finished ceiling for ceiling mounted equipment.

    9.4.3. Provide equipment and infrastructure for treatment of patients using radioactive rays.

9.5. The radiotherapy unit should include the following functional areas, but not limited to:

    9.5.1. CT simulation room with an adjacent control area and changing room

    9.5.2. Treatment planning room for physicist/dosimetrists

    9.5.3. Film processing and storage area

    9.5.4. Physics laboratory/dosimetry equipment area (if thermoluminescent dosimetry (TLD) and film dosimetry are available, an area shall be designed for these activities)

    9.5.5. Film processing room, storage areas

    9.5.6. Radiotherapy room/bunkers to house the equipment to deliver treatment with an adjacent computer control area and changing rooms

    9.5.7. Holding area/recovery area

    9.5.8. Hypothermia room

    9.5.9. Mould room (optional)

    9.5.10. Exam room

9.6. If intra-operative therapy is proposed, the radiation oncology unit shall be only hospital based and located close to the operating unit or with a direct link.

9.7. Areas requiring specific protection measures (controlled areas) include:

    9.7.1. Irradiation rooms for external beam

    9.7.2. Therapy and remote afterloading brachytherapy

    9.7.3. Brachytherapy rooms

    9.7.4. Simulator room

    9.7.5. Radioactive source storage and handling areas

9.8. These areas shall maintain define controlled areas by physical boundaries such as walls or other physical barriers marked or identified with "radiation area" signs.

9.9. The area of the control panel shall be considered as a controlled area, to prevent accidental exposure of patients by restriction of access to non-related persons, and distraction to the operator of a radiotherapy machine.

9.10. Supervised areas may involve areas surrounding brachytherapy patients' rooms or around radioactive source storage and handling areas.

9.11. Certain staff members need to be monitored with individual dosimeters. Individual external doses can be assessed by using individual monitoring devices such as thermoluminescent dosimeters or film badges, which are usually worn on the front of the upper torso. These shall include:

9.11.1. Radiation oncologists

9.11.2. Radiotherapy physicists

9.11.3. Radiation protection officer

9.11.4. Radiotherapy technologists

9.11.5. Source handlers

9.11.6. Maintenance staff

9.11.7. Nursing or other staff who must spend time with patients under treatment with brachytherapy

9.12. Indications for radiation must undergo quality control and auditing

9.13. Healthcare professional requirements for a radiation therapy unit shall be according to the table in Appendix BG.

10. Standard 6: Chemotherapy Requirements

10.1. The chemotherapy unit can be:

10.1.1. A part of a hospital

10.1.2. A satellite unit—on a hospital campus, but not in the hospital.

10.1.3. Integrated cancer care—a part of an oncology center that provides diagnostic services, radiation therapy, and/or surgical facility.

10.1.4. Freestanding unit—in case a chemotherapy unit is a freestanding facility it shall:

(a) Maintain a contract with the closest hospital with inpatient services to manage emergencies or complications.

(b) Provide an in-house ambulance service.

10.2. The chemotherapy unit shall be designed to provide designated, discreet, and easy access for patients who may arrive by public transport or vehicles, with families and children or those who arrive on a wheel chair, ambulance stretcher, or patient trolley.

10.3. Chemotherapy can be provided in an outpatient service except in the case of acute leukemia patients where the patients shall be treated in a multispecialty health facility with inpatient, outpatient, and ICU services.

10.4. The chemotherapy unit can have inpatient services only with an Internal Medicine Consultant/Specialist present at the facility at all times and provide a minimum of 5–6 inpatient beds.

10.5. The chemotherapy unit shall have the following functional areas:

　10.5.1. Reception/waiting area

　10.5.2. Consultation room

　10.5.3. Sterile preparation room/buffer area

　10.5.4. Anteroom/pharmacy

　10.5.5. Aseptic chemotherapy preparation area

　10.5.6. Patient treatment areas/procedure room with treatment chairs or beds

　10.5.7. Isolation room(s)

　10.5.8. Clean utility/dirty utility

　10.5.9. Medication preparation room with a 100% exhaust Class II B2 safety cabinet

　10.5.10. Staff areas

　10.5.11. Support areas

　10.5.12. Storage areas for clinical, non-clinical and bulk items storage, e.g., fluids, equipment including infusion/syringe pump storage

　10.5.13. Waste disposal room

10.6. Patient treatment areas shall consist of treatment bays to provide chemotherapy to patients.

10.7. Patient privacy shall be considered in the design.

10.8. Special consideration given to patients with special needs.

10.9. Nurse call and emergency call facilities shall be provided in all patient areas (e.g., bed/chair spaces, toilets, etc.) and clinical areas in order for patients and staff to request for urgent assistance. The alert to staff members shall be done in a discreet manner.

10.10. Provision of duress alarm system shall be provided for the safety of staff members who may at times face threats imposed by clients/visitors. Call buttons shall be placed at all reception/staff station areas and consultation/treatment areas where a staff may have to spend time with a client in isolation or alone. The combination of fixed and mobile duress units shall be considered as part of the safety review during planning for the unit.

10.11. Inclusion of medical gases (oxygen and suction) units of one (1) per two (2) chairs shall be provided.

10.12. Hand washing facilities with liquid soap dispenser, disposable paper towels, and personal protection equipment (PPE) shall be readily available for staff within the unit.

10.13. The chemotherapy unit shall maintain an easily accessible chemotherapy work flowchart for high quality and standardized care.

10.14. The chemotherapy unit shall maintain a crash cart to deal with emergencies.

10.15. Services that support and are linked with chemotherapy may include:

　10.15.1. Physiotherapy (lymph oedema management)

　10.15.2. Occupational therapy

　10.15.3. Dietetic/nutrition services

10.15.4. Clinical psychology

10.15.5. Social work services

10.15.6. Community and outreach cancer services

10.15.7. Palliative care and hospice

10.15.8. Complementary therapies (e.g., relaxation, stress management, and massage)

10.15.9. Wig and prosthesis services

10.16. All cytotoxic drug waste shall be separated from general waste.

10.17. Cytotoxic waste shall be destroyed in an incinerator approved for the destruction of cytotoxic drugs.

10.18. Breakable contaminated needles, syringes, ampoules, broken glass, vials, intravenous sets and tubing, intravenous and intravesical catheters, etc. shall be placed into designated leak-proof; puncture proof sharps containers that clearly and visibly display the cytotoxic hazard symbol.

10.19. Non-breakable contaminated materials including disposable gowns, gloves, gauzes, masks, intravenous bags, etc. shall be placed in thick sealed plastic bags, hard plastic or cytotoxic containers that clearly and visibly display the cytotoxic hazard symbol. When full, the bags and containers shall be placed in the oncology waste container.

10.20. Clearly marked chemotherapy waste receptacles shall be kept in all areas where cytotoxic drugs are prepared or administered.

10.21. If access to an appropriately licensed incinerator is not available, the acceptable alternative shall be transportation to and burial in a licensed hazardous waste dump.

10.22. Special written protocol shall be maintained for:

10.22.1. Management of an incident in case a patient/family member is contaminated with a cytotoxic agent.

10.22.2. Management of cytotoxic spill in or outside the BSC.

10.22.3. Safe transportation of cytotoxic agents.

10.23. All chemotherapy protocols and deliveries must be audited by oncologists and oncology pharmacists. A flag system should be in place for excessive use of chemotherapy in the last 2 week before death.

11. Standard 7: Surgical Oncology

11.1. All oncology/suspected cancer surgeries must be approved by MDT except emergency surgeries and this must be audited regularly.

11.2. Surgical oncology procedures must be done by surgeons with specialized training in oncology.

11.3. Rectal surgeries must be done by colorectal surgeons.

11.4. Sarcoma surgeries must be done by orthopedic surgeons with special training in oncology.

11.5. For detailed information on operating theater, critical care, airborne infection isolation, emergency area, and inpatient services refer to the "Hospital Regulation" on www.dha.gov.ae.

12. Standard 8: Pediatric Oncology

12.1. The pediatric facility shall:

12.1.1. Be a part of a multidisciplinary hospital.

12.1.2. Have accessible and fully staffed, on-site pediatric intensive care unit (PICU).

12.1.3. Have access to an up-to-date diagnostic imaging facilities to perform radiography, computed tomography, magnetic resonance imaging, ultrasonography, radionuclide imaging, and angiography; positron-emission tomography (PET CT) scanning and other emerging technologies are desirable.

12.1.4. Have an up-to-date radiation therapy equipment with facilities for treating pediatric patients shall be available.

12.1.5. Have an access to hematopathology laboratory capable of performing cell-phenotype analysis using flow cytometry, immunohistochemistry, molecular diagnosis, and cytogenetic and access to blast colony assays and polymerase chain reaction-based methodology shall be available.

12.1.6. Have access to hemodialysis and/or hemofiltration and apheresis for collection and storage of hematopoietic progenitor cells.

12.1.7. Have a clinical chemistry laboratory with the capability to monitor antibiotic and antineoplastic drug levels.

12.1.8. Have an access to blood bank capable of providing a full range of products including irradiated, cytomegalovirus negative, and leucodepleted blood components.

12.1.9. The facility shall have a pharmacy capable of accurate, well-monitored preparation and dispensing of antineoplastic agents and investigational agents.

12.1.10. Have the capability of providing sufficient isolation of patients from airborne pathogens, which can include high-efficiency particulate air (HEPA) filtration, or laminar flow and positive/negative pressure rooms.

13. Standard 9: Patient Care

13.1. All clinical trials should have all regulatory approvals and a designated principal investigator with experience in conducting clinical oncology trials. Patients should be fully aware and consented to unlicensed treatments.

13.2. Palliative care:

13.2.1. The availability of palliative care services is an essential component of cancer care, beginning at the time of diagnosis and being "continuously available" throughout treatment, surveillance, and when applicable.

13.2.2. Palliative care must be available in all cancer centers.

13.2.3. Palliative care services shall be available to patients either on-site or by referral.

13.2.4. An interdisciplinary team of medical and mental health professionals, social workers, and spiritual counselors shall be available or accessible to provide palliative care services.

13.2.5. Palliative care services on-site will vary depending on the scope of the program, staff expertise, and patients treated.

13.2.6. The palliative service team consists of:
   (a) Physician: Hospice and palliative medicine physician is strongly encouraged
   (b) Nurse: Trained in hospice and palliative care is strongly encouraged
   (c) Pharmacist
   (d) Social worker
   (e) Chaplain or spiritual care counselor
   (f) Trained volunteer

13.2.7. Palliative care services include, but are not limited to, the following:
   (a) Team-based care planning that involves the patient and family
   (b) Pain and symptoms management
   (c) Communication among patients, families, and healthcare team
   (d) Continuity of care across a range of clinical settings and services
   (e) Attention to spiritual comfort
   (f) Psychosocial support for patients and families
   (g) Bereavement support for families of patients who die and team members who provided care to the person who died
   (h) Hospice care: Hospice care is one aspect of palliative care and is a service delivery system that provides palliative care for patients who have a limited life expectancy
   (i) Hospice is presented as an option to patients and families when the prognosis is limited and death will not be surprising

13.3. Psychological support.

13.4. Psychosocial services:

13.4.1. Ensure patient access to psychosocial services either on-site or by referral.

13.4.2. These services address physical, psychological, social, spiritual, and financial support needs that result from a cancer diagnosis and help ensure the best possible outcome.

13.4.3. A policy or procedure is in place to ensure patient access to psychosocial services.

13.5. Rehabilitation services:

13.5.1. Ensures access to rehabilitation services and identifies the rehabilitative services that are provided either on-site or by referral.

13.5.2. Rehabilitation services help patients cope with activities of daily living affected by the cancer experience and enable them to resume normal activities.

13.5.3. A policy or procedure is followed to access rehabilitation services.

13.6. Nutrition services:

13.6.1. Nutrition services are essential components of comprehensive cancer care and patient rehabilitation. These services provide safe and effective nutrition care across the cancer continuum (prevention, treatment, and survivorship) and are essential to promote quality of life.

13.6.2. An adequate spectrum of services shall be available (screening and referral for nutrition-related problems, comprehensive nutrition assessment, nutrition counseling, and education) either on-site or by referral, with a procedure in place to ensure patient awareness of and access to services.

13.6.3. A policy or procedure in place to access nutrition services.

13.7. Critical care services:

13.7.1. Every freestanding oncology center must have a contract/agreement with a hospital with an Intensive Care Unit (ICU), which must be accessible (less than 10 min response time) to receive patients in case of emergency.

13.7.2. There must be a competent and DHA licensed RN with suitable training and experience in critical care on duty to provide the critical care services if required. The evidence of competency and training shall include, but not limited to the following:

(a) Recognizing arrhythmias

(b) Infection control principles

(c) Training in using defibrillator

(d) Life support

(e) Airway management

13.7.3. Critical care equipment must be immediately available at the oncology center for immediate and safe provision of care if required.

13.8. Emergency services:

13.8.1. It is the responsibility of the healthcare facility management in addition to the oncologist in charge to ensure that there are facilities for emergency resuscitation, as well as documented protocols/procedures to deal with cardiopulmonary collapse and urgent medical treatment as patients may develop hypotension, fits, or collapse during treatment.

13.8.2. In addition, the healthcare facility management under the supervision of the oncologist in charge must:
   (a) Ensure that there are prior arrangements made for patients receiving treatment to be admitted in a nearby hospital in case of a freestanding facility, shall the need arise, within 10 min driving time.
   (b) Ensure oncology group practice by having standing arrangements with other healthcare professionals to provide immediate medical care in the event that the physician in charge is not available.
   (c) Ensure there is an ambulance available at any given time to transfer the patient to a hospital in case of any medical emergency.
   (d) Ensure that the ambulance service is accessible and at close proximity.
   (e) In case the oncology center has its own ambulance service, the ambulance services shall be ready with licensed, trained, and qualified Emergency Medical Technicians (EMT) for patient transportation if required, this service can be outsourced with a written contract with an emergency service provider licensed in Dubai. Clear patient transport protocol shall be maintained.
13.8.3. The ambulance shall maintain the following, but not limited to:
   (a) Sets of instruments, which shall include suturing set, dressing set, foreign body removal set or minor set and cut down set.
   (b) Disposable supplies which shall include suction tubes (all sizes), tracheostomy tube (all sizes), intravenous cannula (different sizes), IV sets, syringes (different sizes), dressings (gauze, sofratulle, etc.), crepe bandages (all sizes), and splints (Thomas splints, cervical collars, finger splints).
   (c) Portable vital signs monitor (ECG, pulse oximetry, temperature, NIBP, and EtCO2).
   (d) Portable transport ventilator with different ventilation mode (IPPV, SIMV, spontaneous, and PS).
   (e) Suction apparatus.
13.8.4. Emergency drugs, devices, equipment, and supplies must be available for immediate use in the emergency area for treating life-threatening conditions. Minimum emergency medication requirements shall be available as per the DHA Emergency Medication Policy, available on this link: https://www.dha.gov.ae/uploads/112021/3f5565de-9eb7-46c9-9480-17190a531903.pdf.
13.8.5. Storage areas for general medical or surgical emergency supplies, medication, and equipment shall be under staff control and out of path of normal traffic.
13.8.6. A record must be kept for each patient receiving emergency services and must be integrated into the patient's health records, the

record shall patient name, date, time and method of arrival, physical findings, care and treatment provided, name of treating doctor, and discharging/transferring time.

13.9. Transfer planning:

13.9.1. The oncology center shall maintain policies and procedures concerning patient transfer which reflect acceptable standards of practice and compliance with applicable regulations in Dubai.

13.9.2. If patient is transferred to another health facility and in order to ensure continuity of patient care, the other facility shall be informed about the case and approval for transfer shall be documented in the patient file.

13.9.3. The duty manager present at the oncology center is responsible for the coordination of the timely transfer of appropriate information and discharge notice from the oncology center to a hospital or another health facility.

13.9.4. A transfer sheet shall be prepared for all patients being transferred requiring further treatment.

13.9.5. A referral letter shall be given to the patient or family/patient representative. Patient shall not be sent under any circumstances to another facility without prior approval.

13.9.6. Mode of transport shall be decided based on the condition of the patient, the treating physician and the ambulance team shall decide who shall accompany the patient, e.g., physician present or trained nurse.

13.10. Patient assessment:

13.10.1. An effective patient assessment process aims to be comprehensive, includes multidisciplinary teams, and is based on clinical and priority needs of each individual patient. Such assessment shall result in identification and decisions regarding the patient's condition and continuation of treatment as the need arise. The oncology center shall have policies and procedures on patient assessment:

(a) On admission

(b) Following a change of health status

(c) After a fall

(d) When patient is transferred from one level of care to another

13.10.2. The patient assessment shall include, but not limited to, medical history, physical, social, and psychological assessment and identification of patients at risk.

13.10.3. Patients conveying personal health information during any assessment shall be accommodated in an area where privacy is assured.

13.10.4. Discharge preparation starts at admission and includes various persons, information, and resources like:

(a) The pickup person after treatment.

(b) Travel distance to the patient's house.

(c) Post discharge transport.

(d) The carer's contact details and their awareness of possible issues and requirements following discharge.

(e) Contact numbers after discharge in case of an emergency.

(f) Discharge arrangements regarding home care where it is identified.

13.10.5. Healthcare professionals shall use a formal risk assessment process to assess skin integrity and risk of falls of patients.

13.10.6. A comfortable care environment shall be provided in the facility with focus on patient privacy.

13.10.7. The plan of care must be determined and delivered in partnership with the patient and when relevant, patient's family/patient representative/legal guardian, to achieve the best possible outcomes.

13.10.8. The patient has the right to refuse the plan of care but this has to be documented and signed by the patient.

13.10.9. Patient's participation may include:

(a) Procedure date and admission/discharge time

(b) Physician selection

(c) Treatment preparation

13.10.10. Care shall be delivered by DHA licensed and competent healthcare professionals and competent multidisciplinary teams and based on the best available evidence.

13.10.11. A comfortable treatment environment is provided in the facility with focus on patient privacy.

14. Standard 10: Pharmacy and Medication Requirements

14.1. Pharmacy services should ensure adequate stocking, storage, and dispensing mechanisms for medications in a proper storage unit adhering to local laws, DHA pharmacy guidelines and DHA emergency medication policy.

14.2. The facility shall have a pharmacy capable of accurate, well-monitored preparation and dispensing of antineoplastic agents and investigational agents.

14.3. Pharmacy must have an oncology pharmacist available/a pharmacist with an oncology background.

15. Standard 11: Pathology Requirements

15.1. Only an accredited oncology designated lab can diagnose cancer. All specimens suspected of malignancy must be examined and reported independently by two pathologists.

15.2. The oncology healthcare facility must have a designated pathology laboratory for cancer diagnosis.

15.3. Pathology department must be in-house or an accredited outsourced lab.

16. Standard 12: Multidisciplinary Team

16.1. A multidisciplinary team (MDT) recommendation is mandatory for management of all newly diagnosed cancer cases, and prior to initiating treatment.

16.2. The major challenge is that many cancer cases are being misdiagnosed or inaccurately treated leading to poor outcome. MDT must be officially recognized by DHA and must consist of consultant (not specialist) medical oncologist, consultant radiation oncologist, consultant general surgeon, and consultant pathologist with training in oncology and preferred surgical oncologist.

16.3. All hospitals that do not have a full oncology service and do not have a DHA approved MDT must have an agreement with a DHA approved oncology MDT in order to treat cancer patients.

16.4. It is the responsibility of the Chief Medical Officer of each healthcare facility to ensure strict adherence to the protocol: No cancer surgery or cases of suspected cancer shall be scheduled in the operating room without prior recommendation and approval by the Multidisciplinary Team (MDT). This rule is mandatory in all health facilities that provide cancer treatment.

Acknowledgment

HRS developed this document in collaboration with subject matter experts whose contributions have been invaluable. HRS would like to gratefully acknowledge these professionals especially mentioning Emirates Oncology Society and thank them for their dedication toward improving quality and safety of healthcare services in the Emirate of Dubai.

Health Regulation Sector
Dubai Health Authority

References

1. DHA Regulation for Oncology Services. 2016. Available at: https://www.dha. gov.ae/uploads/112021/3e270bc1-7e7e-468d-80da-6a5c4bc3ac73.pdf. Accessed Oct 2023.

2. MOHAP Oncology Center Regulations. 2018. Available at: https://mohap.gov. ae/assets/download/d1451/Oncology%20Services%20Regulations.pdf.aspx. Accessed Oct 2023.

3. A guide to chemotherapy day unit, redesign measures for improvement. Department of Health, State Government Victoria; 2014. p. 1–15. Available at: https://www.healthfacilityguidelines.com/ViewPDF/ViewIndexPDF/iHFG_ part_b_oncology_medical_chemotherapy. Accessed 10 Oct 2023.

4. Abdul Khader MA. Planning a clinical pet centre, vol 11. International Atomic Energy Agency; 2010. p. 1–146. Available at: https://www-pub.iaea.org/MTCD/Publications/PDF/Pub1457_web.pdf. Accessed 10 Oct 2023.

5. American Cancer Society. Chemotherapy drugs. Available at: https://www.cancer.org/cancer/managing-cancer/treatment-types/chemotherapy/how-chemotherapy-drugs-work.html. Accessed 10 Nov 2023.

6. Cancer Center of Excellence. Performance measures, rating system, and rating standard. Florida Health; 2015. p. 1–20. Available at: https://www.floridahealth.gov/provider-and-partner-resources/research/_documents/final-cancer-center-of-excellence-manual.pdf. Accessed Oct 2023.

7. Carey J. Radiation safety officer qualifications for medical facilities. Report of AAPM Task Group 160; 2010. p. 1–33. Available at: https://www.aapm.org/pubs/reports/RPT_160.PDF. Accessed Oct 2023.

8. Center for Disease Control and Prevention. Basic infection control and prevention plan for and prevention plan for oncology settings. 2011. Available at: http://www.cdc.gov/hai/pdfs/guidelines/basic-infection-control-prevention-plan-2011.pdf. Accessed Oct 2023.

9. Cruz A. Standard treatment guidelines oncology. Ministry of Health & Family Welfare, Government of India; 2015. p. 1–137. Available at: http://clinicalestablishments.nic.in/WriteReadData/329.pdf. Accessed Oct 2023.

10. Healthcare Quality International LLC. Cancer Diagnosis and Treatment Center. Sultan Qaboos Cancer Diagnosis and Treatment Center. 2014. Available at: https://cccrc.gov.om/. Accessed Oct 2023.

11. https://hemonc.medicine.ufl.edu/files/2013/07/ChemoPrinciples.pdf.

12. Multi-Disciplinary Team. https://www.frontiersin.org/journals/oncology/articles/10.3389/fonc.2020.00085/full.

13. AlShamsi H. The state of cancer care in the United Arab Emirates in 2022. https://www.mdpi.com/2039-7283/12/6/101.

14. The Comprehensive Cancer Center. 2022. https://link.springer.com/book/10.1007/978-3-030-82052-7.

15. Interventional radiology and the care of the oncology patient. 2011. https://www.ncbi.nlm.nih.gov/pmc/articles/PMC3196980/.

16. Kash B, Tan D. Physician group practice trends. 2016. https://hospital-medical-management.imedpub.com/physician-group-practice-trends-a-comprehensive-review.php?aid=9343.

17. Second opinion in breast pathology: policy, practice and perception. 2014. https://www.ncbi.nlm.nih.gov/pmc/articles/PMC4521120/.

Appendix BG: Healthcare Professionals Requirements for Clinical Radiation Therapy

Consultant radiation oncologist-in-chief	1 consultant and 1 specialist as a minimum per radiation therapy unit
Staff radiation oncologist/physician	1:200/250 patients treated annually No more than 25–30 patients under treatment by a single physician at any one time
Radiation physicist	1:400 patients annually
Treatment planning staff: Dosimetrists or physics assistant	1:300 patients treated annually
RTT (Radio Therapy Technologist)	2:25 patients treated daily
RTT-Simulator	2:500 patients simulated annually
RTT-Brachytherapy	As needed
Registered nurses	1:300 patients treated annually
Social worker	As needed to provide service
Dietician	As needed to provide service
Physiotherapist	As needed to provide service
Biomedical engineer	If equipment serviced "in-house"

Note: If advanced or special techniques are to be undertaken, staff additional to the above will be required

Appendix BH: Standards for Autologous Haematopoietic Stem Cell Transplantation: Version 1

STANDARDS FOR AUTOLOGOUS

HAEMATOPOIETIC STEM CELL

TRANSPLANTATION

Version 1

Issue date: 14/10/2021

Effective date: 14/12/2021

Health Policies and Standards Department

Health Regulation Sector (2021)

800342 (DHA) | dha.gov.ae | @dha_dubai | Dubai Health Authority

Introduction

Health Regulation Sector (HRS) plays an essential role in regulating the health sector and is mandated by the Dubai Health Authority Law No. (6) of 2018 to undertake several functions:

- Developing regulation and standards to improve patient safety and quality and also support the growth and development of the Dubai health sector.

- Licensure and inspection of health facilities and healthcare professionals.
- Managing patient complaints and upholding patient rights.
- Regulating the use of narcotics, controlled and semi-controlled medications.
- Strengthening health tourism and assuring ongoing growth.
- Assuring the management of e-health and innovation.

The Standard for Autologous Haematopoietic Stem Cell Transplantation Stem Cells aims to fulfil several overarching Strategic Objectives and Programs within the Dubai Health Strategy (2016–2021):

- **Objective 1**: Position Dubai as a central medical tourism destination through a comprehensive, integrated, value-based, and high-quality service delivery system.
- **Objective 2**: Direct resources to assure a happy, healthy, and safe environment for Dubai population.
- **Objective 4**: Foster innovation throughout the continuum of patient care.
- **Strategic Program 1**: Care model innovation, care model innovation program. The ambition is to promote innovation and efficiency and ensure residents and visitors in Dubai to have access to high-quality services.
- **Strategic Program 10**: Excellence and quality. The ambition is to promote excellence in healthcare service delivery and enhance patient experience and satisfaction.

Acknowledgement

The Health Policy and Standards Department (HPSD) would like to acknowledge experts in the field for their continued dedication and support to develop the standard and improve patient safety and quality of care in the Emirate of Dubai.
Health Regulation Sector
Dubai Health Authority

Executive Summary

Haematopoietic Stem Cell Transplant (HSCT) or Bone Marrow Transplant (BMT) is a life-saving intervention that has been practiced for over five decades and was historically used to treat bone malignancies. Due to technological advances in medicine, treatment has become possible across many blood cancers (hematologic malignancies) and age groups. HSCT has also expanded into the treatment of solid tumour malignancy, hereditary disorders, and immune deficiency syndromes. Future indications for HSCT therapy include Stroke, CHD, diabetes, neurological and auto-immune diseases.

The purpose of the Standards for Autologous Haematopoietic Stem Cell Transplantation is to maximise quality and patient safety within DHA licensed health facilities. The standard is confined to autologous (same person) treatment with predetermined inclusion and exclusion criteria. The first part of the standard set out the health facility and professional requirements to operationalise an effective AHSCT transplantation unit. The transplant unit shall be led by a Clinical Program Director who has the necessary experience and competencies to supervise the day-to-day operations of the service. The second part of the standard sets out the indications, requirements for the service, stem cell collection, processing, storage, and transportation. The final part of the standard provides the service quality and safety requirements, and the expectations for stem cell preparation, infusion, and post-follow-up care and the requirements for documentation to demonstrate improvement.

Definitions

Adverse event: Any unintended or unfavourable symptom or condition that is temporary and associated with an intervention may have a causal relationship with the intervention, medical treatment, or procedure.

Adverse reaction: An unintended response directly or indirectly caused by the administration of cellular therapy.

Allogeneic: The biological relationship between genetically distinct individuals of the same species.

Apheresis: A medical technology in which blood is separated into parts. The required component is removed, and the remaining components are returned to the donor.

Autologous haematopoietic stem cell transplant: A clinical procedure where one's healthy stem cells are collected from mobilised peripheral blood, cord blood, or bone marrow then processed and stored. The patient then undergoes chemotherapy and/or radiation followed by infusion of the stem cells to treat an array of blood cancers or diseases that affect the bone marrow.

Autologous: Derived from an individual and intended for the same individual.

Clinical program: An integrated medical team housed in a defined location. The program includes a Clinical Program Director who can demonstrate sufficient staff training, adoption of protocols, written Standard Operating Procedures, implementation of quality management systems, clinical outcome analysis, and regular interaction among clinical sites.

Engraftment: Is the process when the transplanted stem cells begin to grow to produce new healthy cells (The reconstitution of recipient haematopoiesis with blood cells and platelets from a donor). It is typical for engraftment to occur between 10 and 15 days, but there are instances where this may take longer. Engraftment is identified through blood analysis of the white blood cells, neutrophil count, haemoglobin, and platelets.

Graft versus host disease: The condition occurs when donated bone marrow stem cells (the graft) identify the host with healthy tissues as alien and leads to an immune response. Graft versus host disease can also occur after an organ transplant or within the first few months of a transplant (acute) or, much later (chronic), damaging human tissue and organs. The signs and symptoms may be severe and life-threatening.

Haematopoietic progenitor cells (HPC): A cellular therapy product that contains self-renewing and/or multipotent stem cells. The cells can mature into haematopoietic lineages, lineage-restricted pluripotent progenitor cells, and committed progenitor cells, regardless of tissue source (bone marrow, umbilical cord blood, peripheral blood, or another tissue source).

ISBT 128: A global standard for identifying, labelling, and transferring human blood, cell, tissue, and organ products.

Peripheral blood stem transplant: Also known as peripheral stem cell support, in which a procedure is undertaken to replace blood stem cells. Medication is used to move cells out of the bone marrow, followed by centrifugation and collection of cells for use or storage. It is the most common of two main types of haematopoietic stem cell transplantation.

Preparative (conditioning) regimen: A treatment used to prepare a patient for stem cell transplantation (e.g. chemotherapy, monoclonal antibody therapy, radiation therapy).

Standard operating procedure (SOP): A written document that describes the process or steps taken to accomplish a specific task.

Stem cell mobilisation: A process whereby certain drugs are used to initiate the movement of bone marrow stem cells into the blood.

Transplantation: The administration of cells to provide transient or permanent engraftment in support of therapy of disease.

Abbreviations

AHSCT	Autologous haematopoietic stem cell transplant
ASTCT	American Society for Transplantation and Cellular Therapy
BMT	Bone marrow transplantation
CIBMTR	The Center for International Blood and Marrow Transplant Research
CMV	Cytomegalovirus
EBMT	European Society for Blood and Marrow Transplantation
FACT-JACIE	The Foundation for the Accreditation of Cellular Therapy and the Joint Accreditation Committee of ISCT-EBMT
GCSF	Granulocyte colony stimulating factor
GvHD	Graft-versus-host disease
HPC	Haematopoietic progenitor cells
HSV-1 or 2	Herpes simplex 1 or 2
ICU	Intensive care unit
ISCT	International Society for Cellular and Gene Therapy
PBSCT	Peripheral blood stem transplant
PCP	Pneumocystis carinii pneumonia
PPE	Personal protective equipment
PTLD	Post-transplant lymphoproliferative disease
QMS	Quality management system

Background

Haematopoietic stem cell transplant is a therapeutic intervention used to treat several malignant and non-malignant disorders. There are two categories of stem cells, allogeneic stem cells and autologous stem cells. Allogeneic stem cells involve cells from a matching donor, which typically involves a member of the family. Autologous stem cells are extracted from the individual, purified and then administered back to the same individual. Autologous stem cell transplantation accounts for the majority of global stem cell transplantation. Autologous Haematopoietic Stem Cell Transplantation (AHSCT)/Bone Marrow Transplantation (BMT) offers life-saving treatment for many haematological malignancies. Haematopoietic stem cells are capable of destroying tumour cells and forming new cells. Haematopoietic stem cell extraction is achieved from two sources: the bone marrow to produce functional cells (after engraftment) to replace diseased cells, or by priming the blood with granulocyte colony-stimulating factor generate new stem cells known as Peripheral Blood Stem Transplant (PBSCT). Once priming is completed, the extraction of stem cells is performed, followed by chemotherapy and/or radiotherapy to destroy blood-forming cells. New cells are infused back into the body intravenously. There are several advantages for Peripheral Blood Stem Cell Transplant (PBSCTs), including rapid engraftment rate, lower infection rate, and lower haemorrhagic morbidity and mortality. Due to the possible indications for stem cells, practice is based on the published case series and clinical consensus.

1. Purpose

1.1. To maximise the quality and patient safety for autologous haematopoietic stem cell transplantation services in DHA licensed health facilities.

2. Scope

2.1. Autologous Haematopoietic Stem Cell Transplantation (AHSCT) services.
2.2. Autologous Haematopoietic Stem Cell Transplantation (AHSCT) cell banking facilities.

3. Applicability

3.1. DHA licensed health facilities and professionals providing Autologous Haematopoietic Stem Cell Transplantation (AHSCT) services.

4. Standard 1: Health Facility Requirements

4.1. All hospitals opting to provide AHSCT services shall apply to the Health Regulation Sector (HRS) https://www.dha.gov.ae for inspection and licensure.
    4.1.1. AHSCT services shall only be performed in a hospital setting that fulfils the requirements set out in the standard.
        (a) Institutions providing AHSCT treatment should be affiliated with a clinical trial approved by the Dubai Health Authority Ethics Committee 12–18 months from service commencement.

4.1.2. Comply with DHA facility design and administrative provisions for inspection and licensure of clinical labs.

    (a) Ensure designated inpatient unit with adequate space that minimises airborne microbial contamination (isolated-positive pressure room).

       (i) A high-efficiency HEPA filter is required for procedures involving immune-compromised patients.

    (b) There is a written plan for monitoring electrical and mechanical equipment for safety, with monthly visual inspections for apparent defects.

    (c) The lighting and utilities are adequate, including temperature controls, water taps, medical gases, sinks and drains, lighting, electrical outlets, and communications.

4.1.3. The unit should only use the equipment required to provide the AHSCT services following the manufacturer's specifications.

4.1.4. The health facility should ensure easy access to the health facility and treatment areas for all patient groups.

4.1.5. The health facility design should provide assurance of patient and staff health and safety.

4.1.6. The health facility should have the appropriate equipment and trained healthcare professionals to manage critical and emergency cases.

4.1.7. To establish an autologous stem cell transplant service, the health facility should have a clear and defined clinical program that includes protocols for stem cell collection, processing, storage, and transportation before the commencement of AHSCT services.

4.2. Scope of services

4.2.1. Written AHSCT scope of services shall be in place, including but not limited to:

    (a) Donor identification, evaluation, selection, eligibility determination, and management.

    (b) Stem cell collection and apheresis.

    (c) Stem cell mobilisation.

    (d) Administration of the preparative regimen.

    (e) Administration of blood products.

    (f) Central venous access insertion and device care.

    (g) Administration of HPC as well as other cellular therapy products, such as products under exceptional release.

    (h) Management of cytokine release syndrome and toxicities of the central nervous system.

    (i) Transfusion blood products and monitoring of blood counts.

    (j) Infection control and sterilisation for AHSCT.

    (k) Communicable disease testing and management.

    (l) Monitoring infections and use of antimicrobials.

    (m) Disposal of medical and biohazard waste.

    (n) Cellular therapy product storage.

   (o) Safe administration of cellular therapy products.

   (p) Monitoring organ dysfunction or failure and institution of treatment.

   (q) Monitoring graft failure and institution of treatment.

   (r) Management of side effects such as vomiting, nausea, pain, and other discomforts.

   (s) Post-transplant clinic follow-ups.

   (t) Patient education (pre- and post-op procedure and graft failure).

   (u) Medication management.

   (v) Clinical laboratory services.

   (w) Nutrition management.

   (x) Medical equipment management and maintenance.

   (y) Patient safety for radiology and chemotherapy.

   (z) Long-term follow-up, treatment, and plans of care.

  (aa) Palliative care.

  (bb) Rehabilitation.

  (cc) Patient transportation and emergency management.

  (dd) Morbidity and mortality management.

4.3. Laws and regulations

   4.3.1. Compliance with laws and regulations including but limited to:

   4.3.2. Comply with DHA requirements (Regulations, Policy, Standards and Guidelines) and Federal Laws:

   4.3.3. Federal Law No. (14) of 2014—Concerning the prevention of communicable diseases.

   4.3.4. Federal Decree-Law No. (5) of 2016—On the regulation of human organs and tissues transplantation.

   4.3.5. Cabinet Resolution No. (33) of 2016—The Executive Regulations of The UAE Federal Law No. 14/2014—On combating communicable diseases.

   4.3.6. Cabinet Resolution No. (67) of 2020—On concerning the implementing regulation of Federal Law No. (5) of 2019—On the practice of human medicine profession.

   4.3.7. Council of Ministers' Decision No. (6) of the Year 2020—On the endorsement of the regulations for cord blood and stem cells storage centres.

   4.3.8. Cabinet Resolution No. (25) 2020—On regulation for human organs and tissue transplantation.

   4.3.9. Cabinet Resolution No. (28) 2020—On the National Cancer Registry.

  4.3.10. Data and the register must not be held outside the UAE as per IC Law No. 2 of 2019, except in cases mentioned in Article no. (2) of the Ministerial Decision no. (51) of 2021.

  4.3.11. Compliance with the Ministry of Health and Prevention for medical devices, consumables, medication, and medical advertisements.

4.4. Accreditation

4.4.1. The hospital must be accredited as per DHA Policy for Hospital accreditation before the commencement of the service.

4.4.2. The hospital lab must be accredited as per DHA Policy for Clinical Lab before the commencement of service.

4.4.3. The health facility should have a Quality Management System (QMS) as 'an organization's comprehensive quality assessment, assurance, control, and improvement system'.

    (a) An action plan for improvement shall be submitted to DHA for review before the commencement of service.

4.4.4. The service shall achieve and comply with FACT-JACIE International Standards for cellular therapy, product collection, processing and administration, storage and collection accreditation 24 months from licensure activation.

    (a) Center for International Blood and Marrow Transplant Research (CIBMTR), FACT clinical inspectors should audit the clinical programs.

    (b) Adhere to FACT-JACIE for personnel, quality management, policies and SoPs, equipment supplier, reagents, coding, and labelling of cellular therapy, process controls, cellular therapy product storage, transportation, shipping, distribution and recipient, disposal.

4.5. In house lab setup and diagnostics

4.5.1. Equipment and supplies for a stem cell processing lab are set out in Appendices BI and BJ.

    (a) Storage of cells in sealed vials, cryobags, or cryopreserved containers for haematopoietic progenitor cells shall meet UAE Ministry of Health and Prevention (MoHaP) requirements.

    (b) Backup equipment shall be identified where there is only one device is in use.

    (c) All essential equipment shall be connected with an uninterruptible emergency power supply.

    (d) All product contact reagents should be sterile and infusion-grade, and disposable.

    (e) Reagents should be dispensed into single-use containers before use to minimise waste.

    (f) All reagents and supplies must be inspected, and lot numbers recorded before use and stored in a controlled environment, separate from non-clinical, potentially harmful research reagents.

4.5.2. Tests, diagnostics, and procedures required for AHSCT include but are not limited to:

    (a) Tissues culture.

    (b) Immunophenotyping.

    (c) Special stains to evaluate iron storage in the marrow for abnormal erythroid (RBC) precursor with iron particles surrounding the nucleus, chromosomal analysis, and fluorescence in situ hybridisation analysis.

       (d) Necessary molecular and cytogenetic tests as per international guidelines such as T-cell receptor gene rearrangement, B-cell immunoglobulin gene rearrangement, JAK2 mutation, BCR-ABL, PML-RARA.

       (e) Routine blood tests.

       (f) Bone marrow aspirate and biopsy.

       (g) Blood transfusion.

       (h) Apheresis.

       (i) Bronchoscopy.

       (j) CT scan, MRI scan, X-ray, and ultrasound.

       (k) Electrocardiogram (ECG) and echocardiogram

       (l) Endoscopy.

       (m) Hickman® Line Insertion.

       (n) Liver biopsy.

       (o) Lumbar puncture.

       (p) Pulmonary function test.

       (q) Urine test.

       (r) Positron emission tomography scan.

       (s) Sperm and ova banking (if not done previously).

4.6. There should be a mechanical freezer capable of storing a liquid nitrogen tank equipped with an audible alarm.

    4.6.1. Self-pressurising dewars should be in place for a regular supply of liquid nitrogen from the main storage tank.

    4.6.2. The space containing the liquid nitrogen storage tanks and supply dewars should be separated from the processing laboratory needs.

    4.6.3. The tanks should have sufficient air handling capacity to maintain safe oxygen levels when the liquid nitrogen (N2) tanks are filled.

    4.6.4. An oxygen sensor alarm to indicate when oxygen levels are dangerously low.

    4.6.5. A temperature sensor should be fitted to track and temperature at least twice a day.

    4.6.6. Adequate backup liquid (or vapour) nitrogen storage capacity should be in place.

5. Standard 2: Healthcare Professional Requirements

5.1. The Privileging Committee and Medical Director of the health facility shall privilege clinical staff in line with his/her education, experience, training, and competencies.

    5.1.1. The privileges shall be granted or removed as per DHA policy for clinical privileging.

5.2. Only a DHA licensed consultant trained to provide AHSCT shall lead the AHSCT service as the Clinical Program Director.

5.2.1. The Consultant shall be in trained in haematology, immunology, or medical oncology with specialty training in Autologous Haematopoietic Stem Cell Transplant (AHSCT):

    (a) The training program should include a minimum of ten (10) successfully completed cases during training.

5.2.2. A consultant with specialty training shall have documented evidence and experience in the field of Haematopoietic Stem Cell Transplantation (HPC) transplantation for a minimum of (5) years post-training.

    (a) The Clinical Program Director must submit evidence of a minimum of ten (10) successful completed cases per year.

5.2.3. The Clinical Program Director must submit evidence of forty (40) CME credits for Autologous Haematopoietic Stem Cell Transplant (AHSCT) per year as per UAE PQR requirements for consultants.

5.3. The Clinical Program Director shall take responsibility for the direct clinical management of HPC transplant patients in inpatient and outpatient settings.

5.4. The Clinical Program Director shall take responsibility for the design, service, and elements of the clinical program. This includes quality management, the selection and care of recipients and donors, and cell collection and processing, whether internal or contracted services including:

5.4.1. All technical procedures.

5.4.2. Performance of the marrow collection procedure.

5.4.3. Supervision of staff.

5.4.4. Administrative operations.

5.4.5. The medical care of autologous donors undergoing marrow collection.

5.4.6. Pre-collection evaluation of autologous donors at the time of donation.

5.4.7. Care of complications resulting from the collection procedure.

5.4.8. The Quality Management Program, including compliance with federal and local regulations.

5.4.9. Evaluations of competence shall be performed before the independent performance of assigned activities and at specified intervals.

5.5. The Clinical Program Director will be responsible for the clinical supervision of physicians and nursing staff and ensure they have a valid and up-to-date certification and training to fulfil the service, including the minimum CME requirements as per UAE prequalification requirement in the past 12 months.

5.6. The Clinical Program Director will ensure all attending physicians:

5.6.1. Have a minimum, 1 year of supervised training. The training shall include the management of transplant patients in both inpatient and outpatient settings.

5.6.2. Clinical training and competency should include the management of autologous transplant recipients.

5.6.3. Evaluations of competence shall be performed for independent performance of assigned activities and at specified intervals.

5.7. The Clinical Program Director shall ensure all AHSCT staff hold written evidence that they have met the service's expected training and competency requirements (Appendix BK).

5.8. Nurses shall be trained on:

5.8.1. Haematology/oncology patient care and cellular therapy process.

5.8.2. Administration of preparative regimens.

5.8.3. Administration of growth factors, blood products, cellular therapy products, and other supportive therapies.

5.8.4. Care interventions to manage cellular therapy complications. This includes and may not be limited to respiratory distress, cardiac dysfunction, tumour lysis syndrome, cytokine release syndrome, neurologic toxicity, macrophage activation syndrome, hepatic and renal failure, disseminated intravascular coagulation, anaphylaxis, neutropenic fever, infectious and non-infectious processes, mucositis, pain management, and nausea and vomiting.

5.8.5. Recognition of emergencies and cellular therapy complications requiring rapid notification of the transplant team.

5.8.6. Palliative and end of life care.

5.9. There shall be written standard operating nursing procedure, including but not limited to:

5.9.1. Care of immunocompromised recipients.

5.9.2. Age-specific considerations.

5.9.3. Administration of preparative regimens.

5.9.4. Administration of cellular therapy products.

5.9.5. Administration of blood products.

5.9.6. Central venous access device care.

5.9.7. Detection and management of immune effect or cellular therapy complications.

5.9.8. Trained to operate the apheresis machine and collection of stem cells and storage.

5.10. Pharmacists shall be trained on:

5.10.1. Haematology and oncology patient care, including the process of cellular therapy.

5.10.2. Adverse events including neurological toxicities and cytokine release syndrome.

5.10.3. Therapeutic drug monitoring shall include but not be limited to anti-infective agents, immunosuppressive agents, anti-seizure medications, and anticoagulants.

5.10.4. Monitoring and recognition of drug/drug and drug and food interactions and necessary dose modifications.

5.10.5. Recognition of medications that require amendment for organ dysfunction.

5.10.6. Conditioning regimens (chemotherapy, monoclonal antibody therapy, and radiation to the entire body).

5.11. AHSCT services shall have the minimum number of healthcare professionals for set up of the service detailed below:

    5.11.1. A Clinical Program Director.

    5.11.2. Facility Medical Director.

    5.11.3. Attending physician (Consultant and specialists in haematology, immunology, oncology, or genetics).

    5.11.4. Multidisciplinary support team.

    5.11.5. A case manager.

    5.11.6. An administrator.

    5.11.7. Two registered nurses.

    5.11.8. Two lab technicians/technologists.

    5.11.9. A clinical pharmacist.

    5.11.10. A ward manager.

    5.11.11. Nurse patient care coordinator.

    5.11.12. Health educator.

    5.11.13. A quality assurance manager.

    5.11.14. Infection control lead.

5.12. Other medical consultants and specialists for a multidisciplinary team shall be available as per patient need: Critical Care, Surgery, Haematology, Oncology, Radiology, Gastroenterology and Histopathology, Pathology, Transfusion Medicine, Dermatology, Dentistry, Internal Medicine, Endocrinology, Nephrology, Cardiology, Pulmonology, Reproductive Medicine, Infectious Diseases, Dietetics, Occupational Therapy, Psychology, Psychiatry and Palliative Care.

6. Standard 3: Permitted Indications for Autologous HSCT

6.1. Inclusions

    6.1.1. Autologous transplant patients for indications within the 'Standard of Care' and 'Clinical Option' as per established clinical practice such as American Society for Transplantation and Cellular Therapy (ASTCT) guidelines on Indications for Haematopoietic Cell Transplantation and Immune Effector Cell Therapy, The European Group for Blood and Marrow Transplantation (EBMT), and the British Society of Blood and Marrow Transplantation (BSBMT).

    6.1.2. Patient health status and overall benefit versus harm should be considered for:

    (a) Repeat transplant patients for failure to engraft.

    (b) Repeat autologous transplants for relapsed disease.

    6.1.3. Non-urgent cases.

    6.1.4. Patients aged 18 years or above.

    6.1.5. Planned tandem transplants (sequential or double transplant) following patient risk score assessment, functional status, and prognosis on using chemotherapies such as bortezomib, lenalidomide, and thalidomide and approval by the Clinical Program Director.

6.2. Exclusions

6.2.1. Allogeneic transplants.

6.2.2. Transplants for indications within the category of 'Developmental' and 'Generally Not Recommended'.

6.2.3. Patients under the age of 18 years.

6.2.4. Emergency cases.

6.3. Use of non-Autologous Haematopoietic Stem Cells.

6.4. Use of double or multiple umbilical cord cells that are not from the same individual.

6.5. Sale, storage, or use of autologous stem cells for any other person(s) who is not the same patient/individual is not permitted.

6.6. Transfer of autologous haematopoietic stem cell in or out of the health facility or Dubai is not permitted. Written approval shall be sought by the competent regulator (DHA or MoHaP).

7. Standard 4: Autologous HSCT Service Requirements

7.1. The service shall adhere to the following:

7.1.1. Written scope of service that is kept up to date.

7.1.2. Documented roles and responsibilities of all staff.

7.1.3. Adherence to ISBT128 standards terminology, identification, coding and labelling (https://www.iccbba.org/home) or Eurocode.

7.1.4. Ensure there is a register for autologous haematopoietic stem cell transplantation that is maintained.

7.1.5. Commencement of a clinical trial within 12–24 months.

7.1.6. Minimum expected number of procedures per year for the quality and safety of the AHSCT Clinical Program:

(a) Five (5) procedures in year one (1).

(b) Five (5) procedures in year two (2).

(c) Ten (10) procedures in year three (3) and thereon.

7.2. Data management and record keeping.

7.2.1. There shall be policies and procedures for all critical electronic record systems to assure the accuracy, integrity, security, and confidentiality of all records.

7.2.2. The Clinical Program shall collect all the data necessary to complete the transplant essential data forms as per the standards set by the Center for International Blood and Marrow Transplant Research (CIBMTR) or the Minimum Essential Data-A requirements of the European Society for Blood and Marrow Transplantation (EBMT).

7.2.3. The Clinical Program shall have in place records for facility maintenance, facility management, complaints, or other general facility issues, quality control, personnel training, and competency.

7.2.4. Patient records, including, but not limited to consent and records of care, should be maintained confidentially as per UAE Law.

7.3. The service should have policy and procedures supported by documentation for the following:

7.3.1. Patient acceptance criteria.

7.3.2. Investigational treatment protocols.

7.3.3. Patient assessment and admission.

7.3.4. Pregnancy testing.

7.3.5. Patient education and informed consent (Appendix BL).

7.3.6. Patient health record.

7.3.7. Pre and post collection care.

7.3.8. Cell collection, processing storage, transportation, and banking.

7.3.9. Conditions and duration of cellular therapy product storage as well as the indications for disposal.

7.3.10. Good tissue manufacturing practice and cell processing.

7.3.11. Use of equipment, supplies and reagents.

7.3.12. Coding, labelling, verification, and tracing of cellular therapy products.

7.3.13. Available therapies and treatment protocols.

7.3.14. Medication management.

7.3.15. Incident reporting.

7.3.16. Patient privacy.

7.3.17. Post-transplant vaccination schedules and indications.

7.3.18. Emergency action plan.

7.3.19. Patient discharge/post-op care/transfer.

7.3.20. Transfer of critical/complicated cases when required.

7.3.21. Quality improvement and control (including outcome at 100 days, 1 year and 5 years).

7.3.22. Cellular therapy emergency and disaster plan, and the clinical program response.

7.3.23. Patient complaint management.

7.3.24. Sentinel, adverse events, and adverse reaction reporting.

7.3.25. Disposal of biological and medical waste as per Dubai Municipality (DM) requirements.

7.4. Infection control program for monitoring and managing infectious processes, including immune-deficiencies and opportunistic infections, central venous catheter infection, and potential patient infections. The program shall assure:

7.4.1. Monitoring of infections and use of antimicrobials.

7.4.2. Blood samples for testing for evidence of clinically relevant infection shall be drawn, tested and reported within timeframes required by local and federal regulations.

7.4.3. Implement post-procedure infection control measures.

7.4.4. Document infection control measures and hazardous waste management.

7.4.5. Compliance with hygiene and use of attire for personal protective equipment.

7.5. The service should maintain the Charter of Patient Rights and Responsibilities at the facility entrances in two languages (Arabic and English).

    7.5.1. Patients have the right to know the percentage of viable cells in the collected samples and estimated success rate over the short- and long-term basis.

8. Standard 5: Stem Cell Collection, Processing, Storage, Transportation, and Banking

8.1. Stem cells shall be collected in a sterile environment.

    8.1.1. Infection control measures should include but not be limited to:

        (a) Processing in clean areas and thorough microbiologic monitoring of all stages of the stem cell preservation procedure as per best practice.

        (b) Screening for microbiologic contamination before cell collection and infusion.

        (c) There should be separate or protected cellular storage to avoid cross-contamination where an infectious graft has been detected.

8.2. Processing of cells should be undertaken within 48 h at a controlled temperature as per the latest evidence-based practice.

    8.2.1. Centrifugation shall be used to achieve the minimum number of cells required for the patient.

    8.2.2. Cells shall be counted (CD34+ cell count), assessed for viability and sterility, and preliminary stored continuously in the recommended controlled temperature (initially -4 °C).

8.3. The sample can be frozen in a controlled manner down to the target temperature of -156 °C (vapour phase) to -196 °C (liquid phase) for longer term storage.

    8.3.1. Cells should be cryopreserved by methods and reagents detailed in FACT-JACIE International using reagents approved for human use in the UAE.

    8.3.2. Assessment of the frozen cells should be performed after 72 h.

    8.3.3. The sample can be thawed in a 37 °C water bath.

    8.3.4. After thawing, cryopreservatives should be washed using a two-step approach through centrifugation to reduce the toxicity of reagents.

    8.3.5. Reassessment of cell viability should be performed to ensure its integrity before stem cell infusion.

8.4. Cell collection, processing, and administration should fulfil the FACT-JACIE International Standards for haematopoietic cellular therapy product collection, processing, and administration requirements.

8.5. Cells that require transportation shall:

    8.5.1. Have an agreement and clear process between the sender and receiver.

        (a) A person (courier) shall be available to accompany the stem cells between the sender and receiver.

    (b) The courier shall be trained for stem cells transportation and verified by the sender or receiver.

        (i) The cell must not be placed in cargo for transportation and should be transported as hand luggage.

    (c) Stem cells must not be exposed to X-ray machines or metal detectors.

    (d) Cells must be placed in the best practice optimal medium to maintain cell viability and ensure cell characteristics are not altered.

8.5.2. Have in place a courier tracking mechanism to determine the status of the cells being transported.

8.5.3. Ensure cells are placed in a credo box that is prepared to 4 °C.

    (a) The credo box should be checked prior for integrity to maintain a controlled temperature of 2–8 °C for 100 h.

    (b) There should be two temperature loggers, and temperature readings should be taken every 15 min.

    (c) The credo box shall be sealed to prevent tampering during transportation.

    (d) Have in place a tracking mechanism to determine the status and position of the cells being transported.

    (e) The credo box shall include labels identifying the product being transported.

8.5.4. Cell transportation should not exceed 72-h to prevent an adverse event.

8.5.5. Transported cells must be documented and coded at both the sending and receiving sites and confirmed by both sites before infusion.

8.6. For stem cell banking, the health facility shall adhere to best practices such as the FACT-JACIE international standards for haematopoietic cellular therapy product collection, processing and administration, and NetCord-FACT International Standards for Cord Blood Collection, Banking, and Release for Administration.

8.6.1. The cell banking system should have written documentation for:

    (a) Cell banking procedures to include reagents, temperature controls, and maintenance of medical equipment and devices.

    (b) Cell types and sizes are being managed.

    (c) Containers, vessels, and closure system used.

    (d) Methods of cell preparation, cryopreservation technique.

    (e) Safe use of reagents and protectants.

    (f) Cell storage and thawing technique.

    (g) Transportation and disposal of medical waste.

    (h) Procedures used to prevent microbiological contamination and cross-contamination and tracing.

    (i) Documentation and labelling procedures.

    (j) Back up and business continuity and recovery from catastrophic events.

    (k) Cell testing technique.

(l) Testing for mycoplasma and sterility before the transfer of cells into the facility.

   (i) Bacteriostasis and fungistasis testing should be performed before sterility testing to assess the sample matrix for inhibition.

(m) Testing program and the schedule should include but not be limited to testing for:

   (i) Species-specific virus (2 weeks).

   (ii) Sterility (2.5 weeks).

   (iii) Mycoplasma testing (3.5 weeks).

   (iv) Retroviruses and animal viruses (5 weeks).

   (v) Adventitious virus (6 weeks).

   (vi) Antibody production (7 weeks).

8.7. The cell banking facility shall ensure:

   8.7.1. Patient consent is obtained, and patients are informed of all costs and timelines to reaffirm consent.

   8.7.2. Patients are informed of the cell quality controls, validation, viability, sterility, count, and cell typing when cells are needed.

   8.7.3. Patients are informed of the site for storing stem cells and any third party agreements.

   8.7.4. Patients are informed of protocols to ensure data confidentiality and privacy.

9. Standard 6: Safety and Quality Requirements

To assure quality and patient safety, the service shall ensure the following:

9.1. A multidisciplinary team is available to manage the patient needs.

9.2. Patient escort and access to emergency services.

9.3. Supply of immunosuppressants is available for the duration of planned therapy.

9.4. Intensive Care Unit (ICU) beds and isolation room are available for patients undergoing AHSCT transplantation.

9.5. Written agreements with suppliers, blood banks, and tertiary hospitals to ensure patient safety and quality of care are not compromised.

   9.5.1. Twenty-four-hour availability of appropriate and irradiated blood products needed to care for cellular therapy recipients.

   9.5.2. Irradiated blood products for patients should be given (if needed) 7 days before transplant and up to 3 months after (unless there are other reasons to continue).

   9.5.3. Patients should be given written information and alert card (if available), and the Dubai Blood Bank should be informed.

9.6. Chemotherapy and radiation are managed in line with the minimum international thresholds to assure reduced-intensity transplantation.

9.7. Infection control measures are robust and monitored regularly.

9.8. Patients and their close family members should take the PCR test 72 h before admission.

9.9. Patients and their close family members should take the Covid-19 vaccine (or booster) post auto graft.

9.10. Patients undergoing autologous transplant should be vaccinated but live vaccines should not be given. Vaccination schedule (doses and months between doses) should be followed as per the latest international guidance: (CDC/WHO):

9.10.1. Influenza A and B inactivated seasonal vaccine.

(a) Recipients aged 65 years and over should receive the adjuvanted trivalent influenza vaccine (aTIV).

(b) The live-attenuated influenza vaccine (Fluenz Tetra®) must NOT be given to transplant recipients. Household members should also receive an inactivated influenza vaccine as there is theoretical potential for transmission of live-attenuated influenza virus in Fluenz Tetra® to immunocompromised contacts for 1–2 weeks following the vaccination 2.

9.10.2. Diphtheria/tetanus/pertussis/inactivated polio/Haemophilus influenzae type b/Hepatitis B (DTaP/IPV/Hib/HepB) hexavalent vaccine.

9.10.3. Meningococcal Group B (Men B) multicomponent protein vaccine.

9.10.4. Meningococcal Groups A, C, W & Y (Men ACWY) quadrivalent conjugate vaccine.

9.10.5. Pneumococcal (Streptococcus pneumoniae) Prevenar 13®, 13 valent conjugate vaccine (PCV13) and for the subsequent dose Pneumovax II®, 23-valent, polysaccharide vaccine (PPSV23).

9.10.6. Measles/Mumps/Rubella (MMR) live-attenuated vaccine should not be given to autologous transplant recipients.

9.11. Appropriate sedation is provided for iliac crest bone marrow harvest and to manage post-transplant complications.

9.12. Medications to manage symptoms subject to patient profile and risk.

9.13. Growth factors for neutrophils should be used to prevent infection and fungus during the low count and engraftment phase and early and late convalescence.

9.14. Adequate anticoagulants should be in place to avoid cell aggregation for storage and transportation for long periods (24–72 h).

9.15. Cellular processing and storage/cryopreservation are controlled in the laboratory does not compromise the quality, quantity, and efficacy of AHSCT.

9.15.1. Cryopreservation initial temperature −4 °C.

9.15.2. −156 °C when stored in the vapour phase.

9.15.3. −196 °C when stored in the liquid phase, depending on where the specimen is stored in the container.

9.16. Cell typing is confirmed before infusion.

9.17. Pre-care, treatment, and aftercare program is comprehensively aligned to best practice to meet patient needs.

10. Standard 7: Pre-transplant Period

The pre-transplant period forms an essential part of identifying suitability for patients to benefit from ASHCT. Pre-transplant workup will include assessing eligibility for transplantation, tissue investigations, and assessing the patient's fitness (Appendices BM and BN).

10.1. A detailed medical history of the patient and testing should be taken for all patients indicated for AHSCT, and the European Medical Blood and Marrow Transplant (EMBT) scoring system should be adopted to inform clinical decisions and protocol for treatment. The findings from EBMT should be discussed with the transplant team and recorded in the patient's medical file.

10.2. The test should include but not be limited to:

10.2.1. The patient's age, fitness status, previous and current disease status, therapies, relapse, drug intake, and prior surgical procedures should be taken.

10.2.2. Patient profile and suitability for AHSCT should be considered as per the available evidence base and consensus.

10.2.3. Disease criteria for bone marrow transplant should be met as per clinical best practice.

(a) Screening for infectious disease shall be undertaken as per the health facility infectious disease protocols.

10.2.4. The intensity of treatment required and stem cell source.

10.2.5. Contraindication and their absence should be considered.

10.3. The patient should undergo several pre-diagnostic tests before admission, including but not limited to a dental exam, cardio pulmonary exam, thyroid, and dietary changes. Computed Tomography (CT) scan and gynaecological exam should be done where indicated.

10.4. Blood work and urine tests should be performed to assess the blood cells' status, infectious disease status, and liver and kidney function.

10.5. Referral to reproductive medicine (for storage of ova or sperm) should be done as chemotherapy and radiation may affect family planning.

10.6. Counselling and psychological services should be offered to the patient to prepare the patient and manage emotional stress.

10.7. Treatment options and duration should be discussed with the patient (and next of kin where available), including risks and recorded in the patient's medical file.

10.8. Care coordination and the medical care plan should be discussed and agreed upon with the transplant team and approved by the Clinical Program Director.

10.9. Preparation for stem cell collection should be undertaken once CD34 levels have been achieved.

10.10. Use of a central line or Hickman line insertion for peripheral blood stem cell collection/harvesting should be done before chemotherapy and/or irradiation.

10.10.1. Patients should be managed for toxicities, symptoms, and side effects.

10.11. The conditioning regimen should be done for 4 days with Grannis Colony Stimulating Factor (GICSF).

    10.11.1. Apheresis machine should be utilised for stem cell collection only, and the volume should align with the patients' weight calculation (750–1000 mL). The buffy coat with white cells (haematopoietic stem cells) should be separated and placed into a collection bag.

        (a) The plasma and red blood cells should be counted (CD34+ cell count) for a viable transplant and returned to the patient to minimise blood loss.

10.12. Peripheral blood stem cells in the collection bag shall be labelled, processed (typing, nucleic sub count, culture), weighed, processed, and cryopreserved.

10.13. If bone marrow harvest is pursued, it should be prepared and conditioned for transplantation as per best practice protocols (immunosuppression, growth factors, and myeloablation).

    10.13.1. Sedation and aseptic techniques must be met for bone marrow harvest.

    10.13.2. Use of anticoagulation should be administered to prevent clotting.

    10.13.3. Bone marrow harvest (iliac crest aspiration) should align to weight calculations and required stem cell volume (10–20 mL/kg).

    10.13.4. Disposable needles should be used for the punch biopsy.

    10.13.5. Imaging should be used to guide the biopsy needle.

10.14. All bone marrow stem cells that are collected from the patient shall be maintained in a collection bag, labelled, weighed, processed (typing, nucleic sub count, culture), and labelled in a laboratory according to clinical need within five (5) to ten (10) day turnaround for all patients.

10.15. Media such as Normasal-R (electrolytes and glucose) should be used to maintain cell metabolism.

11. Standard 8: Transplant Period (Infusion)

11.1. Stem cells should be thawed at the bedside in a water bath and intravenously infused to transplant and engraft the stem cells.

    11.1.1. Stem cell infusion should be done slowly to minimise reactions.

        (a) Side effects such as vomiting, abdominal cramp, nausea, chills, chest pain, and passing red urine should be managed and documented.

    11.1.2. Patients should be monitored during the recovery period to ensure sufficient neutrophils are in place to minimise the risk of infection.

    11.1.3. An aftercare program should be developed with the patient and their primary care practitioner should be updated on the treatment and aftercare plan.

    11.1.4. Patients and/or next of kin should be updated regularly and provided with the required aftercare information.

12. Standard 9: Post-transplant Period

Engraftment is expected 10–15 days post-transplant and may vary according to the transplant, patient, complications, and late effects. The transplant team should ensure:

12.1. The timeframes for anticipated engraftment and follow-up are documented.
12.2. There is a dedicated registered nurse in the transplant inpatient and outpatient areas trained (knowledgeable and skilled) to monitor vital signs, fluid, and electrolyte balance and implement the treating physicians' instructions to manage potential complications. Moreover, side effects related to infection, drugs, or stem cell transplantation.
    12.2.1. The nurse should monitor the patients' health status and follow emergency procedures issued by the treating physician.
12.3. Blood tests are undertaken to verify engraftment and graft-versus-host disease status.
12.4. Patient discharge is done once written approval is issued by the treating physician and clinical director.
    12.4.1. Patients who have been approved for discharge should be issued with a discharge plan in a non-technical manner, supported by verbal explanation to assist the patient and their nominated caregiver in understanding the care plan, and the availability of outpatient services to meet the patients' needs.
    12.4.2. The discharge plan should include:
        (a) Drug management to manage potential complications.
        (b) Key contact numbers to seek advice on symptoms or side effects.
        (c) Precautionary measures and advice to prevent community infections should be issued by the treating physician and infection control lead for common infections:
            (i) Month 1—Herpes Simplex (HSV1/2), bacterial and fungal infections.
            (ii) Months 2–3—Cytomegalovirus (CMV), fungal infection, pneumocystis and carinii pneumonia (PCP).
            (iii) Months 0–12—Varicella-Zoster Virus (VZV) infection.
            (iv) Months 3–6—Home infection control measures, e.g. replacing air condition filters, removing plants, hand hygiene, dental hygiene, healthy lifestyles, Personal Protective Equipment (PPE), and avoidance of public places.
        (d) Advice on vaccines and use of over the counter medications.
        (e) Follow-up appointments at regular intervals to assess:
            (i) The efficacy of the treatment and relapse.
            (ii) Potential second malignancies such as organ dysfunction or myelodysplastic syndrome.
            (iii) Post-AHSCT vaccination protocol (including close family vaccination).
            (iv) Long-term post-AHSCT complication follow-up.

13. Standard 10: Key Performance Indicators

13.1. The health facility should capture performance measures for each patient and for the AHSCT program (Tables BH.1, BH.2, and BH.3).

13.2. Performance measures should be readily available upon request

13.2.1. The provider is required to report on any additional performance requirements or measures issued by DHA.

13.3. Reports should reflect outcomes achieved in the previous quarter.

13.4. The Clinical Director should ensure that all treating physicians maintain an up-to-date log of treatment and patient outcomes using validated tools.

13.4.1. The service should follow up with patients at frequent intervals to determine patient outcomes and success rates, and remission status (1 and 5 years).

13.4.2. Follow up of patient outcomes and reporting should be done as soon as patient complications have been resolved.

13.4.3. Adverse and sentinel events should be logged and reported to the Medical Director.

Table BH.1 Baseline assessment

Patient ID	Diagnosis	Date of intervention	Tool for Assessment (e.g. pain score, biological testing, cardiology, pulmonology and renal physicians and tests including echo, pulmonary function tests and renal function tests, ADL, patient experience)	Baseline Assessment	Treatment Method, dosage and start date Transplant protocol for intravenous infusions of chemotherapy and stem cells)	Primary outcome measure

Name and signature of Treating Physician:

Name and signature of Clinical Program Director:

Table BH.2 Post-intervention assessment

Patient ID	One-month outcome	Adverse Event at one month (No, if yes provide details)	Three-month outcome	Adverse Event at three months (No, if yes provide details)	Six-month outcome	Adverse Event at six months (no, if yes provide details)	12-month outcome	Adverse Event at 12 months (no, if yes provide details)

Name and signature of Treating Physician:

Name and signature of Clinical Program Director:

Table BH.3 Service performance measures for AHSCT program (adopted from Aljurf et al. 2021 and NHS 2017. Specialised services quality dashboards—blood and infection metric definitions for 2017/18)

- Clinical program:
  - Clinical indicator collection indicator processing indicator
  - Number of SCT-certified physicians
  - Number of SCT-certified nurses
  - Number of oncology certified nurses
  - Number of publications
  - Cancellations
  - Incidents reports
  - Number of medication errors
  - Patient volume
  - Bed capacity
  - Outpatient clinic capacity
  - The average length of hospital stay for inpatient transplants
  - Indication of AHSCT
  - Overall survival and mortality
    - Survival rate at day 100
    - Survival rate at 1 year
    - Survival rate at 5 years
    - Treatment-related (non-relapse mortality)
  - Engraftment outcome
    - Engraftment by type of HCT and source of stem cells, ANC, and platelet count
    - Median time to engraftment
    - Graft failure outcome
  - Infections
    - Central venous catheter site infections
    - Percentage of microbial contaminations
  - Outcome readmission rate
  - Number of HCT patient ED visits
  - Staff satisfaction
  - Patient satisfaction
- **Stem cell collection program**
  - Number of trained stem cell collection and apheresis staff
  - Number of autologous products
  - Number of stem cell infusion
  - HCT complications during the collection procedure
- **Processing laboratory program**
  - Number of trained cell processing staff
  - Quality of collected product (CD34 quantitation)
  - SC processing turnaround time
  - Number of acceptable HPC viability cells post-cryopreservation
  - Number of available SC processing reagents

References

1. Agency for Healthcare Research and Quality. Hematopoietic stem-cell transplantation in the pediatric population. Effective Healthcare Program. Comparative Effectiveness Review Number 48. 2012.
2. Aljurf M, et al. Quality management and accreditation in hematopoietic stem cell transplantation and cellular therapy: the JACIE guide. 2021. Available on: https://link.springer.com/book/10.1007/978-3-030-64492-5#toc. Accessed 2 Aug 2021.
3. Australian Government. Stem cell treatments: a quick guide for medical practitioners. National Health and Medical Research Centre; 2013.
4. Barban JB, et al. Brazilian nutritional consensus in hematopoietic stem cell transplantation: adults. Einstein (Sao Paulo, Brazil). 2020;18:AE4530. https://doi.org/10.31744/einstein_journal/2020AE4530.
5. Bone Marrow Transplant Program. Guidelines. A component of the haemopoietic progenitor cell program. Technology Assessment and Access Division. Department of Health; 2018.
6. British Society of Blood and Marrow Transplantation and Cellular Therapy. BSBMT transplant trainee fellow curriculum. 2012. Available on: https://bsbmtct.org/documents-publications/. Accessed 28 Jun 2021.
7. Cabinet Decision No. (6) of 2020 Approving controls for the work of umbilical cord blood and stem cells storage centres.
8. Cabinet Resolution No. (25) of 2020 on Regulation of Human Organs and Tissue Transplantation.
9. Careras E, Dufour C, Mohty M. The EBMT handbook. Hematopoietic stem cell transplantation and cellular therapies. Springer Open; 2010.
10. Daniele N, Zinno F. Quality controls of cryopreserved hematopoietic stem cells. J Bone Marrow. 2015;3:1. https://doi.org/10.4172/2329-8820.1000157.
11. Cord Blood Association. Model criteria for regulation of cord blood banks and cord blood banking: The Cord Blood Association, Board of Directors, January 29, 2019. Stem Cells Transl Med. 2019;8(4):340–3. https://doi.org/10.1002/sctm.cbmc.
12. European Society for Blood and Marrow Transplantation. EBMT book. 2019. Available on: https://www.ebmt.org/education/ebmt-handbook. Accessed 28 Jun 2021.
13. European Society for Blood and Marrow Transplantation. EBMT textbook for nurses 2018. https://www.ebmt.org/ebmt-nurses-textbook. Accessed 28 Jun 2021.
14. Federal Decree No. (5) of 2016—Concerning regulating the transfusion and transplantation of human organs and tissues.
15. Foundation for the Accreditation of Cellular Therapy. International standards for hematopoietic cellular therapy product collection, processing, and administration. 8th ed. University of Nebraska Medical Centre; 2018.

16. Foundation for the Accreditation of Cellular Therapy. NetCord-FACT international standards for cord blood collection, banking, and release for administration. 7th ed. 2020.

17. Healthcare Costing for Value Institute. Introduction to healthcare outcomes. Healthcare Costing for Value Institute; 2016. Available on: https://www.hfma. org.uk/docs/default-source/our-networks/healthcare-costing-for-value-institute/institute-publications/introduction-to-health-outcomes. Accessed 28 Jul 2021.

18. Henig I, Zuckerman T. Hematopoietic stem cell transplantation-50 years of evolution and future perspectives. Rambam Maimonides Med J. 2014;5(4):e0028. https://doi.org/10.5041/RMMJ.10162.

19. Hołowiecki J. Indications for hematopoietic stem cell transplantation. Polskie Archiwum Medycyny Wewnetrznej. 2008;118(11):658–63.

20. Federal Law No. (2) of 2019, Concerning the use of ICT in health fields.

21. Snowden JA, et al. (2018). Autologous haematopoietic stem cell transplantation (aHSCT) for severe resistant autoimmune and inflammatory diseases—a guide for the generalist. Clin Med. 2018;18(4):329–34.

22. Snowden J, et al. Benchmarking of survival outcomes following haematopoietic stem cell transplantation. A review of existing processes & the introduction of an international system. The European Society for Blood and Marrow Transplantation (EBMT) and the Joint Accreditation Committee of ISCT & EBMT (JACIE). Bone Marrow Transplant. 2020;55(4):681–94. https://doi. org/10.1038/s41409-019-0718-7.

23. Kanate AS, et al. Indications for hematopoietic cell transplantation and immune effector cell therapy: guidelines from the American Society for Transplantation and Cellular Therapy. Biol Blood Marrow Transplant. 2020;26(7):1247–56. https://doi.org/10.1016/j.bbmt.2020.03.002.

24. Kumar S. Stem cell transplantation for multiple myeloma. Curr Opin Oncol. 2009;21(2):162–70. https://doi.org/10.1097/CCO.0b013e328324bc04.

25. Khaddour K, Hana CK, Mewawalla P. Hematopoietic stem cell transplantation. StatPearls Publishing; 2020.

26. Leemhuis T, et al. Essential requirements for setting-up a stem cell processing laboratory. Bone Marrow Transplant. 2014;49:1098–105. https://doi. org/10.1038/bmt.2014.104.

27. Leukaemia and Lymphoma Society. Blood and marrow stem cell transplantation guide. Providing the latest information for patients and caregivers. 2019.

28. Majhail NS, et al. Indications for autologous and allogeneic hematopoietic cell transplantation: guidelines from the American Society for Blood and Marrow Transplantation. Biol Blood Marrow Transplant. 2015;21(11):1863–9. https://doi.org/10.1016/j.bbmt.2015.07.032.

29. Malaysian Medical Council. Stem cell research & stem cell therapy. Guideline of the Malaysian Medical Council. MMC Guideline 002/2009. 2009.

30. Ministry of Malaysia. National organ, tissue and cell transplantation policy. Surgical and Emergency Services Unit, Medical Services Section, Medical Development Division, Ministry of Malaysia; 2007.

31. Ministerial Decision No. (51) of 2021—Concerning the health data and information which may be stored or transferred outside the country.
32. Ministry of Health Malaysia. National standards for stem cell transplantation: collection processing, storage, and infusion of haemopoietic stem cells and therapeutic cells. Clinical Support Services Unit, Medical Development Division; 2009.
33. NHS England. National health service contract for haematopoietic stem cell transplantation (adult). Service Specification B04/S/a. 2014.
34. NHS England. Clinical commissioning policy haematopoietic stem cell transplantation (all ages), revised. NHS England B04/P/a. 2015.
35. NHS. Specialised services quality dashboards—blood and infection metric definitions for 2017/18. 2017. Available on: https://www.england.nhs.uk/publication/specialised-services-quality-dashboards-blood-and-infection-metric-definitions-for-201718/. Accessed 2 Aug 2021.
36. Oxford University Hospitals. Immunisation schedule for autologous and allogeneic blood and marrow transplant recipients. Department of Clinical Haematology Oxford BMT Programme; 2021.
37. Passweg JR, et al. Hematopoietic stem cell transplantation. A review and recommendations for follow-up care for the general practitioner. Swiss Med Wkly. 2012;142:w13696. https://doi.org/10.4414/smw.2012.13696.
38. Research Australia. What is autologous stem cell transplant (AHSCT)? 2021. Available on: https://msra.org.au/ahsct/autologous-haematopoietic-stem-cell-transplant/. Accessed 2 Aug 2021.
39. Rebolj K, Veber M, Drobnič M, Maličev E. Hematopoietic stem cell and mesenchymal stem cell population size in bone marrow samples depend on the patient's age and harvesting technique. Cytotechnology. 2018;70(6):1575–83. https://doi.org/10.1007/s10616-018-0250-4.
40. Santosa D, et al. Establishing the hematopoietic stem cell transplant (HSCT) in a developing country. The journey of HSCT in Semarang, Indonesia. Bone Marrow Transplant. 2021;56(1):270–3. https://doi.org/10.1038/s41409-020-0973-7.
41. Standards for Blood Banks and Transfusion Services. The standards for blood banks and transfusion services. 32nd ed. 2020.
42. The Foundation for the Accreditation of Cellular Therapy and The Joint Accreditation Committee of ISCT-EBMT. FACT-JACIE accreditation manual. 8th ed. 2021. Available on: https://www.ebmt.org/sites/default/files/2021-05/STS_5_2_042_FACT-JACIE%20AccreditationMANUAL%20Eighth%20Edition_8.1_R2_05302021_for%20web.pdf. Accessed 2 Aug 2021.
43. The Foundation for the Accreditation of Cellular Therapy and The Joint Accreditation Committee of ISCT-EBMT. FACT-JACIE standards. 8th ed. 2021. Available on: https://www.ebmt.org/sites/default/files/2021-05/STS_5_2_041_FACT-JACIE%20Standards%20Eighth%20Edition_R1_05172021_for%20web.pdf. Accessed 4 Aug 2021.

44. Wagner JE, et al., Blood and Marrow Transplant Clinical Trials Network. One unit versus two-unit cord-blood transplantation for hematologic cancers. N Engl J Med. 2014;371(18):1685–94. https://doi.org/10.1056/NEJMoa1405584.
45. Welsh Health Specialised Services Committee. Specialised services policy position. Haematopoietic stem cell transplantation (HSCT) for adults. 2019.

Appendix BI: Equipment Needed to Start a Cell Processing Lab (Adopted from Leemhuis et al., 2014. Essential requirements for setting-up a stem cell processing laboratory. Bone Marrow Transplant 49, 1098–1105)

Required equipment		
Biosafety cabinet (or equivalent)	Refrigerator	Balance (scale)
Water bath	Centrifuge (with carriers to hold 600 mL blood bags)	Freezer ($\leq$−70 °C)
Plasma extractor	Tubing sealer	Tubing stripper
Cryo-transporter (−80 °C) or liquid nitrogen dry shipper	Micropipettes (100 and 1000 μL)	Reference thermometer
Pipette aid	Hemostats	
Desired equipment		
Sterile connecting device	Controlled rate freezer	LN_2 storage freezer
Label printer	CO_2 incubator	Haemocytometer
Microscope	Personal computer	
Shared equipment		
Flow cytometer	Automated instrument for cell processing	Microbiology lab for bacterial and fungal culture
Haematology analyser		

Abbreviation: LN_2 liquid nitrogen

© The Editor(s) (if applicable) and The Author(s), under exclusive license to Springer Nature Singapore Pte Ltd. 2024
H. O. Al-Shamsi (ed.), *Cancer Care in the United Arab Emirates*,
https://doi.org/10.1007/978-981-99-6794-0

Appendix BJ: Essential Requirements for Setting-Up a Stem Cell Processing Laboratory (Adopted from Leemhuis et al., 2014. Essential requirements for setting-up a stem cell processing laboratory. Bone Marrow Transplant 49, 1098–1105)

Miscellaneous laboratory supplies		
Cryobags (for example: 50; 250; 500 mL)	Transfer packs (300; 600 mL)	Syringes (1, 3, 10, 30, 60 mL)
Safety needles; couplers	Spike to needle, spike to spike adapters; stopcocks	Alcohol swabs, iodine swabs, syringe caps, sterile swabs
Labels, laminating tags; zip ties	15, 50, 175 mL conical tubes	Pipettes (1–50 mL)
Biohazard sample bags	Tube racks	Pipette tips
Cryovials, microtubes	Biohazard bags; sharp containers; garbage bags; trash can	Dry ice
Sterile overwrap bags		
Sample reagent list (will vary depending on products and services offered)		
DMSO	Plasmalyte (or equivalent)	ACD-A
Human serum albumin	Hetastarch	Heparin
70% IPA; bleach; bactericidal and fungicidal detergent	Flow cytometry reagents	Trypan blue

Abbreviations: *ACD-A* acid citrate dextrose solution A, *DMSO* dimethyl sulfoxide, *IPA* isopropyl alcohol

Appendix BK: Training for Clinical Program Directors and Attending Physicians (Adopted from BSBMT 2012)

Knowledge	Skills
Indications for • Autologous transplant • Allogeneic transplant	• Understands the use of indication tables (S, CO, D, GNR) • Understand the outcome of alternative treatment strategies
Patient selection and pre-transplant assessment • Co-morbidity • Choice of conditioning regimens	• Understands how to assess co-morbidities and how they affect TRM and overall outcome • Understands the factors implicated in deciding between FI/RIC • Knowledge of organ assessment methods and interpretation of results
Conditioning regimens • Full intensity • Reduced-intensity	• Understands the side effects of specific chemo/radiotherapy • Understands the long-term effects of specific chemo/radiotherapy • Competent at prescribing conditioning chemo/radiotherapy
Administration of high-dose therapy • Radiotherapy	• Knowledgeable about the principles of TBI • Recognises acute toxicities • Knowledge of long-term toxicities (screening and treatment)
Administration of high-dose therapy • Chemotherapy	• Understands the mechanism of action of chemotherapy conditioning • Understands the use of prophylactic agents (e.g. mesna)

Knowledge	Skills
Stem cell mobilisation (PBSC-autologous) • Cytokine alone • Chemo/cytokine • Target cell doses	• Understand the indications, benefits and side effects of different harvesting regimens • Knowledgeable about the principles and practice of apheresis procedures • Competent at prescribing GCSF (or another mobilising agent) and understands side effects • Knowledge of cell dose targets and pre-collection CD34 counts • Competent at prescribing chemotherapy for stem cell mobilisation
Stem cell harvest (BM-autologous)	• Competent at bone marrow harvesting
Identification and selection of HPC source • Sibling • Haploidentical/another relative • UD/cord	• Understands selection algorithms and is knowledgeable of risks and benefits associated with different sources
Identification and selection of UD/cord	• Competence in requesting an unrelated donor/cord blood search, including understands of donor registries • Competence in donor selection and suitability • Understands the methodology and implications of HLA typing
Donor issues	• Competence in taking informed consent from donors, including the safety of GCSF • Understands the implications of different donation methods (BM/PBSC) • Knowledgeable about infectious diseases testing
Stem cell processing/lab	Knowledgeable about the principles and practice of: • Stem cell processing, including cell counts and cryopreservation • Basic knowledge of techniques to determine CD34+ cell counts • Positive and negative selection of CD34 positive cells, red cell depletion, and plasma depletion
Stem cell infusion	• Competent at requesting/prescribing cells (stem cells or DLI) from donor registries • Competent at prescribing cells (and pre-medication) for infusion • Proficient in HPCP infusion (including cryopreserved products)
Post transfusion, non-haemolytic complications like TRALI, TACO, GvHD	
Use of post-transplant growth factors	

Knowledge	Skills
Management of early transplant-related toxicity	Able to recognise and treat: • Neutropenic sepsis • Nausea and vomiting • Pain and mucositis • Veno-occlusive disease (SOS) • TTP • Haemorrhagic cystitis • Bleeding • Pulmonary toxicity • Multi-organ failure • Renal impairment
Blood product support	• Knowledge on the safe and appropriate use of blood products, including granulocytes • Understands the implications of ABO incompatibility (patient/donor) and group switching
Graft failure	• Understand the risk, cause, and outcome of graft failure • Knowledge of strategies to manage graft failure • Understands of methods and interpretation of chimerism analysis
Infections in the transplant setting • Prophylaxis • Treatment	Competent in: • Diagnosis, prevention, and management of fungal disease • Diagnosis and management of viral disease • Diagnosis and management of viral reactivations, including CMV and EBV • Diagnosis and management of PTLD
Graft-versus-host disease (GvHD) • Acute and chronic	• Competent in the diagnosis and management of acute and chronic GvHD, including novel therapies (e.g. mesenchyma cells, Tregs, ECP)
Disease relapse post-transplant	• Understands the risks, management, and outcomes of relapse post-transplant • Knowledgeable about the utility of second transplants of donor leukocyte infusions • Knowledgeable about methods to monitor patients at risk of relapse (e.g. MRD monitoring)

Knowledge	Skills
Late effects of transplant	• Understands the long-term effects of chemo/radiotherapy, including screening for secondary malignancies • Knowledge of the diagnosis and management of post-transplant immuno-deficiencies and organ toxicity • Knowledge about the long-term anti-infective prophylaxis and vaccination • Recognises the need for a multidisciplinary approach, especially in patients with chronic GvHD
Psychological issues	• Competence in breaking bad news • Understands the management of terminal care patients and referral to palliative care professionals
Ethical issues	• Understands the importance of ethics in all aspects of patient care, including • Donor rights and care • Cord blood donation and banking • Advanced directives • Research • Minority group issues
Quality/governance	• Knowledge of the regulatory bodies pertinent to transplantation and legal requirements • Knowledge of the national and international societies and their roles • Understands the importance of a quality management plan • Understands the function and importance of the MDT
Funding/commissioning	• Understands the funding streams within the NHS, including tariffs
Data collection	• Knowledgeable about data submission (e.g. Med-A and Med-B) • Understands the principles and use of the promise database • Knowledgeable about data protection
Research	• Understands the importance of research in the transplant environment • Understands GCP • Understands documentation and reporting for patients on investigational protocols

Appendix BL: Minimum Requirements for Consent

Patient Declaration: The approval of the treatment does not mean it has been evaluated by DHA. I voluntarily request (**insert physician names**) as my physician(s), and such associated deemed necessary, to diagnose and treat my condition, which has been explained to my satisfaction in a non-technical language. I (**insert patient name**) know the potential benefits and risks of this (**insert procedure name**) and have talked to my treating physician(s) before participating. I understand clearly that the evidence for Autologous Haematopoietic Stem Cell Transplant (AHSCT) is limited. There is no guarantee that the procedure will be successful. I understand that a positive infectious disease status may render the possibility for AHSCT transplantation. I have also consented to the appropriate treatment to be administered to carry out the procedure. The specific risks for this (**insert procedure name**) treatment have been explained to me and include (**list all risks**):

    Patient Name and Signature: Date: Time:

    The legal guardian of the patient

    If unable of consent (name and signature):   Date: Time:

Treating Physician(s) Declaration: I (**insert names**) have explained the diagnosis, prognosis, alternative options, and the stem cell procedure (**insert name of procedure and site**) to be performed and the pertinent contents to the patient. I have answered all the questions from the patient to the best of my knowledge, and the patient has been adequately informed of the potential benefits and risks, complications, and the patient has consented to the (**insert name of procedure and site**). I have explained to the patient that the success of the treatment can vary from case to case. I have explained how anaesthesia/sedation will be administered and the associated risks. I will adhere to best practices and have ensured compliance with the health facilities written protocols for this treatment and agree to assess the treatment's progress and advise the patient accordingly.

    Physician(s) name (s) and Signature: Date: Time:

    Witness Name and Signature: Date: Time:

    Relationship and/or Designation:

Appendix BM: Patient Pathway for Haematopoietic Stem Cell Transplantation (Adopted from Welsh Health Specialised Services Committee 2019)

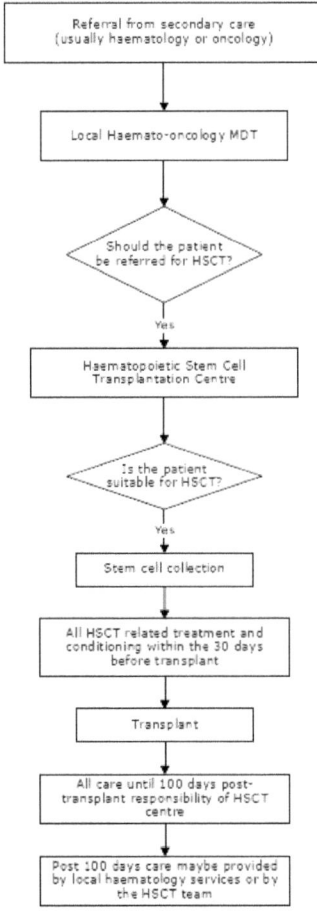

Appendix BN: Steps for Haematopoietic Stem Cell Transplantation (AHSCT) (Adopted from Research Australia 2021)

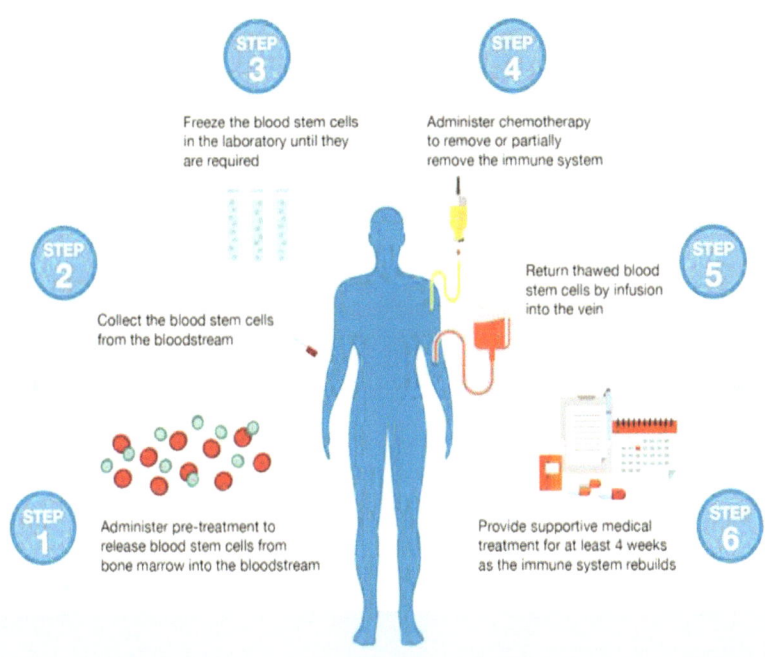

H. O. Al-Shamsi (ed.), *Cancer Care in the United Arab Emirates*,
https://doi.org/10.1007/978-981-99-6794-0